I0754314

ELEMENTS

OF

PHYSICS;

OR,

NATURAL PHILOSOPHY,

GENERAL AND MEDICAL:

WRITTEN FOR UNIVERSAL USE,

IN

PLAIN OR NON-TECHNICAL LANGUAGE;

AND CONTAINING

NEW DISQUISITIONS AND PRACTICAL SUGGESTIONS.

COMPRISED IN FIVE PARTS:

1. SOMATOLOGY, STATICS AND DYNAMICS.
2. MECHANICS.
3. PNEUMATICS, HYDRAULICS AND ACOUSTICS.
4. HEAT AND LIGHT.
5. ANIMAL AND MEDICAL PHYSICS.

BY NEILL ARNOTT, M. D.

OF THE ROYAL COLLEGE OF PHYSICIANS.

A NEW EDITION, REVISED AND CORRECTED FROM THE LAST ENGLISH EDITION.

WITH ADDITIONS.
BY ISAAC HAYS, M. D.

COMPLETE IN ONE VOLUME.

PHILADELPHIA:
BLANCHARD AND LEA.
1856.

ADVERTISEMENT

OF THE

AMERICAN PUBLISHERS.

THE very valuable and popular work of Dr. Arnott, has passed through several editions in this country, in the form in which it was originally published by the author, in separate parts. A new edition being now called for, the work has been carefully revised and corrected, and the whole condensed into one volume. In this form it cannot fail to be more acceptable to the public, and rendered more convenient and useful for the purposes of instruction in the various Colleges and Seminaries of Learning that have adopted it as a Class Book for their pupils. This volume embraces *all* that has been prepared or published by the author.

INTRODUCTION.

To appreciate the importance of PHYSICS or NATURAL PHILOSOPHY, as an object of study not only to all persons engaged in scientific pursuits, but, in the present day, to all who pretend to a moderately good education, we must take a rapid glance at the nature of human knowledge generally, and at its bearings on the existing condition of mankind.

WHILE the inferior races of animals on earth seem to have changed as little in any respect since the beginning of human records, as the trees and herbs of the thickets, which gave many of them shelter, the condition of man himself has fluctuated, but, on the whole, progressed in a very remarkable manner. The inferior animals were formed by their Creator such, that within one life or generation they should attain all the perfection of which their nature was susceptible. Their wants were either immediately provided for—as instanced in the clothing of feathers to birds, and of furs to quadrupeds; or were so few and simple, that the supply was easy to very limited powers—except in a few cases where considerable art was required, as by the bee in making its honey-cell, or by the bird in constructing its beautiful nest, and there, a peculiar aptitude or instinct was bestowed. Thus a crocodile which issues from its egg in the warm sand, and never sees its parent, becomes as perfect and knowing as any crocodile that has lived before or that will appear after it.—But how different is the story when we turn to man? He comes into the world the most helpless of living beings, long to continue so; and if deserted by parents at an early age, so that he can learn only what the experience of one life may teach him,—as to a few individuals has happened who yet have attained maturity in woods and deserts,—he grows up in some respects inferior to the nobler brutes. Now as regards many regions of the earth, history exhibits the early human inhabitants in states of ignorance and barbarism, not far removed from this lowest possible grade, which civilized men may shudder to contemplate. But these countries, occupied formerly by straggling hordes of miserable savages, who could scarcely defend themselves against the wild beasts that shared the woods with them, and the inclemencies of the weather, and the consequences of want and fatigue, and who to each other were often more dangerous than any wild beasts, unceasingly warring among themselves, and destroying each other with every species of savage and even cannibal cruelty—countries so occupied formerly, are now become the abodes of peaceful, civilized and friendly men, where the desert and the impenetrable forest are changed into cultivated fields, rich gardens and magnificent cities.

It is the strong intellect of man, operating with the faculty of language as a means which has gradually worked this wonderful change. By language, fathers communicate their gathered experience and reflections to their children, and these to succeeding children, with new accumulation; and when, after many generations, the precious store had grown until simple memory could retain no more, the arts of writing and then of printing, arose, making language visible and permanent, and enlarging illimitably the repositories of knowledge. Language thus, at the present moment of the world's existence, may be said to bind the whole human race of uncounted millions into one gigantic rational being, whose memory reaches to the beginnings of written records, and retains imperishably the important events that have occurred; whose judgment analyzing the treasures of memory, has discovered many of the sublime and unchanging laws of nature, and has built on them all the arts of life, and through them piercing far into futurity, sees clearly many of the events that are to come, and whose eyes and ears, and observant mind at this moment, in every corner of the earth, are watching and recording new phenomena, for the purpose of still better comprehending the magnificence and beautiful order of creation, and of more worthily adoring its beneficent author.

It might be very interesting to show here, in minute detail, how the arts and civilization have progressed in accordance with the gradual increase of man's knowledge of the universe; but to do so would lead too far from the main subject. We deem it right, however, to make evident to the student the arousing truths, that the progress is not yet at an end; that it has been vastly more rapid in recent times than ever; and that it seems still to proceed with increasing celerity:—and we know not where the Creator has fixed the limits of the change! Although there are thousands of years on the records of the world, our BACON, who first taught the true way to investigate nature, lived but the other day. NEWTON followed him, and illustrated his precepts by the most sublime discoveries which one man has ever made. HARVEY detected the circulation of the blood only two hundred years ago. ADAM SMITH, DR. BLACK and JAMES WATT were friends, and the last, whose steam-engines are now changing rapidly the condition of empires, may be said to be scarcely cold in his grave. JOHN HUNTER died not long ago; HERSCHEL'S accounts of newly-discovered planets, and of the sublime structure of the heavens, and DAVY'S account of chemical discoveries not less important to man, are in the late numbers of our scientific journals;—illustrious Britons these, and who have left worthy successors treading in their steps. On the continent of Europe during the same period, a corresponding constellation of genius has shone; and LAPLACE was lately the bright star shining between the future and the past.

But there is a change going on in the world, connected closely with the progress of science, yet distinct from it, and more important than a great part of the scientific discoveries;—it is the *diffusion of existing knowledge* among the mass of mankind. Formerly, knowledge was shut up in convents and universities, and in books written in the dead languages—or in books which, if in the living languages, were so abtruse and artificial, that only a few persons had access to their meaning; and thus the human race being considered as one great intellectual creature, a small fraction only of its intellect was allowed to come into contact with science, and therefore into activity. The progress of science in those times was correspondingly slow, and the evils of general ignorance prevailed. Now, however, the strong barriers which confined the stores of wisdom have been thrown down, and a flood is overspreading the earth; old establishments are adapting themselves to the spirit of the age; new establishments are arising; the inferior schools are introducing improved systems of instruction; and good books are rendering every man's fireside a school. From all these causes there is growing up an *enlightened public opinion*, which quickens and directs the progress of every art and science, and through the medium of a free press, although overlooked by many, is now rapidly becoming the governing influence in all the affairs of man. In Great Britain, partly perhaps as a consequence of its insular situation, which lessened among its inhabitants the dread of hostile invasion, and sooner formed them into a united and compact people, the progress of enlightened public opinion had been more decided than in any other state. The early consequences were more free political institutions; and these gradually led to greater and greater improvements, until Britain became an object of admiration among the nations. A colony of her children, imbued with her spirit, now occupies a magnificent territory in the new world of Columbus; and although it has been independent as yet for only half a century, it already counts more people than Spain, and will soon be second to no nation on earth. The example of the Anglo-Americans has aided in rendering their western hemisphere the cradle of many other gigantic states, all free, and following, although at a distance, the like steps. In the still more recently discovered continent of Australasia, which is nearly as large as Europe, and is empty of men, colonization is spreading with a rapidity never before witnessed; and that beautiful and rich portion of the earth will soon be covered with the descendants of free-born and enlightened Englishmen. Thence, still onward, they or their institutions will naturally spread over the vast archipelago of the Pacific Ocean, a tract studded with islands of paradise. Such, then, is the extraordinary moment of revolution, or transit, in which the world at present exists! And where, we may ask again, has the Creator predestined that the progress shall cease? Thus far at least we know, that he has made our hearts rejoice to see the world filling with happy human beings, and to observe that the increase of the sciences can make the same spot maintain thousands in comfort and godlike elevation of mind, where with ignorance even hundreds had found but a scanty and degrading supply.

The progress of knowledge which has thus led from former barbarism to present civilization, has gone on by certain remarkable steps, which it is easy to point out; and which it is very useful to consider, because we thereby discover the nature of human knowledge, with the relations and importance of its different branches; and we obtain great facilities for studying science, and for quickening its farther progress.

The human mind, when originally directed to the almost infinity of objects in the universe around it, must soon have discovered that there were resemblances among them; in other words, that the infinity was only a repetition of a certain number of kinds. Among animals, for instance, it would distinguish the sheep, the dog, the horse; among vegetables, the oak, the beech, the pine; among minerals, lime, flint, the metals, and so forth. And becoming aware that by studying an examplar of each kind, its limited power of memory might acquire a tolerably correct knowledge of the whole, while this knowledge would enable the possessors more easily to obtain what was useful to them, and to avoid what was hurtful, the desire for such knowledge must have arisen with the first exercise of reason. Accordingly, the pursuit of it has been unremitting, and the labour of ages has at last nearly completed an arrangement of the constituent materials of the universe, under three great classes of MINERALS, VEGETABLES and ANIMALS; commonly called the *three kingdoms of Nature*, and of which the minute description is termed NATURAL HISTORY: and museums of natural history have been formed, which contain a specimen of almost every object included in these classes, so that now, a student within the limits of an ordinary garden, may be said to be able to examine the whole of the material universe.

While men are examining the *forms* and other qualities of the bodies around them, they could not avoid noticing also the *motions* or changes going on among bodies; and here, too, they would soon make the grand discovery that there were resemblances in the multitude. Self-interest, as in the case of the bodies themselves, having prompted to careful classification, in the present day, as the result of countless observations and experiments made through the series of ages, we are enabled to say, that all the *motions*, or *changes*, or *phenomena* (words synonymous here) of the universe, are merely a repetition and mixture of a few simple manners, or kinds of motion or change, which are as constant and regular in every case as where they produce the returns of day and night, and of the seasons. All these phenomena are referable to four distinct classes, which we call *Physical*, *Chemical*, *Vital* and *Mental*. The simple expressions which describe them are denominated *General Truths* or *Laws of Nature*, and as a body of knowledge, they constitute what is called SCIENCE OF PHILOSOPHY, in contradistinction of NATURAL HISTORY, already described. Now as man cannot, independently of a supernatural revelation, learn anything but what respects, 1st, the momentary state, past or present, of himself and the objects around him; and 2d, the manner in which the states are changed; *Natural History* and *Science*, in the sense now explained, make up the whole sum of his knowledge of nature.

To exemplify the process by which a general truth or law of nature is discovered, we shall take the physical law of *gravity* or *attraction*. 1st. It was observed that bodies, in general, if raised from the earth, and left unsupported, fell towards it; while flame, smoke, vapors, &c., if left free, ascended away from the earth. It was held, therefore, to be a very general law, that things had *weight;* but that there were exceptions in such matters, as now mentioned, which were in their nature *light* or ascending. 2d. It was discovered that our globe of earth is surrounded by an ocean of air, having nearly fifty miles of altitude or depth, and of which a cubic foot taken near the surface of the earth, weighs about an ounce. It was then perceived that flame, smoke, vapor, &c., rise in the air only as oil rises in water, *viz.*, because not so heavy as the fluid by which they are surrounded; it followed, therefore, that nothing was known on earth naturally *light*, in the ancient sense of the word. 3d. It was found that bodies floating in water, near to each other, approached and feebly cohered; that any contiguous hanging bodies were drawn towards each other, so as not to hang quite perpendicularly; and that a plummet suspended near a hill was drawn towards the hill with force only so much less than that with which it was drawn towards the earth, *viz*, the weight of the plummet, as the hill was smaller than the earth. It was then proved that weight itself is only an instance of a more general *mutual attraction*, operating between all the constituent elements of this globe; and which explains, moreover, the fact of the rotundity of the globe, all the parts being drawn towards a common centre, as also the form of dew-drops, rain-drops, globules of mercury, and of many other things; which, still further, is the reason why the distinct particles of which any solid mass, as a stone or piece of metal, is composed, cling together as a mass, but which, when overcome by the repulsion of heat, allows the same particles to assume the form of a liquid or air. 4th. It was further observed, that all the heavenly bodies are round, and must, therefore, consist of material obeying the same law. 5th. And lastly, that these bodies, however distant, attract each other; for that the tides of our ocean rise in obedience to the attraction of the moon, and become *high* or *spring-tides*, when the moon and sun operate in the same direction. Thus the sublime truth was at last made evident, and by the genius of the immortal Newton, that there is a power of attraction connecting together the bodies of this solar system at least, and probably limited only by the bounds of the universe.

Acquaintance with the laws of nature has been very slowly obtained, owing to that complexity of ordinary phenomena, which is produced by several laws operating together, and under great variety of circumstance. With respect to many laws of Chemistry and Life, men seem to be yet little farther advanced than they were with respect to the physical law of *attraction*, when they knew only that heavy things fell to the earth. But we have learned enough to perceive that the great universe is as simple and harmonious as it is immense; and that the Creator, instead of interposing separately, or miraculously, in the common sense of the word, to produce every distinct phenomenon, has willed that all should proceed according to a few general laws. There is nothing in nature so truly miraculous and adorable as that the endless and beneficent variety of results which we see, should spring from such simple elements. In times of ignorance, men naturally regarded every occurrence which they did not understand, that is to say, which they could not refer to a general law, as arising from a direct interference of supreme power; and thus, for many ages, among some nations still, eclipses and earthquakes, and many diseases, particularly those of the mind, and the winds and weather, were or are accounted miraculous. Hence arose, among heathens, many ceremonies, and sometimes even barbarous sacrifices, for propitiating or appeasing their offended deities; but founded on expectations no more reasonable than if we should now pray to have the day of the year made shorter, or to have a coming eclipse averted. They had not yet risen to the sublime conception of the one God, who said, "Let there be light," and the light was; and who gave the whole of nature permanent laws, which he allows men to discover for the direction of their conduct in life—laws so unchanging, that by them we can calculate eclipses backward or forward for thousands of years, almost without erring, by the time of one beat of a pendulum; and as our knowledge of nature advances, we can anticipate and explain other events with equal precision. Even the wind and the rain, which, in common speech, are the types of uncertainty and change, obey laws as fixed as those of the sun and moon; and already, as regards many parts of the earth, man can foretell them without fear of being deceived. He plans his voyages to suit the coming monsoons, and he prepares against the floods of the rainy seasons.

The general laws of Nature, divisible, as stated above, into the four classes of, 1st, *Physics*, often called *Natural Philosophy;* 2d, of *Chemistry;* 3d, of *Life*, commonly called *Physiology;* and 4th, of *Mind*, may be said to form the pyramid of Science, of which Physics is the base, while the others constitute succeeding layers in the order now mentioned; the whole having certain mutual relations and dependencies well figured by the parts of a pyramid. We must describe them more particularly, to show these relations.

Physics.—The laws of *Physics* govern every phenomenon of nature in which there is any sensible change of place, being concerned alone in the greater part of these phenomena, and *regulating* the remainder which originate from chemical action, and from the action of life. The great physical truths, as comprehended in the present day by man, are reduced to four, and are referred to by the words *atom*, *attraction*, *repulsion* and *inertia*. It gives an astonishing, but true idea of the nature and importance of methodical *Science*, to be told that a man, who understands these words, *viz.*, how the ATOMS of matter by mutual ATTRACTION approach and cling together to form masses, which are solid, liquid, or aeriform, according to the quantity or REPULSION of heat among them, and which, owing to their INERTIA or stubbornness, gain and lose motion, in exact proportion to the force of attraction or repulsion acting on then,—understands the greater part of the phenomena of nature; but such is the fact! *Solid* bodies existing in conformity with these truths, exhibit all the phenomena of *Mechanics; Liquids* exhibit those of *Hydrostatics* and *Hydraulics; Airs*, those of *Pneumatics;* and so forth, as seen in the table of heads given below, at page xii. And the whole of this work is merely a list of the most interesting physical phenomena, arranged in classes under these heads.

Chemistry.—Had there been only one kind of substance or matter in the universe, the laws of Physics would have explained all the phenomena; but there are *iron*, and *sulphur*, and *charcoal*, and about fifty others, which to the present state of science, appear essentially distinct. Now these, when taken singly, obey the laws of Physics; but when two or more of them are placed in contact, under certain circumstances, they exhibit a new order of phenomena. Iron and sulphur, for instance, brought together and heated, disappear as individuals, and unite into a yellow metallic mass, which, in most of its properties, is unlike to either:—under other new circumstances, the two substances will again separate, and assume their original forms. Such changes are called *chemical*, (from an Arabic word signifying *to burn*, because so many of them are effected by means of heat,) but during the changes, the substances are not withdrawn from the influence of the physical laws,—their weight or inertia, for instance, is not altered; and indeed the phenomenon is merely a modification of general *attraction* and *repulsion.* Many chemical changes, besides, are only the beginnings of purely mechanical changes, as when the new chemical arrangement produced by heat among the intimate atoms of gunpowder, causes the mechanical or physical motion of the sudden expansion or explosion. And all the manipulations of Chemistry, as the transferring of gases from vessel to vessel, the weighing of bodies, pounding, grinding, &c., are directed to Physics alone. Chemistry, then, is truly, as figured above, a superstructure on Physics, and cannot be understood or practised by a person who is ignorant of Physics. The chief departments of study involving the consideration of Chemical in conjunction with Physical laws, are enumerated in the table below, under the head of CHEMISTRY.

Life.—The most complicated state in which matter exists, is where, under the influence of life, it forms bodies with a curious internal structure of tubes and cavities, in which fluids are moving, and producing incessant internal change. These are called *Organized Bodies*, because of the various distinct parts or *organs* which they contain; and they form two remarkable classes, the individuals of one of which are fixed to the soil, and are called *Vegetables;* and of the other, are endowed with power of locomotion, and are called *Animals.* The phenomena of growth, decay, death, sensation, self-motion, and many others, belong to life, but from occurring in material structures which subsist in obedience to the laws of Physics and Chemistry, the life is truly a superstructure on the other two, and cannot be studied independently of them. Indeed the greater part of the phenomena of organic life are merely chemical and physical phenomena, modified by an additional principle. The science of *Life* is divided into *animal* and *vegetable Physiology* (*see the table below*).

Mind.—The most important part of all science is the knowledge which man has obtained of the laws governing the operations of his own MIND. This department stands eminently distinct from the others, on several accounts. Unlike that of *organic life*, which could not be understood until physics and chemistry had been previously investigated, this had made extraordinary advances in a very early age, when the others, as methodical sciences, had scarcely begun to exist. In proof of this assertion we need only refer to the writings of the Greek philosophers. The most brilliant discoveries and applications, however, were reserved for the moderns, as will occur to many readers, on perusing, in the table below, the several divisions of the subject, and recollecting the honoured names which are now associated with each. It is truly admirable to see the modern analysis, deducing from a few simple laws of mind, act the subordinate departments, just as it deduces mechanics, hydrostatics, pneumatics, &c., from the laws of physics: and let us hope that sound opinions on this subject, ensuring human happiness, and therefore, beyond comparison, more important than any other knowledge, will soon be widely spread. The crowning science of Mind, although in certain respects independent of the science of Matter, is still closely allied to them in the following ways. The faculties of the mind are originally awakened or called into activity solely by the impressions of matter or external nature; all the language used in speaking of mind and its operations, is borrowed from matter; and many mental emotions are entirely dependent on bodily conditions. The science of Mind, therefore, cannot be studied until after knowledge acquired of an external nature; and cannot be studied extensively until that knowledge be extensive.

Quantity.—To express most of the facts and laws of Physics, Chemistry and Life, terms of QUANTITY are required, as when we speak of the magnitude of a body, or say, that the force of attraction between two bodies diminishes, in a certain proportion, as their distance increases. Hence arises the necessity of having a set of fixed measures or standards, with which to compare all other quantities. Such measures have been adopted; and they are, for NUMBERS, the fingers, *fives* and *tens;* for LENGTH, *the human foot, cubit, pace, &c.;* and lately the *second's pendulum* and the French *metre,* (taken from the magnitude of our globe); for SURFACES, the simplest forms of *circle, square, triangle, &c.*, compared among themselves by the lengths of their diameters or other suitable lines; and for SOLID BULK, the corresponding simple solids, of *globe, cube, pyramid, cone, &c.*, similarly compared by the lengths of diameters or of other lines of dimension. The rules for applying these standards to all possible cases, and for comparing all kinds of quantities with each other, constitute a body of science, called the *Science of Quantity*, the *Mathematics.* It may be considered as a subsidiary department of human science, created by the mind itself, to facilitate the study of the others.

Supposing *description of particulars*, or *Natural History*, to be studied along with the different parts of the *System of Science* sketched in the table, there will be included in the scheme the whole knowledge of the universe which man can acquire by the exercise of his own powers: that is to say, what he can acquire independently of a supernatural *Revelation.* And on this knowledge all his arts are founded,—some of them on the single part of Physics, as that of the machinist, architect, mariner, carpenter, &c.; some on Chemistry, (which includes Physics,) as that of the miner, glass-maker, dyer, brewer, &c.; and some on Physiology, (which includes much of Physics and Chemistry,) as that of the scientific gardener or botanist, agriculturist, zoologist, &c. The business of teachers of all kinds, and of governors, advocates, linguists, &c., &c., respects chiefly the science of Mind. The art of medicine requires in its professor a comprehensive knowledge of all the departments.

TABLE OF SCIENCE AND ART.

1. Physics.

Mechanics,
Hydrostatics,
Hydraulics,
Pneumatics,
Acoustics,
Heat,
Optics,
Electricity,
Astronomy,
&c.

2. Chemistry.

Simple substances,
Mineralogy,
Geology,
Pharmacy,
Brewing,
Dyeing,
Tanning,
&c.

3. Life.

Vegetable Physiology.
Botany,
Horticulture,
Agriculture,
&c.

Animal Physiology,
Zoology,
Anatomy,
Pathology,
Medicine
&c.

4. Mind.

Intellect,
Logic,
Mathematics,
&c.

Motives to action,
Emotions and Passions,
Morals,
Government,
Political Economy,
Theology,
Education.

In the first stages of education, *viz*., during the years of childhood and youth, the learning acquired is necessarily of the most mixed kind, and much of it is determined by what is called accident; but from the mutual dependence of the different departments of science, as explained in the preceding paragraphs, it follows that with a view to complete erudition, the order exhibited in "The Table," is that in which they should afterwards be studied, so as to prevent repetitions and anticipations, and to diminish, as much as possible, the labor of acquirement.

Every man may be said to begin his education, or acquisition of knowledge, on the day of his birth. Certain objects, repeatedly presented to the infant, are, after a time, recognized and distinguished. The number of objects thus known gradually increases, and from the constitution of the mind, they are soon associated in the recollection, according to their resemblances, or obvious relations. Thus, sweetmeats, toys, articles of dress, &c., soon form distinct classes in the memory and conceptions. At a later age, but still very early, the child distinguishes readily between a *mineral* mass, a *vegetable*, and an *animal;* and thus his mind has already noted the three great classes of natural bodies, and has acquired a certain degree of acquaintance with *Natural History*. He also soon understands the phrases "a falling body," "the force of a moving body," and has therefore a perception of the great physical laws of gravity and inertia. Then having seen sugar dissolved in water, and wax melted round the wick of a burning candle he has learned some phenomena of Chemistry. And having observed the conduct of the domestic animals, and of the persons about him, he has begun his acquaintance with Physiology, and the science of mind. Lastly, when he has learned to count his fingers and his sugar plums, and to judge of the fairness of the division of a cake between himself and brothers, he has advanced into Arithmetic and Geometry. Thus within a year or two, a child of common sense has made a degree of progress in all the great departments of human science; and in addition has learned to name objects, and to express feelings, by the arbitrary sounds of language. Such, then, are the beginnings or foundations of knowledge, on which future years of experience, or methodical education, must rear the superstructure of the more considerable attainments which befit the various conditions of men in a civilized community.

In the course of the preceding disquisition, we have seen that *Physics*, or *Natural Philosophy*, the subject of the present volume, is fundamental to the other parts, and is therefore that of which a knowledge is indispensable. Bacon truly calls it "the root of the sciences and arts." That its importance has not been marked by the place which it has held in common systems of education, is owing chiefly, 1st, to the misconception that a knowledge of technical mathematics was a necessary preliminary; and, 2d, to an opinion, also erroneous, that the degree of acquaintance with Physics which all persons acquire by common experience, is sufficient for common purposes; now it is true, that the toys of childhood, as the windmill, ball, syphon, tube, and a hundred others, furnish so many exemplifications of the laws of Physics, and may well be called a philosophical apparatus; but they give information which is exceedingly vague, and not at all such as is absolutely requisite in the practice of many of the arts. If, then, the study of Physics be so easy as now appears, and so important as we shall try still farther to show, there can be no excuse for neglecting it.

The greatest sum of knowledge acquired with the least trouble is, perhaps, that which comes with the study of the few simple truths of Physics. To the man who understands these, very many phenomena, which, to the uninformed, appear prodigies, are only beautiful illustrations of his fundamental knowledge, and this he carries about with him, not as an oppressive weight, but as a charm supporting the weight of other knowledge, and enabling him to add to his valuable store every new fact of importance which may offer itself. With such a principle of arrangement, his information, instead of resembling loose stones or rubbish thrown together in confusion, becomes as a noble edifice, of correct proportions and firm contexture, and is acquiring greater strength and consistency with the experience of every day. It has been a common prejudice, that persons thus instructed in general laws, had their attention too much divided, and could know nothing perfectly. But the very reverse is true; for general knowledge renders all particular knowledge more clear and precise. The ignorant man may be said to have charged his hundred books of knowledge, to use a rude simile, with single objects, while the informed man makes each support a long chain, to which thousands of kindred and useful things are attached. The laws of Philosophy may be compared to keys which give admission to the most delightful gardens that fancy can picture; or to a magic power, which unveils the face of the universe, and discloses endless charms of which ignorance never dreams. The informed man, in the world, may be said to be always surrounded by what is known and friendly to him, while the ignorant man is as one in a land of strangers and enemies. A man reading a thousand volumes of ordinary books as agreeable pastime, will receive only vague impressions; but he who studies the methodized *Book of Nature*, converts the great universe into a simple and sublime history, which tells of God, and may worthily occupy his attention to the end of his days.

We have said already, that the laws of Physics govern the great *natural* phenomena of Astronomy, the tides, winds, currents, &c. We will now mention some of the *artificial* purposes to which man's ingenuity has made the same laws subservient.

Nearly all that the civil engineer accomplishes, ranges under the head of Physics. Let us take, for instance, the admirable specimens scattered over the British Isles:—the numerous canals for inland traffic; the dock to receive the riches of the world, pouring towards us from every quarter; the many harbours offering safe retreat to the storm-driven mariner; the magnificent bridges which every where facilitate intercourse; hills bored through to open ways for commerce by canals, common roads and rail-roads, the canals in some places being supported, like the roads, on arches across valleys or above rivers, so that here and there the singular phenomenon is seen of one vessel sailing directly over another; vast tracts of swamps or fen-land drained and now serving for agriculture; the noble light-house, rearing its head amidst the storm, while the dweller within trims his lamp in safety, and guides his endangered fellow-creature through the perils of the night, &c., &c.

In Holland, great part of the country has been won and is now preserved from the sea, by the same almost creating power, and now rich cities and an extended garden smile, where, as related by Cæsar, were formerly only bogs and a dreary waste.

As a general picture, it is interesting to consider, that in many situations on earth where formerly the rude savage beheld the cataract falling among the rocks, and the wind bending the trees of the forest, and sweeping the clouds along the mountain's brow, or whitening the face of the ocean, and regarding these phenomena with awe and terror, as marking the agency of some great but hidden power, which might destroy him; in the same situations now, his informed son, who works with the laws of nature, can lead the waters of the cataract, by sloping channels, to convenient spots, where they are made to turn his mill-wheel, and to do his multifarious work; the rushing winds, also, he makes his servant, by rearing in their course the broad-vaned wind-mill, which then performs a thousand offices for its master, man; and the breezes which whiten the ocean are caught in his expanded sails, and are made to waft their lord and his treasures across the deep, for his pleasure or his profit.

In Architecture, also, Physics in supreme, and has directed the construction of the temples, pyramids, domes and palaces which adorn the earth.

In respect to machinery, generally, Physics is the guiding light. There are, for instance, the mighty steam-engine; machines for spinning and weaving, and for moulding other bodies into various shapes, yea, even iron itself, as if it were plastic clay; wind-mills and water-mills, and wheel carriages; the plough and implements of husbandry; artillery and the furniture of war; the balloon, in which man rides triumphantly above the clouds, and the diving-bell, in which he penetrates the secret caverns of the deep; the implements of the intellectual arts, of printing, drawing, painting, sculpture, &c.; musical instruments, optical and mathematical instruments, and a thousand others.

But Physics is also an important foundation of the healing art. The medical man, indeed, is the engineer pre-eminently; for it is in the animal body that true perfection and the greatest variety of mechanism are found. Where, to illustrate *Mechanics*, is to be found a system of levers and hinges, and moving parts, like the limbs of an animal body; where such an *hydraulic* apparatus, as in the heart and blood-vessels; such a *pneumatic* apparatus, as in the breathing chest; such *acoustic* instruments, as in the ear and larynx; such an *optical* instrument, as in the eye; in a word, such variety and perfection, as in the whole of the visible anatomy? All these structures, then, the medical man should understand, as the watchmaker knows the parts of a time-piece about which he is employed. The watchmaker, unless he can discover where a pin is loose, or a wheel injured, or a particle of dust adhering, or oil wanting, &c., would ill succeed in repairing an injury; and so, also, of the ignorant medical man in respect to the human body. Yet will it be believed, that there are many medical men who neither understand mechanics, nor hydraulics, nor pneumatics, nor optics, nor acoustics, beyond the merest routine; and that systems of medical education are set forth at this day which do not even mention the department of *Physics!* That such is the case, furnishes an illustration of what is stated in the beginning of this essay, *viz.*, that the sciences and arts are progressive, and that perfect methods of education must arise gradually, like all other things of human contrivance. It is within the recollection of persons now living, that political economy was discovered to be a grand foundation of the art of government, indicating means of security against many national misfortunes common in former times, yea, even against famine and war. And the day is not distant, when the members of the medical profession generally will understand how much the correct knowledge of animal structure and function, and of many remedies, must depend on precise acquaintance with Physics.—Besides the more strictly professional matters contained in the medical sections of the present work, there are many others scattered through it which greatly interest the medical man; such are the subjects of *meteorology, climate, ventilation* and *warming* of dwellings, *specific gravities*, &c., &c.

The laws of Physics having an influence so extensive as appears from these paragraphs, it need not excite surprise that all classes of society are at last discovering the deep interest they have to understand them. *The lawyer* finds that in many of the causes tried in his courts, an appeal must be made to Physics,—as in cases of disputed inventions; accidents in navigation, or among carriages, steam-engines, and machines generally; questions arising out of the agency of winds, rains, water-currents, &c.: the *statesman* is constantly listening to discussions respecting bridges, roads, canals, docks, and the mechanical industry of the nation: the *clergyman* finds ranged among the beauties of nature, the most intelligible and striking proof of God's wisdom and goodness; the *sailor* in his ship has to deal with one of the most admirable machines in existence: *soldiers*, in using their projectiles, in marching where rivers are to be crossed, woods to be cut down, roads to be made, towns to be besieged, &c., are dependent chiefly on their knowledge of Physics: the *land-owner*, in making improvements on his estates, building, draining, irrigating, road-making, &c.: the *farmer* equally in these particulars, and in all the machinery of agriculture; the *manufacturer*, of course; the *merchant* who selects and distributes over the world the products of manufacturing industry—all these are interested in Physics; then also the *man of letters*, that he may not, in drawing his illustrations from the material world, repeat the scientific heresies and absurdities which have heretofore prevailed, and which, by shocking the now better-informed public, exceedingly lower the estimation in which such specimens of the Belles Lettres are held, and lessen their general utility; and lastly, *parents of either sex*, whose conversation and example have such powerful effect on the character of their children, who when grown up, are to fill all the stations in society; all should study Physics, as one important part of their education.

And it is for such reasons that Natural Philosophy is becoming daily more and more a part of common education. In our cities now, and even in an ordinary dwelling-house, men are surrounded by prodigies of mechanic art, and cannot submit to use these, regardless of how they are produced, as a horse is regardless of how the corn falls into his manger. A general diffusion of knowledge, owing greatly to the increased commercial intercourse of nations, and therefore to the improvements in the physical departments of astronomy, navigation, &c., is changing every where the condition of man, and elevating the human character in all ranks of society. In remote times the inhabitants of the earth were generally divided into small states or societies, which had few relations of amity among themselves, and whose thoughts and interests were confined very much within their own little territories and rude habits. In succeeding ages, men found themselves belonging to larger communities, as where the English heptarchy was united; but still distant kingdoms and quarters of the world were of no interest to them, and were often totally unknown. Now, however, every one feels that he is a member of one vast civilized society, which covers the face of the earth; and no part of the earth is indifferent to him. In England, for instance, a man of small fortune may cast his looks around him, and say with truth and exultation, "I am lodged in a house which affords me conveniences and comforts which some centuries ago, even a king could not command. Ships are crossing the seas in every direction, to bring me what is useful to me from all parts of the earth. In China, men are gathering the tea-leaf for me; in America, they are planting cotton for me; in the West Indies, they are preparing my sugar and my coffee; in Italy, they are feeding silk-worms for me; in Saxony, they are shearing the sheep to make me clothing; at home, powerful steam-engines are spinning and weaving for me, and making cutlery for me, and pumping the mines that minerals useful to me may be procured. Although my patrimony was small, I have post-coaches running day and night on all the roads to carry my correspondence; I have roads and canals, and bridges, to bear the coal for my winter fire; nay, I have protecting fleets and armies around my happy country, to secure my enjoyments and repose. Then I have editors and printers, who daily send me an account of what is going on throughout the world, among all these people who serve me. And in a corner of my house I have BOOKS! the miracle of all my possessions, more wonderful than the wishing-cap of the Arabian Tales: for they transport me instantly, not only to all places, but to all times. By my books I can conjure up before me, into vivid existence, all the great and good men of antiquity; and for my individual satisfaction I can make them act over again the most renowned of their exploits, the orators declaim for me; the historians recite; the poets sing; and from the equator to the pole, or from the beginning of time until now, by my books, I can be where I please." This picture is not overcharged, and might be much extended, such being God's goodness and providence, that each individual of the civilized millions dwelling on the earth, may have nearly the same enjoyment as if he were the single lord of all.

Reverting to the importance of Natural Philosophy as a general study, it may be remarked that there is no occupation which so much strengthens and quickens the judgment. This praise has usually been bestowed on the Mathematics, although a knowledge of abstract Mathematics existed with all the absurdities of the dark ages; but a familiarity with Natural Philosophy which comprehends Mathematics, and gives tangible and pleasing illustrations of the abstract truths, seems incompatible with the admission of any gross absurdity. A man whose mental faculties have been sharpened by acquaintance with these exact sciences in their combination, and who has been engaged, therefore, in contemplating *real relations*, is more likely to discover truth in other questions, and can better defend himself against sophistry of every kind. We cannot have clearer evidence of this than in the history of the sciences, since the Baconian method of *reasoning by indication* took place of the visionary *hypotheses* of preceding times. Until then, even powerful minds did not recoil from the most absurd theories on all subjects. Astronomy was mixed with Astrology; Chemistry with Alchemy; Physiology with the singular hypotheses which preceded the discovery of the circulation of the blood; Politics with the errors of monopolies, prohibitions, balance of trade, &c. Even Religion itself, in various ages and countries, has felt the influence of the state of the public mind as to solid attainments. To a man with the knowledge of nature which we now possess, the fables and licentious abominations of the Greek and Roman theologies are shocking indeed; as are the religions of the God of Fire in China, of Vishnoo in India, of Mahomet's imposture and pretended miracles, &c. But the enlightened Christian minister earnestly recommends the study of nature; first because from contemplating the beauty of creation, with the wisdom and benevolent design manifest in all its parts, there spring up in every undepraved mind those feelings of admiration and gratitude, which constitute the adoration of natural religion, and which form, as shown by many estimable writers on Natural Theology, a fit foundation for the sublime doctrine of immortality, and secondly, because a Revelation being probable only by the miracles occurring at its establishment; to enable men to distinguish between miracles and the usual course of nature, a perfect knowledge of that course, or of Natural Philosophy, is essential: all the false religions of antiquity were founded on, and upheld by pretended miracles. As regards the question of immortality, even independently of Revelation, no man who contemplates the order and beauty of the material world, and then thinks on the hideous deformities of the moral world—where vice so often triumphs, and modest virtue pines and dies—can for a moment believe that they are the work of the same author, unless there be a hereafter of retribution; and feeling thus that eternal justice requires another state for man, he embraces with delight the cheering promises of immortality. There have been, however, at various times, even among Christians, sincere, but weak-minded or ill-informed men, who decried the study of the natural sciences, as inimical to true religion; as if God's ever-visible and magnificent revelation of his attributes in the structure of the universe could be at variance with any other revelation. But such prejudices are now quickly passing away. Wherever considerable knowledge of nature exists, debasing and gloomy superstition must cease. It is not the abject terror of a slave which is inspired by contemplating the majesty and power of our God, displayed in his works, but a sentiment akin to the tender regard which leads a favored child to approach with confidence a wise and indulgent parent.

It remains for the author now only to say a few words with respect to the present work. He was originally led to the undertaking with the view of supplying the desideratum in medical literature, of a treatise on *Medical Physics;* but soon perceiving that the preliminary investigation of *General Physics*, necessary to adapt the work to medical readers, would require to be nearly as extensive as it would for general readers, and reflecting that every person of liberal education must now possess such a book, not to be read once and then thrown aside as a novel is, but to be frequently consulted as a manual, he determined to make his book as complete and as extensively useful as possible. He has been encouraged during his labor, by the belief that the growing light of science, which now exhibits more clearly the natural relations of the different departments of study, as attempted to be portrayed in the preceding pages, might enable him to avoid some of the defects of former elementary treatises, and to add features of novelty and improvement to his own. The sections on *Animal Physics* were, of course, written for medical men; and a great service will be rendered by the work, if it only awakens them to a just sense of the importance of Physics as one of the foundations of their art. But even for general readers there are few parts of these sections which the author would exclude. There is nothing more admirable in nature than the structure and functions of the human body, and there are many reasons why no liberal mind should be careless of the study. The details here given are not more anatomical than the illustrations from the animal economy contained in the common treatises on *Natural Theology*. From the attempt in this work to compress into the smallest possible space the greatest possible sum of scientific information, few historical details have been admitted, whether relating to the distinguished men who have benefitted the world as authors or inventors, or to the history of the progress of science:—such details form an interesting, but distinct branch of study.

The author must not conclude without observing, that no treatise on Natural Philosophy can save, to a person desiring full information on the subject, the necessity of attendance on experimental lectures or demonstrations. Things that are seen, and felt, and heard, that is, which operate on the external senses, leave on the memory much stronger and more correct impressions than where the conceptions are produced merely by verbal description, however vivid. And no man has ever been remarkable for his knowledge of Physics, Chemistry, or Physiology, who has not had practical familiarity with the objects. With reference to this familiarity, persons who take a philanthropic interest in the affairs of the world, must observe, with much pleasure the now daily increasing facilities of acquiring useful knowledge, afforded by the scientific institutions formed and forming, not only through this kingdom, but through most civilized nations.

Bedford Square, 1st *March*, 1827.

ELEMENTS

OF

NATURAL PHILOSOPHY.

SYNOPSIS, OR GENERAL REVIEW.

If it excite our admiration that a varied edifice, or even a magnificent city can be constructed of stone from one quarry, what must our feeling be to learn how few and simple the elements are out of which the sublime fabric of the universe, with all its orders of phenomena, has arisen, and is now sustained. These elements are general facts and laws which human sagacity is able to detect, and then to apply to endless purposes of human advantage.

Now the four words, *atom*, *attraction*, *repulsion*, *inertia*, point to four general truths, which explain the greater part of the phenomena of nature. Being so general, they are called *physical* truths, from the Greek word signifying *nature* as also "truths of Natural Philosophy," with the same meaning, and sometimes "mechanical truths," from their close relation to ordinary machinery. These appellations distinguish them from the remaining general truths, namely the *chemical* truths, which regard particular substances, and the *vital* and mental truths, which have relation only to living beings. And even in the cases where a chemical or vital influence operates, it modifies, but does not destroy, the physical influence. By fixing the attention, then, on these *four fundamental truths*, the student obtains, as it were, so many keys to unlock, and lights to illuminate the secrets and treasures of nature.

1st. Atom. Every material mass in nature is divisible into very minute indestructible and unchangeable particles,—as when a piece of any metal is bruised, broken, cut, dissolved, or otherwise transformed, a thousand times, but can always be exhibited again as perfect as at first. This truth is conveniently recalled by giving to the particles the name *atom*, which is a Greek term, signifying *that which cannot be further cut or divided*, or an exceeding minute resisting particle.

2d. Attraction. It is found that the atoms above referred to, whether separate or already joined into masses, tend towards all other atoms or masses,—as when the atoms of which any mass is composed are, by an invisible influence, held together with a certain degree of force; or when a block of stone is similarly held down to the earth on which it lies; or when the tides on the earth rise towards the moon. These facts are conveniently recalled by connecting with them the word *Attraction* (a drawing together) or *gravitation.*

3d. REPULSION. Atoms under certain circumstances, as of heat diffused among them, have their mutual *attraction* countervailed or resisted, and they tend to separate;—as when ice heated melts into water, or when water heated bursts into steam, or when gunpowder ignited explodes. Such facts are conveniently recalled by the term *Repulsion* (a thrusting asunder.)

4th. INERTIA. As a fly-wheel made to revolve, at first offers resistance to the force moving it, but gradually acquires speed proportioned to that force, and then resists, being again stopped, in proportion to its speed, so all bodies or atoms in the universe have about them, in regard to motion, what may be figuratively called a *stubborness,* tending to keep them, in their existing state, whatever it may be—in other words, they neither acquire motion, nor lose motion, nor bend their course in motion, but in exact proportion to some force applied. Many of the motions now going on in the universe with such regularity—as that turning of the earth which produces the phenomena of day and night—are motions which began thousands of years ago, and continue unvarying in this way. Such facts are conveniently recalled by the term *inertia* applied to them.

A person comprehending fully the import of these four words, that is to say, having present to his mind numerous good types or exemplars of the facts referred to them, may predict or anticipate correctly, and may control very many of the facts and phenomena which the extended experience of a life can display to him; and such a person is commonly said to know the causes or reasons of things and events. Now it is important here to observe, that when a person gives a reason or explanation of any fact, other than that it is a fact, or than that the Creator has willed it, he is merely, although he may not be aware of this, showing its resemblance to many other facts, no one of which he understands better than itself—and what he calls a general truth, or law, or principle, is merely an expression for the observed but unaccountable resemblance of the facts. Thus, when a man says that a stone falls because of *attraction* or *gravitation,* he only uses a word which recalls thousands of instances which he has witnessed of one body approaching another; but by any cause of the approach, other than that God has willed it, is to him utterly unknown. Should men, in the course of their researches, discover that the phenomena now classed by them under the heads of *attraction* and *repulsion,* although apparently opposite, are really as closely allied as they already know the rising of a balloon and the falling of a stone to be (the balloon rises like a cork in water, being pushed up by the fluid air around it, heavier than it, and seeking to descend,) they will not have discovered a new cause, but a new resemblance, (new to them) among phenomena, and will only have advanced one step farther in perceiving the simplicity of creation. In accordance with these views, it will be found that this volume is chiefly an extensive display of the most important phenomena of nature and art, classified so as to be explained by the four physical truths, and mutually to illustrate one another. They will be distributed under the following heads or divisions:

PART I.

CONSTITUTION OF MASSES, MOTIONS AND FORCES.

The four fundamental truths extensively examined, and used to explain generally, in
Section
1. The *nature or constitution of the material masses* which compose the universe; (a department technically called SOMATOLOGY, from Greek words signifying a *discourse* on *body*.)
2. The *motions* or *phenomena* going on among the masses;—a department including the common divisions of STATICS (things stationary or at rest,) and DYNAMICS (what relates to *force* or *power*.)

PART II.

PHENOMENA OF SOLIDS.

The four truths explaining the peculiarities of state and motion among *solid* bodies:—a department called, in a restricted sense, MECHANICS, (from the Greek, and signifying *machine*.)

PART III.

PHENOMENA OF FLUIDS.

The truths explaining the peculiarities of state and motion among *fluid* bodies:—a department called HYDRODYNAMICS (from Greek words signifying *water* and *force*.)
Section
1. HYDROSTATICS (*water at rest or in equilibrium*.)
2. PNEUMATICS (*air phenomena*.)
3. HYDRAULICS (*water or fluid in motion*.)
4. ACOUSTICS (*phenomena of sound and hearing*.)

PART IV.

PHENOMENA OF IMPONDERABLE SUBSTANCES.

The truths aiding to explain the more recondite phenomena of IMPONDERABLE SUBSTANCES, under the heads of
Section
1. HEAT or *Caloric*.
2. LIGHT or *Optics*.

PART V.

ANIMAL AND MEDICAL PHYSICS.

In this part will be ranged the most interesting illustrations afforded by the animal economy, constituting—ANIMAL AND MEDICAL PHYSICS.

As no man can well understand a subject of which he does not carry a distinct outline in his mind, it is recommended to the reader of this work to study the general *synopsis*, and the *analysis* placed at the heads of the *chapters* and *sections*, until the memory be well impressed with them.

PART I.

THE FOUR FUNDAMENTAL TRUTHS MINUTELY EXAMINED, AND USED TO EXPLAIN GENERALLY, FIRST, THE NATURE OR CONSTITUTION OF THE MATERIAL MASSES WHICH COMPOSE THE UNIVERSE, AND SECONDLY, THE MOTIONS OR PHENOMENA GOING ON AMONG THEM.

SECTION I.—THE CONSTITUTION OF MASSES.

ANALYSIS OF THE SECTION.

The visible universe is built up of very minute indistructible ATOMS *called matter, which by mutual* ATTRACTION, *cohere or cling together in masses of various form and magnitude. The atoms are more or less approximated, according to the quantity or* REPULSION *of heat among them, and hence arise the three remarkable forms in the masses, of solid, liquid and air, which mutually change into each other with change in the quantity of heat. Certain modifications of attraction and repulsion produce the subordinate peculiarities of state called crystal, dense, hard, elastic, brittle, malleable, ductile and tenacious.*

"*Minute Indestructible* ATOMS."*

THAT the smallest portion of any substance which the human eye can perceive, is still a mass of many ultimate atoms or particles, which may be separated from each other, or newly arranged, but which cannot individually be hurt or destroyed, is deduced from such facts as the following:

A particle of powdered marble, hardly visible to the naked eye, still appears to the microscope a block susceptible of indefinite division; and, when it is broken by fit instruments, until the microscope can hardly discover the separate particles of fine powder, these may be yet further divided, by solution in an acid; the whole becoming then absolutely invisible, as part of a transparent liquid.

A small mass of gold may be hammered into thin leaf, or drawn into fine wire, or cut into almost invisible parts, or liquefied in a crucible, or disolved in an acid, or dissipated by intense heat into vapour; yet, after any and all these changes, the atoms can be collected again to form the original mass of gold, without the slightest diminution or change. And all the substance of

* The different heads or titles, which appear thus, throughout the work, between inverted commas, are the successive portions of the *Analysis*, detached for separate consideration. The reader is particularly requested to re-peruse the analysis at the several interruptions, that he may have constantly before him that clear view of the general relations among the different parts of the subject, which is essentially to a perfect understanding of it.

elements of which our globe is composed, may thus be cut, torn, bruised, ground, &c., a thousand and a thousand times, but are always recoverable as perfect as at first.

And with respect to delicate combinations of these elements, such as exist in animal and vegetable bodies, although it be beyond human art, originally to produce, or even closely to imitate many of them—for we cannot build up a feather or a rose—still, in their decomposition and apparent destruction, the accomplished chemist of the present day does not lose a single atom. The coal which burns in his apparatus, until only a little ash remains behind, or the wax-taper that seems to vanish altogether in flame, or the portion of animal flesh which putrefies, and gradually dries up and disappears—present to us phenomena which are now proved to be only changes of connection and arrangement among the indestructible ultimate atoms; and the chemist can offer all the elements again, mixed or separate as desired, for any of the useful purposes to which they are severally applicable. When the funeral piles of the ancients, with their charge of human remains, appeared to be wholly consumed, and left the idea with survivors that no base use could be made, in after time, of what had been the material dwelling of a noble or beloved spirit, the flames had only, as it were, scattered the enduring blocks of which a former edifice had been constructed, but which were soon to serve again in new combinations.

Facts to be stated under the heads of "chemical composition" and "crystal," will prove, that the ultimate particles of any substance must be, among themselves, perfectly similar.

"*Minute.*" (Read the Analysis, page 22.)

The following are interesting particulars in the arts or in nature, helping the mind to conceive how minute the ultimate atoms of matter must be.

Goldbeaters, by hammering, reduce gold to leaves so thin, that 360,000 must be laid upon one another to produce the thickness of an inch. They are so thin, that if formed into a book, 1,800 would occupy only the space of a single leaf of common paper; and an octavo volume an inch thick would have as many pages as the books of a well-stocked ordinary library containing 1,800 volumes of 400 pages each; yet those leaves are perfect, or free from holes, so that one of them laid upon any surface, as in gilding, gives the appearance of solid gold.

Still thinner than this is the coating of gold, upon the silver wire of what is called gold lace; and we know not that such coating is of only one atom thick. If we place a piece of this wire in nitric acid, so as to dissolve the silver within, the gold coating remains as a metallic tube of exquisite tenuity.

Platinum can be drawn into wire much finer than human hair.

A grain of blue vitriol or carmine, will tinge a gallon of water, so that in every drop the colour may be perceived.

A grain of musk will scent a room for twenty years, and will have lost but little of its weight.

The carrion crow seems to smell its food at a distance of many miles.

The thread of the silk worm is so small, that many folds have to be twisted together to form our finest sewing thread; but that of the spider is smaller still, for two drachms of it by weight would reach from London to Edinburgh, or 400 miles.

In the milk of a cod-fish, or in water in which certain vegetables have been infused, the microscope discovers animalcules, of which many thousands together do not equal in bulk a grain of sand; yet these have their blood and other subordinate parts like larger animals; and, indeed, nature, with a singular prodigality, has supplied many of them with organs as complex as those of the whale or elephant. Now the body of an animalcule consists of the same elementary substances, or ultimate atoms, as the body of man himself. In a single pound of matter, it thus appears, that there may be more living creatures than of human beings on the face of this globe, What scenes has the microscope laid open to the admiration of the philosophic inquirer.

Water, mercury, sulphur, or, in general, any substance, when sufficiently heated, rises as invisible vapour or gas; in other words, is made to assume the aeriform state. Great heat, therefore, would cause the whole of the material universe to disappear, the previously most solid bodies becoming as invisible and impalpable as the air we breathe. Utter annihilation would seem but one stage beyond this.

"*Matter.*"

The inconceivable minuteness of ultimate atoms, as shown above, has led some inquirers to doubt whether there really be *matter;* that is to say, whether what we call substance or matter have existence or not. In answer to this it has been usual to adduce, besides the weights of the substances, and the proofs of indestructibility already mentioned, which seems conclusive, the fact that every kind or portion of matter obstinately occupies some space to the exclusion of all other matter from that particular space. This occupancy of space is the simplest and most complete idea which we have of material existence. The awkward word *impenetrability* has been used to express it, with reference of course to the individual atoms. The following are elucidations:

We cannot push one billiard-ball into the substance of another, and then a second, and then a third, and so on; or the material of the universe might be absorbed in a point.

A mass of iron on a support will resist the weight of thousands of pounds laid upon it and pressing to descend into its place; and although a very great weight might crush or break it into pieces, still one particle would not be annihilated. In a forcing pump, or in Braham's water-press, millions of pounds can not push the piston down, unless the water below it be allowed to escape.

A weight laid upon bladders full of air, or on the piston handle of a closed air-pump, is supported in the same manner.

A quantity of air escaping from a vessel under water ascends through the water as a bubble displacing its bulk of water in its way.

A glass tube, left open at bottom, while the thumb closes the top, if pressed from air into water, is not filled with water, because the air contained in it resists; but if the air be allowed to escape by removing the thumb from the top, the tube becomes filled immediately to the level of the water around it. In a goblet or basin pushed into water, with the mouth downwards, the entrance of water is resisted for the like reason; and if the goblet be inverted over a floating lighted taper, this will continue to float under it, and to burn in the contained air, however deep in the water it may be carried—exhibiting the curious phenomenon of light below water, and being an emblem of the living inmate of a diving bell, which is merely a larger goblet holding a man instead of a candle.

"*Mutual attraction.*" (See the Analysis, page 22.)

Any visible mass of matter, then, as of metal, salt, sulphur, &c., we know to be really a collection of dust, or minute atoms, by some cause made to cohere or cling together; yet there are no hooks connecting them, nor nails, nor glue; and the connection may be broken a thousand times, by processes of nature or art, but is always ready to take place again; the cause being no more destroyed in any case by interruption, than the weight of a thing is destroyed by frequent lifting from the ground. Now the cause we know not, but we call it *attraction*. The phenomena of attraction and its contrary, repulsion, particularly when occurring between bodies at considerable distances from each other, are as inexplicable as any subjects which the human mind has to contemplate; but the manner or laws of the phenomena are now well understood. The general nature and extensive influence of attraction may be judged of from the following facts:

Logs of wood floating in a pond, or ships in calm water, approach each other, and afterwards remain in contact. When the floating bodies are very small, or can approach very near to each other at the water's edge—as glass bulbs in a tea-cup—an additional force is called into play, as will be explained under the head of "capillary attraction."

The wreck of a ship, in a smooth sea after a storm, is often seen gathered into heaps.

Two bullets or plummets suspended by strings near to each other, are found by the delicate test of the torsion balance (which will be described afterwards) to attract each other, and therefore not to hang quite perpendicularly.

A plummet suspended near the side of a mountain inclines towards it, in a degree proportioned to its magnitude; as was ascertained by the well-known trials of Dr. Maskelyne near the mountain Schehallion, in Scotland.

And the reason why the plummet in such a case tends much more strongly towards the earth than towards the hill, is only that the earth is larger than the hill

At New South Wales, which is situated on our globe nearly opposite to England, plummets hang and fall towards the centre of the globe, as they do here; so that in respect to England, they are hanging and falling upwards, and the people there, like flies on the opposite side of a pane of glass, are standing with their feet towards us,—hence called our antipodes. Weight, therefore, is merely general attraction acting everywhere.

But it is owing to this general attraction that our earth itself is a globe:—all its parts being drawn towards each other, that is, towards a common centre, the mass assumes the spherical or rounded form.

And the moon also is round, and all the planets; nay, the glorious sun, too, so much larger than these, is round;—suggesting the inference that all must at one time have been a certain degree fluid, and that all are subject to the same law.

Descending again to the earth and observing minuter masses, we have many interesting instances of roundness from the same cause; as—the particles of a mist or fog floating in air—these, mutually attracting and coalescing into larger drops, and so forming rain—dew-drops—water trickling on a duck's wing—the tear dropping from the cheek—drops of laudanum—globules of mercury, like pure silver beads, coalescing when near, and forming larger ones—melted lead allowed to rain down from an elevated sieve, and by cooling as it descends so as to retain the form of its liquid drops, becoming the spherical shot-lead of the sportsman, &c.

The cause of this extraordinary phenomenon which we call attraction, acts at all distances.—The moon, though 240,000 miles from the earth, by her attraction, raises the water of our ocean under her, and forms what we call the tide.—The sun, still farther off, has a similar influence; and when the sun and moon act in the same direction, we have the spring tides.—The planets, so distant that they appear to us little wandering points in the heavens, yet, by their attraction, affect the motion of our earth in her orbit, quickening it when she is approaching them, retarding it when she is receding.

The *attraction is greater* the nearer the bodies are to each other; as the light of a taper is more intense near to the taper than at a distance.

A board of a foot square, represented in fig. 1 by A B at a certain distance from a light supposed at C, just shadows a board of two feet square, as E D, at double distance; but a board with a side of two feet has four times as much surface as a board with a side of one foot, for it is not only twice as high or long, which would make it double, but twice as broad also, which

Fig. 1.

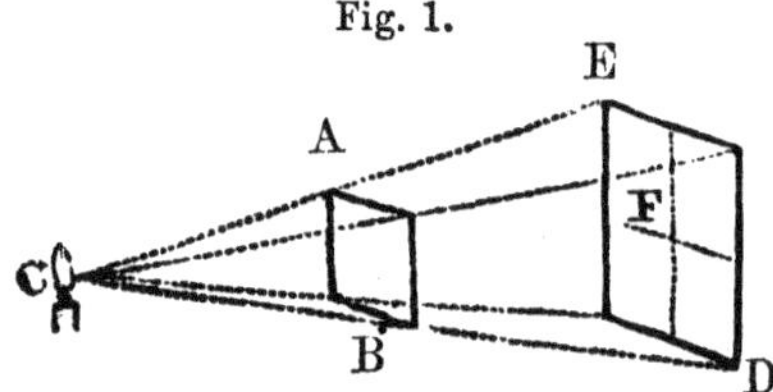

makes it quadruple—as a globe of two feet in diameter requires just four times as much paper to cover it as a globe of one foot,—and the corner, or fourth part E F, of the larger square here shown is just equal to the whole of the smaller square A B. Light, therefore, at double distance from its source, being spread over four times the space, has only one-fourth of the intensity; and for a similar reason, at thrice the distance it has only a ninth part, at four times a sixteenth part, and so on. Now light, heat, attraction, sound, and indeed every influence from a central point are found to decrease in the proportion here illustrated, *viz.*, as the surface of squares which shadow one another increases. The technical expression is, "*the intensity is inversely as the squares of the distance;* (the distances being estimated from the centres of attraction or radiation) or one-fourth part as strong at double distance, four times as strong at half distance, and in a corresponding manner for all other distances.

Accordingly, what weighs 1,000 lbs. at the sea-shore, weighs five lbs. less at the top of a mountain of a certain height, or when raised in a balloon—as is proved experimentally by a spring balance, or other means;—and at the distance of the moon, the weight, or force towards the earth, of 1,000 lbs., is diminished to five ounces, as is proved by astronomical test.

ATTRACTION has received different names as it is found acting under different circumstances. The chief distinctions are *Gravitation, Cohesion, Capillary* and *Chemical attraction.*

Gravitation is the name given to it when acting at sensible distance, as in the cases of the moon lifting the tides—the sun and earth attracting each other—a stone falling, &c. Most of the facts enumerated at page 25, belong to this head.

Cohesion is the name given when it is acting at very short distances, as in keeping the atoms of a mass together.

It might appear, at first sight, that it cannot be the same cause which draws a piece of iron to the earth with the moderate force called its weight, and which contains the constituent atoms of the iron in such strong cohesion; but when we recollect that attraction is stronger as the substances are nearer to each other, the difficulty is met. Atoms very nearly in contact may be a million times nearer to each other than when only a quarter of an inch apart, and therefore, when the heat among the atoms of any mass allows them to approach very near, they should attract mutually with great force.

If, then, the surfaces of the bodies were not in general so very rough and irregular, that, when applied to each other, they can touch only in a few points of the million, perhaps, which each surface contains, bodies would be invariably sticking together or cohering by any accidental contact. The effect of artificially smoothing the touching surfaces is seen in the following examples:—we may remark, however, that besides irregularity of surface, there is another reason, explained a little farther on, which prevents the cohesion.

Similar portions being cut off with a clean knife from two leaden bullets, and the fresh surfaces being brought in contact with a slight turning pressure, the bullets cohere, almost as if they had been originally cast in one piece.

Fresh-cut surfaces of India-rubber or caoutchouc, cohere in a similar way. We may hence make elastic air-tight tubes, by cutting off the edges of a strip of India-rubber and bringing the cut surfaces into contact by winding the strip spirally round any small rod or cylinder, and fixing it there for a time by tape or cord.

Two pieces of perfectly smooth plate glass or marble, laid upon each other, adhere with great force: and so, indeed, do most well-polished flat surfaces.

Cohesion between a solid and liquid, and between the particles of a liquid among themselves, is seen in the following instances:

A flat piece of glass, balanced at the end of a weighing beam, and then allowed to come into contact with water, adheres to the water, and with much more force than the weight of water remaining upon it when again forcibly raised! If there were not cohesion or attraction of the water particles among themselves, as well as to the glass, the latter could only be held down by the weight of the water which directly adhered to it. In pouring water from a mug or bottle-lip, the water does not at once fall perpendicular, but runs down along the inclined outside of the vessel; chiefly in consequence of the attraction between this and the water; hence the difficulty of pouring from a vessel which has not a projecting lip.

The particles of water cohere among themselves in a degree which causes small needles gently laid on the surface to float:—the weight of the needles is not sufficient to overcome the cohesion of the water surface.

For the same reason, many light insects can walk upon the surface of water without being wetted.

It is chiefly the different force of the attraction of cohesion in different liquids that causes their drops or gutts from the lip of a phial to be of different magnitude. Sixty drops of water fill the same measure as 100 drops of laudanum from a lip of the same size.

In a larger mass of liquid, the attraction which, if acting alone, would draw the particles into the form of a distinct globe, yields to that which draws them towards the centre of the earth, and therefore the liquid assumes more or less completely, what is called the level surface, that is to say, a surface corresponding with the general surface of the earth.

Attraction is called *capillary* when it acts between a liquid and the interior of a solid, which is tubular or porous.

When an open glass tube is partially immersed in water, the water within it stands above the level of that on the outside; and the difference of level is greater as the tube is less, because in small tubes, the glass all round being nearer to the raised water, attracts it more powerfully.

Between the two plates of glass standing near to each other, with their lower edges in water, a similar rising of water will occur; and if they are closer at one perpendicular edge than at the other, the surface of the suspended water will be higher there. The two plates of glass in such a case are found to be drawn towards each other by the interposed waters with a certain force, as happens also to glass beads, or other small bodies, floating in water with their surfaces so near to each other at the water's edge, that the water may rise between them,—and the nearer they approach, the higher the water rises, and the more strongly it attracts.

Water, ink, or oil, coming in contact with the edge of a book, is rapidly absorbed far inwards among the leaves.

A piece of sponge or a lump of sugar touching water by its lowest corner, soon becomes moistened throughout.

The wick of a lamp lifts the oil to supply the flame, from two or three inches below it.

A mass of cotton thread hanging over the edge of a glass from the water within it, will empty it as a syphon would. A towel will empty a basin of water in the same way.

Dry wedges of wood driven into a groove formed round a pillar of stone, on being moistened, will swell so as to rive off the portion from the block. In some portions of Germany, mill-stones are thus cut from the rock.

An immense weight or mass suspended by a dry rope may be raised a little way, by merely wetting the rope;—the moisture imbibed by capillary attraction in the substance of the rope causes it to swell laterally, and to become shorter.

At one time, the small vessels of vegetables were supposed to raise the sap from the roots, by capillary attraction; but this is known now to be chiefly an action of vegetable life.

Attraction has received the name of *chemical attraction*, or *affinity*, when it unites the atoms of two or more distinct substances into one perfect compound.

There are about fifty substances in nature which appear, in the present state of science, distinct from each other, and are therefore called *kinds of matter;* such as the various metals, sulphur, phosphorus, &c.; but whether these are in truth, originally and essentially different, or only one simple primordial matter, modified by circumstances as yet unknown to us, we cannot at present positively determine. Diamond and pure black carbon are the same substance only with different arrangement of atoms; and steel, which in the soft state the graver cuts as it would copper or silver, is exactly the same substance as when, after being tempered by heating and sudden cooling, it has become as hard nearly as diamond itself. Yet these differences are more striking than appear between some substances, which we now account essentially distinct.

It is found, however, that the atoms of what we call different substances will not cohere and unite indifferently, to form masses, as atoms of the same kind do,—there being singular preferences and dislikes among them, if it may be so expressed, or affinities, as the chemists term it: and when atoms of two kinds do combine, the resulting compound generally loses all resemblance to either of the elements.—Thus:

Sulphuric acid will unite with copper and form a beautiful transcendent blue salt; with iron it will form a green salt; and if a piece of iron be thrown into a solution of the copper salt, the acid will immediately let fall the copper, and take up or dissolve the iron.—Sulphuric acid will not unite with or dissolve gold at all.—Quicksilver and sulphur unite in certain proportions and form the paint called vermillion; in other proportions they form the black mass called Ethiops Mineral.—Lead, with oxygen absorbed from the atmosphere or other source, forms what is called red lead, used by painters. —Sea-sand, or flint and the substance called soda, when heated together, unite and form that most useful substance called glass.—Certain proportions of sulphur and of iron combine and produce those beautiful cubes of pyrites or gold-like metal which are seen in slate. Chemical attraction operating thus, does not, in the slightest degree, interfere with general attraction or gravity, for every chemical compound weighs just as much as its elements taken separately.

The history and classification of such facts connected with the combinations and analysis of different substances, constitute the science of chemistry, so attractive and so useful. It explains how the fifty kinds of matter above alluded to, by variously combining, form the endless diversity of bodies which constitute, as far as it has yet been explored, the mass of our globe. The reasons of these various modifications of attraction are yet much hidden from us.

It is a remarkable truth, that when different substances combine in the way now described, the proportions of the ingredients are always uniform, and such as to lead to the conclusion, that for every atom present, of one substance, there is exactly one, or two, or three, &c., of the other; so that, if there be ten atoms of one substance, there are exactly ten, or twenty, &c., of the other, but never an intermediate number, as 13 or 23 to 10, for then a particle of the compound would consist of one atom of the first, and of one and three-tenths, or two and three-tenths, &c., of the second substance, an absurdity if the atom be indivisible. For instance, a certain number of atoms of quicksilver, which weigh twenty-five grains, combine with a certain number of atoms of sulphur, weighing two grains, and form a black compound called Ethiops Mineral, or black sulphur of mercury; and if a little more of either ingredients be added, it lies as a foreign mixture in the sulphuret of mercury; but if just as much more sulphur be added as at first, so that there may be two atoms of it, instead of one, in every particle of the compound, a perfect combination of the whole will take place, and a new substance will appear, which we call vermilion. Many elementary substances will only unite with each other in one proportion, so that any two such substances form only one compound; but others unite in several proportions, so that several distinct compounds arise out of the same two elements.

It thus appears, that although we do not know the exact number of atoms in a given quantity of any substance,—whether, for instance, a grain of sulphuret of mercury has more or less than a million of them; still, as we know that in that grain there are just as many atoms of sulphur as of mercury, and that the weight of the whole sulphur to that of the whole mercury is as two to twenty-five, we know that the single atoms must have the same relation, or that the atom of mercury is 12½ times as heavy as that of sulphur.

Tables have been formed exhibiting the relative weights of the atoms of different substances; and the number standing opposite to each substance is called its *chemical equivalent*,—that is to say, the weight of its atom in relation to the weight of the atom of some other substance chosen as a standard. The *equivalent* of a compound substance depends, of course, both on the equivalents of the ingredients, and on the number of atoms existing in one integrant particle of the compound.

There is no such thing as an *atom* of vermilion, or of any other compound, for the ultimate molecule or particle must contain at least one atom of the respective ingredients.

The facts of the peculiarities and constancy of chemical unions are among the strongest arguments for the existence of similar ultimate atoms.

Besides the simple cases of attraction now explained, there are two curious modifications, called *electrical* and *magnetical* attractions, which from their peculiarities are reserved for consideration in a future division of this work.

"*Atoms are more or less close, according to the quantity or* REPULSION *of heat among them; hence the forms of solid, fluid, air, &c.*" (Read the Analysis, p. 22.)

Were there in the universe only atoms and attraction, as hitherto explained, the whole material of creation would rush into close contact, forming one huge solid mass of stillness and death. But there is also heat or caloric, which counteracts attraction, and singularly modifies the results. It has been described by some as a most subtle fluid, pervading all things, somewhat as water pervades a sponge: others have accounted it merely a vibration among the atoms. The truth is, that we know little more of heat as a cause of repulsion than of gravity as a cause of attraction; but we can study and classify most accurately the phenomena of both.

When a continued addition of heat is made to any body, it gradually increases the mutual distance of the constituent atoms, or dilates the body. A solid thus is first enlarged and softened; then melted or fused, that is to say, reduced to the state of liquid, as the cohesive attraction is overcome; and lastly, the atoms are repelled to still greater distances, so that the substance is converted into elastic fluid or air. Abstraction of heat from such air causes return of states in the reverse order.

Thus ice, when heated becomes water, and the water when farther heated becomes steam; the steam when cooled again becomes water as before, and the water when cooled becomes ice. Ice, water and steam, therefore, are three forms or states of the same substance—one of the most common in nature, being the material of the ocean.

Other substances are similarly affected by heat, but as all have different relations to it, some requiring much for liquefaction, and some very little, we have that beautiful variety of solids, liquids and air, which constitutes our external nature.

Dilitation.—A rod of iron, which, when cold, will pass through a certain opening, and will lie lengthwise between two fixed points, when heated, becomes too thick and to long to do either.—For accurate mensuration, therefore, rods or chains used as the measure, must either be at a given temperature, or due allowance must be made for the difference.

The walls of a building, under the pressure of a heavy roof, had begun to bulge out so as to threaten its stability. No force tried was sufficient to restore them to perpendicularity, until the idea occurred of using the contracting force of cooling iron. The opposite walls were then connected by a number of iron bars, passing through both, and having nuts to screw close to the wall upon their projecting ends, of which bars one-half were heated at a time, *viz.*, every second or alternate bar, by lamps placed under them, and while lengthened in consequence, and projecting farther beyond the wall, their nuts were again screwed close up; so that on cooling and contracting, they pulled the walls in a degree back to its place. The nuts of the second set of bars being then screwed home, the others were again heated, and advanced the object as much as the first; and so on, until the object was accomplished.

The iron rim of a coach wheel, when heated, goes on loosely and easily, but when afterwards cooled, it binds the wheel most tightly, giving remarkable firmness and strength.

Iron hoops on masts and casks, are made to bind in a similar manner.

The common thermometer for measuring degrees of heat, is a glass bulb, filled with mercury or other fluid, and having a narrow tube rising from it, into which the fluid, on being expanded by heat, ascends, and so marks the degree.

A bladder not quite full of cold air, on being heated, becomes tense, and if weak, may even be burst.

Liquid and Air.—A piece of gold, lead, pitch, ice, sulphur, or of other thing, if sufficiently heated, melts or becomes liquid; each substance, however, requiring a different degree of heat—gold requires 5,000 degrees, lead 600, ice, 32, and so forth; and if the heating be afterwards continued, most things at certain higher temperatures suddenly expand again to many time the liquid volume, and becomes aeriform fluids.

The conversion of water into steam is familiarly known to all. One pint of water driven off as steam from the boiler of a low-pressure steam-engine, fills a space of nearly 2,000 pints, and raises the piston through this, with a force of many thousands of pounds: it immediately afterwards appears again in the cold condenser as a pint of water.

Six times as much heat is required to convert a pint of water into steam, as to raise it from an ordinary temperature to that of boiling but the steam, by occupying nearly 2,000 times the space of the water, proves that heat merely produces a revulsion among the particles, and by no means fills up the interstices. The steam rising from boiling water does not appear to the thermometer hotter than the water itself; and hence it was that Dr. Black, whose genius shed so much light on this part of knowledge, gave the excess of heat the name of *latent heat.*

The latent heat of common air is made sensible in the *match syringe.* In this, which is close at the bottom, the piston is driven down quickly and strongly, so as to compress very much the air which is underneath it, and the heat then condensed with the air is sufficiently intense to light a small piece of tinder attached to the bottom of the piston.

Not only are spirits, æthers, oils, &c., convertible, as water is into aeriform fluid, but also sulphur, phosphorus, mercury, and, indeed, all the metals and elementary substances;—some of them, however, requiring heats of great intensity.

The varieties of form, then, in the bodies on the face of this earth, may be considered accidental, as dependent on the temperature of the earth, and do not mark the permanent nature ef the substances.

In the planet Mercury, which is near the sun, resin, tallow, wax and many vegetable substances deemed by us naturally solid, would all be liquid, as oil is with us; and a certain mixture of tin, zinc and lead, which with us is solid at common temperatures, but melts in boiling water, would there be always liquid like our quicksilver. Our water, oils, and spirits, would there be in a state of steam or air, and could not be known as liquids, except by cooling processes and compression, such as we have lately learned to use for reducing our different airs to the form of liquids.

Again, in the cold planet Herschel, which is nineteen times farther from the sun than our earth is, water, if it exist, can be known only as rock crystal, which fire would have to melt as it does glass with us: our oils would be as butter or resins, and quicksilver might be hammered as lead or silver is with us.

On our own earth, near the equator, common sealing-wax will not retain impressions; butter is oil in the day, and a soft solid at night; and tallow candles cannot be used. And near our pole, in winter, the quicksilver from a broken thermometer is solid metal; water must be melted by fire for use; oils are solid, &c.

To judge, then, of the constitution of nature aright, we must always take extended surveys, and not allow prejudice to mislead us, as it did that Eastern potentate, who put a traveller to death for saying he had visited remote northern countries, where water was sometimes to be seen solid like crystal, and sometimes white and fleecy, like feathers.—The ancients believed that there were just four elements concerned in forming our globe, with all upon it, viz., *earth, water, air* and *fire.* What a contrast between former and present knowledge!

Repulsion without sensible Heat.

As we stated in a former paragraph that besides general attraction, under names *gravitation, cohesion, capillary* and *chemical attraction*, there are modifications which have the names of *electrical* and *magnetical* attractions; so we have now to remark, that, besides the general repulsion of heat just described, there are peculiarities which we call *electrical* and *magnetical repulsions.* Whether these depend altogether on different causes, or are only modifications of effect from the same cause, we cannot yet positively decide.

And it is a curious fact connected with the subject, that there seems to be a film of repulsion, so to express it, covering the general surfaces of all bodies, and preventing their meeting in absolute contact, even when they appear to the human eye so to meet. Were it not for this, things would be constantly approaching so closely to each other, that they would stick or cohere, in a way to disturb the common operations of nature. The following facts illustrate this superficial repulsion, and the means which art uses to overcome it for particular purposes.

Newton found that a ball of glass, or a watch-glass, laid upon a flat surface of glass does not really touch it and cannot be made to touch it by a force of even 1,000 pounds to the inch.

In like manner, when glass, stone, porcelain, or indeed almost any body is broken, we cannot make the parts cohere again by simply pushing them together in their former position. Where a union, therefore, between separate masses is desired, we are compelled to have recourse to various artifices

A few cases in which cohesion is easily affected, were enumeratéd at page 27 : the following are other instances of a different kind.

Gold leaf laid upon clean steel, and then forciby struck by a hammer, coheres to the steel, and gilds it permanently.

But iron can be made to cohere to iron, only by rendering both pieces red hot before hammering :—the process is called welding. Iron and platinum are the only metals that can be welded.

Tin and lead, in sheets, pressed together between the strong rollers of a flatting-mill, cohere.

The other metals require to be melted before the superficial repulsion gives way so as to allow separate quantities to cohere or run into one mass. It is thus, for instance, that gold, silver, lead, &c., are treated.

In many cases the substances are not such as can be melted, (wood or marble, for instance,) and then it is necessary to use some soft glue or cement. Cements must have strong attraction for both substances, and, when dry or cool, must be tenacious in themselves; solder, paste, common glue, motar, &c., are the principal substances of this kind.

"*Certain modifications of attraction produce the subordinate states, called crystal, porous, dense, &c.*" (Read the Analysis, page 22.)

" It is a remarkable circumstance, that attraction, in causing the atoms to cohere so as to form solid masses, seems not to act equally all around each atom, but between certain sides or parts of one, and corresponding parts of the adjoining one; so that when atoms are allowed to cohere according to their natural tendencies, they always assume a certain regular arrangement and form, which we call crystalline. Because in this circumstance they seem to resemble magnets, which attract each other only by their poles, the fact has been called the polarity of atoms. It is the cause of several of the peculiarities above enumerated, as elasticity, &c.

"*Crystallization*" is exemplified in the following particulars :

Water beginning to freeze, shoots delicate needles across the surface; these thicken and interweave until the whole mass has become solid, but the crystalline arrangement always remains. In most substances, this arrangement is remarkably proved, by the forms of the surfaces left, when the mass is broken.

Moisture, freezing on the window-pane in winter, exhibits a beautiful variety of arborescence.

A flake of snow viewed in the microscope, is seen to be as symmetrically formed as a fern-leaf or a swan's feather.

If a piece of copper be thrown into a solution of silver in nitric acid, it is preferred by the acid to silver, and is dissolved accordingly: the silver in the mean time, during its precipitation or separation, assumes the form of a singularly beautiful shrub or tree, resting on the remaining copper as its root. This appearance is call the *arbor Dianæ*.

Any metal which has been melted, when allowed to cool again, slowly and at rest, becomes solid first on the outside of the mass. If, before the cooling be completed, the remaining liquid be poured from within, a curious internal crystalline structure, like grotto work, is seen. What is called the grain of a metal is the result of this crystallization.

Saltpetre, glaubler salt, copperas (to use popular names,) or any other of the many neutral salts, being dissolved in water, and the water being then allowed slowly to evaporate, reappears in beautiful regular crystals, each salt having its peculiar forms, bounded by perfectly plane and polished surfaces. If any such crystal be broken in any part, the broken surface appears to the microscope as if regular layers of particles had been disturbed, (as we see on a larger scale in a broken stack of bricks, or broken pile of shot in a battery yard,) and the defect of the crystal will be exactly filled up by replacing it in the evaporating solution—proving that the ultimate particles are all of the same size.

All the precious stones are crystals, and can be well cut only parallel to their natural surfaces.

The basaltic pillars of the Giant's Causeway in Ireland, and of the Isle of Staffa, which appears like a garden supported on magnificent columns in the midst of the ocean, are natural crystalline arrangements of particles, equalling in regularity and beauty any human work, and in granduer so far surpassing even the Egyptian pyramaids, that superstitious conjecture naturally supposed them the work of giant architects.

It would be endless to go on enumerating crystalline masses, for nature's forms generally, in the inanimate creation, as well as in organized bodies, are regular and symmetrical; and what we see on earth of broken continents, and islands, and rocks, and wild Alpine scenery, are the effects of subsequent convulsions, which have deranged a primitive and natural order.

Much ingenuity has been employed to account for the specific forms which different crystalline bodies assume; but the subject is not yet reduced to a state fitting it to be a part of this elementary study. A familiarity with the various figures which the exact *science of measures* treats of, is required in the person who expects to pursue it with pleasure or advantage. The facts are extremely curious, and the scientific investigation of them may ultimately give important information respecting the intimate constitution of material nature.

"*Porous.*"—The crossing of the constituent crystalline needles or plates in bodies, causes them to be porous or full of small vacant spaces. In some cases these are visible to the eye, in many more cases, they are visible to the microscope, and in all, they are to be proved in some way.

Owing to the porosity arising from the new arrangement of atoms of solidifying, water and a very few other substances become more bulky in the change from the liquid to the solid state. Water then dilates with such force as to burst the strongest vessels which art can provide, and in winter to split even rocks, where it has been retained in their crevices;—freezing water thus curiously producing effects which surpass those of exploding gunpowder. This agency of water contributes to the gradual breaking down of our Alpine summits, and the falling of their destructive fragments into the valleys.

The stone called hydrophane (agate) is opaque, until dipped into water, when it absorbs into its pores one-sixth of its weight of the water, and afterwards gives passage to light.

Into crystallized sugar, and various stones, much water will enter without increasing the bulk.

A kind of sandstone, suitably shaped, forms an excellent filter or strainer for water.

Pressure will force water through the pores of the most solid gold :—as was seen in the famous Florentine experiment, where a hollow, thick, golden ball, being filled with water and squeezed, to try the compressibility of water, was found to perspire all over.

The examples of porosity in animal and vegetable bodies, are, however, the most remarkable.

Bone is a tissue of cells and partitions, as little solid as a heap of empty packing-boxes.

Wood is a congeries of parallel tubes, like bundles of organ pipes. It has lately been proposed to prepare wood for certain purposes, as for making the great wooden pins or nails used in ship-building, by squeezing it to half its lateral bulk between very strong rollers, and thus making its density approach to that of metal.

A piece of wood sunk to a great depth in the ocean, and exposed to the pressure there, has its pores soon filled with water, and becomes nearly as heavy as stone. Thus it was with the boat of a whale-fishing ship, which had been dragged far under water by a whale, and which, on being afterwards drawn up, was supposed by the crew to be bringing a piece of rock with it.

A piece of cork in a strong, close glass vessel, nearly full of water, may be seen floating at the top; but if more water be then forcibly pumped into the vessel, the cork will be squeezed and reduced in size, until at last it becomes heavier than water, and sinks. On water being afterwards allowed to escape, the cork will resume its bulk and will rise. A cork sunk 200 feet under water will never rise again of itself.

A bottle of fresh water, corked and let down thirty or forty feet into the sea, often comes up again with the water saltish, although the cork be still in its place: the explanation being, that the cork, when far down, is so squeezed as to allow the water to pass in or out by its sides, but on rising, resumes its former size.

"*Density*," or the quantity of atoms which exist in a given space, is very different in different substances.

A cubic inch of lead is forty times heavier than the same bulk of cork Mercury is nearly fourteen times heavier than an equal bulk of water.

The density must depend on, first, the size or weight of the individual atoms; secondly, the degree of porosity just now explained; and thirdly, the proximity of the atoms in the more solid parts which stand between the pores.

From many circumstances it appears, that the atoms even of the most solid bodies are nowhere in actual contact, but are retained in their places by a balance between attraction and repulsion—thus,

A body dilates or contracts, according as heat is added or taken away from it.

A weight placed on any upright rod or pillar, shortens it and lessens its bulk, and if suspended from the bottom, lengthens it and increases its bulk, —the rod in both cases returning to its former dimensions when the weight is removed.

When a plank or rod is bent, the atoms on the concave side are, for the time, approximated, and those on the convex side are drawn more apart. It is remarkable in solid bodies, not only how precisely the balance between attraction and repulsion determines the relative position of the particles, but also how strongly; for any farther separation of the particles is resisted by all the force which we call the tenacity or cohesion of the substance, and any nearer approach by all the force which we call the hardness or incompressibility.

Tin and copper, when melted together, to form bronze, occupy less space by one-fifteenth than when separate: proving that the atoms of the one are partially received into what were vacant spaces in the other. A similar condensation is observed in many other mixtures. A pound of water and a pound of salt, when mixed, form two pounds of brine, but which has much less bulk than the ingredients apart. So also of a pound of sugar dissolved in a pound of water.

Water and liquids generally resist compression very powerfully, but yield enough to show that the particles are not in contact. It is found that at 1,000 fathoms down in the sea the water is compressed by the superincumbent water so as to have bulk about a hundredth part less than it would have at the surface.

In aeriform masses the atoms are very distant, and hence the masses are more easily compressed. A pint of water, on assuming the aeriform state, in which it is called steam, under ordinary pressure, acquires nearly 2,000 times its former bulk. A hundred pints of common air may be compressed into a pint vessel, as in the chamber of an air-gun; and if the pressure be much farther increased, the atoms will at last collapse and form a liquid. The heat which was contained in such air, and gave it its form, is squeezed out in this operation, and becomes sensible all around.

From these proofs of the non-contact of the atoms, even in the most solid parts of bodies; from the very great space obviously occupied by pores—the mass often having no more solidity than a heap of empty boxes, of which the apparently solid parts may still be as porous in a second degree, and so on; and from the great readiness with which light passes in all directions through dense bodies like glass, rock crystal, diamond, &c., it has been argued that there is so exceedingly little of really solid matter, even in the densest mass, that the whole world, if the atoms could be brought into absolute contact, might be received into a nut-shell. We have as yet no means of determining exactly what relation this idea has to truth.

The comparative *weights* of *equal bulks* of different bodies are called their *specific gravities.*

In thus comparing bodies, it was necessary to choose a standard; and water, as being the substance most easily procurable at all times and in all places, has been generally adopted.

The metal called platinum, the heaviest of known substances, is about twenty-two times as heavy as an equal bulk of water, and is therefore said to have specific gravity of 22—gold is nineteen times as heavy— mercury thirteen and a half—lead eleven—iron eight and a half—copper eight—common stones about two and a half—woods from half to one and a half—cork one-quarter, &c.

"*Hardness,*" is not proportioned, as might be expected, to the density of the different bodies, but to the polarity of the atoms in them, that is, to the force with which the atoms hold their places in some particular arrangement.

Hardness is measured generally by the circumstance of one body being capable of scratching another. It is here worthy of notice, however, that the powder or dust of a softer body will often, through an effect of motion to be described below, aid in wearing down or polishing one that is harder.

Gold, though soft, is four times heavier than the hard diamond; and mercury, which is fluid, is nearly twice as dense as the hardest steel.

Diamond is the hardest of known substances. It cuts or scratches every other body, and is generally polished by means of its own dust.

Glass-cutters use a point of diamond as a glass-knife, for dividing and shaping their panes.

Common flint also cuts glass, as is proved by the frequent scribblings on windows.

It is remarkable, that the preparation of iron, called steel, may either be soft like pure iron, or from being heated and suddenly cooled, in the process called tempering, may become nearly as hard as diamond. The discovery of this fact is, perhaps, second in importance to few discoveries which man has made; for it has given him all the edge tools and cutting instruments by which he now moulds every other substance to his wishes. A savage will work for twelve months, with fire and sharp stones, to fell a great tree, and to give it the shape of a canoe; where a modern carpenter, with his tools, could accomplish the object in a day or two.

The project has lately been realized, of engraving on plates of soft steel, instead of copper, and afterwards tempering the steel to such hardness, that it may be used as a type or die to make its impression, not on paper, but on other plates of soft steel or of copper; each of which then is equal in value to an original and distinct engraving. By this means the beautiful productions of art, instead of being limited to a comparatively small number of copies and of persons, may be multiplied almost to infinity, becoming the cheap delight of all.

"*Elasticity*" is present in a mass when the atoms, cohering in a particular arrangement only, yield, however, to a certain extent, when force is applied, but move back or regain their natural positions on the force being withdrawn.

Elastic bodies vary much as to the extent to which they yield without breaking, and as to the degree of perfection with which, after the bending, or displacement of atoms, they regain their former state. India rubber is extensively elastic, for it yields far; but it is not perfectly elastic, for when stretched much or often, it becomes perfectly elongated. Glass, again, is perfectly elastic, for it will retain no permanent bend; but, unless in very thin plates indeed, or in fine threads, it will not bend far without breaking.

All hard bodies are elastic, as steel, glass, ivory, &c., and many soft ones, as caoutchouc, silk, a harp string, &c. The aeriform bodies are all perfectly elastic, as is rudely seen in a bladder filled with air, when squeezed, and allowed to expand again; and they will change volume to a very great extent. Liquids also are perfectly elastic, but to a small extent.

A good steel sword may be bent until its ends meet, and yet, when allowed, will return to perfect straightness.

A rod of bad steel, or of other metal, will be broken in bending, or will retain a bend.

An ivory ball, let fall on a marble slab, rebounds, owing to the great elasticity of both bodies, nearly to the height from which it fell, and no mark is left on either. If the slab be wet, it is seen that the ivory or marble, or both, had yielded considerably at the point of contact, for a circular surface of some extent on the slab is found dried by the blow. The sudden expulsion of air from between the meeting surfaces might contribute to the effect, but the result is very nearly the same when the experiment is made in a vacuum. Billiard balls scarcely lose even their polish by long wear, although the touching parts yield at every stroke.

A marble chimney-piece, long supported by its ends, is found at last to be bent downwards in the middle; and the bend is permanent.

A steel watch-spring, although so much and so constantly bent, resumes its original form when freed at the end of a century; but occasionally, without evident cause, while in action, it will suddenly give way.

Elasticity is a property of bodies of great utility to man, as in his time-pieces, carriage-springs, gun-locks, &c., &c.

"*Brittleness,*" designates that constitution of a body where, with hardness, and elasticity perfect as far as it goes, the cohesion among the atoms exists within such narrow limits that a very slight change of position or increase of distance among them is sufficient to produce a rupture. A comparatively slight force, therefore, if sudden, breaks them. It belongs to most very hard bodies.

Glass scratches an iron hammer, proving that it is harder than iron—yet glass is the very type of fragility; yielding to the stroke of soft wood, or, indeed, of almost any thing which can give a blow.

Steel, when tempered so as to be very hard, becomes brittle also. The steel chisels and tools with which artificers now shape the stones and metals as they formerly did wood, require, of course, to be exceedingly hard; but they thereby lose in regard to the *extent* of their elasticity, and hence are frequently broken. Cast iron, which is much harder than malleable or wrought iron, is very brittle, while soft iron and steel are the toughest things in nature.

"*Malleable,*" or reducible into thin plates or leaves by hammering. This property, in opposition to elasticity and brittleness, belongs to bodies whose atoms cohere equally in whatever relative situations they happen to be, and therefore yield to force, and shift about among each other, without fracture or change of property, almost like the atoms of a fluid.

Gold is remarkably malleable, for it may be reduced to leaves of the thinness of 360,000 to the inch, or of 1,800 to a sheet of common paper. For gold-beaters the metal is first formed into rods, these are afterwards rolled or battened into ribbons; the ribbon is cut into portions, which are extended, by hammering, to great breadth and thinness, and which being again divided into portions, are hammered and extended to the thinness described.

Silver, copper and tin may also be hammered until very thin. Most other metals crack or are torn before the operation is carried far; and some, on being struck, are broken at once, almost like glass.

"*Ductile,*" or susceptible of being drawn into wire. One might expect malleability and ductility to belong to the same substances and in the same degrees—but they do not. In ductile substances, as in malleable, the atoms seem to have no more fixed relation of position than in a liquid, but yet they cohere very strongly.

One end of a rod of iron, or other ductile metal, being reduced in size so as to pass through an opening in a plate of steel, is seized by strong nippers on the other side of the plate, and the whole rod is drawn through. It is thus reduced, of course, to the size of the opening, and is lengthened in a like proportion. By repeating the operation through smaller holes successively, a wire may at last be obtained to the size of a hair.

Dr. Wollaston's ingenuity produced platinum wire finer than spider's thread. He filled a space in the axis of a silver wire with small platinum wire. He then drew or reduced the compound piece to the smallest wire possible, and on dissolving the silver from the outside, he exposed to view the delicate filament of platinum.

The order in which metals may be ranged according to their ductility is, platinum, silver, iron, copper, gold, &c.

Melted glass has great ductility. The workers draw or spin it into threads by merely attaching a point, pulled out from the mass, to the circumference of a turning-wheel. A uniform thread then continues to be drawn out and wound upon the wheel, at a rate of 1,000 yards or more per hour. This glass thread, when lying together in quantities, resembles beautiful white hair, and when cut in bunches, it serves as an ornament to the female head, waving in the air like the delicate plume of a bird of paradise.

"*Pliant.*" In bodies distinguished by this title, the cohesion is not destroyed by considerable change of direction among the particles, but there is little elasticity, and unlike what happens in a ductile mass, the same atoms always remain together.

Of all pliant things, the chief are animal and vegetable fibres and membranes—as silk, bladder, lint, hemp, &c., &c.

"*Tenacity*" means the force of cohesion among the atoms of any mass. It belongs more or less to all solids, and even to liquids.

This property varies much in different substances. Iron and its modification called steel possess it in the most remarkable degree.

The following table shows the comparative tenacity, or strength to resist pulling of certain metals and woods. Supposing similar wires or rods of each to be used, and of such a size that the surface of a broken end or cross-section would be the one-thousandth of a square inch, the weights supported would be nearly as follows:

METALS.	
Cast Steel	134 lbs.
Best wrought iron	70
Cast Iron	19
Copper	19
Platinum	16
Silver	11
Gold	9
Tin	5
Lead	2
WOODS.	
Teak	13
Oak	12
Beech	12½
Ash	14
Deal	11

Iron compared in this way, is five or six times stronger than oak.

Steel wire will support about 39,000 feet, that is, 7½ miles of its own length.

Certain animal substances have great tenacity; as—the silk-worm's thread, which is our strongest connecting or sewing material, and has such flexibility united with its strength—the ligaments and tendons of the animal body, possessing at once such admirable strength, elasticity and pliancy: these when dried, and otherwise prepared, constituted the tough bow-strings of our remote forefathers—the hair or wool of animals twisted into threads, and worked into strong and beautiful textures of the loom—strips of animal intestines prepared and twisted, forming the cords of harp and violin, and in strength and uniformity rivaling the steel wires of keyed instruments.

The gradual discovery of substances possessed of strong tenacity and which man could yet easily mould to his purposes, has been of great importance to his progress in the arts of life. The place of the hempen cordage of European navies is still held in China by twisted canes and strips of bamboo; and even the hempen cable of Europe, so great an improvement on former usage is now rapidly giving way to the more complete and commodious security of the iron chain—of which the material to our remote ancestors existed only as a useless stone or earth. And what a magnificent spectacle is it, at the present day, to behold chains of tough iron stretched high across a channel of the ocean, as at the Menai Strait, between Anglesea and England, and supporting there an admirable bridge-road of safety along which crowded processions may pour, regardless of the deep below, or of the storm; while under it, ships with full sails spread pursue their course, unmolesting and unmolested!

APPENDIX

TO PART I.—SECTION I.

BY THE AMERICAN EDITOR.

If the reader has studied the preceding section with attention he is prepared to understand the following *propositions.*

Prop. 1—Matter is endowed with properties.

Prop. 2.—The properties of matter are distinguishable into two classes, first, those which are *general* or belong to all kinds of matter, and second, those which are *peculiar* or belong only to particular kinds of matter.

Prop. 3.—The *general* properties of matter are, indestructibility (*p.* 22;) extension or the property of occupying a portion of space (*p.* 24;) divisibility (*p.* 23;) impenetrability (*p.* 24;) and inertia, (*p.* 42.)

Prop. 4.—Every particle of matter, and also all masses, have a mutual *attraction* for one another, or endeavor to get near each other; and this attraction is inversely as the squares of the distances.

Attractions may be primarily distributed into two classes: one consisting of those which exist between the molecules or constituent parts of bodies, and the other between the bodies themselves. The former are called molecular or atomic attractions, the latter gravitation (*p.* 26:) of the former there are several varieties, 1st, cohesion (*p.* 27;) when this variety of molecular attraction is exhibited by liquids pervading the interstices of porous bodies, ascending in crevices or in the pores of small tubes, it is called capilliary attraction (*p.* 28.) The other varieties of molecular attractions are affinity or chemical attraction (*p.* 28,) and electric and magnetic attraction, (*p.* 30.)

Prop. 5.—Attraction of gravitation, or that force by which all the masses of matter tend towards each other, is exerted at all distances.

Prop. 6.—Attraction of cohesion acts only within certain limits, and where its sphere of attraction ends, a *repulsive* force begins.

Prop. 7.—Repulsion, except when dependent on electricity or magnetism, is owing to the presence of heat, which latter pervades all matter.

Prop. 8.—The particles of matter are more or less close, according to the quantity of heat among them; but they are never in actual contact (*p.* 30–31,) and hence *porosity* is usually considered as one of the properties of matter.

Prop. 9.—The *peculiar* properties of matter are density (*p.* 35,) hardness (*p.* 36,) elasticity (*p.* 37,) brittleness (*p.* 38,) malleability (*p.* 38,) ductility (*p.* 38,) pliability (*p.* 39,) tenacity, (*p.* 39,) &c.

SECTION II.—THE MOTION OR PHENOMENA OF THE UNIVERSE.*

ANALYSIS OF THE SECTION.

The bodies or masses composing the universe may be at rest or in motion, and to change any present state, force proportioned to the quantity of the body and to the degree of change, is equally required, whether to give motion, to take it away, or to bend it:—a truth expressed by saying that matter has INERTIA, *or figuratively, a stubbornness. Uniform straight motion, then, is as naturally permanent as rest. And the motion in any body, measured by its velocity, quantity of matter and direction, is the measure of the amount and direction of any single force or of any combination of forces, which has produced it, as also of the force or momentum which the body can exhibit again when opposed or made to act itself as a cause of some new motion.*

The great forces of nature, referred to by the two words ATTRACTION *and* REPULSION, *acting upon* INERT *matter, produce the* equable, accelerated, retarded *and* bent *motions which constitute the great phenomena of the universe.—Tides, currents, winds, falling bodies, &c., exemplify* attraction.—*Explosion, steam collision, &c., exemplify* repulsion. *And as in every case of attraction or repulsion two masses at least must be concerned, there is no motion or* action *in the universe, without an equal and opposite motion or* re-action.

"Motion"

Is the term applied to the phenomenon of the changing of place among bodies.

Were there no motion in the universe it would be dead. It would be without the rising or setting sun, or river flow, or moving winds, or sound, or light, or animal existence.

To understand the nature and laws of the motions or changes which are going on around him, is to man of the greatest importance, as it enables him to adapt his actions to what is coming in futurity, and often to interfere so as to control futurity for his special purposes.

Motion, in any particular case, is described by referring to certain objects to mark place, and to some other motion chosen as the standard of velocity. —A man sitting on the deck of a sailing ship, has *common* motion with the ship: if walking on the deck, he has *relative* motion to the ship: but if he be walking towards the stern, just as fast as the ship advances, he is at rest relatively to the bottom or shore. A ship sailing against the tide, just as fast as the tide runs, is as much at rest relatively both to the earth and water as if she were at anchor. *Absolute* motion is that which is relative to the whole universe, or rather to the space in which the universe exists. We have no means of ascertaining such: for although we know how fast our globe whirls upon its axis and wheels round the sun, we have no measure of the motion of the sun himself—revolving possibly round some

* The reader should here re-peruse the title and Analysis at page 22.

more distant centre, but almost certainly having a progress in space, and carrying all the planets along with him.

Motion is called *rapid*, as that of lightning—*slow*, as that of the sun-dial shadow; both terms have reference to the ordinary intermediate velocities observed upon earth. It is called *straight* or *rectilineal*, in the apparent path of a failing body—*bent*, or *curvilinear*, in the track of a body thrown obliquely—*accelerated*, in a stone falling to the earth—*retarded*, in a stone thrown upwards while rising to the point where it stops before again descending.

" *Owing to the* INERTIA *of bodies, force is equally required to impart motion and to take it away.*" (Read again the last Analysis.)

If a man put his hand to the crank of a heavy fly-wheel or grindstone, to turn it, he experiences a certain resistance, which, however, gradually yields to his effort, and he leaves the wheel whirling with velocity proportioned to the effort. If he then puts out his hand again to stop the wheel, he experiences an opposite but similar resistance, which however, as before, gradually yields, and he brings the wheel to rest. In the second case the effort required of him is less than the first, by reason of the friction of the turning axle, and the resistance of the air in which the wheel moves,—obstructions which, when he was giving motion, opposed him, but when taking it away assisted him. That these obstructions caused the whole difference in such a case, and that they are the great reasons why all ordinary motions on earth seem to tend of themselves to cease, will be shown in subsequent pages. It is the resistance overcome in moving the wheel or in stopping it, and occasioning an expenditure of force proportioned to the mass and to the degree of change of state, which is called the INERTIA of the mass, or the *vis inertiæ*, and sometimes, to help the conception of the student, the *stubbornness*, *sluggishness*, or *inactivity;* but none of these words can originally suggest to the mind all that is intended to be conveyed.

An exact measure of the amount of inertia is contained in the familiar fact that a body let fall near the surface of the earth, falls rather more than 16 feet in the first second of time,—the well-known weight of the body, or force of terrestrial attraction acting upon it for one second, being just sufficient to overcome its inertia to the extent stated. Were the inertia of matter only half of what it is, a body near the earth would fall 32 feet in the second, instead of 16, as it equally would, if, with present inertia, the attraction of the earth were doubled. And were there no inertia, it would fall or pass through any height, however great, in one instant. As the amount of inertia thus determines the amount of other force required to give motion to a mass, so does it determine the amount of force required to destroy motion in a mass. A heavy cannon-ball, if wanting inertia, might be dispatched with the speed of lightning by the slightest force, but then the stiffness of a stalk of corn would suffice to arrest it; and while the ball, with the inertia now existing, takes the force of pounds of gunpowder to give it its usual motion, it may not be stopped, even by the cohesion of a block of granite, which accordingly it shivers to pieces. The numerous examples now to follow will prove the immense importance of *inertia* in the general operations of nature.

When the sails of a ship are first spread to receive the force or impulse of the wind, the vessel does not acquire her full speed at once, but slowly, as the continuing force gradually overcomes the inertia of her mass. When the sails are afterwards taken in, she does not loose her motion at once, but slowly again, as the continued resisting force of the water destroys it.

Horses must make a greater effort at first to put a carriage in motion than to maintain the motion afterwards. And a strong effort is required to stop a moving carriage. When a carriage, of which the body hangs from springs, is first moved, the body appears to fall back, and a person within seems to be suddenly forced against the back cushion. When the carriage is stopped again the body swings forward, and if the stoppage be very sudden, a careless passenger may unwittingly pop his head through a front glass. These particulars prove the inertia, first of rest, and secondly of motion.

A man standing carelessly at the stern of a boat, when the boat begins to move, falls into the water behind; because his feet are pulled forward while the inertia of his body keeps it where it was, and therefore behind its support. The stopping of a boat, again, illustrates the opposite inertia of motion, by the man's falling forward.

An awkward rider on horsback may be left behind, when his horse starts forward suddenly: or may be thrown off on one side by the horse starting to the other. A horse at speed, stopping suddenly, often sends his cavalier over his ears—as was mortifyingly experienced by a coxcomb, who, on an old cavalry horse, chose to canter along a foot-path, to the annoyance of the company, and whose horse on hearing the word *halt* loudly addressed to it by a waggish officer of the regiment, who happened to be there and to recognize it, suddenly stood and got rid of its load. The mind or will of the beau had sinned against the law of propriety, but his body very perfectly obeyed the laws of inertia and gravity, by shooting forward in a parabolic curve to the earth.

A young man not yet accustomed to the whip, drove his phaeton against a heavy coach on the road, and then to his father foolishly excused his awkwardness, in a way which led to the prosecution of the coachman for furious driving. At the trial, the youth and the servant both deposed that the shock of the coach was such as to throw them over their horses' heads, and thus lost the case, by unconsciously proving, that the faulty velocity was their own.

A man jumping from a carriage at speed is in great danger of falling forward, when his feet reach the ground; for his body has as much forward velocity as if he had been running with the speed of the carriage and unless he advance his feet like a running man, to support his advancing body, he must as certainly be dashed to the ground, as a runner whose feet are suddenly arrested. A man racing who receives a signal to stop, and a man jumping from a flying vehicle, must check their motion nearly in the same way.

A person wishing to leap over a ditch or chasm, first makes a run, that the motion thereby acquired may help him over. A standing leap falls much short of a running one.

An African traveller saw himself pursued by a tiger, from which he could not escape by running; but perceiving that the animal was watching an opportunity to seize him by its usual spring or leap, he artfully led it to where the plain terminated in a precipice hidden by brush-wood, and he had just time to transfer his hat and cloak to a bush, and to retreat a few paces when the tiger sprung upon the bush, and by the moral inertia of its body, was carried over the precipice and destroyed.

From a glass of water suddenly pushed forward on the table, the water is spilt or left behind; but if the glass be already in motion, as when carried by a person walking, and if it then be suddenly stopped by coming against an impediment, the water is thrown or spilt forward.

A servant carrying a tray of glasses or china in the dark, and coming suddenly against an obstacle, hears all his freight slipping forward and crashing at his feet: and a too hurried departure with such a load causes equal destruction, on the opposite side.

The actions of beating a coat or carpet with a cane, to expel the dust; of shaking the snow from one's shoes, by kicking against a door-post; of cleaning a dusty book by knocking it against a table, or shutting it violently—all illustrate the same principle.

If a guinea be laid on a card which is already balanced on the point of the finger, a small fillip or blow to the edge of the card will cause it to dart off, but the guinea, owing to its inertia, will remain resting on the finger,—its inertia being greater than the friction on it of the card passing from underneath it.

When we desire a person, with suspected disease of the brain, to shake his head and tell whether and where he feels pain, we are doing nearly as if we touched the naked brain with the finger to find the tender part; for the inertia of the brain, when the skull is moved, causes a momentary pressure between it and the skull, almost equivalent, for the purpose desired, to such a touch.

This kind of pressure is sufficient to break and destroy tender wares—as glass or eggs—in packages which are too suddenly moved or stopped.

A weight suspended by a spring on ship-board is seen vibrating up and down as the ship pitches with the waves. It seems to fall as the ship rises, and to rise as the ship falls: but the motion is really in the ship, and the comparative rest is in the weight. A heavy weight so supported, and connected with a pump-rod, would work the pump.

Like the weight last mentioned, the mercury of a common barometer on ship-board is seen rising and falling in the tube; and until the important improvement was lately made, of narrowing one part of the tube to prevent this, the mercurial Barometer was useless at sea. The explanation is, that the tube rises and falls with the ship, from being connected with it; but the mercury, which plays freely in the tube, and is supported by the atmospheric pressure, tends, by its inertia, to remain at rest, and thus makes the motion of the ship apparent.

What happens to the mercury in the barometer-tube on ship-board, indicates what happens to the blood in the vessels of animals under similar circumstances. In any long vein below the heart, when the body falls, the blood, by its inertia and the supporting action of the vessels, does not fall so fast, and therefore really rises in the vein: and as there are valves in the veins preventing return, the circulation is thus quickened without any muscular exhaustion on the part of the individual. This helps to explain the effect of the movement of carriages, of vessels at sea, of swings, &c., and of passive exercise generally, on the circulation, and leaves it less a mystery why these means are often so useful in certain states of weak health.

If a cannon ball were to break to pieces in its flight, its parts would still advance with the previous velocity. And thus, in the deadly contrivance of the Shrapnell-shell, which is in a case containing hundreds of musket bullets, when these are scattered at the desired distance from the devoted body of men, they retain the forward velocity of the shell, and spread death around like the near discharge of a whole battalion of musketry.

On the awful occasion of a ship in rapid motion being suddenly arrested by a sunken rock, all things on board, men, guns, and furniture, start from their places and dash forwards; while the onward inertia or moral obstinacy of the hinder parts of the ship, suffices to crush her bow against the rock.

"Motion as naturally permanent as rest."

From the instances now given, it is seen that a body at rest would never move if force were not applied, and that a body put in motion retains motion, at least for a time, after the force has ceased; but there is a feeling from common experience, that motion is an unnatural or forced state of bodies, and that all moving things, if left to themselves, would gradually come to rest. It is recollected that a stone projected comes to rest, or a wheel left moving, or a bowl rolled on the green, or the waves heaving after a storm—and in a word, that there is no perpetual motion on the earth.

On more attentive consideration, however, it may be perceived that there are prodigious differences in the duration of motions, and that the differences are always exactly proportioned to evident causes of retardation, and chiefly to *friction* and the *resistance of the air.*

Friction is the resistance which bodies experience when rubbing or sliding upon each other; and however much it may be diminished by art, it can in no case be annihilated. Air-resistance, again, to motions going on in air, is of the same nature as water-resistance to motions going on in water, only less in degree: and as advancing science has shown the true nature of our atmosphere, the amount of this resistance is perfectly ascertained.

A smooth ball rolled on the grass soon stops—if rolled on a green cloth over a smooth plank it goes longer—on the bare plank, longer still—on a smooth and level sheet of ice, it hardly suffers retardation from friction, and, if the air be moving with it, will reach a distant shore.

Two little wind-mill wheels set in motion together with equal velocity, but of which one has the flat sides of the vanes turned to their course, and the other the edges, if moving in the air, will stop at very different times, but if tried in a vessel from which the air has been removed, they will both go much longer, and will then stop exactly together.

As it is to facilitate the motion of fishes in the water, that they are of sharp form before and behind; so it is to facilitate the motion of birds in the air that they have somewhat of a similar form.

A large spinning-top, with a fine hard point, set in motion in a vacuum, and on a hard, smooth surface, will continue turning for hours.

A pendulum moving in a vacuum has only to overcome slight friction at its point of suspension, and, therefore, if once put in motion, will vibrate for a day or more.

But it is in the celestial spaces that we see motions completely freed from the obstacles of air and friction—and there they seem eternal.

Had the human eye, unassisted, been able to descry the four beautiful moons of Jupiter, wheeling around him for these thousands of years, with such unabated regularity, and which now form, to the telescope of the astronomer, a perfect and magnificent time-piece in the sky, or had science long proved that the velocity imparted to our globe, when first launched into its present orbit, still wheels it along as swiftly as in the days of the first man, this error or prejudice, that motion is always tending to rest, would never have arisen.

Indeed, had these or other such truths, been long familiar to the common mind, the opposite prejudice might as well have obtained, that motion is the natural state, and rest a forced or unknown state. We know of nothing which is absolutely at rest. The earth is whirling round its axis and round the sun; the sun is moving round its axis and round the centre of gravity of the solar system, and possibly, round some more remote centre in the great universe, carrying all its planets and comets about his path.

If there were any natural tendency in moving bodies to stop, a thing floating in a trough of water, on board a sailing ship, should always be found at the end of the trough nearest the stern; and in all the seas and lakes of the earth, the floating things should be accumulated on the western shores, because the surface of the earth is always turning towards the east. We know that neither of these suppositions is truth. A man on board a moving ship can throw any body just as far towards the bow as towards the stern; although in the two cases the velocity, as regards the earth is so different.

Ignorance of the law of moral inertia led a story-telling sailor to assert, as a proof of the speed of his favourite ship, that when a man one day fell from the mast-head, the ship had passed from under him before he reached the deck: the fact in such a case, being, that he must have fallen on the same part of the deck, whether the ship were in motion or at rest, because his body had just the motion or rest which belonged to the ship.

Another equally sapient man, reflecting that the earth turned round once in twenty-four hours, proposed rising in a balloon, and waiting aloft, until the country which he desired to reach should be passing under him.

"*Motion naturally uniform.*" (See the Analysis.)

It is only repeating that a body can neither acquire motion nor lose motion without a cause, to say that free motion must be uniform.

The perfect uniformity of undisturbed motion is proved by every fact observed in the universe. If any continued motion, as of a planet, for instance, be found at one time to have certain relative velocity to some other continued motion, the same relation is found always to hold: or deviations from perfect uniformity are exactly proportioned to the disturbing causes. Thus we can foretell the exact time of an eclipse, a thousand years before its occurrence.

Had motion not been in its nature uniform, a man could have formed no rational conjecture or anticipation as to future events; for it is by assuming, for instance, that the earth will continue to turn uniformly on its axis, that he speaks of *to-morrow* and of *next week*, &c., and that he makes all his arrangements for future emergencies: and were the coming day, or season, or year, to arrive sooner or later than such anticipation, it would throw such confusion in all his affairs, that the world would soon be desolate.

To calculate futurities, then, or to speak of past events, is merely to take some great uniform motion as a standard with which to compare all others; and then to say of the remote event, that it coincided or will coincide with some described state of the standard motion. The most obvious and best standards are the whirling of the earth about its axis, and its great revolution round the sun. The first is rendered very sensible to man by his alternately seeing and not seeing the sun, and it is called *a day;* the second is marked by the succession of the seasons, and it is called *a year.* The earth turns upon its axes nearly 365 times while it is performing one circuit round the sun, and thus divides the year into so many smaller parts, and the day is divided into smaller parts, by the progress of the earth's whirling being so distinctly marked, in the constantly varying direction of the sun, as viewed from any given spot on the face of the earth. When advancing civilization made it of importance for men to be able to ascertain with precision the very instant of the earth's revolution, connected with any event, various contrivances were introduced for the purpose; as,—sun-dials, where the shadow travels progressively round the divided circle;—the uniform flux of water through a prepared opening—the flux of sand in the common hour-glass, &c. But the great triumphs of modern ingenuity are those astronomical clocks and watches, in which the counted equal vibrations of the pendulum, or balance-wheel, have detected periodical inequalities even in the motion of the earth itself, and have directed attention to unsuspected disturbing causes, important to be known.

It is the natural uniformity of undisturbed motion which causes any number of bodies moving together, as the furniture of a sailing ship, to appear among themselves as if at rest,—no one tending to pass before, or to fall behind, or to move to one side or another. For the same reason a person who is moving with such bodies is absolutely insensible of his uniform progression, and knows it only by reasoning from such facts as the changing appearance of other objects around which do not share the motion, the rushing of the waves or wind, &c. When a ship is becalmed at sea, she may, as numberless sad accidents have proved, be carried by rapid currents in any direction, without one of the crew suspecting that she has motion at all; and if the suspicion do arise, the truth can be come at only by such means as the sounding line, where the bottom can be reached, or careful observation of the heavenly bodies where it cannot. A man in the hold of a ship in a river or tides-way cannot say whether the rushing of water, which he hears from without, be a rapid tide passing the ship at anchor, or the effect of the ship's advance in the river. A man in a balloon going 80 miles an hour, knows not in what direction he is moving, nor indeed that he is moving at all, but by observing the objects below.

This explains why men are not sensible of the motion of the earth itself, which they know, however, to be turning around its axis once in twenty-four hours, and therefore to have its surface near the equator moving with a speed of more than 1,000 feet per second; and as in the case of a ship or balloon, there will be no difference of sensation whether the speed were of one mile per hour or of 10 or 100, so in the case of the earth, there would be none whether it turned as now, once in twenty-four hours; or, like the planet Jupiter, once in ten. A hunter among the hills, who during the heat of noon, rests and contemplates around him a sublime scene of solitude and silence, may little think that if, amidst that apparent repose of nature, he were for a moment lifted up from the earth and held *at rest* above its surface, he would see its face of hill and dale sweeping past beneath him at the prodigious rate of 1,000 miles an hour, on account solely of the whirling of the earth.

The fact that a cannon-ball can be shot just as far upon the surface of the earth, eastward, in the direction of the earth's motion, as westward, against it, illustrates the truth, that whatever *common* motion objects may have, it does not interfere with the effect of a force producing any new relative motion among them. All the motions seen on earth are really only slight differences among the common motions: as in a fleet of sailing ships, the apparent changes of place among them are in reality only slight alterations of speed or direction, in their individual courses.

A man continuing to throw upwards a ball or orange, or several of them at once, and to catch and return them alternately, uses no difference of art as regards them, whether he be standing on the earth and whirling with it, or on a sailing ship's deck, or in a moving carriage, or on a galloping horse's back. He and the oranges have always the same forward common motion. And when a man, standing on a galloping horse, leaps through a hoop held across his course, he does not leap forward—for this would throw him over the horse's ears—but merely jumps up and allows his moral inertia to carry him through.

The reason why a lofty spire or obelisk stands more securely on the earth, than even a short pillar stands on the bottom of a moving wagon, is, not that the earth is more at rest than the wagon, but that its motion is uniform.—Were the present rotation of our globe to be arrested but for a moment, imperial London, with its thousand spires and turrets, would, by the moral inertia, be swept from its valley towards the eastern ocean, just as loose snow is swept away by a gust of wind.

"*Force is required to bend motion.*"

If a body moving freely cannot vary its velocity without a cause, neither can it vary its course without a cause; and free motion, therefore, is *straight* as well as *uniform*.

A ball shot directly up or down gives men their simplest idea of straight motion.

A bullet or arrow, projected horizontally, is gradually drawn downwards by the attraction of the earth, but it deviates neither to the right nor to the left.

William Tell, trusting to the natural straightness of motion, obeyed the tyrant's order, and shot an apple placed on his child's head.

And the right eye of Philip of Macedon is said to have been destroyed by an arrow which brought a label on it, telling its destination.

Riflemen shooting at a target, hit the very spot they choose to aim at.

A stone in a sling, the moment it is set at liberty, darts off as straightly as an arrow from the bow-string or a bullet from a gun-barrel, and it is only because the point of its circle, from which it should depart, cannot in practice be accurately determined, that the same sure aim cannot be taken with it.

A body moving in a circle, then, or curve, is constrained to do what is contrary to its inertia. A person on first approaching this subject, might suppose that a body, which for a time has been constrained to move in a circle, should naturally continue to do so when set at liberty. But on reflecting that a circle is as if made up of an infinite number of little straight lines, and that the body moving in it has its motion bent at every step of the progress, the reason is seen why constant force becomes necessary to keep it there, and force just equal to the inertia with which the body tends, at every point of the circle, rather to pursue the straight line, called a tangent, of which that point, as seen in Fig. 2, is the commencement, than the circle itself. The force required to keep the body in the bent course, is called *centripetal* or centre-seeking force; while the inertia of the body tending outwards, that is, to move in a straight line rather then in a curve, is called the *centrifugal* or centre-flying force; and the term *central forces* is applied to both.

Fig. 2.

A sling-cord is always tight while the cord is whirling: and its tension is of course the measure both of the centripetal and centrifugal force. A means, then, of measuring the tension of a sling-cord would experimentally demonstrate the amount of centrifugal force; and such a means we possess in the contrivance called the "whirling table," upon which is a leading sling, or any mass with a string attached to it, may be placed to revolve, at any desired distance from the centre, and with any desired velocity, while the string passing over a pulley at the centre, is made to lift weights proportioned to the outward dragging of the revolving mass. By this apparatus it is found, as would be expected, that centrifugal force—in other words the force with which the inertia of moving matter resists the bending of its course from straight to circular, is proportioned, first, to the quantity of matter moved—every separate particle having its own inertia; second, to the size of the circle or orbit described in the same time—a body moving in a circle of double diameter for instance, having to be forced inwards from the tangent, at every departure, twice as far in a given time; third, that with a double revolution in the same time, the centrifugal force is not double but quadruple (a corresponding proportion existing for other velocities,) because, not only are there twice as many bendings or angular departures from the tangent for the two circles as for one, requiring, as may be said, twice as many tugs or impulses of the centripetal force, but every impulse must be made with double energy, for it has to drive the mass inwards through the required distance in half the time; and twice as many impulses, every one being twice as strong, make a quadruple amount of force on the whole; fourthly and lastly, it is found, agreeing with the relation between inertia and terrestrial gravity described at page 43, that a body revolving, for instance, in a circle of four feet diameter, that it may have centrifugal force just equal to its weight, required to complete its revolution in one second and a half of time. This and similar facts will be more particularly considered when we come to treat of the motions of the planets round the sun. This analysis of central forces will suffice to excite in the student a due interest touching the kindred phenomena now to be described.

Bodies laid on a whirling horizontal wheel, are readily thrown off.

In a corn-mill, the grain, after being admitted between the stones through an opening in the centre of the upper stone, is then kept turning round between them, and is, by its centrifugal force, always tending and travelling outwards until it escapes as flour from the circumference.

A man, if he lie down on a turning millstone with his head near the edge, falls asleep, or dies of apoplexy, from the new pressure of blood on the brain.

A wet mop, or bottle-brush, made to turn quickly on its handle as an axis, throws the water off in all directions, and soon dries itself.

Sheep, in wet weather, thus discharge the water from their fleeces, by a semi-rotatory shake of the skin. Water-dogs, on coming to land, dry themselves by the same action.

A tumbler of water placed in a sling, may be made to vibrate like a pendulum with gradually increasing oscillation, and at last to describe the whole circle, and continue revolving about the hand, without spilling a drop:—the water, by its inertia of straightness, or centrifugal force, tending more away from the centre of motion towards the bottom of the tumbler, even when that is uppermost, than towards the earth by gravity.

As solid bodies laid on a whirling table are thrown off, so water in a vessel caused to spin round in any way, as on the centre of a horizontal wheel, instead of lying at the bottom, is raised up all round, against the sides of the vessel.

Water, poured obliquely into a funnel, runs round the interior of it, and often leaves an open passage of air all the way down through it, as if there were merely a lining of water to the funnel. The centrifugal force of the turning water is a chief reason of this phenomenon: another reason will be considered farther on, under the head of atmospheric pressure.

Great whirlpools at sea, and smaller ones, or eddies in rivers, occur whenever a current is obliged suddenly to bend, as in rounding a point of land or a rock, or in meeting and mingling with a contrary current. The water, by tending to continue its straight motion, falls in behind the obstruction, reluctantly as it were, and leaves there a pit surrounded by a liquid revolving ridge. Charybdis, in the Mediterranean, and the great whirlpool off the Norwegian coast, are noted examples.

It is owing to the centrifugal force in any bending part of a stream of water, that is to say, the tendency away from the centre of the curvature, that when a bend has once commenced, it increases, and is soon followed by others, until that complete serpentine winding is produced, which characterizes most rivers in their course across extended plains. The water being thrown by any cause to the left side, for instance, wears that into a curve or elbow, and, by its centrifugal force, acts constantly on the outside of the bend, until rock or higher land resists the gradual progress; from this limit being thrown back again, it wears a similar bend to the right hand, and after that, another to the left, and so on.

Carriages are often overturned in quickly rounding corners. The inertia carries the body of the vehicle in the former direction, while the wheels are suddenly pulled round by the horses into a new one. A loaded stage-coach running south, and turning suddenly to the east or west, strews its passengers on the south side of the road. Where a sharp turning in a carriage-road is unavoidable, the road towards the outside of the bend should always be made higher than at the inside, to prevent such accidents.

A man or a horse turning a corner at speed, leans much inward, or towards the corner, to counteract the centrifugal force, that would throw him away from it.

In skating with great velocity, this leaning inwards at the turnings becomes very remarkable, and gives occasion to the fine variety of attitudes displayed by the expert; and if a skater, in running, finds his body inclined to one side and in danger of falling, he merely makes his skate describe a slight curve towards that side, when the tendency of his body to move straightly, or its centrifugal force, refusing to follow in the curve, allows the foot to push itself again under the body, and to restore the perpendicularity. Skating becomes to the intelligent man an intellectual as well as a sensitive or bodily treat, from its exemplifying so pleasingly the law of motion.

The last example explains, also, why a hoop rolled along the ground goes so long without falling: if it incline to one side, threatening to fall, by that very circumstance, the part touching the ground is made to bend its course to that side, and as in the case of the skater who turns his foot, the supporting base is again forced directly under the mass of the body.

A coin dropped on the table or floor often exhibits the same phenomenon. It is said to run and hide itself in the corner. Just before falling, if not obstructed, it describes several turns of a decreasing spiral, the minute examination of which is a pleasing mathematical exercise.

The reason also why a spinning top stands, will be understood here.

While the top is quite upright, the extremity of its peg, being directly under its centre, supports it steadily, and although turning so rapidly, and with much friction, has no tendency to move from the place: but if the top incline at all, the *edge* or *side* of the peg, instead of its very *point*, is in contact with the floor, and the peg then becoming as a turning little roller, advances quickly, and describes a curve somewhat as a skater's foot does, until it come directly under the body of the top as before. It thus appears that the very fact of the top inclining, causes the point to shift its place, and to continue moving until it comes again directly under the centre of the top. It is remarkable that even in philosophical treatises of authority the standing of a top is still vaguely attributed to *centrifugal force*. And some persons believe that a top spinning in a weighing scale, would be found lighter than when at rest; and others most erroneously hold that the centrifugal force of the whirling, which of course acts directly away from the axis, and quite equally in all directions, yet becomes, when the top inclines, greater upwards than downwards, so as to counteract the gravity of the top. The way in which centrifugal force really helps to maintain the spinning of a top is, that when the body inclines or begins to fall in one direction, its motion in that direction continues until the point describing its curve, like the foot of a skater, has forced itself under the body again.

By reason of centrifugal force also, it is easier to do feats of horsemanship in a small ring as at our theatres, than if the animal were running on a straight road. We see the man and the horse always inclining inwards to counteract centrifugal force, and if the rider tend to fall inwards, he has merely to quicken the pace; if to fall outwards, he has to slacken it, and all is right again.

If a pair of common fire-tongs, suspended by a cord from the top, be made to turn by the twisting or untwisting of the cord, the legs will separate from each other with force dependent on the speed of rotation, and will again collapse when the turning ceases. Mr. Watt adapted this fact most ingeniously to the regulation of the speed of his steam-engine. His *steam-governor* may in truth be described as a pair of tongs with heavy balls at the ends, to make their opening more energetic, attached to some turning part of the machine. If the engine move with more than the assigned speed the balls open or fly asunder beyond their middle station, and by a simple contrivance are then made to act on a valve which contracts the steam tube; on the contrary, with too slow a motion, they collapse and open the valve.

A half-formed vessel of soft clay, placed in the centre of the potter's table,—which is made to whirl and is called his wheel,—opens out or widens merely by the force of its sides and thus assists the worker in giving its form.

A ball of soft clay, with a spindle fixed through its centre, if made to turn quickly, soon ceases to be a perfect ball. It bulges out in the middle, where the centrifugal force is great, and becomes flattened towards the ends, or where the spindle issues.

This change of form is exactly what has happened to the ball of our earth. It has bulged out seventeen miles at the equator, in consequence of its daily rotation, and is flattened at the poles in a corresponding degree.—A mass of lead that weighs one thousand pounds at our pole, weighs about five pounds less at the equator, by reason of the centrifugal force.

In the planets Jupiter and Saturn, of which the rotation is much quicker than of our earth, the middle or equator bulges out still more—even so as to offend an eye which expects a perfect sphere.

If the rotation of our earth were seventeen times faster than it is, the bodies or matter at the equator would have centrifugal force equal to their gravity, and a little more velocity would cause them to fly off altogether, or to rise and form a ring around the earth like that which surrounds Saturn. Saturn's double ring seems to have been formed in this way, and is now supported chiefly by the centrifugal force of the parts. Were it to crumble to pieces, the pieces might still revolve, as so many little satellites. His true satellites are only more distant masses sustained in the same manner. And our earth and the other primary planets have the same relation to the sun that these satellites have to Saturn—all being sustained by an admirable balance between centrifugal force and gravity.

"*The quantity of Motion in a body measured by the velocity and quantity of matter.*

If a single atom of matter were moving at the rate of one foot per second it would have a definite quantity of motion expressed by these words; and if it were moving ten feet per second, it would have ten times the quantity. Again, in a mass consisting of many atoms, the quantity of motion would be still as much greater as there were more atoms in it than one.

By experiment it is found, that if a ball of soft clay of one pound, suspended by a cord as a pendulum, be allowed to fall with a velocity of ten feet per second, against a ball of nine pounds suspended in the same way, but at rest, the two, after contact, will start together at the rate of one foot per second, the original quantity of motion being then diffused through ten times the quantity of matter, and therefore exhibiting only one tenth of the velocity.

A cannon ball of a thousand ounces, moving one foot per second, has thus the same quantity of motion in it as a musket-ball of one ounce, leaving the gun-barrel with a velocity of a thousand feet in the second.

"*The quantity of motion in a body is the measure of the force which produced it.*"

• The experiment of the balls of clay mentioned above furnishes one instance of this truth. Again, a body falling for ten seconds, acquires ten times as much velocity as by falling for one second; its motion thus measuring the force of gravity which has been exerted upon it.

When a large body or mass of many atoms falls, it of course has as much more motion than a smaller body, as there are more atoms in it than in the smaller; but as gravity acts equally on every atom, the force causing either body to fall is still exactly indicated by the quantity of motion in it.

A large body or mass of many atoms falls where there is no impediment, with the same velocity as a smaller body or a single atom; for gravity pulls equally at each atom, and must overcome its inertia equally, whether it be alone or with others.

This remark contradicts the popular opinion, that a large and heavy body should fall to the earth much faster than a small and light one; an opinion which has arisen from our constantly seeing such contrasts, as the rapid fall of a gold coin, and the slow descent of a feather. The true cause of the contrast is, that the atoms of the feather are much spread out, so as to be more resisted by the air than those of the gold. If the two be let fall together in a vessel from which the air has been extracted—as in the common air-pump experiment, they arrive at the bottom in exactly the same time;

and even in the air, if the coin be hammered out into gold leaf, it will fall still more slowly than the feather. One brick dropped from a height, because its motion is not much affected by the air, reaches the earth very nearly as soon as ten bricks let fall near it, whether they be connected or separate—as a single horse may reach the goal as soon as ten horses galloping abreast.

A man's force will move a small skiff quickly, a loaded barge very slowly, and a large ship in a degree scarcely to be perceived. In each case, however, the quantity of motion may be the same, and a true measure of the force which produced it.

A ball of one pound weight, impelled by a given force, moves twice as fast as a ball of two pounds impelled with the same; yet, although the velocities are different, the quantities of motion, as ascertained by the rule already given, are equal, and indicate an equality of producing force.

"*The quantity of motion in a body is the measure also of the force or momentum which it can exhibit again.*" (See the Analysis, 42.)

Bodies, owing to their inertia, may be regarded as passive reservoirs of force or motion, always ready to return as much as they have received. *Momentum* is the name given to the motion in a body; with reference to the production by it of new motions or the overcoming of resistances, and is but another term for the *quantity of motion*

A cannon ball, according to the quantity of motion in it, may have only the force or momentum that will bruise a plank, or it may have enough to penetrate a tree, or even to shoot its rapid way through a block of the hardest stone.

A block of wood floating against a man's leg with moderate velocity, would be little felt; but a loaded barge, coming at the same rate and pressing it against the quay, might break the bones; a large ship, again, although moving no faster, would crush his body against any fixed obstacle; and an island of ice, opposed in its approach to another, even by a first-rate man-of-war, would destroy it, as meeting barges destroy a floating egg-shell.

A hail-stone falling, strikes rudely; a stone rolled from a height, as of old, by the besieged against besiegers, may carry death with it to many; an avalanche, breaking from its hold on a mountain steep, may sweep away a village.

To meeting bodies, the shock is the same, whether the motion be shared between them or be all in one.

If a running man come against a man who is standing, both receive a certain shock. If both be running at the same rate in opposite directions, the shock is doubled. In some such cases, as where swift skaters have met, the shock has proved fatal.

The meeting fists of boxers not unfrequently dislocate or break bones.

A man's skull is fractured as certainly by its being dashed against a tree or beam, while he is on a galloping horse, as by a blow of a similar beam coming upon him with the velocity of the horse.

When two ships in opposite courses meet at sea, although each may be sailing at a moderate rate, the destruction is often as complete to both as if with a double velocity they had struck on a rock. Many melancholy instances of this kind are on record. In the darkness of night a large ship has met one smaller and weaker, and in the lapse of a few seconds, have followed the shock of the encounter, the scream of the surprised victims, and the horrible silence when the waves had again closed over them and their vessel for ever.—In November, 1825, on the coast of Scotland, the Comet steamboat was thus destroyed, and carried to the bottom with her about seventy passengers, into whose ears the drowning water rushed before the sound of arrested music and joy had died away.

"Direction of the force or forces producing motion."

When only one force acts on a body, the body obeys in the exact direction of the force.

A ball floating in water, or lying on smooth ice, is driven exactly south by a wind blowing to the south. A bullet issues from the mouth of a cannon, in the direction of the axis of the cannon—which is, as the force impels it.

When two or more forces, not in the same direction, act upon a body at the same time, as it cannot move two ways at once, it holds a middle course between the directions. This course is called the *resulting direction*, viz., resulting from the *composition of the forces.*

A ball or ship moving south by a direct wind, may, at the same time, be carried east, just as fast, by a tide or current moving east; every instant, therefore, it will go a little *south* and a little *east*, and really will describe a middle line pointing *south-east.*

These particulars may be well represented on paper, as by fig. 3: where *b* is the original place of the ball or ship, *e* the east, *s* the south, and *b a* the middle line pointing to the south-east, and showing the true course of the vessel. This figure is called the *parallelogram of forces*, and is an important help to the understanding of many facts in natural philosophy. The minute investigation of the subject belongs to the *science of measures*, or technical mathematics; but the general truths are quite intelligble to common sense, or the mathematics of common experience.

Fig. 3.

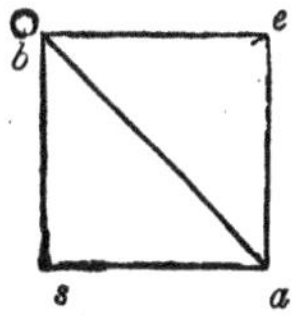

When two forces act upon a body, like the wind and tide in the last example, the result is the same, whether they act together or one after the other. For instance, if the wind drive a vessel one mile south, as from *b* to *s*, fig. 3, and immediately afterwards the tide drive it one mile east, as *s* to *a*, the vessel will be in the same place at last, viz., at *a*, as if she had been driven at once south-east, in the line *b a*, by the simultaneous action of the two. Therefore, by drawing the lines *b s* and *b e* to represent the force and direction of the two causes of motion, and by then adding one of them, or an equivalent, to the end of the other as *s a* to *b s*, or *e a* to *b e*, the square or parallelogram is sketched, of which the middle line or *diagonal*, as it is called, shows the *resultant* of the forces, and the true course of the body obeying them.

What is thus true of the effect of continued forces like wind and tide is

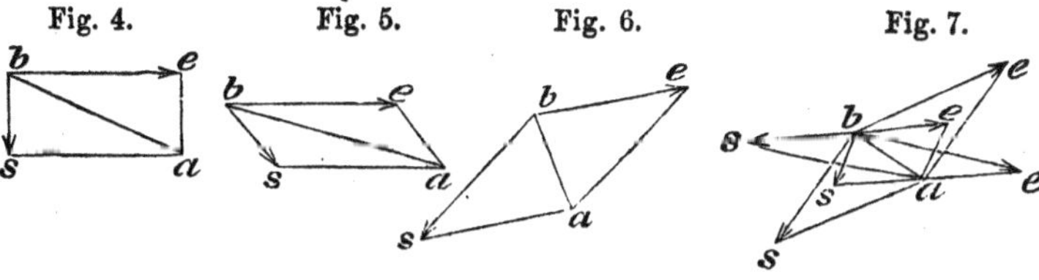

Fig. 4. Fig. 5. Fig. 6. Fig. 7.

true also of momentary impulses, like the blows of clubs simultaneously striking a ball, or of two billiard-balls striking a third.

When the forces exactly cross each other, and are equal, as in the case of the ship above supposed, the figure becomes a square, at fig. 3; but if one of the forces be greater than the other, the figure becomes oblong, as at fig. 4; if the forces cross obliquely, the fig. becomes as at fig. 5; and if they cross in an opposing direction, it will be as at fig. 6. In all the cases, however, the diagonal still shows the *result*. It is evident that the same line may be the diagonal of many figures, as seen in *b a* at fig. 7; and therefore, that very different degrees and directions of combined forces may produce the same *result*.

Forces crossing each other so obliquely as to be represented by lines drawn in almost opposite directions, would form a parallelogram having scarcely any breath, that is to say, the diagonal would approach to nothing; showing thus, that opposing forces neutralize or destroy each other. In fig. 6, by reason of this crossing, the *resultant* is less than either of the constituents. And for the same reason, when forces cross so acutely as to advance nearly parallel to each other, the *resultant* is longer than either, as seen in fig. 5. Forces directly opposed, or entirely agreeing in direction, give as their resultant their difference or their sum.

Forces crossing each other directly, or at right angles, as is true of the exactly eastward force *b e*, and the exactly southward force *b s*, in figures 3 and 4,—do not in the slightest degree neutralize or alter each other, for the body, when arrived at *a*, is just as far east as it would be at *e*, and as far south as it would be at *s*. This explains why the progressive motion of the planets in their orbits is not at all affected by the directly crossing centripetal force of gravity which keeps them at their due distances from the sun.

In all cases where the two crossing forces are equal, with whatever obliquity they cross, the resulting direction must be midway between them.—Thus a boat impelled by oars, goes straight, although the direction in which the oar acts is constantly changing; because the changing obliquity of the force is always the same on both sides.—This explains also why a bird flying, or a man swimming, holds a perfectly straight course, although in both cases the direction of the impelling forces is constantly varying.—And it explains why a body suspended, as a plummet, or falling to the earth as an apple does from a tree, is always in a line towards the centre of the earth: for, while the part of the earth immediately under the body is pulling it straight down to the centre, the action of parts on any one side of the perpendicular is exactly counterbalanced by the action of corresponding parts on the opposite side; and the perpendicular is still the diagonal or middle line of every pair of attracting parts. In fig. 8, *b a* represents the common diagonal. In speaking of the attraction of our earth, therefore, which really is the united attraction of all the individual atoms, we may always consider it as a single force acting towards the centre of the earth.

Fig. 8.

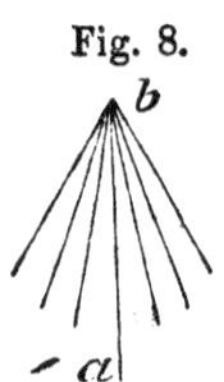

When a body is carried below the surface of the earth, its weight becomes less, because the matter then above it is drawing it up, instead of down, as before. A descent of a few hundred feet makes a sensible difference, and at the centre of the earth, if man could reach it, he would find things to have no weight at all; and there would be neither up nor down, because bodies would be attracted equally in all directions.

When more than two forces act on a body, the resulting direction may be found, first of two, and then of the last *resultant* with each of the others successively :—or the forces may be represented on paper by lines tacked together, of which one denotes the strength and direction of each : the extremity of the last line will mark the place of the body after being acted upon by the combined forces. A sailor, to know the true place of his ship and the course which she has steered, considers, first, the forward progress as found by the log, then the leeway or sideward motion produced by a cross wind, and then the effect of any tide or current in which he may be sailing.

Resolution of Forces is a phrase pointing to another important use of such parallelograms or figures as have just been described, *viz.*, the enabling us when force or motion is given, to find the forces or motions in any other directions of which it may be the *resultant*, and those into which it may itself be resolved.

Thus, if a line *b a* (in any of the preceding figures 4, 5, 6, &c.) represent a force or motion, and the line *b s* represent one of two elements composing it, we have but to complete the parallelogram *b s a e* to obtain the other line, *b e* representing the only other force or motion which, combined with the first element, can produce the given resultant.—If a ship pass from *b* to *a* (fig. 5) while sailing through the water eastward, a distance expressed by *b e*, she must at the same time have been carried by a tide current to the distance and in the direction marked by the line *b s*.

Fig. 9.

Again, if a line be given representing a single force or motion as *b a*, and if it be desired to know how much there is in this capable of acting in another direction as *b d;* it is only necessary to draw a line in the direction, as *b d*, from the commencement of *b a*, and to cut such line by another drawn directly upon it—or at right angles to it, as the term is, from the other end of *b a:* the length of *b d*, so cut off, *viz.*, *b s*, shows the proportion required.

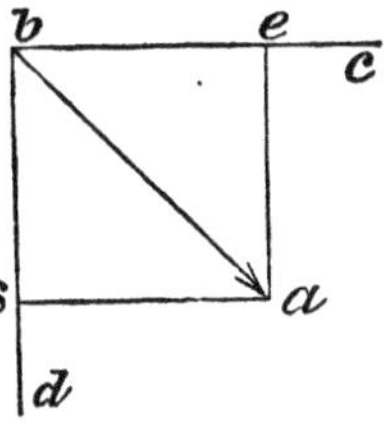

It is thus that a sailor who knows how far he has sailed in an oblique direction finds out how much he has gone north and east or south and west; in other words, finds out the difference in latitude and longitude between his present place and a former one. In the above figure, *b a* may represent the course and distance sailed, *b s* the difference of latitude, and *b e* the difference of longitude.

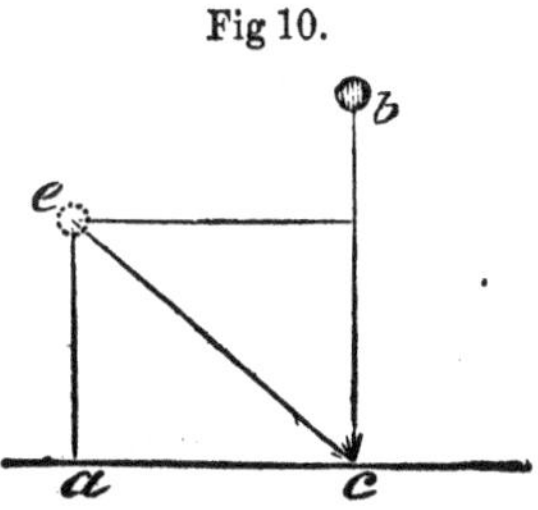

Fig 10.

Thus again, if a ball *b* strike a table *a c*, with velocity and direction, both represented by the line *b c;* and if the ball be supposed afterwards with the same velocity to approach the table in the oblique direction *e c*, it will then strike with as much less force than before, as the line *e a* is shorter than *e c*. For *e a* is found according to the rule for decomposing a force, given above; and, to common sense, it is obvious, that if the whole velocity of the ball be represented by *e c*, the rate of approximation towards the table, or merely downward velocity and therefore the downward force is marked by the line *e a*. The body only *falls* through the distance *e a* while *moving* all the way from *e* to *c*.

Figure 10, explains the important cases of the force of wind upon ships sails, windmill vanes, &c.; and the force of water upon float-boards, water-wheels, &c.; showing that the moving mass exerts force upon a surface, not in proportion to the speed with which it may be passing along or near the surface, but to the rate of perpendicular approximation. It explains also, why the slanting blow of a club or ball is so light, compared with the direct blow.

"*The two great forces of Nature are* Attraction *and* Repulsion." (Read the Analysis.)

A person, on first approaching this subject, is far from supposing that the beautiful and almost endless variety of phenomena exhibited in the universe around, all are referrable to the two principles, *attraction* and *repulsion*, examined in the first section:—but such is the truth.—It will first be shown here, how the great classes of accelerated, retarded, and bent motions arise from them.

Attraction.—Until Newton said, that what we call *weight* of bodies is merely an instance of that universal attraction of matter which diminishes with increasing distance, it was never suspected that weight was less, high up in the air than on the ground; or on a lofty mountain than on the sea-shore. But this we now know to be the case. However, in studying what goes on in obedience to gravity near the surface of the earth, except in a few very nice cases, gravity may be considered as a uniform power; for man has neither approached the centre of the earth in mines, nor receded from it in balloons, by more than about a thousandth part of his distance from it; and weight has relation to the distance from the centre, not to the distance from the surface.

"*Accelerated Motion from Gravity.*"

Owing to the inertia of matter, any force *continuing* to act on a mass which is free to obey it, produces in the mass a quickening or accelerated motion; for as the motion given in the first instance, continues afterwards without any farther force, merely on account of the inertia, it follows, that as much more motion is added during the second instant, and as much again during the third, and so on. A falling body, therefore, under the influence of attraction, is, as it were, a reservoir, receiving every instant fresh velocity and momentum.

It is said that Newton's sublime genius read the nature of attraction in the simple incident of an apple falling before him from a lofty branch in his garden.—The eye which perceives an apple beginning to fall, can follow it for a time and mark the gradual acceleration of its descent, but soon sees its path only as a shadowy line.

A boy letting a ball drop from his hand, can catch it again in the first instant, but after a litttle delay his hand pursues it in vain.

A fragment of rock, detached from the brow of a hill by the lightning stroke, begins its motion slowly; but once fairly launched, it gathers fresh speed and momentum with every instant, and bounds from steep to steep driving every obstacle before it.

Any liquid falling from a reservoir, forms a descending mass or stream, of which the bulk diminishes from above downwards, in the same proportion as the velocity of the particles increases. This truth is well exemplified in the pouring out of molasses or thick syrup; if the height of the fall be considerable, the bulky sluggish mass, which first escapes, is reduced, before it reaches the bottom, to a small thread; but the thread is moving proportionately faster, and fills the receiving vessel with surprising rapidity. The same truth is exhibited on a vast scale in the falls of Niagara; where the broad river is seen first bending over the precipice a deep slow moving mass, then becoming a thinner and a thinner sheet as it descends, until at last, surrounded by its foam or mist, it flashes into the deep below, apparently with the velocity of lightning.

When velocity becomes considerable in any case of falling; it cannot be measured accurately by the eye, but its effects ascertain it. A man leaps from a chair with impunity, from a table with a shock, from a high window with fracture of his bones, and in falling from a balloon his body is literally dashed to pieces.

The force of gravity or general attraction is such at the surface of this earth, that, in the first second of time, it gives to a body allowed to fall a velocity of 32 feet nearly per second, that is, a velocity which, remaining uniform from the end of the second, would carry it, without farther action of gravity, through 32 feet in the next second Yet the body falls only 16 feet in the first second; and the reason is, that the velocity of 32 feet possessed at the end of the second is gradually acquired, the body having only half of it at the half second, and as much less than half at any distance before that time, as it has more than half at the same distance afterwards; and the average, therefore, is only half of the 32, or 16 feet in the whole second. In the next second, it falls, of course, through the whole 32 feet, with 16 additional, from the new action of gravity, in all three times as much as in the first second; and in two seconds, therefore, it falls altogether four times as far as in one second. At the end of two seconds the velocity is doubled or is 64 feet per second, so that in the third second the body falls 64, and other new 16, in all five times as much as in the first second; and in three seconds, therefore, it has descended nine times as far as in one second, &c. Knowing this progress, the velocity acquired by a falling body, and the distance through which it falls, in any given time, are easily calculated; and the height of a precipice, or the depth of a well, may be ascertained by marking the time required for a body to fall through the space.

The doctrines of falling bodies are of such importance in the minute examination of many of the phenomena of nature, that much attention has been bestowed upon them. Mr. Atwood's ingenious contrivance by which the motion of falling bodies may be retarded in any desired degree, without the character of the motion being otherwise altered, has enabled experimenters to render evident to the senses all that abstract calculation had anticipated. A pound weight, left quite free, falls towards the ground, sixteen feet in the first second, proving that *attraction* of one pound is just sufficient to overcome the *inertia* of one pound at that rate. But if the inertia were doubled, or tripled, or increased in any other degree, the fall of course would be just so much slower. Now Mr. Atwood's machine in effect increases it, by causing falling weights to overcome not only their own inertia, but also that of other weights; fig. 11. Thus *a* and *b*, being weights of two pounds each, balancing each other over the very easily turned pulley *c*, are moved by a weight of one pound *d*, hooked to one of them; and gravity in pulling this down, with force of one pound, has to overcome, not the inertia of one pound, but of five, for the other two weights must move as fast as the one pound does; and thus, the velocity being reduced to one-fifth of what is natural to a falling body, the descent can be minutely observed. The experiments with Atwood's machine may be varied exceedingly, and they are most interesting.

Fig. 11.

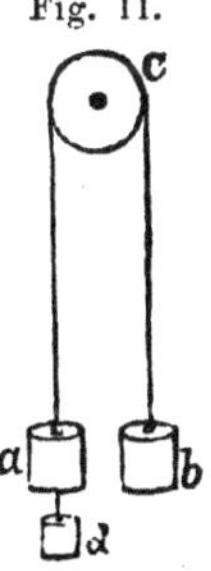

"Retarded Motion," from gravity.

What has been said of the changing velocity of a falling body, from gravity, is exactly true, in a reversed way, respecting a rising body exposed to the same influence.

A bullet shot directly upwards, every instant loses a part of its velocity, until at last it comes to rest in the sky,—where a soaring eagle might see the messenger of death motionless and harmless for a moment by his side: the ball then descends again, and so that, at corresponding points of the ascent and descent, but for the resistance of the air, the velocities would be equal; and, on reaching the ground, it would have acquired exactly the velocity with which it first departed.

It is explained in a preceding paragraph, that a body falls four times as far in two seconds as in one, although the velocity at the end of two seconds is only doubled. For the same reason, a body shot upwards with double velocity, rises four times as far as if shot with a single velocity; if shot with triple velocity, it rises nine times as far, and so forth.

In aiming for amusement at bodies thrown up into the air, it is easy to hit them near their point of turning, and more difficult always as they are nearer to the ground, whether rising or falling.

An upward jet of water is small below, where it issues from the pipe with great velocity, but it becomes more bulky as the water loses velocity in ascending, and at the top it often spreads a little like a palm tree, and any light round solid will continue supported and playing upon its summit.

The rise of a pendulum from the bottom of its arc, is an exact copy, reversed, of its previous descent to that point.

" The Pendulum"

exemplifies well both accelerated and retarded motion. The name is applicable to any body so suspended, that it may swing freely backwards and forwards. When such a body is made of certain form and length, although so simple, it is one of the most admirable contrivances of man's ingenuity.

Galileo having observed the hanging chandeliers of lofty ceilings to continue vibrating long and with singular uniformity, after any accidental cause of disturbance, was led to investigate the laws of the phenomenon; and out of what, in some shape or other, had been before men's eyes, but uselessly, from the beginning of the world, his powerful genius extracted the most important results. Independently of the light which the theory of the pendulum has thrown on various branches of physics, the instrument itself, with a few wheels attached, to record its vibrations, has now become the perfect time-keeper, regulating many of the affairs of men.

A common pendulum consists of a ball, fig. 12, as *a*, suspended from a rod from a fixed point as *b*, and made to swing backwards and forwards, or to vibrite under this point. Being raised to *c*, and then set at liberty, it falls back to *a* with an accelerating motion, like a ball rolling down a slope, and when arrived there, it has just acquired momentum enough to carry it to *d*, at an elevation on the other side; from this it falls back again, again to rise; and would so go on for ever, but for the impediments of air and friction. The pendulum is strictly an object of mathematical study;

Fig. 12.

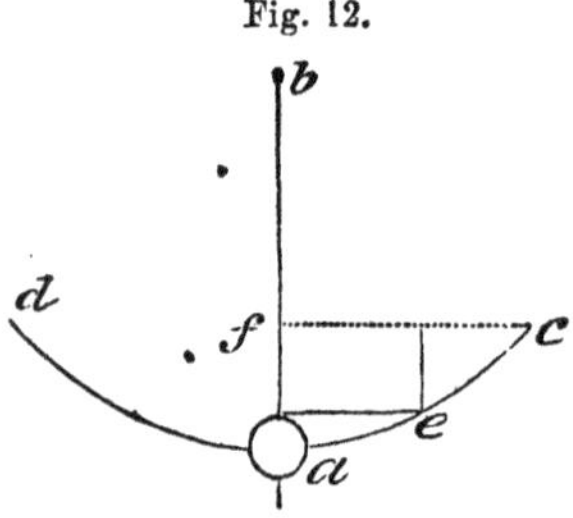

but we shall give a general idea of its important characteristics in common language.

1. The *times of the vibrations* of a pendulum are very neary equal, whether it be moving much or little, that is to say, whether the arc described by it be large or small. This remarkable property is what makes a time-keeper. The reason that a large vibration is performed in the same time as a small one, in other words, that the pendulum always moves faster in proportion as its journey is longer—is, that in proportion as the arc described is more extended, the steeper are its beginning and ending, and the more rapidly, therefore, the pendulum falls down at first, sweeps along the intermediate space, and stops at last. It is evident, for instance, that the portion *c e* of the arc (fig. 13) is much more steep than the equal portion *e c.*—A pendulum made to vibrate in the curve called a *cycloid*, which, in the central part, very nearly coincides with a circular arc, but towards the extremity rises a little more steeply, has its beats perfectly *isochronous*, or in equal times, whatever their extent.

A common clock is merely a pendulum with wheel-work attached to it, to record the number of the vibrations, and with a weight or spring having force enough to counteract the retarding effects of friction and the resistance of the air. The wheels show how many swings or beats of the pendulum have taken place, because at every beat, a tooth of the last wheel is allowed to pass. Now if this wheel has sixty teeth, as is common, it will just turn round once for sixty beats of the pendulum, or seconds, and a hand fixed on its axis projecting through the dial-plate, will be the second hand of the clock. The other wheels are so connected with the first, and the numbers of teeth on them so proportioned, that one turns sixty times slower than the first, to fit its axis to carry a minute hand, and another, by moving twelve times slower still, is fitted to carry an hour hand.

2. The *length of a pendulum* influences the time of its vibration.—Long pendulums virbrate more slowly than short ones, because, in corresponding arcs or paths, the bob or ball of the long pendulum has a greater journey to perform, without having a steeper line of descent. If a pendulum *b a* be twice as long as another reaching from *b* to *e*, it has twice as much to fall in its descending arc *c a*, as the other in its arc *d e*, while in corresponding parts of the two paths, the slope or inclination is always equal:—the ball of the long pendulum may be considered as having rolled twice as far down a given slope as the ball of the short pendulum. Now as a body falls four times as far, either directly or on any uniform slope, in two seconds, as in one, a pendulum must be four times as long, to beat once in two seconds, as to beat every second. A pendulum of a little more than 39 inches beat seconds; one of four times the length is required to beat double seconds, and one of one-fourth the length to beat half seconds.—As a pendulum to answer its purpose must be of invariable length, one which beats seconds constitutes an easily found standard of measure.

Fig. 13.

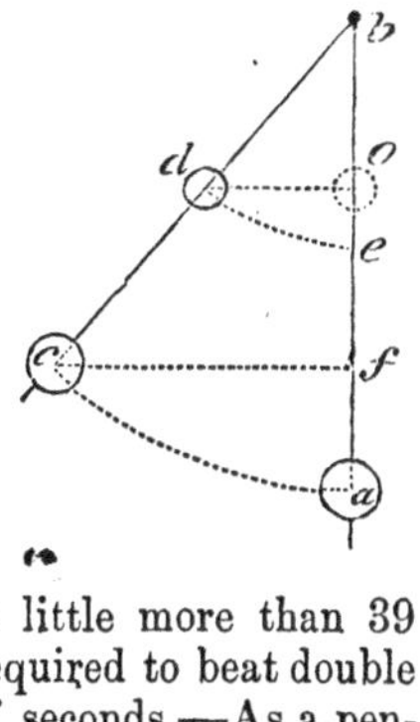

Because the smallest change in a length of a pendulum alters the rate of going of the clock, it is important to be able to counteract the dilatation or contraction of pendulums caused by the changing heat of the seasons; and for this purpose various ingenious means have been contrived. One of the best of these is the *gridiron pendulum*, as it is called, from consisting of various rods of metal. It renders the different dilatability by heat of two metals composing it, the cause of unchanged length in the whole. The adjoining sketch may show that if the central rod of brass represented by the *strong* line from *b* to *c*, dilate alone just as much as the two rods of steel, represented by the *weaker* lines on either side of the other, dilate together (the expansion of brass by heat is about double that of steel,) it will exactly counteract the lengthening of these, and will keep the ball *d* always at the same distance from the point of suspension *a*. Some astronomical clocks in the present day are so perfect that they do not err one beat of the pendulum in a year. Common clocks are regulated by a screw which lifts or lets down the ball of the pendulum, and so changes the effective length, that is, the distance between the point of suspension and what is called the *centre of oscillation*, treated of in the next chapter.

Fig. 14.

3. The *force* of gravity, of course, is what determines how long the pendulum shall be in falling to the bottom of its arc, and how long in rising, for the ball of the pendulum, as already stated, may be considered as a body descending by its weight on a slope; a change in the force of gravity, therefore, would at once alter the rate of all the clocks on earth. At the equator of our earth, where the gravity of bodies is counteracted in a small degree by the centrifugal force arising from the earth's motion (as explained at *page* 53,) a pendulum vibrates more slowly than elsewhere, and must therefore be made shorter to answer the same purpose. Corresponding results take place when a pendulum is carried to a mountain top, and therefore farther away from the centre of the earth, which is the centre of attraction—or when carried to the bottom of a mine, where it is attracted by the matter above it, as well as by the matter beneath.

The popular prejudice refuted at page 53, that a large or heavy body should fall to the earth, even in a vacuum, more quickly than a small or light body attaches itself also to the case of a heavy and a light pendulum. Now there is no difference for pendulums of the same length, whatever their weight or material, but what depends on the resistance of the air. It is a very remarkable fact thus proved, that in all substances the gravity and inertia perfectly agree.

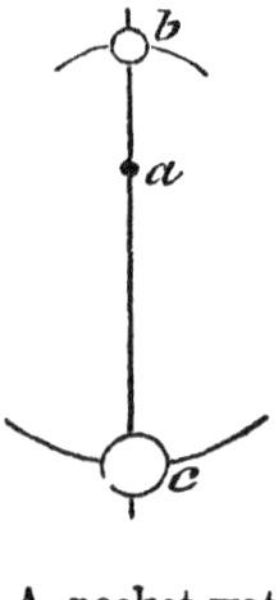

There is a small pendulum called a *metronome*, used by musicians for marking time; which, although very short, may still be made to beat whole seconds, or even longer intervals. The reason of its slow motion is, that its rod is prolonged beyond its axis of support, at *a*, upwards, to *b*, and has a ball upon the top at *b*, as well as on the bottom at *c*; which upper ball prevents the under one from moving so fast as it otherwise would, just as a small weight attached to one end of a weighing-beam, prevents a greater weight attached to the other end from falling so fast as it would if there were no counterpoise. The rate of motion changes with any change in the distance of the ball *b* from the centre of motion *a*; and to allow of such change, the ball is made to slide.

Fig. 15.

A pocket-watch differs from a clock in having a vibrating wheel instead of a vibrating pendulum; and as, in a clock, gravity is always pulling the pendulum down to the bottom of its arc, which is its natural place of rest, but does not fix it there, because the momentum acquired during its fall from one side is just sufficient to carry it up to an equal height on the other—so in a watch, a spring, generally spiral, surrounding the axis of the balance-wheel is always forcing this towards a middle position of rest, but does not fix it there, because the momentum acquired during its approach from either side to the middle position, carries it just as far past on the other side, and the spring has to begin its work again. The balance-wheel at each vibration allows one tooth of the adjoining wheel to pass, as the pendulum does in a clock, and the record of the beats is preserved by the wheels which follow, as already explained for the clock. A main-spring is used to keep up the motion of a watch, instead of the weight used in a clock; and as a spring acts equally, whatever be its position, a watch keeps time although carried in the pocket or in a moving ship.

As the rate of a clock is influenced by the length of its pendulum, so is the rate of a watch by the size or diameter of its balance-wheel; and heat, which retards the motion of a common clock by lengthening the pendulum, retards the motion of a common watch by dilating the balance-wheel. Ingenuity, however, has found a remedy for the latter case as for the former, *viz.*, the contrivance called the *expansion balance-wheel.* Of this the circumference, instead of being a continuous ring, is made up of two half-rings, each attached by one end only, to a cross bar, and which half rings being of brass on the outside and of steel within, bend or curl inwards by heat—as a sheet of damp paper bends when held to the fire—and thus diminish the size of the wheel at their loose extremities, so as just to counter balance its increase by the expansion of the cross bar.

As the motion of a pendulum has relation to the *force of gravity*, so has the motion of a balance-wheel to the *stiffness of the balance-spring;* and the regulator of a watch is merely a pin which bears against the balance-spring, and by sliding backwards or forwards, so as to shorten or lengthen the part of the spring left free to act, changes the degree of its stiffness. A change produced by the variation of temperature is compensated for by the expansion-wheel described above.

It would be exceeding the limit marked out for this general work, to speak more particularly here of those admirable watches which have been produced within the last thirty years under the name of *chronometers*, for the pupose of ascertaining the longitude at sea; but the author may perhaps be excused for mentioning a moment of surprise and delight which he experienced, on first seeing their singular perfection actually proved. After months spent in a passage from South America to Asia, his pocket chronometer, with others on board, announced one morning that a certain point of land was then bearing east from the ship at a distance of fifty miles; and in an hour afterwards, when a mist had cleared away, the looker-out on the mast gave the joyous call of "Land a-head!" verifying the report of the chronometers almost to a mile after a voyage of thousands. It is natural, at such a moment, with the dangers and uncertainties of ancient navigation before the mind, to exult in contemplating what man has now achieved. Had the rate of the wonderful little instrument in all that time been changed even a little, its announcement would have been worse than useless,—but in the night and in the day, in storm and in calm, in heat and in cold, while the persons around it were experiencing every vicissitude of mental and bodily condition, its steady beat went on, keeping exact account of the rolling of the earth and of the stars; and in the midst of the trackless waves it was always ready to tell its magic tale of the very spot of the globe over which it had arrived. The mode of using a chronometer for so valuable a purpose will be explained in the section on astronomy.

Bent or curvilinear motion from attraction.—This takes place whenever attraction is acting across the path of an existing free motion. The flying cannon-ball or stone, drawn down by gravity, is an example, for the projectile force ceases with the first impulse, but the bending force is acting every instant and by every instant producing a new effect, causes a curvilinear path.

An oblique jet of water is to the eye a permanent exhibition of the curve described by a body thus projected. The particles of the liquid move in the line which they would describe if projected singly, and the continued succession of them marks the line of situations through which each passes in its course to the earth.

A cannon or musket-ball, shot quite horizontally over a level plain, will touch the ground or plain just as soon as another ball dropped at the same instant directly from the cannon's mouth; for the forward or projectile motion does not, in such case at all interfere with the action of gravity. This result, which most persons, before consideration, would be disposed to doubt, makes strikingly sensible the extraordinary speed of the cannon-ball; *viz.*, that it has already moved, perhaps, six hundred feet forward, during the half second that a ball dropped from the hand of a standing person requires to reach the earth only four feet beneath. This fact also explains why, for a long range, the gun must be pointed more or less upwards.

A dozen marbles swept horizontally from off a table by a stick, all reach the floor at the same instant, how different soever the distance to which they may respectively be driven.

The particular study of the subject projectiles is very important to military engineers; and we know how successfully they have pursued it by the precision with which they now direct their shot and shells to objects at very great distances.

Fig. 16.

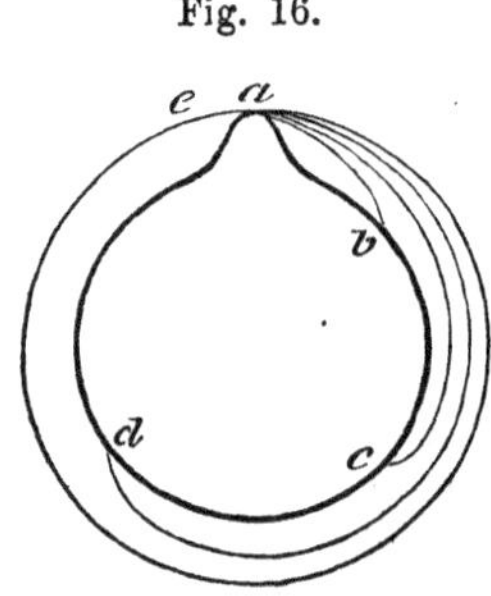

A cannon-ball shot horizontally from the top of a lofty mountain, would go three or four miles. (The mountain is here represented on an enlarged scale, as standing on the globe *b*, *c*, *d*, at *a*.) If there were no atmosphere to resist its motion, or if the mountain top were above the surface of the atmosphere, the same original velocity would carry it thirty or forty miles before it fell, as to *b*: with more force still, it would reach to *c*, and with still more to *d*. And if it could be dispatched with about ten times the velocity of a common cannon-shot, it would not have approached nearer to the earth than at first, even when it had again reached round *e* or to *a*; and its velocity being undiminished, it would perform a second similar tour, and then a third, and so forth: it would, in fact, have become a little satellite, or planetary body, revolving round the earth. In the successive ranges represented in the figure, it is seen that the centrifugal force of the ball, or its tendency to move in a straight line becomes more and more nearly a counterbalance to gravity, and at last is exactly equal to it. If the force given to the ball were more than sufficient to bring it round again to the level of *a*, it would for a time fly off, or increase its distance from the earth, acquiring somewhat the eccentric motion of a comet. There may really be such revolving masses above our atmosphere, although invisible to us, owing to their smallness. It has been supposed by some, that the meteoric stones, which fall to the earth every now and then, came from such bodies, or are the entire masses, having become entangled in our atmosphere, so as to lose their forward velocity. The four little planets discovered lately beyond the orbit of Mars, are not larger than a six-thousandth part of our earth.

REPULSION,—produces *accelerated*, *retarded* and *bent* motions, like attraction, but it acts only at minute distances, while *attraction* draws from the sun, or from the very limits of the universe; *repulsion* acts, for instance, between the adjoining atoms of an elastic fluid. Yet repulsion plays a part in the economy of nature, not at all inferior to its sister attraction. We have already seen, when considering the constitution of masses, in section first, that repulsion prevents or modifies the contact of the atoms of all bodies; that with increase of temperature, it causes these atoms to separate, and of a solid forms a liquid, or even an air; that it operates around all masses as if it were a film or covering, preventing their mutual cohesion, &c., &c.

Accelerated motion from repulsion is seen when the atoms of gunpowder explode and propel the bullet from the bottom of a piece to the muzzle with such rapidly increasing velocity. The strength of this repulsion of gunpowder is so much greater than the strength of gravity or common attraction, that its action on a bullet, during the passage along a barrel of five or six feet in length, may not be overcome by gravity, during an ascent of a mile or more.

A visible *retarded* motion from repulsion is exemplified by a moving body coming against a spring or a bladder full of air, or against the piston-handle of an air-syringe, so as to compress the air beneath it.

Any elastic body striking against another body and recoiling, exhibits in conjunction the phenomena of retardation, acceleration, and often also of bending, chiefly from repulsion; for instance:

An ivory ball, driven forcibly against a marble slab, does not stop at the instant that apparent contact takes place, but still advances and compresses that part of the substance which is against the marble,—as is proved by the facts mentioned at page 37. While this compression of the ivory is going on, the resistance made by the increasing repulsion of the particles gradually retards, and ultimately destroys the forward motion of the ball; and at the instant of its final arrest, the parts in contact, both of the ball and of the marble, being in their greatest degree of compression, act on the ball, and repel it again with gradually accelerating motion, until it leaves the marble with the same velocity which it had on approaching. The retardation and acceleration take place here within so small a space, and in so short a time, that they are not apparent to sense, but the mind perceives the nature of the phenomenon as distinctly as if the ball had rolled against the end of a long steel spring. If the ball strike the marble obliquely, as from *a* to *c*, in a path forming the angle *a c d*, with a perpendicular line, it does not rebound in the same line by which it approached, but just as obliquely towards the other side, *viz.*, from *c* to *b*; and it then exhibits a bent motion from repulsion. This case illustrates also the "resolution of motions," for the oblique descent *a c* being composed of a direct downward motion from *a* to the table, and a horizontal or forward motion from *a* towards the perpendicular, the table destroys the downward motion, and converts it into an opposite directly upward motion, but it does not affect the forward motion, which immediately combines again with the up-

Fig. 17.

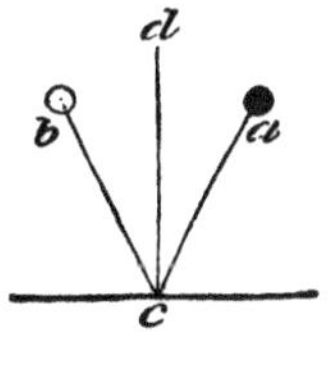

d
b
a
c

ward, and carries the ball as far beyond the perpendicular at *b* as it was distant from it at *a*. The important law in physics, of which this case is an example, is usually expressed—"The angles of incidence and of reflection are equal." It applies to all reflected bodies, as balls, waves, sound, light, &c.

If the ivory ball and marble, in the above case, were supposed to be both perfectly hard, and without elasticity, still the repulsion which surrounds all bodies, as a thin covering, preventing their cohesion, (see page 32,) would act exactly as the real elasticity of the ivory, and would cause a retarded motion until perfect rest came, and then an accelerated motion back again, until the ball recovered its primitive velocity.

Collision between hard bodies always exhibits more or less of the truth now described; when it occurs between soft bodies, as lumps of lead or of moist clay, the approaching parts mutually displace each other, and there is no recoil.

When a straight steel plate, of which the end is fixed in a block, is bent as by a ball rolling against it, the particles on the side which becomes concave are made to approximate, and there is a resistance or repulsion gradually increasing among them; the particles on the convex side, again, are drawn a little more from each other, and are therefore exerting attraction to return: the recoil of the spring is thus owing to both forces trying to replace the particles in their former relative situations.

"*Tides, Winds, &c., exemplify* ATTRACTION." (Read the Analysis, page 42.)

Until we reflect attentively on this subject, we are far from perceiving that all the phenomena of nature are only instances of *attraction* and *repulsion*, acting under a variety of circumstances.

ATTRACTION.—*Tides* are raised by the attraction of the moon and sun, and fall again by the general attraction of the earth; producing in many of the shallower parts of the ocean very rapid horizontal currents. They do a great deal of work for man. They carry his ships along the coasts, and up and down the rivers; they turn water-wheels for him; they fill his docks and canals at convenient times; they rise to receive his ships, launched from elevated building-yards, &c. What a busy scene is a great sea-port river, during the rising and falling of the tide—with the thousands of people along its bank, borrowing assistance in their various occupations!

Winds are produced chiefly by the fluid atmosphere seeking its level, in obedience to the attraction of the earth, after the action of disturbing causes, such as the heat of the sun, &c. They help man in the important business of navigation; they turn his windmills, &c.

The currents of rivers are water constantly descending on slopes, that is, regaining its level, in obedience to the earth's attraction. Water-mills and inland navigation are among the advantages which they afford to man.

All falling and pressing bodies exhibit attraction in its simplest form.

REPULSION—is instanced in *explosion*, *steam*, the action of *springs*, *&c.*

Explosion of gunpowder is repulsion among the particles when assuming the form of air.

Steam, by the repulsion among its particles, moves the piston of the steam-engine. In our days it performs half the labour of society.

Accidental explosion of fire-damp, or hydrogen, in mines, and the tremendous evolutions of elastic fluid in volcanoes and earthquakes, are other instances of the same class.

Elasticity, as seen in springs, collision, &c., belongs chiefly to repulsion; as seen in India-rubber, and other substances resuming their usual length after extension, it belongs chiefly to attraction.

A spring is often, as it were, a reservior of force, kept ready charged for a purpose; as when a gunlock is cocked, a watch wound up, &c.

It will be remarked, with respect to many of the phenomena now and hereafter to be mentioned, that it is not the original Attraction or Repulsion which man uses as his servant, but the momentum gradually accumulated in masses by the exertion of such attraction or repulsion; in other words, the *inertia* is used as a great working power or force.

Electrical, galvanic, magnetical, and *optical*, phenomena, are also in great part peculiar attractions and repulsions, as will be seen in the chapters devoted to the explanation of them. And even the *actions of animals*, so infinitely varied, are all results of a shortening of the fleshy threads called muscular fibres, which is produced by the mutual *attraction* of their component particles;—just as the varied motion of a telegraph, or of a ship's yard, are produced by the shortening of certain ropes of connection.

However closely allied the last-mentioned particular attractions and repulsions may be to the general attraction and repulsion formerly treated of, it is found convenient to consider them apart.

In the remarkable phenomena of nature and art, all the motions being caused, as now shown, by Attraction and Repulsion, these forces do not operate by a single impulse, but through a repetition of impulses, or a continued action, of which the effect is gradually accumulated in the inertia of matter. Thus all great velocities and momenta are the terminations of an accelerated motion.

Meteoric stones falling from great heights, bury themselves deep in the earth by the force of their gradually acquired velocity.

When the wood-cutters among the Alps launch an enormous tree from high on the mountain side, along the smooth wooden trough or channel prepared for it and in fewer minutes than it traverses miles, it is seen plunging into the lake below; it acquires its frightful velocity, not at once, but through the action of gravity continued during the whole of its descent.

The shock or blow of the ram of a pile-engine, is not the effect of momentary attraction between it and the earth, but of that attraction accumulating motal inertia or power, during the descent of the ram through a space of twenty or thirty feet.

A common hammer, in its instantaneous shock, has the condensed effect of the arm and of gravity, as accumulated through its whole previous course; and when a powerful blow is intended, the hammer, or hatchet, or club, or fist in boxing, is lifted high, or carried far back, that there may be time and space for imparting greater power.

The inferior animals, by many of their actions, illustrate the same truth, and prove their experimental or instinctive acquaintance with it.

Sea-birds carry shell-fish up into the air, and drop them on smooth stones to break them, and to obtain the food. It is related in Grecian story, that a bird once mistook the venerable bald head of a sage meditating on the sea-shore for a smooth stone, and by the same act killed an oyster and the philosopher.

There are some long-necked birds, that fight and kill their prey by a blow of their beak. They draw back the head, bending the neck like a swan or serpent, and then dart it forward, with a continued effort, until the strong wedge-like beak reaches its destination, almost with the velocity of a pistol bullet. One snake in darting its fangs at another passing swiftly across its coil, has been known to miss its aim and inflict a mortal wound on its own flesh.

Bulls, rams and goats, in fighting, alternately recede and run at each other, that the shock may be great when their foreheads meet.

A horse in kicking, from the great length of his leg, and the consequent space through which he can be adding velocity to his foot, drives at last against the object almost like a cannon shot.

A bow-string propelling an arrow, follows it through a considerable space, and so gives the great velocity at last produced.

A sling gives to the hand the power of adding velocity to the stone through a long path; for the hand moves in a small circle while the stone moves in a larger, and the hand being kept always in somewhat advance of the stone, pulls at it without intermission, until the moment of discharge.

The battering-rams of the ancients allowed those about them to accumulate in them the efforts of many hands, and of a considerable duration of action, so as to give at last one great and sudden shock.

Even the gentle action of the human breath, exerted for a time on a pea or small hard ball of clay while passing through a long smooth tube, gives a velocity which will inflict a sharp and painful stroke on a distant animal. In Borneo and other of the Eastern Islands, poisoned arrows are thrown in this way with great force and precision.

The action of gunpowder or bullets, although appearing so sudden, is still not an instantaneous, but a gradual, and therefore accelerating action; and accordingly we find the effect to depend on the length of the piece along which the force pursues the ball. A small fast sailing vessel with a single long gun, has often compelled a very superior vessel, whose guns were shorter, to yield.

For the same reason that all great velocities require continued action or repeated impulse to produce them, so do they also destroy them; the inertia of motion and of rest being equal.

A vast mass of rock suspended like a pendulum, and allowed to sweep down its curve to a considerable elevation, would arrive at the bottom like a battering-ram, with force sufficient to shake a thick wall or rampart to its foundation. The continued action of gravity would have given this force, and if, instead of the solid resistance supposed, and which would scarcely be sufficient to take the whole momentum away, the mass were merely allowed to continue its course as a pendulum, and to ascend on the other side, the continued action of gravity then opposing its motion, would bring it to powerless rest again, by the time when it had reached an elevation equal to that from which it fell.

Soft air expanding gives gradually the death-carrying velocity to the cannon-ball; and soft air, or cotton, or wood, resisting in a close strong tube,—if the bullet could be directed exactly into it—would again gradually annihilate the motion. Were the attempt made, however, to stop the ball suddenly, by a block of the hardest granite, the block would instantly be riven by its force.

Bales of cotton or thick masses of cork, attached round a ship, will receive cannon-balls, and bring them to rest, without themselves suffering much, while the naked firmer side of the ship would be penetrated. The cotton or cork offers an increasing resistance through a considerable space, while the oak opposes its hard front at once, and must instantly suffice or be destroyed. A hard body, that it may at once destroy such a motion as we are supposing, must be able to oppose as much force in perhaps the space of one-hundredth of an inch, that is, in the extent to which its elasticity will let it yield without breaking, as the moving cause gave, through a much greater space (a plate of steel will thus oppose a pistol-bullet;) and when it cannot do this, it must be broken or penetrated by the moving body. It is to be remarked, however, that the continued opposition of a thick mass of wood, stone, or earth, to an entered bullet, brings it to rest at least as an elastic unbroken opposition would. Gunners have ascertained the exact depth in each substance to which a ball will penetrate; and they call buildings *bomb-proof* or *ball-proof*, which have a thickness or depth exceeding that.

A hempen or silken rope, supporting the scale of a weighing beam, would resist a greater weight *falling* into the scale than would be resisted by an iron chain which were even stronger than the rope for the purpose of bearing a *quiescent* weight; because the hemp or silk would yield by its elasticity, and continue its resistance through a considerable space and time, and thus would at last gradually overcome the momentum; while the iron, by scarcely yielding at all, would require to be strong enough to stop the mass suddenly or would break.

Yet for the same reason that iron is weakest in such a case as the last, it is stronger than hemp or rope when used as a cable for a ship, to withstand the sudden force of waves.

This will be understood on considering, that the chain by its weight hangs as a curve or inverted arc in the water, while the rope, being nearly of the weight of water, is supported in it almost as a straight line from the anchor to the ship; therefore, when a great wave dashes against the ship, the bent chain will yield until it be drawn nearly straight, by which great extent of yielding, and consequent length of resistance, it will withstand a great shock; whereas, the straight rope, as it can yield only by the elasticity of its material, and comparatively, therefore, a little way, will resist much less.

A heavy ship moving quickly with the tide or wind, could not be stopped instantly by a short rope or chain of any magnitude: if the attempt were made to destroy at once so vast a momentum, something would certainly give way; but a rope of very moderate size, kept tight between the shore and the ship, and from time to time allowed to slip a little round a wooden block, when the tightness threatened its breaking, would accomplish the end very soon and easily.

The following are farther proofs that forces are to be measured as much by the time or space through which they act, as by their difference of intensity or momentary power.

A door standing open, and which would yield readily on its hinges to the gentle push of a finger, is not moved by a cannon-ball piercing through it. Now the ball really overcomes the whole force of cohesion among the atoms of tough wood: but that force is allowed to act or resist for so short a time, owing to the rapid passage of the ball, that it is not sufficient to affect the inertia of the door, in a degree to produce a sensible motion. The cohesion of the circle in the door, cut out by the ball, would have borne a weight of more than a hundred pounds laid quietly upon it, but supposing the bullet to fly twelve hundred feet in a second, and the door to be one inch thick, the cohesion being allowed to act for only the 14,400th part of a second, its influence is not perceived. The following are other examples of the same kind.

A leaden bullet pressed slowly against a pane of glass, breaks it irregularly, where the strength happens to be least; but the same bullet shot at it from a pistol, makes only a small round hole. It has been amusingly said of such a case, that the particles struck and carried away have not time to warn their neighbours of what is happening.

A cannon-ball, having very great velocity, passes through a ship's side, and leaves but a little mark; while one with less speed splinters and breaks the wood to a considerable distance around. A near shot thus often injures a ship less than one from a greater distance.

A sheet of paper standing edgeways on a table, is not driven down by a pistol-ball fired through it.

The truth at present under consideration explains, with respect to gun-shot wounds, why the man often remains ignorant for a time of his misfortune, and why a rapid bullet only kills the parts which it touches, while a spent ball may bruise and injure all around. In many cases of injury popularly attributed to the *wind of a ball*, the ball itself has really reached the part.

A man lying down and receiving the blow of a great hammer on his chest would be killed by it; but if a heavy anvil be first laid upon the chest, and the blow then received upon the anvil, the man bears it with impunity. Here the quantity of motion in the hammer being diffused through the great mass of the anvil, produces but a trifling velocity, which the elasticity of the chest, in its slow yielding, easily overcomes.

A circular plate of soft iron, made to turn with extreme rapidity, will cut through the hardest steel file, almost as a knife cuts through a carrot. In cases where a soft powder suffices to polish a hard body, it acts partly like this plate, by the motion or velocity given to the wearing particles.

"*There is no motion or action in the universe, without a concomitant and opposite action of equal amount.*" (See the Analysis.)

This truth has otherwise been expressed—"action and reaction are equal and contrary." It is evident, that if no action or movement takes place on earth but in consequence of either Attraction or Repulsion—and this has now been shown—there must always be two objects or masses concerned, and each must be *attracted* or *repelled* just as much as the other, although one will have less velocity than the other, as it may be itself greater, or fixed to another mass.

If a man in one boat pull at a rope attached to another, the two boats will approach. If they be of equal size and load, they will both move at the same rate, in whichever of the boats the man may be; and if there be a difference in the sizes, and resistances, there will be a corresponding difference in the velocities, the smaller boat moving the fastest.

A magnet and a piece of iron attract each other equally, whatever disproportion there is between the masses. If either be balanced in a scale, and the other be then brought within a certain distance beneath it, the very same counterpoise will be required to prevent their approach, whichever be in the scale. If the two were hanging near each other as pendulums, they would approach and meet; but the little one would perform more of the journey in proportion to its littleness.

A man in a boat pulling a rope attached to a large ship, seems only to move the boat: but he really moves the ship a little, for supposing the resistance of the ship to be just a thousand times greater than that of the boat, a thousand men in a thousand boats, pulling simultaneously in the same manner would make the ship meet them half way.

A pound of lead and the earth attract each other with equal force, but that force makes the lead approach sixteen feet in a second towards the earth, while the contrary motion of the earth is of course as much less than this as the earth is weightier than one pound,—and is therefore unnoticed. Speaking strictly, it is true, that even a feather falling lifts the earth towards it, and that a man jumping kicks the earth away.

A spring unbending between two equal bodies, throws them off with equal velocity; if between bodies of different magnitudes, the velocity of the smaller body is greater in proportion to its smallness.

On firing a cannon, the gun recoils with even more motion or momentum in it than the ball has, for it suffers the reaction of the expelled gunpowder as well as of the ball; but the momentum in the gun being diffused through a greater mass, the velocity is small, and easily checked.

The recoil of a light fowling-piece will hurt the shoulder, if the piece be not held close to it.

A ship in chase, by firing her bow guns, retards her motion; by firing from her stern she quickens it.

A ship firing a broadside, heels or inclines to the opposite side.

A vessel of water suspended by a cord hangs perpendicularly; but if a hole be opened on one side, so as to allow the water to jet out there, the vessel will be pushed to the other side by the reaction of the jet, and will so remain while it flows. If the hole be oblique, the vessel will constantly turn round.

A vessel of water placed upon a floating piece of plank, and allowed to throw out a jet, as in the last case, moves the plank in the opposite direction.

A steamboat may be driven by making the engine pump or squirt water from the stern, instead of making it, as usual, move paddle-wheels. There is a loss of power, however, in this mode of applying it, as will be explained under the head of "Hydraulics."

A man floating in a small boat, and blowing strongly with a bellows towards the stern, pushes himself onwards with the same force with which the air issues from the bellows-pipe.

A sky-rocket ascends, because, after it is lighted, the lower part is always producing a larger quantity of aeriform fluid, which, in expanding, presses not only on the air below, but also on the rocket above, and thus lifts it. The ascent is aided also by the recoil of the rocket from the part of its substance, which is constantly bursting downwards.

He was a foolish man who thought he had found the means of commanding always a fair wind for his pleasure-boat, by erecting an immense bellows in the stern. The bellows and sails acted against each other, and there was no motion: indeed, in a perfect calm, there would be a little backward motion, because the sail would not catch all the wind from the bellows.

A man supported on a floating plank, by walking towards one end of it gives it a motion in the direction opposite.

A man using an oar, or a steam-engine turning paddle-wheels, advances exactly with the force that drives the water astern.

A swimmer pressing the water downwards and backwads with his hands, is sent forwards and upwards with the same force, by the reaction of the water.

And a bird flying, is upheld with exactly the force with which it strikes the air in the opposite direction.

A man pushing against the ground with a stick, may be considered as compressing a spring between the earth and the end of his stick, which spring is therefore pushing him up as much as he pushes down; and if, at the time, he were balanced in the scale of a weighing beam, he would find that he weighed just as much less as he was pressing with his stick.

Thus an invalid, on a spring plank or chair, who, by a trifling downward pressure of his hand on a staff or on a table, causes his body to rise and fall through a great range, and thus obtains the advantange of almost passive exercise, is really lifting himself while he presses downward.

When a boy cries on knocking his head against a table or pane of glass, he is commonly told, and truly, that he has given as hard a blow as he has received; although his philosophy probably, looking chiefly to results, blames the table for his head hurt, and his head for the glass broken.

The difference of momentum acquired in a fall of one foot or of several, is well known: the corresponding intensities of reaction are unpleasantly experienced by a man who sits down in an easy chair, or who, in sitting down where he supposed a chair to be, unexpectedly reaches the floor.

What motion the wind has given to a ship it has itself lost, that is to say, the ship has reacted on the moving air: as is seen when one vessel is becalmed under the lee of another.

When one billiard-ball strikes directly another ball of equal size, it stops, and the second ball proceeds with the whole velocity which the first had—the action which imparts the new motion being equal to the reaction which destroys the old. Although the transference of motion, in such a case, seems to be instantaneous, the change is really progressive, and as follows. The approaching ball, at a certain point of time, has just given half of its motion to the other equal ball, and if both were of soft clay, they would then proceed together with half the original velocity; but, as they are elastic, the touching parts at the moment supposed are compressed like a spring between the balls, and by then expanding, and exerting force equally both ways, they double the velocity of the foremost ball, and destroy altogether the motion of that behind.

If a billiard ball be propelled against the nearest one of the row of balls equal to itself, it comes to rest as in the last case described, while the farthest ball of the row darts off with its velocity,—the intermediate balls having each received and transmitted the motion in a twinkling, without appearing themselves to move

As farther illustrative of the truths, that action and reaction are equal and contrary, and that in every case of hard bodies striking each other, they may be regarded as compressing a very small strong spring between them, we may mention, that when any elastic body, as a billiard-ball, strikes another body larger than itself, and rebounds, it gives to that other, not only all the motion which it originally possessed, this being done at the moment when it comes to rest, but an additional quantity, equal to that with which it recoils

—owing to the equal action in both directions of the repulsion or spring which causes the recoil. When the difference of size between the bodies is very great, the returning velocity of the smaller is nearly as great as its advancing motion was, and thus it gives a momentum to the body struck nearly double of what it originally itself possessed. This phenomena constitutes the paradoxical case of an effect being greater than its cause, and has led persons, imperfectly acquainted with the subject, to seek from the principle, a *perpetuum mobile.* A hammer on rebounding from an anvil has given a blow nearly double the force which it had itself, for the anvil felt its full original force while stopping it, and then, equally with itself, was affected by the repulsion which caused its return.

Many other interesting facts might be adduced as examples of equal action and reaction, but these will suffice.

This second section of the work has now explained the nature of INERTIA in matter, and has shown that the infinitely varied phenomena of motion, which the universe exhibits, are only *attraction* and *repulsion*, acting on *inertia* of *atoms* separate or conjoined, under diversified circumstances. And such is the sublime simplicity of the whole scheme of nature.

APPENDIX

TO PART I.—SECTION II.

BY THE AMERICAN EDITOR.

THE attentive perusal of the preceding section will prepare the reader to understand the following propositions.

Definitions.

Prop. 1.—When a body is successively changing its place, it is said to be in *motion*, p. 42.

The idea of motion involves those of *space, time, velocity, direction*, the quantity of matter and momentum.

Prop. 2.—The *space described* is the distance passed over by a body during its motion; and is measured by the number of units of length, as a foot, a yard, a mile, &c., contained in this distance.

Prop. 3.—The *time* consists of a certain number of units of time adopted as its measure, as a second, a minute, &c., which have elapsed during the motion of a body.

Prop. 4.—The *velocity* of a body is the rate at which it moves, or the number of these assumed units of space that it passes over during the assumed unit of time.

All the above measures may be represented graphically by lines that are proportioned to them, p. 65.

Prop. 5.—The *direction* of a body may be straight or curved; when straight or rectilinear, it is the angle which its path makes with any straight line in the same plane, adopted as an axis; when the path of a body is a curve, its direction at any point is the angle which the tangent to the curve at the point makes with the fixed axis.

Prop. 6.—The momentum of a body is its quantity of motion, both the mass and velocity being taken into consideration, and its proper measure is the product of the mass into the velocity, pp. 53, 54.

Prop. 7.—A body is said to have a *uniform motion* when its velocity remains constant, that is, when it describes equal spaces in equal successive intervals of time, p. 47.

Prop. 8.—Every motion that is not uniform is said to be *varied*, and is called *accelerated* or *retarded* as the velocity increases or decreases.

Prop. 9.—When the velocity constantly increases or decreases in the direct ratio of the time that the body has been moved, the motion is said to be *uniformly accelerated* or *retarded*, pp. 43, 58, 59, 60.

Prop. 10.—Whatever is capable of producing or destroying the motion of a body is called *force*.

Prop. 11.—A force that produces its effect instantaneously, and then ceases to act, is called an *impulsive* force.

Prop. 12.—A force that acts continually and equally is termed a *constant force.*

Prop. 13.—When the constant force acts in lines directed towards a single point or centre, it is called *centripetal,* and the path of the body its *orbit*, p. 59.

Prop. 14.—That part of the impulsive force which tends to make a body move directly from the centre, is termed the *centrifugal force*, p. 49.

Prop. 15.—A force that is capable of destroying motion without being able, under any circumstances, to produce motion, is termed a *passive force.*

Prop. 16.—The state of rest produced by the action of opposite forces is termed equilibrium.

Prop. 17.—When a body is struck, its particles yield to the impulse, and the form of the body is changed. When the body possesses the inherent power when thus changed, of restoring its form, it is said to be *elastic;* when it has not this power, it is called non-elastic, p. 37.

Prop. 18.—A body oscillating below a point to which it has in any way attached, is termed a pendulum, p. 60.

Laws of Motion.

Prop. 19.—1st. If a body be at rest it will continue at rest, and if in motion, it will continue to advance uniformly in a right line unless compelled to change its state by some external force, pp. 47, 49.

Prop. 20.—2d. The motion of a body is in the direction of the force that produces it, and is proportional to that force, pp. 53, 55.

Prop. 21.—3d. Action and reaction are always equal and opposed to each other; or when a body communicates motion to another, it loses of its own momentum as much as it gives to the other body, pp. 70, 72.

Of Impulsive force and Rectilinear motion.

Prop. 22.—The effect of an impulsive force is to produce *uniform rectilinear motion*, p. 49.

For during the moment of its action on any body, it must set it in motion with a certain velocity; and by the first law of motion, the body must continue to advance in a straight line with that velocity.

Prop. 23.—In rectilinear motion the *space* is as the velocity multiplied into the time.

For if a body move with the velocity of three feet per second, it is evident, that it will move over 6 feet in two seconds, i. e. × 32; and 9 feet in seconds, i. e. × 33, and 12 feet in four seconds, &c. &c.

Prop. 24.—The *time* is as the space divided by the velocity.

For if a body passes over 12 feet for instance, when its velocity is 3 feet per second, it is evident, that in order to find the number of seconds, which the body has employed in passing over 12 feet of space, we need only divide 12 by 3, (i. e., the space by the velocity) and the quotient 4 is the *time* sought.

Prop. 25.—The *velocity* is as the space divided by the time.

For if a body move over 12 feet in 4 seconds, its velocity is evidently 3 feet per second or 12 ÷ 4.

The velocities of two bodies may be compared, in the same manner: the velocities of two bodies A and B, for instance, of which A moves over 54 feet in 9 seconds, and B, 96 feet in 6 seconds; their velocities will be as 6 (54÷9) to 16 (96÷6.*)

Of a constant force and uniformly accelerated motion.

Prop. 26.—The effect of a constant force acting upon a body, is to produce in it a uniformly accelerated motion, p. 58.

For since the effect of force is to produce velocity, a constant force must, in successive instants of time, afford continual and equal additions to the velocity of the body it has set in motion; that is, the velocity will increase in the direct ratio that the body has been moving, which is the definition of *uniformly accelerated motion.*

Prop. 27.—In uniformly accelerated motion the space described is as the square of the time, pp. 58, 59.

Thus it is found by experiment, that if a body move with a gradually and constantly increasing velocity that would carry it through a mile in one minute, that at the end of this time it has acquired such a velocity as would carry it through two miles the next minute, if the force that communicated its motion ceased to act at the end of the first minute; but if the force continues to act, it acquires a velocity that would carry it over an additional mile, so that it will pass over three miles the second minute, or four miles in two minutes. At the end of the second minute it has acquired a velocity that will carry it over double the space in the third minute, that it moved over in the first two minutes, or a velocity of 8 miles in 2 minutes, or 4 miles a minute. But the force still continuing to act, it will move a mile farther or five miles in the third minute. Hence, if a body acted upon by a continued force move a mile the first minute, it will move 3 miles the second, 5 the 3d, 7 the 4th, 9 the 5th, &c.

Thus the spaces described in successive equal parts of time, by uniformly accelerated motion, are always as the odd numbers 1, 3, 5, 7, 9, &c., and consequently the whole spaces are as the squares of the times or of the last-acquired velocities. For the continued addition of the odd numbers yields the squares of all numbers from unity upwards. Thus 1 is the first odd number and the square of 1 is 1; 3 is the second odd number, and this added to one makes 4, the square of 2; 5 is the third odd number and this added to 4 makes 9, the square of 3; and so on for ever. Since, therefore, the times and velocities proceed evenly and constantly as 1, 2, 3, 4, &c., but the spaces described in equal times are as 1, 3, 5, 7, &c., it is evident that the space described,

In 1 minute will be	- -	1=square of	1
In 2 " "	- -	1+3=4= "	2
In 3 " "	- -	1+3+5=9= "	3
In 4 " "	- -	1+3+5+7=16= "	4

* For the benefit of those who are acquainted with algebra, we subjoin the following equation, which expresses all the circumstances of uniform motion.

Let t = the time of motion,

s = the space described in the time t,

v = the velocity;

Then, $s = vt$ from which we obtain

$$v = \frac{s}{t}$$

$$\text{and } t = \frac{s}{v}$$

Of Gravity.

Prop. 28.—The force which causes bodies to fall to the earth is of the kind named constant, and is called gravity, p. 58.

Prop. 29.—The direction of gravity is in lines perpendicular to the earth's surface.

Prop. 30.—The force of gravity is directly proportional to the mass of the body.

For however small the parts into which we divide a body, we find them all affected by gravity, since this force must act upon all the particles of a body.

Hence, in an unresisting medium, all bodies setting out from a state of rest, fall through the same space in the same time, because the force of gravity acting upon them increases in proportion to the mass to be moved.

Prop. 31.—The force of gravity decreases, as the square of the distance from the attracting body increases.

This is proved by astronomical observations.

Motion produced by joint forces.

Prop. 32.—When a body is acted upon at the same moment by a plurality of forces, each of these forces produces its full effect: and the place of the body at the end of any given time is the same as it would have been if the forces had acted in succession each during that time, pp. 55, 56, 57.

Fig. 18.

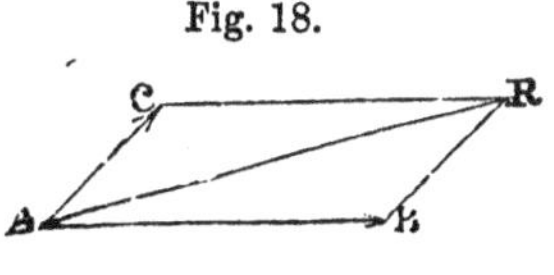

Thus let A B represent the direction of a force that would move a body, A the distance from A to B in a certain interval of time, (a second for example,) and A C, the direction of a force that would propel the same body from A to C in the same interval of time. Suppose the first force acted alone, it would move the body from A to B in one second; if the force A C then acted at B, by drawing B R equal and parallel to A C, B R will represent the direction and velocity of the force A C, and R the position in which the body would be in at the end of the second interval of time. Unite A and R and the line A R will represent the course of the body A if acted upon at the same moment by the two forces A B and A C, and R the position of the body at the end of the first interval of time.

Fig. 19.

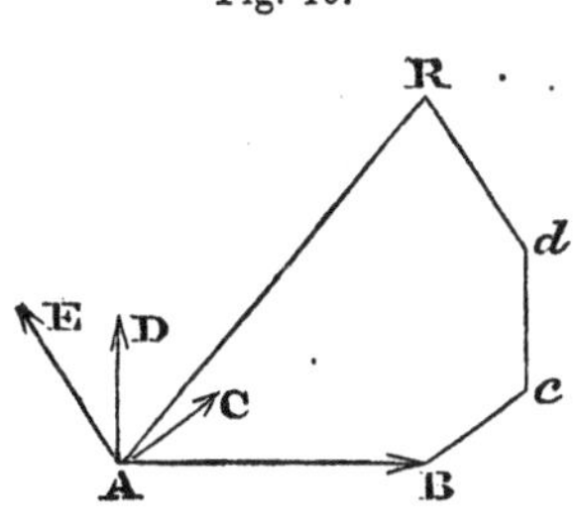

In the same manner the action of any number of forces may be represented. Thus let A B, A C, A D, A E, represent the separate effects of four different forces acting in the same plane, capable of moving a body the distances A B, A C, A D, A E, in a given interval of time. Draw B c, c d, d R, equal and parallell to A C, A D, A E, respectively, and join A R, A B, c d R, will represent the path of the body if these forces had acted successively each during one interval of time, and A R the path of the body if they all act together, and R the position of the body at the end of the first interval of time.

Prop. 33.—The line A R in figures given to illustrate the preceding propositions represents the direction and measure of a single force equivalent to all the others in each figure; and hence the process by which it is determined called the *composition of forces*, pp. 55, 56, 57.

Prop. 34.—Any force may be decomposed into any number of other forces, that shall be equivalent to it, by the reverse of the foregoing operation. This process is called the *Resolution of forces*, p. 57.

Thus the force A R fig. 18, may be separated into two forces A B, A C, and the force A R, fig. 19, into four forces, A B, A C, A D, and A E.

Prop. 35.—When the forces act in the same right line, we have only, in order to ascertain the spaces described by their combined action, to add or substract the spaces which would be described by their separate action, according as these forces act, in the same or opposite direction.

Equilibrium.

Prop. 36.—A body acted upon by a plurality of forces, in opposite directions, will remain at rest, or in *equilibrio;* when these forces were supposed to act in succession each during the same interval of time, the body would arrive at its point of departure.

The simplest and most evident case of *equilibrium* is that in which a body is acted upon by two equal and opposite forces.

On the joint action of an impulsive and a constant force.

A. *When these forces act in the same manner.*

Prop. 37.—When the forces act in the same direction, the place of the body at the end of any given time, may be determined, as in the problem of the composition of forces, by supposing, first, that the impulsive force acts during that time, and then that the action of the constant force commences and acts alone during the same time: the spaces added altogether will give the space passed over by the joint action of these forces during the assumed time.

Prop. 38.—When the forces act in opposite directions, the place of the body may be ascertained by a similar process; in this case, however, the spaces are to be substracted one from the other, pp. 58, 59.

When a constant force is acting in a direction contrary to that of a moving body set in motion by an impulsive force, the retardation that the former produces may be determined by comparing the motion with that of a body moved by the same force.

The degrees by which an ascending body loses its motion, are the same as those by which it is again accelerated at the same points, when it has acquired its greatest height and again descends, for the velocity at the corresponding parts of the ascent and descent are equal. Thus we may calculate to what height a body will rise when projected upwards by an impulsive force, gunpowder, for instance, and retarded by the force of gravity. Since the force of gravitation produces or destroys a velocity of 32 feet in every second, a velocity of 320 feet will be destroyed in 10 seconds; and according to what has been premised, a body will fall in 10 seconds through a hundred times 16 feet or 1600 feet, which is therefore the height to which a velocity of 320 feet in a second will carry a ball projected, without resistance from other cause than gravity, in a vertical direction, p. 60.

B. *When these forces act in different directions.*

* *When the successive directions of the constant forces are parallel.*

Prop. 39.—If the constant force be that of gravity, the successive directions of which are assumed to be parallel, the investigation of the effects produced constitutes the doctrine of projectiles; a projectile being a body thrown in any direction by an impulsive force and at the same time acted upon by the force of gravity, pp. 59, 60.

Prop. 40.—The place of a projectile at the end of any given time may be determined as in the problem of the composition of forces, by supposing first that the impulsive force alone has acted during that time, and then that the action of gravity commences, and acts alone during the same time.

Thus let A H represent a horizontal plane, and A B the initial direction and velocity of a body projected from the point A in the same plane. If the impulsive force alone acted on the body it would describe the path A B B′ B″ B‴ &c. with uniform velocity. But as the force of gravity acts from the moment of projection, the body will be drawn downwards from the line A B‴ so as to be found after the successive intervals of time, at the points g g′ g″, &c., and as the force of gravity produces a velocity which increases as the squares of the distances, if the distances A B, B B′, B′ B″, B″ B‴ be equal, B g, B′ g′, B″ g″, B‴ g‴, &c., will be as the squares of these distances, and the path of the projectile through the points g g′ g″ g‴ will be a curve, and this curve mathematicians have called a parabola.

Fig. 20.

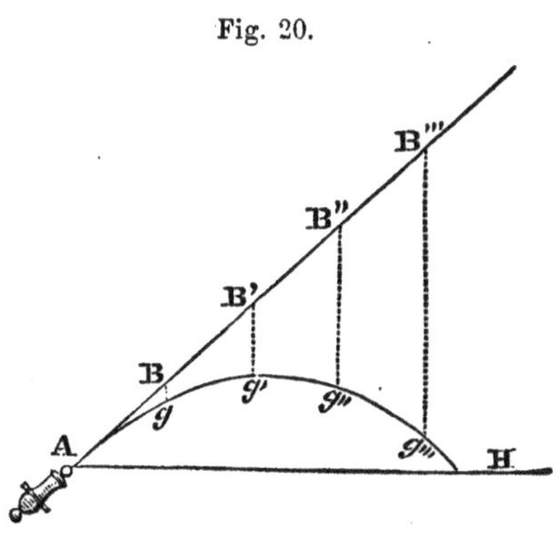

*** When the successive directions of the constant force tend to a common centre.*

Prop. 41.—This case constitutes the doctrine of central forces, see prop. 13, p. 75.

Prop. 42.—The place of the body at the end of any given time may be determined here also by the problem of the resolution of forces.

Thus, suppose A represent a body impelled towards H with such a force, as, by itself, would enable it to run over the equal spaces A B, B F, F G, &c., in equal portions of time: suppose likewise that it is acted upon the same time by constant force, which would enable it to pass over the unequal spaces A I, I K, K L, &c., in the same equal portions of time. It is evident, that the joint action of both these forces would compel the body A to pass over the curvilinear path A N O P, &c. Through B draw the line B C, (viz., in the centre of attraction); through I draw I N parallel to A B; and at the end of the first portion of time the body will be found at N, whence it would proceed in the straight direction N R (by the first law of motion), if the constant force then ceased

Fig. 21.

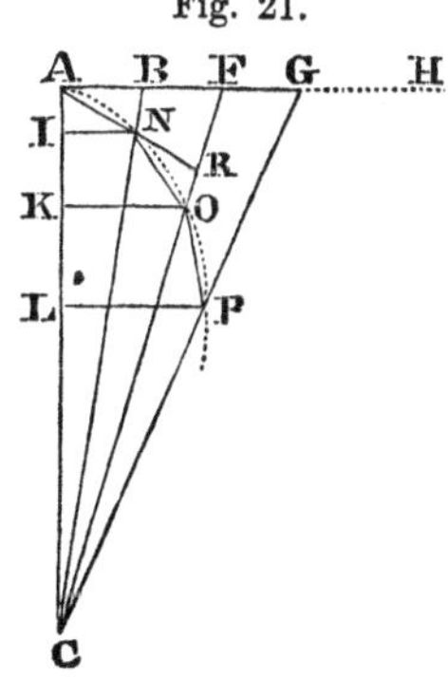

to act. But as this force continues to act, the body at the end of the second portion of time will be found in O; for the like reason, at the end of the third portion of time, it will be found in P, and so on. The course then A N O P is not straight, but consists of the lines A N, N O, O P, forming certain angles with each other. Now it will not be difficult to conceive that, because the attractive force acts not by intervals but constantly and unremittedly, the real path of the body must be a polygonal course, consisting of an infinite number of sides; or more justly speaking, a continuate curved line, which passes through the points A, N, O, P, &c., as is shown by the dotted line.

Prop. 43.—Should the action of the centripetal force cease at any instant, the body would proceed straight forward, p. 49.

The portion of the impulsive force by which this is affected is called the *centrifugal*, prop. 14.

Prop. 44.—Whilst the distance from the centre remains unchanged, as when the body moves in a circular orbit, the centripetal and centrifugal forces are equal.

Laws of Central forces.

Prop. 45.—When bodies revolve in equal circles, their centrifugal forces are proportional to the squares of their velocities.

Prop. 46.—When two bodies revolve with equal velocities at different distances, the centrifugal forces are inversely as the distances.

Consequently (prop 45, 46) the centrifugal forces are in all cases directly as the squares of the velocities, and inversely as the distances.

Prop. 47.—When two bodies revolve in equal times at different distances, their centripetal forces are simply as their distances.

In general, the centripetal forces are as the distances directly and as the squares or the times of revolution inversely.

Prop. 48.—When the forces vary inversely as the squares of the distances, as in the case of gravitation, the squares of the times of revolution are proportional to the cubes of the distances.

Thus, if the distance of one body be four times as great as that of another, the cube of 4 being 64, which is the square of 8, the times of its revolution will be 8 times as great as that of the first body.

Prop. 49.—Where the orbit deviates more or less from a circular form, a right line joining the revolving body and its centre of attraction, always describes equal areas in equal times, and the velocity of the body is therefore always inversely as the perpendicular drawn from the centre to the tangent; and the velocity at any point less than three-eighths, greater than that necessary to make the body describe a circle.

Prop. 50.—To propel a body in an elliptical orbit, the force directed to its focus must be inversely as the square of the distance.

This is proved by astronomical observations, but we have no other proof of it.

The motion of the planets round the sun in the solar system is governed by the laws of central forces, the centripetal force in this case being that of gravity.

On the joint effect of active and inactive forces.

A. *When they have opposite directions.*

Prop. 51.—The effect of passive forces is to restrain and modify the action of other forces so as to confine the motion of a body to a particular course or path, and the direction of the passive force affecting a body at any moment s the line perpendicular to that part of this path at which the body is found at this moment. If the direction of the active force be also perpendicular to this path, the body must evidently remain at rest, since no part of this force can be resolved into the direction of the path in which alone the body can move.

B. *When they have different directions.*

General rule.

Prop. 52.—Resolve the active force into two, one perpendicular, and the other a tangent to the path of the body, the effect of the former force will be entirely destroyed (*prop.* 51,) and the body will advance by the latter alone.

** On the motion of a body impelled obliquely against a plane.*

Prop. 53.—Let M N represent the plane, and A B the direction and velocity from the impulsive force, resolve A B into the forces A C perpendicular to the plane, and C B in its direction, then by the general rule (*prop.* 52) the body will move along the plane with a velocity of which C B is the measure.

Fig. 22

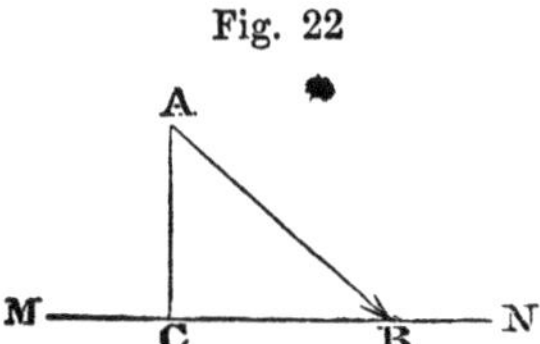

*** On the motion of a body impelled obliquely against a curved surface.*

Prop. 54.—Let M N represent the curve and A B the direction and velocity from the impulsive force. Resolve A B into two forces, C B perpendicular to the curve at B, and B D (equal to A C) a tangent to the curve at the same point. Then B D will represent the velocity at the point B.

Fig. 23.

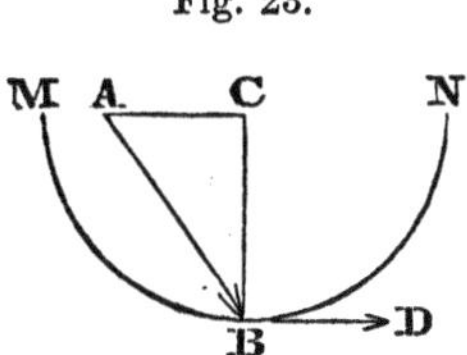

Prop. 55.—If the curve be interrupted at any point, or change the direction of its concavity, the body will advance with its last velocity in a tangent to the curve at that point.

**** On the descent of a body along an inclined plane.*

Prop. 56.—Let M N represent an inclined plane and A B (perpendicular to the horizontal base H N) the force of gravity as measured by the distance which would cause a body to descend in the first second of time. Resolve A B into two, A C, perpendicular to the plane, and C B in its direction, then the body will be urged down the plane by the constant force measured by C B.

Fig. 24.

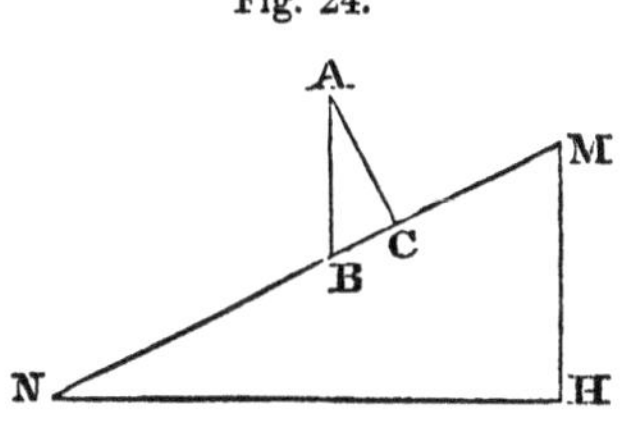

Laws of the descent of bodies down inclined planes.

Prop. 57.—1st. The motion of a body drawn down an inclined plane is uniformly accelerated.

Prop. 58.—2d. The velocity acquired is proportional to the perpendicular descent, so that a body falling from M to H has the same velocity at H as one descending the whole length of the plane at N.

Prop. 59.—3d. The times of descent down planes of the same heights are as their lengths.

Prop. 60.—4th. The times of descent down all planes which are cords drawn to the lowest point of the same circle, are equal.

Fig. 25.

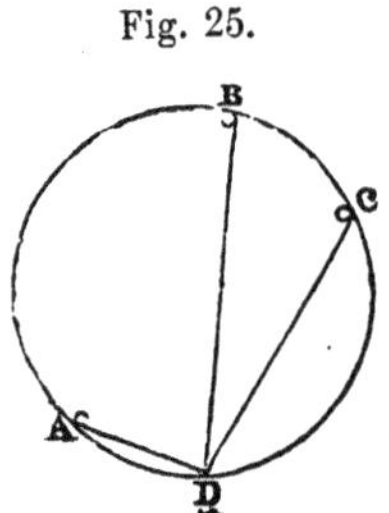

Thus, if the balls A, B, C, be placed at different points of the circle and suffered to descend at the same instant along as many planes which meet at the lowest point of the circle, they will arrive there at the same time.

Or it may be enunciated in the following terms: the times of descent down all the cords drawn from the same point or circumference of a circle will be the same.

This will be made evident by supposing the above figure inverted, D being made the upper point and the balls allowed to fall from that point to A, B, and C.*

**** *On the descent of a body down a vertical curved line.*

Prop. 61.—The times of descent down the cords of different circles are to each other as the square roots of their diameters.

Prop. 62.—If a body fall from a state of rest down a curve, the velocity acquired is equal to that which it would have by falling through the same perpendicular height.

For if the curve be considered as made up of an infinite number of contiguous planes, it is evident that the angle of inclination of any two of these adjacent planes is infinitely small, or nothing, and consequently there is no velocity lost by a change of direction in passing from one to the other. Therefore, as the effect of gravity is not impeded, the truth of the proposition becomes evident.

Prop. 63.—If a body be projected up a curve, the perpendicular height to which it will rise is equal to that through which it must fall to acquire the velocity of projection. For the body in its ascent will be retarded in the same degree that it was accelerated in its descent.

Fig. 26.

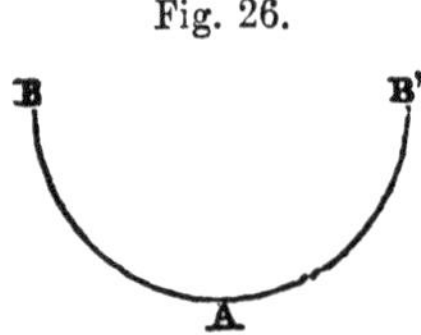

Thus, let B A B' be a curve in which the lowest point is A, and the parts A B, A B' are similar; a body in falling down B A will acquire a velocity that will carry it to B', and since the velocities in all equal altitudes in the ascent and descent are equal, the times of ascent and descent are equal.

The foregoing proposition is equally true whether the body actually move over a solid surface or be retained in its path by a string which is in every part perpendicular to it.

Of the simple Pendulum.

Prop. 64.—The simple pendulum is conceived to be a mere material point suspended by an imponderable and inextensible thread, p. 60.

Prop. 65.—If the simple pendulum vibrates through very small arcs, these may, without sensible error, be conceived to coincide with their chords, and we may derive from this consideration the following theorems:

1st. As the times of descent of the body down different chords of the same verticle circle are equal (*prop.* 60.) the vibrations of the same pendulum, although performed through unequal arcs, will be very nearly equal, p. 61.

2d. The times of vibrations of different pendulums will be to each other as the square roots of the lengths of these pendulums, or, which is the same thing, their lengths are proportioned to the squares of the times of vibration, p. 61.

The times of descent down the chords of different circles are the same as would be occupied in descending vertically through their diameters, and are consequently proportional to the square roots of these diameters.

Of the impact of bodies.

Prop. 66.—When a body in motion strikes directly another body, it always communicates motion to the second body, and loses part of its own, and from the third law of motion it is evident that the momentum gained by the second body is exactly equal to that lost by the first.

Prop. 67.—When one non-elastic body strikes against another, the two bodies will move on together since there is no force to separate them; and as one of the bodies gains all the momentum which the other loses, the momentum after impact will be equal to the sum of the momentum before impact.

Prop. 68.—When an elastic body strikes against another, the second is impelled forward with double the momentum which it would have received under the same circumstances if non-elastic.

For at the moment of impact the form of the body struck is changed by a force equivalent to the momentum which it receives from the striking body, and if this body be perfectly elastic, its form will be restored to it by a force exactly equal to that by which it was changed, and this force (which we have just seen to be equal to the original impulse,) will be exerted in driving the body forward. The body thus receives, besides its original impulse, the equal force of the re-bound.

Prop. 69.—The striking body, when elastic, is also acted upon by the rebound, and loses twice as much momentum as it would have lost if non-elastic.

In this case, as in the former, the sum of the momenta is the same after impact as before it; but the bodies after impact do not move on together.

Prop. 70.—If an elastic body strike against a firm plane, the angle of reflection will be equal to the angle of incidence, p. 66.

PART II.

PHENOMENA OF SOLIDS.

THE FOUR FUNDAMENTAL TRUTHS USED TO EXPLAIN THE PECULIARITIES OF STATE AND MOTION WHICH DEPEND ON THE SOLID FORM OF BODIES; A DEPARTMENT COMMONLY CALLED MECHANICS.

ANALYSIS OF THE CHAPTER.*

A force, which moves part of a solid body, must effect the whole or break off the part.

If the force be directed towards a certain central point in the mass, it will effect the whole equally, whether simply to support the mass, or to move it or to stop it when in motion. The point, according to circumstance, is called THE CENTRE OF GRAVITY OF INERTIA, *or* OF ACTION.

In solid bodies moving about an axis, as exemplified in a wheel or weighing beam the various parts describe circles or move through spaces which are greater in proportion to their respective distances from the centre of motion. Hence forces differing as to speed, may still, through a solid medium, be brought exactly to co-operate or to oppose one another; a slow force counter-balancing or being equivalent to a quicker one, provided that it be more intense in proportion as it is slower. The SIMPLE MACHINES, *or* MECHANICAL, MEDIA *called* LEVER, WHEEL AND AXLE, PULLEY, INCLINED PLANE, WEDGE, SCREW, *&c., are so many arrangementsof solid parts, by which forces of different velocities and intensities maybe thus connected or opposed, or may be conveniently substituted one for another.*

By solid connecting parts, also, the direction of any existing motion or force may be changed, as when the straight motion of running water is converted into the rotary motion of a water-wheel, &c. Hence arises an endless variety of COMPLEX MACHINES.

In all machines, an important circumstance to be considered is the resistance among moving parts which arises from FRICTION :—*and in solid structures generally, the forms and positions of parts have to be adjusted to the* STRENGTH OF THE MATERIALS, *and to the strains which the parts have to bear.*

"*Solid*" is the term applied to a mass in which the mutual attraction of the atoms is so strong, that the mass may be moved about as one body, without the relative positions of the component parts being thereby disturbed.

"*Force moving part of a solid must effect the whole or break off the part.*"

This is a necessary consequence of the description or definition of a solid just given. And it follows that in all cases of breaking, the cohesion of the

* The reader should here re-peruse the general table or synopsis at page 19.

atoms at the fractured part must have been less strong than the weight of the remaining mass, or its inertia resisting the degree of change attempted, or the force fixing it to its place, or than some combination of these particulars.

The sharp blow of a hammer given to an ivory ball, causes it to dart off swiftly, but does not injure it, because the cohesion among the atoms struck is stronger than the opposing inertia of the mass, even under a rapid change; but the blow of a hammer on a large elephant's tusk indents or breaks the part because the opposing inertia of the larger mass is stronger than the cohesion of the atoms which receive the blow.

A vessel of pottery-ware may be safely suspended by its handle; proving that the cohesion which fixes the handle to it is stronger than the weight of the vessel; but if the attempt be made to lift the vessel quickly, the handle may rise and leave the vessel behind; because then the weight and inertia are acting together to destroy the cohesion. Thus servants attempting to lift too quickly the loaded stone-ware dishes at a dinner-table, often break off the part by which they take hold.

Centre of Gravity or Inertia.

If any uniform beam or rod be supported by its middle, like a weighing beam, the two ends will just balance each other. This is in accordance with the general truth or law of *attraction* already explained; for as there is just as much similarly situated matter on one side of the support as on the other, there will also be just as much attraction, and therefore no reason why the matter on one side should overpower that on the other. If equal weights be afterwards attached in corresponding situations on the two arms of the beam the balance will not be thereby disturbed; and the operation of adding weights that counterpoise, above and below, and near and far from the centre may be continued, until a bulky mass is built up upon the beam—and instead of a beam a wheel may be used—yet the whole will remain perfectly supported and in equilibrium about the original centre. In the pages now to follow, it will be shown that, in every body or mass, or system of connected masses, in the universe, there is a point of this kind about which all the parts balance or have equilibrium, and it is this point which is called the *centre of gravity* or of *inertia*. Although in any mass, therefore, every atom has its separate gravity and inertia, and the weight and inertia of the whole are really diffused through the whole, still by supporting this one point, either from above or from below, the whole mass is equally supported; by lifting it, the whole is lifted; by stopping it, the whole is brought to rest; and when it rises or falls, the general mass is really rising or falling. Thus for many purposes, a body, however large, may be considered as compressed into or existing only in the single point called its *centre of gravity* or of *inertia*.

This centre in a mass of regular shape and of uniform substance, as a ball or cube of metal, is easily found, because it is the evident centre of the form; but in bodies that are irregular, either as to density or form, it must be found by rules of calculation hereafter explained.

To say that the centre of gravity will always take the lowest situation which the support of the body will allow, is only to repeat, that bodies tend by their gravity towards the centre of the earth. In a suspended body, therefore, as the lowest situation which the centre of gravity can find is, when it is immediately under the point of suspension, all bodies hanging freely must have their centre of gravity directly under that point. A plummet is an interesting example of this; and the truth furnishes, in many cases of irregular masses, a very simple practical mode of finding the centre.

Thus if an irregular piece of plank or of pasteboard, represented here by the figure *a e b d*, be suspended from any point, as *a*, and the cord of a plummet *a g* be attached at the same point, the centre of gravity of the board must be somewhere in the direction of the plummet, and a chalk line left on the board where the cord touched it, must pass over the centre of gravity. If the board be then suspended by another point, as *d*, and another chalk line *d e* be made in the same manner, the place *c*, where the two lines cross or cut each other, will indicate the centre of gravity; and the board when supported by a cord attached there, will hang evenly balanced.

Fig. 27.

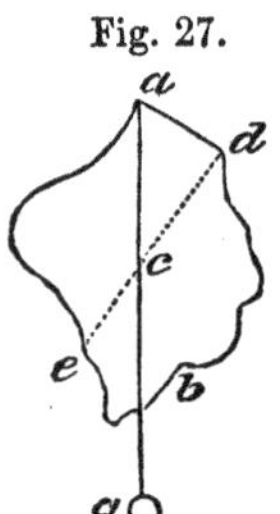

The following cases further illustrate the truth, that the centre of gravity always seeks the lowest place. They seem at first to be exceptions to the law; but when more fully considered, are interesting proofs of it.

Fig. 28.

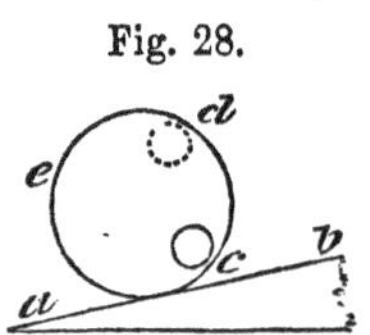

A wooden cylinder or roller *e d c*, placed on a slope or inclined plane *a b*, will naturally descend, because its centre of gravity is thereby approaching the earth; but if there be a heavy mass of lead *c* introduced at one side, which must rise before the roller can descend, the rise of the mass being contrary to gravity, the motion will be arrested. Indeed if the roller were placed on the plane with the lead in the position *d*, the lead would fall down to the position *c*, and so would move the roller towards *b*, exhibiting the singular phenomenon of a body rolling up hill by the action of its weight.

If a billiard-ball be placed upon the small ends of two billiard sticks or cues *a b* and *c d*, laid on a table with their points *c* and *a* in contact, but with the larger ends *b* and *d* so far apart that there may be just room for the ball to touch the table between them, the ball will roll along between the cues, sinking gradually from its high situation near their points, to its lower situation near *b*. To a careless observer, it would then have the appearance of rolling upwards, because the cues on which it rests are thicker towards the end *d* and *b*; but it would really be descending in obedience to gravity. If a double cone, as represented at *f*, were substituted for the ball, it would similarly roll from *c* to *e*, and with still more of the fallacious appearance of rolling upwards, because its ends would always be resting on the upper and rising surfaces of the cues.

Fig. 29.

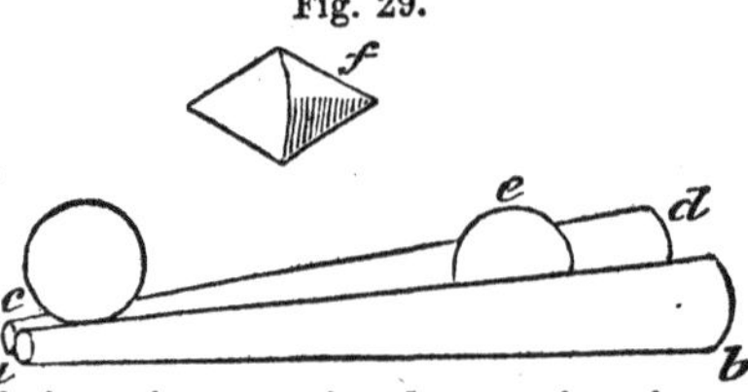

The board or stick *c d* resting on the edge of the table *a b* would naturally fall if left to itself, because more than half of it is beyond the edge of the table; but strange to say, an additional weight *e* attached to its projecting part as at *b* by the cord *b e*, instead of pulling it down faster, shall fix or steady it on the table, provided the weight be pushed inwards a little by a rod *d e* resting against it and against a niche in the stick at *d*. It is evident that the stick *c d*, in falling, must turn round the edge of the table at

Fig. 30.

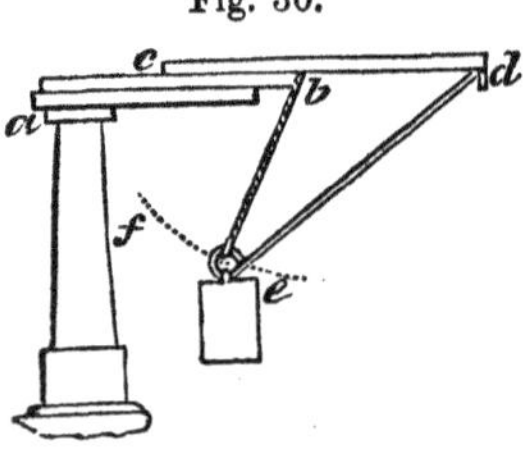

b; but in so doing, after the arrangement now supposed, it must lift the weight *e* along the path *e f*—which rise, as the weight is heavier than the stick (that is to say, as the common centre of gravity of the connected objects is near *e*,) gravity forbids, and therefore the stick and weight will both remain supported by the table. An umbrella or walking cane, hanging on the edge of a table by a crooked handle, is another instance of the same kind. And the common toy of a little man standing on tiptoe upon the top of a pillar, and supporting two leaden bullets by wires descending from his hands, is another combination of parts which places the centre of gravity of the whole the support, making the combination a kind of pendulum.

By attending to the centre of gravity of the bodies around us on earth, we are enabled to explain why, from the influence of gravity, some of them are stable or firmly fixed, others tottering, others falling.

If we find that a body, from its form or position, cannot be overturned, without its centre of gravity being lifted,—knowing now that the general mass is then lifted in the same degree, we see why a weak cause cannot effect the change. The rise of the centre of gravity, or body, in any case of falling over when the centre of gravity is over the middle of the sustaining base, will be proportioned to the breadth of the base of the body, compared with the height of the centre of gravity above the base. This is shown in the annexed figures of which the two particulars of *base* and *height* are combined

Fig. 31.

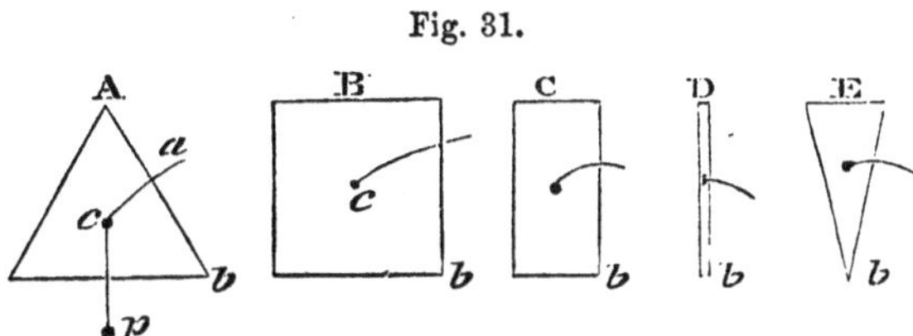

in a series of proportions. In the figures, the dot *c* marks the place of the centre of gravity, and the curved line beginning from the dot marks the path of the centre of gravity, when the body is overturned. This curved line is a portion of a circle which has the edge or extremity of the base (*b*, in fig. A.) as a centre, because the body in turning must rest upon such extremity or corner as the centre of its motion. The farther inwards, therefore, from this extremity that the centre of gravity is, as marked by where a plumb-line as *p*, hanging from it, crosses the base, the farther, of course, is the centre of gravity from the top of the circle which it has to describe in moving, and the steeper, consequently, will be its commencing path; and as in the case of bodies made to roll up slopes, the steeper the ascent, the greater will be the force necessary to give motion.—The line of a plummet hanging from the centre of gravity is called the *line of direction* of the centre, or that in which it tends naturally to descend to the earth.

In fig. A, which has a broad base and little height of the centre of gravity, we see that the centre must rise almost perpendicularly before it can fall over, and the resistance to overturning is therefore nearly equal to the whole weight of the body. Hence the firmness of a pyramid.

In figures B, C, and D, progressively, the commencing path of the centre is less steep, because the base is narrower, and hence the bodies are so much the less stable. B, may represent an ordinary house, C, a tall narrow house, and D. a lofty chimney.

Fig. E, shows a tottering position, for the centre of gravity being directly over a base which is a mere point, the least inclination places it on a descending slope, and the body must fall.

Fig. 32.

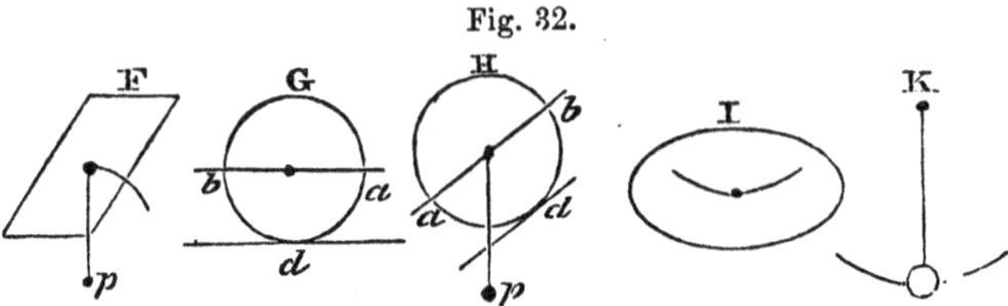

In F, the position is tottering on one side, and stable on the other. This explains how the least inclination of a standing body virtually narrows, in one direction, its sustaining base.

In G, which represents a ball upon a level plane, the whole mass is supported on a single point, as in E, yet the body has no tendency to move, because, in any other possible position, the centre would still be as far from the sustaining plane. In moving, the centre describes the straight level line *a b*.

In H, the ball is on an inclined plane, and rolls down, the centre of gravity describing the oblique line *b a*.

In I, which is an oval body resting on a level plane, when the body is moved to either side, the centre of gravity must rise, as in the case of a pendulum. Hence an oval body on a level will rock or vibrate like a pendulum.

K, is a true pendulum, whose centre of gravity describes the curve here shown, as explained in Section II., at page 60.

The importance of the subject of the centre of gravity will be farther judged of by the facts which are now to be reviewed.

A cart loaded with metal or stone may go safely along a road of which one side is higher than the other, as here shown, but were the same cart loaded with wool or hay it would be overturned; because, although the sustaining base be the same in the two cases, the *line of direction* falls much within it from the low centre of gravity of the metal at *c*, but falls very near the wheel at P, or altogether on the outside, from the high centre of the wool at *a*, and in the latter case the centre has offered to it a descending path.

Fig. 33.

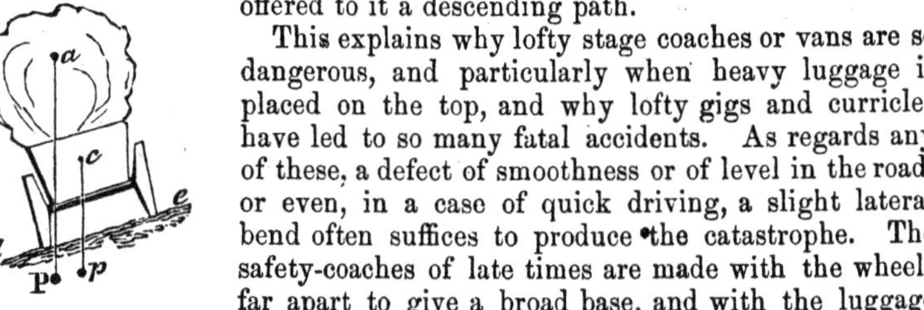

This explains why lofty stage coaches or vans are so dangerous, and particularly when heavy luggage is placed on the top, and why lofty gigs and curricles have led to so many fatal accidents. As regards any of these, a defect of smoothness or of level in the road, or even, in a case of quick driving, a slight lateral bend often suffices to produce the catastrophe. The safety-coaches of late times are made with the wheels far apart to give a broad base, and with the luggage receptacles and seats for outside passengers placed low down before and behind the body of the carriage, instead of on the top, as formerly.

The feet of tripods are generally expanded below to give a broad base. The same is true of our common chairs; but a thoughtless child often leans so far over the back of the chair, that he causes the line of the general centre of gravity to fall beyond the base, and the chair with its load is overturned.

The small lofty chairs made to raise children to the parents' elbow at the dinner-table, are very dangerous if the feet are not made to spread much. Pillar-and-claw tables, candle-sticks, table-lamps, and many other articles of household furniture, have stability given in the same manner.

The least inclination of a standing body virtually narrows the supporting base.

This truth is explained by *fig.* F. It shows the necessity of building the thin walls and tall chimneys of modern houses perfectly upright. And hence the extreme importance and utility of that simple instrument, the *plummet* or *plumb-line,* which, when applied to a body, is a visible indication of the line of its centre of gravity. The mason and many other workmen cannot proceed a step without their guiding plummet.

The brick walls of ordinary houses are so thin, that, to have standing strength, they require to rest against one another; and hence they occasionally exhibit the kind of stability which belongs to a child's house built of cards. As contrasted with the masses of masonry which remain to us from antiquity, resting on firm-spreading basements, they are examples of what is truly ephemeral, in comparison with that which has partaken of the permanency of nature's own works, covering regions with mighty ruins. What magnificent illustrations of strength and durability, dependent on proportions, are those ancient pyramids and temples, which still give such interest to the banks of the Nile, and to the valleys and plains of Asia!

There are many remarkable structures on earth which lean or incline a little; yet so long as the line of their centre of gravity remains within the base, and the parts of the mass have tenacity among themselves sufficient to hold together the structure will stand. The famous tower of Pisa was built intentionally inclining, to frighten and surprise: with a height of one hundred and thirty feet, it overhangs its base sixteen feet, and assumes nearly the air of fig. F. in page 88.

The tall monument near London Bridge inclines so much, that in high winds, from a particular quarter, timid minds have doubted of its stability.

And many of the most lofty and beautiful of our cathedral spires or towers, as that of Salisbury, have lost something of their perpendicularity.

An oval body on a flat level surface, as already explained by fig. I, page 88, oscillates somewhat like a pendulum, because, when disturbed from its middle position, its centre of gravity has risen and seeks to return. The same is true of any regular slice or portion of a solid globe, which will consequently always come to rest with its plane face turned directly upwards.

The rocking-horse of children and the common cradle are exemplifications of the same class.

But perhaps the most curious instances are those rocks called Loggan or Laggan stones, of which there are several among the picturesque barriers of the British coast. An immense mass, loosened in some convulsion of nature, is found with a slightly rounded base resting on a flatter surface of rock below; and is so nearly balanced, that the force of a man suffices to move it. Some of these have been objects of much superstitious veneration to their neighbourhood.

There is an amusing Chinese toy, made in obedience to the same principle. It has the appearance of a little fat, laughing man, sitting on the ground with his feet concealed under him; but where the feet should be, there is only a rounded smooth surface, with heavy lead ballast placed in it, so low, as always, when allowed, to raise the body to the erect or sitting attitude. A child pushes the little fellow down again and again, and would persuade him to be still, but is surprised to see him always up the moment after, shaking about and as lively as ever.

The vibratory motion of a pendulum, as dependent upon the circumstance of the centre of gravity having been moved from its lowest place, which it again constantly seeks, was so fully considered in the last chapter, that it need not be again dwelt upon here; but we have to enumerate the following phenomena as being of the same class:

—The vibrations of a common swing.

—The rocking of a balloon when it first ascends.

—The spontaneous shutting of those gates or doors of which the upper hinge overhangs or projects beyond the lower, causing the gate, when in the shut position, to have its lock lower than when in any other. Such a gate always returns of itself, from either side, to the shut position, just as a penpulum returns to the lowest part of its arc:—the gate, in fact, is but a sloping pendulum.

Of the same nature also is the rocking or rolling of a ship, in particular states of wind and sea. When the centre of gravity of a ship is too low, owing to all the heavy load being placed near the keel, this pendulum-motion, in rough weather, becomes excessive and dangerous.

The actions and postures of animals, and particularly of man, illustrate beautifully the observations made above with respect to the centre of gravity.

A body, we have seen, is tottering in proportion as it has great altitude and narrow base—but it is the noble prerogative of man to be able to support his towering figure with great firmness, on a very narrrow base, and under constant change of attitude. This faculty is acquired slowly because of the difficulty. A child does well who walks at the end of ten or twelve months; while the young of quadrupeds, which have a broad supporting base, are able to stand and even to move about almost immediately after birth.

The supporting base of a man is the space occupied by and included between the feet. The advantage of turning out the toes is, that without taking much from the length of the base, it adds considerably to the breadth.

If there be much art in walking on two perfect feet, there is still more in walking on two slender wooden legs, with rounded extremities:—which, however, we often see done, by mutilated soldiers and sailors.

All the ladies of the empire of China have to acquire nearly the same talent as these victims of war; for barbarous custom has crippled them, by confining their feet for life in such shoes as fitted them in infancy.

But surpassing in difficulty any of these instances is the practice, which is general among the inhabitants of the sandy plains, called the *Landes*, in the south-west of France, of walking on stilts. The *Landes* afford tolerable pasture for sheep; but during one portion of the year are half covered with water, and during the remainder are still very unfit walking ground, by reason of their deep loose sand and thick furze. The natives meet the inconveniences of all seasons by doubling the length of their natural legs, through the addition to them of the stilts mentioned, which they call *des echasses*. Mounted on these, which are wooden poles, put on and off as regularly as the other parts of dress, they appear to strangers a new and extraordinary race of long-legged beings, marching over the loose sand, or through the water, with steps of eight or ten feet in length, and with the speed of a trotting horse; their moderate journeys being of thirty or forty miles in a day. While watching their flocks, they fix themselves in convenient stations by means of a third staff which supports them behind, and then with their rough sheep-skin cloaks and caps, like thatched roofs over them, they appear like little watch-towers, or singular lofty tripods, scattered over the face of the country.

Still beyond the art of walking on stilts is that which some persons attain of walking and dancing on a single rope or wire; or even of keeping the centre of gravity above the base, while standing on the moveable support of a galloping horse.—A rope-dancer usually carries a long pole in his hand, to balance him; it is loaded at each end, and when he inclines, he throws it a little towards the side required, that the reaction may restore his perpendicularity.

Much art of the same sort is shown in the attitudes and evolutions of the skater; in the amusements of supporting a stick upright on the end of the finger; and many other feats of a like kind.

Attitudes generally depend on the necessity of keeping the centre of gravity of the body over the base under variety of circumstances, as in the straight or upright part of a man who carries a load on his head;—the leaning forward of one who carries it on his back; the hanging backwards of one who bears it between his arms;—the leaning to one side of him who is carrying a weight on the other side;—the habitual carriage of very fat people, whose head and shoulders are thrown back, giving a certain air of self-satisfaction,—an air which belongs also to the expectant mother, and even to the dropsical patient, although producing in the latter so sad an incongruity.

When a man walks or runs, he inclines forward, that the centre of gravity may overhang the base: and he must then be constantly advancing his foot to prevent his falling. He makes his body incline just enough to produce the velocity which he desires.

A man, in pulling horizontally at a load, is merely causing his body to overhang its base, so that its tendency to fall may become a force or power applicable to the work.

When a man rises from a chair, he is seen first to bend the body forward, or to draw the feet backward, so as to bring the feet or base under the centre of gravity, and then he lifts the body up. If he lifts too soon, that is, before the body be sufficiently advanced, he falls back again.

A man standing with his heels close to a perpendicular wall, cannot without falling, bend forward sufficiently to pick up any object that lies before him on the ground; because the wall prevents him from throwing part of his body backward, to counterbalance the head and arms which must project forward. A person little versed in such matters, might agree to give ten guineas for permission to possess himself, if he could, of a purse of twenty, laid on the ground before him: he of course would lose his stake.

When a man walks at a moderate rate, his centre of gravity comes alternately over the right and over the left foot. This is the reason why the body advances in a waving line, and why persons walking arm in arm shake each other, unless they make the movements of their feet to correspond, as soldiers do in marching.

Sea Sickness is a subject closely related to the present. Man requiring, as now explained, so strictly to maintain his perpendicularity, that is, to keep the centre of gravity always over the supporting part of his body, ascertains the required position in various ways, but chiefly by comparing the perpendicularity, or other known position of things about him, with his own position. Vertigo and sickness are the consequences of depriving him of his standards of comparison, or of disturbing them.

Hence on shipboard, where the lines of the masts, windows, furniture, &c. are constantly changing, sickness, vertigo and other affections of the same class are common to persons unaccustomed to ships. Many persons experience similar effects in carriages, and in swings; or on looking from a lofty precipice, where known objects being distant, and viewed under a new aspect, are not so readily recognized; also in walking on a wall or roof; in looking directly up to a roof, or to the stars in the zenith, because then all standards disappear; on entering a round room, where there are no perpendicular lines of light and shade, as when the walls and roof are covered with a paper which has no regular arrangement of spot; on turning round, as in waltzing, or if placed on a wheel; because the eye is not then allowed to rest long enough on any standard, &c.

People when in the dark, and therefore blind people, always use standards belonging to the sense of touch; and it is because, on board of a ship, the standards both of sight and touch are lost, that the effect is so very remarkable.

But sea sickness also partly depends on the irregular pressure of the bowels among themselves and against the containing parts, when the influence of their inertia and weight varies with the rising and falling of the ship.

From the nature of sea sickness, as discovered in these facts, it is seen why persons unaccustomed to the motion of a ship, often find relief by keeping their eyes directed to the fixed shore, where visible; or by lying down on their backs and shutting their eyes; or by taking such a dose of exhilarating drink as shall diminish their sensibility to all objects of external sense.

As no condition or form of matter escapes from the great laws of nature, we find the attitudes and general condition of vegetable as well as of animal bodies, characterized by the necessity of having the centre of gravity supported over the base. With what admiration may we contemplate the pine and other trees in the forests or nature, springing up to heaven as perpendicularly as if the plummet had been at work to direct them; and no less on the brows of precipitous hills than in the level plains. On a smaller scale, we see the grasses and corn-stalks of our fields illustrating the same truth. And whenever, in tree or shrub, accident or peculiar nature causes a deviation from perpendicularity, additional strength and support are provided.

Beauty of form or position is often felt to exist in bodies, merely because they possess the shape and support required, that the centre of gravity may be stable.

In architecture, how displeasing is a wall or pillar that is not quite upright; or a column with too small a base; or a very tall narrow house; or a long slender chimney. On the other hand, how beautiful in a lofty edifice is the suitable succession of columns, from the massive Doric of the basement, supporting the whole superstructure, to the light Corinthian or kindred forms seen above. The Chinese pagoda is a fine example of the union of certain requisites for stability, *viz.*, perpendicularity and expanding base, with the other qualities of perfect symmetry, graceful proportion, and fanciful ornament. When seen crowning a rising ground in a wooded island, or springing up from the centre of a rich garden, it forms, perhaps, one of the most beautiful objects which fancy has ever designed.

Beauty of attitude and *grace of carriage* in the human individual are in great part referable to the same principle.

The postures of opera dancers might pass as intentional illustrations of the number of ways in which the centre of gravity may be kept above a narrow base, by counteracting one disturbing motion or extension of a limb by some opposite and corresponding motion. The common statute of the god Mercury on tip-toe is a permanent familiar illustration of such a beautifully balanced attitude.

Grace of carriage includes not only a perfect freedom of motion, but also a firmness of step, or steady bearing of the centre of gravity over the base. It is usually possessed by those who live in the country, taking much and varied exercise, or who make gymnastics a part of their discipline. What a contrast is there between the gait of the active mountaineer, enjoying the consciousness of perfect nature, and that of the mechanic or shop-keeper, whose confinement to the cell of his trade soon produces in his body a shape and air corresponding to it—and in the softer sex what a difference is there, between that active and graceful fair one who recalls to us the fabled Diana of old, and that other sedentary being who, having scarcely trodden but on smooth pavements and carpets, under any new circumstances, carries her person as if it were a load quite new and foreign to her.

The *centre of gravity* is also the *centre of inertia.* When a person lifts a uniform rod by its middle, the inertia of both ends being equal, he overcomes it equally, and raises them evenly together. When he lifts by a part nearer to one end, the shorter and lighter portion having less inertia will rise the first, and there will be a turning motion of the rod round the finger as a centre, proportioned to the excess of inertia in the greater side.

The *centre of gravity*, or *inertia*, however, is not necessarily in the centre of the mass; for if a weight of three pounds, *a*, be affixed to one end of a rod, and a weight of only one pound, *b*, be affixed to the other, the two will still be balanced, if supported or lifted by a point of the rod, *c*, three times nearer to the centre of the large weight than to the centre of the small one. This fact is explained under the head of *lever*, a few pages hence. For the sake of simplicity, in describing such experiments, the weight of the connecting rod itself is neglected.

Fig. 34.

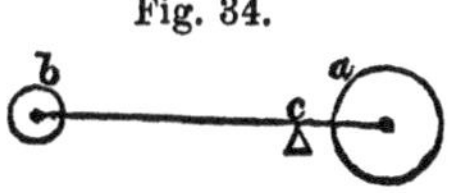

The *centre of gravity* or *inertia* is also the *centre of centrifugal force*:—for if the balls *a* and *b* of the last figure were made to spin round a common centre, as by making the connecting rod rest and turn upon a point or pivot at *c*; unless the point *c* were the centre of inertia of the two, the pivot would always be drawn in the direction of that end of the rod at which there was the greatest centrifugal force. It is on this account that in the case of a mill-stone, or great fly-wheel, or of the balance wheel of a watch, the axis must pass through the centre of inertia, to prevent its being more worn on one side than on the other.

When we say in astronomy, that the earth revolves round the sun, or that the moon revolves round the earth, we do not speak with absolute correctness, for in all such cases, both bodies are revolving round the common centre of inertia of the two. In the case of the sun and earth, as the former is about a million times larger than the latter, the common centre of inertia of the two is a million times nearer to its centre than to the centre of the earth, and is therefore within its body or circumference.

The *centre of inertia* in a body moving evenly is also its *centre of action* or *percussion;* because, if such centre come against an obstacle, the whole momentum of the body acts there and is destroyed; while if any other part than the centre hit, the body loses only part of its momentum, and revolves round the obstacle as a pivot or centre of motion, to pass it on the side towards which the greater inertia happened to be.

In a hammer, or a bar of iron used as a hammer, or in a pendulum, the motion is not said to be *even*, because the velocity of the different parts is different, being greatest far from the hand or centre of motion, and the centre of all the motal inertia is nearer to the fast moving end than to the other. Its exact place, in many cases, is easily ascertained by calculation. In a uniform rod moving as a pendulum, for instance, it is at a distance of one-third from the lower end. In the pendulum it is called the *centre of oscillation.*

If a man use a bar or rod of iron as a hammer, he must take care to make it strike the object by its centre of action, or his own hand will receive a part of the shock. A very heavy mass thus carelessly used will seriously strain the wrist. In a common hammer, as the chief part of the matter is at the end, the centre of percussion is there too, and no precaution of the kind mentioned is required.

If a rod or small log of wood be suspended horizontally by a string tied to its middle, or be floating in water, and if a forward blow be given directly across it near to one end, the other end will be found, in the first instant to have moved a little backward, or in a direction contrary to the blow, as if the rod had been fixed upon an axis. The inertia of the general mass, by resisting the motion becomes in effect a fixed axis. This fact is amusingly illustrated by laying the ends of a long stick on two wine-glasses, and then breaking the stick by a smart downward blow of a poker on its centre. Instead of breaking the glasses also, as by such a blow might be expected, the ends of the stick rise at the instant of the stroke, to turn round certain *centres of resistance* in the fragments, as at *a* and *b* , and then fall harmless on the table.

Fig. 35.

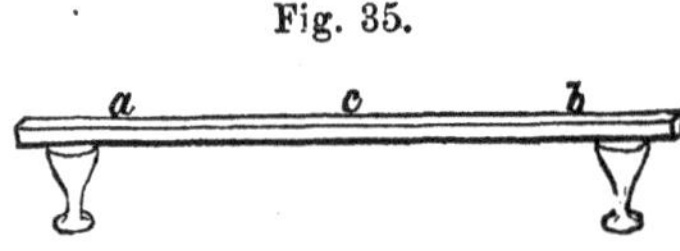

In this *section* we have seen what admirable simplicity is given to many of our reasonings and operations, by considering bodies in reference only to their *centre of inertia*, under one or other of its names.

"*In a solid body moving about an axis, like a wheel or weighing-beam, the different parts have different velocities, according to their respective distances from the axis or centre.*" (Read the Analysis.)

The truth of this proposition is perceived at once on comparing the motion in the rim of a wheel, or near the ends of a weighing-beam, with that in parts nearer the centres. Suppose *a d* to be a line drawn across a wheel, or, along a weighing-beam, the centre of motion in either case being at *c*; then the outer circular line or path, *a c*, which a point in *a* describes when moving, is longer than the corresponding inner line, *b f*, which a point at *b* describes in the same time, as *a* is farther from the centre than *b*. This admits of easy mathematical demonstration, and is indeed merely an instance of the truth, that the proportions existing between any parts or lines in one circle, hold with respect to the corresponding parts and lines in all circles.

Fig. 36.

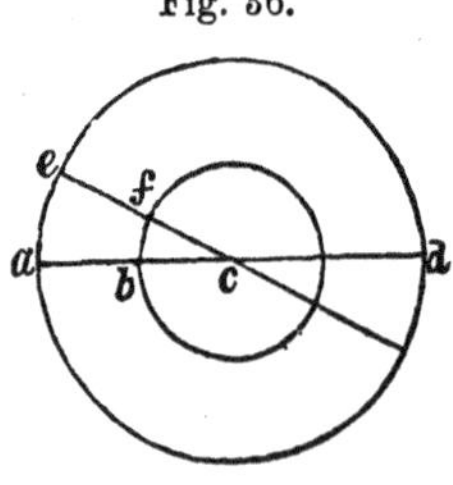

"*Hence forces with different speed may still be placed in continued connection or opposition; and they will balance or be equivalent, if the one be as much more intense than the other as it is slower.*" (Read the Analysis.)

This is the important truth upon which the whole of mechanics may be said to hinge. It gives to man the *simple machines* or *mechanical powers*, as they have been called,—the Lever, Wedge, Pulley, &c., which enable him to adapt any species and speed of power which he can command to almost any work which he has to accomplish: and the discovery of it and of means to apply it may be said to have subjected external nature to his control. His works are of a thousand kinds, from the displacing of a rock to the spinning of a delicate thread; while the natural powers or forces at his command are chiefly wind, waterfalls, fire and animal effort—and of which in any particular case, he may have only one kind at his service;—still, being able to connect together his power and resistance by solid media, of which different parts move with any desired difference of velocities, he can employ any force for a purpose of almost any kind.

There is, however, a false and most pernicious prejudice very generally existing with respect to the *simple machines*, which we must begin by removing, *viz.*, that they increase the quantity of power or force applied to them. For instance, when one pound, as *a* at the end of a beam or lever, is seen balancing two pounds, as *b*, at half the distance on the other side of the axis, or four pounds as *c*, at a quarter of the distance, many persons believe that the lever itself gives or begets a force equal to the difference of the weights so balanced. But we shall now show, that levers and all the other *mechanical powers* (as from the erroneous idea above mentioned they have been called,) merely enable us to make such substitutions, so that of a small weight descending far, in place of a greater weight descending a little way, or of an inferior force working long, instead of a superior force working for a shorter time,—and thus often to accomplish ends to which the force possessed would be quite unsuited if applied directly. In other words the simple machines enable us to concentrate or divide any kind or quantity of force which we possess, so as to suit it to our various purposes, just as mill-ponds and branching channels enable us to accumulate or divide the force of a stream of water; but they no more increase the *quantity of power* than a mill-pond increases the quantity of water. When any slender force is caused through a machine, to produce some effect which seems proportioned to an intense force, it has always to act longer, or through more space than the other, just in proportion as it is more slender; as a small stream of water acting for ten minutes, may produce the same effect as a greater gush in one minute. Twenty feet of the action of a small horse near the circumference of a great wheel, may be rendered, by intervening machinery, equivalent to ten feet of the action of a heavy ox or elephant nearer the centre. And one horse in drawing through six hundred feet, or a hundred horses in drawing through six feet, or the piston of a great steam-engine, in raising one from the bottom to the top of its cylinder, &c., may all be made to do the same work.

Fig. 37.

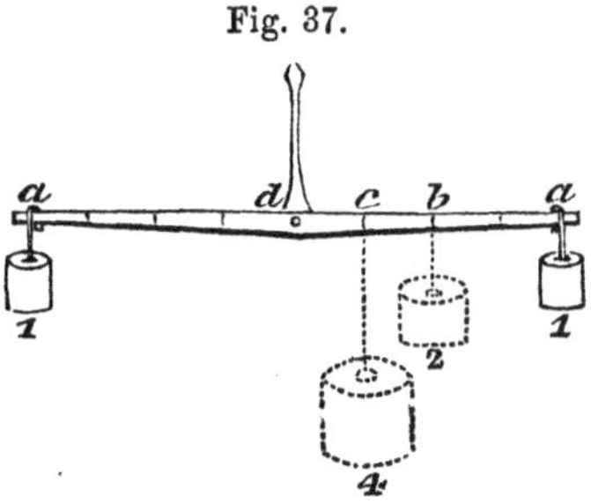

To illustrate this subject farther; we shall suppose a weighing-beam $x\,y$, with a weight of one pound hanging at the end x; then if a spring issuing from the fixed box at E, with uniform force of one pound, be made to push at the other end of the beam y, it will just balance the weight; and if it be in the slightest degree stronger than the weight, it will push the end of the beam y down to B, and will raise the weight to F. If, instead of the single spring of one pound at the end of the beam, two such springs be applied at half way from the centre to the end so as to press at A, where there is just half the extent of motion, or room to act, as at B, exactly the same effect will follow. Now because one spring at the end of the beam is seen here doing the same work as two similar springs, or a single spring of double strength at the middle, it might at first appear that there was a saving of power by using the single spring and longer lever; but let it be observed, that the two middle springs have each issued from their box only one inch, while the single spring at the end has issued two inches: in both cases, therefore, exactly two inches of one-pound spring have been used.

Fig. 38.

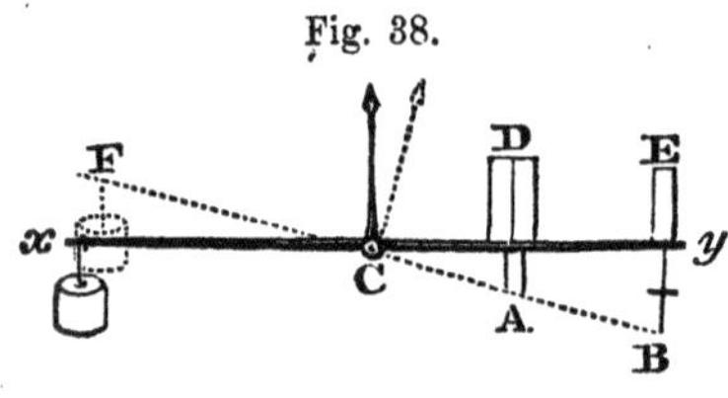

In the last experiment, pound weights or little buckets of water might be used instead of the springs, and with exactly the same result—one pound or pint at the end of the arm producing the same effect as two pounds or pints at the middle of it; but it would be observed that the single quantity fell two inches, while the double quantity at half distance fell only one inch; and to replace them after they had done their work, there would evidently be the same labour, whether a person had to lift the single quantity first one inch, and then another, or had to lift, first, one half of the double equipoise an inch, and then the other half as much.—Each atom of matter may be considered as held to the earth by its thread of attraction, and if one atom rise or fall ten inches, just as much of the supposed thread of attraction will be drawn out or returned as if ten atoms rise or fall one inch. And so where a weight of one pound is made to do any work, instead of a weight of two pounds, there is no more saving than in giving away two yards of single rope instead of one yard of double rope; and in like manner for all other differences of intensity.

If a man were to exert a force of one hundred pounds at A, in the above figure, to lift the weight, a boy at B, with force of fifty pounds, might do the same work; but the man would only have worked or pressed down through one foot, while the boy would have worked through two; and therefore, although the boy, with the assistance of the lever, seemed to become as strong as the man, the case would merely be, again, that of the one-pound spring unbending two inches, to produce an effect equal to that of the two-pound spring unbending one inch. The boy would be using two feet of his smaller force, where the man used one foot of his greater force; and if the work had to be long continued, the boy would have completely exhausted himself, when the man remained yet fresh.

A case of the lever, exhibited in this diagram, serves well to explain the nature of *mechanical powers* in general. Suppose A to be a weight of four pounds at the end of a rod or lever A B, (p. 97) made to turn on c as an axis or fulcrum, and having the arm c B four times as long as the arm c A, (but the two arms of the lever being equipoised so as not to conceal the action of weights subsequently attached to them;) then one pound at the end B, would balance the four pounds at the end of A, and with the slightest additional weight would preponderate. Now let us suppose the arc B *b* to have been fixed to the long arm of the lever with the four projections or shelves here shown on which balls of one pound might rest; then if one of the four balls from the plane *d* were to roll upon the first shelf, it would just balance A, and, with one grain more, would descend to the plane *e*, one inch below; then a second ball of one pound would occupy the second shelf, and would descend in the same way, to be followed by a third, and afterwards by a fourth; and when the whole four had fallen from *d* to *e*, they would just have lifted the four pound mass, at the other end of the lever, one inch. So that, although one pound was seen here lifting four pounds, it would only have lifted them one-fourth part as far as it fell itself, and the sum of the phenomena would be, that four pounds by falling one inch at the long end of the lever, had raised four pounds through the same distance of one inch at the short end. No *mechanical power* or *machine* generates force more than the lever does in this case.

Fig. 39.

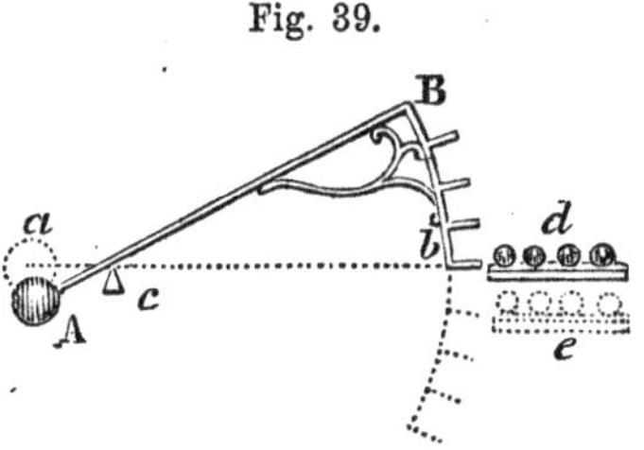

It appears, then, from all this, that as the *quantity of motion* in a body is measured by its velocity and the number of atoms in it conjointly, so the *quantity of force* exerted in any case, is measured by the *intensity* of the force conjointly with the *space* through which it moves. A clear mode, therefore, of comparing forces, is to state the *lengths* and the *intensities*—for instance, to speak of ten feet of one-pound force, as equal to one foot of ten-pound force, &c.

A horse pulling with the force of fifty pounds goes generally at the rate of six miles an hour; the steam-engine piston is generally made to move at the rate of two hundred feet per minute, bearing a pressure of steam of about twenty pounds to each square inch of its surface; a particular mill-stream may have a force of one hundred pounds, with a velocity of a hundred and fifty feet per minute:—now it is easy, by simple arithmetic, and the rule of *length* and *intensity* above explained, to compare all these and other forces as applicable to any given work. We must warn the reader, however, that there are many important considerations connected with the practical employment of forces, according to their respective nature and that of the resistance to be overcome, which cannot be entered upon in this elementary work. In very many cases there is a great waste or unavoidable loss of force, because the resistance, in yielding, runs away or escapes from the force; as when a ship runs away from the wind which is driving her, or the floats of a quick moving water-wheel, from the stream which turned it. Horses drawing boats, or carriages at the rate of five miles an hour, might exert great force, but to have a speed exceeding twelve miles they might require their whole effort to move their own bodies. As a general rule, although *equal quantities* of force balance each other when applied to parts of a lever or wheel altogether or nearly at rest, still when a force is made to act near its axis or fulcrum, to produce considerable velocity in a more distant part of the machinery, much of it is wasted in pressure against the fixed fulcrum.

What an infinity of vain schemes—yet some of them displaying great ingenuity—for perpetual motion, and new mechanical engines of power, &c.,

would have been checked at once, had the great truth been generally understood, that no form or combination of machinery ever did or ever can increase, in the slightest degree, the quantity of power applied. Ignorance of this is the hinge on which most of the dreams of mechanical projectors have turned. No year passes, even now, in which many patents are not taken out for such supposed discoveries; and the deluded individuals, after selling perhaps even their household necessaries to obtain the means of securing the expected advantages, often sink into despair, when their attempts, instead of bringing riches and happiness to their families, end in disappointment and ruin. The frequency, and eagerness, and obstinacy, with which even talented individuals, owing to their imperfect knowledge of this part of natural philosophy, have engaged in such undertakings, is a remarkable phenomenon in human nature. Examples of such schemes will be noticed in different parts of this work, where they may serve to illustrate points under consideration.

"*Lever, wheel and axle, &c.*" (Read the Analysis, at page 84.)

These are the simplest of the contrivances which the circumstances of solidity in masses has enabled man to adopt for the purpose of connecting or opposing forces and resistances of different intensities. We proceed to describe them, and to explain some of their useful applications.

"*Lever.*"

A beam or rod of any kind, resting at one end of a prop or axis, which becomes its centre of motion, is a lever; and it has been so called, probably, because such a contrivance was first employed for lifting weights.

This figure represents a lever employed to move a block of stone; *a*, is the end to which the *power* or *force* is applied, *f*, is the *prop* or *fulcrum*, and the mass *b*, is the *weight* or *resistance*. According to the rule already given and explained at page 96, the power may be as much less intense than the resistance as it is farther from the fulcrum, or moving through a greater space. A man at *a*, therefore, twice as far from the prop as the centre of gravity of the stone *b*, will be able to lift a stone twice as heavy as himself; but he will lift it only one inch for every two inches that he descends: and two men would be required, acting at half the distance, to do the same work.

Fig. 40.

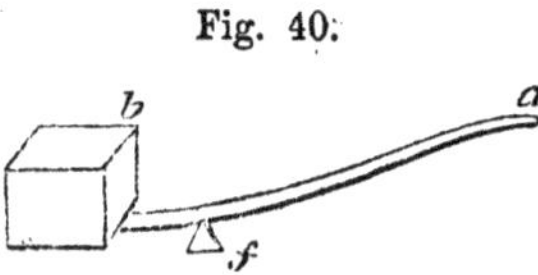

There is no limit to the difference, as to intensity, of forces which may be made to balance each other by the lever, except the length and strength of the material of which levers have to be formed. Archimedes said, "Give me a lever long enough, and a prop strong enough, and with my own weight I will lift the world." But he would have required to move with the velocity of a cannon-ball for millions of years, to alter the position of the earth by a small part of an inch. As stated in a former part of the volume, this feat of Archimedes is, in mathematical truth, performed by every man who leaps from the ground; he kicks the world away from him when he rises, and attracts it again when he falls back.

To calculate the effect of a lever, in practice, we must always take into account the weight of the lever itself, and the fact of its bending more or less; but in expounding the theory of the lever, it is usual to consider, first, what would be the result if the lever were a rod without weight and without flexibility.

The rule for the lever, that the opposing forces, to balance each other, must be more or less intense, exactly as they act nearer to or farther from the centre, holds in all cases, whether the forces be on different sides of the prop or both on the same side, and whether the force nearest to the prop have the office of power or of resistance; it holds, also, whether the lever be straight or crooked.

The following are examples of levers with the prop between the forces.

The *handspike*, represented in page 98, is a lever moving a block of stone. The same form, when made of iron, with the extremity formed into claws, is called *crow-bar*. Both kinds are used by gunners, in working cannon during battle: they are also used generally for lifting and moving heavy masses through small spaces, as the materials of the mason, the ship-builder, the warehouse-man, &c. A short crow-bar is the instrument of house-breakers, for wrenching open locks or bolts, tearing off hinges, &c.

The common *claw-hammer*, for drawing nails, is another example. A boy who cannot exert a direct force of fifty pounds, may yet, by means of this kind of hammer, extract a nail to which half a ton might be quietly suspended,—because his hand moves through, perhaps, eight inches, to make the nail rise one-quarter of an inch. The claw-hammer also proves, that it is of no consequence whether the lever be straight or crooked, provided it produces the required difference of velocity between power and resistance. The part of the hammer resting on the plank is the fulcrum.

A *pincers* or *forceps* consists of two levers, of which the hinge is the common prop or fulcrum. In drawing a nail with steel forceps or nippers, we have a good example of the advantages of using a tool: 1, the nail is seized by the teeth of steel instead of by the soft fingers: 2, instead of the griping force of the extreme fingers only, there is the force of the whole hand conveyed through the handles of the nippers: 3, the force is rendered, perhaps, six times more effective by the lever-length of the handles: and 4, by making the nippers, in drawing the nail, rest on one shoulder as a fulcrum it acquires all the advantages of a lever or claw-hammer for the same purpose.

Common scissors are also double levers, and those stronger *shears* with which, under the power of a steam-engine, bars and plates of iron are now cut as readily as paper is cut by the force of the hand.

The common *fire-poker* is a lever. It rests on the bar of the grate as its prop, and displaces or breaks the caked coal behind as the resistance.

The *mast of a ship*, with sails set upon it, may be regarded as a long lever, having the sails as the power, turning upon the centre of buoyancy of the vessel as the fulcrum, and lifting the ballast or centre of gravity as the resistance. For this reason lofty sails make a ship heel or lean over greatly, and if used in open boats, are dangerous. In some of the islands in the Eastern and Pacific Oceans, for the sake of sailing swiftly, boats are used so extremely narrow and sharp, that to counteract the overturning tendency of their large sails, they have an *outrigger* or projecting plank to windward, on the extremity of which one or more of the crew may sit as a balance.

Perhaps no instance of the lever, with the prop between the forces, is more interesting than the *weighing-beam;* whether with equal arms, forming the common *scale-beam;* or with unequal arms, forming the *steel-yard*.

We have seen why quantities of matter attached at equal distances from the prop, must be equal to each other in order to balance. A lever, therefore, which enables us to place quantities thus exactly in opposition to each other, and which turns easily on its axis, becomes a weighing-beam. Of this the annexed figure shows a common form. The axis or pivot at *c* is sharpened below, wedge-like, that the beam may turn easily, and that its centre of motion may be nicely determined; in a delicate balance for philosophical purposes, the axis is almost as sharp as a knife edge, and rests on some hard smooth surface of support, so as to turn with the weight of a small part of a grain. The scales also of a weighing-beam are suspended on sharp edges to facilitate motion, and to determine nicely the points of suspension. If the two arms of a beam be not of perfectly equal length, a smaller weight at the end of the longer will balance a greater weight at the end of the shorter. An excess of half an inch in the length of a beam-arm, to which merchandise is attached, where the arm should be eight inches long, would cheat the buyer of exactly one ounce in every pound. This case might be detected instantly, by changing the places of the two things balanced; for so, the lightest would be at the short arm, and would then appear doubly too light. A beam intended for delicate purposes, and required, therefore, to turn easily, must have its centre of gravity very near the axis on which the beam turns; for if otherwise, the beam will be in the predicament of a ship with the ballast too high or too low: in the former case, when once inclined, it would fall over, and not to recover itself: in the latter, it would tend to remain horizontal, and therefore would be less free to move. The proper situation of the centre of gravity is a little below the axis or line of support, that the beam may return with sufficient readiness from any state of inclination, to its horizontal position of rest.

Fig. 41.

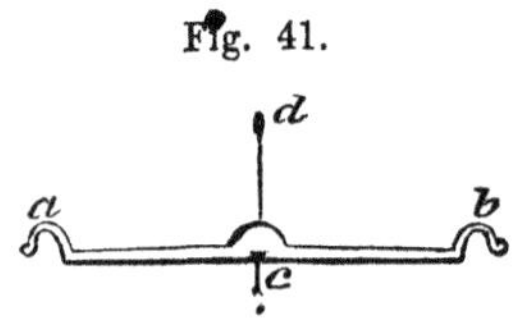

There is a mode of arriving at very accurate results, even with a weighing-beam which is not itself accurately made, provided it has very free motion, *viz.*, first, very nicely to balance in one scale the substance to be weighed, and then to remove it, and to put weights into the same scale, until a perfect balance is produced. Such weights must be the exact equivalent or weight of the substance, however unlike to each other the arms of the balance may be. A projecting rod, or plank, or branch of a tree, may thus be made to answer the purpose of a weighing-beam, by attaching any substance to its extremity and observing minutely how far such substance bends it, and then trying what weights would bend it as much.

The *steel-yard* is a lever with unequal arms, and any weight as *b*, on the long arm, will balance as much more weight as *a* on the short arm, as the former is supported farther from the fulcrum than the latter. Thus, if the hook at the short end be one inch from the centre of support, *c*, a pound weight *b*, on the long arm at four inches, will balance four pounds *a*, at the short arm. This supposes, however, that the steel-yard when bare, hangs horizontally, from having a greater mass of matter in the short arm to counterbalance the long slender

Fig. 42.

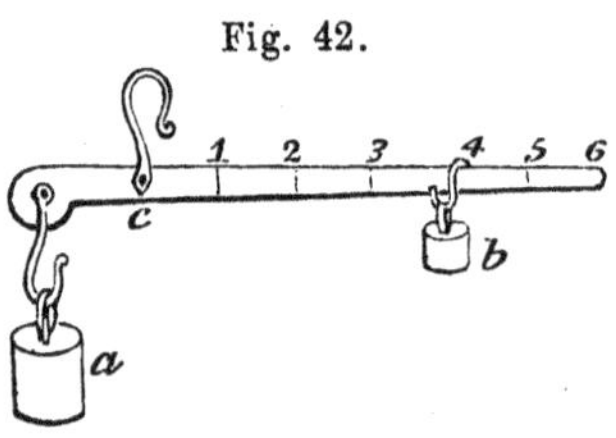

arm from which the shifting weight hangs. When this is not the case, a corresponding allowance has to be made.

The Chinese, who are so remarkable for the simplicity to which they have reduced all their common implements, weigh any small objects by a delicate pocket steel-yard. It is a rod of wood or ivory, about six inches long, with a silk cord passing through it at a particular part, to serve as a fulcrum, and with a sliding weight on the long arm, and a small scale attached to the short one.

The following are examples of levers with both forces on the same side of the prop, and where the more distant force acts as the power.

A common wheel-barrow is a lever, in using which a man bears as much less than the whole weight of the load as the centre of gravity of the load is nearer to the axle of the wheel than to his hands.

When two porters carry a load placed midway between them, on a pole, they share it equally, that is to say, each bears a half, for the pole becomes a lever, of which each porter is a fulcrum, as regards the other; but if the load be nearer to one end than to the other, he to whom it is nearest bears proportionably more of its weight. A load at *c* is equally borne by a porter at *a* and by one at *b*; but a load at *d* gives three-quarters of its weight to the man at *a* and only one-quarter to him at *b*.

F g 43.

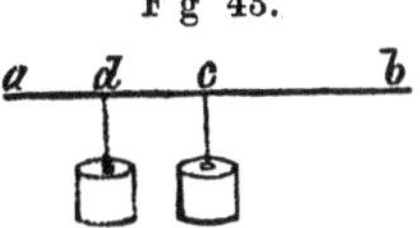

Two horses drawing a plough, act from the ends of a cross bar, of which the middle is usually hooked to the plough. The horses must thus pull equally, to keep the bar directly across. When on heavy land, three horses are yoked, and two of them are made to draw from one end of the bar, it must be attached to the plough by a hook, not at its middle, but half as far from one end of it as from the other.

The oar of a boat is a lever of this kind, where singularly the purpose of fulcrum is served by the unstable water.

The common nut-crackers furnish another instance, by the lever-power of which a person can break a shell many times stronger than he could break with the bare fingers.

The consideration of this kind of lever explains why a finger caught near the hinge of a shutting door is so much injured. The momentum of the door acts by a comparatively long lever, upon a resistance placed very near the fulcrum. Children pinching their fingers near the hinge of a door,—or of the fire-tongs, which furnishes a similar case—wonder why the bite is so keen.

The phenomenon of the branch of a tree giving way, when in autumn overloaded with fruit, or in winter with snow, also exhibit the action of this kind of lever. The resistance is the cohesion of the upper side of the branch to the tree, and the fulcrum is the part below which is last broken.

The following are examples of the lever, where the two forces are on the same side of the pivot, but where that nearest to the pivot acts as the power. In this kind, the power is more intense than the resistance.

The hand of a man who pushes open a gate while standing near the hinges, moves through much less space than the end of the gate, and hence must act with great force.

When a man uses the common fire-tongs, the ends move much further than his fingers, and therefore with less strength. No one fears a pinch with the ends of the fire-tongs.

The most beautiful and remarkable instances of this modification of lever are in the limbs of animals. The object in these was to give to the extremities great range and freedom of motion, without clumsiness of form; and it has been attained most perfectly by the tendons or ropes which move them, being attached near to the joints, which are the pivots or fulcra of the bone levers.

In the human arm, the deltoid muscle, which forms the cushion of the shoulder, by contracting its fibres less than an inch, raises the elbow twenty inches, and of course, if it overcome a force of fifty pounds at the elbow, it must itself be acting with a force at least twenty times as intense, or of one thousand pounds.—What extraordinary strength of muscle, then, is displayed by a man who lifts another at the end of his extended arm; yet this feat is frequently accomplished, and even on both sides of the person at once.

How powerful again must be the wing-muscles of birds, which, by this kind of action, sustain themselves in the sky for many hours together. The great albatros, with wings extended fourteen feet or more, is seen in the stormy solitudes of the Southern Ocean, accompanying ships for whole days, without ever resting on the waves.

A little contraction of the glutæi muscles of the hips gives to the human step a length of four feet.

While the erroneous opinion prevailed, that machines *increased* power, instead of, as they do, merely *accommodating* forces to purposes, this last kind of lever, where a great force acting through a short distance is made to gain great extent of motion and other benefits, was regretted by many as a most unprofitable, contrivance and was called the *losing lever*.

It is almost unnecessary to say, that the same rule of comparative velocities ascertains the relations required between power and resistance, where a combination of levers is used, as where there is only one. If a lever which makes *one* balance *four*, be applied to work a second lever which does the same, *one* pound at the long arm of the first will balance *sixteen* pounds at the short arm of the second, and would balance *sixty-four* at the short arm of a third such, &c.

The general rule for the lever, that a force may be less intense the farther it is from the pivot, supposes always that the force acts at right angles, or directly across the lever; for if there be any obliquity, there is a corresponding diminution of effect, as explained under the head of *resolution of forces*, at page 57. For instance, one pound at b on the end of the long arm of the bent lever $b\ d\ a$, because its weight does not act directly across $b\ d$, has influence only as if it were acting directly at the end of a shorter horizontal arm $d\ f$; and the two-pound weight at a acts only as if it were on a horizontal arm at e; now e being only half as far from the centre as f, two pounds at a, in the position of the lever here shown, would just balance the one pound at b. In every case, the exact influence of weights is known by referring them to places directly above or below them, on a supposed horizontal lever $e\ f$. What is called a *bent-lever balance*, is made on the principle here explained. It has on one side a heavy weight as at a, and on the other side a scale attached at b; and the weight of any thing put into the scale is indicated by the position

Fig. 44.

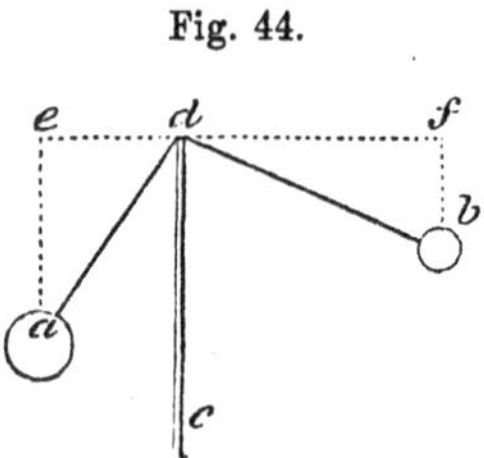

then assumed by the lever, marked by the point at which it cuts an arc of divisions placed behind it. In any common weigh-beam, the point of suspension of the scales being a little below the axis of motion of the beam, there is a degree of the property of the bent-lever balance, and enough to require notice in very nice experiments.

" The Wheel and Axle"

is the next to be mentioned of the *simple machines*. The letter *d* here marks a wheel, and *e* an angle affixed to it; and we see that in turning together, the wheel would take up or throw off as much more rope than the axle, as its circumference or diameter were greater than that of the axle. If the proportions were as four to one, one pound at *b* hanging from the circumference of the wheel, would balance four pounds at *a*, hanging from the opposite side of the axle. The proportions are equally indicated, and are usually expressed by comparison of the diameters of the wheel and the axle.

Fig. 45.

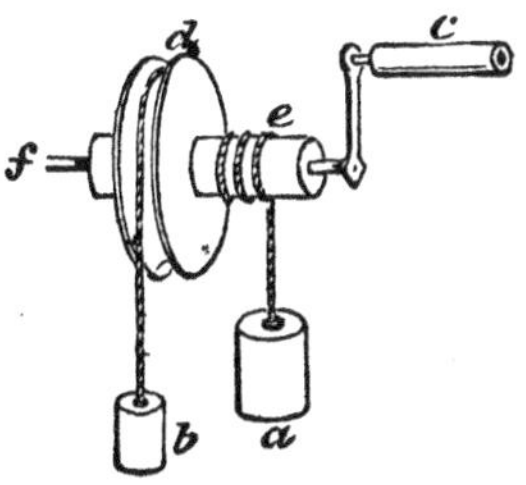

This figure represents the same object as the last, viewed endways. It explains why the wheel with its axle has been called a perpetual lever; for the two weights hanging in opposition, on the wheel at *a*, and on the axle at *b*, are always as if they were connected by a horizontal lever at *a c b*, of which the arms are respectively the diameters of the wheel and the axle turning on the centre *c* as the prop; and while a simple lever could only lift through a small space, it is evident that this construction will lift as long as there is rope to be wound up.

Fig. 46.

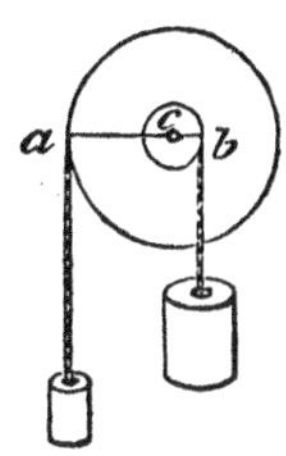

A common crane for raising weights consists of an axle to wind up or receive the rope which carries the weight, and of a large wheel at the circumference of which the power is applied. The power may be animal effort exerted on the rim or outside of the wheel, or the weight of a man or beast walking within it, and moving it as a squirrel moves the cylinder of his cage.

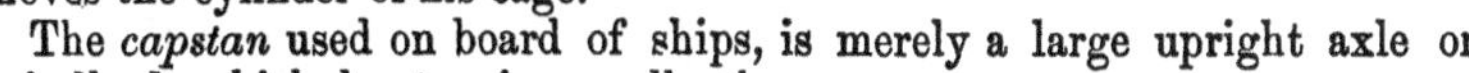

The *capstan* used on board of ships, is merely a large upright axle or spindle *b*, which by turning, pulls the cable or rope *a b c*; and it is moved by the men pushing at the capstan-bars *d*, *e*, *f*, &c., which for the time are stuck into holes made for them in the broader part or drum, usually appearing above the deck at the top of the spindle. These bars may be considered as the spokes of a large wheel, and the effect produced by a man working at one of them is in proportion to his distance from the centre. The capstan is chiefly used on board ships for lifting the anchor, and for doing any other very heavy work; but it is also applied for certain purposes on shore.

Fig. 47.

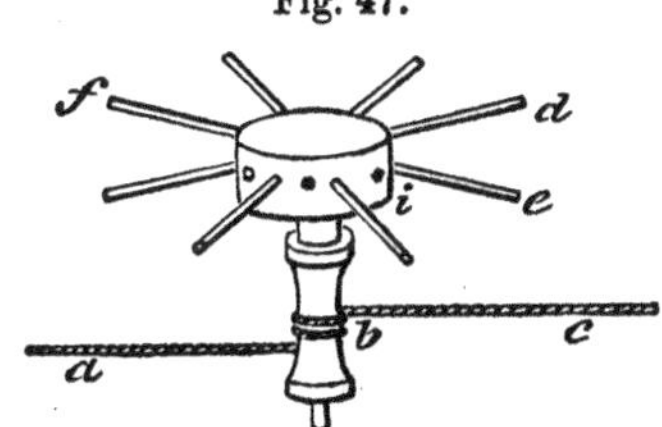

The common *winch* (represented as attached to the wheel and axle at the letter *c*,) with which a grindstone is turned, or a crane worked, or a watch wound up, is really in principle a wheel: for the hand of the worker describes a circle, and there is no difference in the result whether an entire wheel be turning with the hand or only a single spoke of a wheel.

That part of a common watch called the *fusee* is as beautiful an illustration of the principle of the wheel and axle now under consideration, as it is a useful and ingenious contrivance. The spring of a watch, immediately after winding up, being more strained, is acting more powerfully than afterwards when slacker, and if there were no means of equalizing its action, it would destroy the wished-for uniformity in the motion of the time-piece. The fusee is this means. It may be considered as a barrel or spindle, gradually diminishing from its large end *b*, to its small end *a*, with the surface cut into a spiral grove to receive the chain, by pulling at which the spring in the box *c* moves the watch. Now when the watch has been wound up, by a key applied on the axis of the fusee, the fusee is covered with the chain up to the small end *a*, and the newly bent and strong spring begins to pull by this small end or short lever; and afterwards, exactly as the spring becomes relaxed and weaker, it is pulling at a larger and larger part of the fusee-barrel, and so keeps up an equal effect on the general movement.

Fig. 48.

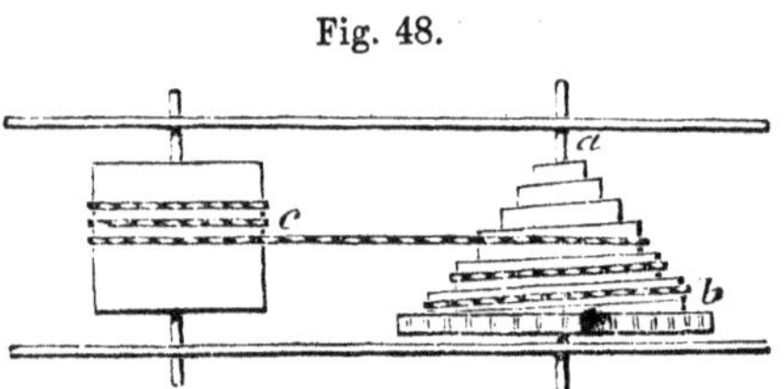

A large fusee in place of a common cylindrical axle, is often used with a winch, for drawing water by bucket or rope from very deep wells. When the bucket is near the bottom of the well, and the laborer has to overcome the weight of the long rope, in addition to that of the bucket and water, he does so more easily by beginning to wind the rope on a small axle, that is to say, on the small end of the fusee; and in proportion as the length of rope diminishes, he lifts by a larger axle.

By the double axle *a b*, very unequal intensities of force may be balanced. We see that in turning it, a rope unwinding from the small end *a* is taken up by the large end *b*, turn for turn, and that the rope below must be shortened at each turn by the difference between the circumference of the ends *a* and *b*. If the weight rise half an inch only, while the handle of the winch describes a circle of fifty inches, one pound force at the winch would balance one hundred pounds at *d*.

Fig. 49.

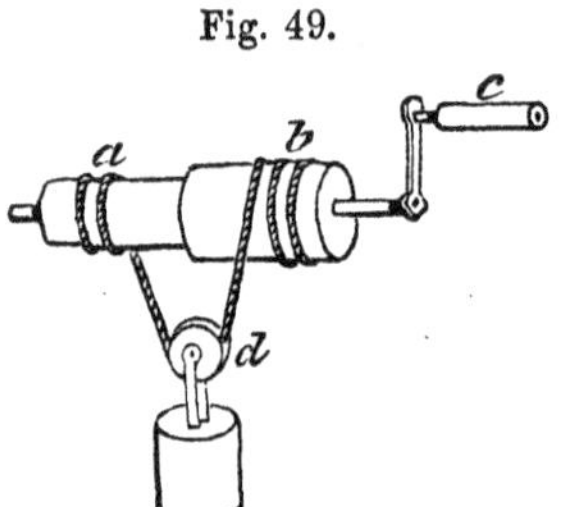

By means of a wheel, which is very large in proportion to its axle, forces of very different intensities may be balanced, but the machine becomes of inconvenient proportions. It is found preferable, therefore, when such an end is desired, to use a combination of wheels of moderate size. In the adjoining figure, three wheels are seen thus connected. Teeth on the axle *d*, of the first wheel *c*, acting on six times the number of teeth in the circumference of the second wheel *g*, turn it only once for every six times that *c* turns; and in the same manner the second wheel, by turning six times, turns the third wheel *h* once; the first wheel, therefore, turns thirty-six times for one turn of the last; and as the diameter of the wheel *c*, to which the power is applied, is three times as great as that of the axle *f*, which has the resistance, three times thirty-six, or one hundred and eight, is the difference of velocity, and therefore of intensity, between weights or forces that will balance here.—An axle with teeth upon it, as *d* or *e*, is called a pinion.

Fig. 50.

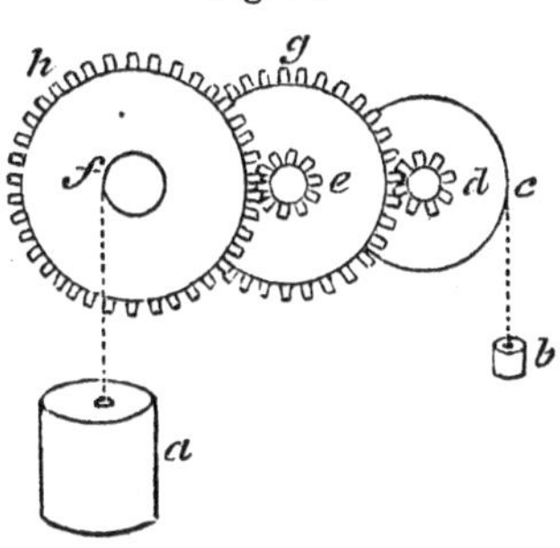

On the principle of combined wheels, *cranes* are made, by which one man can lift many tons. It is even possible to make an engine, by means of which a little windmill, of a few inches in diameter, should tear up the strongest oak by the roots; but of course it would require a very long time for its work.

The most familiar instances of wheel-work are in our clocks and watches. One turn of the axle on which the watch-key is fixed, is rendered equivalent, by the train of wheels, to about four hundred turns or beats of the balance-wheel; and thus the exertion during a few seconds, of the hand which winds up, gives motion for twenty-four or thirty hours. By increasing the number of wheels, time-pieces are made which go for a year; if the material would last, they might easily be made to go for a hundred or a thousand years.

Wheels may be connected by bands as well as by teeth. This is seen in the common spinning-wheel, turning-lathes, grindstones, &c. &c. A spinning-wheel, as *a c*, of thirty inches in circumference, turns by its band a pirn or spindle of half an inch, *b*, sixty times for every turn of itself.

Fig. 51.

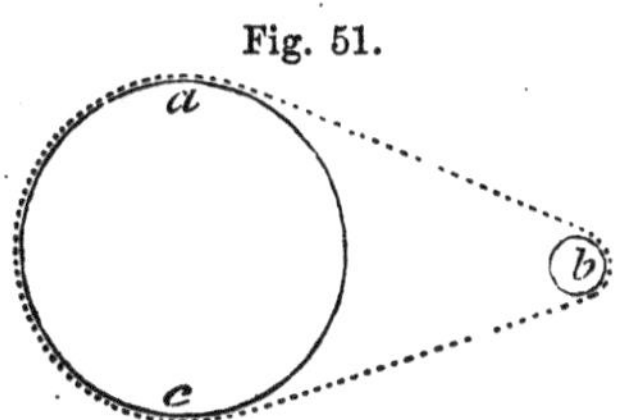

"*The Inclined Plane*"

is the third means which we shall describe, of balancing, by solid media, forces of different intensities. A force pushing a weight from *c* to *d*, only raises it through the perpendicular height *e d*, by acting along the whole length of the plane *c d;* and if the plane be twice as long as it is high, one pound at *b* acting over the pulley *d* would balance two pounds at *a*, or anywhere on the plane: and so of all other quantities and proportions, as already explained under the head of "Resolution of forces," at page 86.

Fig. 52.

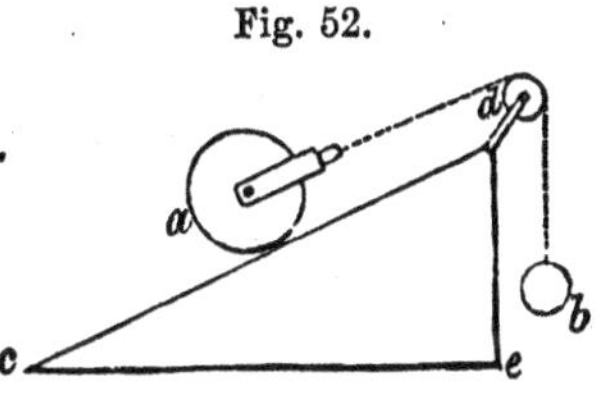

A horse drawing on a road where there is a rise of one foot in twenty, is really lifting one-twentieth of the load, as well as overcoming the friction and other resistances of the carriage. Hence the importance of making roads as level as possible; and hence our forefathers often erred in carrying their roads directly over hills, for the sake of straightness considered vertically, where by going round the bases of the hills they would scarcely have had greater distance, and would have avoided all rising and falling. Hence, also, a road up a very steep hill must be made to wind or zig-zag all the way; for to reach a given height, the ease of the pull to the horses is greater exactly as the road is made longer. This rule of road-making is exhibited remarkably in various parts of the world, where hills with almost perpendicular faces have very safe and commodious roads upon them, leading to forts or residences near their summits. An intelligent driver, in ascending a steep hill on which there is a broad road, winds from side to side of the road all the way to save his horses a little.

The railways of modern times offer a beautiful illustration of this subject. They are made generally quite level, so that the drawing horse or steam-engine has only to overcome the friction of the carriage; or where heavy loads are passing only in one direction, as from mines, they are made to slope a very little, leaving to the horse or other power only the office of regulating the movement.

A hogshead of merchandize, which twenty men could not lift directly, is often seen moved into or out of a wagon, by one or two men, who have the assistance of an inclined plane. In some canals, or rather particular situations on canals, the loaded boats are drawn up by machinery or inclined planes, instead of being raised by water in locks, as is the usual mode.

It is supposed that the ancients (the Egyptians particularly) must have used the inclined plane, to assist in elevating and placing those immense masses of stone, which still remain from their times, specimens of their gigantic architecture.

Our common stairs are inclined planes in principle; but being so steep, are cut into horizontal and perpendicular surfaces, called steps, that they may afford a firm footing.

We may here recall, that a body falling freely, in obedience to gravity, descends about sixteen feet in the first second, and that if made to descend on an inclined plane, it moves just as much less quickly (besides the loss from the friction and the turning produced) as the length of the plane is greater than the height. On a plane sloping one foot in sixteen of its length, a body would descend only one foot in the first second.

The descent of a pendulum in its arc is investigated mathematically by the laws of the inclined plane. And the laws of the inclined plane itself are mathematically explained by the principle of the *resolution of forces*, explained at p. 57.

"The Wedge"

is merely an inclined plane forced in between resistances to separate or overcome them, instead of, as in the last case, being stationary while the resistance is moved along its surface. The same rule as to mechanical advantage has been applied to the wedge as to other simple machines; the force acting on a wedge being considered as moving through its *length* $c\ d$, while the resistance yields to the extent of its breadth $a\ b$. But this rule is far from explaining the extraordinary power of a wedge. During the tremor produced by the blow of the driving-hammer, the wedge insinuates itself, and advances much more quickly than the above rule anticipates.

Fig. 53.

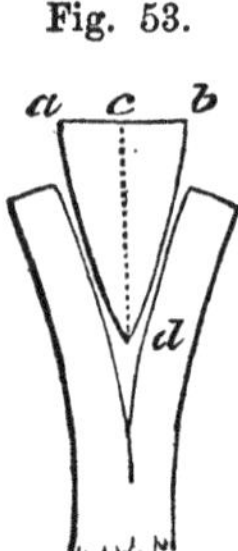

The wedge is used for many purposes; as for splitting blocks of stone and wood: for squeezing strongly, as in the oil-press; for lifting great weights, as when a ship of war, in dock is raised by wedges driven under the keel, &c.

An engineer in London, who had built a very lofty and heavy chimney, common to all his steam-engines and furnaces, found after a time that, owing to a defect in the foundation, it was beginning to incline. However, by driving wedges under one side of it, he succeeded in restoring it to perfect perpendicularity.

Nails, awls, needles, &c., are examples of the wedge; as are also all our cutting instruments, knives, razors, the axe, &c. These latter are often used somewhat in the manner of a saw—which is a series of small wedges—by pulling them lengthwise at the same time that they are pressed directly forward against the object. They themselves, indeed, when viewed through a microscope, are seen to be but finer saws. It appears that the vibration of the particles produced by the drawing of a saw, enables its edge to insinuate itself more easily. The sharpest razor may be pressed directly against the hand with considerable force, and will not enter, but if then drawn along ever so little, it starts into the flesh.

"The Screw"

is another of the simple machines. It may be called a winding wedge, for it has the same relation to a straight wedge that a road winding up a hill or tower has to a straight road of the same length and acclivity.

Fig. 54.

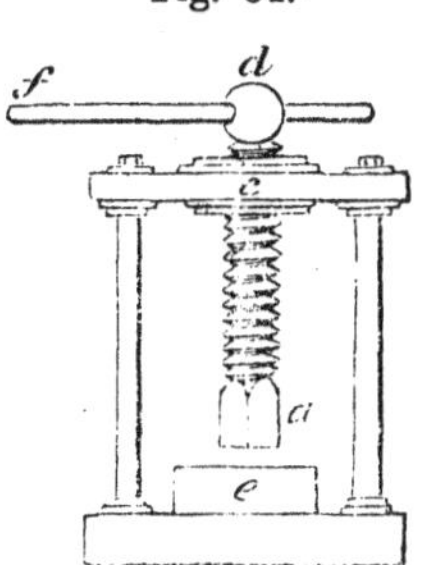

A screw may be described as a spindle *a d*, with a thread wound spirally round it,—turning or working in a nut *c*, which has a corresponding spiral furrow fitted to receive the thread. The nut is sometimes called the female screw. Every turn of the screw carries it forward in a fixed nut, or draws a moveable nut along upon it, by exactly the distance between two turns of its thread: this distance, therefore, is the space passed through by the resistance, while the force moves in the circumference of the circle described by the handle of the screw, as at *f*, in the figure. The disparity between these lengths or spaces is often as a hundred or more to one; hence the prodigious effects which a screw enables a small force to produce.

Screws are much used in presses of all kinds: as in those for squeezing oil and juices from such vegetable bodies as linseed, rapeseed, almonds, apples, grapes, sugar-cane, &c.: they are used also—in the cotton press, which reduces a great spongy bale, of which a few, comparatively, would fill a ship, to a compact package, heavy enough to sink in water;—also, in the common printing-press, which has to force the paper strongly against the types:—a screw is the great agent in our coining machinery,—and in letter-copying machines:—it is a screw which draws together the iron jaws of a smith's vice, &c. The screw, although producing so much friction as to consume a considerable part of the force used in working it, is an exceedingly useful contrivance.

As a screw can easily be made with a hundred turns of its thread in the space of an inch, at perfectly equal distances from each other, it enables the mathematical instrument maker to mark divisions on his work, with a minuteness and accuracy quite extraordinary. If we suppose such a screw to be pulling forward a plate of metal, or pulling round the edge of a circle, over which a sharp-pointed steel marker can be let down perpendicularly, always in the same place, the marker, if let down once for every turn of the screw, will make just as many lines on the plate as the screw makes turns; but if made to mark at every hundredth or a thousandth of a turn of the screw, which it will do with equal accuracy, it may draw a hundred thousand distinct lines in one inch.

The instruments called micrometers, by which the sizes of the heavenly bodies, and of microscopic objects, are ascertained, are worked by fine screws.

A *perpetual screw* is the name given where a screw acts on the teeth of a wheel, so as to produce a continued rotation of the wheel.

· A common cork-screw is the thread of a screw without the spindle, and is used, not to connect opposing forces, but merely to enter and fix itself in the cork Complicated cork-screws are now made, which draw the cork by the action of a second screw, or of a toothed rod or rack and pinion.

"*The Pulley*"

is another *simple machine*, by which forces of different intensities may be balanced. A simple pulley consists of a wheel as *a b*, which rests with its grooved circumference on the bend of a rope, *c a b d*, and to the axis of which the weight or resistance is attached, as at *e*.

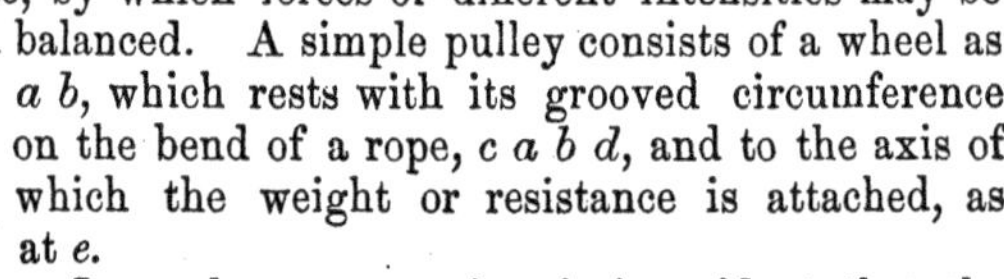

Fig. 55.

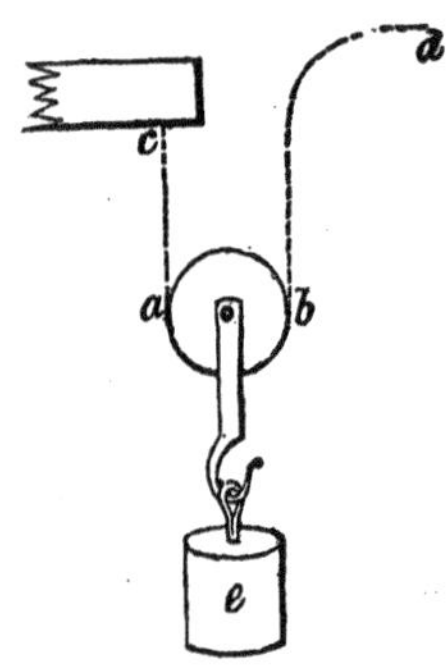

In such a construction, it is evident that the weight (let it be supposed ten pounds) is equally supported by each end of the rope, and that a man holding up one end, only bears half of the weight, or five pounds; but to raise the weight one foot, he must draw up two feet of rope; therefore, with the pulley, he is as if lifting five pounds two feet, where, without the pulley, he would have to lift ten pounds one foot.

Many wheels may be combined together, and in many ways to form compound pulleys. Where-ever there is but one rope running through the whole, as shown here, the relation of power and resistance is known by the number of folds of the rope which support the weight. Here there are four supporting folds, and a power of one hundred pounds would balance a resistance of four hundred.

Fig. 56.

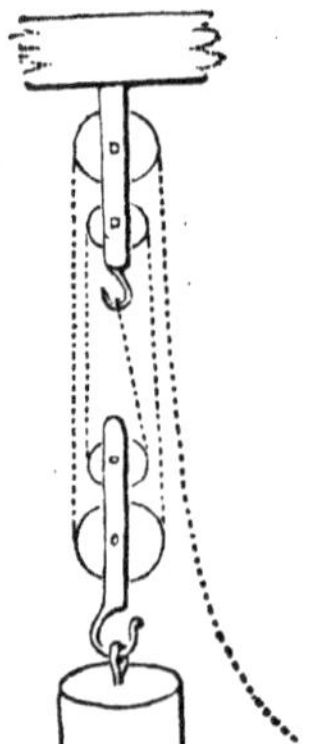

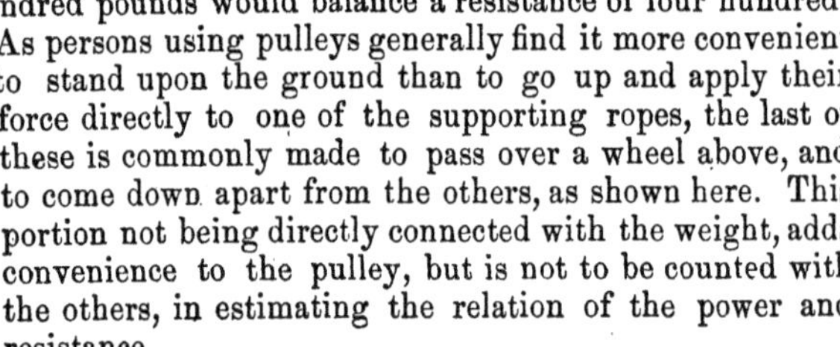

As persons using pulleys generally find it more convenient to stand upon the ground than to go up and apply their force directly to one of the supporting ropes, the last of these is commonly made to pass over a wheel above, and to come down apart from the others, as shown here. This portion not being directly connected with the weight, adds convenience to the pulley, but is not to be counted with the others, in estimating the relation of the power and resistance.

In *fixed pulleys*, like those shown at *a* and *c*, p. 109, there is no mechanical advantage, for the weight just moves as fast as the power; yet such pulleys are of great use in changing the direction of forces. A sailor without moving from the deck of his ship, by means of such a pulley, may hoist the sail or the signal-flag to the top of the loftiest mast. And in the building of lofty edifices, where heavy loads of material are to be sent up every few minutes, a horse, trotting away with the end of the rope from *d*, in a level courtyard, causes the charged basket *b* to ascend to the summit of the building as effectually as if he had the power of climbing, at the same rate, the perpendicular wall.

Fig. 57.

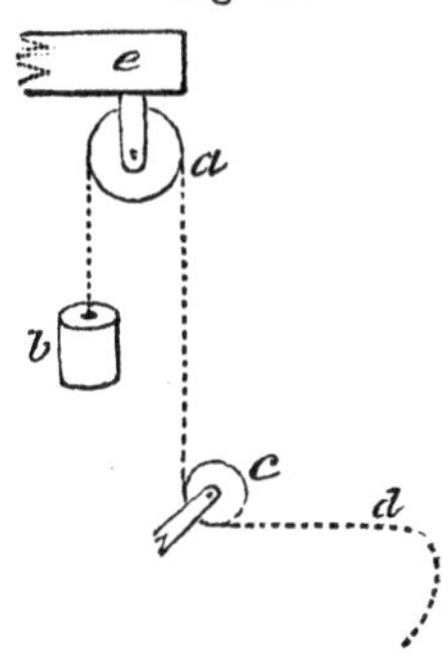

There is a case, however in which a fixed pulley may seem a balancer of different intensities of force; *viz.*, where one end of a rope is attached to a man's body, and the other is carried over a pulley above, and brought down again to his hands;—for safety this end also should be attached to his body. By using the hands then to pull with force equal to half his weight, he supports himself, and may easily raise himself to the pulley. A man, by a pulley thus employed, may let himself down into a deep well, or from the brow of a cliff, with assurance of being able easily to return, although no one be near to help him; and cases have often occurred where, by such means, a fellow-creature's life might have been saved, or other important objects attained. How easily, for instance, might persons either reach or escape from the elevated windows of a house on fire, by such a pulley, which might readily be found and used where ladders could not be obtained! This kind of pulley furnishes a convenient means of taking a bath from a ship's stern windows, &c.

The chief use of the pulley is on ship-board. It is there called a block, although, strictly speaking, the block is only the wooden mass which surrounds the wheel or wheels of the pulley. It aids so powerfully in overcoming the heavy strains of placing the anchor, hoisting the masts and sails, &c., that, by means of it, a smaller number of sailors are rendered equal to the duties of the ship. Pulleys are also used on shore, instead of cranes and capstans, for lifting weights, and overcoming other resistances.

Surgeons, in former days, when they trusted rather to force than to the address which better information gives, used pulleys much to help in the reductions of luxations,—but often hurtfully, from not understanding the force of the pulley. A good surgeon now rarely needs a pulley, and he who should ignorantly stretch his patient on the rack, would be well requited by similar treatment.

The cranks of bell-wires, seen in the corners of our rooms, are bent levers nearly equivalent to fixed pulleys.

There is no reason, but old usage, why the appellation of *mechanical power* should be confined to the six contrivances now explained, for those of which the account is yet to follow equally deserve it; and, as will be seen under hydrostatics and pneumatics, the most powerful mechanical engines do not belong to solids at all.

Engine of oblique action, is a title which may include a considerable variety of contrivances for connecting different velocities.

Suppose *c a* and *c b* to represent two strong rods connected together, like a carpenter's folding rule, by a hinge or joint at *c*. If the distant ends be made to bear against notches in two obstacles, at *a* and *b*, and by force then applied to *c*, either to push or to pull, the joint *c* be straightened or carried towards *d*, the joint *c* will move through a much greater space than the simultaneous increase of distance produced between *a* and *b*; and, in proportion to the disparity, the power applied at *c* will overcome a more intense resistance at the extremities. The mechanical power of this contrivance increases rapidly, the nearer the jointed rods approach to straightness.

Fig. 58.

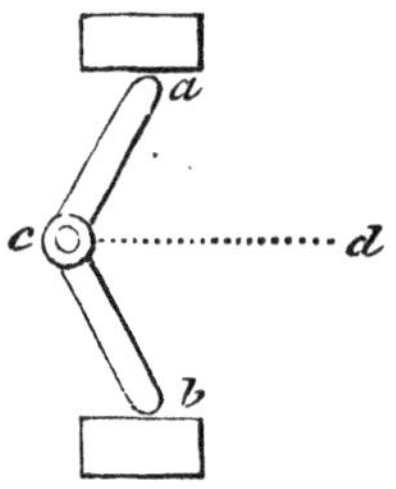

If we suppose the end *a* to be steadied by a hinge on frame-work, and the end *b* to bear upon that part of a printing-press which carries the paper against the types, we have imagined the simple press called, from its contriver, the Russell-press. A man's force at *d*, at the moment when the rods are drawn nearly to a straight line, becomes equivalent to a pressure of many tons.

For the same reason, that by urging *c* toward *d*, in the last figure, the extremities *a* and *b* are separated with great force, so by urging *c* in the contrary direction, the extremities would be drawn together with corresponding force: and if we suppose *a c b* to be part of a rope coming through pulleys at *a* and *b*, to one end of which rope, beyond *a*, great resistance is attached, one man, by pulling at *c*, may move a weight or resistance many times greater than he could move by his direct power.

The following is another mode of connecting an oblique and a direct force so as to balance them, although of different intensities. If to turn a wheel (represented here by the circle) a weight be suspended from *d*, it is acting directly, for it descends just as fast as the circumference of the wheel moves, and would, therefore, be impelling with its whole strength: but if it were suspended from the point *e*, it would then be acting obliquely to the motion of that part of the wheel, and from not descending so fast as if at *d*, it would have as much less effect on the wheel than if there, as the line *e b* is shorter than the line *d c*. The reason of this will be understood by referring to the subjects of *resolution of forces* and of *bent levers*, in former parts of the work. For the same reason, if such a wheel were used in lifting weights, a man turning it could lift as much more attached at the point *e* than at the point *d*, as the line *d c* is longer than *e b*. A man turning this wheel in the direction from *e* to *a*, with a weight hanging at *e*, would be lifting that weight exactly as if he were rolling it up the inclined plane or curve *e a*. This figure is useful in explaining the varying intensity of the action of a crank or winch, in different parts of its revolution, and of the combination of levers used in the *Stanhope printing-press*, in their different positions: it explains also the degrees of strength and support afforded by oblique stays in buildings and in ships' rigging, and many other kindred matters.

Fig. 59.

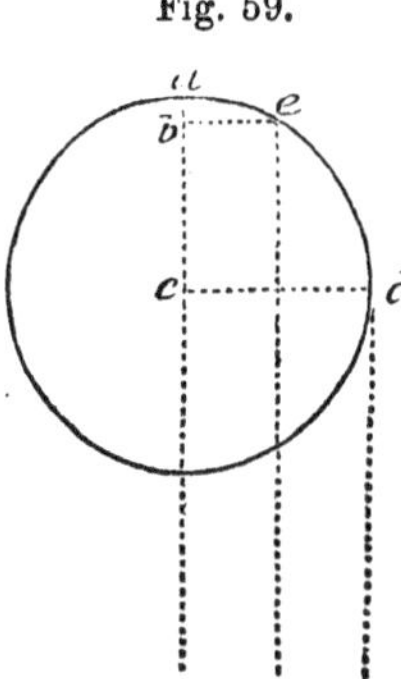

Fig. 60.

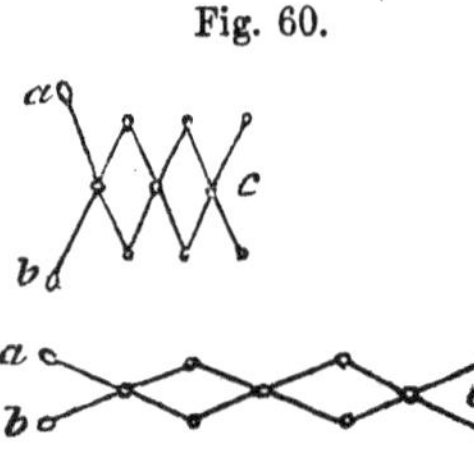

The arrangement of cross-jointed wires, represented here, connects different velocities, and therefore is really a mechanic power. It has been applied to some curious purposes, but to none of much utility. By pressing the ends *a* and *b* towards each other, the wires, from being in the position represented in the upper figure, immediately assume the position represented in the lower; so that the end *c* darts outward much farther than the ends *a* and *b* approximate.

Different intensities of force are balanced, although not simultaneously, by the following means; which, therefore, according to the old idea, have some claim to the name of *mechanic powers.*

A man may have a purpose to effect, which a forcible downward push would accomplish: but his body being too weak to give that push directly, he may employ a certain time in carrying a weight to such an elevation, above his work, that when let fall its momentum may do what is required. Here the continued effort of the man in lifting the weight, to a height of perhaps thirty feet, may be just sufficient to produce a blow which will cause a stake or pile to sink into the earth one inch; and the contrivance has therefore balanced forces, of which the relation as to intensity is marked by the spaces thirty feet and an inch.

So also *hammers, clubs, battering-rams, slings, &c.*, are machines which enable a continued moderate effort to overcome a great but short resistance.

The *fly-wheel,* which by persons ignorant of natural philosophy, has often been accounted a positive power, in common cases merely equalizes the effect of an irregular force.

In using a winch to turn a mill, for instance, a man does not act with equal force all around the circle: but a heavy wheel fixed on the axis moderates acceleration, and receives or absorbs momentum, while his action is above par, and returns it again, giving to the machine, while his action is below par, thus equalizing the movement. And in the common instances of circular motion produced by a crank, as when, by the pressure of the foot on a treadle, we turn a lathe or grind-stone, or spinning-wheel, the force is only applied during a small part of the revolution, or in the form of interrupted pushes; yet the motion goes on steadily, because the turning grindstone, or wheel, or lathe, becomes a fly and reservoir, equalizing the effect of the force. In a steam-engine which moves machinery by a crank, the upward and downward pushes of the piston are converted, by means of a heavy fly-wheel into a very steady rotatory motion.

A heavy wheel, however, has sometimes been used as a concentrator of force or a machanic power. By means of a winch, or a weight, or otherwise, motion or momentum being gradually accumulated in the wheel, is then made to expend itself in producing some sudden and proportionally great effect. Thus, a man may lift a very heavy weight by first, in any way giving motion to a fly-wheel, and then suddenly hooking a rope from the weight to the axle of the wheel, which rope being wound upon the axle, lifts the weight.

A fly-wheel moved in the same manner, and containing the result of a man's action during perhaps one hundred seconds, if made to impel a screw-press, will, with one blow or punch, stamp a perfect medal, or from a rough flat plate of silver will form a finished spoon, or other utensil.

A spring, in the same sense, may become a mechanical power. A person may expend some minutes in bending it, and may then let fly its accumulated energy in an instantaneous blow. A gun-lock shows this phenomenon on a small scale. The slow bending of a bow, which afterwards shoots its arrow with such velocity, is another instance.

These, then, are the principal means which the solid state of bodies affords us of balancing forces of different intensities. We shall find other such means or mechanic powers belonging to liquids and airs. All of them are of inestimable value to man, by enabling him to accommodate the forces which he can command to any kind of work which he has to perform. Thus he makes his millstone turn with the same velocity, whether it be moved by the slow exertion of a horse or bullock, walking in a ring, or by the quicker motion of a river gliding under the wheel, or by the rapid gush of a water-fall, or by the invisible swiftness of the wind. And again, each of these forces he can equally apply to turn the heavy millstone or to twist a cotton thread.

The wants of men seem first to have led them to use the *simple machines*, for the purpose of raising great weights, or overcoming great resistances and hence the name long used of *mechanic powers*,—particularly for the Lever, Wheel and Axle, Plane, Wedge, Screw, and Pulley: but the term conveys to the uninformed a false idea of their real nature, and has begotten the common prejudice with respect to them, that they *generate* force, or have a sort of innate power for saving labour. Now so far is this from being true, that in using them in any case, even more labour or bodily exertion is expended than would suffice to do the work without them. This assertion is intentionally rendered paradoxical to arrest, attention, but its truth will appear from the following considerations.

One man may be able, with a tackle of pulleys having ten plies of the rope, to raise a weight which it would require ten men to raise at once without pulleys. But if the weight is to be raised a yard, the ten men will raise it by pulling at a single rope and walking one yard, while the one man at his tackle must walk until he has shortened all the ten plies of rope of one yard each; that is, he must walk ten yards, or ten times as far as the ten men did. In both cases, therefore, to accomplish the same end, we have just the same quantity of man's work expended, in the first, performed by ten men in one minute, in the second, by one man in ten minutes; and if the work were of a nature to continue longer, let us say a whole day for the ten men, it would last ten days for the single man, and there would be ten days' wages of a man to pay in both cases. There is, therefore, no direct saving of human effort from using pulleys; indeed, there is a loss, because of the great friction which has to be overcome. Now exactly the same is true of all other simple machines, or mechanic powers; none of them save labour, in a strict sense of the phrase; they only allow a small force to take its time to produce any requisite magnitude of effect, at the expense of additionality overcoming a certain amount of friction or other such risistance.

The real advantage of these machines are such as the following:

That one man's effort, or any small power, which is always at command, by working proportionally longer, will answer the purpose of the sudden effort of many men, even of hundreds or thousands, whom it might be most inconvenient and expensive, or even impossible, to bring together.

A ship's company of a few individuals easily weighs a heavy anchor by means of the capstan.

A solitary workman, with his screw or other engine, can press a sheet of paper against types, so as to take off a clear impression; to do which without the press, the direct push of fifty men would scarcely be sufficient: and these fifty men would be idle and superfluous except just at the instants of pressing, which occur only now and then. In this way the screw may be said to do the work of fifty men, for it is as useful.

A man with a crow-bar may move a great log of wood to a convenient place, where twenty men would have been required to move it without the crow-bar; and although the single man takes twenty minutes, perhaps, to do what the many men would have done in one minute, as the twenty might not have been wanted again for the rest of the day, the crow-bar may really be as useful as the twenty men.

It is so important to have correct notions on the subject of the simple machines or mechanical powers, that more space has been here allotted to the explanation of the general principle, than has been usual in such works. After the examination which it has now undergone, however, the author hopes that none of his readers will have difficulty in conceiving clearly, that "whatever, through a machine, is gained in power, is lost in speed or in time, and *vice versa*"—or will have difficulty in detecting immediately any common fallacy connected with the subject;—as that of supposing, for instance, that a lever, or great pendulum, or spring, or heavy fly-wheel, &c., can never exert more force than has passed into it from some source of motion.

"*By solid connecting parts, also, the direction of any existing motion or force may be changed. Hence the endless variety of* COMPLEX MACHINES." (Read the Analysis at p. 84.)

It is this power of changing the direction of motion, added to the power of connecting and adjusting various intensities of force and resistance by the simple machines last described, which has enabled man to make complex machines, rivaling in their performances the nicest work of human hands. It would be endless to attempt the enumeration of the modes in which the directions of motions may thus be changed, for it would be to enumerate and describe the whole apparatus of the arts and sciences; but we shall advert to a few as specimens.

Straight motion changed into rotatory.—The straight motion of wind or water becomes rotatory in wind or water-wheels.—The straight downward pressure of the human foot, acting at intervals on a treadle and crank, turns round the grindstone, and common lathe, and spinning-wheel. The alternate rising and falling of the piston of a steam-engine is made, by means of a crank, to turn the great fly-wheel and any other wheels which a steam-engine may move.

Rotatory motion into straight.—An axle in turning will wind up a rope, and lift a weight in a straight line.—A crank on a turning axle, if connected with a pump rod, will work the piston up and down; or it will work a saw. Pallets or teeth on a turning-wheel act on the handle of a great forge hammer, so that every one in passing lifts the hammer and produces a blow.

We need not multiply instances. By a visit to great manufacturing towns, or, indeed, by simply directing the eyes to what is passing around, in any part of the civilized world, we discover miracles of mechanic art:—machines driven by wind, water or steam for grinding corn;—machines for sawing wood and giving it various forms;—machines in which rods of metal are seized between great rollers, and are flattened at once into thin plates, as if they were of clay, and these plates again are slit into bars or ribbons—spinning machines, which perform their delicate office even more uniformly than human hands, forming thousands of threads at once, in obedience to the impulse of a single steam-engine;—weaving machines, which accomplish their difficult task with the most admirable perfection;—paper-making engines, which convert worn-out and apparently useless remnants of our apparel, into the uniform and beautiful texture of paper, a texture which, with the farther assistance of the pen, or types, or engraved plate, becomes a magic conservatory of mind, shutting up among its folds the brightest effusions of genius, and ready, at any instant, to disclose them again to the delighted student, nothing changed after revolving centuries;—coining machinery, which form a bar or plate of metal cuts out and stamps thousands of beautiful medals in an hour, and keeps an exact record of its work;—cranes,—pile engines,—turning-lathes,—time-pieces,—all the implements of agriculture, of mining, of navigation, &c., &c. If Aristotle deemed the title or definition of *tool-using animal* appropriate to man two thousand years ago, what title should be given now?

In many of the complex machines, several of the simple ones are found as elements; and in the same machine may be comprised many of the means of changing the direction of motion.

"*Friction.*" (Read the Analysis, p. 84.)

In estimating the effects of mechanical contrivances, by the rule of comparative velocities of the power and resistance, there is an important correction to be made, on account of the mutual friction of the moving parts. In the steam-engine, where the rubbing parts are numerous, the loss of power, from friction often amounts to one-third of the whole.

Impediment from friction seems to be owing to two causes; 1st, a degree of cohesive attraction between the touching substances; 2nd, the roughness of these surfaces, even where, to the naked eye, they appear smooth.

It is supposed to be, because the roughness, or little projections and cavities, in pieces of the same or of homogeneous substances mutually fit each other, as the teeth of similar saws would, so as to allow the bodies, in a degree to enter into each other, that the friction is greater between such than between pieces of different or of heterogeneous substances with dissimilar grain.

The friction of one piece of iron, wood, brick, stone, &c., on another piece of the same substance, has been measured by using the second piece as an inclined plane, and then gradually lifting one end of it until the upper mass began to slide,—the inclination of the plane, just before the sliding commences, is called the angle of repose. This angle, different for different substances, is found to be, for metals, generally such as to mark that the force required to overcome the friction between small pieces of them is equal to about a fourth of the weight of the moving piece, and for woods it is about a half. But for large pieces or great pressures, the friction is proportionably much less.

It is this angle in the substances concerned, which determines the degrees of acclivity which can exist in the sides of hills composed of sand, gravel, earth, &c., in the banks of canals, rivers, &c.

If the thread of a screw winds round the spindle with an angle less than this, the screw can never recoil or slide back from force acting against its point.

But for friction, men walking on the ground or pavement would always be as if walking on ice; and our rivers, that now flow so calmly, would all be frightful torrents. Friction is, therefore, in these cases of great use to men.

Friction is useful, also, when it enables men, out of the comparatively short fibres of cotton, flax, or hemp, to form their lengthened webs and cordage,—for it is friction alone, consequent upon the interweaving and twisting of the fibres and threads, which keeps the material of these fabrics together.

The following means are used to diminish friction between rubbing surfaces; and they are used singly or in combination, according to the circumstances.

1. Making the rubbing surfaces smooth;—but this must be done within certain limits, for great smoothness allows the bodies to approach so near that a degree of cohesion takes place.

2. Letting the substances which are to rub on each other be of different kinds. Axles are made of steel, for instance, and the parts on which they bear are made of brass; in small machines as time-keepers, the steel axles often play in agate or diamond. The swiftness of a skater depends much on the great dissimilarity between steel and ice.

3. Interposing some lubricating substance between the rubbing parts; as oils for the metals, soap, grease, black-lead, &c., for the woods. There is a laughable illustration of this in the holiday sport of soaping a lively pig's tail, and then offering him as the prize of the clever fellow who can catch and hold him fast by his slippery appendix.

4. Diminishing the extent of the touching surfaces; as in making the rubbing axis of a wheel very small.

5. Using wheels as in wheel-carriages, instead of dragging a rubbing load along the ground. Casters on household furniture are miniature wheels.

6. Using what is called friction-wheels;—which still farther diminish the friction even of a smooth axis, by allowing it to rest on their circumferences, which turn with it. Here *a* represent the end of an axis, resting on the exteriors of two friction-wheels, *b* and *c*.

Fig. 61.

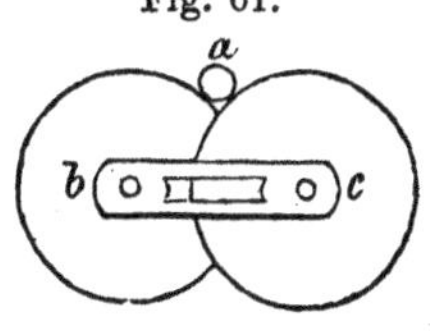

7. Placing the thing to be moved on rollers or balls as when a log of wood is drawn along the ground upon rounded pieces of wood; or when a cannon with a flat circular base to its carriage, turns round by rolling on cannon-balls laid on a hard level bed. In these two cases there is hardly any friction, and the resistance is merely from the obstacles which the rollers or balls may have to pass over.

Of all rubbing parts the joints of animals, considering the strength, frequency and rapidity of their movements, are those which have the least friction. The rubbing surfaces in these are covered, first, with a layer of elastic cartilage, and then with an exceedingly smooth membrane, over which there is constantly poured from the glands around, a fluid called synovia, more emollient and lubricating than any oil, and which is renewed constantly as may be required. We study and admire the perfection of animal joints, without being able very closely to imitate it.

Wheel-carriages merit notice here, as illustrating many of the circumstances connected with friction; and moreover as being among the most common of machines.

Wheel-carriages have three advantages over the sledges for which they are the substitutes:

1. The rubbing or friction, instead of being between an iron shoe and the stones and irregularities of the road, is between the axle and its bush, of which the surfaces are smoothed and fitted to each other, and well lubricated.

2. While the carriage moves forward, perhaps fifteen feet, by one revolution of its wheel, the rubbing part, *viz.*, the axle, passes over only a few inches of the internal surface of its smooth greased bush.

3. The wheel surmounts any abrupt obstacle on the road by the axle describing a gently rising slope or curve, — as shown in this figure, where *a* represents an obstacle, and where the curve from *c*, of which the beginning has the direction shown by the line *c e*, represents the path of the axle in surmounting it. The wheel is as if rising on an inclined plane, and gives to the drawing animal the relief which such a plane would bring. This kind of advantage is greater in a large wheel, for evidently the smaller wheel here represented, in having to surmount the same size of obstacles, has to rise in the steeper curve beginning at *d*,—but the difference of advantage, in this respect, is not so great as the difference of size. It is true again, that a small wheel would sink to the bottom of a hole, where a larger one would rest on the edges as a bridge, and would sink less. The fore wheels of carriages are usually made small, because such construction, by allowing the wheel to go under the body of the carriage, facilitates the turning of the carriage. It is not true, however, according to the popular prejudice, that the large hind wheels of coaches, wagons, &c., help to push on the little wheels before them, as if the carriage were on an inclined plane resting on the wheels; but there is the accidental advantage, that in ascending a hill, when the horses have to put forth their strength, the load rests chiefly on the hind wheels, and in descending, when an increased resistance is desirable, the load falls chiefly on the fore-wheels.

Fig. 62.

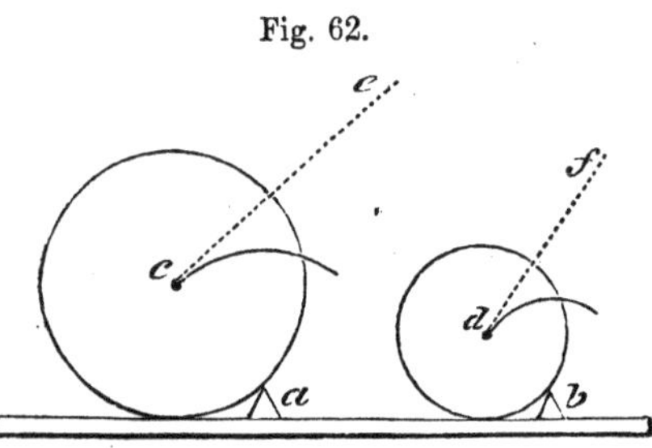

From the causes mentioned in the last paragraphs, the difference in performing the same journey of a mile, by a sledge and by a wheel carriage, is that while the former has to rub over every roughness in the road and to be jolted by every irregularity, the rubbing part of the latter, the axle, glides very slowly over about thirty yards of a smooth oiled surface, in a gently waving line. Thus, by wheels, the resistance is reduced to about the hundredth part of what it is for a sledge.

On hilly roads, in descending, it is common to *lock* or fix one of the wheels of a carriage, and the horses have then to pull nearly as much as on a level road with the wheel free; showing the effect of a little increase of friction.

The wheel of a carriage, simple as from our extreme familiarity with it, it now appears to us, is a thing of very nice workmanship, and which has exercised much ingenuity.—It acquires astonishing strength, indeed that of the arch, from what is called its *dished* form, seen here in the wheel *c*, as contrasted with the flat wheel *a*. In a wheel of this form, the extremity of a spoke cannot be displaced inwards, or towards the carriage, unless the rim of the wheel be enlarged, or all the other spokes yield at the same time, and it cannot be displaced outwards, or away from the carriage, unless the rim be diminished, or the other spokes yield in the opposite direction:—now the

Fig. 63.

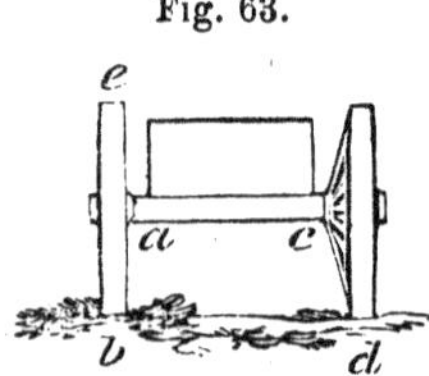

rim being strongly bound by a ring, or tire of iron, cannot suffer either increase or diminution, and the strength of all the spokes is thus by it conferred on each individually. In a *flat* wheel a given degree of displacement outwards or inwards of the extremities of a spoke, would less affect the magnitude of the circumference, and therefore the rim of such a wheel, secures much less firmly. A watch-glass and a round piece of egg-shell are stronger than flat pieces of like substances, for the same reason that a dished wheel is stronger than a flat wheel.—The dished form of a wheel is farther useful by leaving more room between the wheels for the body of the carriage, and is useful, also, in this, that when the carriage is on an inclined road, and more of the weight consequently falls upon the wheel of the lower side, the inferior spokes of that wheel become nearly perpendicular, and thereby support the increased weight more safely. The strongest form of wheel is the *doubly dished*, that is, a wheel having half of the spokes passing from within to the rim as from *c* to *d*, fig. 63, and the other half similarly from without. This form is adopted in the wheel recently constructed entirely of iron, in which there is a farther peculiarity that the load is supported more by hanging by the upper spokes than by resting on the lower. When wheels, instead of standing upright, like *b* and *d* shown, fig. 63, are made to incline outwards, as is common, owing to the ends of the axletrees being bent down a little to give a security against the accident of the wheels falling off, the pull to the horses in deep or sandy roads is much increased; for an inclining wheel would naturally describe a curved and outward path, as is seen when a hoop or wheelbarrow inclines; and the horses, therefore, in drawing straight forward; have constantly to overcome the deviating tendency in all inclining wheels. This cause of resistance is still more remarkable when the wheels have broad rims. Such wheels must be conical, that is, of smaller diameter at the outer than at the inner edge, as the end of a cask is smaller than its middle, and then, as the iron hoops or tires which cover the different parts cannot all, by an equal number of turns, truly measure the same length of road, there will be a constant rubbing or grinding forward of the lesser rings, and a grinding backward of the larger, injuring the road, rapidly wearing the iron, and exhausting the strength of the pulling animals. Such wheels rolling free would describe a circular path, as is exemplified when a thimble, or drinking glass, or sugar-loaf, which also are conical, is pushed forward on any plane surface.

The application of springs to carriages, which is an improvement of comparatively recent date, not only renders them soft moving vehicles on rough roads, but much lessens the pull to the horses. When there is no spring, the whole load must rise with every rising of the road, and if time be given, must sink with every depression, and the depression costs as much labour as the rising, because the wheel must be drawn up again from the bottom of it; but in a spring-carriage moving rapidly along, only the parts below the springs are moved in correspondence with the road-surface, while all above, by the inertia of the matter, have a soft and even advance. Hence arises the superiority of those modern carriages, furnished with what are called *under-springs*, which insulate from the effect of shocks, all the parts excepting the wheels and axletrees themselves. When only the body of the carriage is on springs, the horses have still to rattle the heavy frame-work below it over all irregularities, and then the wheels as well as the structure generally require to be of much greater strength and weight to bear the consequent shocks.

The subject of wheel carriages is interesting to medical men, from their having often to direct in transporting the sick or wounded.

It is perhaps difficult to conceive anything more elegant and perfect than the carriages of modern refinement; and, therefore, a man, who sees them gliding swiftly along the prepared levels and slopes of our present landscapes, and thinks of the clumsy vehicles on the bad roads of former times, may readily imagine that absolute perfection is at last attained. Yet we are perhaps now on the eve of a farther change which, for many purposes, will be of greater importance than all that has yet been achieved—*viz.*, the general adoption of rail-roads, with new-fashioned carriages to suit them. To all who study such subjects, it is now known, that to draw a loaded wagon up one inconsiderable hill, costs more force than to send it thirty or forty miles along a level rail-way; and the conclusion is obvious, that although the original expense of forming the level line might considerably exceed that of making an ordinary road, still, in situations of great traffic, the difference would soon be paid for by the savings, and when once paid, the savings would be as a profit for ever. To readers conversant with political economy, it would be superfluous to speak here of the advantages of any great facility of intercourse, but to those who are not, the following reflections may be interesting.

In reviewing the history of the human race, we find that every remarkable increase in civilization has taken place very much in proportion to the facilities of intercourse offered in the particular situation. First, therefore, civilization, grew along the banks of great rivers, as the Nile, the Euphrates and the Ganges; or along the shores of inland seas or archipelagoes, as in the Mediterranean and the numerous islands of Greece; or over fertile and extended plains, as in many parts of India. When the situation thus bound a great number of individuals into one body, the useful new thought or action of any one unusually gifted, and which, in the insulated state, would soon have been forgotten and lost, extended its influence immediately to the whole body, and became the thought or action of all who could benefit by it, besides that it was recorded for ever, as part of the growing science of art of the community. And in a numerous society, such useful thoughts and acts would naturally be more frequent, because persons feeling that they had the eyes of a multitude upon them, and that the rewards of excellence would be proportionally great, would be excited to emulation in all the pursuits that could contribute to the well-being of the society. Men soon learned to estimate aright these and many other advantages of easy intercourse, and after having possessed themselves with avidity of the stations naturally fitted for their purposes, they began to improve the old and to make new stations. They created rivers and shores, and plains of their own, that is, they constructed canals and basins, and roads; and so connected, artificially, regions which nature seemed to have separated forever.—In the British isles, whose favoured children have taken so remarkable a lead in showing what prodigies a wise policy may effect, the advantages arising from certain lines of canal and road first executed, soon led to numberless similar enterprises, and within half a century the empire has been thus bound together in all directions: and it seems as if the noble work was now to be crowned by the substitution of level railways for many of the common roads and canals.* Several rail-

* These observations were first published (the substance had been written long before,) soon after the Darlington rail-road, the first of any note intended for passengers, was opened. The Manchester and Liverpool rail-road has since then admirably verified the anticipations.

roads of short extent have already been established, and although they and the carriages upon them are far from having the perfection which philosophy says they will admit, the results have been very satisfactory. If we suppose the progress to continue, and the price of transporting things and persons to be thus reduced to a fourth of the present charge—and in many cases it may be less—and if we suppose the time of journeying with safety also to be reduced in some considerable degree,—of which there can be as little doubt—the general adoption of such roads would operate an extraordinary revolution and improvement in the state of society. Without in reality changing the distances of places, it would in effect bring all places nearer to each other, and would give to every spot in the kingdom the conveniences of the whole,—of town and country, of sea-coast and of highland district. A man, wherever residing, might consider himself virtually near to any other part, when, at the expense of time and money now expended in travelling a short way, he might travel very far, and he would thus find remarkably extended, the sphere both of his business and of his pleasures. The over-crowded and unhealthy parts of towns would scatter their inhabitants into the country; for the man of business could be as conveniently at his post from a distance of several miles, as he is now from an adjoining street. The present heavy charges for bringing distant produce to market being nearly saved, the buyer everywhere would purchase cheaper, and the producer would be still better remunerated.

In a word, such a change would be affected, as if by magic the whole of Britain had been compressed into a circle of a few miles in diameter, yet without any part losing aught of its magnitude or beauties. All this may appear visionary; but it is less so than seventy years ago it would have been to anticipate much of what, in respect to travelling, has really come to pass,—as, that the common time of passing from London to Edinburgh would be forty-six hours. At the recent opening (in 1825) of the railroad near Darlington, a train of loaded carriages was dragged by one little steam-engine a distance of twenty-five miles within two hours; and in some parts of the journey the speed was more than twenty miles an hour: the load was equal to a regiment of soldiers, and the coal expended was not of the value of a crown. An island with such roads would be an impregnable fortress; for in less time than an enemy would require to disembark on any part of the coast, the forces of the country might be concentrated to defend it.

"*Strength depends on the magnitude, form and position of bodies, as well as on the degree of cohesion in the material.*" (Read the Analysis, page 84.)

The minute details connected with this branch of the subject belong to the practical engineer, but there are some of the general truths which should be familiar to every body.

Of similar bodies the largest is proportionally the weakest.

Suppose two blocks of stone left projecting from a hewn rock, of which blocks one, as *d*, p. 120, is twice as long, and deep, and broad, as the other, *b*. The larger one will by no means support at its end as much more weight than the smaller, as its mass is greater, and for two reasons. 1st. In the larger, each particle of the surface of attachment at *c*, in helping to bear

Fig. 64.

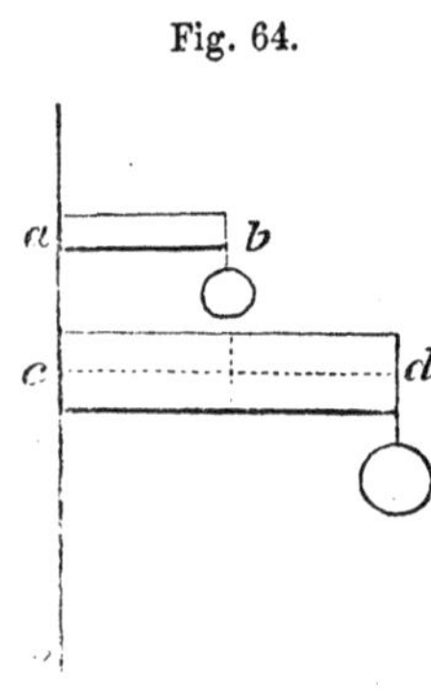

the weight of the block itself, has to support by its cohesion twice as many particles beyond it in the double extent of projection, as a particle has to support in the shorter block at *a*; and 2ndly, both the additional substance, and anything appended at its outer extremity, are acting with a double lever advantage to break it, that is, to destroy the cohesion at *c*. Hence, if any such mass be made to project very far, it will be broken off or will fall by its own weight alone. And what is thus true of a block supported at one end, is equally true of a block supported at both ends, and indeed of all masses, however supported, and of whatever forms, if they have projecting parts. It is to be observed, also, that masses, like an absolutely perpendicular cliff, which have no projecting or overhanging parts, are still limited, as to size, by the degree of cohesive force among their particles, for the upper part of such a mass tends to crush or break down the lower. A lofty pillar cannot be formed of soft clay.

That a large body, therefore, may have proportionate strength to a smaller, it must be still thicker and more clumsy than it is longer: and beyond a certain limit, no proportions whatever will keep it together, in opposition merely to the force of its own weight.

This great truth limits the size and modifies the shape of most productions of nature and art;—of hills, trees, animals, architectural or mechanical structures, &c.

Hills. Very strong or cohesive material may constitute hills of sublime elevation, with very projecting cliffs and very lofty perpendicular precipices; and such accordingly are seen where the hard granite protrudes from the bowels of the earth, as in the Andes of America, the Alps of Europe, the Himalayas of Asia, and the Mountains of the Moon in Central Africa. But material of inferior strength exhibits more humble risings and more rounded surfaces. The gradation is so striking and constant, from granite mountains, down to those of chalk, or gravel, or sand, that the geologist can often tell the substance of which a hill is composed by observing the peculiarities of its shape.

Even in granite itself, which is the strongest of rocks, there is a limit to height and projection; and if an instance of either, much more remarkable than now remains on earth, were by any chance to be produced again, the law which we are considering would prune the monstrosity. The grotesque figures of rocks and mountains, seen in the paintings of the Chinese—or actually formed in minature for the gardens, to express their notions of perfect sublimity and beauty—are caricatures of nature for which originals can never have existed. Some of the smaller islands in the Eastern Ocean, however, and some of the mountains of the chains seen in the voyage towards China, along the coasts of Borneo and Palawan, exhibit, perhaps, the very limits of possibility in singular shapes. In the moon, where the weight or gravity of bodies is less than on earth, on account of her smaller size, mountains of a given material might be many times higher than on earth — and observation proves that the lunar mountains are in fact very high.

By the action of winds, rains, currents, and frosts, upon the mineral masses around us, there is unceasingly going on an undermining and wasting of supports, so that every now and then immense rocks, or almost hills, are torn by gravity from the station which they have held since the earth received its present form, and fall in obedience to the law now explained.

The size of vegetables, of course, is obedient to the same law. We have no trees reaching a height of three hundred feet, even when perfectly perpendicular, and sheltered in forests that have been unmolested from the beginning of time: and oblique or horizontal branches are kept within comparatively narrow limits by the great strength required to support them. The truth, that to have proper strength, the breadth or diameter of bodies must increase more quickly than the length, is well illustrated by the contrast existing between the delicate and slender proportions of a young oak or elm, yet in the seedman's nursery, and the sturdy form of one which has braved for centuries all the winds of heaven, and has become the monarch of the park or forest.

Animals furnish other interesting illustrations of this law.

How massive and clumsy are the limbs of the elephant, the rhinoceros, the heavy ox, compared with the slender forms of the stag, antelope, and greyhound! And unless the bones were made of stronger material than now, an animal much larger than the elephant would fall to pieces owing to its weight alone. The whale is the largest of animals, but feels not its enormous weight because lying constantly in the liquid support of the ocean. A cat may fall with impunity from a greater height than would suffice to dash the bones of an elephant or an ox to pieces.

For the reason which we are now considering, the giants of the heathen mythology could not have existed upon this earth; although, on our moon, where, as already stated, weight is much less, such beings might be. In the planet Jupiter, again, which is many times larger than the earth, an ordinary man from hence would be carrying in the simple weight of his body a load sufficient to crush the limbs which supported him. The phrase *a little compact man*, points to the fact that such a person is stronger in proportion to his size than a taller man.

The same law limits the heighth and breadth of architectural structures. In the houses of fourteen stories, which formerly stood for protection close under the Castle of Edinburgh, there was danger of the superincumbent wall crushing the foundation.

Roofs. Westminster hall approaches the limit of width that is possible without either very inconvenient proportions or central supports; and the dome of the church of St. Peter in Rome is in the same predicament.

Arches of a bridge. A stone arch, much larger than those of the magnificent bridges in London, would be in danger of crushing or splintering its material.

Ships. The ribs or timbers of a boat have scarcely a hundredth part of the bulk of the timbers of a ship only ten times longer than the boat. A ship's yard of ninety feet contains, perhaps, twenty times as much wood as a yard of thirty feet, and even then is not so strong in proportion. If ten men may do the work of a three-hundred-ton ship, many more than three times that number will be required to manage a ship three times as large. Very large ships, such as the two built in Canada, in the year 1825, which carried each nearly ten thousand tons of timber, are weak from their size alone; and the loss of these first two specimens of gigantic magnitude will not encourage the building of others.

The degree in which the strength of structure is dependent on the *form* and *position* of their parts, will be illustrated by considering the two cases of *longitudinal* and *transverse* compression. And the rule for giving strength to any structure will be found to be, to cause the force tending to destroy it, to act, as equally as possible, on the *whole* resisting mass at once, and with as little mechanical advantage as possible.

In *longitudinal compression*, as produced by a body *a*, on the atoms of the support *b*, the weight, while the support remains straight, can only destroy the support, by crushing it in opposition to the repulsion and impenetrability of all its atoms. Hence a very small pillar, if kept perfectly straight, supports a very great weight; but a pillar originally crooked, or beginning to bend, resists with only part of its strength; for, as seen in *c d*, the whole weight above is supported chiefly on the atoms of the concave side, which are therefore in greater danger of being oppressed and crushed, while those on the convex side, separated from their natural helpmates, are in the opposite danger of being torn asunder. The atoms near the centre in such a case are almost neutral, and might be absent without the strength of the pillar being much lessened.

Fig. 65.

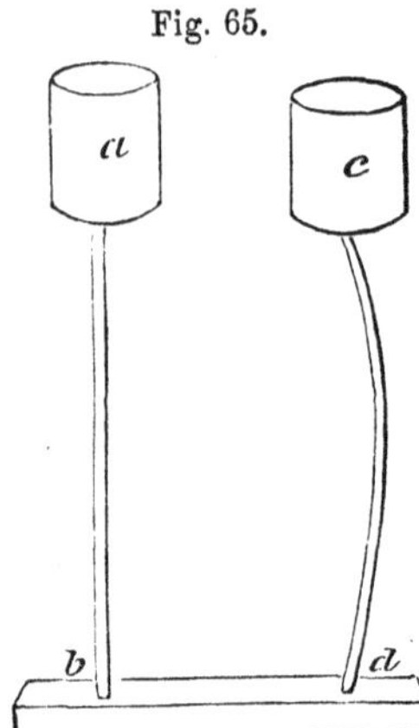

Long pillars or supports are weaker than short pillars of the same diameter, because they are more easily bent; and they are more easily bent because a very inconsiderable, and therefore easily effected yielding between each adjoining two of their many atoms, makes a considerable bend in the whole; while in a very short pillar there cannot be much bending without a great change in the relation of proximate atoms, and such as can be effected only by great force. The weight resting on any pillar, and bending it, may be considered as acting (with obliquity dependent on the degree of bending) at the end of a long lever which reaches from the extremity to the centre of the pillar, against the strength resisting always directly at a short lever reaching from the side *d* to the centre; the strength of the pillar, therefore, has relation to the difference between these levers and to the degree of bending. Shortness, then, or any stay or projection as *a e b*, which, by making the resisting lever longer, opposes bending, really increases the strength of a pillar.

Fig. 66.

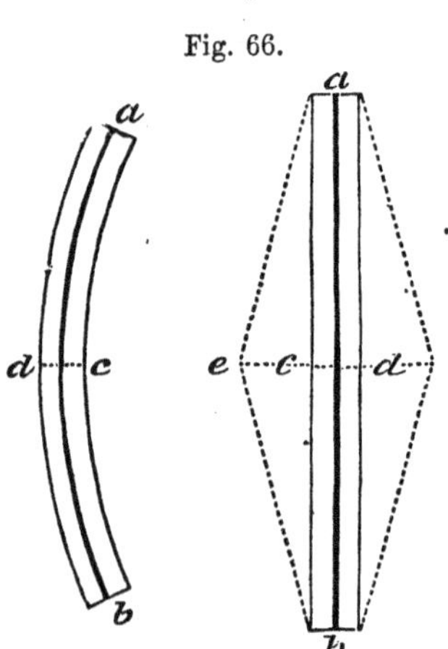

A column with ridges projecting from it, is on this account stronger than one that is perfectly smooth.

A hollow tube of metal is stronger than the same quantity of metal as a solid rod, because its substance standing farther from the centre resists bending with a longer lever. Hence pillars of cast-iron are generally made hollow, that they may have strength with as little metal as possible.

In the most perfect weighing-beams for delicate purposes, that there may be the least possible weight with the required strength, the arms, instead of being of solid metal, are hollow cones, of which the substance is not much thicker than writing paper.

Masts and yards for ships have been made hollow in accordance with the same principle.

In Nature's work we have to admire numerous illustrations of the same kind.

The stems of many vegetables, instead of being round externally, are ribbed or angular and fluted, that they may have strength to resist bending. Many also are hollow, as corn-stalks, the elder, the bamboo of tropical climates, &c., thereby combining lightness with their strength."—A person who has visited the countries where the bamboo grows, cannot but admire the almost endless uses to the inhabitants, which its straightness, lightness and hollowness, fit it to serve. Being found of all sizes, it has merely to be cut into pieces of the lengths required for any purposes, and nature has already been the turner, and the polisher, and the borer, &c. In many of the Eastern Islands it is the chief material, both of the dwellings, and of the furniture; there are the bamboo huts and bungallows, and then the fanciful chairs, couches, beds, &c.; flutes and other wind instruments there, are merely pieces of the reed with holes bored at the requisite distance: conduits for water are pipes of bamboo; bottles and casks for preserving liquids are single joints of larger bamboo with the natural partitions remaining; and bamboo split into threads is twisted into rope, &c.

From the animal kingdom also we have illustrations of our present subject:—as in the hollow stiffness of the quills of birds; the hollow bones of birds; the bones of animals generally—strong and hard, and often angular externally, with light cellular texture within, &c.

Transverse Pressure.

When a horizontal beam is supported at its extremities, as at *a* and *b*, its weight bends its middle down more or less, as here shown, the particles on the upper side being compressed, while the parts below are distended; and the bending and tendency to break are greater, according as the beam is longer and its thickness and depth is less.

Fig. 67.

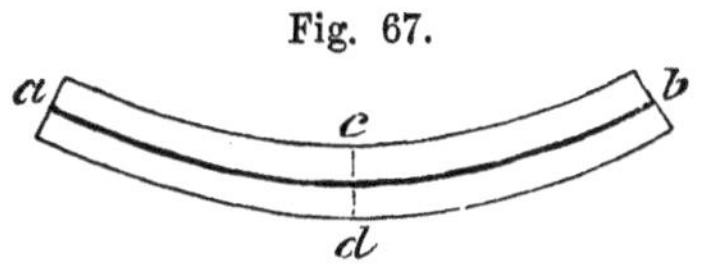

The danger of breaking in a beam, so situated is judged of, by considering the destroying force as acting by a long lever reaching from an end of the beam to the centre, and the resisting force or strength as acting only by a short lever from the side *d* to the centre: while only a little of the substance of a beam on the under side is allowed to resist at all. This last circumstance is so remarkable, that the scratch of a pin on the under side of a plank resting, as here supposed, will sometimes suffice to begin the fracture.

Because the resisting lever is small in proportion as the beam is thinner, a plank bends and breaks more readily than a beam, and a beam resting on its side bears less weight than if resting on its edge. Where a single beam cannot be found deep or broad enough to have the strength required in any particular case, as for supporting the roof of a house, several beams are joined together, and in a great variety of ways, as is seen in house-rafters, &c., which although consisting of three or more pieces, may be considered as one very broad beam, with those parts cut out which would contribute least to the strength.

The arched form, resting against immoveable abutments, bears transverse pressure so admirably because by means of it the force that would destroy, is made to compress not one side only, but all the atoms or parts of both sides nearly in the same degree. By comparing this figure with the last, we see that the atoms on the under side of an arch, must be compressed about as much as those on the upper side, and are therefore in no danger of being torn, or overcome separately. The whole substance of the arch therefore resists, nearly like that of a straight pillar under a weight, and is nearly as strong.

Fig. 68.

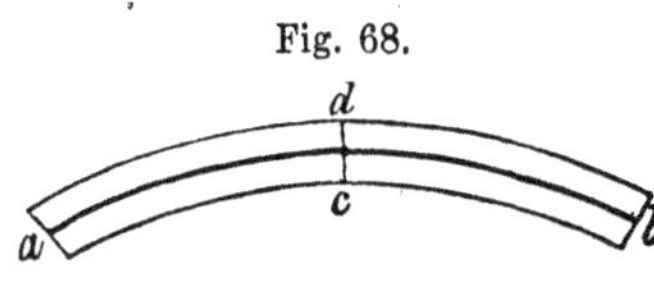

An error, which has been frequently committed by bridge-builders, is the neglecting to consider sufficiently the effect of the horizontal thrust of the arch on its piers. Each arch is an engine of oblique force (see page 56,) pushing the pier away from it. In some instances, one arch of a bridge falling, has allowed the adjoining piers to be pushed down towards it, by the thrust, no longer balanced, of the arches beyond, and the whole structure has given way at once like a child's house or bridge built of cards.

It is not known at what time the arch was invented, but it was in comparatively modern times. The hint may have been taken from nature, for there are instances in Alpine countries of natural arches, where rocks have fallen between rocks, and have there been arrested and suspended, or where burrowing water has at last formed a wide passage under masses of rock, and has left them balanced among themselves as an arch above the stream. Nothing can surpass the strength and beauty of some modern stone bridges; —those, for instance, which spans the Thames as it winds through London.

Iron bridges have been made with arches twice as large as those of stone; the material being more tenacious and easily moulded, is calculated to form a lighter whole. The bridge of three fine arches lately built between the city of London and Southwark, is a noble specimen, and compared with those erected in the preceding century, appears almost a fairy structure of lightness and grace.

The great domes of churches, as those of St. Peter's in Rome and St. Paul's in London, have strength on the same principle as simple arches. They are in general strongly bound at the bottom with chains and iron-bars, to aid the masonry in counteracting the horizontal thrust of the superstructure.

The Gothic arch is a pointed arch, and is calculated to bear the chief weight on its summit or key-stone. Its use, therefore, is not properly to span rivers as a bridge, but to enter into the composition of varied pieces of architecture. With what effect it does this, is seen in the truly sublime Gothic structures which still adorn so many parts of Europe.

The following are instances, in smaller bodies, of strength obtained by the arched form.—A thin watch-glass bears a very hard push; a dished or arched wheel for a carriage is many times stronger to resist all kinds of shocks than a perfectly flat wheel;—a full cask may fall with impunity, where a strong square box would be dashed to pieces;—a very thin globular flask or glass, corked and sent down many fathoms into the sea, will resist the pressure of water around it, where a square bottle, with sides of almost any thickness, would be crushed to pieces.

We have, from the animal frame, an illustration of the arched form giving strength, in the cranium or skull, and particularly in the skull of man, which is the largest in proportion to its thickness :—the brain required the most perfect security, and in the arched form of the skull has obtained it with little weight.—The common egg-shell is another example of the same class : what hard blows of the spoon or knife are often required to penetrate this wonderful defence of a dormant life! The weakness of a similar substance not arched, is seen in a scale from a piece of freestone so readily crumbling between the fingers.

To determine, for particular cases, the best forms and positions of beams and joists, and of arches, domes, &c., is the business of strict calculation, and belongs therefore to mathematics, or the *science* of *measures.*

It was a beautiful problem of this kind, which Mr. Smeaton, the English engineer, solved so perfectly, in the construction of the far-famed Eddystone light-house. He had to determine the form and dimensions of a building, which would stand firm on a sunken rock, in the channel of a swift ocean tide, and exposed to the fury of tempests from every quarter. Only the man who has himself been driven before the irresistible storm in the darkness of night, and in the midst of dangers, and whose eyes have watched the steady ray from the light-house which saved him, can appreciate fully the importance of the studies which bring such useful results; or can feel how happy he is to have fellow-men, whose talents, although exerted usually for individual good, are yet, by God's providence, made to accomplish the most philanthropic ends, and to bind the whole of human kind into one great society of helping brotherhood.

[For Animal and Medical Mechanics, see Part V. Sect. 1.]

PART III.

THE PHENOMENA OF FLUIDS.*

SECTION I.—HYDROSTATICS.

ANALYSIS OF THE SECTION.

The particles of a fluid mass are freely moveable among one another, so as to yield to the least disturbing force; and if bearing force at all, can be at rest only when equally forced in all directions. Hence:

1. *In a mass of fluid submitted to compression, the whole is equally affected, and equally in all directions. A given pressure, for instance, made by a plug forced inwards upon a square inch of the surface of a fluid filling a vessel, is suddenly communicated to every square inch of the vessel's surface, however large, and to every inch of the surface of any body immersed in the fluid.*
2. *In any fluid, the particles that are below bear the weight of those that are above, and there is, therefore, within the mass, a pressure increasing exactly with the perpendicular depth, and not influenced by the size, or shape, or position of the containing vessel.*
3. *The open surface of a fluid is level; and if various pipes or vessels communicate with each other, any fluid admitted to them will rise to the same level in all.*
4. *A body immersed in a fluid displaces exactly its own bulk of it, which quantity having been just supported by the fluid around, the body is pressed upwards, or supported, with a force exactly equal to the weight of the fluid displaced, and must sink or swim according as its own weight is greater or less than this. By comparing, therefore, the weight of a body with the force which holds it up in a fluid, the comparative weight or* specific gravities *are found.*

"Fluid."

It was explained in Part I., that the same atoms may exist in the form of a solid or of a fluid; and as a fluid, they may either constitute a dense liquid like water, or a light elastic mass like air. A pound of ice, or a pound of water, or a pound of steam, differs only in the particles being more or less distant from each other, owing to the different quantities of heat among them. In the ice, they are comparatively near, and are held together by attraction, as if they were spitted or glued to each other; in the water, the repulsion of heat seems nearly to balance attraction, and to leave the particles at liberty to glide about among each other almost without

* Read again the Synopsis, page 20.

friction; and in the steam, the repulsion altogether overcomes the attraction, and the particles separate to a great distance, as if held apart by some bulky elastic medium. The few facts not evidently reconcilable with the simple and satisfactory explanation of so many phenomena,—as that water in freezing, and even in cooling down from forty degress to the freezing point, increases in volume, instead of contracting, like things in general, and like itself in cooling at other temperatures,—and that baked clay, in proportion as it is more heated, contracts instead of dilating,—are treated of in other parts of our work.

Whether matter be in the solid or fluid form, the properties of the individual atoms remain unchanged, that is, the atoms always exist in accordance with the "general truths;" but as, in the chapter on Mechanics, we found so many important modifications of effect produced by the circumstance of the attraction being in the degree which produces solid cohesion among the particles, in this chapter on fluids we shall find as many important results springing from the circumstances of non-cohesion or fluidity.

In a liquid the particles, although comparatively near to one another, seem not to be in actual contact; for the mass may be condensed indefinitely by pressure. The force required, however, to change the volume of a liquid in any sensible degree, is so great, that until improved means of experiment, recently contrived, liquids were accounted absolutely incompressible. In aeriform fluids, on the contrary each particle, under common circumstances, has about two thousand times as much space to itself as when forming part of a liquid or solid; and hence it is that these fluids are so extensively compressible and dilatable—or elastic, as they are called. On account of this elasticity, they exhibit so many important phenomena, in addition to those of mere fluidity, that the consideration of them requires to be gone into apart, and forms the branch of the subject called *pneumatics*, from a Greek word, signifying "spirit" or "breath!"

"*In a quantity of fluid submitted to compression, the whole mass is equally affected, and similarly in all directions. A given pressure, therefore, made upon an inch of the surface of a fluid confined in a vessel, as by a plug forced inwards, is suddenly borne by every inch of the surface of the vessel, however large, and by every inch of the surface of any body immersed in the fluid.*"

This truth is of great importance, both from its explaining so many remarkable phenomena of nature, and from the useful applications of it in the construction of machinery.

When a man compresses in his hands a bladder full of air, he readily conceives that the air immediately under his fingers is not at all more compressed than that in every other part of the bladder; and of course that every part of the bladders's surface must be pressing the air as much as those parts of it on which his fingers rest, and must be bearing a reaction or resistance of the air in an equal degree; and that every single particle of air must be acted upon equally on every side, so that if a small opening were made in the bladder anywhere, the air would issue from it with equal readiness. This is in accordance with the characteristic of fluidity, "that the particles glide about among one another almost without friction, so that a particle can never be at rest unless when equally urged in all directions."

In like manner, if a close vessel B be filled with water, and into the top of it a tube *a c* be screwed, and if then, by means of a cork or moveable plug in the tube at *c*, the surface of the water in the vessel be pressed upon with a force of one pound, the water throughout the whole will be squeezed or condensed in proportion to the pressure, and every other portion of the vessel B, of equal surface with *c*, will be keeping up the condensation just as much as *c*, and will be bearing the resistance or elasticity of the water to the extent of one pound. And if there were another similar tube, *b*, also with a plug, screwed into the top of the box B, the force of one pound depressing the plug *c* would be pushing up the plug *b*, with the same force. And if there were many other similar tubes and plugs, by acting on one, all would be equally affected; and a plug or piston of double size would be twice as much affected as the smaller one; and a plug *d*, of ten times the size, would be lifted with a force of ten pounds. Hence it appears that, through the medium of confined fluid, a force of one pound, acting upon an inch square of the fluid surface in a vessel, may become a bursting force of ten, or a hundred, or a thousand pounds, according to the size of the vessel, or may be used as a mechanical power to overcome a force much more intense than itself. It will be explained below that the well-known hydrostatic press is merely a large plug or piston as here described, forced up against the substance to be pressed by the action of a smaller piston in another barrel.

Fig. 69.

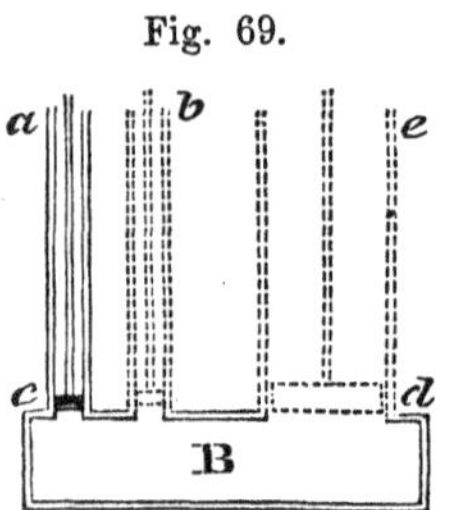

If, in the above figure, the tube *a* were such as to contain just one pound of water, on the plug *c* being withdrawn from it, and water being poured in to fill it, the same pressure or condensation would take place in the box B as when the plug was pressed with the force of one pound; and of course exactly the same effects would follow on the sides of the vessel and on the other pistons; and if, in the other tubes also, water were substituted for the pistons, it is evident that, to effect a balance in all, it would require to stand as high in every one as in the tube *a c*, producing the same level in all, whatever their size.

Fig. 70.

The fact that the weight of one pound of water, or any other force of one pound similarly applied, may be made, through the medium of extended fluid surface to produce a pressure of hundreds or of thousands of pounds, has been called the "hydrostatic paradox," yet there is nothing in reality more paradoxical in it than that one pound at the long end of the lever should balance ten pounds at the short end: indeed it is but another means, like the contrivances usually called mechanical powers, and described in the last chapter, of balancing different intensities of force, by applying them to parts of an apparatus moving with different velocities. Here the tube *a* being ten times smaller than the tube *e*, the piston *a* must descend ten inches to raise the greater piston in *e* one inch.

This law of fluid pressure is rendered very striking in the experiment of bursting a strong cask by the weight or action of a few ounces of water. Suppose a cask *a* already filled with water, and that a long small tube *b c* is screwed tightly into its top, which tube will contain only a few ounces of water; by pouring these few ounces into the tube, the cask will be burst. In explanation it is unnecessary to say more than that if the tube have an area of a fortieth of an inch, and contain, when filled, half a pound of water, that water would produce a pressure of half a pound upon every fortieth of an inch all over the interior of the cask, or nearly 2,000 lbs. on the square foot,—a pressure greater than a common cask can bear.

A similar effect is seen in what is called the *hydrostatic bellows.* This consists of a long small tube *a b*, into which water is poured to enter the body of the apparatus at *c*, which resembles the common bellows, in having wooden boards above and below, and strong leather connecting them. If the tube *a b* holds an ounce of water, and has itself only one-thousandth of the area of the top of the bellows, an ounce of water in it will balance weights of a thousand ounces placed on the top of the bellows at *d.* If mercury were substituted in this machine for water, the effect would be fourteen times greater, because mercury is fourteen times heavier in the same bulk. And if a man stand on a large bellows of the kind, he may raise himself by blowing into the tube with his mouth.

Fig. 71.

The annexed cut will give an idea of Mr. Braham's singularly powerful and useful *hydrostatic* or *hydraulic press;* which, if compared with the bellows, exhibits merely a strong forcing pump instead of the lofty tube, and a barrel with its piston instead of the leather and boards. The letter *e* points out the piston of the forcing pump worked by the handle *d*, and driving water along the horizontal tube into the space *f* under the large solid piston *c*, which last, with its spreading top, is urged against the object to be compressed. If the small pump have only one-thousandth of the area of the large barrel, and if a man by means of its lever-handle *d*, press its piston down with a force of five hundred pounds, the piston of the great barrel will rise with a force of one thousand times five hundred pounds, or more than two hundred tons. Scarcely any resistance can withstand the power of such a press; with it the hand of an infant can break a strong iron bar; and it is used to condense substances, as cotton or hay for sea voyages, to raise great weights, to uproot trees, to tear things asunder, &c.

Fig. 72.

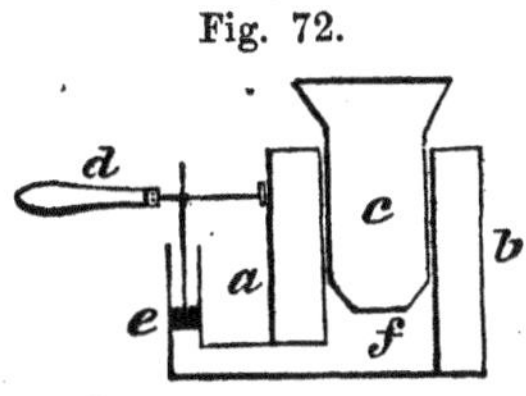

The Dilater is a surgical instrument of extensive applicability, of which the action depends on the principle of the communication of fluid pressure. It was proposed by the author some years ago, and was brought to great practical perfection by his brother Dr. James Arnott, (now superintendent surgeon in the service of the Hon. East India Company,) in whose publication it is minutely treated of. Many professional men in this country doubted of its power, from not being aware of the nature of fluid nature; but it is in reality a kind of hydraulic press, allowing the operator to act with the most gentle or most energetic force. Further remarks are made upon it in the medical section which follows this chapter.

"*In any fluid, the particles that are below, bear the weight of those that are above, and there is therefore a pressure among them increasing in exact proportion to the perpendicular depth, and not influenced by the size, or shape, or position of the containing vessels.*"

The atoms of matter having gravity, it is evident that the upper layer of any mass of fluid must be supported by the second, and this with its load by the third, and the third with its double load by the fourth, and so on. This truth is experimentally proved by putting different heights of liquid into an upright tube, of which the bottom is closed by a flap having a spring or lever to support it, and to indicate the force acting on it. And what is true of the entire column of water in the tube, may be considered true of any single line of atoms; just as it would be true of a line of bricks piled one above another.

A tube of which the area is an inch square, holds, in two feet of its length, nearly a pound of water; hence, the general truth, well worth recollecting, that the pressure of water, at any depth, whether on the side of a vessel or on its bottom, or on any body immersed, is nearly one pound on the square inch for every two feet of depth.

The striking effects from the increase of pressure in a fluid, at great depths, are, of course, most commonly exhibited at sea. The following instances will illustrate them.

If a strong square glass bottle, empty, and firmly corked, be sunk in water, its sides are generally crushed inwards by the pressure before it reaches a depth of ten fathoms.

A chamber of air similarly let down with a man it, would soon allow him to be drowned by the water bursting in upon him; as really happened to an ignorant projector.

When a ship founders in shallow water, the wreck, on breaking to pieces, generally comes to the surface, and is cast upon the beach; but when the accident happens in deep water, the great pressure at the bottom forces water into the pores of the wood, and makes it so heavy that no part can ever rise again to reveal her fate.

A bubble of air or of steam, set at liberty far below the surface of water, is small at first, and gradually enlargens as it rises.

A man who dives deep, suffers much from the compression of his chest, as the elastic air within him yields under the strong pressure. This limits the depth to which divers can safely go.

It is not known whether there is a limit to the pressure which fishes can bear to impunity, but they are chiefly found living in the shallower waters on coasts, or on banks in the midst of the ocean, such as the banks of Newfoundland, the Dogger-bank, and other fishing stations out at sea. In rounding the Cape of Good Hope, at a considerable distance from land, ships pass over the bank of Lagullas, where a hook let down with a bit of red rag, or almost any thing as a bait, immediately secures its codfish.

By sending a vessel prepared for the purpose, down into the deep sea, we can readily prove the compressibility of water. Suppose the vessel to be made with only one entrance, and that a small round opening, into which, instead of cork, a sliding rod has been closely fitted. If, then, when filled with water, and having the rod inserted into the opening, it be allowed to sink into the sea, the pressure around will push the rod inwards, in a degree proportioned to the yielding or compression of the water within; and if there be on the road a stiff sliding-ring, or other contrivance to indicate on the return of the vessel how far the rod has been driven inwards, the apparatus will show the degree of compression at the greatest depth to which it has descended. Water a thousand fathoms below the surface is less bulky by about one-twentieth part than when at the surface.

The following are proofs of the pressure of weight in an open fluid, operating in all directions, as any pressure does in the case of a confined fluid.

A bottle-cork carried far under water, is not flattened as if it were pressed unequally, but is reduced in all its dimensions so as to appear a phial-cork of the usual form.

If a corked empty bottle be sent down into the sea, the cork is generally forced inwards at a given depth, and equally so in whatever direction the mouth of the bottle may happen to point.

If a vessel containing water have an opening in the side, covered by a valve or flap so contrived as to tell the force required to keep it shut, we find that the water tends to escape just as powerfully through such an opening as it would through one in the bottom, with the same elevation of water over its centre. And different equal openings in the side of a vessel require to be closed with forces exactly proportioned to the heights of liquid above their centres.

In an open square sided vessel full of water, the whole pressure on any upright side is just half the pressure on an equal extent of horizontal bottom; for the centre of the side being just half as deep as the bottom, the pressure on any point there is only half as great as on a point at the bottom, and on points above the level of the centre is just as much less than half, as, at corresponding distances below, it is more than half, and so it amounts to an exact half in the whole. Considering that the pressure on every point below the central level is greater than on every point above it, we see the reason why, to support a sluice or flood-gate by a single stay on the outside, the point at which the pressure has to be made is below the central level. Calculation discovers that this point, called the centre of pressure, is at one-third from the bottom. The knowledge of such facts furnishes rules for the construction of large vessels for liquids, canal embankments, &c.

The pressure on a given extent of the side of a narrow vessel is just as great as on the same extent of the side of a wide vessel, having the same depth of fluid; because, as now explained, it depends entirely on the extent of surface acted upon and the depth of liquid.

Hence a flood-gate or sluice which shuts out the ocean, as in docks opening to the sea, bears no more pressure than if it stood only against an equal depth of lake or river; or than if it were one of two such flood-gates become the sides of a very narrow vessel, made to contain only a few hogsheads of water.

Hence, again, the fear is unfounded which has been expressed with reference to the formation of a canal between the Red Sea and the Mediterranean,—that because the former, owing to the effects of easterly winds at its mouth, &c., is twenty feet higher than the later, it might burst through the flood-gates, and carry devastation along its course.

A deep crevice in a rock, when filled by a shower, is often the cause of the rock being torn asunder, and of part being precipitated.

Extensive walls or faces of masonry, intended to confine banks of sand or earth, if no openings were left for water to escape from behind them, would be burst after a rain unless they had the strength of flood-gates of the same size. Ignorance of this danger has led to some extraordinary catastrophies.

Other examples of the pressure in fluids being in all directions, and proportioned to the depth, are :—the swelling and bursting of leaden pipes when filled from a very elevated source:—the tearing up of the coverings of subterranean drains or water courses, when, during a flood, any accident chokes them near their lower openings:—the violence with which water escapes by an opening near the bottom of any deep vessl, or enters by an opening or leak near the keel of a deep-floating ship:—the great strength required in the lower hoops and securities of those enormous vessels of porter-brewers, called vats, some of which contain many thousand barrels of liquid.

In speaking of the pressure of a fluid in all directions, some persons have difficulty in conceiving that there is an upward as well as a downward and a lateral pressure. But if, in a fluid mass, the particles below had not a tendency upwards equal to the weight or downward pressure of the fluid over them, they could not support that fluid, which entirely rests upon them. Their tendency upward is owing to the pressure around them from which they are trying to escape. Accordingly, if a long tube, open at both ends, and with a sliding plug or piston in it near one end, be partially plunged into water by the plugged end, the water is found to press the plug upwards with force proportioned to the depth to which it is carried, and exactly equal to the force with which water presses upon an equal extent of the bottom or side of any other vessel having in it the same depth; or, with which, in the same vessel, it would press other plugs in other branches of the tube projecting in various directions. On removing such a plug altogether, the upward pressure is visibly proved and measured by the column of water pushed into the tube from below, and supported there to the level of the water around.

The pressure in a mass of fluid is proportioned to the perpendicular depth, and is not at all influenced by the size, shape or position of the containing vessel.

A body immersed in the water of a lake, one foot under the surface is just as much pressed upon as if it were one foot under the surface of the sea, and no more than if it were one foot under the surface of a small cistern.

Suppose vessels differing from each other in form and capacity, as sketched here at *a*, *b*, and *c*, but all having flat bottoms, of exactly the same area; if fluid be poured into all to the same level or perpendicular height, as represented here by the dotted lines, although the quantity be very different in each, the pressure on the bottom will be the same in all. This truth is easily proved experimentally, by having the bottoms moveable, and held to their places by weights or springs capable of measuring the pressure: or by letting the three vessels all communicate with the same vessel of water below them, and then observing that the water in all has still the same level. These results are other exemplifications of the truths, "*pressure equal in all directions,*" "*pressure as depth,*" and "*pressure as the extent of surface.*" For as a column of the fluid, resting on the middle of each bottom, just presses with its whole weight, and therefore, according to its altitude, this column could not remain at rest if there were any greater or less pressure than its own near it; then as the fluid really is at rest in all the cases, and in all a central column is of the same height, the pressure must be equal on all the bottoms. The case of the largest vessel *a*, is in a degree illustrated by

Fig. 73.

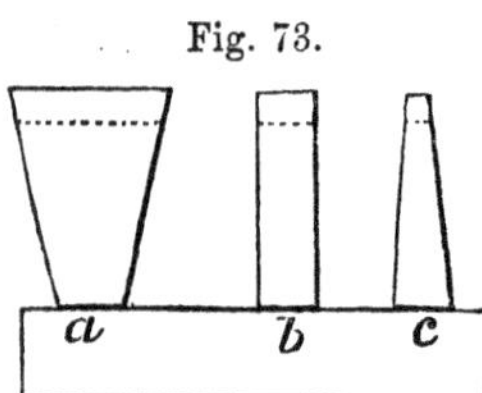

supposing the water in it to be suddenly converted into smooth upright small columns or rods of ice or glass; Then evidently only those pieces which rested on the bottom, could press on it while the others would be supported by the oblique sides of the vessel, and by the lateral resistance of the pieces around them.

"Level surface of a Fluid." (Read the Analysis.)

That the surface of a fluid must be level, follows from the facts of all the particles being equally attracted towards the centre of the earth, and being perfectly moveable among themselves. The particles forming the surface may be regarded as the tops of so many columns of particles, supported at any given level below, by a uniform resistance of pressure;—for no particle of an inferior level can be at rest unless equally urged in all directions, and therefore all the particles at such a level, and which, by equally urging one another, keep themselves at rest, must all be bearing the weight of equal columns: thus a higher column must sink and a lower one must rise, until just balanced by those around; that is, until all become alike. Besides, just as a ball rolls down a slope or inclined plane, so do the particles of a fluid slide or move from any higher situation among themselves to any lower unoccupied situation near them. The account now given explains why any accidental elevation or depression of a fluid surface, usually called a wave, continues to rise and fall, or to oscillate, for some time with gradually diminishing forces;—for when the mass is raised above the general level it is not quite supported, and therefore soon sinks, but in sinking, like, a falling pendulum, it acquires momentum which carries it below the general level, until opposed and arrested by a resistance greater than its weight, it then rises again, by acquiring a new momentum in its rise, it has to fall again, again to rise, and this alteration continues, until the lateral sliding of the particles, and the friction among them, gradually destroy it.

A perfectly level surface on earth really means one in which every particle is equi-distant from the centre of the earth, and it is therefore truly a spherical surface; but so large is the sphere, that if a slice of it of two miles in diameter were cut off, and laid on a perfect plane, the centre of the slice would only be four inches higher than the edges. Any small portion of it, therefore, for all common purposes, may be accounted a perfect plane.

So truly smooth does a fluid surface become, that it forms a perfect mirror; that is, it reflects or throws back the rays of light, which fall upon it so exactly in the order which they had on leaving the object, that an eye which receives them may fancy the object to be placed in the direction of the mirror.—It was over the glassy surface of the fountain or the lake, that the shepherdess of the young world bent themselves, to learn the charms which nature had bestowed on them. And a child contemplates with wonder and delight, through the window of a still pool or gliding stream, another sky below the ground, with its cloulds, and sun or stars; and another landscape, with inverted wood and mountains, the supposed dwelling of fairy beings.

In the cutting of canals, the making of railways, and in many other operations of engineering, it is of essential importance to determine the level or horizontal direction at any place; and this is usually done by a tube or glass, *a c*, filled with spirit except one bubble of air *b*, and called a spirit level. When this tube is horizontal, the bubble has no tendency to move to either

Fig. 74.

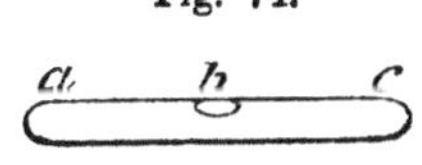

end; but if the tube inclines ever so little, the bubble rises to the end which is highest; or to speak more correctly, the denser spirit falls down to the lower end, and forces the light bubble away from it. Such a tube properly fixed in a frame, with a telescope attached to it, or simply with sight-holes to look through, becoming the engineer's guide in many of his most important operations.

A hoop surrounding the earth would bend away from a perfectly straight line four inches in a mile. In cutting a level canal, therefore, which may be considered as part of a hoop, there must be everywhere a falling from the straight line, called by geometers *a tangent*, in the proportion now described. All rivers also have the curvature of hoops applied to the surface of the earth.

Canals leading from sea-ports to the interior of countries have generally to ascend; but as water cannot become stagnant in any channel which is not level, the canal is divided, by gates or sluices, into portions at different levels, like steps of a stair, the rising at the joinings being generally from six to twelve feet. The boats are raised or lowered from one level to another by the contrivance called a lock, which is merely a portion of the canal, of sufficient capacity for the boat to lie in, furnished with high walls, and with flood-gates at both ends; so that when the gates below are shut, and water is gradually admitted from above, it becomes part of the high level, ready as such to deliver a boat, or receive one; and when the upper flood-gates are shut, and the water is gradually allowed to escape from the lock, it becomes a part of the low level, and a boat may enter it, or leave it by its lower gates.

The cutting of canals is one of the great items in the mass of modern improvement, which both mark and hasten the progress of civilization. Adverting to the importance of easy intercouse, as explained in a former section, we need only say here, that a horse which can draw only one ton on our best roads, can draw thirty with the same speed in a canal-boat.

And what a glorious triumph to science and art it is, to be able to conduct vessels of all kinds, even those originally intended for the ocean surge alone, through the quiet valleys of an interior country! In Scotland, now, along the Caledonian canal, a noble frigate may be seen, wandering as it were among the inland solitudes, and displaying her grace and majesty to the astonished gaze of the mountain shepherd; and when she has traversed the kingdom, and visited the lonely lakes, whose waters until lately had borne only the skiff of the hunter, she descends again by the steps of her liquid stair, and safely resumes her place among the waves.

It was lately in contemplation to lead a ship canal across the isthmus which joins North and South America. The elevation to which the canal must reach, to surmount the central ridge, is considerable, and will increase the difficulty; but such important consequences would follow the accomplishment of the object, that, with the continuance of general peace, and the increase of political wisdom, it will probably be attained. If so, the loaded vessel, rising from the Atlantic, would soon be descried among the mountain heights, and, a few hours after would be safely lodged in a port of the opposite sea; having performed, by a near cut, a voyage which at present costs months of delay and hazard, in a tedious navigation round the whole southern continent.—And if the Red Sea and Mediterranean were joined in the same way, as has also been proposed, the operation would, in effect, bring India nearer to Europe, and would more and more strengthen the bonds of mutual utility and brotherhood among the nations of the earth. Then, indeed, might it be said with truth, that the world is a vast garden, given to man for his abode, of which every spot has its peculiar sweets and treasures; but because the cultivator of each may exchange a share of its produce for shares in return, the same general result follows as if every field or farm contained within itself the climates and soils and capabilities of the whole.

In a canal, the least deviation from the true level would immediately cause any water admitted into it to flow towards the lower end. This flux to a lower situation is what is going on in the myriads of streams, which render the face of the earth a scene of such varied beauty and incessant change.

As in the animal body, from every the minutest point a little vein endowed with living power, takes the blood which has just brought life and nutriment to the part, and delivers it into a larger vein, whence it passes into a larger still, until at last, in the great reservoir of the heart, it meets the blood returned from every part of the body, so, in this terraqueous globe, where the magic moving power is simply fluid seeking its level, does the rain which falls to sustain vegetable and animal life, and to renovate nature, glide from every point of the surface into a lower bed, and from thence into a lower still, until the countless streams, so formed, after every variety of course combine to form the swelling rivers, which return the accumulated waters into the common reservoir of the ocean. In the living body, the arteries carry back the blood with renewed vitality to every point whence the veins had withdrawn it, and so complete the circulation; and in what may be called the living universe, the circulation is completed by the action of heat and of the atmosphere, which, from the extended face of the ocean raise a constant exhalation of watery vapour of invisible purity, which the winds then carry away and deposit as rain or dew on every spot of the earth.

A very slight declivity suffices to give the running motion to water. Three inches per mile, in a smooth straight channel, gives a velocity of about three miles per hour. The Ganges, which gathers the waters of the Himalaya mountains, the loftiest in the world, is, at eighteen hundred miles from its mouth, only eight hundred feet above the level of the sea—that is, above twice the height of St Paul's Church in London; and to fall these eight hundred feet, in its long course, the water takes nearly a month. The greater river Magdalena, in South America, whose channel, for a thousand miles, is between two ridges of the Andes, falls only five hundred feet in all that distance. Above the commencement of the thousand miles, it is seen descending in rapids and cataracts from the mountains. The gigantic Rio de la Plata has so gentle a descent to the ocean, that, in Paraguay, fifteen hundred miles from its mouth, large ships arrive which have sailed against the current all the way, by the force of the wind alone: that is to say, which on the beautifully inclined plane of the stream, have been gradually lifted by the soft wind, and even against the current, to an elevation greater than that of our loftiest spires.

A small lake or extensive mill-pond, with uneven bottom, if suddenly emptied by a sluice or opening at its lowest part, would exhibit a number of pits or pools of various size and shape left among the inequalities. But supposing rain to fall, and frequently to recur, the water seeking its level would soon effect a very remarkable change. In consequence of each pool discharging over its lowest part, that is, sending out a streamlet either into another lower pool, or into a channel leading directly to the sluice or opening, there would be a constant wearing down of the part or side of the pool over which the water was running, that is to say, a deepening of a breach or channel there, and the surface of water in the pool would be consequently becoming lower, while, at the same time, the bottom would be rising, owing to the deposit of sand or mud washed down by the rain from the elevations around: and these two operations continuing, the pool would at last altogether disappear. And by this change going on in every pool through the whole of the emptied mill-pond, the general bottom would at last exhibit only a varied or undulated surface of dry land, with a beautiful arrangement of ramifying water channels, all sloping with a precision unattainable by art, to the general mouth or estuary.—The reason that, in the supposed case, and in every other, a water course soon becomes so singularly uniform, both as to dimensions and descent, is, that any pits or hollows in it are filled up by the sand and mud carried along in the stream, and deposited where the current is slack; while any elevations are worn away by the action of the more rapid current which accompanies shallowness.

The above paragraph describes, in miniature, what has been going on over the general face of our earth ever since that convulsion of nature which produced its present form. In many places the phenomenon is already complete; in others it is only in progress. The whole of what is now dry land, has at some period been under water, and much of it has evidently been a gradual deposition from water. By some extraordinary convulsion, therefore, our present continents and islands must have been thrown up from the bottom of an ocean, or an ocean must have subsided away from them; and in either case the land must have merged as checkered and unsightly as the bottom of the emptied lake above supposed. And it is the gradual operation of *water seeking its level* which has at last converted the earth into the paradise which we now behold.

The marks of the former state of the world, and of the progressive change, are every where most strikingly evident to the enlightened eye of philosophy. The present kingdom of Bohemia, for instance, is the bottom of one of the great lakes formerly existing over Europe. It is a basin or amphitheatre, formed by a wall of mountains, and the only gate or opening to it, is that remarkable one by which the water now escapes from it, and which evidently has been gradually cut or formed by the action of the running stream. As the bottom became uncovered, owing to the sinking of the water, and the formation of a regular sloping channel from every part, the former lake was converted into a fine and fertile country, a fit habitation for man; and the continued drain from it of the rains which fall over its surface, and either pass rapidly away, or sink into the earth, and ooze again more gradually in the form of springs, is the beautiful river which we now call the Elbe.

In Switzerland, many of the valleys which were formerly lakes, have the opening for the exit of water so narrow, that, as happened in one of them a few years ago, a mass of snow or ice falling into it, converts the valley once more into a lake. On the occasion alluded to, the accumulation of water within was very rapid; and although, from the danger foreseen to the country below, if the impediment should suddenly give way, every means was tried to remove the water gradually, the attempt had not succeeded when the frightful burst took place, and involved the inferior country in common ruin.

The magnificent Danube is the drain of a chain of basins or lakes, which must, at one time, have discharged or run over one into another; but owing to the continued stream cutting a passage at last low enough to empty them all, they are now regions of fertility, occupied by civilized man, instead of the fishes which held them formerly. This operation is still going on in all the lakes of the earth. The lake of Geneva, for instance, although confined by hard rock, is lowering its outlet, and the surface has consequently fallen within the period of accurate observation and records; and as, at the same time, the wearing of the neighbouring mountains, brought down by the winter torrents, are filling up its bed, if the town of Geneva last long enough, its inhabitants may have to speak of the river in the neighbouring valley, instead of the picturesque lake which now fills it. Already several towns and villages, which were close upon the lake a century ago, have fields and gardens spreading between them and the shore.

Illustrating this subject, it is very interesting to observe the contrast between the pure blue water of the Rhone issuing from the lake of Geneva, and the turbid streams which join its course a little farther down. The torrents which fall into the lake all around, are equally charged with the *debris* or wearings of the mountains; but having deposited all their load in the still bosom of the lake, the pure water alone escapes to form the river. The streams, however, coming to the Rhone directly from the Alps, and bringing with them their charge of broken-down earth, even after they have joined it, are long distinguishable by their muddy waters. It is the mud deposited as here described, which is gradually filling up all lakes, and which has formed the vast regions of flat country seen about the mouths of great rivers. The greater part of Holland is deposition of this kind, the whole of lower Egypt, a great part of Bengal, &c., &c.

There are some lakes on the face of the earth which have no outlet towards the sea,—all the water which falls into them, being again carried off by evaporation alone—and such lakes are never of fresh water, because every substance, which, from the beginning of time, rain could dissolve in the regions around them, has necessarily been carried towards them by their feeding streams, and there has remained. The great majority of lakes, however, being basins with the water constantly running over at one part towards the sea, although all originally salt, have, in the course of time, become fresh, because their only supply being directly from the clouds, or from rivers and springs fed by the clouds, is fresh, while what runs away from them must always be carrying with it a proportion of any substance that remains dissolved in them. We thus see how the face of the earth has been gradually washed to a state of purity and freshness fitting it for the uses of man, and why the great ocean necessarily contains in solution all the substances which originally existed near the surface of the earth, soluble in water:—*viz.*, all the saline substances. The city of Mexico stands in the centre of a vast and beautiful plain, 7,000 feet above the level of the sea, and surrounded by sublime ridges of mountains, many of them snow-capped. One side of the plain is a little lower than the other, and forms the bed of the lake, which is salt for the reasons stated above;—but the lake will not long be salt, for it now has an outlet. About 150 years ago, owing to unusual rains, an extraordinary increase of the water took place, and covered the pavements of the city. An artificial drain was then cut from the plain, at the distance of about sixty miles from the city, to the lower external country. This soon freed the city from the water, and since then, becoming every year deeper by the wearing effects of the uninterrupted stream, it is still lowering the surface of the lake, is daily rendering the water less salt, and is converting the vast salt marshes, which formerly surrounded the city, into fresh and fertile fields.

The vast continent of Australasia, or New Holland, (as large as Europe,) is supposed by some to have been formed at a different time from what is called the Old World, so different and peculiar are many of its animal and vegetable productions; and the idea of a later formation receives countenance from the existence of immense tracts of marshy or imperfectly drained land discovered in the interior, into which rivers flow, but seem not yet to have worn down a sufficient outlet or discharging channel toward the ocean.

Where the soil or bed of a country through which a water-track passes is not of a soft consistence, so as to allow readily the wearing down of higher parts, and the filling up of hollows by deposited sand, lakes, rapids and great irregularities of current remain. We have, for instance, the line of the lakes in North America, the rapids of the St. Lawrence, and the stupendous falls of Niagara, where at one leap the river gains a level lower by a hundred and sixty feet. A softer barrier than the rock over which the river pours, would soon be cut through, and the line of lakes would be emptied.

The contemplation of the fact, that water in seeking its level is constantly wearing where it rubs, and carrying the abraided portions down to lower level, and ultimately to the bed of the ocean, brings irresistibly the awful idea, that this earthly abode of ours, owing to natural causes already in operation, can have but a limited existence in its present state. No shower falls that does not send portions of mountains and plains into the depths of the ocean, and thus cause a corresponding encroachment on the shores by the rising water; and with revolving ages, unless new convulsions of nature disturb the progress, or art succeed, as in Holland and elsewhere, in shutting out the ocean from extensive low tracts by means of sea dykes or embankments, the dry land must at last disappear, and another gradual deluge embrace the globe.

There is, perhaps, nothing which illustrates in a more striking manner the exact resemblance among nature's phenomena, or their accordance with the few general expressions or laws which describe them all, than the perfect level of the ocean as a liquid surface. The sea never rises nor falls in any place, even one inch, but in obedience to fixed laws, and therefore its changes may generally be foreseen and allowed for. For instance, the eastern trade-winds and other causes force the water of the Indian Ocean towards the African coast, so as to keep the Red Sea about twenty feet above the general ocean level; and the Mediterranean is a little below that level, because the evaporation from it is greater than the supply of its rivers, causing it to receive an additional supply by the Strait of Gibralter;—but in all such cases, the effect is as constant as the disturbing cause, and therefore can be calculated upon with confidence.

Were it not for this perfect exactness, in what a precarious state would the inhabitants exist on the sea shores, and on the banks of low rivers! Few of the inhabitants of London, perhaps, reflect, when standing by the side of their noble river, and gazing on the rapid flood-tide pouring inland through the bridges, that although sixty miles from the sea, the water there is, at the moment, lower than the surface of the sea, which may at the time be heaving moreover, in lofty waves, covered perhaps with wrecks and the drowning.

The horrid destruction that would follow any alteration of the level of the ocean, may be judged of by the effects of occasional floods, produced by rains and melting snow in the interior of countries, or by these combined with winds and high tides on the coasts. The flood at St. Petersburg, in 1825, was dreadful, in which strong westerly winds had retarded the flow of the Neva so much, that the water rose forty feet (the height of an ordinary house) above its usual mark, covered all the low parts of the town, and drowned thousands of the people.

In Holland, which is a low flat, formed chiefly by the mud and sand brought down by the Rhine and neighboring rivers, much of the country is really below the level of the common spring-tides, and is only protected from daily inundations by artificial dykes or ramparts, made strong enough to resist the ocean. On one occasion the water broke into such an enclosure, and drowned more than sixty thousand people. What awful uncertainty then would hang over the existence of the Dutch, if the level of the sea were subject to change; for while we know that its waters, owing to the centrifugal force or of the earth's rotation, are seventeen miles higher at the equator than at the poles, if the level, as now established, were from any cause to be suddenly changed but ten feet, millions of human beings would be the victims.

Where inundation is regularly periodical, as in the Nile and many other rivers, the hurtful effects can be guarded against, and the occurrence may even become useful, by fertilizing the soil.

Tracts of land in contact with rivers, of which land, the surface lies between the levels of ebb and flood-tide, if surrounded with dykes, may be kept constantly covered with water, by opening the sluices only at high water; or may be kept constantly drained, by opening the sluices only at low water. A vast extent of rice fields, near the mouths of rivers in India and China, is managed in this way, the admission or exclusion of water being regulated by the age of the rice plant. A great part also of the rich sugar plantations of Demerara, Esequibo, &c., on the coast of South America, are in the same predicament; and another advantage which these have over the plantations on the West-India Islands, is the saving of the labour of transport effected by the canals which intersect all the fields.

"*If various tubes and vessels communicate with one another, fluid admitted into them will rise to the same level in all.*" (Read the Analysis, p. 84.)

The following sketch may represent a variety of tubes and vessels, fixed upon and opening into the cistern or box G. Water poured into any one would fill the box, and would then rise to the same level in all. The dotted lines from a to f may represent the surfaces of the fluid in the different vessels. In the figure at p. 128, it was seen why, in all upright cylindrical vessels, as a, b and c, the fluid rises to the same level; and the figure at p. 132 explained why shape of the vessel cannot affect the level. Although in the oblique vessel c, represented here, there is more water than in a, still there is the same pressure at the bottom of both, because e supports part of the weight of its contained fluid on the principle of the inclined plane.

Fig. 75.

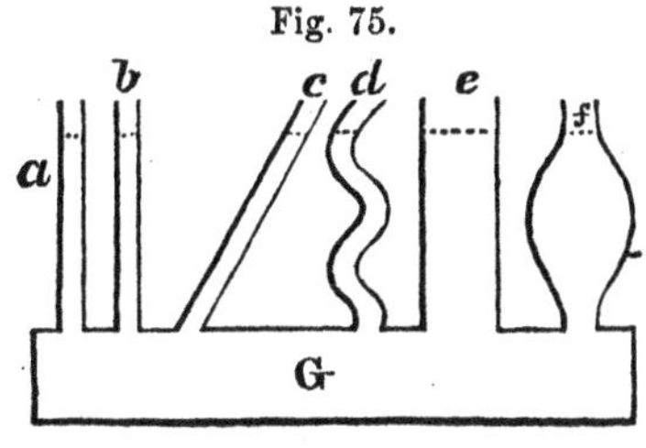

If a tube twenty miles long, and rising and descending among the inequalities of a country, were filled with water, and could have its ends brought together for comparison, it would exhibit two liquid surfaces having precisely the same level; and on either end being raised, the fluid would sink in it to overflow from the other.

An easy mode of determining a level line at any spot is to have an open tube, bent up at its ends a and b, and nearly filled with liquid: by then looking along the two liquid surfaces, or through floating *sights* resting on them, an observer looks in a line which is quite horizontal at the middle point between them.

Fig. 76.

If there were two lakes on adjoining hills of different heights, a pipe of communication descending across the valley and connecting them, would soon bring them to the same level; or if one were much higher than the other, would empty that one into the other.

A projector thought that the vessel of his contrivance, represented here, was to solve the renowned problem of the perpetual motion. It was goblet-shaped, lessening gradually towards the bottom until it became a tube, turned upwards at c, and putting with an open extremity into the goblet again. He reasoned thus: A pint of water in the goblet a must more than counterbalance an ounce which the tube b will contain, and must therefore be constantly pushing the ounce forward into the vessel again, and keeping up a stream or circulation, which will cease only when the water dries up. He was confounded when a trial showed him the same level always in a and in b.

Fig. 77.

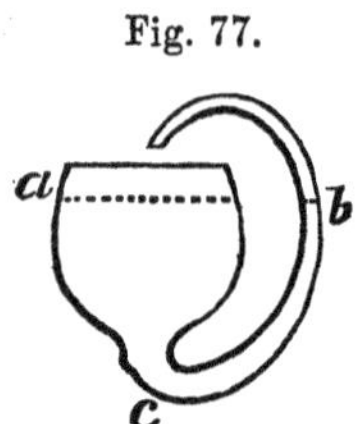

A glass tube inserted near the bottom of a cask or cistern of any sort, not air-tight above, which tube is then bent upwards, to appear on the outside like a barometer tube, shows by the elevation of a fluid in it, the height of the greater mass of fluid within.

In like manner a tube brought from a river into a neighbouring cellar or pit, will indicate the height of water in the river.

A knowledge of the truth, that water in pipes will always rise again to the height or level of its source, has enabled men in modern times to construct those admirable systems of iron pipes, which distribute water in great towns. The water brought to any elevated site, in or near the town, may be delivered from a reservoir there, by the effect of gravity alone, to every cistern which is under the level of the reservoir, the result not being affected by the pipes having to rise over heights and to descend into valleys many times in their course.

On the hill north of London, on which Pentonville stands, there is such a reservoir to which water is brought from Hertfordshire, by a channel cut for the purpose, upwards of thirty miles in length, and called the New River. Another reservoir has lately been constructed by the West Middlesex Water Company, at Primrose Hill, higher than any house in town. It is filled by operation of steam-engines at the Company's works, near Hammersmith, five miles off. It will supply water to the summits of all the houses connected with it, and is exceedingly useful in cases of fire.

Many persons have believed that the ancients were ignorant of the law, that fluid in pipes rises to the level of its source, because, in all the ruins of their aqueducts, the channel is a regular slope. Some of the aqueducts, as works of magnitude, are not inferior to the great wall of China, or the Egyptian Pyramids; yet, at the present day, a single pipe of cast-iron is made to answer the same purpose, and even more perfectly. It is now ascertained, however, that it was not ignorance of the principle, but want of fit material for making the pipes, which cost our forefathers such enormous labour.

The supply and distribution of water in a large city, particularly since the steam-engine has been added to the apparatus, approach closely to the perfection of nature's own work in the circulation of blood through the animal body. From the great pumps or a high reservoir, main pipes issue to the chief divisions of the town; these then send suitable branches to the streets, which branches again divide for the lanes and alleys; and at last subdivide until every house has its small leaden conduit carrying its precious freight, if required, even into the separate apartments, and yielding it anywhere to the turning of a cock A corresponding arrangement of drains and sewers, most carefully constructed in obedience to the law of level, receives the water again when it has answered its purposes, and sends it to be purified in the great laboratory of the ocean. And so admirably complete and perfect is this counter-system of sloping channels, that a heavy shower may fall, and after washing and purifying every superficial spot of the city, and sweeping out all the subterranean passages, may, within the space of an hour, form part of the river passing by. It is the recurrence of this almost miracle, of extensive, sudden, and perfect purification, which makes modern London the most healthy, while it is the largest city in the world.

English citizens have now become so habituated to the blessing of a supply of pure water, more than sufficient for all their purposes, that it no more surprises them than the regularly returning light of day or warmth of summer. But a retrospect into past times may still awaken them to a sense of their obligation to advancing art. How much of the anxiety and labour of men in former times had relation to the supply of this precious element! How often, formerly, has periodical pestilence arisen from the deficiency of water; and how often has fire devoured whole cities, which a timely supply of water might have saved! Kings have received almost divine honours for constructing aqueducts, to lead the pure streams from the mountains into the peopled towns. In the present day, it is he who has travelled on the sandy plains of Asia or Africa, where a well is more prized than mines of gold, or who has spent months on ship-board, where the fresh water is often doled out with more caution than the most precious product of the still, or who, in reading history, has vividly sympathized with the victims of siege or shipwreck, spreading out their garments to catch the rain from heaven, and then with mad eagerness, sucking the delicious moisture —it is he who can appreciate fully the blessing of that abundant supply which most of us now so thoughtlessly enjoy. The author of this work will long remember the intense momentary regret with which, at once approaching a beautiful land after months spent at sea, he saw a stream of fresh water gliding over a rock into the salt waves—it appeared to him as if a most precious essence, by some accident were pouring out to waste.

The subject of *fluid level* leads to a consideration of springs or wells, and of the operation of boring for water.

The water which falls from the clouds, and which must all ultimately return to the sea, may find its way to the rivers, either by running directly along the surface of soils which refuse it admittance; or by first sinking into porous earth, and again oozing out at lower situations in the form of springs. If a spring be as low as the bottom of the porous earth from which it issues, that is to say, as low as the surface of the impermeable clay or rock on which at some depth all such earth rests, it may drain the whole; but if not, the water will stand at a certain level among the earth as it would among bullets in a water-tight vessel. If a hole or pit be then dug in such earth, reaching below the level of the water lying in it, the pit will soon be filled with water up to the level, and will be called a well. In many places this water-level is very far below the surface of the ground; and in some places, by reason of the water having an easy drainage towards the sea, or of the superficial soil being altogether impermeable to it, there is none to be found within an accessible depth.

A remarkable illustration of this subject occurred a few years ago, in Kent, on the occasion of cutting between Rochester and Gravesend the canal called the Thames and Medway canal. This canal consists of but one cut or level, seven miles long, of which two are in a tunnel through the hill—which level is that of high water in the connected rivers; the intention having been to let the canal be filled always from the rivers at high water;—but as the level of the subterranean water in the surrounding land, and therefore of all the inhabitants' wells there, is, as might be anticipated, half-way between the levels of high and low tides, the salt water from the rivers was no sooner admitted to the canal than it spread into the land on either side, where the resisting internal water-level was lower, and destroyed all the wells. If the canal had been dug a few feet lower, the mischief would not have occurred, and the company would have escaped paying the heavy damages, which rendered their undertaking a very ungainful speculation.

All the wells and springs in the world are merely the rain water which has sunk into the earth, appearing again, and gradually escaping at lower places: nature thus admirably making the bowels of the earth an ever-stored reservoir of the substance most indispensable to the comfort and existence of man, and of all living creatures. It is worthy of remark here, that high cultivation or agricultural improvement of a country has a great effect on the quantity of spring water in it. While the face of a country is rough the rain-water remains long among its inequalities, slowly sinking into the earth to feed the springs, or slowly running away from the surface as from bogs and marshes towards the rivers. The rivers, hence, have a comparatively uniform and regular supply, even when rain has not fallen for a long time:—but in a well-drained country, the rain, by a thousand prepared channels, finds its way to the brooks and rivers almost immediately, producing often dangerous floods or inundations of the neighboring low grounds. A friend of the author had a waterfall and mill in Surrey, which he formerly let for a rent of £1,200 a year; but after agricultural improvements in the district from which the water came, the supply of water was generally either superabundant or deficient, and the value of the mill was reduced to one-half.

The surface of our globe is formed of different strata or layers, as of clay, chalk, sand, gravel, &c., &c., which appear all to have been at former periods horizontal, formed under water, and to have been afterwards thrown up, by some convulsion or convulsions of nature, into every variety of position. In particular situations, the upper surface is now concave or basin-shaped, the different strata or layers, when water-tight, being like cups or basins placed one within another; and as water poured in, to fill the space between two basins so placed, would spring out to the height of its upper or level surface, through any hole made in the side of either, so on boring for water, through an innermost or superior water-tight stratum or basin of earth, the water often springs out and rises far above the surface of the ground. London stands in a hollow of which the first-met layer is a basin of clay, placed over chalk, and on boring through the clay (sometimes of three hundred feet thickness,) the water issues and in many places will form a jet considerably above the surface of the ground; showing that there is a higher source or level somewhere—as among the hills of Surrey, or those north of London.

When fluids of different kinds and of different weights under the same bulk, are made to oppose or to balance each other in communicating vessels—as water, for instance, in one leg of the bent tube *b d c*, and oil in the other—the surfaces will not all rest or settle at the same height or level, but that of the lighter fluid will be just as much higher than that of the other as it is lighter. Thus a column of oil must be of a length as *d o*, to balance a column of water *d w:* and alcohol, because lighter than oil, to balance the same water, would have to stand higher still, as at *a;* while mercury, because thirteen times weightier than water, would stand only about *m*. The shape, size, or position of the vessels in which the opposing fluids might stand, would have no influence on the relative heights of the surfaces; for if we suppose a larger vessel, such as is represented here by the dotted lines between the letters *e f m*, to be substituted for the leg *c d* of the tube, the various fluids to balance the water in *b d*, would have to stand just as high in it as in the smaller tube.

Fig. 78.

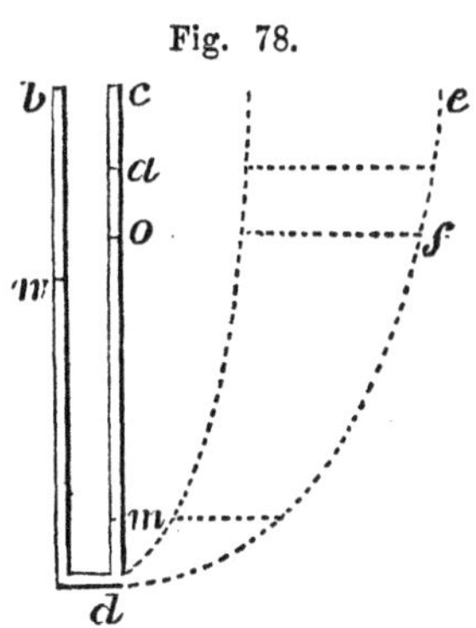

" *A body immersed in a fluid, displaces exactly its own bulk of it, which quantity having been just supported by the fluid around the body is held up with force exactly equal to the weight of the fluid displaced, and must sink or swim according as its own weight is greater or less than this.*"

A bladder full of air, and maintaining the bulk of a pound of water, requires a force of one pound (except a few grains, the weight of the air,) to plunge it under the water. The same bulk of gold is held up in water with exactly the same force; so that, if previously balanced at the end of a weighing beam, it appears on immersion to have lost one pound of its weight.

And a piece of wood, ivory, or any other substance, having exactly the same bulk, is opposed on entering the fluid by the same resistance.

The reason of this is obvious, for the immersed body takes the place of water which weighed one pound and yet was supported, and whose pressure was necessary for the equilibrium of the rest. In a vessel of water represented here by the figure *a b*, let us append to any portion of the water, a single column of particles, for instance, represented by the line *c d:* we know that each column is steadily supported in its place, because the particle of the liquid immediately under it is tending upwards to escape from the surrounding pressures, with force exactly equal to the weight of the column; and what is true of a column of single particles, is true of any other portion, such as the larger column represented by the figure *f h g*. If such portion weighed exactly a pound, the surface under it would be tending upward with the force of a pound; and if the portion, without changing its bulk or form, were to become ice, it would still be exactly supported by the surface below pressing upwards with a force of a pound; and farther, if a similar column of wood, or stone, or metal, were there, the surrounding pressure would still be the same. Again, if we suppose only half the column to be solidified, the portion *h g* for instance, it, would be pressed upwards with a force of one pound at *g;* but its own weight of half a pound, and the weight of the half pound of water above it, would produce an exact balance and maintain rest.

Fig. 79.

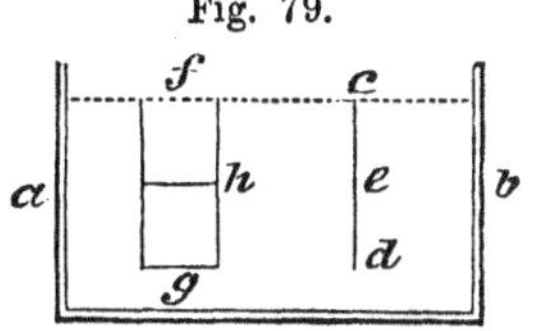

It is very important to have clear notions on this subject; and as different minds apprehend such matters with different degrees of facility, and in different ways, we shall state the same general truth in other words.

Let us consider a mass of fluid as consisting of a vast number of extremely minute columns of single particles standing side by side, where every particle supports those above it by tendency upwards which it requires through the pressure of the fluid surrounding it. Now if we suppose the particles of a portion of a fluid mass, of any shape, to stick together, or to become ice without change of bulk or weight, that portion when solid would still be between the same forces as when fluid, and therefore would be equally supported, and would remain at rest. And if gold, or silver, or glass, or wood, having the same bulk, were substituted for the supposed ice, such new substance would still be sustained with the same force; so that a substance of exactly the same weight as the ice or water displaced, would have no tendency either to rise or to fall more than the water itself had; but a substance heavier would sink, and one lighter would swim, and in either case with force exactly proportioned to the difference between its weight and that of an equal bulk of water.

Few persons, in now reading the statement of this truth—in appearance so simple and obvious—would imagine that it had remained so long unknown and that the discovery of it may be accounted one of the most important which human sagacity ever made,—but such is the case. We owe the discovery to one of the master-minds of antiquity—that of Archimedes. He caught the idea one day while his limbs were resting on the liquid support of a bath: and as his God-like intellect darted into futurity, and perceived many of the important uses to which the knowledge was applicable, he is said to have become so moved with admiration and delight, that he leapt from the water, and unconscious of his nakedness, pursued his way homewards, calling out "ευρηκα, ευρηκα," I have found it. He was thinking chiefly of the ready means, thus obtained, of ascertaining in all cases what has since been called the *specific gravity* of bodies, *viz.*, the comparative weights of equal bulks of different substances; as of gold, or silver, or copper, or iron, compared with water; and in the case of mixtures, as of gold with silver for instance, of declaring at once the proportion present of each—important problems, which, until then, could not be correctly solved.

The hydrostatic law now explained, has since led to great advances in various arts. It may be regarded as a chief foundation of chemistry, for by it the chemist distinguishes one substance from another, distinguishes a pure from an impure substance, and discovers the nature of many mixtures or compounds. The merchant often judges by it of the worth of his merchandize. In any case it enables an inquirer to ascertain at once the exact size or solid bulk of a mass, however irregular—even of a bundle of twigs. It has become the cause of improvements in navigation, in marine architecture, and in many other arts.

We shall now discuss more particularly the subject of *comparative weights or specific gravity.*

"*The force with which a body is held up in a fluid, being the exact weight of its bulk of that fluid, by ascertaining this force and comparing it with the weight of the body itself the comparative weights or* SPECIFIC GRAVITIES *are found.*" (Read the Analysis, p. 126.)

If any body, *c*, a mass of gold, for instance, be suspended by a thread or hair from the bottom of one scale *b* of a weighing-beam, and be balanced by weights put into the other scale *a*, and if a vessel of water be then lifted under it so that the water shall surround it, the body is pushed up or supported by the water with force equal to the weight of the water which it displaces; the weights, therefore, then required in the scale *b* to restore the balance, show truly the exact weight of the water displaced; or of water equal in bulk to the body; and the weights in the two opposite scales show the comparative weights of the body and of its bulk of water. In the supposed case, whatever weight the gold had in the air, it would seem to lose when the water surrounded it, about a nineteenth part of such weight; that is, the water would support it with this force; and gold would thus be proved to be about nineteen times as heavy as water.

Fig. 80.

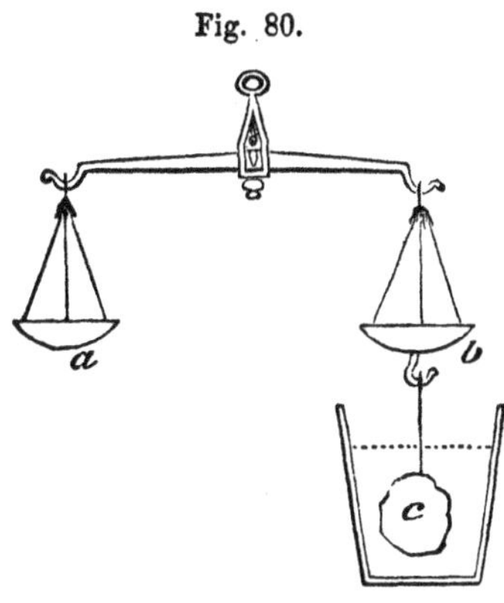

In making a table of specific gravities, it was necessary to select a common standard with which all other substances should be compared, and this has been done in choosing water; the reason of preference being, that water can be so easily procured in a state of purity, and therefore of uniformity, in all situations. When we say, therefore, that gold is of the specific gravity 19, and copper 9, and cork $\frac{1}{7}$, we mean that these substances are just so much heavier or lighter than their bulk of pure water in its densest state, *viz.*, at the temperature of 40 degrees of Fahrenheit's thermometer.

As the substances in nature differ as to form and other qualities, corresponding differences have to be made in the manner of ascertaining their specific gravities: the following cases are most important.

Solid bodies insoluble in water and *heavier* than it—as the metals, &c., are merely suspended by a thread or hair, having nearly the specific gravity of water, to one scale of the *hydrostatic balance* (simply a good weighing-beam with a water-vessel below one of the scales;) and the body being first balanced or weighed in the air, and then in water, as already described, the weight and the loss, represented, if the operator chooses, by the weights in the opposite scales, are the weights of equal bulks of the two substances; and by finding, through the arithmetical operation of *division*, how often the weight of the water is contained in the weight of the solid, we find the specific gravity of the solid, or how much it is weightier than its bulk of water.—It is almost superfluous to remark, that putting weights into the scale, *b* or taking them out of the scale *a*, are equivalent operations. We shall explain afterwards, that for very delicate purposes bodies must be weighed first in a vacuum, instead of in air, or a suitable allowance must be made; for air itself supports a little any body immersed in it.

Solids lighter than water, as cork, are weighed in it by attaching to them a mass of metal or glass heavy enough to sink them, and already balanced in water for the purpose; or by making the line which connect them with the weighing beams pass under a small pully fixed at the bottom of the vessel, so that the rising of the end of the beam to which they are attached shall draw them down.

A solid soluble in water, as a chrystal of any salt, may be protected during the operation of weighing in water, by previously dipping it in melted wax, so as to leave a thin covering on it; or it may be weighed in some liquid which does not dissolve it, allowance being afterwards made for the difference between the weight of such liquid and of water.

Powders insoluble in water, such as gold dust, are weighed in a glass cup which has previously been balanced in water for that purpose.

Powders soluble in water, must be weighed in some other liquid.

Mr. Leslie, the highly endowed professor of natural philosophy in the University of Edinburgh, has lately suggested a novel and ingenious mode of acertaining the specific gravity of pulverized or porous bodies; but as it can be understood only by persons acquainted with the doctrines of *pneumatics*, the consideration of it must come under that head.

Other liquids may be compared with water in several ways. 1st. If a phial be made to hold exactly one thousand grains of distilled water, at the temperature of 40°, the weight of the same measure of any other liquid is found, by simply filling the phial, and weighing it. Of sulphuric acid, for instance, such a phial will contain nearly nineteen hundred grains, while of alchohol it will receive only about eight hundred. 2d. A bulb of glass, which loses one thousand grains when weighed in water, (which thousand grains is therefore the weight of its bulk in water,) may be weighed in other liquids, and the difference of loss marks the specific gravity, as in the last case. The bulb for this purpose may be of any size, but one which loses in water exactly one thousand grains, is preferable, from the simplicity thereby given to the calculations :—This remark applies also to the phial last mentioned. 3d. A contrivance which renders the beam and scales altogether unnecessary, is a hollow floating bulb of glass or metal *a*, with a slender stalk rising from it to support the little scale or dish *b*, and with another stalk descending to carry the weight or weights at *c*, which serve as ballast to it. The whole is so adjusted that when displacing one thousand grains, or other known quantity of pure water, it shall float with a certain mark upon the upper stalk just at the surface of the water. By then immersing it in other liquids and finding how much weight must be added to, or taken from it above or below, to make it float in them at the same elevation, the comparative weights of these other liquids and of water are found :—or the difference of weight which makes it float at different elevations in water, having been previously ascertained, it will only be necessary, in any other case, to note exactly its elevation; an inch of the slender stalk may be equivalent to a difference of ten grains. This instrument is called an *hydrometer*. There are generally printed tables and directions, accompanying all forms of it, telling the exact import of the several indications, and the allowances to be made for temperature, &c. It may be used for weighing solids as well as liquids, for if any mass be put into the saucer *b*, weights exactly equal to the mass must be taken out of the saucer *b*, or from below at *c*, to restore the equilibrium of the instrument. The mass may be afterwards placed at *c*, and weighed in water. 4th. The shortest mode of ascertaining the specific gravities of liquids, is to have a set or series of small glass bubbles of different specific gravities, so that when they are thrown into any liquid, those heavier than it will sink, and those lighter will swim, while that one which marks its specific gravity will remain merely suspended.

Fig. 81.

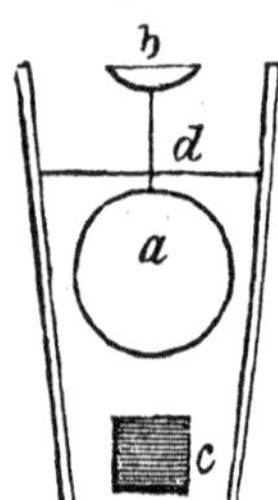

The bubbles must, of course be numbered, and the specific gravity of each be previously known.

A common use of hydrometers is to ascertain the quality of the distilled spirits brought to market, as of rum, brandy, gin, &c. All these consist of alcohol more or less diluted with water; and duty or tax is levied upon them in proportion to their strength, or the quantity of alcohol which they contain. A delicate hydrometer discovers this at once.

A shop-keeper in China sold to the purser of a ship, a quantity of distilled spirit according to a sample shown; but not standing in awe of conscience, he afterwards, in the privacy of his store-house, added a certain quantity of water to each cask.

The spirit having been delivered on board, and tried by the hydrometer, was discovered to be wanting in strength. When the vendor was charged with the intended fraud, he at first denied it, for he knew of no human means which could have made the discovery; but on the exact quantity of water which had been mixed being specified, a superstitious dread seized him, and having confessed his roguery, he made ample amends. On the instrument of his detection being afterwards shown to him, he offered any price, for what he foresaw might be turned to great account in his trade.

The specific gravity of *aeriform substances* is ascertainnd by means of a glass flask of known size, furnished with a stop-cock. It is first weighed when emptied by the air-pump, and afterwards when filled successively with water and with different airs or gases. Comparison of the weights gives the specific gravities, as already described.

The following table shows, in round numbers, the comparative weights or specific gravities of some common substances. Water is the standard kept in view, and any equal bulk of another substance is heavier or lighter than water, according to the numbers severally attached to them.

Platinum	$22\frac{3}{4}$	Common Salt, . . .	2
Gold	$19\frac{1}{3}$	Brick	2
Mercury	$13\frac{1}{2}$	Alcohol	$\frac{8}{10}$
Copper	$8\frac{3}{4}$	Æther	$\frac{3}{4}$
Steel and Iron	8	Cork	$\frac{1}{7}$
Diamond	$3\frac{1}{2}$	Atmospheric Air . .	$\frac{1}{800}$
Glass	3	Hydrogen Gas . . .	$\frac{1}{12000}$
Common stones . .	$2\frac{1}{2}$		

Complete tables are found in systems of Dictionaries of Chemistry.

A cubic foot of water happens to weigh very nearly one thousand ounces avoirdupois, or $62\frac{1}{2}$ pounds. Hence, in the foregoing table, the figures denoting the specific gravities tell how many times a thousand ounces of the different substances a cubic foot contains. Of gold, for instance, a cubic foot contains more than nineteen thousand ounces, being worth in money about £63,000 sterling. A cubic foot of common air contains only a little more than one ounce; and of hydrogen gas, the lightest of ponderable things, a cubic foot contains less than a drachm.

The following facts are also illustrations of the truth, that a body immersed in a fluid is held up, or has its entrance resisted, with force equal to the weight of the quantity of fluid which it displaces.

A stone which on land requires the strength of two men to lift it, may be lifted and carried in water by one man. There are cases, therefore, where the support of water thus rendered useful, is equivalent to the assistance of additional hands. A boy will often wonder why he can lift a certain stone to the surface of water, but no farther.

The invention of the diving-bell in modern times, having enabled men, in the building of piers, bridges, &c., to work under water almost as freely as above, many have experience of this influence of water: but workmen are generally surprised at first, to find that below they can move much larger and heavier stones than they can in the air. Some had supposed the fact accounted for by saying that the denser air of the diving-bell, when received into the lungs gave greater strength. In recovering property from a sunken ship by the diving-bell, everything is found to be lighter in the proportion now stated.

This law explains also why stones, gravel, sand and mud, are so easily moved by waves and currents. Many people expressed astonishment, in March, 1825, to learn that at the Plymouth Breakwater, the storm had displaced blocks of stone of many tons weight; but we now see that the moving water had only to overcome about half the weight of the stone.

When a person lies in a bath, the limbs are so nearly supported by the water as to require scarcely any exertion on the part of the individual. When this softest of all beds has been indulged in for half an hour or more, the person, on first lifting a limb out of the water, feels surprise at its great apparent weight. The workers about diving-bells always experience the sensation now spoken of, on returning to the air.

The bodies of most fishes are nearly of the specific gravity of water, and, therefore, if lying in it without making exertion, they neither sink nor rise very quickly. When this subject was less understood, many persons believed that fishes had no weight in water; and it is related as a joke at the expense of philosophers, that a king having once proposed to his men of science to explain this extraordinary fact, many profound disquisitions came forth, but not one of the competitors thought of trying what really was the fact. It was beneath the dignity of science in those days to make an experiment. At last a simple man balanced a vessel of water in scales, and on putting a fish into the water, showed its scale preponderating just as much as if the fish had been weighed alone.

In the sense now explained, water is said to have no weight in water. The least force will raise a bucket of water from the bottom of a well to the surface; but if the bucket be lifted at all farther, its weight is felt just in proportion to the part of it which is above the surface.

"*A body lighter than its bulk of water will float, and with force proportioned to the difference.*" (Read the Analysis, p. 126.)

The reason of this is clear. If any body, the cylinder *a b c d* for instance, be partially immersed in water, we know that the upward pressure of the water on the bottom *c d*, is exactly what served to support the water displaced by the body, *viz.*, water of the bulk, *e f c d*. The body, therefore, that it may remain out as far as here represented, must have exactly the weight of the water which the immersed part of it displaces; and if it be lighter than this, it will rise farther; if heavier, it will sink farther until the exact balance be produced.

Fig. 82.

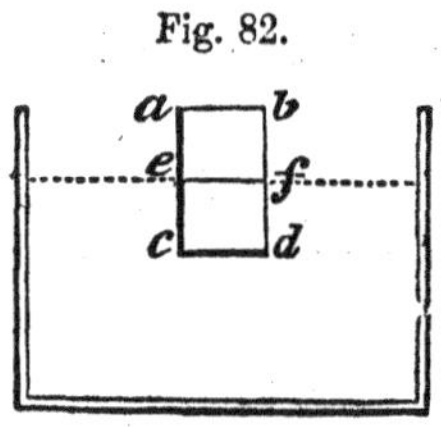

Hence of any body which floats in water, a pound weight displaces just a pound of water, whether the body be very light in proportion to its bulk, as cork, or heavier, as a piece of dense wood. This is experimentally shown by putting such bodies to float in a vessel originally full of water. The water displaced by each must run over the sides of the vessel, and may be caught and measured.

Hence a porcelain basin weighing four ounces will sink in water only as far as a similar wooden basin or bowl of the same weight; and the weight of either basin may be in the substance of which it is formed, or in anything else put into it as a load.

Hence a boat made of iron floats just as high out of water as a boat of similar form and size made of wood, provided the iron be proportionately thinner than the wood, and therefore not heavier on the whole. An empty metallic pot or kettle is often seen floating with a great part of it above the surface of the water.—Prejudice for a long time prevented iron boats from being used, although, for various purposes, they are superior to others: and there are still people who would fear to go on board of a ship built of the strong and singularly durable Indian teaks, because it is heavier than water, and, in the form of a log, therefore, sinks in water. Many fine ships of the line, however, and East-Indiamen of fifteen hundred tons or more, are now built of teak.

Hence a ship carrying a thousand tons weight will draw just as much water, or float to the same depth, whether her cargo be of cotton or of lead:—and the exact weight of any ship and her cargo may be determined by finding how much water she displaces. In canal boats, which are generally of a simple form, this truth affords a ready rule for ascertaining the quantity of their load.

The human body, in an ordinary healthy state with the chest full of air, is lighter than water.

If this truth were generally and familiarly understood, it would lead to the saving of more lives, in cases of shipwreck and in other accidents, than all the mechanical life-preservers which man's ingenuity will ever contrive.

The human body with the chest full of air naturally floats with a bulk of about half the head above the water,—having then no more tendency to sink than a log of fir. That a person in water, therefore, may live and breathe it is only necessary to keep the face uppermost. The reason that in ordinary accidents so many people are drowned who might easily be saved, are chiefly the following:—

1st. They believe that the body is heavier than water, and therefore, that continued exertion is necessary to keep it from sinking; and hence, instead of lying quietly on the back, with the face upwards, and with the face only out of the water, they generally assume the position of a swimmer, in which the face is downwards, and the whole head has to be kept out of the water to allow of breathing. Now, as a man cannot retain this position but by continued exertion, he is soon exhausted, even if a swimmer, and if he is not, the unskilful attempt will scarcely secure for him even a few respirations. The body raised for a moment by exertion above the natural level, sinks as far below it when the exertion ceases; and the plunge, by appearing the commencement of a permanent sinking terrifies the unpractised individual, and renders him an easier victim to his fate.—To convince a person learning to swim of the natural buoyancy of his body, it is a good plan to throw an egg into water about five feet deep, and then desire him to bring it up again. He discovers that instead of his body with the chest full of air naturally sinking towards the egg, he has to *force* his way downwards, and is lifted again by the water as soon as he ceases his effort.

2d. They fear that water entering by the ears may drown, as if it entered by the nose or mouth, and they make a wasteful exertion of strength to prevent it; the truth being, however, that it can only fill the outer ear, as far as the membrane of the drum, where its presence is of no consequence. Every diver and swimmer has his ears thus filled with water, and cares not.

3d. Persons unaccustomed to the water, and in danger of being drowned, generally attempt in their struggle to keep their hands above the surface, from feeling as if their hands were imprisoned and useless while below; but this act is most hurtful, because any part of the body held out of the water, in addition to the face which must be out, requires an effort to support it, which the individual is supposed at the time ill able to afford.

4th. They do not reflect, that when a log of wood or a human body is floating upright, with a small portion above the surface, in rough water, as at sea, every wave in passing must cover it completely for a little time, but again leave its top projecting in the interval. The practiced swimmer chooses this interval for breathing.

5th. They do not think of the importance of keeping the chest as full of air as possible; the doing which has nearly the same effect as tying a bladder of air to the neck, and without other effort, will cause nearly the whole head to remain above the water. If the chest be once emptied, while from the face being under water the person cannot inhale again, the body remains specifically heavier than water, and will sink.

When a man dives far, the pressure of deep water compresses, or diminishes the bulk of the air in his chest, so that, without losing any of that air, he yet becomes really heavier than water, and would not again rise, but for the exertion of swimming. The author of this work once saw a sailor (a fine-bodied West India negro) fall into the calm sea from a yard-arm eighty feet high. The velocity on his reaching the water was so great, that he shot deep into it, and, of course, his chest was compressed as now explained: probably also the shock stunned him, for although he was an excellent swimmer, he only moved his arms feebly once or twice, and was then seen gradually sinking for a long time afterwards, until he appeared only as a black and distant speck, descending towards the unknown regions of the abyss.

Every person need not learn to swim; but every one who makes voyages should have practiced the easy lesson of resting in the water with the face out. The head, from the large quantity of bone in it is a heavy part of the body, yet, owing to its proximity to the chest, which is comparatively light, a little action of adjustment with the hands, easily keeps it uppermost; and there is an accompanying motion of the feet, called *treading the water*, not diffcult to learn, which suffices to sustain the entire head above the surface. Many of the seventy passengers who were swallowed up on the sudden sinking of the Comet steam-boat near Greenock, in November, 1825, might have been saved by the boats, which so soon went to their assistance, had they known the truth which we are now explaining.

A man having to swim far, may occasionally rest on his back for a time, and resume his labor when he is somewhat refreshed.

So little is required to keep a swimmer's head above water, that many individuals, although unacquainted with what regards swimming or floating, have been saved after shipwreck, by catching hold of a few floating chips or broken pieces of wood. An oar will suffice as a support to half a dozen people, provided no one of the number attempts by it to keep more than his head out of the water; but often, in cases where it might be thus serviceable, from each person wishing to have as much of the security as possible, the number benefitted is much less than it might be.

The most common contrivances, called *life-preservers*, for preventing drowning, are strings of cork put round the chest or neck, or air-tight bags applied round the upper part of the body, and filled, when required, by those who wear them blowing into them through valved pipes.

On the great rivers of China, where thousands of people find it more convenient to live in covered boats than in houses upon the shore, the younger children have a hollow ball of some light material attached constantly to their necks, so that, in their frequent falls overboard, they are not in danger.

Life-boats have a large quantity of cork mixed in their structure, or of air-tight vessels of thin copper or tin plate: so that, even when the boats are filled with water, a considerable part still floats above the general surface.

Swimming is much easier to quadrupeds than to man, because the ordinary motion of their legs in walking and running is that which best supports them in swimming. Man is at first the most helpless of creatures in water. A horse while swimming can carry his rider with half the body out of the water. Dogs commonly swim well on the first trial.—Swans, geese, and water-fowls in general, owing to the great thickness of feathers on the under part of their bodies, and the great volume of their lungs, and the hollowness of their bones, are so bulky and light, that they float upon the water like stately ships, moving themselves about by their webbed feet as oars.

A water-fowl floating on plumage half as bulky as its naked body, has about half that body above the surface of the water; and similarly a man reclining on a floating mattrass, as in the hydrostatic bed afterwards to be described, has nearly as much of his body above the level of the water-surface, as he forces of the mattrass under it. His position, therefore, depends on the thickness of the mattrass.

A man walking in deep water may tread upon sharp flints or broken glass with impunity, because his weight is nearly supported by the water.

But many men have been drowned in attempting to wade across the fords of rivers, from forgetting that the body is so supported by the water, and does not press on the bottom sufficiently to give a sure footing against a very trifling current. A man, therefore, carrying a weight on his head, or in his hands held over his head, as a soldier bearing his arms and knapsack, may safely pass a river, where, without a load, he would be carried down the stream.

There is a mode practised in China of catching wild ducks, which requires that the catcher be well loaded or ballasted. The light grain being first strewed upon the surface of the water to temp them, a man hides himself in the midst of it, under what appears a gourd or basket drifting with the stream, and when the flock approaches and surrounds him, he quickly obtains a rich booty by snatching the creatures down one by one—adroitly making them disappear as if they were diving, and then securing them below. Each bird becomes as a piece of cork attached to his body.

Fishes can change their specific gravity, by diminishing or increasing the size of a little air-bag contained to their body. It is because this bag is situated towards the under side of the body, that a dead fish floats with the belly uppermost.

Animal substances, in undergoing the process of putrefaction, give out much aeriform matter. Hence the bodies of persons drowned and remaining in the water, generally swell, after a time, and rise to the surface, again to sink when the still increasing quantity of air shall burst the containing parts.

A floating body sinks to the same depth whether the mass of fluid supporting it be great or small:—as is seen when a porcelain basin is placed first in a pond, and then in a second basin only so much larger than itself that a spoonful or two of water suffices to fill up the interval between them. One ounce of water in the latter way may float a thing weighing a pound or more, exhibiting another instance of the *hydrostatic paradox* :—And if the largest ship of war were received into a dock, or case, so exactly fitting it that there were only half an inch of interval between it and the wall or side of the containing space, it would float as completely, when the few hogsheads of water required to fill this little interval up to its usual water-mark were poured in, as if it were on the high sea. In some canal locks, the boats just fit the place in which they have to rise and fall, and thus the expense of water at the lock is diminished.

The preceding examples of floating are all illustrations also of the truth that the pressure of a fluid on any immersed body is exactly proportioned to the depth and extent of the surface pressed upon. The lateral pressures just balanced one another, and the upward pressure has to be balanced by the weight of the body.

Similar reasoning to that which proves that the whole weight of a body acts as if lodged in the point called its centre of gravity, proves that the whole buoyancy of a body, or the upward push of the fluid in which a body is immersed, acts as if lodged in the point which was the centre of gravity of the fluid displaced. This point, consequently, is called the "centre of buoyancy."

A floating body, to be stable in its position, either must have its *centre of gravity* below the *centre of buoyancy*—in which case it resembles a pendulum ; or it must have a very broad bearing on the water, so that any inclination may cause the centre of gravity to ascend—in which case it resembles a cradle or rocking-horse.

Hence arises, in the stowing of a ship's cargo, the necessity of putting the heavy merchandise underneath, and generally of putting iron ballast under all the merchandise. Hence, also, the danger of having a cargo or ballast which is liable to shift its place. A ship loaded entirely with stones, is sometimes lost by a wave making her incline for a moment so much that the load ships to one side, which is then kept down. For a similar reason, a cargo of salt or sugar has a peculiar danger attached to it, for if the ship leak, the cargo may be dissolved, and then pumped out with the bilge water, leaving her with altered trim. In a fleet coming home from India, in 1809, four fine ships disappeared during a hurricane off the isle of France, and from what happened to the other ships that were saved, the cause of the destruction was supposed to be, that the saltpetre of the cargoes had been dissolved and pumped out, and that the ships in consequence became unmanageable.

Bladders used by beginners in swimming are dangerous, unless secured so as not to shift towards the lower part of the body.

A great inventor (in his own estimation) published to the world, that he had solved the important problem of walking safely upon the water ; and he invited a crowd to witness his first essay. He stepped boldly upon the wave, equipped in bulky cork boots, which he had previously tried in a butt of water at home; but it soon appeared that he had not pondered sufficiently on the centres of gravity and of floatation, for in the next instant all that was to be seen of him was a pair of legs sticking out of the water, the movements of which showed that he was by no means at his ease. He was picked up by help at hand, and, with his genius cooled and schooled by the event, was conducted home.—Some soldiers once finding a few cork *jackets*, among old military stores, determined to try them; but mistaking the shoulder straps for lower fastenings, they put them on as *drawers*, and on then plunging in, with the hope of being able to sit pleasantly on the water, their heavy heads went down, and they were nearly drowned.

When, on the return of summer, the ice breaks up in the polar regions, immense islands of it are set afloat, rising high into the air and sinking deep into the sea. The melting process, in most cases, does not go on equally in the water and in the air, and from the mass, consequently, changing form, its stability is often lost, and one of the grandest phenomena in nature follows—the overturning of a mountain—the sudden subversion of an island—producing a tumult in the ocean around, felt often at the distance of many leagues.

The phenomena of pressure, floating, &c., in fluids, vary in proportion to the weight or specific gravity of the fluid.

A ship draws less water, or swims lighter, by one thirty-fifth, in the heavy salt-water of the sea than in the fresh water of a river: and for the same reason a man swimming supports hmiself more easily in the sea than in a river.

Many kinds of wood that float in water will sink in oil.

A man floats on mercury as the lightest cork floats on water, and with practice he might be able to walk upon mercury.

Had the water of our ocean been but a little heavier than it is, men after shipwreck might have died of famine and cold, but would not have been drowned.

Oil floats on water, but sinks in alcohol or æther. The term *proof spirit* means spirit light enough for oil to sink in it. The strength of spirit is proportioned to its lightness.

Cream rises in milk, and forms a covering to it.

Blood, allowed to rest after flowing from the living body, separates into parts or layers, which arrange themselves according to their specific gravities. The buffy coat of inflammation (where this exists) is uppermost, forming the surface of the general coagulum: towards the lower part of the coagulum there is an accumulation of red globules; and the whole of the solid part floats in the serum, which is therefore lowest of all. When the red globules escape from the coagulum, they fall to the bottom even of the serum.

Wine, if slowly and carefully poured on water, will float upon it. In a vessel shaped like a common sand-glass, only with a larger opening between the chambers at *c*, if wine be put into the under chamber, and water into the upper, the two liquids will gradually change places: and if the lower half of the glass be covered, so as to leave the upper half with the appearance of a simple goblet, the water will seem to have been changed into wine. The liquids are less mixed, and change places sooner, when there is a tube *b* to carry the water down to the bottom without touching the wine, and a tube *a* to carry the wine directly to the top.

Fig. 83.

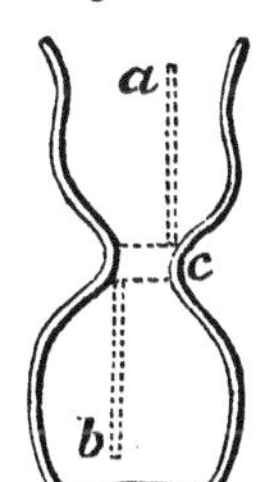

Mercury, water, oil, air, and some other fluids may all be shaken together in the same vessel, and on standing will separate again and arrange themselves in the order of their specific gravities.

When, in a mass of water, part of it is heated more than the rest, that part, by its expansion, becomes specifically lighter than the rest, and rises to the surface. Hence, when heat is applied to the bottom of a vessel containing water, a circulation is established, which goes on from the first moment until the operation of heating finishes:—water is always rising from the hotter parts of the vessel, and descending over the colder parts.

In like manner, when a tall glass containing hot water is dipped into cold water, a downward current takes place within the glass near the sides all round, and there is an upward current in the middle. This motion may be rendered very obvious by small portions of amber thrown into the water, for these being nearly of the specific gravity of water, rise and descend with it. On account of the current established in such cases, heat applied to the bottom of a vessel of liquid is soon equally diffused over it; but heat applied at the top is there confined, because the heated and lighter fluid does not descend. Water may be made to boil at its surface, while a piece of ice lies at the bottom. The converse is impossible.

The current in a fluid, produced by local change of temperature, is an important part of the following process, which the author deems applicable to various useful purposes.—Heat may be transferred from one liquid to another, without mixing them, by making the hot liquid descend in a very thin metallic tube, through the cold liquid rising around it in a larger tube, Boiling water from the vessel *e*, for instance, may descend slowly by the small tube *e a b f*, which is surrounded from *a* to *b* by cold water ascending through the tube *c g*. Then, as the temperature of two liquids, brought so nearly into contact with each other, will not, after a very short time, differ, in any one place more than a few degrees, it follows that the water lately cold, will on leaving the part of the tube *g*, which is in contact with the boiling water descending directly from *e*, be nearly boiling, while the water lately hot will, on leaving the tube at *b*, which is in contact with cold water just arrived from *h*, be itself nearly cold; and thus equal quantities of hot and cold water will have exchanged temperatures. The flux of the hot water is to be regulated by a cock *b*, and that of the cold water by a cock *h*. The water in the part of the tube *c g d* rises, because it is hotter and therefore specifically lighter than that in the part *h c*.—The author believes that an apparatus made on this principle, with an arrangement of many thin flat tubes instead of a single large tube, for the descending fluid, and a spacious box *c g* to contain these and the rising fluid, would be an excellent refrigerator in a distilling apparatus, and for cooling the wort of brewers; or would serve as a means of diminishing the expense of warm baths, by transferring the heat from the water lately used to pure water. In distilling, the *wash* or *low wines*, about to enter the still, might be used as the cold condensing fluid to surround the warm or vapor tubes, and thus, without expense, would be

Fig. 84.

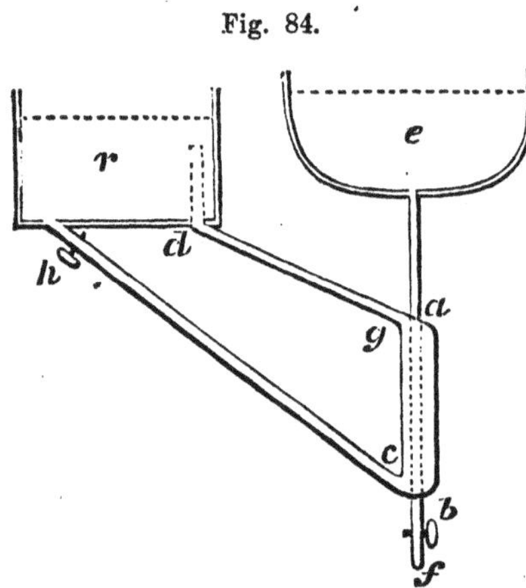

heated in its progress to the still. Half the original expense of a great porter brewery is in the construction of the numerous water-tight floors on which the hot wort is thinly spread to cool. The practice of warm bathing, so conductive to health, is less common in this country, because the present expense is so great.

It is a general truth in nature, that substances contract in size as they cool. There is, however, in water, a curious exception to this rule, which, operating through the principle of specific gravities, effects most important purposes in the economy of nature. Water contracts only down to the temperature of 40 deg., below which, towards 32 deg., or the freezing point, it goes on dilating again, and as ice is much lighter than as a fluid. Ice, therefore, floats on the surface of water, and being a very slow conductor of heat, defends the water underneath from the cold air, and preserves it liquid, and a fit dwelling for the finny tribe, until the return of the mild season. And not only is the extreme of cold below thus prevented, but because very cold water remains floating on the surface of a wintry lake, as cream floats on milk, it preserves underneath that warmth which is agreeable to the fishes, just as very hot water in summer remains uppermost, preserving underneath an agreeable coolness. By the dilation of very cold water, then, and the formation of ice, nature has prepared a winter garb for the inhabited lakes and rivers, as complete and effectual as for terrestrial animals, by the periodical thickening of their wool or fur. Had ice become heavier than water, so that it must have fallen to the bottom, and have left the surface without protection, a deep lake in European winters, would have been frozen into a solid lifeless mass, which summer suns would no more have melted than they now do the glaciers of Switzerland. But for this important exception, therefore, to a general law of nature, many of the now most fertile and lovely portions of the earth's surface would have remained for ever barren and uninhabited wastes.

PART III.

THE PHENOMENA OF FLUIDS.

SECTION II.—PNEUMATICS.

ANALYSIS OF THE SECTION.

In aeriform fluids, that is, in such as have their particles held far apart by mutual repulsion, which yields, however, to any force applied, so that the mass suffers great change of volume under different degrees of compression.—the phenomena are modified by the GREAT LIGHTNESS *and* ELASTICITY *of the fluids, but are still in strict accordance with the general properties of fluids already explained, viz.,* PRESSURE EQUAL IN ALL DIRECTIONS—PRESSURE AS THE DEPTH—LEVEL SURFACE, *and* FLUID SUPPORT. *The pressure of air, in all directions, and as the depth, may be studied in the effects of our atmosphere*—on solids—on liquids:—*or when it concurs with heat, in producing the phenomena of* boiling, evaporation, clouds, rain, dew, &c.; *or when, by varying in degree, it allows certain substances to exist sometimes in the* liquid *and sometimes in the* aeriform *states. The fluid support in air is exemplified by* ballons, *the* ascent of flame, *and* smoke, winds, &c.

WHAT a change has taken place in the degree of man's knowledge of nature, since philosophers thought that air was one of four primary elements, viz., *air, fire, water*, and *earth*, of which all things were composed, and each of which was for ever distinct from the others. We now know that air or gas is merely an accidental state, in which any body may exist, according to the quantity of heat pervading it: the body being solid when the absence of heat allows its atoms to obey freely their mutual attraction, and to cohere—as in ice, for instance; being liquid, when so much heat is present as nearly to balance the attraction, and to let them slide freely among each other—as they do in water; and being aeriform when still more heat is added, causing the atoms mutually to repel and dart asunder to a great distance—as they do in steam. But in any one of these three states, the various substances are as much themselves as in the others, and at the command of the chemist will assume any of the forms which he desires. As most substances in nature have a different relation to heat, there are some which, at the medium temperature of our earth are solid, some which are liquid, and some aeriform. The solids, in general, are the heaviest under a given volume, and therefore sink down and form the great mass or centre of the earth; the liquids follow next in order, and float upon this solid centre, filling up its inequalities with a level surface, so as to constitute the ocean; while the airs are the lightest of all, and as a second ocean, rest above the sea and above the highest mountain, to an elevation of about fifty miles. Among the substances whose relation to heat causes them, when not restrained in certain combinations, to assume the form of air at very low temperatures, there are two in particular, viz., *oxygen* and *nitrogen*, which are very abundant in nature in such uncombined state, and of these, therefore, the atmosphere chiefly consists; but smaller portions of almost every other substance are found in it. Water, among the supplementary matters, is much more abundant than any of the others, and in various states of cloud, mist, rain, dew and snow, it answers a thousand useful purposes, and serves beautifully to vary the scenes of nature. The atmosphere is about fifty miles high or deep, and therefore, in relation to the bulk of the earth, is as a covering of one-tenth of an inch in thickness to a common library globe of a foot in diameter.

The atmospheric ocean is the great laboratory in which most of the actions of life go on, and on the composition of which they depend. A human being requires for breathing a gallon of fresh air every minute, dying equally if deprived of air, or if confined to the same. All other animals also require fresh air, but in various proportions. And in the vegetable creation, the beautiful green leaf and delicate flower are merely broad and tender expansions of surface for the contact of the vivifying air. Animals give out to the atmosphere a substance which vegetables absorb, and vegetables, by the absorption, fit the air again for the use of animals; so that, upon the whole, in the various changes of nature, there is a perfect balancing of actions, which preserves the atmospheric mass in a uniform state, constantly fit for its admirable purposes.

While the ancients had that notion of air, which made them apply to it vaguely, and almost indifferently, the names of *air*, *ether*, *spirit*, *breath*, *life*, &c., they never dreamed of making experiments upon it, with a view to prove its relation to common matter:—and one of the most beautiful portions of the modern history of man's progress in knowledge, is that which exhibits the light gradually breaking in upon this most interesting subject. Galileo was led to conclude that air made a definite pressure upon things at the surface of the earth; Torricelli and Pascal proved that this was occasioned by its weight, and hence, moreover, they deduced the height of the aerial ocean; Priestly, Black, Lavoisier, and others, discovered that air might be united with a metal, so as to increase its weight, and to produce a compound of totally new qualities, for they showed that many of the ores of our mines are merely metals concealed, by being thus united with a substance which, when set free, ascends as one of the ingredients of the atmosphere. They at last analyzed the atmosphere itself, and exhibited its two ingredients as distinct substances. And within a few years the nature of air or gas has been so thoroughly investigated, that we can now take a little of many a light, invisible, impalpable fluid such as we breathe, and squeezing the heat out of it by strong pressure, can make its particles collapse from their aeriform distances to assume the state of a tranquil fluid; which may then be retained as such for ever, or may be decomposed and made solid in combination with other bodies, or may be again set at liberty.

The suspicion once excited, that air was as much a material fluid as water, only much less dense, by reason of a greater separation and repulsion of the particles, it was easy to follow out the parallel, and to confirm the supposition by reference to the commonest facts. Thus, a leathern sack or pouch, opened and dipped into water so far as to become full, if its mouth be then carefully closed, retains the water, and its sides cannot afterwards be pressed together: a similar sack or bladder, opened out, and then closed in air, is found to remain, in a corresponding way, bulky and resisting, and forms what is called an air-pillow. The motion of a flat board is resisted in water: the motion of a fan is resisted in the air. Masses of wood, sand, and pebbles, are rolled along or floated by currents of water: chaff, feathers, and even rooted trees are swept away by currents of air. There are mills driven by water; and there are mills driven by the wind. Oil set free under the surface of water, or placed there in a bladder, rises to the surface: hot air or hydrogen gas placed in a balloon, rises in the air. A fish moves itself by its fins in water: a bird moves itself by its wings in the air; and as on taking the water from a vessel in which a fish swims, the creature falls to the bottom, gasps a few moments, and dies, so, on exhausting the air from a vessel in which birds or butterflies are enclosed, their useless wings may flap; but they sink to the bottom, and if the cruel experiment be continued, they soon become motionless and forever.

We proceed now to prove that air or gas, as a fluid, differs from the other fluids, which we call liquids, only in the two circumstances of great lightness or rarity, and of being very extensively elastic, that is to say, the particles being so related, that pressure brings them much more nearly into contact, and on ceasing, allows them to regain their former distance.

Lightness of Air.

The lightness or rarity of atmospheric air, as it is found on the general surface of the earth, is such, that if, by the action of a pump, a bag of it holding a cubic foot be emptied into the copper ball of an air-gun, the ball weighs about an ounce and a quarter more than before. The same volume of water weighs nearly a thousand ounces; so that common air is about eight hundred times lighter than water. Other gases, or substances in the aeriform state, have their various specific gravities, just as the same substances have when liquid or solid. Thus water in the form of air, that is to say, when existing as steam, and of the common density, is little more than half as heavy as the same bulk of common air; hydrogen is only one-fourteenth part as heavy: and carbonic acid gas, which is the air that rises out of soda-water, brisk ale, champagne wine, &c., is so much heavier, that even in the atmosphere it may be poured out of one open vessel into another, as a liquid might, or, more exactly, as water might be poured out under oil.

Elasticity of Air.

A small bladder full of air may be pressed or squeezed between the hands so as to be much reduced in size, but on being relieved from the pressure, it will immediately resume its former bulk.

Fig. 85.

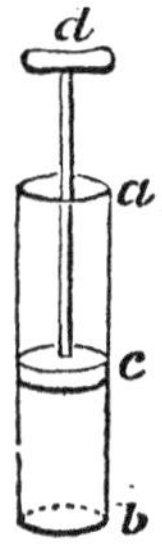

If a metalic tube or barrel of perfectly uniform bore *a b*, be fitted with a moveable plug or piston *c*, which is covered with leather and oiled, so as to slide up and down without allowing the air to pass by its sides, the air between the piston and the close bottom *b* may be compressed to a hundredth or less of its usual bulk; but when allowed, will push the piston back again with the same force as it opposed to the condensation, and will recover the volume which it had before the experiment.

Again, if the plug, at the commencement of the experiment, were only an inch from the bottom, enclosing air of the usual density, on drawing it up to the top, the inch of air beneath it would expand so as to occupy the whole tube, having become, of course, proportionally less dense.

To the question why the air, which admits of such various density, is found to have that certain degree of it met with at the surface of the earth, we answer, that as the water, in any place near the bottom of the ocean, is pressed with force exactly proportioned to the quantity of water above it, so the air at the surface of the earth bears the pressure of the superincumbent mass of air, and on account of its extensive elasticity, suffers, like the lowermost bags of cotton or wool in a great heap, that degree of compression which the superincumbent mass is calculated to produce. We shall see below that the density of the air near the earth is changing with every circumstance which affects the weight of the atmosphere above, as winds, clouds, rain, &c., and that it bears relation to the altitude of the place of observation above the level of the sea.

The tube with its piston, described in the last page, becomes, according to the position of its valves, either a syringe for injecting and condensing air, or a pump for exhausting or removing it from any vessel; both operations depending on the elasticity of air.

A barrel and piston is a *condensing syringe*, when in a passage of communication between the bottom of the syringe and a receiving vessel, there is a flap or *valve* allowing air to pass towards the receiver but not to return. The piston, therefore, at each stroke, forces what the barrel contains of air into the receiver. When the piston is lifted again after the stroke, air re-enters the barrel from the atmosphere, either through a valve in the piston, or through a small hole near the top of the barrel. That useful contrivance, *a valve*, for whatever purpose used, and in whatever way formed, is in principle merely a moveable flap, placed on an opening, against which it is held by its weight, or by some other gentle and yielding force. Such a flap, it is evident, will allow fluid to pass only in one direction, *viz.*, outwards from the opening, for any fluid tending inwards must shut the flap, and press it the closer, the greater the tendency.

To convert a forcing syringe or pump into an exhausting syringe or pump, commonly called an *air-pump*, it is only necessary to reverse the position of the valves; then, on the descent of the piston, all the air between it and the bottom, instead of entering the vessel or receiver, as in the last case, escapes by a valve in the piston itself towards the atmosphere, and on the rising of the piston, a perfect vacuum would be left under it, but that the valve below, then opened by the elasticity of the air in the receiver, allows a part of that air to follow it. Thus, at each stroke, a quantity of the air, proportioned to the size of the pump, is removed from the receiver. In a good air-pump, there are two similar pumping barrels, as *a* and *b*, to quicken the operation of exhausting; and both are worked at the same time by the reciprocating winch or handle *f*, with its pinion *e*, acting on the teeth of the piston rods *d*

Fig. 86.

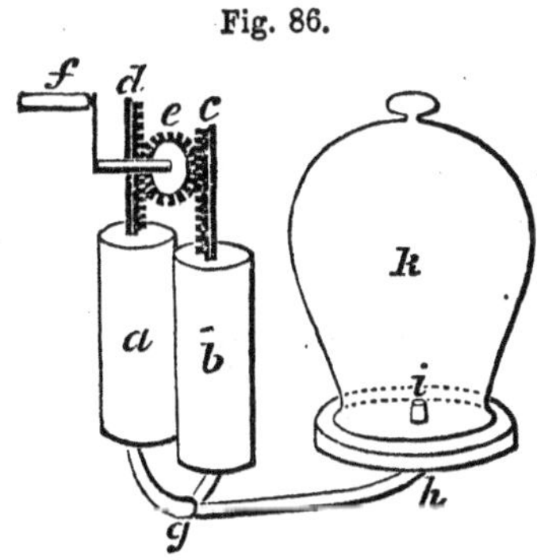

and *c*. This double construction has the farther advantage, that the atmospheric pressure, if fifteen pounds per square inch on the upper surface of either piston, and which for a single piston would have to be overcome by the worker in lifting it, as here balanced always by the corresponding pressure on the other piston. Both pumps communicate with the tube *g h*, which at *h* rises tightly through the round plate of the machine to *i*. This flat plate is so smooth, that a glass bell or *receiver k*, with a smooth ground lip, when placed upon it, forms an air-tight joining. On working the pump, such a bell is exhausted of its air, and fitted for showing the many interesting phenomena which the air-pump can display,—and which will pass under review as we proceed. The supporting frame-work of the pump is not shown here.

The law of the elasticity of air is, that its spring, or resistance to compression, increases exactly with its density or the quantity of it collected in a given space. Hence, by finding in any case either the density of the air, or the spring, or the compressing force, we know all the three.

It has been ascertained by experiments described a few pages farther on, that in the atmospheric ocean surrounding the earth, there are nearly fifteen pounds of air above every square inch of the surface of the earth; and that the air nearest the earth, and bearing this superincumbent weight or pressure, has the density of an ounce and a quarter of weight to a cubic foot of volume. We further find that such air is reduced to half its bulk, or becomes of what is called double atmospheric density, by an additional pressure of fifteen pounds on the inch, and of triple density, by tripple pressure, and so forth; and on the other hand, that it dilates to double bulk, if the pressure be diminished to half, and to any greater bulk, even beyond a thousand-fold, if the pressure be diminished in a corresponding degree; and any air bearing a given force or pressure, is always acting as a spring of that force on whatever it touches.

It is very important to be familiar with this truth or law, for it holds very nearly with respect to all aeriform fluids as well as common air, and throws light, therefore, on the action of steam-engines, air-guns, pneumatic machines generally. It also explains the condition of our atmosphere as to density at various elevations; telling us, for instance, that when a balloon has risen through half of the atmospherical mass, the air around it will be of only half the density which exists at the surface of the earth.

We know not exactly to what extent the rarefaction of air may go on the removal of pressure; in other words, at what distance the gravity of the particles becomes just a balance to their mutual repulsion; and therefore we know not exactly what the degree of rarity is at the top of our atmosphere; but we know that it must be exceedingly great, from the fact that the air left in the receiver of an air-pump has still spring or elasticity enough to lift the valve of the pump, when less remains than the thousandth part of the original quantity. In the most perfect air-pumps, that the exhaustion may be as complete as possible, the machine itself is made to raise the valve.

The expansion of air is well illustrated by a bladder, having a very little air in it, placed under the receiver of an air-pump. On exhausting the receiver, the bladder gradually swells, with force sufficient to lift a moderate weight laid upon it, and at last appears quite full, and may even be burst. A shriveled apple treated in the same way becomes plump. The explanation of such phenomena is, that at first the air in the bladder or apple is in a state of condensation, like all air, at the surface of the earth under the pressure of the superincumbent atmosphere; but that its volume increases as that pressure is diminished by the air-pump:—it is rarefied in the same proportion as the air which remains in the receiver surrounding it.

The curious instrument called the air-gun has a strong globular vessel of copper attached under the lock, into which air is usually forced to be thirty or forty times as dense as the air in the atmosphere around: hence the pressure or elasticity tending outwards is thirty or forty times fifteen pounds on the inch, and when the valve is opened for an instant by the action of the lock, a portion of the air issues and propels the charge with this force. The effect of air thus condensed nearly equals that of gunpowder, and one charge of the ball suffices for many shots, the force, however, becoming less for every successive discharge.

If a bottle or vessel *a b*, partly filled with water, have a tube *c d* passed tightly through the cork to near the bottom of the water; and if more air be then forced through this tube in any way, so as to accumulate in the upper part of the vessel above the water surface *a b*; on turning the cock *c*, which opens the tube, the elasticity of the condensed air will press the water out as a beautiful jet, to a height proportioned to the condensation, and gradually diminishing as the condensation diminishes. Or if such a vessel, with air of common density, be placed under a tall air-pump receiver, on working the pump so as to diminish the density of the air in the receiver, the jet of water will equally rise.—A table-lamp, by the force of condensed air, may be supplied with oil from a reservoir far below the wick: and lately an enema syringe and a shower-bath have been constructed on the same principle.

Fig. 87.

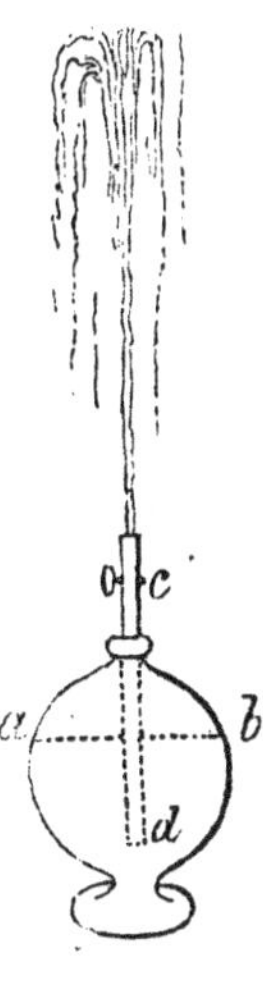

The elasticity of air is rendered very serviceable in connection with great water-pumps, such as those used for the supply of cities. A pump throws its water by a distinct gush at each stroke, while the current through the pipe towards the city should be uniform. Now uniformity is attained by causing the gushes from the pump *a* to enter by the passage *b* at one side of a large vessel *c*, of which the upper part is full of the condensed air, and from the other side of which at *d* the water issues on its way. The air in this vessel (called the *air-vessel*) is condensed, as a spring, by the entering water, and its resisting elasticity, both immediately, and afterwards during the interval of the strokes, forces the water along the pipe *d*. Each entering gush has only the effect of compressing the air a little more for the time, while the flow in the great pipe continues nearly uniform. The pump itself is made to take in a little air at each stroke, so that not only is the vessel always supplied, but some air is constantly passing on with the water, and effecting the highly useful purpose of giving an elasticity to the whole contents of the pipe and its ramifications.

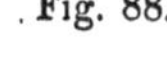

Fig. 88.

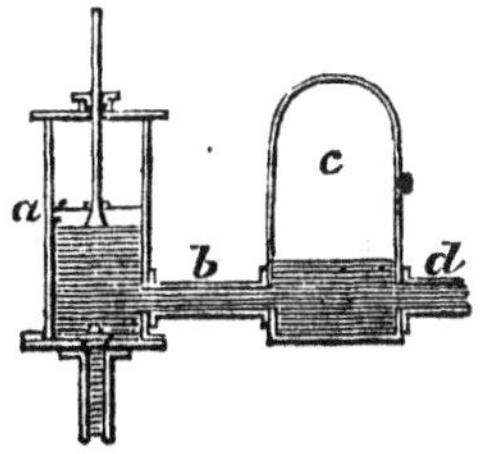

The same object is attained by the same means in the fire-engine used to check conflagration. In it there are generally several water-pumps working together, which throw their interrupted supply into an air-vessel whence it passes in a nearly uniform jet to the point desired.

The compressibility and corresponding spring of air are remarkably exhibited in that singular contrivance of modern times, the *diving-bell*, in which men now descend with safety to considerable depths in the ocean, there to reside and labor, attaining many objects of high importance to them :—they recover sunken treasures,—they are enabled to pursue works of submarine architecture, as in constructing light-houses and noble harbours, where formerly no foundations could have been laid, &c. The diving-bell, in point of utility, has proved a remarkable contrast to its sister invention, the balloon, which, although so wondrously bearing man aloft to the regions of the clouds, has brought him as yet little advantage to compensate for the many fatal accidents which its use has occasioned.

The diving-bell is a large heavy open-mouthed vessel, with accommodation in it for one or more persons. It is let down into the water with its mouth undermost, from a crane to which it is suspended, and which rests on a suitable carriage either on the shore, or on the deck of a ship, or barge fitted for its service. On first entering the water it appears full of air; but air being compressible, according to the law now explained, and the pressure of the water around the descending bell increasing with the depth, the volume of the air gradually diminishes, and at thirty-four feet is reduced to half. The bell, then, unless more air be supplied, will of course be half full of water, and a person breathing in it, at each inspiration will receive twice as much air into the lungs as when breathing at the surface. A constant supply of fresh air is sent down to the bell by a forcing-pump above: and the heated and contaminated air, which has served for respiration, and which rises to the top of the bell, is allowed to escape by a cock placed there for the purpose. The men who work at a distance from the bell have tubes of communication with it, by which they inhale the air required; and they allow the used air to rise through the water above them. A man cannot breathe comfortably by such a tube if he be either much above or much below the level of the water in the bell; for if above, the air in the bell is more compressed than his chest, and is forced towards him so as to require an effort to control its admission; and if below, his chest is bearing greater pressure than the air in the bell, and he must therefore act strongly with the muscles of the ribs to draw the air down to him. A phenomenon similar to this takes place when two bladders of air are connected by a long tube, and immersed in water to unequal depths, the air being always strongly forced from the lower into the upper one, because the lower one is more pressed. The difficulty of pumping air down to the diving-bell increases, of course, with the depth to which it has ascended: for if the bell be so low that the water is pressing on the air in it with a force of fifteen pounds per inch, (which would happen at thirty-four feet,) it is evident that a syringe or pump can not inject more air unless it act with a force greater than this. Divers might often, if not always, more conveniently receive their supply of air through tubes from an air vessel kept charged to the necessary density in a boat over them, or on the shore, than from a bell below. If they would, moreover, dress in India-rubber cloth, and use a hood of metal with windows for the head, they might work under water without wetting any part but their hands.

It is remarkable, when the use of the diving-bell has become so familiar, that a kindred and still more simple contrivance of the same class has not been introduced for certain purposes, particularly of sudden emergency, such as to aid in the recovery of the bodies of drowning persons. A ten-gallon cask, or vessel of any kind, filled with air, and made heavy enough just to sink in water, with a breathing tube from it like that of a diving-bell, would be a provision of air for a man below water for ten minutes; and a man with it under his arm, might instantly descend from a boat or walk from the shore, into water of any depth, to recover the body of a fellow-creature lately sunk, and in time probably to save the life, which a few minutes wasted in waiting or in unsuccessful dragging would suffer to be lost. The author would propose this as an addition to the apparatus of the Humane Society for the recovery of persons apparently drowned.—It shows the remoteness from common trains of thinking of the truths connected with the constitution of our atmosphere and sea, when a means so simple and easily procured should never have been thought of or tried in any way by pearl-fishers, or by persons who gain their bread by diving to recover things dropped overboard in harbours or anchoring stations; all of whom have hitherto been limited to the single gulp of air taken on descending. In the case of a man working under water, cask after cask of air might be sent down, to enable him to remain as long as necessary.

There is an exceedingly beautiful philosophical toy, of which the action depends chiefly on the elasticity of air; and as it moreover illustrates most of the laws of fluidity, it is deemed worthy of description here. It is a small balloon or thin globe of glass *c*, having an opening at the bottom, and its little car or basket hanging to it. If put to float in water while the globe contains air only, it is so light that half the globe remains above the surface; but water may be introduced to adjust the specific gravity of the whole, until it becomes only a little less than that of water. If the balloon be then placed in a tall jar of water *a b*, the mouth of which is closely covered by bladder-skin or India-rubber tied upon it, on pressing such covering with the hand, the balloon will immediately descend in the water, to rise again when the pressure ceases, and will float about, rising, or falling, or standing still, according to the pressure made. The reason of this is, that pressure on the top of the jar first condenses the air between the cover and the water surface; this condensation then presses upon the water below, and by influencing it through its whole extent, compresses also the air in the balloon globe, forcing as much more water into this as to render the balloon heavier than water, and therefore heavy enough to sink. As soon as the pressure ceases, the elasticity of the air in the balloon repels the lately entered water, and the machine, becoming as before, lighter than water, ascends to the top. If the balloon be adjusted to have a specific gravity too nearly that of water, it will not rise of itself after once reaching the bottom, because the pressure of the water then above it will perpetuate the condensation of the air which caused it to descend. It may even then, however, be made to rise again by inclining the water-jar to one side, so that the perpendicular height of water over it shall be diminished.

Fig. 89.

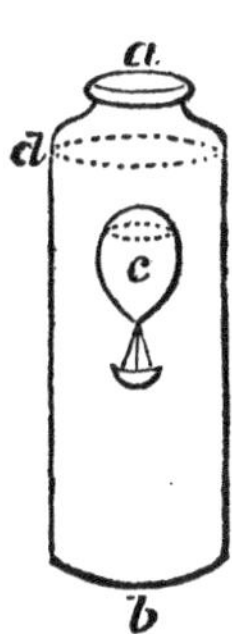

This toy proves many things—the *materiality* of air, by the pressure of the hand on the top being communicated to the water below through the air in the upper part of the jar—the *compressibility* of air, by what happens in the globe just before it descends—the *elastic force* of air shown in expansion, when, on the pressure ceasing, the water is again expelled from the globe—the *lightness* of air, in the buoyancy of the globe:—it shows, also, that in a fluid *the pressure is in all directions*, because the effects happen in whatever position the jar be held—it shows that *pressure is as the depth*, because less pressure of the hand is required the farther that the globe has descended in the water—and it exemplifies many circumstances of *fluid support*. A young person, therefore, familiar with this toy, has learned the leading truths of hydrostatics and pneumatics, and has had much amusement as well as instruction.

Fig. 90.

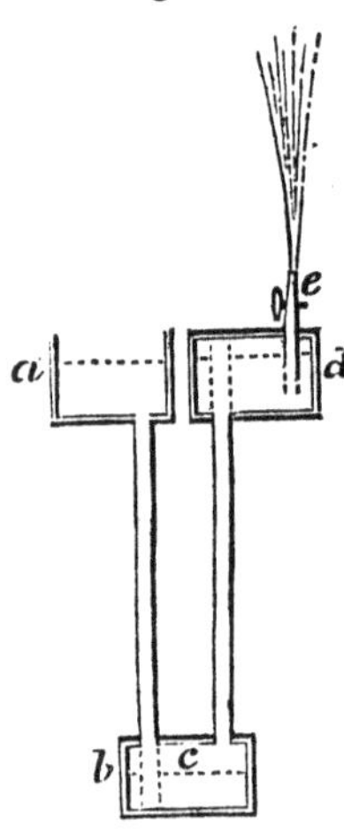

Fig. 91.

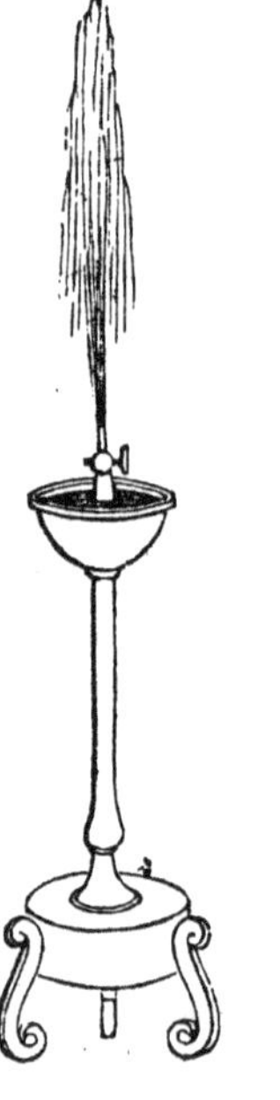

On the same principle as the balloon now described, three or four little figures of men may be formed of glass, hollow within, and having each a minute opening at the heel, by which water may pass in or out. If these be placed in a jar as the balloon was, and be adjusted by the quantity of water admitted into them, so that in specific gravity, they shall differ a little from each other, and if then, a gradually increased pressure be made on the cover of the jar, the heaviest figure will descend first, and the others will follow in succession; and they will stop or return to the surface in reverse order when the pressure ceases. A person while exhibiting these figures to spectators who do not understand them, may appear only carelessly to rest his hand on the cover of the jar while he is making the required pressure, and he will seem to have the power of ordering their movements by his will. If the jar containing the figures be inverted, and the cover be placed over a hole in the table, through which, unobserved, the exhibitor can act by a rod rising through the hole and obeying his foot, he may produce the most amusing and surprising evolutions among the little men, in perfect obedience to his word of command.

The beautiful fountain, called the fountain of Hiero, by which water is made to spout far above its source, depends for its action upon the resisting elasticity of compressed air. The vessel *d* is first filled with water, while *b* and *a* contain air only. On then pouring water into *a*, the water of *d* darts upwards through the jet-pipe *e*, to an elevation nearly equal to the length of the tube from *a* to *b*. The reason is, that the water from *a* descends by the tube to *b*, and compresses the air at *c*; which compression conveyed along the other tube from *c* to *b*, acts on the water in the vessel *d*, and causes it to jet. As the pressure is produced by the column of water *a b*, the jet is proportioned to the length of that column.—This kind of fountain may have its parts concealed under a variety of forms as here exemplified, and may thus become a beautiful ornament among flowers in a summer drawing-room. It may be made of size to play for an hour or more, and it will always recommence on the water being shifted from the low to the high reservoir.—The useful table-lamp, consisting of a simple column or pillar with the oil rising to the flame from far below, is a Hiero's fountain, only the oil, instead of being allowed to jet out, rises in a tube to the flame. The contrivance for maintaining the two columns always of the same length, notwithstanding the expenditure of oil has to be explained some pages hence.

Having now explained the two peculiarities which distinguish aeriform from other fluids, *viz.*, their *lightness* and extensive *elasticity*, we proceed to show that they have the four other properties already described under hydrostatics, as belonging to fluids generally: and first,

"*Pressure in all directions.*" (Read the Analysis, at pages 140 and 172.)

A quantity of air or gas shut up in any vessel and compressed, is equally affected throughout, and its tendency to escape from the pressure is equal in all directions, as is proved by the force necessary to keep similar valves close wherever placed. Hence the hydrostatic press and hydrostatic bellows described in the last section, which depend for their action on this law, may be worked by air or gas as by a liquid.

Owing to this law, air, when allowed, will always rush from where there is more pressure to where there is less. The actions of the common fire-bellows, and of the animal chest in breathing, blowing, sucking, &c., are so many instances.

The suddenness with which any compression made on part of a confined aeriform fluid is communicated through the whole, is strikingly seen in the simultaneous increase or burst of all the gas-lights over an extensive building or even in a long street, at any instant when the force supplying the gas is augmented.

Many very interesting illustrations of the fluid pressure of air being in all directions, will occur under the next head, joined with proofs of the atmospheric pressure being as the depth.

"*Pressure as the depth.*"

On first approaching this subject, a person is naturally surprised to hear the depth or height of the atmosphere spoken of as something perfectly ascertained, although nobody can ever have approached the surface to measure it; but science often furnishes means of reaching precise truth, in cases where ignorance would not even dream of the possibility of making an approximation. It may facilitate the apprehension of this point as regards air, to describe first some parallel cases in which water is concerned.

The bottom of a lake evidently supports all the water in the lake, and each portion bears just the weight of the water directly over it: a means then of ascertaining the weight or pressure of water on any portion of the bottom would tell how much water stood over that portion, and by the known relation of the weight and bulk of water would tell also the depth at that part. In like manner the ocean of air which surrounds the globe rests with its whole weight upon the surface of the globe, and each portion of the surface bears its share: if we ascertain, then, the pressure of the atmosphere on a given extent of the surface, we find how much air is standing directly over it; in other words, the weight of a column of air resting on such surface as its base, and reaching to the top of the atmosphere. Having then the weight of the whole column, and finding the weight of a given bulk of it at the botton (ascertained as described at page 158,) and knowing the law of aerial elasticity (explained at page 158,) we determine the depth or height of the column by a simple calculation. Now accurate experiments show that there are nearly fifteen pounds of air over every square inch of the earth's surface; producing the same pressure as would be made by a depth of water of thirty-four feet, or by a depth of quicksilver of thirty inches; and from this fact and the ascertained lightness and elasticity of air, we know that its depth on earth must be nearly fifty miles, which, as already stated, is about as much in relation to the size of the earth as the tenth of an inch is to a globe of one foot in diameter. The remaining part of this section has chiefly to trace the effects of this mass of matter resting upon the earth's surface, and as a fluid embracing and compressing every object placed there.

Water is a substance much more obvious to the human senses than air, and which is constantly under observation; yet many of its most important agencies escape the notice of common observers. Few persons, for instance, of themselves discover the law explained in the last section, of the pressure in water being proportioned to the depth: but when made to observe that a piece of cork plunged deep into it is compressed to much smaller bulk, and that strong empty vessel of glass, or even of metal under the same circumstances, are crushed or broken inwards, and that pieces of sunken wood are, at great depths, filled with water through all their pores, so as to become nearly as heavy as stone, &c., their minds are roused to a sense of the important fact that a fluid presses, and in proportion to its depth. If the truths of hydrostatics thus escape notice, we need not wonder that those of pneumatics escape still longer.

If a piece of bladder-skin or a pane of glass be laid at the bottom of a vessel, holding water, the bladder or glass exhibits no sign of being pressed upon, although it bears on its upper side the whole weight of the water directly above it: the reason being that water beneath the bladder resists just as strongly as the water above presses, in the same way that one stone in a pillar resists those above it: but if the bladder be tied closely over the mouth of a common drinking glass or tumbler filled with water, and placed at the bottom of the vessel, and if then, by means of a syringe or pump, the water be extracted from within the glass, the bladder itself has to bear the whole pressure of the water above it, (independently of a pressure of air, to be explained afterwards,) and will probably be torn or burst. The degree of pressure, and consequently the depth of the water, in such a case, might be ascertained by placing some support, of which the action could be measured, under the bladder to sustain it after the removal of the interior water.—Now this case may be closely copied in our atmosphere or sea of air. A glass held in the hand is immersed in the fluid air, and is full of it as the other glass was supposed full of water: its mouth may be covered over with bladder, and no external pressure will be apparent, because there is a resistance of the air within, just equal to the pressure of the air on the outside:—but if the air be extracted from under the covering by means of an air-pump, the bladder is first seen sinking down and becoming hollow from the weight of the air over it, and at last bursting inwards with a great noise or crack. By placing a circular piece of wood under the bladder-skin, for it to rest on, and a spring of known force to support the wood, we may ascertain very nearly the weight and pressure of the air over it. This mode, however, of ascertaining the weight of the atmosphere, is not that commonly used, but is described here as a good illustration of the present subject; the problem being solved much more elegantly and accurately by means of the barometer described farther on. The phenomenon of atmospheric pressure is often exhibited by placing the hand on the mouth of a glass so as to cover it closely, and then extracting the air from underneath the hand: the weight of the atmosphere holds the hand down on the mouth of the glass with the force of fifteen pounds to the inch.

As should follow, from the pressure of fifteen pounds per inch thus detected at the surface of the earth, being the weight of our superincubent atmosphere, we find that exactly as we rise from the earth, and leave part of the atmosphere beneath us, the pressure diminishes. This fact now furnishes the readiest means of ascertaining the heights of mountains and of balloon ascents, as will be explained in considering the barometer.

After the many explanations here given of fluid pressure being equal in all directions, it is almost superfluous to remark, that the downward weight of the atmosphere becomes a pressure in all directions. This is seen in the fact of the bladder seen above, being as readily burst if turned sideways as if turned directly upwards. Every body or substance, therefore, on the surface of the earth, dead or living, solid or fluid, is compressed with this force. In general the pressure on one side of a body, is just balanced by the equal pressure on the other, so that no sensible effect follows; and it is on this account that philosophers were so long in discovering it at all, and that half-informed persons are still disposed to doubt its existence; but the proofs offered on all sides to the now awakened attention are irresistible. We shall speak first of

"Atmospheric pressure on solids."

The atmosphere, then, presses on the two sides of a plate of glass or metal, with force of fifteen pounds on the inch. Under ordinary circumstances, no sensible effect follows, because the opposite pressures counterbalance; but if two plates of smooth glass or metal be laid against each other, and the air be prevented from entering between them, they cannot be separated by less force than fifteen pounds per inch of their surface.

In like manner, to draw down the piston of a syringe from the bottom of its barrel, while no air is allowed to enter between them, requires force of fifteen pounds to the square inch of surface of the piston. But if the experiment be made in the exhausted receiver of an air-pump, the piston falls by its own weight. It is pushed back immediately on re-admitting the air. Wherever a vacuum is produced at the surface of the earth, there is an external pressure, of the force stated, seeking admittance all round.

An air-pump receiver of five inches diameter has nearly twenty square inches of surface in its upper part or roof, and bears a weight or pressure of atmosphere, of twenty times fifteen, or three hundred pounds. While it has air within it, this pressure is exactly balanced, and is not sensible; but when exhausted on the plate of the air-pump, it is pressed against the plate with this force. As the atmospheric pressure is in all directions, the pump-plate, of course, is equally pressed upwards against the receiver, and the sides of the receiver are pressed towards each other. This explains why air-pump receivers must be made arched or of dome-shape to withstand the great pressure. A flat piece of glass of great thickness, laid upon the upper mouth of a receiver, so as to form an air-tight cover to it, is broken instantly by exhausting the air beneath; and a bottle or receiver with flat side, when exhausted suffers in the same manner.

Illustrative of this pressure on solids there is the experiment of the Magdeburgh hemispheres, as it is called. Two hollow half globes of metal *a* and

Fig. 92.

b, are fitted to each other, so that their lips when touching may be air-tight. While there is air between them or within, resisting the pressure of the outward air, they can be separated from each other without difficulty; but when the air is exhausted from within by the air-pump, a force is required to separate them of as many times fifteen pounds as there are square inches in the area of the mouth. The air is extracted by unscrewing one of the handles at *b*, and then connecting the remaining stalk (which is hollow and has a stop-cock) with the air-pump.—This experiment merits recollection, because it was one of the first which drew attention to the material nature and properties of the air; and it astonished the world. Otto Guericke, Burgomaster of Magdeburgh, the inventor, had hemispheres made of three feet in diameter, and once when he exhausted them, on the occasion of a public exhibition, twenty coach-horses of the emperor were unable to pull them asunder. There being no air-pump when Guericke began his experiments, although he himself invented it afterwards, he originally emptied the balls of their air by first filling them with water, and then extracting the water by a common pump or syringe applied to the bottom.

It is a phenomenon of the same kind as the last described, when a boy with his foot presses a circular piece of wet leather as *a*, against a flat-faced stone as *b*, and then lifts the stone by pulling at a cord *c*, rising from the centre of the leather. If the leather be so close in its texture that air cannot pass through it, and stiff enough not to be puckered or drawn together, he must exert a force before detaching it, of as many times fifteen pounds as there are square inches of surface covered by it for such is the weight or pressure of the air over it, while there is no counterbalancing pressure underneath nearer than on the other side of the stone. The weight of the stone that may be lifted is thus determined by the size of the leather. The contrivance has been called a *sucker*, or *pneumatic tractor*. A very large *sucker* applied upon a rock or wall, would resist the pull of horses like the Magdeburgh hemispheres.

Fig. 93.

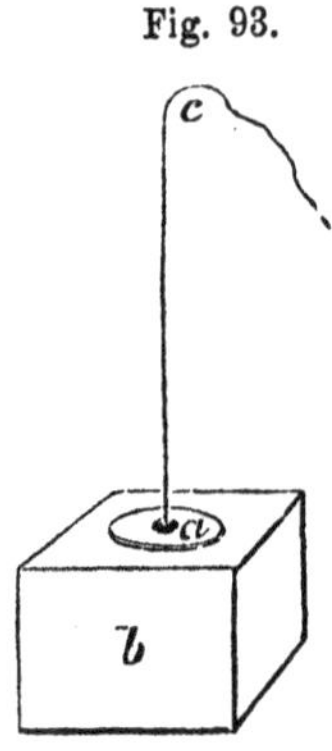

This contrivance seems suited to some purposes of surgery. It might assist, for instance, in raising depressed portions of a fractured skull, and might thus sometimes save the operation of trepanning:—for such a purpose it would be preferable to the small cupping-glass sometimes used, from its being perfectly inactive, except during the instants when pulled at; whereas the cupping-glass, by keeping up a continual flow of blood to the part, might do injury. There is another surgical application spoken of in the last *section* of the present *part*, which the professional reader may consult immediately.

It is from having feet that act on the principle of the tractor, that the common fly and other insects can move along ceilings, and even polished surfaces of glass or metal with their bodies hanging downwards; and there are many marine animals which attach themselves to rocks, or other objects by a similar action.

If two pneumatic tractors be applied to each other, men pulling opposite ways, to separate them, must act with a force of fifteen pounds to the square inch of the surface of contact, as if they were separating the Madgeburgh hemisphere.

The case of the pneumatic tractor may be well illustrated by an experiment made in a vessel containing a liquid. If a body with a flat surface be applied to the bottom of the vessel so as perfectly to exclude the liquid, the body bears the whole weight of liquid directly over it, and cannot be detached without force equal to this. The case is striking when a flat piece of cork is pushed against the smooth bottom or side of a vessel containing mercury, and is found not to rise again when the hand is withdrawn from it, but to be firmly held down by the weight of the mercury. We have to remark that in such experiments made in vessels open to the air, the weight of the atmosphere on the liquids adds a pressure of fifteen pounds on every inch of the surface of a body immersed in it.

"Atmospheric pressure on liquids."

The pressure of the atmosphere on liquids produces many important effects, and now that we comprehend them, we wonder that they should have been so long misunderstood. We have familiar examples of it in the working of pumps and syphons. All such phenomena, in former times, were referred to what was called *nature's horror of a vacuum*, or to an obscurely imagined *principle of suction*. It was not until the time of Galileo that their true nature began to be detected. The discovery has led to many very important results in the arts.

Persons may at first have a difficulty in conceiving that a fluid so rare and subtle as air should affect or resist a dense liquid like water: but the action or resistance of air in contact with water, is familiarly shown in the facts that a glass does not become full of water when plunged, with its open mouth downwards, from the air into water; and that when a tube, open at both ends, has been partially immersed in water, and, therefore, partially filled, the water can be forced out of it by blowing air in at the upper end, to return only when the blowing ceases. Then it may be recollected that a hundred pounds of feathers are as great a load as a hundred pounds of lead.

That there are fifteen pounds of air above every square inch of the earth's surface, is confirmed by the effects above described of the atmospheric pressure on solids; and we now proceed to show that many of the phenomena among liquids, which long appeared so mysterious, are merely the necessary consequences of the same pressure upon them. It will facilitate the comprehension of these effects, if we first view them as they may be produced by more visible agents, *viz.*, by one liquid pressing upon another; and for this purpose the author has contrived the apparatus represented in the next page, in which a layer of oil rests upon a layer of water, or upon a layer of mercury.

It has already been shown, that an ocean of oil, spread over the earth, to have the same weight as our atmosphere, requires to be about thirty-seven feet deep A vessel, then, *a b c*, with water in it up to the level W, and with thirty-seven feet of oil above this, up to the level O, is fitted to illustrate many of the phenomena of atmospheric pressure on liquids. The following are the seven principal cases.

1st. The weight of the oil pressing with a force of fifteen lbs. per inch on the water at W, would not at all disturb the level surface of the water. Neither does the weight of the atmosphere of fifteen lbs. per inch disturb any liquid surface.

Fig. 94.

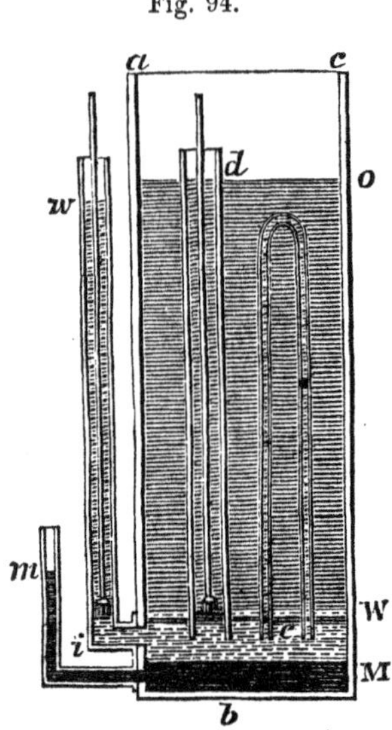

2. If the oil were gradually poured into the vessel *a b c*, over the water, the water would rise in the tube *i w*, as already explained by the figure at page 143; so that when there were thirty-seven feet in height, or fifteen pounds in weight of oil on the inch, the water *i w* would stand thirty-four feet above its level in the large vessel. If these thirty-four feet of water were then lifted out of the tube by a plug or piston drawn up from the bottom of it at *i*, a second equal quantity would be pressed up by the oil, to be removed, if desired, in the same way as the first, and the tube and piston would constitute a pump. Now when the atmosphere, instead of the oil, is allowed to press upon a water surface in such a vessel, but is excluded from the tube, the water rises in the tube thirty-four feet, as in the last case; and if this quantity be lifted out of the tube by a piston, a second equal quantity is pressed up, and the tube and piston become a complete example of the common *lifting or sucking pump.* We have to describe it more particularly hereafter.

3d. If there were a quantity of mercury or of quicksilver at the bottom of the vessel *a b c*, filling it up to the level M, and if a tube *i m* issued from under this level, the mercury pressed upon by thirty-seven feet of oil would rise in this short tube as the water did in the larger; but by reason of its greater specific gravity, it would only reach a height of thirty inches above its level, the water having stood at thirty-four feet. Now thirty inches of mercury is the height of column which the atmospheric pressure, acting in the same way really produces, as is seen in a similar apparatus made expressly for measuring that pressure, and called a *barometer* or *measure of weight.*

4th. If a tube *d*, of an inch square and open at both ends, were plunged into the oil, it would of course always be full up to the level of the oil on the outside of it; and if it were pushed low enough to touch the water at W, it would just contain fifteen pounds of oil resting on an inch square of the water-surface at its mouth; which surface would therefore be bearing a weight of fifteen pounds like every inch of the surface around, but would not yield, owing to the force with which it tended upwards to escape from the pressure corresponding to its depth in the oil. Then if the tube were pushed a little farther down, and if, by a piston or plug in it, the fifteen pounds of oil were lifted out of it, water would rise into it until enough had entered to reproduce the pressure of fifteen pounds on the surface below as before; that is to say, the water would rise thirty-four feet, as in the external tube *w i*. This internal tube and piston again would form a *pump.* In like manner, when a tube open at both ends is plunged from the air into water, the air presses on the surface of the water within the tube, as on the surface around it, with a force of fifteen pounds to the inch, and the two surfaces are not affected by the equal pressures; but if, by a piston, we lift the air out of the tube, as we suppose the oil lifted in the last experiment, the water will then rise, following the piston to the altitude of thirty-four feet. This arrangement of parts is the most usual for the *lifting* or *household pump.*

5th. If a common bottle or vessel of any shape, as the bent tube *e*, were filled with water, and placed under the oil with its mouth or mouths reaching below the water surface at the level W, it would remain full of water, owing to the pressure of the oil surrounding it.—For a similar reason, any such vessel or tube, surrounded only by air, when filled with water, and placed with its mouth or mouths under the surface of water, remains full; and if such a bent tube has one of its ends in another vessel lower than the first, a current is established in it;—the contrivance being then called a *syphon.*

6th. A fish in the water below the level W, would be bearing the pressure of the oil from O to W, as well as the pressure of the water.—So a fish in water open to the air, is bearing the atmospheric pressure of *fifteen pounds per inch*, in addition to that of the water itself. This is proved by extracting the air from over water in which a fish is swimming: for then the air-bag of the fish, situated near its under side, as already described, immediately dilates and turns the fish upon its back.

7th. To separate the Magdeburgh hemispheres, or to produce a vacuum in any way, under the water level W, would require force proportionate to the weight of oil above, in addition to that required on account of the water;—and to separate the Magdeburgh hemispheres under any water-surface pressed upon by the atmosphere, a force is required of *fifteen pounds per inch* beyond what would balance the effect of the water itself.

The following remarks illustrate more minutely some of the objects which we have just been explaining.

The common *lifting-pump* (or *sucking-pump* as it used to be called,) is then merely a barrel *a b*, with a close-fitting moveable plug or piston in it *c*. When the lower end *b*, is plunged into water, and the piston is drawn up from the bottom, the atmosphere being prevented from pressing on the surface of the water within the tube, the pressure on the surface external to the tube, drives the water up after the piston. That the water which thus rises may not fall again, there is a valve or flap at the lower part of the pump-barrel *b*, which opens only to water passing upwards; and that the piston may be allowed to pass downwards through the water in the barrel, to repeat its stroke, there is in it a similar valve. The piston, in rising during a second or succeeding stroke, causes all the water above it to run over at the spout *d*.—Formerly a lifting-pump was said to act by *sucking* the water up from the well beneath it; the true meaning of which phrase we now perceive to be, that the piston merely lifts or holds off the air which was pressing on the water within the barrel, and allows the water to rise in their obedience to the external pressure of the air around. The reason is apparent, then, why, in the lifting-pump, the water will only follow the piston to a certain elevation, *viz.*, until its weight balances the external pressure of the atmosphere.

Fig. 95.

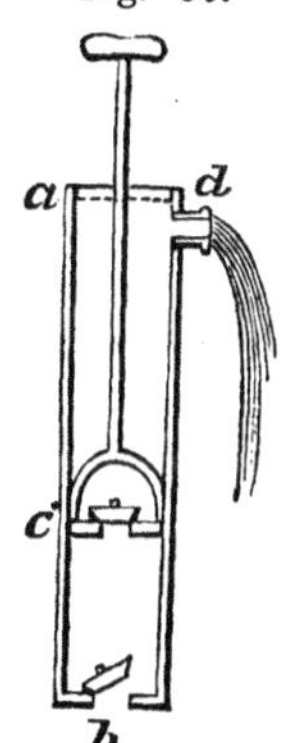

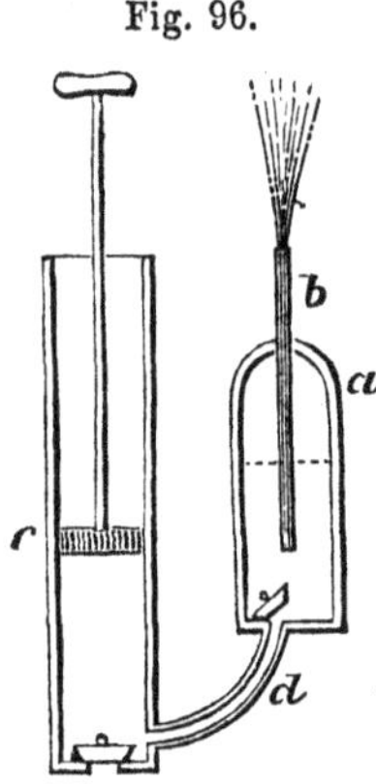

Fig. 96.

When the piston of a pump is solid, or without a valve, as at *c*, the machine is called a *forcing-pump*. The water rises beneath the piston, as already explained for the lifting-pump, but then as it cannot pass through the descending piston, as in the lifting-pump, it is forced into any other desired direction, as to *d*. A forcing-pump can bring water from only thirty-four feet below the piston, but can send it to any elevation. In forcing-pumps, it is usual to make the water enter an air-vessel *d a* (already explained at page 161,) from which it is again urged by the elastic air, through the pipe *b*, in a nearly uniform stream.

The animal action of *sucking* is an approximation to what we have described in the lifting-pump. The difference is that the chest or mouth can make only a partial vacuum, and therefore cannot raise a liquid very far.

A *syphon* remains full of liquid, although partially raised above the general surface of the liquid, as explained above. For common purposes, a syphon is made of the form here represented, *viz.*, a bent tube *c b a*, with one end longer than the other. To use it, the end *c* is first immersed in liquid, and the end *a* being then stopped for the time by the finger or a cock, the air is extracted by the mouth or otherwise, through the small tube *a d*, and the atmosphere immediately fills the whole tube with liquid from *c*. If the instrument be then left to act, the liquid will run from the longer leg, because a long column of liquid overbalances a short one, until the shorter has drunk up all within its reach. Whether the external extremity be in the air only, or immersed in liquid, makes no difference, except that the immersion shortens so much the descending column. If both extremities be immersed in liquid, and in different vessels, by alternately lifting one vessel or the other, the liquid will be made to pass and repass, and will come to rest in the syphon only when the surfaces in the two vessels are at the same level. Thus the same leg becomes alternately the long and the short leg, according to the height of the liquid in which it is immersed. A syphon is sometimes made with both legs equal and turned up, as here represented, so that it remains full of liquid although lifted away from the vessel, and therefore is always ready for action. As it is the same cause which lifts the water in a pump and in a syphon, the top of a syphon must evidently be within thirty-two feet of the water-surface below. In the syphon as the cases of balancing liquids, described at page 131 (which see,) the comparative diameters of the legs are of no importance, nor their oblique length, provided the perpendicular heights of the two columns have the necessary relation :—even an inverted tea-pot may be used as a syphon. This truth is well exemplified in what may be called the *syphon-paradox* an exact

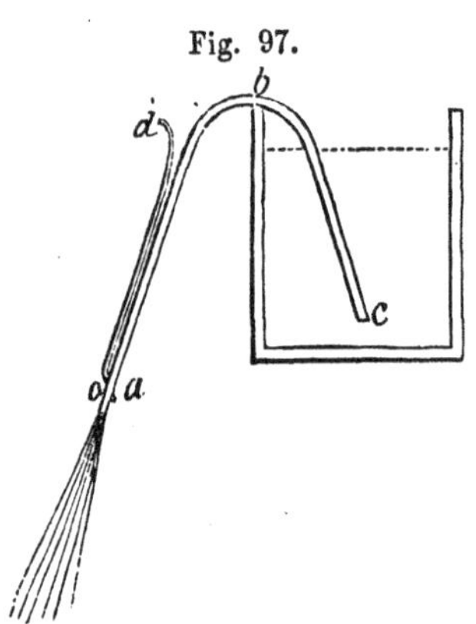

Fig. 97.

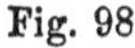

Fig. 98.

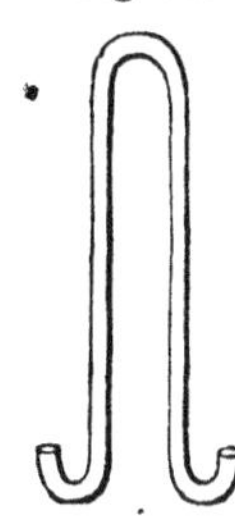

counterpart of the paradox of the "Hydrostatic Bellows," already explained. If the apparatus of the bellows be filled with water in the ordinary way (see page 130,) and be then reversed or turned so that the tube becomes like the long leg of a syphon, the little stream of water issuing from it at *a* will lift as great a weight suspended *from* the board *d*, as the same slender column in the standing position can lift *upon* the board. As farther illustrative of the atmospherical pressure exerted in producing this effect, and in rendering a syphon active, we may advert to the striking fact, that a long small tube of water screwed into the side or bottom of a close cask of water so as to communicate with it, and then allowed to discharge like the long leg of a syphon, will cause the cask to be crushed inwards, just as the same tube screwed into the top of the cask, as represented at page 131, causes the cask to burst outwards.

The syphon is very useful for drawing off liquids, where there is a sediment that should not be disturbed, or where it is desirable not to make an opening in the lower part of the vessel. A large syphon would empty a lake or mill-pond over its bank without injuring the bank. To fill a large syphon that it may act the most convenient way is, instead of pumping out the air from it, to close the two ends for the time, and to pour in water through a cock at the top.

There is a pretty syphon-toy, called a Tantalus-cup, having in it a standing human figure which conceals a syphon. The short branch of the syphhon rises in one leg of the figure to reach the level of the chin, and the long branch descends in the other leg to pierce the bottom of the cup towards a reservoir below. On pouring water into the cup, the syphon begins to act as soon as the water reaches the chin of the figure, and the cup is then emptied as if by magic.

Among the infinitely varied water-drains or courses in the bowels of the earth, some are syphons, and produce what are called intermitting wells or fountains. These may alternately run and cease for longer or shorter periods, according to the comparative magnitudes of the collecting reservoir and the drain. The reservoir may be an internal cave of a mountain, receiving a regular supply of water by a slow filtering of moisture from above, and the drain is a syphon-formed channel, which, like that of the Tantalus-cup, begins to act only when the water in the reservoir has reached the level of the top of the syphon, and then carries off the water faster than it is supplied. There are some fountains that flow constantly, but at regular intervals have a remarkable increase. In them a common spring is joined with a syphon-spring.

The author has suggested an application of the syphon, which obviates a strong objection to the high operation for stone, as explained in the next medical section.

The following facts have close relation to those now explained, as farther illustrative of atmospheric pressure on liquids.

A long glass of jelly, if inverted and placed with its mouth just under the surface of warm water, will soon be found to have lost the jelly, but to be full of water in its stead. The jelly is heavier than water, and when melted by the heat sinks down, and is replaced by water from below, sent up by the atmospheric pressure.

The slaves in the West Indies steal rum, by inserting the long neck of a bottle full of water through the top aperture of the rum-cask. The water falls out of the bottle into the cask, while the lighter rum ascends in its stead.

The common water-glass for bird-cages has its only opening near the bottom through the neck *b*; yet no water can escape from it but when the level of the water at *c*, in the open part, becomes low enough for some air to pass into the body of the glass by the channel *b*. When a bubble of air does pass in, an equal bulk of water comes out, and by raising the water level in *c*, prevents the passage of more.—An ink-glass made on this principle preserves the ink well, because there is so small a surface exposed to the air; if made too large, however, the accidental expansion of the air in it by heat may cause it to overflow.

Fig. 99

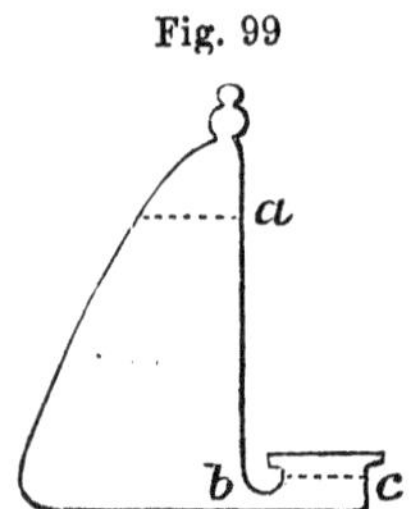

In the common *Argand* or *fountain-lamp*, the provision of oil is in a vessel like an inverted bottle, higher than the flame, and with its mouth immersed in a small reservoir of oil, nearly on a level with the flame, then no oil can escape from above, but as the flame consumes the free oil from the small reservoir, which supply is thus maintained always at the same elevation.—In the Hiero's fountain lamp, mentioned at page 164, that the two balancing columns of oil may be always of the same height, the oil is suppplied to them from high reservoirs, with the mouths dipping in them as above described, and keeping their tops, therefore, always at the same level; and that the descending column may not be shortened by the rising of the oil in the low reservoir *c*. the tube containing it is turned up at the bottom like an end of the "ever ready syphon," and discharges near the top of *c*.

We have hitherto been contemplating only the direct weight or downward pressure of the atmosphere on liquids: in the following instances we have proof of the same pressure acting upon them in all directions.

If a bottle or cask be filled with liquid, and closely corked, and if a small hole be then drilled in the bottom or side, the liquid will not escape by it, because of the existing pressure of the atmosphere, and of there not being room in the opening for a current of air to enter while the current of water escapes: but if a second hole be drilled in the top, a jet from the lower opening will follow immediately, because then the air will press on the upper surface of the liquid as well as on the lower, and the weight of the liquid will be free to act:—thus, a cask, of beer or wine cannot be emptied by a cock near the bottom, unless what is called a *vent-hole* be made at the top. If the lower opening, however, in any case be so large, that the air may enter by one side of it, as is seen in decanting a bottle of wine. In such a case, it is the interrupted entrance of the air which causes that gurgling sound so delightful to the ear of the drunkard, instead of allowing the smooth stream which fall from a funnel.

Even a large opening at the bottom of a vessel which is close above, may be prevented by the pressure of the air, from discharging liquid if any mutual passing of the two currents of air and liquid be rendered difficult. An inverted bottle of water will not discharge, if a piece of paper simply be applied against its mouth. Even a wineglass filled with water may be inverted, and yet will spill none, if the piece of paper, laid loosely upon its mouth, be held to it during the turning,—the pressure of the atmosphere against the paper keeping it in its place, and supporting the water above it. Any vessel or tube of water of less height than thirty-four feet may be kept closed at the bottom in the same way.

The animal body is made up of solids and fluids, and is affected by the atmospheric pressure accordingly.

There is a difficulty at first in believing that a man's body should be bearing a pressure of fifteen pounds on every square inch of its surface, while he remains altogether insensible of it: but such is the fact, and the reason of his not feeling the fluid pressure is its being perfectly uniform all around. If a pressure of the same kind be even many times greater, such, for instance, as fishes bear in deep water, or as a man supports in the diving-bell, it equally passes unnoticed. Fishes are at their ease in a depth of water where the pressure around will instantly break or burst inwards almost the strongest empty vessel that can be sent down; and men walk on earth without discovering a heavy atmosphere about them, which, however, instantly crushes together the sides of a square glass bottle emptied by the air-pump, or even of a thick iron boiler, left for a moment by any accident, without the counteracting internal support of steam or air.

The fluid pressure on animal bodies, thus unperceived under ordinary circumstances, may be rendered instantly sensible by a little artificial arrangement. In water, an open tube partialy immersed becomes full to the level of the water around it, and the water contained in it is supported, as already explained, by that which is immediately below its mouth: now a flat fish resting closely against the mouth of the tube, would evidently be bearing on its back the whole of this weight, perhaps one hundred pounds; but the fish would not thereby be pushed away, nor would it even feel its burden, because the upward pressure of the water immediately under it would just counterbalance the weight, while the lateral pressure around would prevent any crushing effect of the upward and downward forces. But if, while the fish continued in the situation supposed, the hundred pounds of water were suddenly lifted from off its back by a piston in the tube, the opposite upward pressure of one hundred pounds would at once crush its body into the tube. At a less depth, or with a smaller tube, the effect might not be fatal, but there would be a bulging or swelling of the substance of the fish into the mouth of the tube.—In air and on the human body a perfectly analogous case is exhibited. A man without pain or any peculiar sensation, applies his hand closely to the mouth or opening of a tube, or of any vessel containing air, but the instant that the air is withdrawn from within the tube or vessel, the then unresisted pressure of the external air fixes the hand upon the opening, causes the flesh to swell or bulge into it, and makes the blood ooze from any crack or puncture in the skin.

These last lines describe closely the surgical operation of *cupping;* the essential circumstances of which are, the application of a cup or glass, with a smooth blunt lip, to the skin of any part of the body, and the extraction by a syringe, or other means, of a portion of the air from within the cup. To some minds the exact comprehension of this phenomenon may be facilitated, by considering the case of a small bladder or bag of India-rubber full of any fluid and pressed between the hands on every part of its surface except one:—at that one part it would swell, and even burst if the pressure were strong enough. So in cupping, the whole body, except the surface under the cup, is squeezed by the atmosphere, with a force of fifteen pounds to the square inch, while in that one situation the pressure is diminished according to the degree of exhaustion in the cup, and the blood consequently accumulates there. The application of a cup with exhaustion only, constitutes the operation called *dry-cupping*. To obtain blood, the cup is removed and the tumid part is cut into by the simultaneous stroke of a number of united lancets: and the cup is then applied again as before and exhausted, so that the blood may rush forth under the diminished pressure.

The partial vacuum in the cup may be produced either by the action of a syringe, or by burning a little spirit in the cup and applying it while the momentary dilation effected by the heat has driven out the greater part of the air. The human mouth applied upon any part becomes a small cupping apparatus, and formerly, in cases of poisoned wounds, was used as such. Our present perfect cupping glasses, of stronger and more permanent operation, are not yet always used, as they might be, to assist in removing the poison after the bites of rabid or venomous animals.

The author has suggested an extension and modification of the operation of the dry cupping, which he believes will prove an important remedy in the hands of the medical practitioner. It is intended as a substitute for bleeding in certain cases where blood can ill be spared, and as a more sudden and effectual check than even bleeding itself, in certain cases of inflammatory disease. It is explained in the next medical section of this work.

The atmospheric pressure on living bodies produces an effect which is rarely thought of, although of much importance, *viz.*, keeping all the parts about the joints, firmly together, by an action similar to that exerted on the Magdeburgh hemispheres. The broad surfaces of bone forming the knee-joint, for instance, even if not held together by ligaments, could not, while the capsule surrounding the joint remained air-tight, be separated by a force of less than about a hundred pounds; but on air being admitted to the articular cavity, the bones at once fall to a certain distance apart. In the loose joint of the shoulder, this support is of great consequence. When the shoulder or other joint is dislocated, there is no empty space left, as might be supposed, but the soft parts around are pressed in to fill up the natural place of the bone. When a thigh bone is dislocated, the deep socket called thee actabulum instantly becomes like a cupping-glass, and is filled partly with fluid and partly with the soft solids. In all joints, it is the atmospheric pressure which keeps the bones in such steady contact, that they work smoothly and without noise.

The barometer we have seen at page 170, is a column of fluid supported in a tube by the pressure of the atmosphere, and therefore indicating most exactly the degree of that pressure. It is an instrument now of such importance, both in a scientific point of view and in the business of common life, that for the sake of minds which conceive such subjects with difficulty we shall add here the two following farther illustrations of its nature.

If mercury be poured into a bent tube open at both ends, it will stand at the same level in the two legs, as at a and b, and the air will be pressing on the two surfaces at a and b with equal force of 15 lbs. per square inch. If the air be then removed from one leg a, by a piston or otherwise, while it continues to press in the other leg b, the mercury will be pushed down in b, until the growing height of a column in a produces a weight so much greater than that in b, as just to counteract the pressure: now this balance takes place, in fact, when the mercury in a stands about thirty inches higher than in b; that being the height of a column of mercury weighing 15 lbs on the square inch.

If the top of the tube *a* were then closed permanently, the mercury would for ever remain elevated in it, marking most perfectly the atmospheric pressure; now this construction, only with the empty and useless part of the tube above *d* cut off or wanting, forms a common barometer. The exact altitude of the mercury in it is known by observing how much the surface near *c* is higher than that near *d*. Often, in such a barometer, a little mass of metal is placed to float on the mercurial surface at *d*, and as it rises and falls, is caused, by a thread passing from it over a wheel or pulley, to move an index like the hand of a clock connected with the wheel, and this index tells the degree of elevation. This modification is called the *wheel barometer*.

Fig. 100.

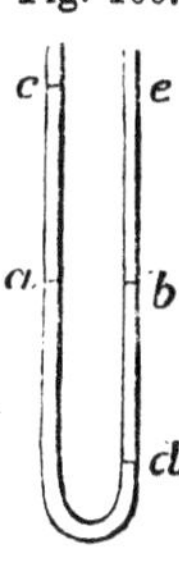

Again, as water at *a*, in the bottom of a closed pump-barrel, if pressed upon by the piston *b c*, of which the rod *d* were hollow or tubular, would rise in the rod to a height proportioned to the pressure made by the piston, so, in a straight exhausted barometer-tube, which is as this hollow piston-rod, the mercury or water rises, because the atmospheric pressure around it is as the piston forcing the fluid up. To make a barometer of this kind it is only necessary to procure a glass tube more than thirty inches long, and close at one end, and then having filled it with mercury, to plunge its mouth (stopped by the finger while turning) into a small cup or basin of mercury;—the fluid falls away a little from the top of the tube, leaving a vacuum there, and stands at the elevation which the atmospheric pressure is fitted to maintain. We know from the law of hydrostatics already explained that it is of no importance, in such a case, what the shape, or inclination, or size of the tube may be, as only the perpendicular height can measure or be measured by the pressure. This fact enables us to construct barometers with the upper part of the tube bent obliquely, so that for one inch rise of mercury in a perpendicular tube, there shall be an advance of several inches in the oblique top, rendering any change of elevation so much more apparent.

Fig. 101.

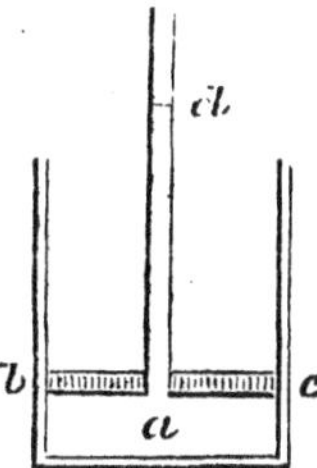

Galileo had found that water would rise under the piston of a pump to a height only of about thirty-four feet. His pupil Torricelli, conceiving the happy thought, that the weight of the atmosphere might be the cause of the ascent, concluded that mercury, which is about thirteen times heavier than water, should only rise under the same influence to a thirteenth of the elevation;—he tried and found that this was so, and the mercurial barometer was invented. Pascal then, to afford farther evidence that the weight of the atmosphere was the cause of the phenomena, carried the tube of mercury to the tops of buildings and of mountains, and found that it fell always in exact proportion to the portion of the atmosphere left below it;—and he found that water-pumps in different situations varied as to sucking power, according to the same law.

It was soon afterwards discovered, by careful observation of the mercurial barometer, that even when remaining in the same place, it did not always stand at the same elevation; in other words, that the weight of the atmosphere over any particular part of the earth was constantly fluctuating; a truth which without the barometer could never have been suspected. The observation of the instrument being carried still farther, it was found, that in serene dry weather the mercury generally stood high, and that before and during storms and rain it fell:—The instrument, therefore, might serve as a prophet of the weather, becoming a precious monitor to the husbandman or the sailor.

The reasons why the barometer falls before wind and rain will be better understood a few pages hence; but we may remark here, that when water which has been suspended in the atmosphere, and has formed a part of it, separates as rain, the weight and bulk of the mass are diminished: and that wind must occur when a sudden condensation of aeriform matter, in any situation, disturbs the equilibrium of the air; for the air around will rush towards the situation of diminished pressure.

To the husbandman the barometer is of considerable use, by aiding and correcting the prognostics of the weather which he draws from local signs familiar to him; but its great use as a weather-glass seems to be to the mariner who roams over the whole ocean, and is often under skies and climates altogether new to him. The watchful captain of the present day, trusting to this extraordinary monitor, is frequently enabled to take in sail and to make ready for the storm, where, in former times, the dreadful visitation would have fallen upon him unprepared. The marine barometer has not yet been in general use for many years, and the author of this work was one of a numerous crew who probably owed their preservation to its almost miraculous warning. It was in a southern latitude; the sun had just set with placid appearance, closing a beautiful afternoon, and the usual mirth of the evening watch was proceeding, when the captain's order came to prepare with all haste for a storm. The barometer had began to fall with appalling rapidity. As yet, the oldest sailors had not perceived even a threatening in the sky, and were surprised at the extent and hurry of the preparations: but the required measures were not completed, when a more awful hurricane burst upon them than the most experienced had ever braved. Nothing could withstand it; the sails already furled and closely bound to the yards, were riven away in tatters: even the yards and masts themselves were in great part disabled; and at one time the whole had nearly fallen by the board. Such, for a few hours, was the mingled roar of the hurricane among the rigging, of the waves around, and of the incessant peals of thunder, that no human voice could be heard, and amidst the general consternation, even the trumpet sounded in vain. In that awful night, but for the little tube of mercury which had given the warning, neither the strength of the noble ship, nor the skill and energies of the commander, could have availed anything, and not a man would have escaped to tell the tale. On the following morning the wind was again at rest, but the ship lay upon the yet heaving waves, an unsightly wreck.

The marine barometer differs from that used on shore, in having its tube contracted in one place to a very narrow bore, so as to prevent that sudden rising and falling of the mercury which every motion of the ship would else occasion.

Civilized Europe is now familiar with the barometer and its uses, and therefore, that Europeans may conceive the first feelings connected with it, they almost require to witness the astonishment or incredulity with which people of other countries still regard it. A Chinese once conversing on the subject with the author, could only imagine of the barometer, that it was a gift of miraculous nature, which the God of Christians gave them in pity, to direct them in the long and perilous voyages which they undertook to unknown seas.

A barometer is of great use to persons employed about those mines in which *hydrogen gas*, or *fire-damp*, is generated and exists in the crevices. When the atmosphere becomes unusually light, the hydrogen being relieved from a part of the pressure which ordinarily confines it to its holes and lurking-places, expands or issues forth to where it may meet the lamp of the miner, and explode to his destruction. In heavy states of the atmosphere, on the contrary, it is pressed back to its hiding-places, and the miner advances with safety.

We see from this, that any reservoir or vessel containing air would itself answer as a barometer if the only opening to it were through a long tubular neck, containing a close-sliding plug; for then, according to the weight and pressure of the external air, the density of that in the vessel would vary, and all changes would be marked by the position of the movable plug. A barometer has really been made on this principle, by using a vessel of glass, with a long slender neck, in which a globule of mercury is the movable plug.

The state of the atmosphere, as to weight, differs at different times in the same situation, so as to produce a range of about three inches in the height of the mercurial barometer; that is to say, from twenty-eight to thirty-one inches. On the occasion of the great Lisbon earthquake, however, the mercury fell so far in the barometers, even in Britain, as to disappear from that portion at the top usually left uncovered for observation. The uncovered part of a barometer is commonly of five or six inches in length, with a divided scale attached to it, on which the figures 28, 29, &c., indicate the number of inches from the surface of the mercury at the bottom to the respective divisions:—on the lower part of the scale the words *wind* and *rain* are generally written, meaning that when the mercury sinks to them, wind and rain are to be expected; and on the upper part, *dry* and *fine* appear, for a corresponding reason: but we have to recollect, that it is not the absolute height of the mercury which indicates the existing or coming weather, but the recent change in its height; a falling barometer usually telling of wind and rain; a rising one of serene and dry weather.

The barometer answers another important purpose, besides that of a *weather-glass*—in enabling us to ascertain readily the height of mountains, or of any situation to which it can be carried.

As the mercurial column in the barometer is always an exact indication of the weight or pressure of air above its level, being, indeed, as explained in the foregoing paragraphs, of the same weight as a column of the air of equal base with itself, and reaching from it to the top of the atmosphere—the mercury must fall when the instrument is carried from any lower to any higher situation, and the degree of falling must always tell exactly how much air has been left below. For instance, if thirty inches barometrical height mark the whole atmospheric pressure at the surface of the ocean, and if the instrument be found, when carried to some other situation, to stand at only twenty inches, it proves that one-third of the atmosphere exists below the level of the new situation. If our atmospheric ocean were of as uniform density all the way up as our watery oceans, a certain weight of air thus left behind in ascending would mark everywhere a change of level nearly equal, and the ascertaining any height by the barometer would become one of the most simple of calculations: the air at the surface of the earth, being between eleven and twelve thousand times lighter than its bulk of mercury, an inch rise or fall of the barometer would mark everywhere a rise or fall in the atmosphere of nearly twelve thousand inches or one thousand feet. But owing to the elasticity of air, which causes it to increase in volume as it escapes from pressure, the atmosphere is rarer in proportion as we ascend, so that to leave a given weight of it behind, the ascent must be greater, the higher the situation where the experiment is made: the rule, therefore, of one inch of mercury for a thousand feet, holds only for rough estimates near the surface of the earth. The precise calculation, however, for any case, is still very easy; and a good barometer, with a thermometer attached, and with tables, or an algebraical formula, expressing all the influencing circumstances, enables us to ascertain elevations much more easily, and in many cases more correctly, than by trigonometrical survey.

The weight of the whole atmospherical ocean surrounding the earth being equal to that of a watery ocean of thirty-four feet deep, or of a covering of mercury of thirty inches, and the air found at the surface of the earth being 828 times lighter than water, if the same density existed all the way up, the atmosphere would be 34 times 828 feet high, equal to about five miles and a half. On account of the greater rarity, however, in the superior regions, it really extends to a height of nearly fifty miles. From the known laws of aërial elasticity, explained at page 158, we can deduce what is found to hold in fact, that one-half of all the air constituting our atmosphere exists within three miles and a half from the earth's surface; that is to say, under the level of the summit of Mont Blanc. A person unaccustomed to calculation, would suppose the air to be more equally distributed through the fifty miles than this rule indicates, as he might at first also suppose a tube of two feet diameter to be only twice as capacious as a tube of one foot, although in reality it is four times as capacious.

In carrying a barometer from the level of the Thames to the top of St. Paul's Church, in London, or of Hampstead Hill, the mercury falls about half an inch, making an ascent of about five hundred feet. On Mont Blanc it falls to half of the entire barometric height, making an elevation of fifteen thousand feet; and in Du Lucts' famous balloon ascent it fell to below twelve inches, indicating an elevation of twenty-one thousand feet, the greatest to which man has ever ascended from the surface of his earthly habitation.

The extreme rarity of the air on high mountains must, of course, affect animals. A person breathing on the summit of Mont Blanc, although expanding his chest as much as usual, really takes in at each inspiration only half as much air as he does below—exhibiting a contrast to a man in a diving-bell, who, at thirty-four feet under water, is breathing air of double density, at sixty-eight feet of triple, and so on. It is known that travellers, and even their practised guides, often fall down suddenly as if struck by lightning, when approaching lofty summits, on account, chiefly, of the thinness of the air which they are breathing, and some minutes elapse before they recover. In the elevated plains of South America, the inhabitants have larger chests than the inhabitants of lower regions—furnishing another admirable instance of the animal frame adapting itself to the circumstances in which it is placed. It appears from all this, that although our atmosphere be fifty miles high, it is so thin beyond three miles and a half, that mountain ridges of greater elevation are nearly as effectual barriers between nations of men, as islands or rocky ridges in a sea are between the finny tribes inhabiting the opposite coasts. The intense cold which appertains to high situations, and forms another obstacle to human approach, remains to be considered in our next division.

A barometer connected with an air-pump, indicates exactly the progress and degree of exhaustion in the receiver. When the mercury falls to half its height, it shows that half of the air is extracted; and so for all other proportions. A barometer then is a necessary appendage to a complete air-pump; but as its chief purpose is to mark when the exhaustion is carried nearly to completion, a very short tube, corresponding to the bottom of a common barometer, is all that is generally provided, and it is usually made of syphon form.

The ingenious method, mentioned at page 145 of ascertaining the specific gravity of the solid material forming any porous mass or powder, includes the agency of a barometer. It proceeds upon this reasoning. The interstices of a porous or pulverized mass are filled with air of the density of the surrounding atmosphere, and if the atmospheric pressure on which that density depends be diminished upon the mass in any given degree, an exactly corresponding proportion of the air will issue from the pores, and if measured will declare the whole quantity, and, therefore, the amount of interstices or pores in the solid mass. Now if the substance were inclosed at the end or bottom of a syringe, the pressure of the atmosphere might be held off from it in any degree by drawing at the piston, and the air would issue from the pores as described, and would follow the piston; but as, owing to the friction or a solid piston, it would be difficult to measure the precise action, the liquid piston of a mercurial column has been substituted, of which the force is always proportioned to the length. The operator takes an open glass tube, *a e*, of known dimensions, and prepare a part of its top, *a b*, as a receptacle for the substance under trial, by affixing a partition at *b*, which shall support the substance, but allow passage to air. Having then filled *a b* with the substance, he gradually immerses the tube in a vessel of mercury *d f*, until the mercury stand both inside and outside of the tube at the level of *b*, the air from the tube having passed out through the substance in *a b*. It is evident that on then closing the tube at *a* in the air-tight manner, and lifting the tube, a column of mercury will remain standing in it, above the level of the external mercury at *d*, and will be acting as a piston pulling down from *b* with force proportioned to its height. If the tube be lifted until such mercurial column *c d* be just of half the length in a column in a common barometer, the air in the pores of the substance will be relieved from half of the atmospheric pressure, and will dilate to double bulk: so that while half of the air will remain in the pores, the other half will have issued forth to occupy a space, as *b c*, between the surface of the mercury at the partition at *b*. This space *b c*, therefore will be exactly equal to the amount of the pores or interstices; and as it may be measured and compared with the whole space *a b*, its ascertained magnitude will solve the problem. It has been found in this way that charcoal, which is usually said to be only half as heavy as its bulk of water, is really formed of matter nearly four times as heavy; proving, in a new way, the identity of charcoal and diamond, and that light pumice-stone consists of matter heavier than granite or marble. This very ingenious application of the barometer may lead ultimately to many useful results: and the contrivance merits consideration here, as exhibiting, under a new and interesting aspect, the rationale of barometric action and the elasticity of air.

Fig. 102.

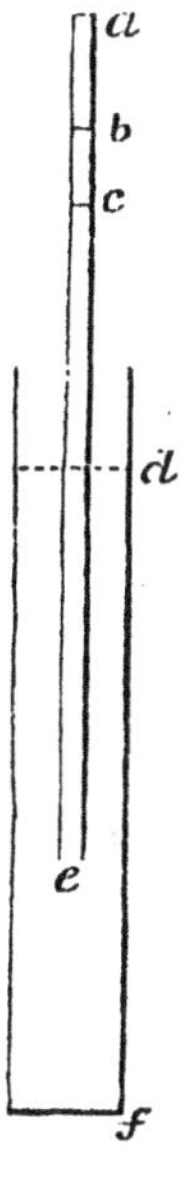

Atmospheric pressure determining the liquid or aeriform state of certain substances. (See the Analysis, page 156.)

It has already been stated that the permanent gases—or substances usually in the aeriform state—may be reduced to the liquid, or even solid form, by simple pressure, and abstraction of the heat which is combined with them in the aeriform state. Carbonic acid, the common coal gas, &c., have been treated in this way. Now it becomes an interesting question whether many of the substances known as liquids on the face of the earth, where they are bearing the pressure of the atmosphere, would not appear as airs if that pressure did not exist.

On investigating this subject by experiment, we accordingly find, that *æther, alcohol* or *ardent spirits, volatile oils, &c.*, and even *water* itself, are known to us here as liquids, only because their particles are kept together by the weight and pressure of a superincumbent atmosphere. Any of these substances, relieved by art from such pressure, quickly becomes an air or gas, just as a common gas, which has been kept in the state of liquid by any great pressure, becomes air again on being relieved.

In our first chapter we explained the dependence of the three forms which any body may assume, *viz.*, of solid, liquid, or air, on the quantity of heat diffused among the particles; we now see, however, that to understand the subject completely, we must consider also the effect of accidental pressure; for, while heat is the power separating the atoms in the changes mentioned, it has to overcome both the mutual attraction of the atoms and the additional force of the atmosphere pressing them together. The combined influence of these forces is fully displayed in the two phenomena called *boiling* and *evaporation*, which exhibit the progress or the change of a liquid into an aeriform fluid. We now proceed to examine these phenomena.

Boiling.—If water be placed in a suitable vessel over a common fire, or over the flame of a lamp, it is gradually heated to a certain degree; and then small bubbles of aeriform matter, *viz.*, water, in the state called steam, are seen forming at the bottom of the vessel; and successively rising to the surface, where they disappear by mixing with the atmosphere; and the operation being continued, the quantity of water diminishes with every bubble, until the whole vanishes under the new form of air.

This change takes place in water, under common circumstances at the degree of heat mark 212° on Fahrenheit's thermometer, and called, on that account, the *boiling point* of water; at which therefore, the repulsive power among the particles is just sufficient to overcome both their natural attraction, and the compressing force of the atmosphere of fifteen pounds on the inch. But a less degree of heat suffices if the pressure of the atmosphere be lessened or removed: and a greater degree is required if pressure be increased. Water on the top of Mount Blanc boils at 180°, because relieved from the pressure of the air that is below the level of the mountain's summit; and at all intermediate heights in descending to the level of the sea, and beyond that into mines, there is a corresponding increase of the boiling temperature. So exactly is this the case, that we now find it to be a good method of ascertaining the heights of places, merely to observe the heat of boiling water at them. To many persons the information here given that boiling water is not equally hot in all places, will appear extraordinary; and they will not understand that even in the same place, at different times, when the barometer is high or low, there will be corresponding differences. Again, near the bottom of a boiler, the water is hotter than above, because it is bearing an additional pressure proportioned to the depth and does not, therefore, give out the steam which it would part with if a little higher up. In very large and deep boilers, therefore, such as are used in great porter breweries, the liquor is much more heated that it can be in smaller vessels; a circumstance which probably has an influence on its ultimate quality.

While water, under common atmospheric pressure, or when the barometer stands at thirty inches, boils at 212°, other substances, with other relations to heat, have their *boiling points* higher or lower: æther, for instance, at 98°; spirit or alcohol at 174°; fish-oil and tallow at about 600°; mercury at 650°.

It is in consequence of the different temperatures at which the paticles of different substances requires repulsion enough to rise against the atmospheric resistance that we are enabled to perform the operation called *distilling*. If a mixture of spirits and water, for instance, be heated up to 180 deg., the spirit will pass off in the aeriform state, leaving the water behind, and may be caught apart and cooled to condensation in any fit receiver. Distillation is the best means we possess of separating many substances from each other, as spirit from wine and other fermental liquor; various acids from water; water itself from its common impurities;—and even the separation of mercury from silver or gold which it has been used to dissolve from among the rubbish of a mine or river-bottom, is merely a distillation.

We must call to mind here what was mentioned in a former part of the work, that a large quantity of heat combines with every substance during the change of form from solid to liquid, or from liquid to air; a quantity which, from not remaining sensible to the thermometor, has received the name of *latent or concealed heat.* The whole of this is given out again in the contrary change. In the conversion of water into steam, the heat which thus disappears is about 1,000 degrees, or six times as much as is required to raise the cold water to the boiling point; this is proved by the time and fuel expended in boiling any quantity to dryness, and by the fact that a pint of water in the form of steam will combine instantly with six pints of cold water, raising the whole to boiling heat.

But for the fact of latent heat, the conversion of a liquid into air would not be the gradual process of boiling which we now see, but a sudden and terrible explosion: for when any quantity of water were raised to the boiling heat one degree more would be sufficient to convert the whole into steam. And but for the same reason, the thawing of winter snow would always be a sudden and frightful inundation; the whole load of a mountain or plain becoming at once as a lake bursting from its enclosing barriers. On the other hand, if water in freezing had not to give out again its latent heat, after any quantity were at once cooled down to the freezing point, the abstraction of one degree more would instantly convert the whole into a solid mass. Thus, then, by an arrangement effecting most important purposes in nature and art, all changes from solid to liquid and from liquid to air, and the converse changes, are very gradual.

If a little heat be abstracted from steam, a part of the steam proportioned to the abstraction is immediately condensed into water. What is called steam, in common language—as the vapour which becomes visible at a little distance from the spout of a boiling kettle or the chimney of a tea-urn—is not truly *steam*, but small globules of water already condensed by the cold air and mixed with it. Steam is as dry and invisible as air itself; but the instant that it comes in contact with air or other bodies colder than itself, it becomes water.

By means of the exhausting air-pump on one hand, and of the condensing syringe on the other, all the above-mentioned phenomena, depending on the atmospheric pressure, and its increase or diminution, may be strikingly shown.

Thus, to exhibit the effect of diminished pressure, water not heated by several degrees to the boiling point of ordinary low situations, but which would be boiling at the top of Mont Blance, is caused to boil instantly by placing it under the receiver of an air-pump, and making a few strokes of the piston; if the exhaustion be rendered nearly complete, the water will boil, even when colder by 20 degrees than the blood of animals; and at degrees of temperature still much lower, it will rapidly assume the form of air, although not with force sufficient to produce the violent agitation of boiling. Other liquids, as spirits, æther, &c., from requiring inferior degrees of heat to separate their particles to aeriform distances, boil under the receiver of an air-pump at very low temperatures; æther, for instance, when as cold as freezing water.

On the other hand, to exhibit the effect of increasing pressure, if we confine the particles of a liquid still more than by a common atmospheric or equivalent pressure, degrees of heat higher than the common boiling point will be required to separate them. In a diving-bell, the boiling point of water is higher than 212 deg. in proportion to the depth which the bell has reached: and if at the surface of the earth, we heat water in a close vessel into which air is forced so as to press thirty pounds on the inch instead of fifteen, as the atmosphere does; or from which we prevent the steam's escaping until it has acquired the force of a double atmosphere,—before making the liquid boil, we shall have to raise the heat, in a corresponding proportion beyond 212 deg. Under a very strong pressure, water may be rendered almost red-hot, but the force with which its particles are then tending to separate is almost that of inflamed gunpowder. Even then, however, if a gradual issue were allowed, only a certain quantity of the water would absorb and render latent the existing excess of heat above 212 deg., and would become common steam, leaving behind a considerable portion as boiling water of the ordinary temperature.

The fact that liquids are driven off, or made to boil at lower degrees of heat when the atmospheric pressure is lessened or removed, has recently been applied to some very useful purposes.

The process of refining sugar is to dissolve impure sugar in water, and after clarifying the solution, to boil off or evaporate the water again, that the dry crystallized mass may remain.

Formerly this evaporation was performed under the atmospheric pressure, and a heat of 218° or 220° was required to make the syrup boil; by which degree of heat, however, a portion of the sugar was discoloured and spoiled, and the whole product was deteriorated. The valuable thought occurred to Mr. Howard, that the water might be dissipated in boiling the syrup in a vacuum, or at least a place from which air was nearly excluded, and therefore, at a low temperature. This was done accordingly; and the saving of sugar and the improvement of quality were such, as to make the patent-right, which secured the emoluments of the process to him and other parties, worth many thousand pounds a year. The syrup, during this process, is not more heated than if in a vessel merely exposed to a summer sun.

In the preparation of many medicinal substances, the process of boiling in *vacuo* is equally important. Many extracts from vegetables, have their virtues impaired, or even destroyed, by a heat of 212°; but when the water used in making the extract is driven off in *vacuo*, the temperature need never be higher than blood-heat, and all the activity of the fresh plant remains in the extract.

In the same manner, in the process of distillation,—which is merely the receiving and condensing again in appropriate vessels the aeriform matter raised by heat from any mass,—substances which are changed and injured by an elevated temperature may be obtained of admirable quality by carrying on the operation in a vacuum. The essential oils of lavender, peppermint, &c., never had the natural flavour and virtues of the plants until within the last few years, since this plan has been adopted.

The influence on the human system of vegetable medicines obtained in the old or in the new way, is so different, that the prescriber should carefully advert to the circumstance.

The apparatus for evaporating and distilling *in vacuo* consists of vessels strong enough to bear, when quite empty, the external atmospheric pressure and which are therefore generally of arched form. The vacuum is produced and maintained by air-pumps driven by a steam engine or otherwise; or by first admitting steam to expel the air, and then condensing the steam into water.

The author has suggested a very simple contrivance to answer, in certain cases, the purpose of such air-pumps and steam-engines or apparatus. It is merely to establish a communication between a close boiler, as *a*, and the vacuum at the top of a water barometer, as *b*, To produce that vacuum, the strong vessel *b* forming the top of the barometer, and thirty-six feet of tube below, reaching to *d* are first filled with water through a cock *c* at the top; this cock being then shut, and another cock *d* at the bottom, which was shut, being opened, the water will sink down out of the vessel *b*, until the column in the tube be only thirty-four feet high as at *f*, that being the height which the atmosphere will support. On then opening a communication between the boiler *a* and the vacuum in *b*, the operation will go on as desired, and the steam rising from *a* may be condensed in *b* by a little stream of cold water allowed constantly to run through from above. This water, it is evident, would always pass downwards to form part of the column below, without filling up or impairing the vacuum. If air should find admittance in any way, the original degree of vacuum could easily be reproduced as at first; and to prevent interruptions, it might be convenient to have two vessels like *b*, of which one could always be in action while the other was being emptied of air. The author planned this as a simple apparatus for the preparation of medical extracts; and it appears well suited also for the manufacture of sugar in the colonies, where air-pumps and nice machinery can with difficulty be either obtained or managed. On many sugar estates there is a fall of water, which would supply the barometer without the trouble of pumping. The tube *d c* need not be perpendicular, provided it be longer in proportion to its obliquity; and it may be very small; some yards of common lead-pipe would answer.

Fig. 103.

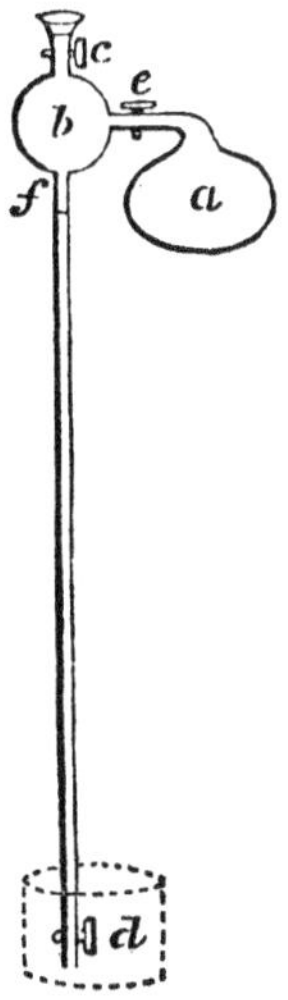

c
e
b
f
a
d

When it was understood that, at common temperatures, water and many other liquids would be existing in the form of air, but for an atmospheric pressure opposing the separation of the particles, it became of great importance in many of the arts, and for comprehending certain phenomena of nature, to ascertain, very exactly, with respect to some of these liquids, the degrees of expansive force belonging to them at different degrees of temperature. The subject, as far as water is concerned, has been investigated with great care, and the following table shows part of the results. The left-hand column marks temperature from 32 deg. of Fahrenheit's thermometer, or the freezing point of water, to 290 deg.; and the right-hand column marks the corresponding degrees of force with which the water tends to expand into the state of steam, and therefore also the force and density of the steam existing in any vessel above the water which it contains. One ounce and a half per square inch, is the force exerted on the sides of any containing vessel by steam rising from freezing water, that is to say, the force with which freezing water seeks to dilate into steam or air; and sixty pounds per inch is the force of water heated to 290 deg. To many readers the idea will be quite new and surprising, that if some freezing water, or even ice, be placed in a bladder containing nothing else, and the bladder be then placed in the exhausted receiver of an air-pump or other vacuum, the bladder will quickly be distended with steam strong enough to support one ounce and a half on every square inch of its upper surface.

At 32° force of steam is	1½	oz. per inch.
50 - - - - -	2¾	oz.
100 - - - - - -	13	
150 - - - - - -	4	lbs.
180 - - - - - -	7½	lbs.
212 - - - - - -	15	lbs.
250 - - - - - -	30	lbs.
275 - - - - - -	45	lbs.
290 - - - - - -	60	lbs.

In this table we have to remark how much more rapidly the tendency to become steam increases than the temperature of the water: for a rise of eighteen degrees, *viz.*, from 32° to 50°, at the beginning of the scale, only increases the dilating force *one ounce and a quarter* on the inch, while an equal rise at the top of the scale, *viz.*, from 272 deg. to 290 deg., increases it *fifteen pounds.* It is most important to distinguish, however, between the *tendency to form steam* at any temperature, and the *bulk or quantity of steam* formed by a given quantity of heat; for the matter imperfectly understood has led to many vain schemes for improving the *steam-engine.* The truth is, that *high-pressure steam* is merely *condensed steam,* as *high-pressure air* is *condensed air;* in other words, the density of steam is greater, or there must be more of it, exactly as its force is greater according to the rule explained at page 160: and the heat absorbed in its formation being proportioned to the quantity of steam in a given space, or the density, the force and the cost in fuel have always nearly the same relation to each other. In one pint of steam, at 290 deg., having an elastic force of sixty pounds on the inch, there are very nearly four times as much water and four times as much latent heat as in one pint of steam at 212 deg., which has a force of fifteen pounds on the inch; indeed, the one pint, at 290 deg., may be changed into four pints at 212 deg., or the contrary, by merely changing the degrees of pressure. It does not accord with the plan of the present work to enter farther into the details of this subject, but they may be found in various modern treatises.

Seeing the rapid increase of the expansive force in the preceding table, we have the explanation of the terrible effects occasionally produced by confined water when overheated. A boiler of any kind completely closed, and having no safety valve, if heated to a certain degree, will explode as if charged with gunpowder. Unhappily the instances are too numerous where the incautious or ignorant use of steam has produced explosions, which have shattered buildings and destroyed whole neighborhoods.

The prodigious force generated by heating water would at first only surprise and terrify men, but in the course of time would lead inventive minds to enquire whether it might not be turned to use; in other words, whether some mechanism, to be called a *steam-engine*, might not be contrived to enable men to make it aid them in their various labours. To this inquiry, after numerous less successful attempts, a glorious answer has been given in our own day by the illustrious WATT;—and to this part of our work it belongs to consider what he has accomplished, *viz.*, to describe

The Steam-Engine,

which, in the few years since the genius of WATT carried it to its present state of perfection, has changed the direction of human industry, and may almost be said to have elevated man in the scale of existence.

Fig. 104.

The name of *steam engine*, to most persons, brings the idea of a machine of the most complex nature, and hence to be understood only by those who will devote much time to the study of it; but he who can understand a common pump, may understand a steam-engine. It is in fact only a pump in which the fluid passing through it is made to impel the piston instead of being impelled by it, that is to say, in which the fluid acts as the *power* instead of being the *resistance.* It may be described simply as a strong barrel or cylinder *c d*, with a closely fitting piston in it, here shown at *b*, which is driven up and down by steam admitted alternately above and below it from a suitable boiler; while to the end of the piston rod *a*, at which the whole force may be considered as concentrated, there is attached in any convenient way the work which is to be performed. The power of the engine is of course proportioned to the size or area of the piston, on which the steam acts with a force, according to its density, of from 15 to 100 or more pounds to each square inch. In some of the Cornish mines there are cylinders and pistons of more than ninety inches in diameter, on which the pressure of the steam equals the efforts of six hundred horses.

In one place this wonderful piston-rod may be seen acting upon the end of a great vibrating beam, to the other end of which capacious water-pumps are attached, whose motion causes almost a river to gush up from the bowels of the earth. In another place, it is seen working a crank, and urging complicated machinery. One steam-engine four miles from London is at the same instant filling all the water reservoirs, and baths, and fountains of the finest quarter of the town. One engine stretching long arms over a great barrack or manufactory, keeps in one quarter, thousands of spinning-wheels in motion, while in another it is carding the material of the thread, and in another weaving the cloth. In like manner, one steam engine, in a great metropolitan brewery, may be seen at the same time grinding the malt, pulling up supplies of all kinds from wagons around the building, pumping cold water into some of the coppers, sending the boiling wort from others up to lofty cooling-pans, over which it is turning the fans, perhaps also working the mash-tub, drawing water from the deep wells under ground, and loading the drays—in a word, performing the offices of a hundred hands. Again, there are manufactories where this resistless power is seen with its mechanic claws seizing masses of iron, and in a few minutes delivering them out again pressed into thin sheets, or cut into bars and ribbons, as if the iron had become to it like soft clay in the hands of the potter. And now for some years, over nearly the whole world, has this wonderful piston-rod, working at its crank, been turning the paddle-wheels of innumerable steam-boats, thereby sitting at defiance the violence of the winds and waves, and the currents of the fleetest rivers, while it carries men and civilization into the remote recesses of all the great continents. To wherever a river leads, the region, although concealed perhaps since the beginning of the world, is now by the steam-engine called as it were from its solitude, to form a part of the great garden which civilized man is beautifying.—Such are a few of the prodigies which this machine is already performing, and every day is witnessing new applications of its utility.

The following account of the parts of the steam-engine is intended, without entering into minute practical details, still fully to explain the principle or general nature of the machine. It should serve to render very interesting to an attentive reader, a visit to any place where a steam-engine is in use: and it should make evident the folly of many of the modern schemes for improving the engine. To avoid complexity in the figure, the parts which the reader can easily conceive are not here sketched.

Fig. 105.

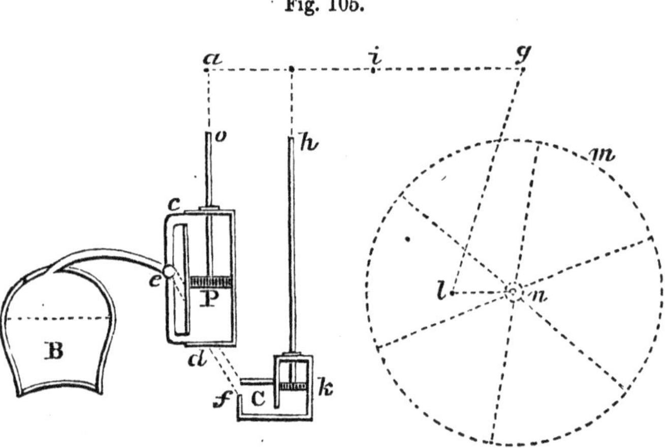

1st. The part which first claims attention is the great *barrel c d*, already spoken of as the centre or main portion of the machine, in which the *piston* P is moved up and down by the action of steam entering alternately above and below it, through the pipes *e c* and *e d*. The barrel or cylinder is bored with extreme accuracy, and the piston is padded round its edge with hemp or other soft material so as to be perfectly air or steam-tight. Lately pistons have been made altogether of metal, and, in some cases, from working with less friction, these answer even better than the others.—2d. The next part to be mentioned is the *boiler* B, which is made of suitable size and strength.—3d. The steam passes from the boiler along the pipe to *e*, and there by any suitable *cock* or *valves*, worked by the engine itself, is directed alternately to the upper and under part of the barrel; and while it is entering to press on one side of the piston, the waste steam is allowed to escape from the other side, either to the atmosphere, for high-pressure engines, or into—4th, the *condenser* at C, for those of low-pressure; the condenser being always kept at a low temperature by cold water running into it and pumped out again by the piston *k*.—5th. The *supply of steam* from the boiler to the cylinder is regulated by a *valve* placed somewhere in the pipe B *e*, and made obedient to what is called—6th, the *governor*, a contrivance not represented here, but already described at page 52, to illustrate centrifugal force. We may recall it by saying, that it consists of two balls hanging by jointed rods like the legs of a tongs, from opposite sides of an upright spindle, which is made to revolve by connection with some turning part of the machinery:—when the spindle turns at all faster than with the desired speed, the balls fly more apart and are made to affect the steam valve so as to narrow the passage; and, on the contrary, when it turns more slowly than is desired, they collapse, and by so doing open the valve wider.—7th. The *supply of water* to the boiler is regulated by a *float* on the surface of the water in the boiler; which float, on descending to a certain point, by reason of the consumption of water, opens the valve to admit more.—8th. There is a *safety valve* in the boiler, *viz.*, a well-fitted flap or stopper, held against an opening by a weight, but loaded so as to open before danger can arise from the over-heating of the water.—9th. The *rapidity of the combustion*, or force of the fire, is exactly regulated by the state of the boiler and the wants of the machine, thus:—there is a large open tube (not represented here) rising from near the bottom of the boiler, through its top, to the height of several feet, and when the water in the boiler is too hot, and the steam therefore too strong, part of the water is pressed up into this tube, and by the agency of a float which rests on its surface, it shuts the chimney valve or *damper:* the draught is then diminished and the fuel saved, until a brisker fire is again required.—10th. In this figure *a i g* marks the place of the *great beam*, turning on an axis at *i*, and transmitting the force of the piston to the remote machinery. When the object is to raise water, the pump rods are simply connected with the end *g* of the beam; but when any rotatory motion is wanted, the end *g* is made to turn.—11th. A *crank l n* by the rod *g l;* and uniformity of motion is obtained by the influence of—12th, the great *fly wheel m*, fixed to the axis of the crank.

The smallest and simplest steam engine, and therefore the cheapest, is that called the *high-pressure engine.* In it steam is used of great density, and consequently of great force, as of 50 lbs. or more to the inch; and while the fresh steam is admitted to press on one side of the piston, the steam which has already worked, is allowed to escape, or is driven out to the air, from the other side. The atmospheric resistance to the issue of the steam diminishes the working force of the piston just 15 lbs. per inch. The simplicity of this form of engine recommends it, but the danger of a large boiler of overheated water, always, like inflamed gunpowder, seeking to escape, has, by numberless fatal accidents, been proved to be so great, that the use of such an engine is limited to certain situations. Notwithstanding all the ingenious securities recently contrived against the danger, and which will suffice for small engines, such as are required for steam carriages, the high-pressure engine is not employed in a single English passage-vessel.*

In the low-pressure engine, the steam is used generally of force not exceeding 20 lbs. on the inch, which force is only 5 lbs. more than the atmospheric pressure, and is insufficient to burst a common boiler, or to do serious mischief:† but as the interior of the low-pressure engine is kept in a state of

* In this country, also, what is *called* the low *pressure* engine, is at present employed in all the steamboats on our eastern waters, but we have reason to believe that few, if any, of them are worked with a less pressure than 25 lbs. on the square inch, and we have seen many of them worked with a much higher pressure. It is stated, that the steam guage on board the Pulaski, just before the fatal explosion, indicated a pressure of 26 lbs., which was considered by the engineer a safe working force. The British Ocean Steamers work with a pressure of only 3½ to 4½ lbs. to the square inch. On our western waters, the low pressure engine has been discarded in favour of the high pressure. AM. ED.

† Our author must be understood as saying, only, that explosions do not take place in low pressure engines whilst working under a pressure of only 20 lbs. to the square inch, But from various circumstances, the elastic force of the steam in a low pressure engine may be greatly increased, and instead of its ordinary power of 20 lbs., it may acquire one of 100 or 200 lbs., and if the boiler, as is often the case, is unable to bear this pressure, it will burst. In this way explosions have repeatedly taken place in low pressure engines. Several of these will be found related in an interesting memoir by M. Arago, originally published in the "*Annuaire du Bureau des Longitudes*," and which has been translated and published in that useful periodical, "*The Journal of the Franklin Institute of Pennsylvania.*" M. Arago, in this memoir, remarks, "I ought not to conclude so long a paper on the subject of the explosion of steam boilers, without explaining why I have not separated the examples of the explosion of high pressure boilers from those of the low pressure; it is because I think there is no reason to make such distinction. Every one must, in fact, admit, that at the time of an explosion, all boilers contain high pressure steam."

The belief that low pressure boilers are not liable to burst, or do mischief, has led, as has been already observed, to their exclusive use in passage vessels; this belief, however, is founded in error. M. Arago, in the memoir already quoted, observes, "it does not appear established by any means, that explosions take place more frequently in high than in low pressure boilers; the contrary has been maintained by different engineers, among whom may be classed Messrs. Perkins, Oliver Evans, &c." Indeed, a little reflection will show that high pressure boilers ought not to be more liable to explosion than the low. Boilers may be made of either iron or copper of sufficient strength to resist a much greater force than that of the steam ever employed in high pressure engines. Now boilers are always constructed of a strength proportionate to the pressure they are to sustain. Thus, in a low pressure engine working with a force of 20 lbs., the boiler is made of strength calculated to support from 3 to 5 times that pressure; in a high pressure engine destined to work with a pressure of 150 lbs., the boiler is constructed so as to resist from 3 to 5 times that pressure. It will immediately be asked, why cannot the boiler of a low pressure engine be made of the same strength as if they were for a high pressure engine? In answer to this it may be remarked, that independent of many difficulties to be overcome before we can exceed certain limits in the thickness of boilers, the weight and cost of the low presure engine are already so great that it would be impossible to persuade owners of steamboats to incur any addition to these particulars—and even would they do so, perfect safety would still not thus be obtained. We have already observed, that at the time of an explosion, all boilers contain high pressure steam, and as we know no limits to the force of this steam, however strong the boiler may be, it may burst, unless this be prevented by other means.

It was long ago known, that if a vessel, however strong it might be, containing water, be placed over a fire, it will burst, unless an opening is provided for the escape of the steam as fast as produced. The temperature which will cause the rending of a vessel, must depend upon its form and dimensions, and upon the tenacity and thickness of the material of which it is made. If we could keep the heat of our furnace below a certain limit, no other precaution would be required to prevent explosions. But it is evident that this cannot be done, we must, therefore, resort to some other expedient, and the safety valve, invented by Papin, would seem to answer this purpose. We must be allowed to anticipate the subject a little, in order to explain the nature of this valve.

The safety valve consists of a hole, say of an inch square, made in the upper part of the boiler, upon which is placed a metal plate loaded with a certain weight. It is evident that the hole will remain closed as long as the pressure of the steam within the boiler is less that the weight of the valve, together with that of the atmosphere, upon the square inch, and that as soon as the pressure within shall exceed this, the valve will be raised and give a free vent to the steam.

It would lead us too far to explain how it has happened that so simple and apparently vacuum, except where the steam is acting, the whole pressure of 20 lbs. is made available, and the engine has the same power, if of equal size, as a high pressure engine working with steam of 35 lbs. on the inch The required vacuum is preserved by means of a separate vessel or box, represented at C. called the condenser, into which cold water is constantly running to condense the steam, and is afterwards pumped out with the condensed steam, and with any little air that may have entered : the pump is represented at *k* in the figure. Steam, on coming into contact with a cold body, is condensed almost with the rapidity of an explosion ; and therefore the instant that the opened valves make a communication between the cold condenser and any part of the engine containing steam, this rushes to the condenser, and becomes water, leaving a vacuum behind. The great merit of Mr. Watt was in the contrivance of this separate condenser, for, until his time, cold water had always been thrown directly into the working cylinder, cooling it so much, that twice or thrice its fill of steam was destroyed at each stroke to warm it again before it could work. This single change saved three-fourths of the quantity of fuel formerly expended.

Before Watt's day, the only steam engine in use was a rude *single-stroke engine* as it was called, in which steam, admitted under the piston, allowed the weight of the pump-rods at the far end of the beam to lift the piston, and the steam being then condensed so as to leave a vacuum in the cylinder, the pressure of the atmosphere pushed the piston down to do its work : on this last account the engine was also called an *atmospheric engine*. It was used almost solely for pumping water ; but it wasted so much fuel, from causes of which the chief is mentioned in the last paragraph, that the expense was not much less than that of employing horses.

In the atmospheric engine, the steam which lifted the piston against the atmospheric pressure, required to be at least as strong as that pressure, to the very end of the stroke. Another of Watt's great improvements was, his

efficient means has not always proved efficacious; since these causes are various and many of them as yet not perfectly understood.

Those who wish to investigate the subject, we refer to the memoir of M. Arago, and to the report of a committee of the Franklin Institute, who have collected an account of all the explosions in this country, and who have instituted a very interesting series of experiments, in order to examine into the causes of the explosion of steam boilers, and devise means for its prevention. This report has been published in the Journal of the Institute.

We must not omit, however, to mention that when a low pressure boiler does explode, it has been found to produce greater destruction than a high pressure one, in consequence of the great size, and, therefore, larger quantity of water contained in the former. It may, perhaps, be supposed that the steam from a high pressure engine would scald more severely than that from a low pressure one. This, however, is not the fact: on the contrary, whilst the steam issuing from a low pressure engine scalds at all moderate distances from the boiler, that from a high pressure one scalds only at certain distances. Thus the hand may be placed an inch from an aperture in a high pressure engine without any inconvenience being felt; at a greater distance, however, it will scald most severely. A friend has informed us that he has placed his hand within an inch of the aperture in a boiler from which the steam was issuing at a time when the force of the steam within the boiler was equal to 300 lbs. on the square inch, without feeling any inconvenience. Some interesting experiments on this subject have been instituted by Peter Ewort, Esq., and an account of which will be found in the fifth volume of the *Journal of the Franklin Institute.*

It must not be supposed from any thing that we have said in this note, that explosions of steam boilers cannot be prevented. But we may be allowed to quote on this subject the following remarks of M. Arago. "No cause of explosion exists which cannot be avoided, by means at once simple and within the reach of every one. As we should not trust fire-arms in the hands of children, so, I think, we should not trust the direction of a steam engine to a man either unskilful, without experience, or wanting in intelligence. It is a mistaken idea, that because steam engines usually move without attention to them, such attention is not required; Watt contended strongly against this error." AM. ED.

excluding altogether the air from his machine, by doing which he not only avoided the cooling effect of the air, but was at liberty to shut off the steam, as it is expressed, or to stop the supply for each stroke before the cylinder was full, and then to make the farther expansion of the quantity admitted impel the piston to the end of the stroke. This principle of causing the mere expansion of steam to do work was afterwards carried to a great extent by Messrs. Hornblower, Woolfe, and others, who constructed engines with two barrels, in the first and smaller of which, the steam was made to act in its dense or strong state, as it issued from the boiler, and when it had finished a stroke there, instead of being at once sent useless to the condenser, it was admitted to a larger piston, which it moved by its continued expansion alone :—the same steam thus doing double work or more. All the advantages of the two cylinders, however, are obtainable from the single cylinder with its condenser, as now used in most of the Cornish mines. Steam of about 60 lbs. pressure on the inch is admitted to the cylinder, until the piston is driven nearly one-third of its way, and the valve being then shut, the same steam is left to finish the stroke by its expansion. The pressure of the expanding steam gradually diminishes, it is true, in proportion as the volume increases : but in pumping water there is a great saving of time, from having the power more intense at the beginning of the stroke, when the vast mass of water and machinery has first to be put into motion. Steam, while doubling its volume by mere expansion, will do about *two-thirds* as much work as while originally rising from the boiler, and by every subsequent doubling it might do as much as by the first: the increasing size of the cylinder, however, and increased friction, confine this mode of using it to narrow limits.

It might be supposed that high pressure engines without condensers would be comparatively wasteful, because in them the steam which has acted must be driven out of the cylinder against the powerful resistance of the atmosphere, while in the low pressure engine it has instant access to the condenser, and leaves effective the whole pressure of the fresh steam on the opposite side of the piston. But as in the low pressure engine, nearly half the power of the steam is expended in overcoming the friction and other impediments of the numerous parts, while in that of high pressure, the parts are so much fewer, and the piston is so much smaller in proportion to the force acting upon it, that the loss from friction is often less than a fourth or even a sixth of the steam power, although the resistance of the air is to be overcome by the high pressure engine, still there is often a saving on the whole. The saving becomes very considerable if the steam be allowed to act by its expansion also, as described in the last paragraph.

From misapprehension of the law of increase of force by increase of heat in water, explained by the table at page 186, some exceedingly false conclusions have been drawn and acted upon at great expense (as lately by Mr. Perkins) in attempts to make engines work with an excessively high pressure. Besides making the error now alluded to, and others, Mr. Perkins overlooked the fact, that we possess no material for cylinders and pistons strong enough to bear the contemplated pressure and friction even for a moderate time. Perhaps more striking examples could not be adduced of the absurdities into which even highly ingenious men may fall, when not sufficiently acquainted with the general truths of nature on which the arts which occupy them are founded, than in the history of supposed inventions and improvements connected with the steam-engine.

The fertile genius of James Watt did not stop at the accomplishment of the two or three important particulars described above, but throughout the whole detail of the component parts, and of the various applications of the engine, he contrived miracles of simplicity and usefulness. We should exceed the prescribed bounds of this work by entering more minutely into the subject; but we may remark that, in the present perfect state of the engine, it appears a thing almost endowed with intelligence. It regulates with perfect accuracy and uniformity the *number of its strokes* in a given time, *counting* or *recording* them, moreover, to tell how much work it has done, as a clock records the beats of its pendulum;—it regulates the *quantity of steam* admitted to work;—the *briskness of the fire;*—the *supply of water* to the boiler;—the *supply of coals* to the fire;—it *opens and shuts its valves* with absolute precision as to time and manner;—it *oils its joints;*—it *takes out any air* which may accidentally enter into parts which should be vacuous;—and when any thing goes wrong which it cannot of itself rectify, it *warns its attendants* by ringing a bell:—yet with all these talents and qualities, and even when exerting the force of hundreds of horses, it is obedient to the hand of a child;—its aliment is coal, wood, charcoal, or other combustible;—it consumes none while idle;—it never tires, and wants no sleep;—it is not subject to malady when originally well made; and only refuses to work when worn out with age;—it is equally active in all climates, and will do work of any kind;—it is a water-pumper, a miner, a sailor, a cotton-spinner, a weaver, a blacksmith, a miller, &c., &c.; and a small engine in the character of a *steam pony*, may be seen draging after it on a rail-road a hundred tons of merchandize, or a regiment of soldiers, with thrice the speed of our fleetest horse coaches. It is the king of machines, and a permanent realization of the *Genii* of Eastern fable, submitting supernatural powers to the command of man.

We need not wonder that the inventor of an engine having such qualities, should be deemed deserving of the highest honours from his fellow-men. In November, 1825, a public meeting was called, to vote a monument to WATT, then not long deceased; and the most distinguished men of the empire, of all parties, philosophers and statesmen, met to vie with each other in speaking his praise. Perhaps a series of such eloquent discourses has rarely been pronounced at one time; but perhaps in the progress of the arts of civilization there can rarely be offered such motive and occasion. The common voice of that assembly scarcely exaggerated, when attributing to WATT's genius and perseverance that increase of our national commerce and riches, which had enabled free Britian, single-handed, at an extraordinary crisis of human affairs, to contend with Europe combined against her, and at last to triumph, so as to secure her own happy destinies, and probably much to influence those of the human race.

As science and the twin sister art are making constant advances, who shall say that even the steam-engine, perfect as we have described it, forms the limit to human discovery of mighty yet obedient force? It is true that the nature of steam, and the laws of its formation and action, are now so well understood, that the intelligent engineer no more hopes for great improvement in steam-engines, than he hopes for it in the mode of using a waterfall to turn a mill; but still there are kindred regions of nature left almost unexplored. We shall have occasion to make a remark on this subject in our chapter on the nature of *heat.*

The explosion of gunpowder and of all fulminating mixtures bears so strong an analogy to the phenomenon of the formation of steam, that the mind may advantageously contemplate the subject in this place.

The ingredients of which gunpowder is formed are chiefly substances which, when separate, exist, at any common temperature, in the form of air; and the combustion sets them loose, with a production of intense heat, causing an increase of volume which is instantaneous, and almost irresistible. By experiment and mathematical deduction, it appears that the exploding particles begin to separate from each other with a velocity as if ten thousand volumes of air had been condensed into one: and this explains the corresponding force and swiftness with which a bullet is propelled.

All the fulminating metals are chiefly combinations of the like substances with the metals; and the ingredients are held together by so slight a tie, that a little friction or elevation of temperature disunites them so as to produce the explosion.

The escape of condensed air from the chamber of an air-gun, is a species of explosion; but is very gentle compared with the shock of discharged gunpowder.

It has lately been shown that a gun-barrel may be connected with a high-pressure steam-boiler, in the same manner as with a chamber of condensed air; and as the steam may be supplied as long as water remains in the boiler, if bullets be allowed to fall into the barrel fast enough, a hundred or more may be thrown out every minute, with the same force and precision as if each issued from a common fire-arm. The rapid succession resembles the issue of water from a jet pipe; and if such an engine were used in a field of battle, its barrel of death, made to point gradually along a line of men, would mow them down like corn-stalks before the scythe—none could escape. The horrible idea and proposal have been excused by saying, that to prove the possibility of such carnage must have the effect of putting an end to war altogether.

The invention of gunpowder, with the consequent change of military tactics, because it gave to a handful of men possessing it the mastery over thousands who had it not, was hailed by the philosophers of the day as a certain security against the relapse of civilized mankind into such a state of barbarism as followed the irruption into Europe of the Goths and Vandals:—none but well-instructed and disciplined armies could then enter a European kingdom. This consideration, however, has lost its interest, since the invention of printing, and other changes in society, have afforded still better and more humane securities.

Besides the interesting instances above cited of the pressure of the atmosphere determining whether certain substances shall or shall not have the form of air, there are others that deserve mention, where the effect is modified by the mutual attraction of substances.

The pressure of the atmosphere at the surface of the earth keeps a certain quantity of air in combination with water, so as to form part of the liquid mass. This air re-appears at once on taking off the pressure. If we place a glass of water under the receiver of an air-pump and then exhaust this, the water is soon crowded with bubbles of air, seen adhering to the glass all round, or rising through the water. This admixture of air in water is necessary to the life of fishes. It is driven off by boiling, and hence the vapid taste of water that has recently been boiled.

In the making of beer, wine, and other fermented liquors, there is formed, during the fermentation, a large quantity of the substance called carbonic acid. Much of it flies off in its usual form of gas, but, because of the pressure of the atmosphere, much still remains in union with the liquid. On removing this pressure suddenly, the liquid appears almost to boil, as when a glass of warm beer is placed in the air-pump vacuum.

A degree of pressure still greater than that of the atmosphere keeps a proportionably larger quantity of this carbonic acid in liquid combination; as in bottled porter or sparkling champagne before the cork is drawn; but as soon as the compression maintained by the cork is removed, the gas escapes, causing the thin champagne to sparkle, and the more viscid beer, which retains the little bubbles as they rise, to be covered with froth. After the sparkling or frothing has ceased under the atmospheric pressure, the phenomenon may be renewed by placing the glass in the air-pump receiver.

Carbonic acid so readily becomes liquid when its attraction for water assists the compression, that enough of it may be united with water to make a pint become a pint and a half. The soda water, or aerated water, now so generally used as a drink in warm weather, is water with several times its bulk of carbonic acid forced into it by pressure; and a part of this is seen escaping always at the instant of the confining cork being drawn.

Carbonic acid forms nearly half of the substance of marble or lime-stone. When an acid with stronger attraction, as vinegar or sulphuric acid, is poured upon marble, it dispossesses the carbonic acid, and unites itself with the pure lime. The carbonic acid in rising, constitutes the effervescence which then appears. Carbonic acid, for the manufacture of the common soda water and other aerated drinks, is obtained in this way.

Many mineral waters contain carbonic acid, which remains in tranquil combination, while the water is bearing a certain pressure under ground, but which in part escapes as soon as the water issues to the air and only the atmospheric pressure remains: such waters are called sparkling waters.

The reason that champagne and aerated waters are so cool when first decanted is, that the carbonic acid, in assuming its gaseous form, absorbs as latent heat, a large proportion of the heat which was previously existing in the liquid.

The atmospheric pressure, by making the density of the air in any place dependant upon the height of the place above the level of the sea, causes corresponding differences of temperature.

The explanation of this is simple. If a gallon of air, at the surface of the earth, contain a certain quantity of heat, this must be diffused equally through the space of the gallon; but if the air be then compressed into one-tenth of the bulk, there will be ten times as much heat in that tenth as there was before; the increase affecting the thermometer to an extent modified by circumstances explained in a future part of this work. In like manner, if by taking off pressure, the gallon be made to dilate to ten gallons, the heat will be in the same degree diffused, and any one part will be colder than before. It is known that air may be so much compressed under the piston of a syringe, that the heat in it, similarly concentrated, becomes intense enough to inflame tinder attached to the bottom of the piston;—this means, under the name of the *match-syringe*, being in common use for obtaining an instantaneous light.

Now, for the reason here explained, the air near the surface of the earth, forming the bottom of the atmosphere, because condensed by the weight of the air above it, is much warmer than if it were suddenly carried higher up, to where, from the pressure being less, it would be more expanded or thin. In many cases the height of mountains may be estimated by the difference of temperature observed at the bottom and at the top. While a thermometer stands at 60° at the bottom of St. Paul's Cathedral, in London, another marks only 58° at the top of the dome; and in the lofty ascent of a balloon, the thermometer soon falls to the freezing point and below it, the cold to the aeronaut becoming almost insupportable.

In every part of the earth, at a certain elevation in the atmosphere, different according to the latitude or proximity to the equator, the thermometer never rises above the freezing point,—and this limit in the atmosphere is called the line or level of perpetual congelation. In Norway it is at five thousand feet above the level of the sea; in Switzerland at six thousand five hundred; in Spain and Italy at seven thousand; farther south, at Teneriffe, at nine thousand; directly under the sun, as in central Africa, and among the Andes in America, it is about fourteen thousand. We see therefore why the snow-capt mountains are not the tenants only of high northern and southern latitudes. In this effect of elevation which renders many of the tropical regions of the earth not only tolerable abodes for man, but as suitable as any others, contrary to the opinion of the ancient philosophers of Europe, who accounted them by reason of the great heat, an everlasting barrier, as regarded man, between the northern and southern hemispheres. Much of the tropical land of America is so raised, that it rivals, as to agreeable temperature, even a European climate; while the lightness and purity of the air, and the brightness of the sun, add delightfully to its charms. The vast expanse of table land forming the empire of Mexico is of this kind, enjoying the immediate proximity of the sun, and yet, by its elevation of seven thousand feet above the level of the ocean. possessing the most healthful freshness.

The land in many parts has the fertility of a cultivated garden, and can produce naturally most of the riches which vegetation offers over the diversified face of the globe. The plains of Columbia in South America, and indeed all along the ridge of the Andes, are similarly circumstanced. The contrast is very striking, after sailing a thousand miles up the level river Magdelena, in a heat scarcely equalled on the plains of India, at once to climb to the table-land above, where *Sante Fe de Bogota*, the capital of the republic, is seen smiling over interminable plains, that bear the livery of the fairest fields of Europe.

Persons not understanding the law which we are now illustrating, will express surprise that wind or air blowing down upon them from a snow-clad mountain, should still be warm and temperate. The truth is, that there is just as much heat combined with an ounce of the air on the mountain top as in the valley; but above, the heat is diffused through a space perhaps twice as great as when below, and, therefore, is less sensible. It may be the same air which sweeps along as a warm gale on a plain at the foot of a mountain,—which then rises and freezes water on the summit—and which in an hour after, or less, is playing among the flowers of another valley, as warm and genial as before.

As the temperature in different parts of the atmosphere depends thus upon the rarity of the air, and therefore upon the height, the vegetable productions of each distinct region or elevation are of a distinct character; and many other peculiarities of place and climate acknowledge the same cause.

Because the atmospheric pressure determines the temperature of the air in different situations, as now explained, it has also a corresponding influence upon the state of aerial humidity, which is modified by the temperature.

It was explained at page 184, that water and other liquids under a vacuum, rise in the form of air or vapour with force, and in quantity having a strict relation to the temperature—heat being in fact the cause of their rising; and the table at page 186 exhibits the force, and therefore the density of watery vapour corresponding to some certain temperatures. Now it is a remarkable circumstance, that vapour in the same quantity and of equal tension rises from any liquid, whether placed under the pressure of air, or under a vacuum; only through a space containing air it diffuses itself more slowly than if the air were not present. As regards the former case, it was for a long time supposed that the air dissolved a liquid as a liquid dissolves a salt: but it now appears that there is merely a mechanical mixture of the two. If the vapour, while rising from a liquid, has not a tension or elastic force equal to the pressure of the atmosphere the process is tranquil, and is called *evaporation*, and it goes on only as the vapour can diffuse itself among the particles of the air, and therefore slowly in air perfectly quiescent, but quicker as the air is moving more, or as the density of the air is less. But when the vapour, owing to a greater heat, is strong enough to overcome the atmospheric pressure of fifteen pounds per inch, and the weight of a certain quantity of liquid over it, the phenomena of boiling arises as already described.

For the reason now explained, the air of our atmosphere contains diffused through it a large quantity of invisible aeriform water; and if there were no intestine motion, and no changes of temperature in the atmosphere, the quantity of water would soon everywhere reach a *maximum*, or would be the greatest that the temperature of the place could support; instead of this, however, from a variety of causes to be explained below, the air is moving about constantly as winds, and the local temperatures are ever fluctuating, and when the temperature is lowered, in situations where a maximum of watery vapour is present, part of this is instantly reduced to the state of water again, and appears, according to circumstance, in the form of *mist, rain, snow* or *hail;* while to supply material for these phenomena, evaporation is going on wherever, over water, there is not a *maximum* of vapour in the air. These opposing operations of evaporation and condensation keep up that constant circulation of moisture which is the life of nature.

When a given quantity of water assumes the aeriform state, it contains the same quantity of latent heat in all cases, whether rising, for instance, from a boiling caldron, or from the surface of the lake. Hence we see why evaporation is so cooling a process to any liquid or moistened solid from which it is arising: and as we have already shown that a rapid passing of dry air, or the substance being placed in a vacuum, quickens evaporation, we now see why both of these conditions accelerate the cooling. Wet linen placed in a strong wind, which does not contain a maximum of moisture, becomes dry almost immediately; a bottle of wine covered with a wet cloth and suspended in a current of air, as is practised in warm climates to prepare wine for the table is quickly cooled; mats hung around the walls of houses in India, and frequently wetted through the day, preserve a delightful freshness in the apartments. Sprinkling water or vinegar over a hot sick room cools and refreshes it; and watering the streets of a city moderates in them the intensity of summer heat. In warm climates water is cooled for drinking by being put into vessels so porous that the external surface is always moist, the vessel being then suspended in a current of air, or during a calm being made to vibrate in the manner of a pendulum. Again, the rapidity of evaporation from water under the exhausted receiver of an air-pump, and particularly when some other substance which powerfully absorbs watery vapour is included in the receiver, is so rapid, and carries off the heat so quickly, that the mass of water freezes before much of it has been carried away. This process is used for making ice in India.

It is partly because air saturated with moisture, that is to say, having as much water diffused in it as can be supported in the invisible or aeriform state of the existing temperature,—lets fall a part on any reduction of the temperature, that air which as a portion of the atmosphere, has been heated by the sun during the day, and has received such moisture, lets it fall again during the night, and exhibits the night fogs of certain seasons, which float upon the surface of the earth, until again acted upon by the beams of the next morning's sun. Fog, when farther condensed, by groups of the minute particles uniting, forms rain; and rain when cooled becomes snow or hail.

The quantity of dew which falls at night is influenced by the quantity of moisture taken up by the atmosphere during the heat of the day; and the immediate cause of the dew is, as was ingeniously proved by Dr. Wells, some years ago, that the temperature of the objects on which it settles has become lower during the night than that of the air around, and than is required to maintain in the invisible state, the moisture in the surrounding atmosphere. There is a tendency in heat to diffuse itself uniformly among bodies, by a constant radiation from one to another, rapid in proportion to the differences of temperature, and which, if continued, would reduce all to the same degree. The earth, therefore, during the day, receives radiated heat from the sun, and becomes comparatively hot, and during the night it gives out heat again by radiation towards the sky, from which there is little or no return. When there are clouds in the atmosphere at night, they receive the heat darted upwards from the bodies on the earth's surface, and they radiate heat back, becoming thus, as it were, a clothing to maintain the warmth of the earth beneath them,—and on cloudy nights there is no dew,—but with a clear sky, the heat radiated upwards, darts into boundless space, and is lost altogether to the objects which emitted it. These objects, therefore, which during the day had the same, or even a higher temperature than the atmosphere around, now become colder, and the aeriform water which comes in contact with them is condensed, and forms what we call dew. This beautiful provision of nature supplies the necessary moisture to vegetables during seasons when rain is deficient. Dew on very cold objects freezes as it settles, and is then called *hoar frost.* A phenomenon which may be classed with dew, is the perspiration, as it is vulgarly called, of massive walls and furniture, occurring on the sudden setting in of warm weather, or on the occasion of a warm moist air of higher temperature than the walls being suddenly introduced, as when a crowd assembles in a cold church:—the wall or other object then, from not having yet acquired the temperature of the surrounding air, condenses upon itself a copious deposition of the atmospheric moisture. For a similar reason a bottle of wine brought from a cold cellar or from an ice-pail, into a room with company, is soon covered with thick moisture or dew; as are the glasses also into which the wine is poured. It is another phenomenon of the same kind when we see the moisture of warm breath condensed on any cold polished surface, as on a mirror's face, or on the glasses of a carriage shut up, or on the windows of a room in winter, when the surface is very cold, the moisture being frozen with the appearance of beautiful aboresence.

Many instruments have been contrived, with the name of *hygrometers,* for indicating the quantity of water in the atmosphere. A prepared human hair is the essential part of one of the best of those formerly used; the lengthening or shortening of the hair, according to the quantity of moisture around it, being caused to move an index like that of a wheel-barometer; to mark the degrees. This, however, and other common hygrometers, are only philosophical toys; but Mr. Daniel (see his excellent work, entitled Meteorological Essays) has lately given to the philosophical world a correct and simple instrument for the purpose, depending on the principle explained above,—that whenever the temperature of a body in the atmosphere is reduced below that at which the quantity of watery vapour in the air around it can be maintained in the aeriform or invisible state, dew forms on the body. His apparatus consists of a bulb of glass, which can be cooled to any desired degree from being connected with another bulb enveloped in an evaporating liquid; and when moisture begins visibly to settle upon the first, its temperature is exhibited on a thermometer enclosed within it; and the proportion of water mixed with the air around is then, as indicated by the table, partially copied here, at page 186.

A great fall of the barometer marks a diminished pressure in the atmosphere around, with a consequent dilatation of the air and fall of temperature, as explained a few pages back; and if the air at such a time hold a maximum of moisture, a part of this must become visible as fog or rain. Thus a fall of the barometer, a fall of temperature, and a fall of rain, often occur as associated phenomena.

Illustrating this by experiment, we find, that on the extraction of air from the receiver of an air-pump, a cloud of mist generally appears in it with the first strokes of the piston :—the reason being that the still remaining air, because cooled by the refraction, absorbs heat from the vapor in combination with it, and renders the water visible. The mist is then removed by the subsequent action of the machine, or is re-dissolved when the usual quantity of air is re-admitted.

We understand from this why rain happens so much more frequently among mountains than on extended plains. When air saturated with moisture approaches the mountain ridge to rise over it, for every foot that it rises it escapes from a degree of the pressure which it bore while lower down, and in then dilating, it becomes colder, and lets fall part of its moisture. It is the rain copiously thus produced in mountainous regions which constitutes the chief supply of their many rivers, and which, with periodical changes of wind bringing more moisture, causes the extraordinary annual overflowing of such rivers as the Nile, the Ganges, &c.

Those who have visited the Cape of Good Hope, will recollect a striking phenomenon illustrative of our present subject, observed there when the wind blows from the south-east. Beyond the city, as viewed from the bay, there is a mountain of great elevation, called from its extended flat summit, the Table Mountain. In general its rugged steeps are seen rising in a clear sky; but when the south-east wind blows, the whole summit becomes enveloped in a cloud of singular density and beauty. The inhabitants call the phenomenon the spreading of the table-cloth. The cloud does not appear to be at rest on the hill, but to be constantly rolling onward; yet to the surprise of the beholder, it never descends, for the snowy wreaths seem falling over the precipice towards the town below, vanish completely before they reach it, while others are formed on the other side to replace them. The reason of the phenomena is this: The air constituting the wind from the south-east having passed over the vast southern ocean, comes charged with as much invisible moisture as its temperature can sustain. In rising up the side of the mountain it is rising in the atmosphere, and is therefore gradually escaping from a part of the pressure lately borne; and on attaining the summit it has dilated so much, and has consequently become so much colder, that it lets go part of its moisture. This then appears as the cloud just described; but it no sooner falls over the edge of the mountain, and again descends in the atmosphere to where it is pressed, and condensed, and heated as before, than it is re-dissolved and disappears:—the magnificent apparition dwelling only on the mountain top.

When the elevation to which moisture is suddenly carried is very great, the fall of temperature is proportioned, and the separating water becomes snow instead of rain. This phenomenon is remarkably illustrated by a great *Hiero's* fountain, used in one of the mines of Hungary; during the play of which, the air in one place is so compressed, that on being suddenly released it expands and cools enough to cause the moisture driven out with it to appear, even in summer, as a shower of snow.

The foregoing reasoning explains why, along the sides of mountain ridges, clouds are generally seen floating at a certain height only, and therefore in horizontal strata. The water is separated from the air at a certain temperature, which is dependent on the height, and above that height the air is at the time too dry and rare to have clouds. Very lofty summits are always seen much above the clouds, and the admirer of nature who climbs towards them, may often contemplate the grand phenomena of the thunder-storm far beneath his feet. Teneriffe soars so sublimely, that the distant sailor not unfrequently mistakes the line of clouds hanging around its sides for the white streak which elsewhere indicates the cliffs and waves of the sea-shore.

Fluid support or floating in air. (Read the Analysis, page 156.)

When it was explained under "Hydrostatics," that any body immersed in a fluid has its downward tendency or weight resisted with exactly the force which supported the quantity of the fluid previously occupying the same space, and therefore that the body will sink or swim, according as it is heavier or lighter than its bulk of the fluid, the reasoning was as applicable to the case of a body immersed in an air or gas as in a liquid.

We hence see why a body weighed in an air appears lighter, by the exact weight of its bulk of the air, than when weighed in an empty space or vacuum;—and why, for the same reason, the jocular question, whether a pound of lead or a pound of cork be the heavier, it is not truly answered by saying that they are of equal weight; the cork being really the heavier, for when balanced in air, bulky cork is more supported than dense lead. A small weighing beam having attached to its opposite ends pieces of cork and lead which equipoise in the air, if placed under the exhausted receiver of an air-pump, quickly exhibits the cork preponderating.

As any liquid lighter than water, such as oil or spirits, on being set at liberty under the surface of water, will rise, while any heavier liquid, such as brine, syrup, or sulphuric acid, will sink; and in both cases with force proportioned to the difference of specific gravities: so we find that in common air, a mass of hydrogen, or hotter air, ascends, because specifically lighter; while oxygen, carbonic acid gas, or colder air, descends, because specifically heavier. This truth is well exemplified in

The Balloon,

which is a thin light bag of varnished silk, generally shaped like a globe or egg, and filled with a fluid lighter than common air. It is made sufficiently large that the difference between its weight when filled and that of an equal bulk of common air, may enable it to carry aloft the material of which it is constructed, with the aeronauts, and their apparatus. It is in principle like a bladder of oil immersed in water. A globe of thirty-five feet diameter has a capacity of nearly twenty-two thousand cubic feet. This quantity of common air weighs about *sixteen* hundred pounds, and the same quantity of hydrogen gas, of easily obtained purity, weighs only one-eighth as much as *two* hundred pounds. Such a globe, therefore, being buoyed up, or supported in common air, with a force of sixteen hundred pounds, while, if filled with hydrogen, it only weighs two hundred, will carry up into the sky fourteen hundred pounds of material and load.

The first balloon was exhibited by a man ignorant of what he was really effecting. Seeing the clouds float high in the atmosphere, he thought that if he could make a cloud and enclose it in a bag, it might rise and carry him with it. Then, erroneously deeming smoke and a cloud the same, he made a fire of green wood, wool, &c., and placed a great bag over it with the mouth downwards to receive the smoke. He soon had the joy to see the bag full, and, when set free, ascending; but he understood not that the cause was the hot and dilated air within, which, being lighter than the surrounding air was buoyed up; while the visible parts of the smoke, which chiefly engaged his attention, was really heavier than the air, and was an impediment to his wishes.

This modification called the *hot air or fire balloon*, was afterwards better understood, and was used by aeronauts, until the more commodious and less dangerous modification, called the *inflammable air balloon*, or balloon of hydrogen gas, was substituted.

Since the modern introduction of gas lights, the *carburretted hydrogen* prepared for them is generally employed for filling balloons. It is considerably heavier than pure hydrogen, but is so much more readily obtained, that aeronauts like better to make a larger balloon to suit it, than a smaller one which obliges them to prepare the other.—A thin paper bag, filled with the hot-air rising from a large lamp, is a miniature *hot air or fire balloon;* and a common soap bubble, filled with hydrogen is a little *inflammable air balloon*, which mounts with great rapidity.

There are, perhaps, few occasions on which a youth is more surprised and delighted than when he first beholds a balloon sailing high in the bosom of the air and bearing a human being to regions far beyond what the soaring eagle has ever reached; while to the intrepid aeronaut himself, the scene of a world displayed beneath him is unquestionably the grandest, except that of the starry heavens, which mortal eye has ever compassed. To him even wide spread London, the queen of the cities of the earth, and a little world within itself, when viewed from a great elevation in the sky, appears but as a dusky patch upon a map, with the far-famed Thames winding there as a silvery line, and the magnificent temples and palaces scattered around appearing but as darker points rising out of the general mist of buildings, in which a million and a half of human beings reside.

The first aeronautic expeditions astonished the world, and endless reveries passed through mens' minds of important uses to which the new discovery might be applied; but more mature reflection, and now frequent trials have shown that the balloon, while furnishing philosophers with the opportunity of making some observations in elevated regions of the atmosphere, is still interesting chiefly as a philosophical toy. The French, under the Directory in 1796, attempted to use it as a military station, from which the position and motions of an enemy might be descried: but the plan was eventually abandoned. It has since been thought of as a means by which travellers might obtain information while penetrating into unknown countries, like the almost interlineable plain of *Australasia*. Although aeronauts, while aloft, have the power of making the balloon rise farther by throwing out part of the sand-ballast which they carry with them, or of making it descend by opening a valve at the top, through which the hydrogen may escape, still they have no power of producing a lateral motion. The idea which yet strongly excites the minds of some projectors, that by wings or other means, a balloon may be directed in the sky nearly as a ship is directed on the sea, is not much more reasonable than to suppose that an insect, suspended to a huge block of wood, driven along at the rate of eight or ten miles an hour by river torrent, should have power to stop or sail against the stream. A man in a balloom would generally have to resist or change a motion exceeding fifty miles in an hour.

A balloon which is only half full at the surface of the earth, becomes quite full when it has risen three miles and a half, because, at that altitude, air from below doubles its volume on account of the diminished pressure. A balloon, therefore, if quite distended on first rising, must let air escape as it ascends, or it will burst: this is true also of the drum of the human ear under the same circumstances, and in a contrary way under the opposite circumstances of descending in a diving-bell.

The downy seeds of plants seen floating about upon the winds of autumn are not lighter than air, but have so much bulk and surface in proportion to their weight, that the friction upon them of the moving air, is greater than their weight, and carries them along.

A sheet of paper made in some degree to resemble a balloon, by its having a little weight, representing the hanging car, attached by threads from its angles, is often seen rising at a street corner, to the delight of the boy who watches it. Its rise depends upon eddy winds or currents which the corner produces.

The ascent of flame and smoke

in the atmosphere, affords other examples of a lighter fluid rising in a heavier; for both these are merely hotter air rising in the midst of colder.

The phenomenon of flame is produced when a burning substance contains some ingredient capable, on being heated, of assuming the form of air or gas, which ingredient, or ascending, burns or combines with the oxygen of the atmosphere, with intensity of action sufficient to produce a white heat. It is because charcoal and coke have nothing in them thus volatile, that they burn without flame, appearing like red-hot stones. The flame of a lamp or candle is merely the oil, wax, or tallow converted into gas, and allowed to burn as it is disengaged and rises. The same gas obtained by heating the oil, &c., in vessels which exclude the atmosphere, so as to prevent immediate combustion, and from which tubes lead to suitable receptacles, is the common oil-gas used for illumination.

Smoke consists of all the dust and visible particles which are separated from the fuel without being burned, and are, moreover, light or minute enough to be carried aloft by the rising current of heated air; but all that is visible of smoke is really heavier than air, and soon falls again as powdered chalk falls in water. In the receiver of an air-pump, where a candle has been extinguished by exhausting the air, the steam of smoke that continues to pour from the wick after the exhaustion, is seen to fall on the pump-plate, because there is no air to support it.

Chimneys quicken the ascent of hot air by keeping a long column of it together. A column of two feet high rises, or is pressed up with twice as much force as a column of one foot, and so in proportion for all other lengths; just as two or more corks strung together and immersed in water, tend upwards with proportionally more force than a single cork; or as a long spear of light wood, allowed to ascend perpendicularly from a great depth in water, acquires a velocity which makes it dart above the surface, while a short piece under the same circumstances rises very slowly. In a chimney where one foot in height of the column of hot air is one ounce lighter than the same bulk of the external cold air, if the chimney be one hundred feet high, the air or smoke in it is propelled upwards with a force of one hundred ounces. In all cases, therefore, the *draught*, as it is called of a chimney, is proportioned to its length. The following facts are consequences of this truth.

In low cottages, and in the upper floors of houses, the annoyance of smoky rooms is much more frequent than were chimneys are longer.

If there are two fires in the same room, or in any rooms open to each other, which have chimneys of different lengths, and of which the doors and windows are very close, so that the air to supply the draught cannot enter by them, the taller chimney will overpower the shorter, and cause it to smoke into the room; just as the long leg of a syphon overcomes the short one, or as a long log of wood, held down in water by a cord passing from it round a pulley at the bottom to a shorter log also floating, will rise, and pull down the shorter log.

A long chimney, for the reasons above explained, causes a current of air to pass through the fire very rapidly, and it has the advantage also of acting more uniformly than any bellows or blowing machine. On these accounts, of fires of steam engines, and many others, it is the means of blowing generally preferred. The importance of length in a chimney explains the remarkable appearance of some mining districts and modern English towns, where steam-engines abound.

When we heap dying embers together, so that the hot air rising among them may become a mass or column of considerable altitude, this column has the effect of blowing them gently, and helps to light them up again. A piece of burning paper thrown upon the top of a half-extinguished fire, often makes it blaze afresh, by causing a more rapid current of air to pass through it from below.

The action or draught of a chimney, influenced as we have seen, by its length, depends also on the degree in which the air in it is heated, because this determines the dilitation, or comparative lightness, which makes the air ascend.

In what are called *open fire-places*, such as those in the sitting-rooms of Britain, a large quantity of air directly from the apartment enters the chimney above the fire, and mixes with the hot air from the fire itself. This mixture ascends more slowly than if hot air alone entered, and in a proportion dependent on the degree of mixture. The effect of excluding a part of this colder air, is seen when a board or plate of metal is suspended across the opening of the chimney, so as to narrow the entrance:—almost instantly a quicker action is produced, and the fire begins to roar as if blown by a bellows. This means is often used to blow the fire instead of bellows, or to cure a smoky chimney by increasing the draught. What is call a *register stove* is a kindred contrivance. It has a flap placed in the throat of the chimney, which serves to widen or contract the passage at pleasure. Because the flap is generally opened only enough to allow that air to pass which rises directly from the fire, the chimney receives only very hot air, and therefore acts well. The register stove often cures smoky chimneys: and by preventing the too ready escape of the moderately warmed air of the room, of which so much is wasted by a common fire-place, it also saves fuel. In what are called *close fire-places*, as those of steam-engines, or brewers' coppers, when the furnace door is shut, no air can enter the chimney but directly through the fire; hence the action of such chimneys is very powerful.

In a room with two fires, or in drawing rooms communicating with each other, although the chimneys be of equal length, that one over the best fire will act the most strongly; and if the doors and windows of the apartment be so close as to prevent a sufficiency of air from entering by them to supply both fires, cold air will enter by that chimney which has the weakest fire, and the smoke from it will spread into the room. How often is an assembling dinner party annoyed by the smoke of a second drawing-room fire just lighted before their arrival, and which had therefore to contend with the antagonist fire already in powerful action all the day. While only one fire was lighted, the cold chimney was admitting the air to feed it, just as an open pane in the window would have done. A room may be so close that no air can find entrance, and in such a case the smoke of its fire must all spread into the room.

When all the windows and doors of a house fit so closely as not to admit air for the acting chimneys, the supply comes down the chimneys that are not in use. Inattention to this fact causes many a good chimney to incur the imputation of being smoky, because on the attempt being made to light a fire at it, the smoke at first is always thrown back. The truth is, that at the time when the servant begins to light the fire, there is a downward current in the chimney, repelling of course, any heated air and smoke that approaches it, and spreading them over the whole house; but were the room door to be shut for a few minutes, so as to cut off communication with the other *drawing* chimneys in the house, while at the same time the windows were opened, the chimney would act at once; and when sufficiently heated, would continue to act in spite of the others, as well as they.

There are some cases of smoky rooms not to be so easily corrected as what we have now mentioned. When a low house adjoins a lofty house, the wind blowing towards the latter, is obstructed and becomes a gathering or condensation of air against the wall; and if the top of a low chimney be there, the compressed air enters it and pours downwards. The same happens occasionally from the proximity of trees or rocks. In such cases, to avoid the influence, the chimneys of the low houses are often made very lofty. Again, whenever, from the nature of buildings, eddies of wind occur, or unequal pressures, as at street corners, &c., the chimneys around do not act regularly. It is proverbial, that corner houses, or those at the end of a row, are smoky houses; and we see the uniformity of architecture in a street often destroyed by the necessity of lengthening the chimneys of the houses at the extremities.

When smoke is found descending into a room where there is no fire, the empty chimney is serving as an inlet for air to the house, while the smoke of a neighboring chimney is passing closely over the top of it.

In summer, when fires are not in use, there is often a strong smell of soot perceived in the apartments during the whole of the day, but which ceases at night. The reason is, that during the day the chimney is colder than the external air, and by condensing the air which enters it, causes a downward current through the soot. During the night, again, when the external air becomes colder, owing to the absence of the sun, the chimney, by retaining the heat absorbed during the day, is hot enough to warm the air in it, and to cause an upward current. These currents, in chimneys left open during the days and nights of summer, are almost as regular as the land and sea breezes of tropical countries.

All these remarks prove how important it is to be able to conceive clearly of the motions going on, according to the simple laws of matter, in the invisible air around us. Were such subjects better and more generally understood, many prevalent errors in the arts of life, influencing much the comforts and health of the community, would soon be corrected.

If we are filled with admiration on discovering how perfectly the simple law of a lighter fluid rising in a heavier, provides a constantly renewed supply of fresh air to our fires, which supply we should else have to furnish by the unremitted action of some expensive blowing apparatus, still more must we admire that the operation of this law should effect the more important purpose of furnishing the ever renewed supply of the same vital fluid to breathing creatures. The air which a man has once respired becomes poison to him; but because the temperature of his body is generally higher than that of the atmosphere around him, as soon as he has discharged any air from the lungs, it ascends completely away from him into the great purifying laboratory of the atmosphere, and new air takes its place. No art or labor of his, as by the use of fans or punkas, could have done half so well what this simple law unceasingly and invisibly accomplishes, and accomplishes without effort or even attention on his part, and in his sleeping as in his waking hours. Truly in this, may he be said to be watched over by a kind Providence.

The warming and ventilating of houses,

is an important art, founded chiefly on the foregoing considerations, and at present too little understood, not only by the public at large, but even by medical practitioners, whose management of disease, though judicious in other respects, is often rendered vain by error or omission in this.

Excellent fuel is so cheap in Britain, owing to the profusion which beds of rich coal are scattered in it, that a careless domestic expenditure has arisen; which, however, instead of securing the comfort and health that might be expected, has led to plans of warming which often prove destructive to both. The mischief lies chiefly in the unsteadiness or fluctuations of our domestic temperature; for in still colder countries, and where fuel is more expensive, as in the north of continental Europe, the necessity for economy has led to contrivances which give steady temperature and impunity.

In cold countries, to retain and preserve the heat once obtained, the houses are made with thick walls, double windows, and nice fittings; and, moreover, with close stoves or fire-places, which draw their supply of air, not from the apartments where they are placed, wasting the temperate air of these, but directly from without. Thus fuel is saved to a great extent, and a uniformity of temperature is produced, both as regards the different parts of the room, so that the occupiers may sit with comfort where they please, and as regards the different times of the day, for the stove being once heated in the morning, often suffices to maintain a steady warmth until night. The temperature can be carried to any required degree, and sufficient ventilation is easily effected.

In England, again, the apartments, with their open chimneys, may be compared to great air funnels, constantly pouring out their warm contents through a large opening, and constantly requiring to be replenished. They thus waste fuel exceedingly, because the chimney being large enough to allow a whole room-full of air to pass away in two or three minutes, the air of the room has to be warmed, not once in the course of the day, but very many times. The temperature in them is made to fluctuate by the slightest causes, as the opening a door, the omitting to stir the fire, &c. The heat is very unequal in different parts of the room, rendering it necessary in general for the company to sit near the fire; where they must often submit to be almost scorched on one side, while they are chilled on the other. There is generally a warm stratum of air above the level of the chimney-piece, surrounding, therefore, the upper part of the bodies of persons in the room, while a cold stratum below envelopes the sensitive feet and legs. As a very rapid current is constantly ascending in the chimney, a corresponding supply must be entering somewhere; and it can only enter by the crevices and defects in the doors, windows, floors, &c. :—now there is nothing more dangerous to health than to sit near such inlets, as is proved by the rheumatisms, stiff necks and catarrhs, not to mention more serious diseases, which so frequently follow the exposure. There is an old Spanish proverb, thus translated,

> "If cold wind reach you through a hole,
> Go make your will and mind your soul,"

which is scarcely an exaggeration.

Consumption is the disease which carries off a fifth or more of the persons born in Britain; owing in part, no doubt, to the changeableness of the external climate, but much more to the faulty modes of warming and ventilating the houses. To judge of the influence of temperature in producing this disease we may consider,—that miners who live under ground, and are always, therefore, in the same temperature, are strangers to it, while their brothers, and relatives, exposed to the vicissitudes above, fall victims,—that butchers and others, who live almost constantly in the open air, so as to be hardened by the exposure, enjoy nearly equal immunity,—that consumption is scarcely known in Russia, where *close* stoves and houses preserve a uniform temperature within doors, while fit clothing gives safety on going out,—and that in all countries and situations, whether tropical, temperate or polar, the frequency of the disease bears relation to the degree and manner of change. We may here remark, also, that it is not consumption alone which springs from changes of temperature, but a great proportion of acute diseases, and particularly of the common winter diseases of England. There are a few cases of these in which the invalid has not to remark, that if he had avoided cold or wet on some certain occasion, he might yet have been well.

While temperature is thus so frequently an original cause of disease, it is also a circumstance of the very highest importance in the treatment,—as is proved by every fact bearing upon the question. We may, therefore, at first wonder that it should be so negligently and unskilfully controlled as we often see it; disease and death being thence allowed to lurk, as it were, undisturbed in the sanctuaries of our homes: but when we reflect on the subtile and invisible nature of air and heat, and that the science which detects their agencies has been hitherto so little an object of general study, and is, indeed, of modern discovery, the fact is accounted for.

In England, the open fire-place is so generally in use for common dwellings, and the cheerful blaze is accounted so essential to the comforts of the winter days and long evenings, that it would be difficult to persuade persons to abandon it: let us hope, then, that when the subject which we are now discussing comes to be better and more generally understood, the open fire, with close flooring, better for double windows, doors that fit well, register stoves, and good general management, may be rendered almost as efficient for warming, and as safe to health, as any other contrivance.

The following considerations present themselves in this place:—Small rooms in winter are more dangerous to health than large ones, because the cold air, entering towards the fire by the doors and windows, reaches the persons in the room before it can be tempered by mixing with the warmer air already around them. Stoves in halls and stair-cases are useful, because they warm the air before it enters the rooms; and they prevent the hurtful chills often felt on passing through a cold stair-case from one warm room to another. It is important to admit no more cold air into the house than is just required for the fires and for ventilation; hence there is a great error in the common practice of leaving all the chimneys that are not in use quite open, each admitting air as much as a hole in the wall, or an open pane in the window would do. Perhaps the best mode of admitting air to feed the fires is through tubes, leading directly from the outer air to the fire-place, and provided with what are called throttle-valves, for the regulation of the quantity; the fresh air admitted by them being made to spread in the room either at once, or after having been warmed during its passage inwards, by coming near the fire. In a very close apartment, ventilation must be expressly provided for by an opening near the ceiling, through which the impure air, rising from the respiration of the company, may pass away. With an open fire the purpose is effected, although less perfectly, by the frequent change of the whole air of the room which that construction occasions.

With a view to have, in rooms intended for invalids, the most perfect security against cold blasts and fluctuations of temperature, and still to retain the so much valued appearance of the open fire, a glazed frame or window may be placed at the entrance to the chimney or stove, so as completely to prevent the passage of air from the room to the fire. The room will then be warmed by the fire through the glass, nearly as a green-house is warmed by the rays of the sun. It is true that the heat of combustion does not pass through glass so readily as the heat of the sun; but the difference for the case supposed is not important. The glass of such a window must, of course, be divided into small panes, and supported by a metallic frame-work to resist the heat; and there must be a flap or door in the frame-work, for the purpose of admitting the fuel and stirring the fire. Air must be supplied to the fire, as described above, by a tube leading directly from the external atmosphere to the ash-pit. The ventilation of the room may be effected by an opening into the chimney near the ceiling; and the temperature may be regulated with great precision by a valve placed in this opening, and made to obey the dilatation and contraction of a piece of wire affixed to it, the length of which will always depend on the temperature of the room. The author contrived the arrangements here described, for the winter residence of a person threatened with consumption, and the happy issue of that particular case, and of others treated on similar principles, has led him to doubt whether many of the patients with incipient consumption who are usually sent to warmer climates, and who die there after suffering hardships on the journey, and distress from the banishment sufficient to shake even strong health, might not be saved by judicious treatment in properly warmed and ventilated apartments, under their own roofs, and in the midst of affectionate kindred. And if a boy be almost certainly secured from consumption by being made a miner or a butcher, may we not hope that, when all the influencing circumstances come to be better understood, something of the same immunity may be obtained for persons in all the professions and conditions of civilized society?

It must not be supposed that the remarks made in this section exhaust even nearly the very important subject of temperature as affecting health. The questions of *clothing*, of *hot and cold bathing*, of *exercise*, and others, equally belong to it, but the consideration of them falls under other departments of study.

Winds or currents in the atmosphere

are also phenomena, in a great measure dependent on the law, that lighter fluids rise in heavier. As oil let loose under water is pressed up to the surface and swims, so air near the surface of the earth, when heated by the sun, rises to the top of the atmosphere, and spreads there, forced up by the heavier air around; this heavier air rushing inwards, constitutes the wind felt at the surface of the earth. The cross currents in the atmosphere arising as now described, are often rendered evident by the motion of clouds or balloons.

If our globe were at rest, and the sun were always beaming over the same part, the earth and air directly under the sun would become exceedingly heated, and the air there would be constantly rising like oil in water, or like the smoke from a great fire; while currents or winds below would be pouring towards the central spot, from all directions. But the earth is constantly turning round under the sun, so that the whole middle region or equatorial belt may be called the sun's place: and therefore, according to the principle just laid down, there should be over it a constant rising of air, and constant currents from the two sides of it, or the north and south, to supply the ascent. Now this phenomenon is really going on, and has been going on ever since the beginning of the world, producing the steady winds of the northern and southern hemispheres, called *trade winds*, on which in most places within thirty degrees of the equator, mariners reckon almost as confidently as on the rising and setting of the sun himself.

The trade winds, however, although thus moving from the poles to the equator, do not appear on the earth to be directly north and south, for the eastward whirling, or diurnal rotation of the earth, causes a wind from the north to appear as if coming from the north-east, and a wind from the south as if coming from the south-east. This fact is illustrated by the case of a man on a galloping horse, to whom a calm appears to be a strong wind in his face; and if he be riding eastward, while the wind is directly north or south, such wind will appear to him to come from the north-east, or south-east:—or again, is illustrated by the case of a small globe made to turn upon a perpendicular axis, while a ball or some water is allowed to run from the top of it downwards; the ball or water will not immediately acquire the whirling motion of the globe, but will fall almost directly downwards, in a track which, if marked upon the globe, will appear not as a direct line from the axis to the equator, that is, from north to south, but as a line falling obliquely. Thus, then, the whirling of the earth is the cause of the oblique and westward direction of the trade winds, and not, as has often been said, the sun drawing them after him.

The reason why the trade winds at their external confines, which are about 30 degrees from the sun's place, appear almost directly *east*, and become more nearly *north* and *south* as they approach the central line, is, that at the confine they are like fluid coming from the axis of a turning wheel, and which has approached the circumference, but has not yet acquired the velocity of the circumference; while, nearer the line, they are like the fluid after it has for a considerable time been turning on the circumference, and has acquired the rotary motion there, consequently appearing at rest as regards that motion, but still leaving sensible any motion in a cross direction.

While, in the lower regions of the atmosphere, air is thus constantly flowing towards the equator and forming the steady trade winds between the tropics in the upper regions, there must, of course, be a counter-current distributing the heated air again over the globe: accordingly, since reasoning led men to expect this, many striking proofs have been detected. At the summit of the Peak of Teneriffe, observations now show that there is always a strong wind blowing in a direction contrary to that of the trade wind on the face of the ocean below. Again, the trade winds among the West India Islands are constant, yet volcanic dust thrown aloft from the Island of St. Vincent, in the year 1812, was found, to the astonishment of the inhabitants of Barbadoes, hovering over them in thick clouds, and falling, after coming more than 100 miles directly against the strong trade wind, which ships must take a circuitous course to avoid. Persons sailing from the Cape of Good Hope to St. Helena, have often to remark that the sun is hidden for days together, by a stratum of dense clouds passing southward high in the atmosphere; which clouds consist of the moisture raised near the equator with the heated air, and becoming condensed again as it approaches the colder regions of the south.

Beyond the tropics, where the heating iufluence of the sun is less, the winds occasionally obey other causes than those we have now been considering, which causes have not yet been fully investigated. The winds of temperate climates are in consequence much less regular, and are called *variable;* but still, as a general rule, whenever air is moving towards the equator, from the north or south poles where it was at rest, it must have the appearance of an east wind, or a wind moving in the contrary direction of the earth itself, until it has gradually acquired the whirling motion of that part of the surface of the earth on which it is found; and again, when air is moving from the equator, where it had at last acquired nearly the same motion as that part of the earth, on reaching parts nearer the poles, and which have less eastward motion, it continues to run faster than they, and becomes a westerly wind. In many situations beyond the tropics, the westerly winds, which are merely the upper equatorial currents of air falling down, are almost as regular as the easterly winds within the tropics, and might also be called trade winds:—witness the usual shortness of the voyages from New York to Liverpool, and the length of those made in the contrary direction. North of the equator, then, on earth, true north winds appear to be north-east, and true south winds appear to be south-west:—which are the two winds that blow in England for three hundred days of every year. In southern climates the converse is true.

While the sun is beaming directly over a tropical island, he warms very much the surface of the soil, and, therefore, also, the air over it; but the rays which fall upon the ocean around penetrate deep into the mass, and produce little increase of superficial temperature. As a consequence of this, there is a rapid ascent of hot air over the island during the day, and a cooler wind blowing towards its centre from all directions. This wind constitutes the refreshing *sea-breeze* of tropical islands and coasts. A person must have been among these, to conceive the delight which the sea-breeze brings after the sultry stagnation which precedes it. The welcome ripple shorewards is first perceived on the surface of the lately smooth or glassy sea; and soon the whole face of the sea is white with little curling waves, among which the graceful canoe, lately asleep on the water, now shoots swiftly along.

During the night a phenomenon of opposite nature takes place. The surface of the earth then no longer receiving the sun's rays, is soon cooled by radiation, while the sea, which absorbed heat during the day, not on the surface only, but through its mass, continues to give out heat all night. The consequence is, that the air over the earth becoming colder than that over the sea, sinks down, and spreads out on all sides, producing the *land-breeze* of tropical climates. This wind is often charged with unhealthy exhalations from the marshes and forests, while the sea-breeze is all purity and freshness. Many islands and coasts would be absolutely uninhabitable but for the sea-breeze.

The peculiar distribution of land in the Asiatic part of the globe, produces the curious effect there of a sea-breeze of six months, and a land-breeze of six months. The great continent of Asia lies chiefly north of the line, and during its summer, the air over it is so much heated, that there is a constant steady influx from the south—appearing south-west, for the reason given in a preceding page; and during its winter months, while the sun is over the southern ocean, there is a constant land breeze from the north—appearing, for a like reason, north-east. These winds are called *monsoons;* and if their utility to commerce were to be a reason for a name, they also deserve the name of trade winds. In early periods of navigation, they served to the mariner the purpose of compass, as well as of moving power; and one voyage outward, and another homeward with the changing monsoons, filled up his year.—On the western shores of Africa and America, also, the trade winds are interfered with by the heating of the land; but much less so than in Asia, and always in accordance with the laws now explained.

The frightful tornadoes, or whirlwinds, which occasionally devastate certain tropical regions, making victims of every ship or bark caught on the waters, and the shore gusts or squalls met with every where, are owing to some sudden chemical changes in the atmosphere, not yet fully understood.

The Pneumatic Trough and Gasometer

of the chemist are contrivances constantly displaying the truth now under consideration, "that a lighter fluid is pushed up and floats on a heavier." They are important parts of the apparatus for operating on substances while in the form of air.

The trough *a* may be made of metallic plate, or of wood lined with metal, and of any convenient size. It is nearly filled with water, and has at one end about an inch under the surface of the water, a shelf on which jars or vessels, as *b* and *c*, may rest. Any particular air or gas is preserved separate from the atmosphere, by being placed in one of these jars with the mouth downwards. The gas is passed into the jar by the operator first immersing the jar in the trough, so as to fill it with water and to expel the common air from it; and then holding its mouth over the gas while rising under the water from another vessel or pipe:—

Fig. 106.

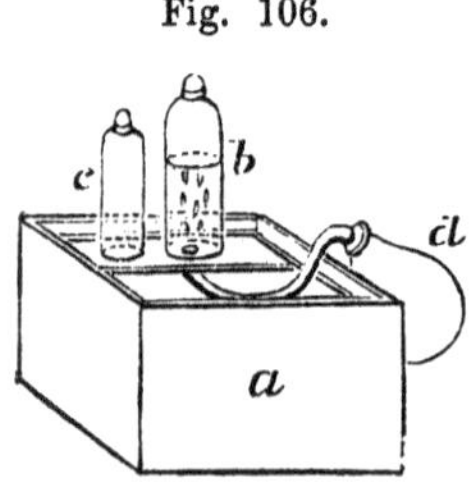

d represents a long-necked vessel, used to contain the ingredients for the production of gases by chemical action. The gas of course rises to the top of the jar *b*, and gradually displaces the water. During the operation of filling, the jar may be supported by the hand or by resting on the shelf;—in the latter case the gas is allowed to rise into it through a hole in the shelf, provided with a small funnel gaping downwards to catch the air more readily. The shelf may have room on it for many jars, and it may have more holes than one; and if the gas under operation be such that water absorbs or changes it, some other liquid, as mercury, may be used instead of water.

A *gasometer* or *gas-holder*, is merely a larger jar or vessel as *a*, dipping into water, with its mouth downwards, in a trough of its own shape, *b c*, and so supported or counterpoised by a weight at *d*, over pullies, that very little force suffices to move it up or down. Air forced into it through a pipe *f* opening under it, causes it to rise or float higher in proportion to the quantity. The air is made to pass from it again when wanted, either through the same tube or through another as *e*.

Fig. 107.

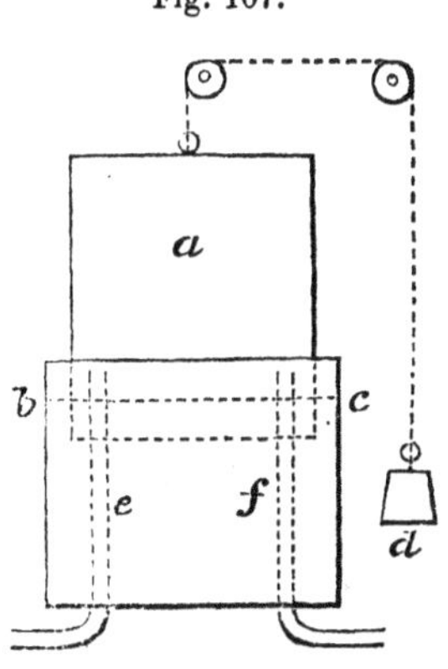

The huge gasometers, exceeding in size an ordinary house, and containing the supply of gas for the lamps of a town, are vessels suspended as above represented, in great pits or troughs, filled with water. The gas issues with force proportioned to the downward pressure of the containing vessel, which may be nicely regulated in a variety of ways, and is generally made to equal the action of a column of water of two inches in height; that is to say, such, that a pipe issuing from the gas holder, and dipping into water at its other end, shall allow gas to escape, if immersed less than two inches perpendicularly.

It would be encroaching on the province of the chemist to treat here particularly of the substances which most generally exist in the aeriform state; but to give an increased interest to the description of the gas apparatus, a few leading facts may be mentioned.

Of about fifty distinct substances known as the materials of our globe, five, when uncombined, and under common circumstances of heat and pressure, exist as air or gases The water used to fill the apparatus above described is a compound of two of the substances, *viz.*, *oxygen* and *hydrogen*. By directing an electrical current through water, it is gradually decomposed, and from one side, a stream of aeriform oxygen may be received, and from the other a stream of hydrogen The two gases may be again united to form water, by mixing them in a proper vessel, and passing an electric spark through them. They combine with explosion.

This *oxygen*, so called from its relation to acids, (the name consisting of two Greek words, signifying *acid* and *to form*,) has been accounted, for many reasons, the most important substance in nature. It forms eight-ninths, by weight of the ocean; one-fourth of the atmosphere; and perhaps one-fourth of the solid matter of the globe: possibly, therefore, although most persons think of it only as an air or gas, there is not a millionth part of the quantity of oxygen in the world existing as air. It unites readily with most other substances, and generally with such intense action as to produce the phenomena of fire or combustion; the word *combustible* chiefly apply to substances that quickly combine with oxygen.

Oxygen assumes a singular variety of character in its different combinations. Thus with *hydrogen*, it forms water; with *lead*, it forms the substance called *red-lead;* with *nitrogen*, in one proportion, it forms *atmospheric air*, in another proportion, the *nitrous oxide*, or what is called the *laughing gas*, in a third, the acid called *aqua fortis;* with sulphur it forms the *sulphuric acid* or *oil of vitriol;* with iron, and all metals it forms their ores called oxides: and so forth. But the most important character in which we know it, is as that ingredient of our atmosphere, without which animals and vegetables cannot live, and fire cannot burn. Oxygen, from this part of its history, was long named *vital or pure air.*

Pure oxygen, in the state of air is a little heavier than common air; but when holding a quantity of charcoal in solution, it forms aeriform *carbonic acid*, which is nearly twice as heavy as common air, and may be poured out of one vessel into another like water. Carbonic acid is what issues from soda-water, brisk ale, champagne, &c., while they sparkle. If drawn into the lungs in breathing, it is fatal to life. A charcoal fire left in a close room with sleeping persons, has often been fatal to them, because carbonic acid gas is the product of the combustion. So likewise, houseless wretches in winter lying down in a brickmaker's field to leeward of a burning-heap of bricks, often fall asleep for ever. The famous *Grotto del Cane*, in Italy, is a cavern always full of carbonic acid, which springs into it from below, as water springs into a well, and runs over like water from a well:—it received its name from the circumstance of dogs dying instantly when thrown into it. Carbonic acid rising in fermentation has often proved fatal to persons leaning over the edge of fermenting vats. It is common to see a rat die instantly, in the attempt to run a plank laid across the mouth of a fermenting tub.

Hydrogen, the other ingredient of water, so called from its relation to *water* (the name consists of the Greek words for *water* and to *form*,) when in the state of air, is sixteen times as light as oxygen. With it balloons are filled. When it holds in solution a certain quantity of carbon or charcoal it becomes the common gas used for illumination, and is the fire-damp of mines, of which the burning and explosion are so terrible. It forms one-ninth of the ocean, and much of the animal and vegetable bodies.

Nitrogen, so called from its relation to *nitric acid*, is the third and last substance which we shall mention. It is what remains of the atmosphere when the oxygen is removed. It forms about four-fifths of the atmosphere, one-fourth of the animal flesh, and is found in small quantities in the other combinations. It will not support life by itself, and therefore formerly was called *azote:* with a larger portion of oxygen, it forms *nitric acid* or the *aqua fortis* of old.

The last few paragraphs may serve to show how many of the manipulations of chemistry are directed by the principles of physics or mechanical philosophy; and therefore, how essential to the chemist the preliminary study of physics becomes.

PART III.

OR

THE PHENOMENA OF FLUIDS.

(CONTINUED.)

SECTION III.—HYDRAULICS—PHENOMENA OF FLUIDS IN MOTION.

ANALYSIS OF THE SECTION.

Whether the particles of matter exist in the form of solid or fluid, the circumstance does not affect their properties of INERTIA *and* GRAVITY.—*Hence liquids and airs, in proportion to their quantity, resist, receive, and impart motion, and have weight and friction, as is true of solids. This is seen in the phenomena of*

1. *Fluids issuing from vessels, or moving in pipes and channels.*
2. *Waves.*
3. *Fluids resisting the motion of bodies immersed in them; or themselves moving against other bodies.*
4. *Fluids lifted, or moved in opposition to gravity.*

"*Fluids issuing from vessels, or moving in channels.*"

WATER admitted to a tube ascending from near the bottom of a reservoir will rise in it, as already explained, to the level of the liquid-surface in the reservoir. If such a tube be afterwards cut off, except a small part at the bottom, then prepared as a jet-pipe, the water will spout from this still to the same height, with a certain deduction for the resistance of the air and friction. Now as a body shot upwards to any height has that velocity in departing, which it again acquires by falling back to the same place or level, (with a certain deduction for the resistance of the air,) as explained at page 60, it follows that fluid issues from any orifice in a reservoir with velocity equal to what a body acquires in falling as far as from the level of the fluid surface in the reservoir to the orifice. By referring then to the law of falling bodies, as explained at page 50, we may learn the velocity of the issue of water in any case, and therefore the quantity delivered by an opening of a given magnitude.

Thus, a body by gravity falls sixteen feet in the first second, with speed gradually increasing, and at the end of the second has a velocity of thirty-two feet per second; therefore a reservoir with an opening of an inch square at sixteen feet below the water's surface, will deliver, in one second of time, with a certain deduction for resistance of air, friction, &c., thirty-two feet of a jet of water of an inch square; and according to the same rule, an opening at four times the depth should deliver a double quantity; at nine times the depth, a triple quantity; and so on, as really happens. An inquirer is at first surprised that the quantity should not be quadruple, where the height of column or pressure forcing it out is quadruple, ninefold when the pressure is ninefold, &c., but on reflection, he may perceive that the real effects, as stated above, are still exactly proportioned to the causes; for when only twice as much water is forced out in the same time, there is still an effect four times as powerful, because each particle of the double quantity issues with twice the force or velocity, and increase of velocity costs just as much force as increase of quantity. Similar reasoning holds with respect to the triple or other quantities. Because a body shot upward with a double velocity gains a quadruple height, (see page 60,) the jet issuing with only double velocity from four times the depth, still reaches the level of the surface of the reservoir.

The knowledge of this rule for discharging orifices is of the greatest importance in the construction of water-works, because when joined with other rules assigning the effects of friction, bending, unequal width, &c., in pipes, it ascertains the quantity of water which a conduit of any magnitude, length and slope, will deliver.

It is a curious fact, that more water issues from a vessel through a short pipe, than through a simple aperture of the same diameter as the pipe; and still more if the pipe be funnel-shaped, or wider towards its inner extremity.

The explanation is, that the issuing particles coming from all sides to escape, cross and impede each other in rushing through a simple opening, as is proved by the narrow neck which the jet exhibits a little beyond the opening; but in a tube, this narrowing of the jet cannot happen without leaving a vacuum around the part, and the pressure of the atmosphere, resisting the vacuum, causes a quicker flow. The funnel-shape again leads the water by a more gradual inclination to the point of exit, and thus considerably prevents the crossing among the particles; besides that, because its mouth surrounds the narrow neck of the jet, it allows that part to be deemed the commencement of the jet.

Another remarkable effect of atmospheric pressure on running liquids is, that in a tube of considerable length, descending from the reservoir, it much quickens the discharge. Water naturally falls like any other body with accelerating velocity, but if it so fall in a tube which it fills like a piston, either portions of it below must outstrip portions above, leaving vacuous spaces between, or water from above must be pressed into the tube by some other force than its weight. Now the atmospheric pressure becomes this force, and it prevents a vacuum, partly by impelling water more rapidly into the top of the tube, and partly by resisting the discharge from below. The forcing in of the water at the top of the tube causes that depression of the water surface in the reservoir over it, which becomes more conspicuous as the depth in the reservoir diminishes, and at last is a deep hole in the water extending far into the tube, and sometimes even as in a common funnel extending quite through.

The friction or resistance which fluids suffer in passing along pipes is much higher than might be expected. It depends on the cohesion of the particles to the surface of the pipe and among one another, and on the particles near the outside being constantly driven from their straight course by the irregularities in the surface of the pipe. An inch tube of two hundred feet in length, placed horizontally, is found to discharge only a fourth part of the water which escapes by a simple aperture of the same diameter. All passing along tubes is still more retarded. A person who erected a great bellows at a water-fall, to blow a furnace two miles off, found that his apparatus was totally useless. When gas lights were first proposed, some engineers feared that the resistance by friction to the passing air would be fatal to the enterprise.

Higher temperature in a liquid increases remarkably the quantity discharged by an orfice or pipe,—apparently by diminishing that cohesion of the particles which exists in certain degrees in all liquids and affects so much their internal movement. The additition of 100 degrees of heat will, in certain cases, nearly double the discharge.

The flux of water through orifices under uniform circumstances is so steady, that before the invention of clocks and watches, it was employed as a means of measuring time. The vessels were called *clepsydræ*. That of Ctesibius is famous, in which the issuing water took the form of tears from the eyes of a figure, deploring the rapid passing of precious time; and these tears being received into a fit vessel, gradually filled it up and raised another floating figure, who pointed to the hours marked on an upright scale. This vessel was daily emptied by a syphon, when charged to a certain height, and its discharge worked machinery which told the month and the day. The common hour-glass of running sand is another modification of the same principle, with this remarkable difference, however, that depth of the sand does not quicken the flux.

The progress of water in an open conduit, such as the channel of a river or acqueduct, is influenced by friction, &c., in the same manner as in close pipes. But for this, a river like the Rhone, drawing its waters from the elevation of 1,000 feet above the level of its mouth, would pour them out, with the velocity of water issuing from the bottom of a reservoir, 1,000 feet deep; that is to say, at the rate of about 170 miles per hour. The ordinary flow of rivers is about three miles per hour, and their channels slope three or four inches per mile.

The velocity of a water current is easily ascertained by immersing in it an upright tube, of which the bottom bent at right angles becomes an open mouth turned towards the stream. The water in the tube will stand above the surface of the stream, as much as would be necessary in a reservoir, according to the explanation given above, to cause a velocity of jet equal to the velocity of the stream. A modification of this contrivance may be made to measure the velocity of the wind. A common mode of telling the velocity of an open stream, is to observe with a stop-watch the progress of a body floating in some part of it from which its medium speed may be known; and knowing that speed and the depth and width of the channel the quantity delivered in a given time becomes a matter of simple calculation. The speed of the wind may be ascertained by observing how long the shadow of a cloud takes to pass across a field of known dimensions.

Fig. 108.

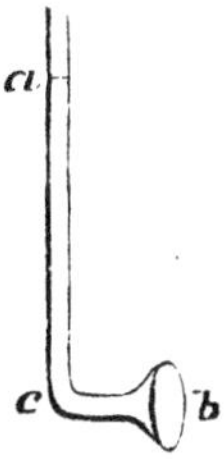

The friction of water moving in water is such, that a small stream directed through a pool, with speed enough to rise over the opposite bank, will soon empty the pool. Extensive fens have been drained on this principle. The friction between air and water is also singularly strong, and is proved on a great scale by the magnitude of the ocean-waves, which is a consequence of it; and on a small scale, by the amusing experiment of making a light round body dance or play upon the summit of a water-jet—a chief cause of its remaining there being, that the current of air which rises around the jet by reason of the friction, presses it inward again, whenever it inclines to fall over. Oil thrown upon the surface of water, soon spreads as a film over it. and defends it from farther contact and friction of the air. If oil be thus spread at the windward side of a pond where the waves begin, the whole surface of a pond becomes as smooth as glass; and even out at sea, where the commencement of the waves cannot be reached, oil thrown upon them smooths their surface to leeward of the place, and prevents their curling over or breaking. It is said that boats having to reach the shore through a raging surf, have been preserved by the crews first spilling a cask of oil in the offing.

The most magnificent examples that ever existed, or probably ever will exist of artificial water-courses, were the acqueducts of ancient Rome, about twenty in number. Several of them exceeded forty miles in length, passing through hills in their way, and resting on tiers of splendid arches across the valleys. They were constructed of such durable materials, and so skilfully, that the principal of them remain perfect to this day. Considered as one object, they rank, in point of magnitude, with any other work of human labour, not excepting the pyramids of Egypt.

While the acqueducts are cited as specimens of grandeur, we may mention the fountains in the gardens of France and Italy as specimens of beauty. Those at Versailles are well known. In them the most magical effects are produced by varying the ways in which water is made to spout from orifices. In one place it is seen darting into the air as a single upright pillar: in others many such pillars rise together, like giant stalks of corn; sometimes an inclination given to the jets makes them bend so as to form beautiful arches, of which a portion appear as the roofs of apartments built of water while others mingle together with endless variety: here and there water-throwing wheels throw out spiral streams, and hollow spheres with a thousand openings are the centres of immense bushes or trees of silvery boughs. Such effects amidst cascades, smooth lakes, and scenes of lovely landscapes, constitute a whole as enchanting, perhaps, as art by moulding nature has ever produced.

"*Waves.*"

The form, magnitude and a velocity of waves, are subjects admitting of deep mathematical research; and are rendered the more interesting, because certain phenomena of *sound* and *light* are of kindred nature. Here, however, they must be treated with great brevity.

A stone thrown into a smooth pond, causes a succession of circular waves to spread from the spot where it falls as a common centre. They become of less elevation as they expand, and each new one is less raised than the preceding, until gradually the liquid mirror becomes again perfect as before. Several stones falling at the same time in different places, cause crossing circles which, however, do not disturb the progress of one another—a phenomenon seen in beautiful miniature at each leap of the little insects which cover the surface of our pools in the calm hours of summer. The rationale of the formation of waves in such cases is as follows: When the stone falls into the water, because the liquid is incompressible, a part of it is displaced laterally, and becomes an elevation or circular wave around the stone. This wave then spreads outward in obedience to the laws of fluidity, already explained, and the circle is seen to widen. In the meantime, where the stone descended, a hollow is left for a moment in the water, but owing to the surrounding pressure, is soon filled up, chiefly by a sudden rush from below. The rising water does not stop, however, at the exact level of that around, but like a pendulum sweeping past the centre of its arc, it rises almost as far above the level as the depression was deep. The central elevation now acts as the stone did originally, and causes a second wave, which pursues the first; and when the centre subsides, like the pendulum still, it sinks again almost as much below the level as it had mounted above: hence it has to rise again, again to fall, and so on for many times, sending forth a new wave at each alternation. Owing to the friction among the particles of the water, each new wave is less raised than the preceding, and at last the appearance dies away.

A wave passing through any gap or opening, spreads from it as a new centre; and a wave coming against a perpendicular surface of wall or rock, is completely reflected from this, and acquires the appearance of coming from a point as far beyond the reflecting surface, as its real origin or centre is distant on the side where it is moving.

So absolutely level is a liquid surface, and so sensitive or mobile, that the effect of any disturbing cause is perceived at great distances. A boat rowed across a still lake, ruffles its surface to a great extent; and although the widening waves become at last such gentle risings as not to be perceptible to the eye, they still produce a rippling noise where they fall among the pebbles on shore. In seas liable to sudden but partial hurricanes, the roar of breakers on distant coasts often tells of the storm which does not otherwise reach them. The author once, in the eastern ocean, had an opportunity of contemplating waves of extraordinary magnitude rolling along during a gloomy calm, and therefore with unbroken surface, appearing like billows of molten lead. At that very time, about a hundred and fifty miles to the north-east, four of the finest ships of the India Company were perishing in a storm.—In the polar seas which are comparatively tranquil, because partially defended from the wind by the floating islands of ice, a few sudden waves are occasionally observed, and quickly all is calm again. Such a phenomenon announces, that the occurrence described at page 153 has happened somewhere, of an island of ice turning over, when the place of its centre of gravity is changed by partial melting.

The common cause of waves is the friction of the wind upon the surface of the water. Little ridges or elevations first appear, which, by continuance of the force, gradually become loftier and broader, until they are the rolling mountains seen where the winds sweep over a great extent of water. The heaving of the Bay of Biscay, or still more remarkably, of the open ocean beyond the southern capes of America and Africa, exhibits one extreme, and the stillness of the tropical seas, which are sheltered by near encircling lands exhibits the other. In the vast archipelago of the east, where Borneo and Java and Sumatra lie, and the Molucca Islands and the Philippines, the sea is often fanned only by the land and sea breezes, and is like a smooth bed in which these islands seem to repose in bliss—islands in which the spice and pefume gardens of the world are embowered, and where the bird of paradise has its home, and the golden pheasant, and a hundred other birds of brilliant plumage, among thickets so luxuriant, and scenery so picturesque, that European strangers find there the fairy land of their youthful dreams.—One who has visited these islands in his early days, may perhaps be pardoned for thus adverting to their beauties.

In rounding the Cape of Good Hope, waves are met with, or rather a swell, so vast, that a few ridges and a few depressions occupy the extent of a mile. But these are not so dangerous to ships, as what is termed a *shorter* sea, with more perpendicular waves. The slope in the former is comparatively gentle, and the rising and falling are much less felt; while among the latter, the sudden tossing of the vessel is often destructive. When a ship is sailing directly before the wind, over the *long swell* now described, she advances as if by leaps; for as each wave passes, she is first descending headlong on its front, acquiring a velocity so wild that she can scarcely be steered; and soon after, when it has glided under her, she appears climbing on its back, and her motion is slackened almost to rest, before the following wave arrives. To a passenger perched at such a time on the extremity of the bowsprit, and looking back on the enormous body of the ship, with perhaps its thousands of a crew, a hundred feet behind him, heaved by those billows as a cork is on a ruffled lake, the scene is truly sublime. When a coming wave lifts the stern and in the same degree depresses the bow, he is deep in the hollow or valley between the waves, and sees only the ship rushing headlong down towards him as if to be engulphed; but soon after, when the stern is down, and the bow is raised, he looks from his station in the sky upon an awful scene beneath him and around.

The velocity of waves has relation to their magnitude. The large waves just spoken of, proceed at the rate of from thirty to forty miles an hour.—It is a vulgar belief that the water itself advances with the speed of the wave, but in fact the *form* only advances, while the *substance*, except a little spray above, remains rising and falling in the same place, with the regularity of a pendulum. A wave of water, in this respect, is exactly imitated by the wave running along a stretched rope when one end is shaken; or by the mimic waves of our theatres, which are generally undulations of long pieces of carpet moved by attendants. But when a wave reaches a shallow bank or beach, the water becomes really progressive, for then, as it cannot sink directly downwards, it falls over and forwards, seeking the level.

So awful is the spectacle of a storm at sea, that it generally biases the judgment; and, lofty as waves really are, imagination pictures them loftier still. Now no wave rises much more than ten feet above the ordinary sea-level, which, with the ten feet that the surface afterwards descends below this, give twenty feet for the whole height, from the bottom of any water-valley to an adjoining summit. This is easily verified by a person who tries at what height on a ship's mast the horizon remains always in sight over the top of the waves—allowance being made for accidental inclinations of the vessel, and for her sinking in the water to considerably below her water line, at the time when she reaches the bottom of the hollow between the two waves. The spray of the sea, driven along by the violence of the wind, is of course much higher than the summit of the liquid wave; and a wave coming against an obstacle, or entering a narrow inlet, may dash to an elevation much greater still. At Eddystone light house, which is about ninety feet high, placed on a solitary rock ten miles from land, when a surge breaks which has been growing under a storm all the way across the Atlantic, it often dashes to 100 feet above the lantern at the summit.

The magnitude of waves is well judged of when they are seen breaking on an extended shore or beach. In the deep sea the wave is only an elevation of the water, sloping on either side; but as it rolls towards the shore, its front becomes more and more perpendicular, until at last it curls over and falls with its whole weight, and when several miles of it break at the same instant, its force and noise may shake the country abroad.

Along the east, or Coromandel Coast of India, at certain season, vast waves are constantly breaking; and as there are no good harbours there, communication between the sea and land is rendered impossible to ordinary boats. The natives of the coast, at Madras, for instance, have hence become almost amphibious. They reach ships beyond the breakers by the help of what are called *catamarans*, consisting of three small logs of wood tied together. On these they secure themselves, and boldly advance up to the coming wall of water, which they shoot into, and rise to the smooth surface beyond it, like water-fowls after diving. Boats unsuited to the breakers often perish in them. The author of this work had gone on shore with a watering party on the coast of Sumatra, and during the hours spent there, a swell had risen in the sea, which on their return was already bursting along the beach and across the river's mouth in lofty breakers. The boat in which he happened to be, regained the high sea in safety, but a larger boat which followed at a short distance was overwhelmed, and an officer and part of the crew perished.

There is a phenomenon observed at the mouths of many great rivers, called the *Boar*, which has resemblance to a wave. When the tide returning from sea meets the outward current from the river, and both have the force which in certain situations belongs to them, the stronger mass from the ocean assumes the form of an almost perpendicular wall, moving inland with resistless sweep. This is called the boar. It is in fact the great sea-wave of the tide, produced twice a-day by the attraction of the moon, rolling in upon the land and inlets, where contracting channels concentrate its mass. In the different branches of the Ganges the boar is seen in a remarkable degree. Its roaring is heard long before it arrives. Smaller boats and skiffs cannot live where it comes; and as it passes the city of Calcutta, even the large ships at anchor there are thrown into such commotion, as sometimes to be torn away from their moorings.—The nature and effects of this boar are strikingly illustrated upon certain coasts where extensive tracts of sand are left uncovered at low water. In such situations, of which there are many on the western shores of Britain, the returning tide is seen advancing with steep front, and with such rapidity, that the speed of a galloping horse can scarcely save a person who has incautiously approached too near. Many, every year, are the victims of temerity or ignorance on these treacherous plains.

In the end of the year 1831, on the low flat coast of the Indian peninsula, north of Madras, one great wave of the kind now described was produced during a very high-spring tide of midnight, by an extraordinary wind, and spread ten miles in upon the inhabited land. It had retired with the ebbing tide before morning, but the next day's sun disclosed a scene of devastation rarely matched. Amidst the total wreck of the villages and fields, there lay the drowned carcases of more than ten thousand human beings, mixed with those of elephants, horses, bullocks, wild tigers and the other inhabitants of the land.

It has been proposed lately to construct *sub-marine boats*, or vessels calculated to swim so deep in the water as to be below the superficial motion of the waves, and therefore beyond the influence of storms at the surface. Such a boat has been tried with considerable success; and man's increasing familiarity with sub-marine matters since the invention of the diving-bell, may ultimately lead to improvements rendering the sub-marine vessel, for certain purposes, commodious and safe.

"*Fluids resisting the motion of bodies immersed in them, or themselves moving forcibly against other bodies.*" (See the Analysis.)

The same force is required to give or to take away, or to bend motion, in a fluid, as in an equal quantity of solid matter. A pound of water enclosed in a bladder is not more easily thrown to a given height than a pound of ice or of lead; nor, if falling into the scale of a weighing beam, does it require less as a counterpoise; nor, if made to revolve at the end of a sling, does it render the cord less tight.

A convenient measure of the force of moving water on an obstacle, or of the resistance of still water to a moving body, exists in the facts already explained, that the pressure of a known height of fluid column produces from an orifice a certain velocity of jet, while conversely, that jet, or a current of equal speed, directed against the orifice supports the column. The impulse given or received, therefore, by a flat surface in water, such as the vane of a water-wheel, whether that of a steam-boat pressing against the water, or that of a corn-mill pressed by it, is measured by the weight of the column alluded to, the height of which is, according to the velocity and the breadth or diameter, according to the breadth or extent of the solid surface concerned. This estimate supposes that the pressure of air upon the surface is direct; if it be oblique, there is a diminution according to the rule given under the head of "resolution of forces."

Many persons looking carelessly at the subject of fluid resistance, would expect that if a body, as a boat, moving through a fluid at a given rate, meets a given resistance, it should just meet double resistance when moving twice as fast. Now the resistance is four times greater with a double rate.

This fact is but another example of a principle already explained, and when more closely examined, is easily understood. A boat which moves one mile per hour, displaces or throws aside a certain quantity of water, and with a certain velocity;—if it move twice as fast, it of course displaces twice as many particles at the same time, and requires to be moved by twice the force on that account; but it also displaces every particle with a double velocity, and requires another doubling of the power on this account; the power then being doubled on two accounts becomes a power of four. In the same manner with a speed of three, three times as many particles are moved and each particle with three times the velocity; therefore, to overcome the resistence, a force of nine is wanted; for a speed of four, a power of sixteen; for a speed of five, a power of twenty-five, and so forth: the relations being that which mathematicians indicate by saying *that the resistance increases as the square of the speed.* The corresponding numbers, up to a speed of ten are as here shown.

Speed . .	1	2	3	4	5	6	7	8	9	10
Corresponding resistance	1	4	9	16	25	36	49	64	81	100

Thus, if even the resistance at the bow of a vessel was all that had to be considered, the force of one hundred horses would only drag the vessel ten times as fast as the force of one horse. But there is another important element in the calculation, *viz*, the lessening, as the vessel's speed quickens, of the usual water pressure on the stern,—which pressure while she is at rest is equal to the pressure on the bow, and the force therefore required to produce an increased velocity is still considerably greater than as noted in the table.

There is not a more important truth in physics than the law of fluid resistance to moving bodies here treated of; it explains so many phenomena of nature, and becomes a guide in so many matters of art. We will now set forth some interesting examples.

It explains at what a heavy expense of coal high velocities are obtained in steam-boats. If an engine of about 50 horse power would drive a boat 7 miles an hour, two engines of 50, or one of 100 would be required to drive it 10 miles, and three such to drive it 12 miles, even supposing the increased resistance to the bow, as already stated, to be the measure of the whole work done, which it is not, and that engines worked to the same advantage with a high velocity as with a low, which they do not.—For the same reasons, if all the coal which a ship could conveniently carry were just sufficient to drive her 1,000 miles, at a rate of 12 miles per hour, it would drive her more than 3,000 at a rate of 7 miles per hour; and more than 6,000 at a rate of 5 miles per hour. This is a very important consideration for persons concerned in steam navigation to distant parts.

The same law shows the folly of putting very large sails on a ship; the trifling advantage in point of speed by no means compensating for the additional expense of making and working the sails, and the risk of accidents in bad weather. The ships of the prudent Chinese have not, for the same tonnage, one-third so much sail as those of Europeans, and yet they move but little slower on that account. A European ship under jury-masts does not lose so much of her usual speed as most people would expect.

This law explains also why a ship glides through the water one or two miles an hour when there is very little wind, although with a strong breeze she would only sail at the rate of eight or ten miles. Less than the 100th part of that force of wind which drives her ten miles an hour, will drive her one mile per hour, and less than the 400th part will drive her half a mile. Thus, also, during a calm, a few men pulling in a boat can move a large ship at a sensible rate.

These considerations show strikingly of what importance to navigation it might be to have, as a part of a ship's ordinary equipment, one or two water-wheels, (or ready means of forming them,) to be affixed upon the ship's side when required, like the paddle-wheels of a steam-boat, and by turning which, the crew might easily deliver themselves from the tedium, or even disastrous consequences of a long calm at sea.—This idea occurred to the author while in a ship completely becalmed for weeks on the Line: during which wearisome periods, the breezes were often seen roughening the water a mile or two farther on; and any means that could have enabled the ship's company to advance her that little distance might have saved the delay. The wheels might be driven by connection with the capstan, at which, under such circumstances, the crew would most willingly work. Delay in a large vessel often costs hundreds of pounds per day, and may retard the execution of important projects.—But the propelling of a ship in a calm seems, by no means, the most important purpose which such wheels might serve. If from disease, fatigue, or other cause, the crew were inadequate to existing necessities, two wheels affixed to the extremities of an axis crossing the ship might be equivalent in many cases to additional hands, or to a steam-engine of great power; for when acted upon by the water as the ship sailed, they would turn with the force of water-wheels on shore, and might be made to move the pumps, to hoist the sails, and to do any work which a steam-engine could perform. Many a gallant vessel has perished because the exhausted crew could not longer labour at the pumps, where such water-wheels as now contemplated, or a wind-mill wheel in the rigging would have performed the duty most perfectly.

The law that resistance to a body moving in a fluid increases in a greater proportion than in speed of the body, applies where the fluid is aeriform, as well as where it is liquid.

A bullet shot through the air with a double velocity, for the reason assigned above, experiences four times as much resistance in front, as with a single velocity: the motion is retarded also by the diminution of the usual atmospheric pressure of 15 lbs. per inch on the posterior surface, which diminution is proportioned to the speed. It is farther true, that when the velocities of bodies moving in the air are very great, the resistance increases in a still quicker ratio than in liquids,—probably because the compressibility of air allows it to be much condensed or heaped up before the quick moving body. It is useless to discharge a cannon-ball with a velocity exceeding 1,200 feet in a second, because the powerful resistance of the air to any velocity beyond that, soon reduces it to that at least.

The rule of reciprocal action between a solid and fluid, now explained, holds equally when the fluid is in motion against the solid, as when the solid moves through the fluid.

If a ship be anchored in a tide's way, where the current is four miles an hour, the strain on her cable is not one fourth part so great as if the current were eight miles.

A wind moving three miles an hour is scarcely felt: if moving six miles, it is a pleasant breeze; if twenty or thirty miles, it is a brisk gale; if sixty, it is a storm; and beyond eighty, it is a frightful hurricane, tearing up trees and destroying every thing.

Supposing the wind to move one hundred miles per hour, there are one hundred times as many particles of matter striking any body exposed to it, as when it moves only one mile per hour, and each particle strikes, moreover, with one hundred times the velocity or force, so that the whole increase of force is a hundred times a hundred, or ten thousand. This explains how the soft invisible air may by motion acquire force sufficient to unroof houses, to level oaks which have been stretching their roots around for a century, and in some West India hurricanes, absolutely to brush every projecting thing from the surface of the earth.

The law of rapidly increasing resistance assigns a limit to many velocities, both natural and artificial.

It limits the velocity of bodies falling through the air. By the law of gravity, a body would fall with a constantly accelerating speed, but as the resistance of the air increases still more quickly than the speed at a certain point, this resistance and the gravity balance each other, and the motion becomes uniform.

The *parachute*, by means of which a person may safely descend to the earth from a balloon at any elevation, furnishes a good example. The contrivance resembles a large flat umbrella. The aeronaut attaches himself underneath it, and when it is let loose from the balloon, he is partly supported by the resistance which its broad expanse experiences in falling through the air, and falls, therefore, in a corresponding degree more slowly. After the first second or two, for the reason stated above, it descends with a uniform motion; and its breadth is generally made such, as to allow a velocity of about eleven feet in a second, or that which a man acquires in jumping from a chair two feet high.

No ship sails faster than fifteen miles in an hour.—And it is because the resistance to be overcome in steam-carriages on rail-ways, *viz.*, their friction, does not increase with their velocity like the fluid resistance to steamboats, that the speed of the former may so much exceed that of the latter.

No fish swims with a velocity exceeding twenty miles an hour; not the dolphin, when shooting ahead of our swiftest frigates, nor the salmon, when darting forward with a speed which lifts him over the water-fall.

And the flight of birds through the thin air has a limited celerity. The crow, when flying homewards against the storm, cannot face the wind in the open sky, but skims along the surface of the earth in the deep valleys, or wherever the swiftness of the wind is retarded by terrestrial obstructions. The great albatross, stemming upon the wing the current of a gale so as to keep company with a driving ship where the air is passing at the rate of a hundred miles an hour, often takes shelter momentarily under the lee-side of the lofty billows. The bird called the *stormy petrel* abides chiefly in the midst of the Atlantic Ocean, but the irresistible violence of the wind occasionally sweeps it from the waves, and causes its appearance on the western shores of Europe. Vessels from the high sea, approaching a coast from which the wind blows, generally become resting places to exhausted land birds driven off the shore by wind which they have not had strength of wing to stem;—sad evidences of the myriads which are constantly perishing where no resting place is found, and where no human eye notes their fate.

The action or resistance between a meeting fluid and solid, is influenced by the shape of the solid.

This follows from what has already been said of direct and oblique impulse. If a flat surface directly opposed to the fluid experience a certain resistance, a projecting surface like that of a sphere or short wedge is resisted in a less degree, and a concave surface in a greater. The explanation is, that a flat or plane surface throws the particles of fluid almost directly outwards from its centre to its circumference, and therefore with greater velocity, while the convex or wedge-like surface, although displacing them just as far, still does so more slowly, and therefore with less expenditure of force, in proportion to the obliquity of surface, or as its point is in advance of its shoulder or broadest part; and a concave surface must give to some of the particles a forward as well as lateral motion. The shape of the hinder part of a solid moving through a fluid is of importance for corresponding reasons.

The following are instances of projecting or wedge-like surfaces, intended to diminish the resistance. Fishes are wedge-like, both before and behind, their form being modified, however, in relation to other objects than mere speed of motion. Birds are so, also; and they stretch out their necks while flying, so as to make their form perfect for dividing the air. In the form of the under part of boats and ships, men have, in a degree, imitated the shape of fishes. The light wherries which shoot about upon the surface of the Thames, appear the very essence of what imagination can picture of form combining utility and grace. There are boats used in China called *snake-boats*, which are only a foot or two broad, but perhaps a hundred feet in length, and when moved, as they often are, by nearly a hundred rowers, their swiftness is extreme. The problem of which it is the object to assign for a ship's hull or bottom the best possible form that she may have speed of sailing, is not yet completely solved; so that a kind of empiricism prevails in the matter, and very unexpected results often arise. Yet the subject merits much attention, for when vessels have to chase and to flee, speed becomes of the greatest importance; and at all times the sailor's heart swells with delight to find his well-beloved vessel performing well.

The following instances exhibit the mutual influence of meeting solids and fluids, where the surface of the solid is plane or concave.—In a water-wheel, whether the water be moving against the wheel, as is the case where a stream acts to drive machinery, or the wheel be moving against the still water, as in the case of the paddle-wheels of a steam-boat, the extended faces of the vanes or float-boards give or receive a powerful impulse. When a wheel with float-boards has its lower part merely dipping into a stream of water, to be driven by the momentum, it is called an *undershot-wheel;* when the water reaches the wheel near the middle of its height, and turns it by falling on the float-boards of one side as they sweep downwards in a curved trough fitting them, the modification is called a *breast-wheel;* and when the float-boards are shut in by flat sides, so as to become the bottoms of a circle of cavities or buckets surrounding the wheel, into which the water is allowed to fall at the top of the wheel; and to act by its weight instead of its momentum, the modification is called the *overshot-wheel.* To have a maximum of effect from wheels moved by the momentum of water, they are generally made to turn with a velocity about one-third as great as that of the water; and wheels moved by the simple weight of water usually have their circumference turning with a velocity of about three feet per second. The subject of water-wheels is one of the most important in practical mechanics; for moving water performs a great deal of labor for man.

Oars for boats are made flat, and often a little concave, that the mutual action between them and water may be as great as possible. The webbed feet of water-fowl are oars; in advancing, they collapse like a shutting umbrella, but open outwards in the thrust backwards, so as to offer a broad concave surface to the water. The expanded wings of birds are in like manner a little concave towards the air which they strike. The sails of ships, when they are receiving a fair wind, are left slack so as to swell and become hollow.

The resistance between a meeting solid and fluid being nearly proportioned to the breadth of the solid, it follows that large bodies, because containing more matter in proportion to their breadth or surface than smaller bodies of similar form, are less resisted in proportion to their weights, than smaller bodies.

The science of measures tells us that a bullet or other solid of two inches diameter, has eight times as much matter in it as a similar solid of one inch diameter, while it has only four time the breadth or surface. Thus, by putting eight dice or little cubes together, as here represented, we have a larger cube, of which compared with a single dice, the edge is evidently *twice* as long, the surface *four* times as great; and the quantity of matter *eight* times as great;—again, *twenty-seven* dice similarly put together form a cube with sides *three* times as long, and the surface *nine* times as great; and sixty-four dice form a cube with sides *four* times as long, and a surface sixteen times as great. All solids similar have to each other this kind of relation, which, in the language of the science of quantity, is called the relation of cubes: they are said to be to each other as the cubes of any of their corresponding lines. Hence if a bullet of eight pounds, and a bullet of one pound be shot off with equal velocity, because that of eight pounds has only half as much surface in proportion to its weight, and therefore to its moral inertia or force, as the other, it will go much farther than the other.

Fig. 109.

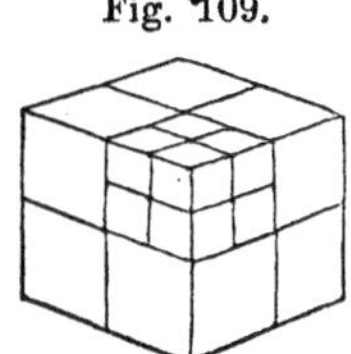

This important rule explains why shells and large shot may be thrown four or five miles, while smaller cannon-balls, musket-bullets, pistol and swan shot, and the common small shot of the sportsman, all of which are generally discharged from their respective pieces with the same commencing velocity, have a shorter range, as the size of the projectile is less. Even water is sometimes thrown from a gun or powerful syringe to stun birds, that they may be obtained with uninjured plumage; but it soon divides in the air so minutely that it reaches only a short distance.

Water falling through the air from a great height, goes on suffering a gradual division into smaller and smaller portions, which at last may be said to be nearly all surface: and then the resistance of the air lets them fall very slowly indeed. The relation of the size and resistance is well shown by the difference of celerity in the descent of a minute fog, a drizzling mist, and common rain. The toy called the *water-hammer*, is merely a little water enclosed in a tube exhausted or empty of air; and when, by turning the tube, the water is made to fall from one end to the other, as there is no air to impede or divide it in its descent, it falls as one mass, and makes a sharp noise like the blow of a hammer.

The same law explains why a spider's thread or a single filament of silk floats so long in the air before it falls;—why there is almost constantly suspended in the air, wherever active man resides, that immense quantity of very minute solid particles, which when rendered visible by the sun's light passing directly through them, are called motes in the sunbeam—particles of which are constantly settling on household furniture, and rendering necessary the daily operation of dusting or cleaning;—why the fine dust sent aloft during the eruption of volcanoes is often carried by the wind to a distance of hundreds of miles;—why in the deserts of Africa the strong winds often transport fine sand from place to place, overwhelming caravans, and forming new mountains, which succeeding blasts are again to lift;—why in the bottom of a river, or in a tides-way, fine mud is found where the current is slow; sand where it is quicker; pebbles, or large stones, where it is quicker still; while in rapids and water-falls, only massy rocks can resist the fluid force. Now rocks, pebble, sand and mud, may all be the same material in portions of different magnitude.

This law explains the operation of *levigating*, by which substances insoluble in water are obtained in the state of a very fine powder. Any such substance is first ground or powdered in the ordinary way, and mixed with water. The grosser parts then soon fall to the bottom, while the fine dust remains longer suspended. This is afterwards obtained separately by pouring the fluid which bears it into another vessel, and allowing more time for the slow subsidence. The fine powder of flint used in the manufacture of porcelain is obtained by levigation; as is also that of calamine stone, and other powders used in medicine and various arts.

This law farther explains how, by means of air or water, bodies of different specific gravities, although mixed ever so intimately, may be easily separated. If pieces of cork and lead be let fall together through the air, the lead will reach the ground first, and may be swept away before the cork arrives; but in a vacuum the whole would reach the ground at the same time as is proved by the common experiment of the guinea and feather falling in the exhausted receiver of an air-pump. Again, when a mixture of corn and chaff, as it comes from any threshing machine, is showered down from a sieve in a current of air, the chaff being longer in falling, is carried farther by the wind, while the heavier corn falls almost perpendicularly. The farmer, therefore, by *winnowing* in either a natural or artificial current of air, readily separates the grain from the chaff; and, if he desire it, may even divide the grain itself into portions of different quality. Similar to the operation of separating chaff from corn by wind, is that of separating sand or mud from gold-dust by water:—the soil containing gold-dust is first spread on a flat surface, over which a current of water is then made to pass; which current carries away the lighter rubbish, and leaves the gold. If a mass of metal be affixed on the end of a rod of wood, the rod then, whether simply falling through the air, or advancing as an arrow, will follow the heavier metal as its points. The cork of a shuttlecock is always foremost for the same reason.

The instances enumerated under this head serve to show how many and varied the results may be which flow from a single principle.

When a fluid and a solid meet each other obliquely, the impulse or effect is still perpendicular to the surface of the solid, as if they met directly, but is less forcible as the obliquity of the approach is greater.

Suppose a b to represent the upper edge of a smooth board or of any flat polished surface, standing in a current, the fluid approaching this surface in whatever direction, must act upon it as if approaching perpendicularly, because on account of its smoothness, the fluid can take no hold of it to push it endways, either towards a or b. But the impulse of a stream acting on the surface will be less forcible if the surface be oblique to the stream, both because less fluid will touch, and because the velocity of the effective approach will be less. The line c d marks the breadth, and therefore force, of the part of a stream reaching the board directly; and the shorter line f c marks the smaller breadth that can touch it, of a stream coming obliquely in the direction c b: in the oblique stream, moreover, if the line c b mark the whole velocity, the shorter line c a marks the slower rate of the *direct* approach of any one particle to the board. (This subject was treated of at page 57, under the head of *Resolution of Forces.*)

Fig. 110.

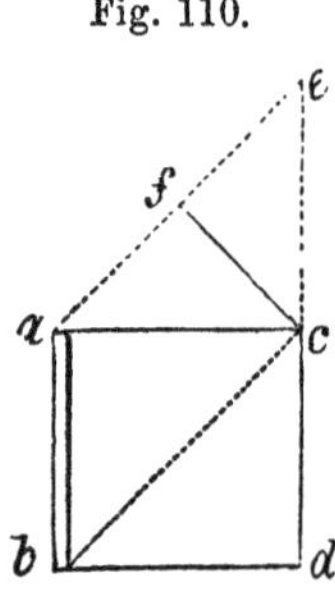

Hence the wind blowing upon the sail of a ship, however obliquely always presses it directly forward or perpendicularly to its surface, but acts less forcibly as the obliquity is greater. If the wind be represented, as to direction and strength, by the line *e d* approaching the sail *a b*, it will act on the sail as if it came from *f*, but with the smaller force *f d*, instead of the whole force *e d*. The effect, therefore, is the same as if the sail were pulled by the rope *d c*. We see in this how a ship can be made to sail in a certain degree against the wind:—for all the sails being adjusted so as to receive the wind in the direction here shown, they all act to produce the same result as if ropes were pulling from each in the direction of *d c;* and a force like *f d*, or a rope like *d c*, urging sideways as well as forwards—as instanced in the tow-rope of a canal boat—makes the vessel advance rapidly forward, but scarcely at all sideways, because the form of vessels causes them to pass forward at least twenty times more easily with their sharp bow than sideways with the long keel; and therefore a force urging equally sideways and forwards makes a ship advance twenty miles in the direction of her keel, that is forwards, for one mile which she deviates sideways. The deviation sideways, which, in sailing vessels, must take place to a certain extent whenever the wind is at all oblique, is called the *lee-way*.

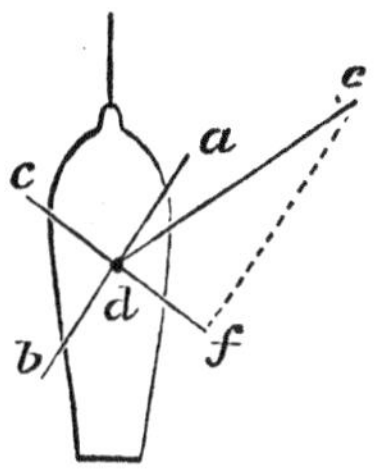

Fig. 111.

Fig. 112.

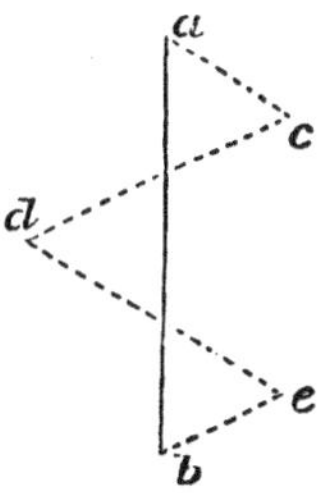

A vessel having to sail from *b* to *a*, while the wind blows directly against her course, or from *a* to *b*, is obliged to sail *close to the wind*, as represented in fig. 112, first perhaps to *e*, as represented by this figure, with the left or larboard side to the wind, then to *tack* as it is called, or turn around, at *e*, and to sail to *d*, with the right or starboard side in the wind; then to go on the larboard tack again to *c*, and thence to the port at *a*.

In making way against a *contrary wind*, the sails of a ship are pointed so nearly edgewise to the wind, that unless very flat, a great portion of their surface becomes useless. The Chinese manner of rigging is, in this respect at least, superior to the European; for in it bamboo reeds attached across the sails, render them as flat as boards. When a Chinese ship has her sails pointed edgeways to a spectator, he only sees the masts which support them.

The reason why a ship of several masts generally sails faster when the wind is more or less from a side, than when directly astern, is, that in the former case all the sails are acting, although individually not to the best advantage, while in the latter the sails in front are becalmed by those behind them. A ship with a side-wind may move faster than the wind itself, as is often true of the outer extremities of the wind-mill's vanes. A corresponding relation of motions is observed when a slippery wedge is forced out two or three inches laterally from its place, by a weight which descends only one inch perpendicularly.

The law now under consideration explains the action of the *rudder* of ships,—that contrivance, by which a single steersman can direct the course of an enormous vessel through rocks and shoals more steadily and safely than an adroit charioteer can guide his tiny vehicle on a common road. The helm or rudder is a flat projection from the stern-post of the ship, turning on strong hinges, in the manner of a door or gate, and moved by a beam or lever

Fig. 113.

called the *tiller*, which proceeds from it forwards to where the steersman stands. In small vessels the tiller is above the deck and the steersman applies his hand directly to it; but in large ships it is below, and is moved by ropes, rising from it to the *wheel* on the deck, where the steersman stands, with the compass before him. While the rudder points directly astern, as to *a*, like a continuation of the keel and stern post, it does not affect the vessel's course; but if it be inclined ever so little to one side, as to *b* on the left or *larboard* side, the water immediately acts on it in the direction of *c b*, perpendicular to its surface, and pushes the stern to the right or *starboard* side,—an action equivalent to pulling to the left or larboard.

It is possible to make a ship or boat steer itself, by placing a powerful vane on the mast-head, and connecting it with the tiller-ropes by two projecting arms from its axis. If it were desired to make the ship sail directly before the wind, the tiller-ropes would be fixed to the arms of the vane so that the helm should be in the middle position, when the vane was pointing directly forward: should the vessel then from any cause deviate from her course, the vane, by its changed position with respect to her, would have produced a corresponding change in the position of her helm, just such as to bring her back to her course. Again, it is evident that, by adjusting such a vane and rudder to each other in different ways, any other desired course might be obtained, and which would alter only with the wind. The vane, to have the necessary power, would require to be of large size; it would be a wide hoop, for instance, with canvas stretched upon it; and the rudder, to turn with little force, might be hung on an axis passed nearly through its middle, instead of, as usual, by hinges at one edge. Cases have occurred where shipwrecked persons might have sent intelligence of their disaster to a distant coast, by a small vessel, or even a block of wood fitted up in this way. The method admits also of other applications, particularly in war.

As fluids act on surfaces, in a direction perpendicular to them, the water on the right side of a ship's bow is always pressing it towards the left side, but owing to the equivalent and contrary pressure there, the ship holds her course evenly between the two, or straight forward. When a ship, however, owing to the side wind, lies over, or *heels*, as it is called, that side of the bow which sinks most is more pressed than the other; and were there not a counteracting inclination of the rudder then made, constituting what is called *weather-helm*, the ship's head would come round to the wind. Now ships so rarely have the wind exactly astern, that to diminish the almost constant necessity for *weather-helm*, the mast or masts, and consequently the mass of the sails are placed more towards the bow than the stern.

Again, because the bow of a ship is oblique downwards as well as sideways, the water, when she moves, is constantly tending to lift the bow; hence when the vessel is dragged by a low horizontal rope, as in the case of a boat attached to a sailing ship's stern, or is moved by paddle-wheels like steam-boats, the bow rises much out of the water, and the stern sinks in the hollow or furrow of the track: but when she is driven by sails, as these are high on the mast, and are acting therefore on a long lever to depress the bow, the two opposite tendencies just balance each other, and the vessel sails evenly along.

The form of the fore part of a ship has less influence upon her speed of sailing than the form of the hind part, called the *run*, from the middle to the stern. When a ship is at rest, there is of course as much forward pressure of water about the stern as of backward pressure on the bow; but when she sails, she is running away from the propelling pressure, and is increasing the resisting pressure. A gradual tapering of the hind part, therefore, or a *fine run*, as it is called, which allows the water to apply itself readily to it, as it passes along, must influence much the rate of sailing. The fore part of any mass drawn through the water, however blunt or square, becomes in effect sharp or rounded by a quantity of water which it pushes on before it. A tree, or the tapering mast of a ship, can be drawn through the water more easily with the large end foremost than in a contrary way.

The *common windmill* furnishes another illustration of the action of fluids on oblique surfaces. The face of the windmill is turned directly to the wind, but the four flat vanes or sails, of which the great wheel consists, are individually oblique. Thus the edge *a* of the vane *a e*, is more forward as regards the coming wind or a spectator in front, than the edge *e*, and the action of the wind, therefore, being perpendicular to the oblique surface *a e*, pushes it in a degree towards *a*. The same remark applies to each of the other vanes where the edges *b c* and *d* are in front and those marked by the fainter lines are behind; so that each vane produces an equal effect in turning the wheel. The law of the "decomposition of forces," explained in page 57, tells in what proportion the force of the wind is exerted to push the wheel backwards against its supports, and to turn it round.

Fig. 114.

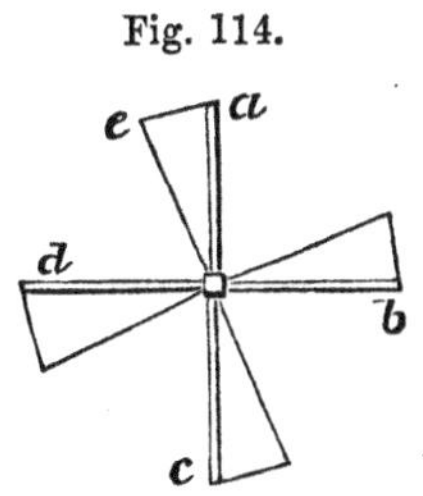

Windmills were first used in Europe in the fourteenth century, and they are still of great importance in countries where there are no waterfalls, and little fuel for steam-engines. In some of the richest European landscapes, every height is crowned by its busy windmill, grinding corn, or sawing wood, or pressing oil-seeds; and over the plains, similar wheels are pumping water for domestic use, or incessantly draining the land.

The *smoke-jack* of our chimneys is a small windmill, driven by the ascending current of air in the chimney.

The *feathering of an arrow* acts in part on the principle of the windmill. The feathery projection from the shaft is not quite straight, but winds round it a little, like the thread of a screw; and the arrow, therefore, constantly turns as it flies, and goes straight to its object although the shaft itself be bent, because any deviation is constantly correcting itself.

The rifle-barrel in fire-arms has spiral furrows or threads along its interior surface, so that the bullet in passing out receives a turning motion corresponding to that of an arrow, and producing similar results. A bullet which receives any other turning motion than round the line of its course—and most bullets from a common barrel do acquire such, owing to the irregularity of their form, or unequal friction at the mouth of the piece—is sure to deviate from its course, because unequally pressed or resisted by the atmosphere. A good rifle fixed to its place will send a succession of shots through the hole made in the target by the first shot, at the distance of 200 yards. Duels have been fought with rifles, and the parties having fired at the same moment, have been corpses the moment after.

It might be supposed that a wheel which the wind turned by *direct* action on the rim, as water turns common water-wheels, would be preferable to the windmill-wheel above described, which is turned by oblique action on the face: accordingly, a wheel like a water-wheel, only with broader vanes, has been placed in a house or cover, so that only one side at a time was exposed to the wind;—but it is a powerless machine. The oblique vane wheel may apply to use only half or less of the force of the air which reaches it, but its wide expanse receives a stream of air of thirty feet in diameter, while an ordinary window would admit that required for a wheel of equal size of the other construction.

There are some situations where it would be an advantage to have water-wheels like the common windmill-wheel, *viz*, where the stream is sluggish, and is deep enough to allow a large wheel to be wholly immersed.

A small wheel of this sort, with broad oblique vanes, has been used as a means of ascertaining the rate of a ship's sailing. It is allowed to drag astern in the water; and the number of revolutions made in a given time marks the ship's speed.

A windmill-wheel made to turn during a calm by force applied to its axle, would be pressed endways, or in the direction of its axle, just as if wind were blowing upon it, owing to the reaction of the still air, through which its oblique vanes were made to sweep. Such a form of wheel fitted to work in water, and called a water-screw, has been applied at the bow or stern of steamboats to propel them in canals where there was no room for side wheels. But as from the obliquity of the surfaces only a part of the applied power becomes propulsive—the remainder being wasted in the lateral strain or twisting of the water—the method is not applicable to general purposes.

Two small windmill-wheels placed horizontally one above the other, on the same axis, and made to turn in opposite ways by springs or otherwise, would rise in the air, carrying a certain load with them, and would constitute, therefore, a flying-machine.

The effect of a single oar projecting from the stern, used to propel a boat or vessel, in the manner called *sculling*, is referable to the law now under consideration. The oar or scull rests on a round-headed prop or nail at the stern, and is made to vibrate from side to side. In all positions it has the surface which pressess the water turned obliquely backwards; hence the reaction of the water drives the boat forward. In China, large vessels are moved by a single sculling oar, which half the ship's company may be urging at the same time. A sculling oar may be regarded as a single vane of such a propelling-wheel or water-screw as above described, made to sweep across, behind the vessel, alternately to the right and to the left.

The action of a fish's tale and of the bending of an eel or snake in water, partly resembles that of the sculling oar. Many people believe that the tail of the fish is only the rudder of the body, and that the fins give it forward motion—as is true of a bird's tail and wings—but the fish's tail is in fact the great instrument of motion, while the fins serve chiefly to steady and direct the motion.

Fig. 115.

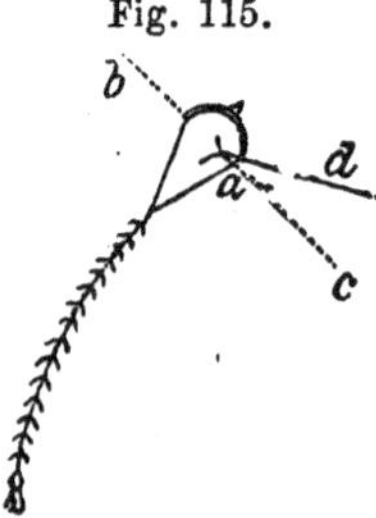

A *paper kite* rising in the air is another example belonging to this place. Its cord *d* is attached to it above the middle of its loop, and therefore so as to make it present always an oblique surface to the wind; and by the action of the wind, perpendicular to its surface, it rises as if pushed up in the direction *c a*, or as if drawn up in the direction of *a b*. A kite might be made large enough to lift a man. Cats have been sent up at kites' tails, and have fallen down safely under parachutes from the greatest elevations. It might be safer for a man to rise at a kite's tail to reconnoitre an enemy's position, or to survey an unknown country, than under a balloon, as was practiced by the French during the revolutionary wars. He might have the security of a parachute, and the power of regulating the obliquity of attachment of the rope, so as to command his ascent or descent at pleasure. An exhibition was made in October, 1827, between Bath and London, of a car drawn along the highway by kites. That they might ascend to a great elevation, where the wind is generally stronger than below, they were attached to each other in a row, so that the second kite mounted as if its cord were held by a hand at the first, the third as if rising from the second, and so forth. The projector of this novelty hoped that he had pointed out a most valuable means of travelling across extensive plains, sandy deserts, tracks of snow, &c., and in all cases, nearly with the speed of the wind.

"*Fluids lifted in opposition to gravity.*" (See the Analysis.)

Water, as we have seen in former parts of this work is to the living universe, in some degree what the blood is to the animal body, and a constant supply and circulation are required. This supply has been provided for to an extraordinary extent, by the operation of natural causes; but for many purposes of human society, water is still required where none naturally exists. A great variety of means have been employed in raising it, some of which, sufficient to illustrate the whole, are now to be considered.

Water may be raised in a bucket which is attached to a rope to be pulled up by the hand.—The rope carrying the bucket may be drawn up more easily by being wound round a barrel or axle turned by a winch.—There may be a succession of buckets on a rope, rising one after the other, and when emptied, descending again on the opposite side of the wheel or axle which lifts them: the rope to which they are attached being a circle or *endless rope*, and constituting with them what is called the *bucket-machine.*—Instead of buckets on such an endless rope or chain there may be a succession of flat pieces of wood, which, on being drawn up through a large tube or barrel, like loose-fitting pistons, will raise a copious stream of water; this is the contrivance called the *chain-pump.*—Or simply an endless rope of hair, very rough, passing round one wheel above, another below, may be whirled quickly by turning the upper wheel, so that a mass of water adhering by friction to its rising half, shall be thrown into a reservoir at the top where it passes over the upper wheel: several such ropes may be joined side by side to increase the effect.—But the most important of all water-raising engines are the *lifting and forcing pumps*, already described at pages 171 and 172. They are used to draw from wells, to drain mines, to send a supply over cities, to pump ships, to throw water for extinguishing fires and for many other purposes.

Fig. 116.

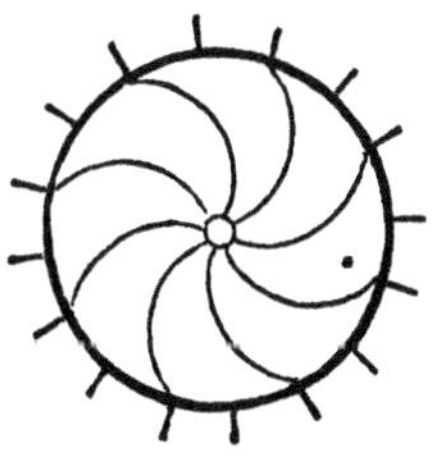

A stream of water passing through a garden, or in the midst of fields, may have beauty with little utility, unless it can be employed to irrigate the vegetable creation around. In the fields and gardens of Persia, where the heat of the sun is very intense, the streams are caused by their own action, to lift a part of their water into elevated reservoirs, from which it again flows in sloping channels to wherever it is required. A large water-wheel is placed so that the stream may turn it, and around its circumference buckets are attached to be filled as they sweep along below, and to be emptied into a reservoir as they pass above—or instead of buckets, the spokes of the wheels themselves are made hollow, and curved as here represented, so that their extremities dip into the water at each evolution, they receive a quantity of it, which runs along them as they rise, and is discharged into a reservoir at the centre. These are usually called *Persian wheels*, but they are as commonly employed on the banks of the Nile and elsewhere as in Persia.

A pipe wound like a screw upon a sloping barrel, and made to dip its lower mouth into water at each revolution of the barrel, will also raise water; the lower portions of the turning pipe will always be full of it, and it will be rising in them to the top, as if on an inclined plane. Archimedes was the inventor of this beautiful water-screw, and his name has remained to it. It may be turned by the hand or by a passing stream which acts on the vanes of a water-wheel fixed upon it.

Fig. 117.

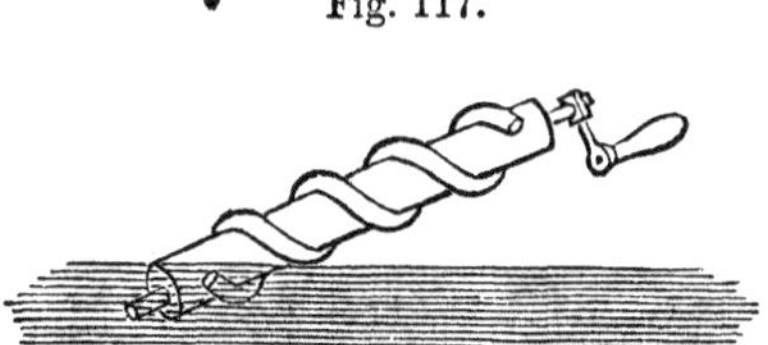

Water may be raised by producing centrifugal force at the upper end of a bent pipe which dips into a reservoir. Supposing the pipe to be bent as here represented, and the horizontal arm a to turn like the spoke of a wheel, round the upright portion as the axis,—if the pipe be once filled with water, and be turned with sufficient speed, it will continue to throw out a constant stream from the end a. To increase the discharge there may be several horizontal arms from one large upright pipe, and emptying themselves into a circular trough or reservoir; and to prevent the necessity of refilling the apparatus after every interruption of its motion, a valve opening upwards must be placed at the bottom. This contrivance has been called the *centrifugal pump*, because the water is raised at b as in a pump, by the pressure of the atmosphere, to supply the place of that which is thrown out from a by the centrifugal force. The velocity of rotation must bear proportion to the height of the discharging aperture a, above the surface of the water in the reservoir.

Fig. 118.

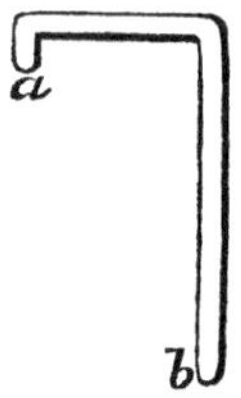

It had long been observed in household experience and elsewhere, that while water is running through a pipe, if a cock at the extremity be suddenly shut, a shock and noise are produced there. The reason is, that the forward motion of the whole water contained in the pipe being instantly arrested, and the momentum of a liquid being as great as if a solid, the water strikes the cock with as much force as if it were a long bar of metal or a rod of wood having the same weight and velocity as the water. Then as a fluid presses equally in all directions, a leaden pipe of great length may be widened, or even burst in this experiment.—Lately this forward pressure of an arrested stream has been used as a force for raising water, and the arrangement of parts contrived to render it available has been called, on account of the shocks produced, the *water-ram*. The ram may be described as a sloping pipe in which the stream runs, having a valve at its lower end, to be shut at intervals to arrest the stream, and having a small tube rising from near that end towards a reservoir above, to receive a portion of the water at each interruption. Now water allowed to run for one second, in a pipe ten yards long, two inches wide and sloping six feet, acquires momentum enough to drive about half a pint, on the shutting of the cock, into a tube leading to a reservoir forty feet high. Such an apparatus, therefore, with the valve shutting every second, raises about sixty half-pints or four gallons in a minute. The valve is so contrived that the steam works it as desired— In this figure, which represents the lower end of the water-ram, *a* is the opening by which the stream escapes from it, and the valve or flap seen below the opening is that which by suddenly shutting arrests the stream. The valve is made so heavy, that the stream must run for a certain time to acquire force enough to shut it: and in the instant of its shutting, a little of the advancing water passes upwards through the valve *b* towards the reservoir. The water in the main pipe then becoming stagnant again, no longer has power, by its weight alone, to keep the valve *a* shut: this, therefore, falls open, and the stream begins again, again to be arrested as before; and as long as the supply of water lasts, the action of the apparatus continues. The action of a water-ram has been compared to the beating of an animal's pulse. The upright tube has usually a reservoir at the bottom, where it first receives the water, constituting there an air-vessel *b*, (described at page 161) which, by the air's elasticity, converts the interrupted jets first received, into a nearly uniform current towards the reservoir. The supply of air to this vessel is manifested by the contrivance called a *shifting-valve*.

Fig. 119.

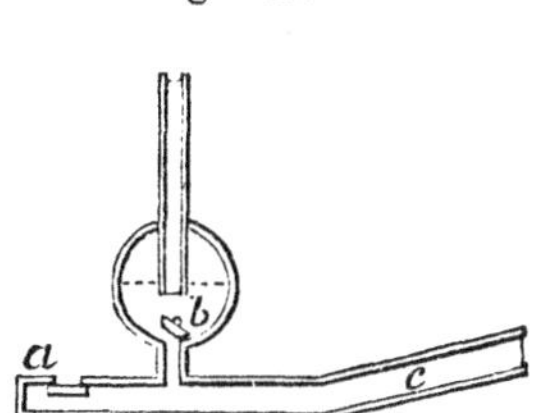

In the preceding examination of the doctrines of fluidity, we have had to touch on many of those phenomena of nature and art which are the most important to man; yet we have seen how beautifully simple and intelligible they are all rendered when referred, by a methodical arrangement, to a few heads dependent on the "fundamental truths." Each one of the many particulars belonging to this department, and which, when now explained, appears so simple and obvious, has yet been a distinct step in the slow progress of discovery or invention, and probably, when first made, has filled some ingenious mind with intense and purest delight.

PART III.

(CONTINUED.)

SECTION IV.—ACOUSTICS, OR PHENOMENA OF SOUND AND HEARING.

ANALYSIS OF THE SECTION.

1. SOUND *is heard when any sudden shock or impulse is given to the air, or to any other body which is in contact directly or indirectly with the ear.*
2. *If such impulses be repeated at very short intervals, the ear cannot attend to them individually, but hears them as a* CONTINUED SOUND, *which is* UNIFORM, *or what is called a* TONE, *if the impulse be similar and at equal intervals, and is* GRAVE *or* SHARP, *according as they are few or many in a given time; and all continued sound is but a repetition of impulses.*
3. *When the number of impulses in a given time producing some uniform continned sound has a simple relation, as of half, third, fourth, &c., to the number producing some other such sound, which is heard either simultaneously with it, or a little before or after, the ear is generally much and pleasingly affected by the circumstance; and the sounds are said to have* MUSICAL RELATION *to each other, or to be* ACCORDANT, *while all others are termed* DISCORDANT.
4. *The shock which causes the sensation of sound* SPREADS *or is propagated in all bodies, somewhat as a wave spreads in water, with decreasing strength as the distance increases, but with a velocity nearly uniform, and which in air is about* 1,142 *feet per second.*
5. *Sound is* REFLECTED *from smooth surfaces, and hence arise many curious and pleasing effects called* ECHOES, *&c.*
6. *The structure of the ear illustrates the law of sound.*

EARLY inquiries into nature had remarked that in most instances of noise or sound there was present a shock or trembling of the sounding body, often visible, but sometimes only sensible to the touch, or discoverable by other means; it was noted for instance, in the string of a harp, in the reed of a hautboy, in the prongs of a tuning-fork, in the lip of a bell; but it was reserved for the moderns to understand fully, that the animal organ called the ear, is merely a structure of parts admirably adapted to be affected by the concussions or tremblings of things around; and that sounds in all their varieties are merely such motions, affecting the ear through the medium of the air which surrounds it, or of some other body, or series of bodies reaching from the trembling thing to the ear.

The delicacy and complexity of an organ destined to feel and to distinguish such slight and varying influences, and the vast importance of it to man, as that which makes him capable of using language, besides being his ever-watchful monitor of surrounding occurrences, and the channel by which the fascination of music enters, render this subject, to all who love to read in nature the attributes of its author, a most favourite study.

Because all the bodies around us are immersed, in common with ourselves, in the ocean of air which covers the earth, we are much more frequently warned of the shocks and tremblings of which we have been speaking, by their effect on the air, than in any other way; hence the early prejudice that air was necessary to sound, and hence, in part, the reason why the doctrines of sound have generally been accounted a part of pneumatics. We shall now find, however, that all bodies are more or less fitted to convey these tremblings, and that air in many cases is neither the quickest nor the best conductor. Although our notions on the subject are thus corrected, it is still convenient to study the theory of sound as a part of *Pneumatics.*

1. "*Sound is heard when any sudden shock or impulse occurs in a body having communication through the air or otherwise, with the ear.*" (Read the Analysis.)

Common instances of a single impulse are—the blow of a hammer—the clap of hands—the crack of a whip—a pistol-shot—any explosion—the thunder-clap.

The loudness of sound conveyed by the air depends on the air's density. A bell enclosed in the receiver of an air-pump is heard less and less distinctly as the air is exhausted, and in a vacuum is not heard at all.—Even the blow of a hammer, in a vacuum, is not heard if care is taken to prevent the shock from being communicated through neighbouring solid bodies.—In the thin air surrounding a lofty mountain-top the report of a pistol is much less loud, and human voices are weaker.—In the condensed atmosphere of a diving-bell a whisper is loud.—When volcanoes and various other resemblances to the constitution of our earth were first discovered in the moon, some persons fancied that during the stillness of night we should hear the thunder there:—but supposing the thunder to happen, and to be ever so loud, it could not be heard on earth, because there is no medium to bear thither the pulses of sound—there is a vacuum between.

2. *Impulses quickly repeated cannot be individually attended to by the ear, and hence they appear as one continued sound, of which the pitch or tone depends upon the number occurring in a given time; and all continued sound is but a repetition of impulses.* (Read the Analysis.)

If a wheel with teeth be made to turn and to strike any elastic plate, as a piece of quill, with every tooth, it will, when moved slowly, allow every tooth to be seen and every blow to be separately heard; but with increasing velocity the eye will lose sight of the individual teeth, and the ear ceasing to perceive the separate blows, will at last hear only a smooth continued sound called a *tone,* of which the character will change with the velocity of the wheel.

In like manner the vibrations of a long harp-string, while it is very slack, are separately visible, and the pulses produced by it in the air are separately audible; but as it is gradually tightened, its vibrations quicken, so that, where it is moving, the eye soon sees only a broad shadowy bellying line; and the distinct sound which the ear lately perceived, seeming now to run together on account of the shortness of the intervals, are felt as one uniform continued tone, which constitues the note or sound then belonging to the string.

Again, if a current of air passing through a tube or opening, be in any way interrupted at regular and very short intervals, as by a little stop-cock placed in the opening, of which cock the plug, instead of being only partially turned by a cross handle, as in a common beer-cock, has a wheel fixed upon it, so that any desired rapidity of rotation may be given to it,—then at every time when the passage for air becomes open, there will be a certain shock given to the air around, and the repetition of such shocks will constitute a musical tone. This apparatus can produce all tones, and it enables us with great precision to ascertain the number of pulses required to constitute any given tone.

It is the elasticity of any string used to produce a tone which causes the repetition of the percussion, and therefore the continuance of the sound, thus:—the string having been pulled at its middle to one side, and then let go, is, owing to its elasticity, carried back quickly to the straight position; but by the time that it has reached this, it has acquired a momentum which, like the momentum of a vibrating pendulum, carries it nearly as far beyond the middle-station as the distance whence it came:—it has to return, therefore, by its elasticity, from this second deviation, in the same way; but still passing the middle as before, it has again to return; and thus continues vibrating uniformly as a pendulum does, until the resistance of the air and friction gradually bring it to rest. A large vibration of any string, like a large oscillation of a pendulum, occupies very nearly the same time as a smaller, because the farther that the string is displaced or bent, the more forcibly, and therefore quickly, is it pulled back again by its elasticity; hence the uniformity of the tone produced by a musical string is not injured by the different force with which the finger of the player may touch the string. According, however, as the vibrations of a string are more extensive or quicker, the impulses given to the air are more sharp or forcible, and hence the sound becomes louder. And this explains why sharp sounds are generally also loud. Vibrations which are comparatively few and slow, strike the ear very gently, as in the flapping of a pigeon's wing, or in the play of a switch.

The most familiar instance of sounding vibration is that of an elastic cord extended between two fixed points, as in stringed instruments of music; but because elastic bodies generally, when by any force their natural form is for a time altered, recover it when allowed, not by a first effort, but like the string of a pendulum, after a series of oscillations, almost all such bodies repeat many times an impulse once given to them, and thus may become the means of producing a continued sound.—If a solid rod of steel, glass, or any other elastic substance, be firmly fixed at one end and left free at the other, and if that other be then pulled a little to one side of its station or rest, and suddenly let go, it will immediately seek its station again, but by the momentum acquired in the approach, will go beyond it: it will then return as before, but again to pass, and so will continue to vibrate with diminishing force for considerable time.—A boy at school, thus, sticks the point of his penknife into the bench, and by one touch makes it produce a continued uniform sound of considerable duration.—The prongs of a tuning-fork, or of the common sugar tongs, vibrate and sound in the same way. In the musical snuff-boxes and chimney-clocks, the sounds are produced by the vibration of little rods of steel, fixed by one end, in a row, like the teeth of a comb, and touched by small pins or points projecting from a turning barrel. Any elastic flap, as of metal or of tough wood, placed over an opening, so as to stand away from it a little when not pressed by passing air, but to close the opening if so pressed, becomes a sounding reed when air is gently forced through the opening: thus, the air pressing on the flap to close them causes a momentary interruption of the current, but the flap immediately recoiling from the blow, as well as by reason of its own elasticity, again opens the passage, and the continued rapid alteration of the shutting and opening produces the tone. The reed of a clarionet is a thin plate of elastic wood, made to vibrate in this way. The drone of the bag-pipe and the common straw-pipe, are reeds of nearly the same kind. The Chinese organ, and the sweet instrument lately introduced under the name of Æolian, have reeds which differ from these, by beating *through* the opening instead of merely *on* its face. Elastic rods simply resting on supports at both ends, or suspended by their middle, will also vibrate; a musical instrument is thus made of pieces of glass laid upon two strings, and struck by a cork hammer: in the island of Java, a rude instrument of the same kind is made of blocks of hard elastic wood. A portion of a hollow sphere of elastic metal very readily takes on a vibration, during which its form is constantly changing from the perfect round to the oval, and conversely; there are consequently repeated percussion of the air, and a continued sound, and the thing is called a *bell.* A bell admits a great variety of shapes, and may be made of any elastic substance, as metal, glass, earthenware, (buyers ring earthenware to ascertain its soundness,) and even of hard wood. The *Chinese gong* is a metallic vessel shaped like a common sieve, having a manner of vibration very peculiar, and producing sounds that are rousing and sublime. The *drum* has a tense elastic membrane on which the blows of the drum-stick are received: its tone ceases quickly, because the motion of so broad a surface is much resisted by the air. In the flute, flageolet, common organ-pipes, &c., the air is forced through narrow passages, and is divided by sharp edges, in such a way as to suffer repeated but perfectly regular condensations or interruptions sufficient to affect the ear; and hence the endless variety of sweet continued sounds which these contrivances are known to produce.

To the production of a tone, it is of no consequence in what way the pulses of the air are caused, provided they follow with sufficient regularity; witness, in addition to some of the instances given above, the pure sound produced by the motion of a fly's wing—supposed by many to be the voice of an insect. The clacking of a corn-mill, and the noise of a stick pulled along a grating, are not tones, because the pulses follow too slowly.

Where a continued sound is produced by impulses which do not, like those of an elastic body, follow in regular succession, the effect ceases to be a clear uniform sound or tone, and is called a *noise.* Such is the sound of a saw or grind-stone—the roar of the waves breaking on a rocky shore, or of a violent wind in a forest—the roar and crackling of houses or of a wood in flames—the mixed voices of a talking multitude—the diversified sounds of a great city, including the rattling of wheels, the clanking of hammers, the voices of street-criers, the noises of manufactories, &c.; which rough elements, however, at last mingle so completely that the combined result has often been called "the hum of men," from analogy to the smooth mingling miniature sounds which constitute the hum of a bee-hive.

" Grave and sharp sounds." (Read the Analysis.)

The difference of sounds, which depends on the different number of vibrations of the sounding body in a given time, divides them into those called *bass*, *low*, or *grave* notes, for comparatively few and slow vibrations; and those called *high*, *shrill*, or *sharp*, for vibrations more numerous and quick.

The frequency of vibrations in strings increases with their *shortness*, *lightness* and *tension*—for if a string be *long* or *heavy*, there is a greater mass of matter to be moved, and hence a slower motion; and if a string be slack, the force of elasticity which pulls it from any deviation back to the straight line is so much the less. It is found that a string taken of half the length or of one-fourth the weight, or of quadruple the tension of another string, vibrates twice as fast on any one of these accounts.

These truths are familiarly illustrated in the violin. The low or bass string is thick and very heavy from being covered with metallic wire, and the others gradually diminish in magnitude and weight, up to the smallest or treble. The strings are tuned to each other by being attached by one end to moveable pins, which, when tuned, increase or diminish their tension; and the sound produced by each may be afterwards varied to a certain extent, by the performer pressing different parts of it with the finger against the board, so as to shorten the vibrating portion.

An analogous law, as to the influence upon tone, of weight and dimensions, holds with respect to bells, glasses, reeds, &c., and enables us to use these also in the construction of musical instruments.

3. *" When the number of impulses producing some continued sound has a simple rotation, as of half, third, fourth, &c., to the number producing some other sound, which is heard either simultaneously, or a little before or after it, the ear is much and pleasingly affected; and the sounds are said to have musical relation to each other, or to be accordant, while all others are termed discordant."* (Read the Analysis.)

Understanding now that all continued uniform sounds are produced by a repetition of similar beats or vibrations, we perceive that in the series from grave to sharp, there must be such as, with respect to the number of beats in a given time, are related to each other, as the numbers 1, 2, 3, 4, &c., or, which is the same thing, as 10, 20, 30, &c. Now as between two sounds, one of which has 20 beats while another has 10, there will be a coincidence by every second beat of the quicker, and between sounds whose beats are to each other as 30 to 20, there must be a coincidence at every third beat of the quicker, and so forth, we should naturally expect the ear to be differently affected by such correspondence than when the coincidence is either less frequent, or is irregular. Accordingly we find that all sounds which have such simple relations to each other, are remarkably agreeable to the ear, either when heard together, or in close succession; while those in which the coincident beats are farther apart, are heard with indifference, or are felt to be positively harsh or disagreeable. It is in fact offering itself to be noticed here, that the coincident or double pulses of any two concordant sounds become the cause of elements of a third sound, perfectly distinct from them, but which is always heard with them, and is called their *grave harmonic* or *resultant:* it is the same as a simple sound having as many vibrations in a given time as there are coinciding beats between the two other sounds.

If a long musical string be made to sound, and the number of its vibrations in a given time be ascertained, we find that if only half of it be allowed to vibrate at a time, as when a finger presses its middle against a board, that half will vibrate twice as fast; and similarly, a third part three times as fast; a fourth part four times as fast; and so on, producing the sounds or tones most nearly related to each other. A fine illustration of this is afforded by the string of a violoncello, when made to vibrate by a bow moved very gently across it, near the bridge; for it often divides itself spontaneously into two, three or four, &c., equally vibrating parts or bellies, with points of rest between them called knots: when this happens, there are heard not only the sound or note belonging to the whole length of the string, but, also, more feebly, the subordinate notes belonging to its half, third, or fourth, &c., according to circumstances, beautifully mingling with the first sound, and forming with it a rich harmony. Often in such a case the subordinate sounds swell with such force as to overpower for a time the fundamental note: but any one such sound is rarely of long duration. The same harmonic sounds. may be produced still more certainly, while drawing the the bow across the string, by touching the string lightly with the finger, at one of the points where we wish it to divide. Even a tune may be so played.

The sounds thus belonging to a single cord or string, and produced by its spontaneous division into different numbers of equal parts, constitute, when heard together or in succession, what may be called the simple music of nature herself. It is produced pleasingly, as just described, by the single string of a violoncello; but in the most perfect manner by the instrument called the Æolian harp.

The Æolian harp is a long box or case of light wood, with harp or violin strings extended on its face. These are generally tuned in perfect *unison* with each other, or to *the same pitch*, as it is expressed, except one serving as a bass, which is thicker than the others, and vibrates only half as fast; but when the harp is suspended among trees, or in any situation where the fluctuating breeze may reach it, each string, according to the manner in which it receives the blast, sounds either entire, or breaks into some of the simple divisions above described; the result of which is the production of the most pleasing combination and succession of sounds that ear has ever listened to, or fancy perhaps conceived. After a pause this fairy harp may be heard beginning with a low and solemn note, like the bass of distant music in the sky: the sound then swells as if approaching, and other tones break forth, mingling with the first, and with each other; in the combined and varying strain, sometimes one clear note predominates and sometimes another, as if single musicians alternately led the band: and the concert often seems to approach and again to recede, until with the unequal breeze it dies away, and all is hushed again.—It is no wonder that the ancients, who understood not the nature of air, nor consequently even of simple sound, should have deemed the music of the Æolian harp supernatural, and, in their warm imaginations, should have supposed that it was the strain of invisible beings from above, come down in the stillness of evening or night to commune with men in a heavenly language of soul intelligible to both. But even now that we understand it well, there are few persons so insensible to what is delicate and beautiful in nature, as to listen to this wild music without emotion; while the informed ear finds it additionally delightful, as affording an admirable illustration of those laws of sound which human ingenuity at last has traced.

And the simple scale of sound, called a *chord*, which nature thus gives by the spontaneous dividing of a single string, has considerable vacancies in it, human taste or feeling, long before there was any theory of music, had joined to it the notes of two additional strings, one sharper or more acute than it, and the other more grave; of which additional notes, while part agreed, or were in unison with certain notes of the principal chord, the remainder just served to fill up its larger intervals, and to complete a scale of nearly uniform interval—as three ladders having unequal intervals between their steps, might still, if placed together, complete a stair of easy ascent. The relation between these strings or chords is such, that the principal beats thrice or twice of the low chord, and the high chord beats thrice for twice of the principal: and in the complete scale of notes, the principal is five notes above the lower and five notes below the higher. So truly natural is the scale thus formed, that it has arisen in all nations, however remote and unconnected; and an untutored individual in attempting to raise his voice by regular steps, falls into it almost as readily as the learned professor. The scale has eight steps or notes between any tone, and the tone above it vibrating twice as fast, or the tone below it vibrating half as fast; these two tones or notes being hence called the *octaves* above and below the *note* with which they are compared, and the intermediate notes which fill up either octave from the fundamental note are distinguished by the names of *second, third, fourth*, &c., in ascending or descending. The numbers which express the relations of beats among the notes of an octave are easily found, from our knowing the relative number of beats in the notes of any one simple chord, and the relation as above described of the three chords forming the compound scale. The following table exhibits these numbers or the arithmetical expression for the notes of an octave, as well as the corresponding lengths of a given string required to produce them, and the English designation of the notes by letters, and the continental designation by names, these names being the first syllables of certain verses sung by learners.

Number of vibrations . . .	1	$\frac{9}{8}$	$\frac{5}{4}$	$\frac{4}{3}$	$\frac{3}{2}$	$\frac{5}{3}$	$\frac{15}{8}$	2
Length of string	1	$\frac{8}{9}$	$\frac{4}{5}$	$\frac{3}{4}$	$\frac{2}{3}$	$\frac{3}{5}$	$\frac{8}{15}$	$\frac{1}{2}$
English characters	C	D	E	F	G	A	B	C
Continental names	ut	re	mi	fa	sol	la	si	ut

The musical scale, however far extended, is a repetition of similar octaves, so that any note in it vibrates just twice as often as the corresponding note in the octave below, and half as often as that in the octave above. The lowest note which is perceptible to the human ear has about thirty beats in a second, and the highest about thirty thousand: and there is included between these two, a range of nearly ten octaves. To certain ears the extremes of this range are totally inaudible, as if their power did not reach so far. Some persons do not hear at all the sharp note of the grasshopper, while some are equally insensible to the lowest tones of an organ or piano; and yet to all the perception of intermediate sounds may be equally perfect. Few musical instruments comprehend more than six octaves, and the human voice in general has only from one to three, the female voice being in pitch an octave higher than the male.

If the intervals in the musical scale were all equal, a performer might choose indifferently any note as a fundamental or key-note, and would only have to attend to the number of intervals above and below it; but, in fact, the relation of the three constituent chords is such that the third and seventh intervals, in ascending from a key-note, are only about half as large as the others. It is owing to this circumstance that in *changing the key* on any instrument, certain notes belonging to other keys are half a note too low or too high, that is, too *flat* or too *sharp*, and must be changed accordingly. And hence, when an instrument is to be used to play on all keys, its larger intervals must be divided into two parts. The fact of these unequal intervals, ill understood, is what gives an appearance of great complexity and difficulty to musical science.

Melody, in music, is when notes, having the simple numerical relations of beat which we have been describing, are played in succession; *harmony* is when two or more such notes are sounded together. The effect of both is delightfully increased by what is called *measure*, viz., making the duration of the notes or strains correspond with certain regular divisions of *time*. This gives to the ear a prescience, to a certain degree, of what is coming, with the pleasure of having expectation realized, as happens similarly from the metre and rhyme of poetry; it moreover enables the memory to retain musical combinations of sound—for the airs of the Æolian harp, which observe no *time*, cannot be learned or repeated. The music of a single drum is that of time only.

Melody, *harmony*, *time* and *varying intensity of sound*, are the four constituents of music, and it seems that almost every state of mind has, in some combination of these, an appropriate expression, intelligible to the general feelings of the human race. The exact relation between the movements of the animal spirits, as it has been expressed, or the fluctuating stream of feeling, and the varying flow of sound in a musical composition is not clearly understood, but the fact of their correspondence and its consequences are most remarkable. Under many circumstances, the association between the feeling and expression is so strong, that the latter is often spontaneously betraying itself;—witness the almost constant humming or low song of some contented beings—the singing and whistling of careless childhood, or of the light-hearted rustic living among the beauties of nature—the heart-rousing strain of the hunter or warrior—and the tender expression of many of the modifications of anxiety and sorrow. The musical sensibilities are by no means limited to the human race, for there is no expression more exquisite than in the song of the nightingale during the evenings of spring, or of the thrush and blackbird, in the same season, amid the quiet retreats of our woodlands,—the music of which untutored songsters is made up of the same elements as our own.

The *accompaniment* of an air afforded to a singer by one or more instruments, and which is so pleasing, is chiefly the sounding, simultaneously, in a subdued manner, some other notes of the chords to which the several vocal notes belong. *Duetts* and more complicated *concert-pieces* have their origin from the same source: and highly cultivated musical sense can even follow and enjoy several melodies played together.

Musical notes, by whatever instrument produced, have to each other the same numerical relations, in the beats or vibrations which constitute them. The different qualities of tone, therefore, from different instruments, can only depend on the peculiarities of the single beat, as to whether they are sharp or soft, strong or weak, &c. Such is the extraordinary nicety of perception which the human ear possesses in this respect, that it cannot only distinguish different kinds of instruments, as a flute and clarionet, playing the same note, but different instruments of the same kind, even to the extent for instance, of recognizing each one of a hundred voices singing the same air. One of the greatest charms of concert music is, that the voice and the different instruments may take up separately, parts of the strain suited to their individual expression—the flute and clarionet, for instance, breathe softness; the trumpet and drum arouse; the harp rolls forth its brilliant chord; the violin leads the flowing sounds through rapid and endless variety; and so of the rest.

That there might be correspondence in instruments when played together and a known pitch when played apart, it became necessary to fix on some tone or number of vibrations as a point of comparison. Hence, *tuning-forks* have been made of steel, with length of prongs calculated to produce a certain note. The note is usually the fourth, A or *la* from the bass of the piano-forte, and vibrates about 430 times in a second;—and when the note of the same name on any instrument is *tuned* in unison with this, the other notes can be easily adjusted according to the harmonic relations above explained.

Almost every substance or contrivance that can produce a uniform continued sound may enter into the composition of a musical instrument; hence the almost endless variety which the world has seen. The chief classes of instruments are *stringed instruments, wind instruments* and *bells* or *rods.*

Of the *stringed instruments* we may mention the *harp*, the *lyre* or *lute*, the *guitar*, the *viol* of all sizes, and *piano-forte.* The harp, lyre and lute were the inventions of antiquity, and have brought down with them to the present times a thousand delightful associations. They awakened to inspiration the bards and poets of the young world, and they were the beloved companions of many of the noblest spirits of succeeding times. Their great charm appears to have been in their power to heighten the emotions produced by music's twin sister, poetry; and the combined effects seem to have been magical.—The other instruments mentioned are of comparatively modern invention, particularly the piano-forte; and their perfection has assisted in carrying the combination of musical sound to degrees of complexity and difficulty of which antiquity dreamt not. It is a question, however, whether the style of much of the music now in vogue does not prove rather a degeneracy, than a desirable refinement of musical taste. Music is a language of nature, intelligible at once to all susceptible minds, and, in a degree even to inferior animals; but modern art is attempting to make of it an artificial and conventional language, in which there may be fashion and change. The ornaments and accompaniments are now often so overwhelming, that the *melody*, in which the idea and sentiment really reside, is masked and almost lost; and an unpractised ear, particularly if listening to an *organ*, often discovers only an unmeaning succession of chords. And when a singer, abandoning the natural simplicity of melody, strains to execute with the voice the complicated movements which belong properly to instrumental accompaniments, the attempt destroys the poetry, by either rendering the words inaudible, or by sacrificing their natural expression to some supposed appropriate expression of the ornamental music. These considerations may account in part for the insensibility of so many highly-endowed persons to *what is now called* excellent music. Some of the tricks on the voice and on instruments, at present so common, are, to natural or graceful music, what tumbling or rope dancing are to natural or graceful gesture. And when we hear noted professors avow their inability to sing a simple ballad, or to play an unadorned melody, must we not conclude that the natural sense of music has left them, as the relish for simple but the most invigorating fare has left the morbid epicure?

The *guitar*, as affording an accompaniment to vocal music, has many advantages. It is not too loud, yet the strains are very distinct; it admits of most touching expression; it is very easily learned by any one who should attempt to learn music; it is portable and cheap. The great facility of accompaniment on it depends on this, that the player is able by one position of the hand to touch the strings so that the sound of all the six shall belong to the same chord:—three positions of the hand, therefore, for one key, produce all the notes and chords which a simple accompaniment requires; and the hand soon falls into these so readily, that the player is hardly sensible of exerting volition.

Among *wind instruments* are the *flute*, the *flageolet*, the *organ*, the *clarionet*, the *hautboy*, the *horn*, the *trumpet*, &c. The pitch or tone of a tubular wind instrument, just as of a musical string, has relation to its length; and the vibrations causing the sound seem to be waves or condensations of air passing from the mouth to the extremity of the tube; being more frequent, therefore, as the tube is shorter;—when the bottom of the tube is closed, the wave has to come back again, and thus renders the note twice as grave. It appears, also, that on blowing more strongly, the air in the tube divides into separate vibrating portions, as a string may divide to produce its harmonic sounds, and produces thus all the harmonic sounds belonging to the fundamental note at the tube. By blowing into a common German flute, for instance, it is possible to produce five ascending harmonics without moving the fingers at all. The music of a trumpet is limited to these five notes of the same chord; but in the flute and other instruments with holes, the effective length of the tube is calculated from the upper end to the nearest hole left open; and each length has its harmonics.—If a tuning fork, Jew's-harp, or any such sounding body, be held at the open end of a tube or other empty space of dimensions calculated to produce a frequency of undulation, in its contained air, according with the pulses of the sounding body, then the tube or space will immediately give out its own beautiful tone; and if the space be enlarged or diminished in a double, treble or any other simple proportion—as a tone may be, by a piston moved up or down in it—then will its note become the fifth, octave, twelfth, &c., above or below the original tone, although that tone continues unchanged. The tones of the Jew's-harp are well known to depend altogether on the varying dimensions of the player's mouth; but to obtain perfect music from it, three harps at least, to be substituted one for the other during the performance, are required to produce the notes of the three constituent chords of the common musical scale.—In wind-instruments with reeds, the tone depends on the stiffness, weight, length, &c., of the vibrating plate or tongue of the reed, as well as on the dimensions of the tube or space with which it may be connected. This truth is well illustrated in that instrument, the Æolian, already mentioned, which, in improved and varied forms, promises to become common, and one of the most expressive wind-instruments.—The sounds of the human voice are the sweetest of all, and are produced by the vibrations of two delicate membranes situated at the top of the windpipe, with a slit or opening, called the *glottis*, left between them, for the passage of the air. The tones of the voice are grave or acute, according to the varying tension of these membranes, and to the size of the opening.—In the *organ* there is a pipe for each note, and wind is admitted from the bellows to the pipes, by the action of the keys, like the keys of a piano-forte. The organ may be played also very perfectly by a barrel, made to turn slowly under the keys, and to lift them in passing, by pins projecting from it at the required situations. Very complicated pieces of music are thus set on barrels, but by a great cost of study and labour, and, therefore, of money; now a plain barrel, made to turn near the keys of an organ during performance on it by the hands, might be made to record, with mathematical accuracy, every touch of the most finished player, by receiving marks of some kind from the keys as they were lifted; and to repeat with absolute accuracy, therefore any performance, however delicate and exquisite, it would only be farther necessary to drive pins into the barrel where the marks remained, and afterwards make these pins lift the keys.

Bells are often conjoined in sets, having the musical relation, and to some persons their music is very agreeable. There are, in the tolling of a single bell, a loudness and a solemnity rendering it a fit accompaniment of funeral rites.

The *Chinese gong* partakes of the nature both of the bell and of a great drum, and has something in its sound which is singularly affecting. In its own country it bears a part in one of the most imposing ceremonies which man has ever imagined. On certain festivals as the sun is sinking in the west, the whole population of China, a host of more than a hundred millions, issues forth under the single canopy of heaven, to testify, amid the thunder of gongs and the continued discharge of fire works, that adoration and gratitude towards the Deity which human nature, in all ages and climes, has felt to be due, and has eagerly sought to express, however blind as to the sublime simplicity of religious truth.

Bells or *goblets* of glass sound still more perfectly than those of metal, and by gentle friction on their edges with a bow or the wetted finger, their tones may be continued for any length of time, and may be made to swell and diminish like a human voice or the notes of a violin. A set of glasses, therefore, attuned to each other, according to the harmonic scale, becomes, for certain species of music, the most perfect of all instruments. It is in fact an Æolian harp at command. Dr. Franklin, who first constructed a set, doubled the long line of glasses upon itself, and placed the half-notes as outside rows. The author of this work, however, during some experiments on sound, found the *zig-zag* arrangement here represented to possess certain advantages. The small open circles represent the mouths of the glasses standing in a box *a b c*, and the relation of the glasses to the written musical notes is shown by the common music lines and spaces which connect them. The learner discovers immediately that one row of the glasses produces the notes written *upon* the lines, and the other row the notes written *between* the lines; and he is mentally master of the instrument by simple inspection. This arrangement also renders the performance easy, for the notes most commonly sounded in succession are contiguous: and the relations of the notes forming a tune are so obvious to the eye, that the theory of musical combination and accompaniment is learned at the same time. The set of glasses here represented has two octaves, and with the additional *flat seventh* and *fourteenth*, seen at *a* and *c*, which, when required, may be substituted for the corresponding glasses in the rows, it is capable of playing the greater part of our simple melodies. All the half-notes, if desired, may be placed in outside rows. The player stands at the side of the box between *a* and *b*, and has the notes ascending towards the right hand, as in a piano-forte.

Fig. 120.

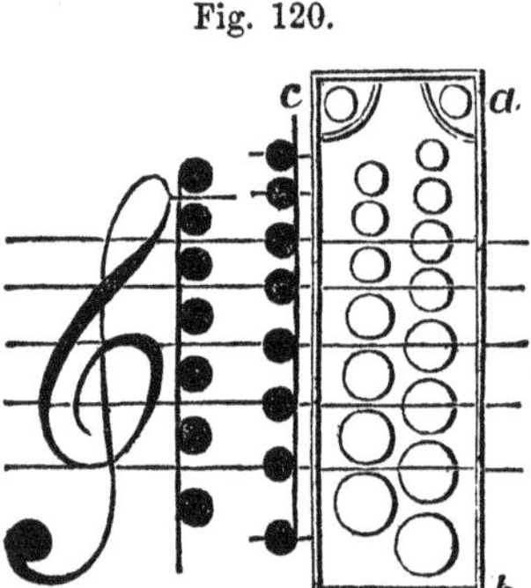

Musical ear.

Philosophers have not yet been able to account for a remarkable difference among individuals, as regards their perception of the musical relations of sounds. Many persons, without understanding any thing of acoustics, or having studied music as a science can tell instantly whether various notes heard together or in succession, have the mutual relations which we call musical—and which we now know to depend on the comparative numbers of beats in a given time; and they quickly recognize and learn to repeat tunes, and to sing a fit second or bass to the performance of another;—while there are persons, again, with an equally perfect sense of hearing, who can neither know if an air be played in tune, nor what air it is, nor can they ever sing alone or accompany. The former class of persons are said to have a *musical ear*, and the latter to want it; and although cultivation will raise mediocrity to considerable expertness, it cannot bestow the faculty where originally deficient. On this subject there is a very common misconception, which becomes a source of great mortification on one side, and of arrogance on the other, *viz.*, that the possession of a musical ear, or the power of distinguishing notes, is the indication of all the finer qualities of the mind while the want of it proves an opposite deficiency; and Shakspeare's opinion of him "that hath no music in himself," is often triumphantly cited as applicable to all who want the distinguishing ear. The truth, however, is, that many who possess this characteristic in a remarkable degree, are deficient in almost all else that humanity reveres,—witness the weak minds and disorderly lives of so many professed musicians,—while many, again, who have it not, are otherwise examples of excellence, and exquisitely sensible to other beauties and harmonies of nature. They may not be deaf, for instance, to the general music of spring, when all nature bursts forth in voice of rejoicing, nor to the awful music of the storm—they may feel as touching music the silence of a lone wood, contrasted with the unceasing din of multitudes—or even the stillness of night in a great city, where the astronomer, contemplating the wondrous spheres above, hears only the tongues of passing time in the church towers, or the call of watchmen, faintly sounding in the distance. In fine, many distinguished poets and philosophers have had no musical ear. —That the charm of music is often as much from early associations as from peculiar aptitude in the individuals, is proved by the effects so well known of the Swiss airs, when heard by native Swiss in foreign lands; and, indeed, of the national melodies of all countries, whose people are happy, and mix song with their usual occupations,—it not being in nature, that at any period of life, or in any clime, a man should cease to deem those modulations lovely, which recall the ecstatic emotions of his infancy and childhood; modulations learned in general from a parent's voice, perhaps an excellant mother's, whose affection was so long around him as a shield; whose tears fell to chide his errors, and to reward where there was promise of virtue; whose steady judgment was his guide, whose faultless life was his example, and who in all things to him was a personification of God's goodness on earth.

It is the prejudice of which we are now speaking with respect to musical ear and musical taste, that in the present day, condemns many young women, possessed of every species of loveliness and talent, except that of *note-distinguishing*, to waste years of precious time in an attempt to aquire this talent in spite of nature; but when they have succeeded as far as they can, they have only the merit of being machines, upon which tunes are set as upon a barrel-organ, and of which the performance is often far from being pleasing to good judges. Such persons when liberty comes to them with age or marriage, generally abandon the offensive occupation; but tyrant fashion will force their daughters to run the same course. The waste of time now spoken of, is only one of many evil consequences which arise from the prevailing false notions with respect to music : a subject which, however interesting, cannot be farther pursued in this place.

"*The trembling which causes the sensation of sounds spreads in all bodies. solid or fluid.*" (Read the Analysis.)

As air consists of material particles held far apart from each other by the repulsion of heat among them, we can conceive how an impulse given to a certain portion of the particles is transmitted to those beyond, by the increase of repulsion as they approximate; and from the second layer in the same manner to the third, and so on. And as in fluids the particles all mutually rest against, or repel each other, we can conceive why a motion produced in any part of a mass should be felt in every direction. The explosion of gunpowder, in which there is a sudden formation of a quantity of air, gives a shock all round which spreads a spherical wave to a great distance.

Although material particles in the form of liquid or solid are much nearer to each other than in the form of air, we still have many proofs, as stated at page 30, that they are not in absolute contact, and we therefore see the reason why the impulses producing sound should be transmitted through a liquid or solid in the same manner as through air, and even, by reason of the greater proximity of the particles, more quickly and forcibly than in air.

Instances of air carrying sound were given at page 235.—As farther examples we may cite the cases of what are called *sympathetic sounds.* Every elastic body being sonorous, that is to say, being fitted to tremble when struck, with a certain frequency of oscillation, depending on its weight and shape, &c., if the air around it be made to tremble by any cause, with the velocity which it is fitted to take on or produce, it immediately begins to tremble in unison with the air; and its motion or sound may continue after the original cause has ceased.—Thus almost any sound produced near a piano-forte whose dampers are raised, finds a responsive string, and if bits of paper are strewed upon the strings generally, those falling on the strings which return unisons or octaves to the sounding body are soon shaken off, while the others remain. A harp or guitar in a room with talking company, is often mingling a note with their conversation.—A wine-glass or goblet may be made to tremble, and if on a table at all inclined, even to fall, by a person sounding on a violoncello near it, the note accordant to its own

Sounding bodies vibrate much more quickly, or have sharper tones, if placed in light hydrogen, than in common air, and more quickly in common air than in any of the heavier gases :—because the lighter the surrounding fluid, the less is the resistance to a body moving in it. Thus also a bell will ring under water, but with a much graver sound than in the air.

That water is a vehicle of sound, is proved by the fact last mentioned,—by the distinctness with which the blows of workers around a diving-bell are heard above,—by the fact that fishes hear very acutely, &c.

And the following are instances of sound conveyed by solids.—A scratch of a pin at one end of a wooden log is distinctly heard by a person applying his ear at the other end, although through the air it is not at all audible even to the person who makes it.—Savages often discover the proximity of enemies or of prey, by applying an ear to the ground and hearing the tread.—The approach of horsemen at night is easily discovered in the same way.—The report of a cannon placed on the ice is carried much farther by the ice, than by the air around.—In the military operation of mining, or cutting away under ground for the purpose of entering a citadel, or blowing up fortifications, the approach of the enemy is often discovered by the subterranean sound of the pioneer's tools.—The awful muttering of earthquakes is merely the sound of subterranean explosions, conveyed from amazing distances, by the solid earth

A superstitious man sleeping in the upper story of a lofty house had for some time heard, during the stillness of the night, a singular beating noise near the head of his bed. There was no adjoining house beyond the wall, nor was there any thing going on near him in his own house to account for it, and he at last deemed it supernatural. Accident at last discovered that in a hovel built at the bottom and outside of the wall against which his bed stood, there was a wooden clock hanging, of which the sound, while all else was still, became audible aloft.

It is easy to ascertain whether a kettle boils, by putting one end of a stick or poker on the lid, and the other end to the ear; the bubbling of the water then appears as loud as the rattling of a carriage in the street.—A slight blow given to a steel poker or common triangle, of which an end is held to the ear, produces a sound which is even painfully strong.

The readiness with which solids receive and transmit sound is farther perceived in the fact, that a small musical box, while held in the hand, is scarcely audible, but when pressed against a table, or a door, will rival a little harp. The vibration communicated from the box pervades the whole of the wood, and the extended surface then acting on the air increases the effect. The construction of violins, harps, guitars, &c., and of sounding boards generally, is governed by the same law. In the dancing-master's *kit* or small fiddle, which he carries in his pocket, there may be the same strings and the same bow as for a violin, but it has very little sound, because the extent of its surface is so small. A heavy piece of metal called a *sourdine*, when fixed upon the bridge of a violin, damps the sound, because it is a dead mass resisting the motion of the elastic wood.

The fact of solids conveying sounds so much more perfectly than air has lately been applied to useful purposes in medicine. Dr. Laennec, of Paris, proposed some years ago to listen to what was going on in the interior of the body, and of the chest particularly, by applying one end of a wooden cylinder, which he called a *stethoscope* or *chest inspector*, to the surface, and resting the ear against the other end. The results of this happy thought have been important.

The actions going on in the chest are, the entrance and exit of the air in respiration, the voice, and the motion of the blood in the heart and blood-vessels;—and so perfectly do all these declare themselves to a person listening through the *stethoscope*, that an ear once familiar with the natural and healthy sounds, instantly detects certain deviations from them. Hence this instrument becomes a means of ascertaining certain diseases in the chest almost as effectually as if there were convenient windows for visual inspection ; and when it is considered that a fourth or fifth part of the inhabitants of Europe die of diseases of the chest, such as inflammations, abscesses, consumption, dropsical collections, aneurism, and various affections of the heart and blood-vessels, each of which requires an appropriate treatment, the importance of such a means may be judged of. By many medical men this instrument was at first ridiculed as quackery and nonsense, and many have yet to learn the use of it. May not both of these facts be attributed to the error which has existed in medical education, of leaving so many practitioners without that knowledge of the general laws of nature, which should enable them to appreciate at once any means likely to be useful in their art, from whatever quarter offered?

"*Velocity of sound.*" (See the Analysis.)

The velocity of light is such, that for any distance on earth its passage may be regarded as instantaneous. The velocity of sound is very much less. —If a woodman be observed at his occupation on the hill, his axe is seen to fall a considerable time before the sound of his blow reaches the spectator's ear.—The flash of a gun fired at a distance is seen long before the report is heard.

Most accurate experiments have been made to ascertain the velocity with which sound travels in the atmosphere; and it is found to be 1,142 feet per second, or a mile in about four seconds and-a-half; varying little either with the density or temperature of the air.

By noting then how long the flash of a gun is seen before the report reaches the ear, we learn the distance of the ship or battery from which the gun is fired. A chasing ship may thus often discover whether she be *nearing* or not the object of her pursuit. In the same manner the distance of thunder may be ascertained: and the reason of the long-continued roll of thunder is, that although the lightning darts instantly through the chains of clouds, perhaps of miles in length, the claps or explosion at each interruption of the chain are only heard successively, as the sound arrives at the ear. The pulse at the wrist of a healthy man is a convenient measure of time for ascertaining distances by the motion of sound,—each beat making nearly a second, and therefore indicating a distance of nearly a quarter of a mile.

A line of muskets fired at the same instant cannot appear a single report to any person who is not in the centre of a circle, of which the line forms a part.

An extended orchestra of musicians cannot be heard equally well from all situations near them.

Wind affects the velocity of sound just as a current in water affects the motion of a sailing ship.

Sound decreases in intensity from the centre where it originates, according to the same law as gravitation or light; that is to say, at double distance it is only one-fourth part as strong, at triple, a ninth, and so on.

By confining it, however, in tubes, which prevent it spreading, its force diminishes much less rapidly, and it will, therefore, extend to much greater distances.—In many manufactories, and even private dwellings now, there are pipes for the conveyance of sound leading to all parts; so that on ringing a bell to attract attention, verbal orders may be given through them to great distances.

Sound travels in water four times quicker, and in solids from ten to twenty times quicker, than in air. The blow of a hammer given to a wall by a person at one end, may be heard twice by a person at the other, *viz.*, almost immediately by an ear applied to the wall, and a little after through the air.

"*Reflection of sound.*" (Read the Analysis.)

As a wave of water turns back at a smooth wall or obstacle, so that at any distance after the reflection, it appears what it would have been at the same distance beyond the wall, only moving in an opposite direction; so the pulses or waves of sound are regularly reflected from flat surfaces, and produce what is called an *echo*. Such flat surfaces of nature's works are found only among the rocks and hills; and hence arose the beautiful fiction of the ancient poets, that echo was a nymph who dwelt concealed among the rocks. Science has now disclosed the secret of the viewless echo; but who does not vividly recollect the wonder and delight with which he has listened in the morning of his days, to his shrill call returned to him from some bold precipice, across the plain or river, or, perhaps, sent down to him again from the vaulted roof of ocean's caves!

The quickness with which an echo is returned to the spot where the sound originates, depends, of course, upon the distance of the reflecting surface; and, as sound travels 1,142 feet in a second, a rock at half that distance returns a sound exactly in one second. The number of syllables that can be pronounced in a second, will, in such a case, be repeated distinctly, while the end of a longer phrase would mix with the commencement of the echo. The breadth of a river may easily be ascertained where there is an echoing rock on the farther shore. A perpendicular mountain's side, or sublime cliffs, such as in many parts skirt the British coasts, return an audible echo of artillery, or of thunder, to a distance of many miles.

If two bold faces of rock or wall be parallel to each other, a sound produced between them is repeated often, playing like a shuttlecock between them, but becoming more faint each time until it is heard no more. In some situations, particularly when the sound plays thus above the smooth surface of water, a pistol-shot may be counted forty times.

The resonance of enclosed spaces depends on this continued reverberation. It often increases the effect of music by converting a simple melody, which is a *succession* of notes, into a harmonized piece, where each note is *accompanied* by some accordant tones; and a young flute-player is often first charmed with his own music when he finds himself performing a duet with echo in a cave or under a spacious arch:—but resonance injures the distinctness of speech, so as even in some ill-contrived halls of assembly or theatres, to render the articulation unintelligible. Small rooms or near surfaces give no perceptible echo, because the interval of time between the original sound and its repetition is too short for the ear to appreciate.

It is worthy of remark, that every apartment or confined space, has a certain musical note proper to it, the pitch of which depends upon the number of pulses or repetitions of a sound produced there in a given time by the returns from its walls. The velocity of sound being uniform, this number must depend on the size of the apartment.

There is a curious effect of echo which both illustrates the nature of the phenomenon, and proves that a tone or musical sound is merely a repetition of pulses following each other very quickly. Iron railings are generally formed of square bars, of which any side is a plane surface, and may produce an echo. Now a sound such as the sharp blow of a hammer, occurring near the end of such a railing, is echoed to a corresponding place on the other side by every bar in it; and as the echoes do not return all at once, but in regular succession, according to the increasing distances of the bars, the consequent regular succession of slight pulses, with uniform and small intervals, affects the ear, not as the echo of a single blow, but as a continued musical tone, the pitch of which depends on the distance of the bars from each other. The writer of this had observed, in passing on horseback along a particular portion of road, where there was first a piece of wall and then two pieces of paling with rails or bars of different width,—that there was from the wall a clear echo of the horse's cantering feet, and afterwards opposite the palings a ringing sound for every step of the horse. He had first concluded that the road there was singularly hard, although it did not appear so, and he slackened the horse's pace to save his feet, until, observing one day that the ringing sound was of different pitch opposite the two pieces of pailing, and so as to correspond with the different width of the bars, the true explanation occurred to him that the sound was an echo of the nature above described.

That an echo may be perfect, the surface producing it must be smooth, and of some regular form; for the wave of sound rebounds according to the same law as a wave of water, or a ray of light, or an elastic ball, &c., as explained at page 65, *viz.*, perpendicularly to the surface, if it fall perpendicular, but if it fall obliquely on one side, departing with an equal degree of obliquity on the other. To express this very important law shortly, we say that "the angle of reflection is equal to the angle of incidence."—According to this law, any irregular surface must break an echo; and if the regularity be very considerable, there can be no distinct or audible reflection at all. A regular concave surface, on the contrary, as *e g*, may concentrate sound, and bring all which falls upon it, as from *a b c d*, to the same centre or *focus*, as at *f*, so as to produce there a very powerful effect.

Fig. 121.

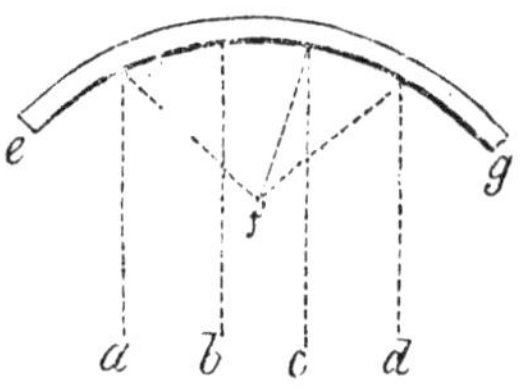

We thus see the reason why echo is much less perfect from the front of a house which has windows and doors, than from the plane end, or any plane wall of the same magnitude,—and why the resonance of a room is so irregular and indistinct when the room contains curtains, carpets and other furniture, or a crowded assembly. Halls for music have generally plane bare walls. Theatres for the drama, again, have boundaries broken in all ways by rows of boxes, and various ornaments.

The concentration of sound by concave surfaces produces many curious effects both in nature and in art.

There are remarkable situations where the sound from a cascade is concentrated by the surface of a neighbouring cave so completely, that a person accidentally bringing his ear into the focus, is suddenly astounded, as if the universe were crushing around him. A chair placed in the cave, so that a person sitting down in it must bring his ear into the focus, insures the success of the sometimes amusing experiment.

Fig 122.

The centre of a circle is the focus in which sound issuing from it is again collected after reflection; hence the powerful echo near the centre of a round apartment. An *oval* has two centres or *foci*—one towards each end, as *a* and *b*—and the nature of the curve is such, that sound, or light, or heat, issuing around from either of the foci, as *a*, by obeying the law of reflection above stated, is all directed from the various points, as at *c d e*, &c., to the other focus, as at *b*. Hence a person uttering a whisper in one focus of an oval room is very audible to the other, although he may not be heard by persons placed between. Such a room may be called a *whispering gallery*. Concave surfaces facing each other, as two alcoves in a garden, or covered recesses on opposite sides of a street or bridge, will enable persons seated in their foci to converse by whispers across louder noises in the space between, and without themselves being overheard in that space.

The reason why a tube conveys sound so far, is, that its sides confine or repress by a continued reflection, the advancing sound which, in the open air, would quickly spread laterally and be dissipated. And the reason that the plane surface of a smooth wall, or of water, &c., also conveys sound so far, is, that it similarly prevents the lateral spreading and dissipation, although only on one side.—Persons far apart may converse along a smooth wall.—The barking of dogs and the clear voice of a street-crier, in a town situated on the border of a lake, may be heard across the water in a calm evening, at a distance of more than five miles—the sound of bells, of course, is audible much farther.—And in the stillness of night, even the splashing oars of a boat will announce its approach to persons waiting at a great distance.

If a sound-reflecting surface be curved inwards, that is, be concave, it not only prevents the spreading of any sound which passes along it, but is constantly condensing the sound by driving the external part inwards. Hence, in a circular space, such as a gallery under a dome, persons close to the wall may whisper to each other at all distances.

An *ear-trumpet* is a tube wide at one end where the sound enters, and narrow at the other where the ear is applied : its sides are so curved that, according to the law of reflection, all the sound which enters is brought to a focus in the narrow end. It thus increases many fold the intensity of a sound which reaches the ear through it, and enables a person who has become deaf to common conversation, to mix again with pleasure in society, The concave hand held behind the ear answers in some degree the purpose of an ear-trumpet, and in a very large theatre is sometimes useful even to persons of quick hearing. A notorious instance of a sound-collecting surface was the *ear of Dionysius*, in the dungeons of Syracuse : the roof of the prison was so formed as to collect the words and even whispers of the unhappy prisoners, and to direct them along a hidden conduit to where the tyrant sat listening. The wide-spread sail of a ship, rendered concave by a gentle breeze, is also a good collector of sound. It happened one day on board a ship sailing along the coast of Brazil, far out of sight of land, that the persons walking on deck, when passing a particular spot, heard very distinctly, during an hour or two, the sound of bells, varying as in human rejoicings. All on board came to listen, and were convinced, but the phenomenon was most mysterious. Months afterwards it was ascertained, that at the time of observation the bells of the city of St. Salvador, on the Brazilian coast, had been ringing on the occasion of a festival: their sound, therefore, favoured by a gentle wind, had travelled over perhaps 100 miles of smooth water, and had been brought to a focus by the concave sail in the particular situation on the deck where it was listened to. It appears from this that a machine might be constructed having the same relation to sound that a telescope has to light.—A friend of the author, on the 18th of June, 1814, while sitting near the wall of his garden, situated near Dover, heard distinctly the firing of the cannon at the battle of Waterloo.

The *speaking-trumpet* is made according to the same law of reflected sound, with the view of directing the strength of the voice to a particular point. The sea captain uses it to hail ships at a distance, or to send his orders aloft, where the unaided voice would be lost in the noise of the wind and waves. A similar form of mouth is used for the *bugle horn* and common trumpet, and fits them to sound the note of command amid the uproar of contending armies.

Some amusing effects have been produced by operating on sounds with tubes and concave surfaces. What was termed the *invisible girl,* was a contrivance where the questions of vistors were caught by a concealed concave, and carried to the director who sat at a distance; and his replies, as in the whispering gallery, become audible to the inquirers alone.

The concave, undulating, and perfectly polished surface of many sea-shells, fits them to catch, mix, and return the pulses of sounds that happen to be trembling about them, so as to produce that curious resonance from within which closely resembles the sound of the distant ocean—so closely, that the spirited boy, after studying the interesting stories of voyagers which paint dangers to be nobly braved, and charm of nature to be seen in distant lands, often feeds his imagination with this voice of a shell, and fancies himself already riding among the billows.

The animal ear,

so admirably adapted to perceive the evanescent tremblings of the air, has of course a structure in nice relation to their nature as now explained. The parts of the ear, and the progress of the sound to the sentient nerve, may be simply described as follows:

1st. There is external to the head, a wide-mouthed tube or ear-trumpet *a,* for catching and concentrating the waves of sound. It is moveable in many animals, so that they can direct it to the place from which the sound comes.

Fig. 123.

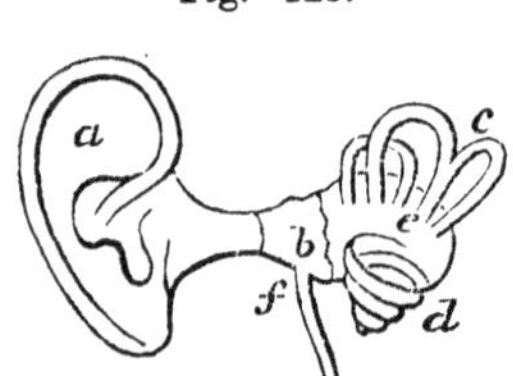

2d. The sound concentrated at the bottom of the ear-tube, falls upon a membrane stretched across the channel, like the parchment of an ordinary drum, over the space called the *tympanum* or *drum of the ear b,* and causes the membrane to vibrate. That its motion may be free, the air contained within the drum has free communication with the external air, by the open passage *f,* called the *Eustachian tube,* leading to the back of the mouth. A degree of deafness ensues when this tube is obstructed, as by wax; and a crack or sudden noise, with immediate return of natural hearing, is generally experienced when, in the effort of sneezing or otherwise, the obstruction is removed.

3d. The vibrations of the membrane of the drum are conveyed farther inwards, through the cavity of the drum, by chain of four bones (not here represented on account of their minuteness,) reaching from the centre of the membrane to the *oval door* or *window* leading into the labyrinth *e.*

4th. The labyrinth, or complex inner compartment of the ear, over which the nerve of hearing is spread as a lining, is full of water; and therefore by the law of fluid pressure (see page 128,) when the force of the moving membrane of the drum, acting through the chain of bones, is made to compress the water, the pressure is felt instantly over the whole cavity, as in a hydrostatic press.—The labyrinth consists of the *vestibule e,* the three *semi-circular canals c,* imbedded in the hard bone, and a winding cavity, called the *cochlea d,* like that of a snail-shell, in which fibres, stretched across like harp-strings, constitute the *lyra.*—The separate uses of these various parts are not yet perfectly known. The membrane of the tympanum may be pierced, and the chain of bones may be broken without entire loss of hearing. Considerable diversity of form and dimension is found in different animals.

The bone containing the cavities of the ear is the hardest in the body, and is the first formed.

The ear has the power of judging of the direction in which sound comes. A person in a thicket, listening to the song of various birds, although they be concealed from his eye by the luxuriance of the vernal foliage, still judges correctly by the ear in what tree every little songster is concealed.—The same truth is strikingly exemplified in the fact that, when horses or mules march in company at night, those in front direct their ears forward; those in the rear, backwards; and those in the centre, laterally or across;—the whole troop seeming to be actuated by one feeling, which watches the common safety.

The intensity of sound is to the ear a measure of distance. In a windy night, the sound of a distant bell may be brought so quickly, that it has not yet had time to spread and be weakened; and a person is often roused from a reverie by its unusual loudness and apparent nearness.—When a stormy wind blows directly upon a coast, and rolls the great waves in upon the sandy beach or among the rocks, the countryman living far inland hears the uproar, as if the ocean had burst its barriers, and were pouring in upon the land. The scene-contrivers at our theatres heighten the illusion of an approaching procession, by letting the accompanying music be first heard from a closed chamber or in a feeble tone, and afterwards with gradually increasing loudness. To the imagination, already excited perhaps to the highest pitch by the drama of some divine mind, the advancing host is thus most vividly portrayed; and when at last, with the thunder of drums and trumpets from the front of the stage, the troop also appears, the effect is complete. It is the varying loudness of the Æolian harp which produces the feeling that the heavenly choir is sometimes approaching and sometimes receding.

[*For an account of the Doctrines of Fluidity in relation to animals*, see Part V. Sec. II.]

PART IV.

DOCTRINES OF IMPONDERABLE SUBSTANCE.

To minds beginning this study, it may facilitate the conception of a substance which is without weight, or at least is imponderable by human art, to consider the nature of *air*. Until lately men were so imperfectly acquainted with the constitution of the universe around them, that a person placed in an apartment which offered to view nothing but the naked walls, would have said that it was empty, meaning literally what he said; and even when advertised that there was *air* in the room, he would still have been far from possessing a clear notion that it was full of aerial fluid just as an open vessel immersed in the sea is full of water, and that if air were not allowed to escape from it, even so small a body as an apple could not be pressed into it additionally by less force than fifty or sixty pounds. This truth, however, is now clearly understood, and daily exemplified in easy pneumatic experiments, and in no way more strikingly than by the recent adoption of the substance of air in place of feathers, as stuffing for beds and pillows. An air-tight bag or sack suspended by its lip in the air, and held quite open by a hoop near its mouth, would appear empty, but if then firmly closed above the hoop, it would have imprisoned its fill of air, just as a bag similarly managed under water would imprison its fill of water; and while in some respects the air would be softer and locally more yielding than feathers, its entire mass would be much less compressible. Now this air, when weighed by means which modern science has furnished, is found in a cubic foot to contain somewhat more than an ounce, and by strongly pressing it, or by causing it to combine chemically with some other substance, we can reduce it to a very small bulk, either with the form of a liquid or of a solid; proving how small a quantity of ponderable matter, under certain circumstances, will occupy great space. And common air is by no means the lightest known substance, which as powerfully resists the intrusion of other bodies where it exists. Hydrogen gas, for instance, of the same space-occupying force, weighs only a fourteenth part as much, and therefore a few drachms of it confined in a bag or bed as broad as the foundation of a house, would support a house or cask as large as a house filled with water to a height of thirty feet, the gas itself being then eighty thousand times lighter than its bulk of gold;—and if the pressure on it were diminished, it would readily expand to a volume a thousand times as great, and would still be exerting a considerable outward elasticity. Again, a mixture of oxygen and hydrogen gases, while uniting with explosive force to form water, dilates for the time, even under the great pressure of the atmosphere, to a bulk about twenty times greater than the gases have while separate.

The mind, pursuing the idea of such expansion or occupancy of space by a small quantity of matter, and reflecting on the wonderful divisibility of matter or minuteness of the ultimate atoms, as explained in Part I. of this work, might almost admit as a possible reality Newton's hypothetical illustration of that divisibility, *viz.*, that even one ounce of substance uniformly distributed over the vast space in which our solar system exists, might leave no quarter of an inch without its particle. Now a fluid in any degree approaching in rarity to this, although it might press, resist, communicate motion, and have other influences in common with more ponderable matter, would have neither weight nor inertia discoverable by means at present known to man. While we are contemplating, then, or modifying the agencies of what causes the phenomena of heat and cold, of light and darkness, of electricity in its forms of thunder and lightning, of galvanism, or of magnetism, in a word, the most striking phenomena of nature, we may be dealing with matter of the subtle constitution now spoken of. And as in the terrestrial atmosphere there are at least two fluids present, *viz.*, oxygen and nitrogen, of distinct nature, so in a more subtle ether, filling all space, there may be various ingredients.

A majority of philosophers now incline to the opinion here sketched, that there is at least one such subtle fluid or ether occupying completely the space of the universe, and tending to uniform diffusion by reason of a strong mutual repulsion of its particles, which fluid pervades denser material substances somewhat as water pervades a sponge or a mass of sand, being attracted in a peculiar way by each substance, and which fluid may or may not have weight and inertia. They believe farther that the phenomena above alluded to, and which human art can exhibit with highest beauty, or with awful intensity, are produced by the motion of other affections of that fluid, as the sensation of *sound* in all its varieties is produced in the delicate structure of the ear by a certain motion in the air, or in any other body having communication with the ear; or as the sensation of *jar* is perceived by a hand held to one end of a log of wood when a blow is given to the other end. Some philosophers again suppose that the causes of the phenomena are material particles projected through space, somewhat as sand might be scattered by an explosion, and which particles are present only when the effects are apparent. Some combine these two hypotheses. And some hold all the phenomena of heat to be mere motions in the common matter of the bodies in which the heat exists.

We mention these hypotheses, not with the view of entering upon a minute examination of their respective merits, or even of asserting that any one of them is true, but merely to make the reader aware of the directions which inquirers' minds have taken in pursuing the investigation. To understand the subjects as far as men yet usefully understand them, and sufficiently for a vast number of most useful purposes, it is only necessary, as in other departments of science, to classify important phenomena, so that their nature and resemblances may be clearly perceived. When, in treating of the human mind, we speak of its *retaining an idea*, or being *depressed*, or being *heated with passion*, &c., we speak of subjects sufficiently definite, although we may have no hypothesis as to the intimate nature of the phenomena:—and in the same manner may we speak of the accumulation, radiation, or other affections of heat and light. We know nothing of the cause even of gravity, the grandest influence in nature, but we can calculate its effects with admirable precision.

PART IV.

(CONTINUED.)

SECTION I.—ON HEAT.

ANALYSIS OF THE SECTION.

Heat (by some called Caloric) may be strikingly referred to as that which causes the difference between winter and summer, between tropical gardens and polar wastes. Its inferior degrees are denoted by the term COLD. *It cannot be exhibited apart, nor prove to have weight or inertia, and the change of its quantity in bodies is most conveniently estimated by the concomitant change of their bulk; any substance so circumstanced as to allow this to be accurately measured constituting a* THERMOMETER.

Heat diffuses itself among neighbouring bodies until all have the same temperature, that is, until all similarly affect a thermometer. It spreads partly through their structure, or by conduction, as it is called, with a slow progress, different for each substance, and in fluids modified by the motion of their particles; and it spreads partly also by being shot or radiated like light from one body to another, through transparent media or space, with readiness affected by the material and state of the giving and receiving surfaces.

Heat, by entering bodies, expands them, and through a range which includes, as three successive stages, the forms of SOLID, LIQUID *and* AIR *or* GAS; *becoming thus in nature the grand antagonist and modifier of that attraction which holds corporeal particles together, and which, if acting alone, would reduce the whole material universe to one solid lifeless mass. Each particular substance, according to the nature, proximity, &c., of its ultimate particles, takes a certain quantity of heat (said to mark its capacity), to produce in it a given change of temperature or calorific tension; undergoing expansion then in a degree proper to itself, and changing its form to liquid and air at points of temperature proper to itself; the expansion in bodies generally increasing more rapidly than the temperature, because the cohesion of their particles lessens with increase of distance; being remarkably greater therefore in liquids than in solids; and in air than in liquids; and the rate of expansion, moreover, being much quickened as the bodies approach their points of changing form to liquid or air, to produce which changes, a large quantity of heat enters them, but in the new arrangement of particles and increased volume of the mass, it becomes hidden from the thermometer, and is therefore called* LATENT HEAT. *For any given substance the changes of form happens so constantly at the same temperature, that they mark fixed points in the general scale of temperature, and enable us to regulate and compare thermometers. Heat, by expanding different substances unequally, influences much their chemical combination.*

*Heat influences also the functions of vegetable and animal life. The great source of heat is the sun; but electricity, combustion and other chemical action, condensation, friction, and the actions of life, are also excitants.**

"*Heat may be strikingly referred to as that which causes the difference between winter and summer, between the gardens of the equator and polar wastes.*" (See the Analysis, page 256.)

In the winter of climates, where the temperature is for a time below the freezing point of water, the earth with its waters is bound up in snow and ice, the trees and shrubs are leafless, appearing everywhere like withered skeletons, countless multitudes of living creatures, owing either to the bitter cold or deficiency of food, are perishing in the snows—nature seems dying or dead; but what a change when spring returns, that is, when heat returns! The earth is again uncovered and soft, the rivers flow, the lakes are again liquid mirrors, the warm showers come to foster vegetation, which soon covers the ground with beauty and plenty. Man, lately inactive, is recalled to many duties; his water-wheels are everywhere at work, his boats are again on the canals and streams, his busy fleets of industry are along the shores; winged life in new multitudes fill the sky, finny life similarly fills the waters, and every spot of earth teems with vitality and joy. Many persons regard these changes of season as if they came like the successive position of a turning-wheel, of which one necessarily brings the next; not adverting that it is the single circumstance of change of temperature which does all. But if the colds of winter arrive too early, they unfailingly produce the wintery scene, and if warmth come before its time in spring, it expands the bud and the blossom, which a return of frost will surely destroy. A seed sown in an ice-house never awakens to life.

Again, as regards climates, the earthly matters forming the exterior of our globe, and, therefore, entering into the composition of soils, are not different for different latitudes,—at the equator, for instance, and near the poles. That the aspect of nature then, in the two situations, exhibits a contrast more striking still than between summer and winter, is merely to an inequality of temperature, which is permanent. Were it not for this, in both situations the same vegetables might grow, and the same animals might find their befitting support. But now, in the one, namely, where the heat abounds, we see the magnificent scene of tropical fertility; the earth covered with luxuriant vegetation in endless, lovely variety, and even the hard rocks festooned with green, perhaps with the vine, rich in its purple clusters. In the midst of this scene, animal existence is equally abundant, and many of the species are of surpassing beauty—the plumage of the birds is as brilliant as the gayest flowers. The warm air is perfume from the spice-beds, the sky and clouds are often dyed in tints as bright as freshest rainbow, and happy

* It is to be remarked here, that many phenomena in which heat plays an important part, have been already described in preceding chapters of this work;—for instance, the action of the steam-engine, the phenomena of winds, many facts in meteorology, &c., under the head of Pneumatics. In a separate treatise on heat, these could not with propriety have been omitted; but in a comprehensive system of science like the present, they find their fit place, where, being surrounded by subjects resembling them in more intricate particulars, they can be more concisely and clearly explained.

human inhabitants call the scene a paradise. Again, where heat is absent, we have the dreary spectacle of polar barrenness, namely, bare rock or mountain, instead of fertile field; water everywhere hardened to solidity; no rain, nor cloud, nor dew; few motions but drifting snow; vegetable life scarcely existing, and then only in sheltered places turned to the sun—and instead of the palms and other trees of India, whose single leaf is almost broad enough to cover a hut, there are bushes and trees, as the furze and fir, having what may be called hairs or bristles, in the room of leaves. In the winter time, during which the sun is not seen for nearly six months, new horrors are added, viz., the darkness and dreadful silence, the cold benumbing all life, and even freezing mercury—a scene into which man may penetrate from happier climes, but where he can only leave his protecting ship and fires for short periods, as he might issue from a diving-bell at the bottom of the ocean. That in these now desolate regions, heat only is wanted to make them like the most favoured countries of the earth, is proved by the recent discoveries under ground of the remnant of animals and vegetables formerly inhabiting them, which now can live only near the equator. While winter, then, or the temporary absence of heat, may be called the sleep of nature, the more permanent torpor about the poles appears like its death; and when we farther reflect, that heat is the great agent in numberless important processes of chemistry and domestic economy, and is the actuating principle of the mighty steam-engine which now performs half the work of society, how truly may heat, the subject of our present chapter, be considered as the life or soul of the universe.

"*Heat cannot be exhibited in a separate state, nor proved to have weight or inertia.*" (Read the Analysis, page 256.)

Although heat is known to be abundant in the sunbeam, and to radiate around from a blazing fire, we cannot otherwise arrest or detect it in its progress than by allowing it to enter, and remain in some ponderable substance. We know hot iron, or hot water, or hot air, but nature no where presents to us, nor has art succeeded in showing us, heat alone.

If we balance a quantity of ice in a delicate weigh-beam, and then leave it to melt, the equilibrium will not be in the slightest degree disturbed. Or if we substitute for the ice, boiling water or red-hot iron, and leave this to cool, there will be no difference in the result. If we place a pound of mercury in one scale of the weigh-beam and a pound of water in the other, and then either heat or cool both through the same number of thermometric degrees, although about thirty times more heat (as will be explained below) enters or leaves the bulky water than the dense mercury, they will still remain equivalent weights.

Again, a sunbeam, with its intense light and heat, after being concentrated by a powerful lens or mirror, may be made to fall upon the scale of a most delicate balance, but will produce no depressing effect on the scale, as would follow if what constitutes the beam had the least forward motal inertia or momentum.

Such are the facts which have led certain inquirers to deny the material or separate existence of heat, and to hold that it is merely motion of one kind among the material particles of bodies generally, as sound is motion of another kind among the same particles. The following facts they consider to have the same bearing in the argument: Heat can be produced without limit by friction, as—when savages light their fires by rubbing together two pieces of wood—when Count Rumford made great quantities of water boil, by causing a blunt borer to rub against a mass of metal immersed in water—when Sir Humphrey Davy quickly melted pieces of ice by rubbing them against each other in a room cooled below the freezing point, &c. Intense heat is produced by the explosion of gunpowder or other fulminating mixture, yet it cannot be conceived to have existed in the small bulk of the powder before the explosion. Other inquirers, on the contrary, have deemed to be proofs of the separate materiality of heat such facts as now follow;—that it is radiated through the most perfect vacuum which we can produce, and even more readily than through air; that it radiates in the same place in all directions, without impediment from the crossing rays;—that it becomes instantly sensible on the condensation of any material mass, as if then squeezed out from the mass; as when, by compressing air suddenly, we inflame a match immersed in it; or when, on reducing the bulk of iron by hammering, we render it very hot, the warming being greater at the first blow (which most changes the bulk) than afterwards,—that when, on mixing bodies which combine so intimately as to occupy less space than when separate, there is a disengagement of heat proportioned to the diminution of the volume:—that the laws of the spreading of heat in bodies do not resemble those of the spreading of sound, or of any other motion known to us;—and that, as to the great and sudden extrication of heat by friction or explosion, it may be as truly a rush of the fluid to the part, as in the case of an electrical accumulation or discharge These facts, moreover, they think, square well with their assumption that the phenomena of heat are produced by an exceedingly subtle fluid, or ether, pervading the whole universe, and softening, or melting, or gasifying bodies, according to the quantity present in each; its own parts being strongly repulsive of each other, and seeking, therefore, widest and most equable diffusion.

"*The change of its quantity in bodies is most conveniently estimated by the concomitant change of their bulk, any substance so circumstanced, as to allow this to be accurately measured, constituting a thermometer.*" (Read the Analysis, page 256.)

If we heat a wire, it is lengthened; if we heat water in a full vessel, a part runs over; if we heat air in a bladder, the bladder is distended; in a word, if we heat any substance, its volume increases in some proportion to the increase of temperature,—and we may measure the increase of volume. The reasons why, in such investigations, a contrivance in which the expansion of mercury may be observed, *viz.*, the mercurial thermometer, is commonly preferred to others, can only be fully understood by the mind which has considered the whole subject of heat; and we touch upon the matter here, only for the purpose of stating that a mercurial thermometer is a small bulb, or bottle of glass filled with mercury, and having a long very narrow stalk or neck, in which the mercury rises when expanded by heat, or falls when heat is withdrawn; the stalk between the points at which mercury standings in freezing and boiling water, being divided into an arbitrary number of degrees, which division appearing on a scale applied to the stalk, is continued similarly above and below these points.

"*Heat diffuses itself among neighboring bodies until all have acquired the same temperature; that is to say, until all will similarly affect a thermometer.*" (See the Analysis.)

An iron bolt thrust in among burning coals soon becomes red hot like them. If it be the heater of a tea-urn, it will, when afterwards placed amidst the water, part with its lately acquired heat to the water, until both are of the same temperature. Boiling water, again soon imparts heat to an egg placed in it, and a feverish head yields its heat to a bladder of cold water or ice. A hundred objects enclosed in the same apartment, if tested, after a time, by the thermometer, will all indicate the same temperature.

" *The inferior degrees of heat are denoted by the term* COLD."

When the hand touches a body of a higher temperature than itself, it receives heat according to the law now explained, and it experiences a peculiar sensation; when it touches a body of lower temperature than itself, it gives out heat for a like reason, and experiences another and very different sensation. The two are called the sensations of *heat* and of *cold*. Now heat and cold, considered as existing in the bodies themselves, although thus appearing opposities, are really degrees of the same object, *temperature*, contrasted by name, for convenience sake, in reference to the particular temperature of the individuals speaking of them—just as any two nearest mile-stones on a road, although merely marking degrees of the same object, *distance*, might receive from persons living between them the opposite names of east and west, or of north and south. It is to be remarked, moreover, that the sensation of heat is producible also by a body colder than the hand, provided it be less cold than a body touched immediately before, or than the usual temperature; and the sensation of cold is produced under the opposite circumstances of touching a comparatively warm body, but which is less warm than something touched just before. This explains the remarkable fact that the same body may appear at the same time, and to the same person, both hot and cold. If a person transfer one hand to common spring-water from touching ice, that hand will deem the water very warm: while the other hand, transferred to it from a warm bath, would deem it very cold. For a like reason; a person from India, arriving in England in the spring, deems the air cold, while the inhabitants of the country are diminishing their clothing, because the heat to them is becoming oppressive. Such facts show how necessary it was for men to discover more correct thermometers than their bodily sensations.

" *Spreading partly through their structure, or by conduction, as it is called, with a progress proper to each substance.*" (Read the Analysis, page 256.)

If one end of a rod of iron be held in the fire, a hand grasping the other end soon feels the heat coming through it. Through a similar rod of glass the transmission is much slower, and through one of wood it is slower still. The hand would be burned by the iron, before it felt warmth in the wood, although the inner end were blazing.

On the fact that different substances are permeable to heat, or have the property of conducting it, in different degrees, depend many interesting phenomena in nature and the arts; hence it was important to ascertain the degrees exactly, and to classify the substances. Various methods for this purpose have been adopted. For solids—similar rods of the different substances, after being thinly coated with wax, have been placed with their inferior extremities in hot oil, and then the comparative distances to which, in a given time, the wax was melted, furnished one set of indications of the comparative conducting powers:—or, equal lengths of the different bare rods being left above the oil and a small quantity of explosive powder being placed on the top of each, the comparative intervals of time elapsing before the explosions gave another kind of measure :—or equal balls of different substances, with a central cavity in each to receive a thermometer, being heated to the same degree and then suspended in the air to cool, until the thermometer fell to a given point, gave still another list. A modification of the last method was adopted by Count Rumford to ascertain the relative degrees in which furs, feathers, and other materials used for clothing, conduct heat, or, which is the same thing, resist its passage. He covered the ball or stem of a thermometer with a certain thickness of the substance to be tried, by placing the thermometer in a larger bulb and stem of glass, and then filling the interval between them with the substance; and after heating this apparatus to a certain degree, by dipping it in liquid of the desired temperature, he surrounded it by ice, and marked the comparative time required to cool the thermometer a certain number of degrees. The figures following the names of some of the substances in the subjoined list, mark the number of seconds required respectively for cooling it 60°.

These experiments have shown, as a general rule, that density in a body favours the passage of heat through it. The best conductors are the metals, and then follow in succession diamond, glass, stones, earths, woods, &c., as here noted:

Metals—silver, copper, gold, iron, lead.	
Diamond.	
Glass.	
Hard stones.	
Porous earths.	
Woods.	
Fats or thick oils.	
Snow.	
Air - - - - - -	576
Sewing silk - - - -	917
Wood ashes - - - -	927
Charcoal - - - - -	937
Fine lint - - - - -	1,032
Cotton - - - - -	1,046
Lamp black - - - -	1,117
Wool - - - - -	1,118
Raw silk - - - - -	1,284
Beavers' fur - - - -	1,296
Eider down - - - -	1,305
Hares' fur - - - - -	1,315

Air appears near the middle of the preceding list, but if its particles are not allowed to move about among themselves, so as to *carry* heat from one part to another, it *conducts* (in the manner of solids) so slowly that Count Rumford doubted whether it conducted at all. It is probably the worst conductor known, that is, the substance which when at rest impedes the passage of heat the most. To this fact seem to be owing, in a considerable degree, the remarkable non-conducting quality of porous or spongy substances, as feathers, loose filamentuous matter, powders, &c., which have much air in their structure, often adherent with the force of attraction which immersion in water, or even being placed in the vacuum of an air-pump, is insufficient to overcome.

While contemplating the facts recorded in the above table, one cannot but reflect how admirably adapted to their purposes the substances are which nature has provided as clothing for the inferior animals;—and which man afterwards accommodates with such curious art to his peculiar wants. Animals required to be protected against the chills of night and the biting blasts of winter, and some of them which dwell among eternal ice, could not have lived at all, but for a garment which might shut up within it nearly all the heat which their vital functions produced. Now any covering of a metallic, or earthy, or woody nature, would have been far from sufficing; but out of a wondrous chemical union of carbon with the soft ingredients of the atmosphere, those beautiful textures are produced called fur and feather, so greatly adorning while they completely protect the wearers;—textures, moreover, which grow from the bodies of the animals, in the exact quantity that suits the climate and season, and which are reproduced when by any accident they are partially destroyed. In warm climates the hairy coat of quadrupeds is comparatively short and thin; as in the elephant, the monkey, the tropical sheep, &c. It is seen to thicken with increasing latitude, furnishing the soft and abundant fleeces of the temperate zones; and towards the poles it is externally shaggy and coarse, as in the arctic bear. In amphibious animals, which have to resist the cold of water as well as of air, the furs grow particularly defensive, as in the otter and beaver. Birds, from having very warm blood, require plenteous clothing, but require also to have a smooth surface, that they may pass easily through the air;—both objects are secured by the beautiful structure of feathers, so beautiful and wonderful that writers on natural theology have often particularized it as one of the most striking exemplifications of creative wisdom. Feathers, like fur, appear in kind and quantity suited to particular climates and seasons. The birds of cold regions have covering almost as bulky as their bodies, and if it be warm in those of them which live only in air, in the water-fowl it is warmer still. These last have the interstices of the ordinary plumage filled up by the still more delicate structure called down, particularly on the breast, which in swimming first meets and divides the cold wave. There are animals with warm blood which yet live very constantly immersed in water, as the whale, seal, walrus, &c. Now neither hair nor feathers, however oiled, would have been a fit covering for them; but kind nature has prepared an equal protection in the vast mass of fat or thick oil which surrounds their bodies—substances which are scarcely less useful to man than the furs and feathers of land animals.

While speaking of clothing we may remark, that the bark of trees is also a structure very slowly premeable to heat, and securing, therefore, the temperature to vegetable life.

And while we admire what nature has thus done for animals and vegetables, let us not overlook her scarcely less remarkable provision of ice and snow, as winter clothing for the lakes and rivers, for our fields and gardens. Ice, as a protection to water and its inhabitants, was considered in Sec. I., in the explanation of why, although solid, it swims on water. We have now to remark that snow, which becomes as a pure white fleece to the earth, is a structure which resists the passage of heat nearly as much as feathers. It, of course, can defend only from clouds below 32° or the freezing point; but it does so most effectually, preserving the roots and seeds and tender plants during the severity of winter. When the green blade of wheat and the beautiful snow-drop flower appear in spring rising through the melting snow, they have recently owed an important shelter to their wintry mantle. Under deep snow while the thermometer in the air may be far below zero, the temperature of the ground is rarely below the freezing point. Now this temperature, to persons some time accustomed to it, is mild and even agreeable. It is much higher than what often prevails for long periods in the atmosphere of the centre and north of Europe. The Laplander, who during his long winter lives under ground is glad to have additionally over head a thick covering of snow. Among the hills of the west and north of Britain, during the storms of winter, a house or covering of snow frequently preserves the lives of travellers, and even of whole flocks of sheep, when the keen north wind catching them unprotected, would soon stretch them lifeless along the earth.

It is because earth conducts heat slowly, that the most intense frost penetrates but a few inches into it, and that the temperature of the ground a few feet below its surface is nearly the same all the world over. In many mines, even although open to the air, the thermometer does not vary one degree in a twelvemonth. Thus also water in pipes two or three feet under ground does not freeze, although it may be frozen in all the smaller branches exposed above. Hence, again, springs never freeze, and therefore become remarkable features in a snow covered country. The living water is seen issuing from the bowels of the earth, and running often a considerable way through fringes of green, before the gripe of the frosts arrest it; while around it, as is well known to the sportsman, the snipes and wild duck and other birds are wont to congregate. A spring in a frozen pond or lake may cause the ice to be so thin over the part where it issues, that a skater arriving there will break through and be destroyed. The same spring water which appears warm in winter, is deemed cold in summer, because although always of the same heat, it is in summer surrounded by warm atmosphere and objects. In proportion as buildings are massive, they acquire more of those qualities which have now been noticed of our mother earth. Many of the Gothic halls and cathedrals are cool in summer and warm in winter—as are also old-fashioned houses or castles with thick walls and deep cellars. Natural caves in the mountains or sea-shores furnish other examples of a similar kind.

When in the arts it is desired to prevent the passage of heat out of, or into any body or situation, a screen or covering of a slow conducting substance is employed. Thus to prevent the heat of a smelting or other furnace from being wasted, it is lined with fire bricks, or is covered with clay and sand, or sometimes with powdered charcoal. A furnace so guarded may be touched by the hand, even while containing within the melted gold. To prevent the freezing of water in pipes during the winter, by which occurrence the pipes would be burst, it is common to cover them with straw ropes, or coarse flannel, or enclose them in a larger outer pipe with dry charcoal, or saw dust, or chaff, filling up the interval between. If a pipe on the contrary, be for the conveyance of steam or other warm fluid, the heat is retained, and, therefore, saved by the very same means. Ice-houses are generally made with double walls, between which, dry straw placed, or saw dust, or air, prevents the passage of heat. Pails for carrying ice in summer, or intended to serve as wine coolers, are made on the same principle—*viz.*, double vessels, with air or charcoal, filling the interval between them. A flannel covering keeps a man warm in winter—it is also the best means of keeping ice from melting in summer. Urns for hot water, tea pots, coffee pots, &c., are made with wooden or ivory handles, because if metal were used, it would conduct the heat so readily that the hand could not bear to touch them.

It is because glass and earthenware are brittle, and do not allow ready passage to heat, that vessels made of them are so frequently broken by sudden change of temperature. On pouring boiling water into such a vessel, the internal part is much heated and expanded (as will be explained more fully in a subsequent page) before the external part has felt the influence, and this is hence riven or cracked by its connection with the internal. A chimney mirror is often broken by a lamp or candle placed on the marble shelf too near it. The glass cylinder of an electrical machine will sometimes be broken by placing it near the fire, so that one side is heated while the other side receives a cold current of air approaching the fire from a door or window. A red hot rod of iron drawn along the pane of glass will divide it almost like a diamond knife. Even cast iron, or the backs of grates, iron pots, &c., although conducting readily, is often, owing to its brittleness, cracked by unequal heating or cooling, as from pouring water on it when hot. Pouring cold water into a heated glass will produce a similar effect. Hence glass vessels intended to be exposed to strong heats and sudden changes, as retorts for distillation, flasks for boiling liquids, &c., are made very thin, that the heat may pervade them almost instantly and with impunity.

There is a toy called a *Prince Rupert's Drop*, which well illustrates our present subject. It is a lump of glass let fall while fused into water, and thereby suddenly cooled and solidified on the outside before the internal part is changed; then as this at last hardens and would contract, it is kept extended by the arch of external crust, to which it coheres. Now if a portion of the neck of the lump be broken off, or if other violence be done, which jars its substance, the cohesion is destroyed, and the whole crumbles to dust with a kind of explosion. Any glass cooled suddenly when first made, remains very brittle, for the reason now stated. What is called the *Bologna jar* is a very thick small bottle, thus prepared, which bursts by a grain of sand falling into it. The process of annealing, to render glass ware more tough and durable, is merely the allowing it to cool very slowly by placing it in an oven, where the temperature is caused to fall gradually. The tempering of metals by sudden cooling seems to be a process having some relation to that of rendering glass hard and brittle.

It is the difference of conducting power in bodies which is the cause of a very common error made by persons in estimating the temperature of bodies by the touch. In a room without a fire all the articles of furniture soon acquired the same temperature; but if in winter, a person with bare feet were to step from the carpet to the wooden floor, from this to the hearth-stone, and from the stone to the steel fender, his sensation would deem each of these in succession colder than the preceding. Now the truth being that all had the same temperature, only a temperature inferior to that of the living body, the best conductor, when in contact with a body, would carry off heat the fastest, and would, therefore, be deemed the coldest. Were a similar experiment made in a hot-house, or in India, while the temperature of every thing around were 98°, viz., that of the living body, then not the slightest difference would be left in any of the substances: or lastly, were the experiment made in a room where by any means the general temperature were raised considerably above blood heat, then the carpet would be deemed considerably the coolest instead of the warmest, and the other things would appear hotter in the same order in which they appeared colder in the winter room. Were a bunch of wool and a piece of iron exposed to the severest cold of Siberia, or of an artificial frigorific mixture, a man might touch the first with impunity, (it would merely be felt as rather cold;) but if he grasped the second, his hand would be frost-bitten and possibly destroyed; were the two substances, on the contrary, transferred to an oven, and heated as far as the wool would bear, he might again touch the wool with impunity (it would then be felt as a little hot,) but the iron would burn his flesh. The author has entered a room where there was no fire, but where the temperature from hot air admitted was sufficiently high to boil the fish, &c., of which he afterwards partook at dinner; and he breathed the air with very little uneasiness. He could bear to touch woolen cloth in this room, but no body more solid.

The foregoing considerations make manifest the error of supposing that there is a positive warmth in the materials of clothing. The thick cloak which guards a Spaniard against the cold of winter, is also in summer used by him as a protection against the direct rays of the sun:—and while in England, flannel is our warmest article of dress, yet we cannot more effectually preserve ice than by wrapping the vessel containing it in many folds of softest flannel.

In every case where a substance of different temperature from the living body touches it, a thin surface of the substance immediately shares the heat of the bodily part touched—the hand generally; and while in a good conductor, the heat so received quickly passes inwards, or away from the surface, leaving this in a state to absorb more, in the tardy conductor the heat first received tarries at the surface, which consequently soon acquires nearly the same temperature as the hand, and therefore, however cold the interior of the substance may be, it does not cause the sensation of cold. The hand on a good conductor has to warm it deeply, a slow conductor it warms only superficially. The following cases further illustrate the same principle. If the ends of an iron poker, and of a piece of wood of the same size, be wrapped in paper and then thrust into a fire, the paper on the wood will begin to burn immediately, while that on the metal will long resist:—or if pieces of paper be laid on a wooden plank and on a plate of steel, and then a burning coal be placed on each, the paper on the wood will begin to burn long before that on the plate. The explanation is, that the paper in contact with the good conductor loses to this so rapidly the heat received from the coal, that it remains at too low a temperature to inflame, and will even cool to blackness the touching part of the coal; while on the tardy conductor the paper becomes almost immediately as hot as the coal. It is because water exposed to air cannot be heated beyond 212°, that it may be made to boil in an egg-shell or a vessel made of paper, held over a lamp, without the containing substance being destroyed; but as soon as it is dried up, the paper will burn and the shell will be calcined, as the solder of a common tin kettle melts under the same circumstances. The reason why the hand judges a cold liquid to be so much colder than a solid of the same temperature is, that from the mobility of the liquid particles among themselves, those in contact with the hand are constantly changing. The impression produced on the hand by very cold mercury is almost insufferable, because mercury is both a ready conductor and a liquid. Again, if a finger held motionless in water feel cold, it will feel colder still when moved about; and a man in the air of a calm frosty morning does not experience a sensation nearly so sharp as if with the same temperature there be wind. A finger held up in the wind discovers the direction in which the wind blows by the greater cold felt on one side, the effect being still more remarkable if the finger is wetted. If a person in a room with a thermometer, were with a fan or bellows to blow the air against it, he would not thereby lower it, because it had already the same temperature as the air, yet the air blown against his own body would appear colder than when at rest, because being colder than his body, the motion would supply heat-absorbing particles more quickly. In like manner, if a fan or bellows were used against a thermometer hanging in a furnace or hot-house, the thermometer would suffer no change, but the air moved by them against a person, would be distressingly hot, like the blasting sirocco of the sandy deserts of Africa. If two similar pieces of ice be placed in a room somewhat warmer than ice, one of them may be made to melt much sooner than the other, by blowing on it with a bellows. The reason may here be readily comprehended why a person suffering what is called a cold in the head, or catarrh from the eyes and nose, experiences so much more relief on applying to the face a handkerchief of linen or cambric than one of cotton :—it is, that the former, by *conducting*, readily absorbs the heat and diminishes the inflammation, while the latter, by refusing to give passage to the heat, increases the temperature and the distress. Popular prejudice has held that there was a poison in cotton.

"Heats spreading in fluids chiefly by the motion of their particles."
(Read the Analysis, page 256.)

Owing to the mobility among themselves of fluid particles, heat entering a fluid anywhere below the surface, by dilating and rendering specially lighter the portion heated, allows the denser fluid around to sink down and force up the rarer; and the continued currents so established, diffuse the heat through the mass much more quickly than heat spreads by conduction in any solid.

Count Rumford's experiments led him at first to conclude that liquids, but for this carrying process, by the particles changing their place, were absolutely impassable to heat. A piece of ice will lie very long at the bottom of water which is made to boil at the top by the contact of any hot body; and when it at last melts, Count R. believed that it did so entirely from the heat which passed downwards through the sides of the vessels containing the water. But an ingenious experiment by Dr. Murray decided the question differently. He made a vessel of ice, which, of course, could not carry downwards any heat greater than 32°, as ice melts at that degree; and having put into the vessel a quantity of oil at 32°, with the bulb of a thermometer being a quarter of an inch under the surface of the oil, he placed a cup of boiling water in contact with the surface of the oil :—in a minute and a half the thermometer rose nearly a degree, and in seven minutes it rose five degrees, beyond which it did not go. The heat then must have passed downwards through the liquid, proving a conducting power ;—unless, indeed, it passed by radiation, as explained in a subsequent page.

The internal currents or circulation produced by heat in fluid masses, and of which there are so many important instances in nature, were more fitly explained in the chapter on *Hydrostatics* and *Pneumatics;* we shall here, therefore, allude to them very shortly.

Perhaps the best experimental illustration of the subject is obtained by placing a tall glass jar, filled with water, in which small pieces of amber are diffused to show its movements, first in a warm bath, and then in a cold bath. In the first case, the water and amber near the outside of the jar where they are heated, will exhibit a rapid upward current, while in the centre of the jar they will form an opposite and downward current. In the second case, or when the jar is placed in a cold bath, the direction of the currents will be reversed.

Consideration of these currents led the author of this work, some years ago, to propose what he deemed a great improvement on the construction and management of boilers and evaporating pans generally; namely, to convert the upward and downward current in the mass of boiling liquid into a lateral current below, constantly and rapidly sweeping the bottom of the vessel. In ordinary boilers, when a portion of liquid is converted into steam in contact with the horizontal bottom, it does not separate from the bottom immediately, but remains until a steam-bubble of considerable size be formed, and in the mean time the part of the boiler defended by it from the contact of the liquid, becomes overheated, and the following evil consequences ensue:—1. Rapid destruction of the boiler, and on that account a rapid expense; 2. Necessity for having originally much thicker, and therefore dearer boilers; 3. The thickness being an impediment to the passage of heat, there is a proportionate waste of fuel; 4. and last, When the liquid is a vegetable juice or extract, as sugar-cane juice, of a nature to be carbonized and blackened when overheated, the quality of the product is often exceedingly deteriorated. The patentees of an apparatus, described at 184, for boiling sugar *in vacuo*, and therefore at a low and steady temperature, gained, it was said, more than $40,000 a year by preventing the injury now spoken of. And when the liquid is a saline solution, like the sea-water used in steam ships, the salt soon encrusts the bottom of the boiler; and powerfully both prevents the passage of heat and destroys the boiler. Now a current sweeping the bottom prevents all these consequences, and may be easily obtained. The most obvious method is, to place in the liquid some upright tubes with open tops at a certain distance under the surface of the liquid, and with the bottoms also open, but latterly, and all in one direction; the consequence will be, that as soon as the liquid begins to boil, the general mass, consisting of liquid mixed with bubbles of steam, becomes of considerably less specific gravity than the liquid in the tubes, remaining unmixed, because steam will not enter the lateral mouths of the tubes, and the columns of heavier liquid will therefore descend rapidly and issuing by the lateral openings of the tubes all in one direction along the bottom of the boiler, will powerfully and uninterruptedly sweep it. In a long wagon-shaped boiler the tubes, instead of being round, should be made flat and broad enough to reach from side to side; and if a very rapid current be desired, they must be made larger than the spaces between them in which the steam has to rise, for thus the steam bubbles, being driven closer together, will make the rising column so much the lighter, and its ascent consequently the more rapid; and there will be a corresponding rapidity of issue of the sweeping current. In a moderate sized pan or boiler of the usual basin or half-globe shape, the simplest method of producing the current is to have a smaller vessel of similar shape made of thin metal, and placed within the other so as to have about an inch space all round between them, and having one large opening at its bottom,—then all the steam mixed with fluid will rise between the outer and inner vessel, while the unmixed liquid will descend through the open bottom of the inner vessel, and spread in every direction, sweeping the bottom of the outer. The sweeping of the bottom of a boiler might also be effected by a wheel kept turning, to cause the liquid to resolve horizontally (as was done with another view in the large Scotch whiskey stills,) or by a frame made somewhat like a rake or gridiron, kept moving backwards and forwards upon the bottom.

As stated in a previous section, it is the heating and dilatation of the fluid air over a tropical island while acted upon during the middle of the day by the powerful rays of the sun, which allow the colder and heavier air from the face of the ocean around to press inwards upon it and force it upwards in the atmosphere—the cold current forming the delightful sea-breeze of the climate. And it is the general heating of the air over the whole equatorial belt of the earth, which, rendering it specifically lighter than the air nearer the poles, allows this to assume the form of cool trade-winds, constantly blowing towards the sun's path, and pressing upwards the hot air, which then spreads away on the top of the atmosphere towards the poles, to mitigate the severity of the northern and southern cold. In the watery ocean also there is a circulatory motion of the same kind, although less in degree, tending to distribute heat and equalize temperature, and contributing to produce some of the great sea currents known to mariners.

The vertical currents produced by heat, in the ocean and in great masses of water generally, preserve in and over them a comparatively uniform temperate freshness, while the rocks and soil on the shores around may be either parched under a burning sun, or bound up in frost. A keen frost chills, and soon hardens in its icy grasp the surface of the ground; but of water similarly exposed, the part first cooled descends to the bottom by its increased density, and forces up a warmer water to take its place; this in its turn is cooled and descends, and a continued circulation is established, so that the surface cannot become ice until the whole mass, of whatever depth, has been cooled down to its greatest density. Hence the very deep sea is not frozen even in the coldest climates, and in temperate climates, the severest winter does not freeze even the ordinary lakes. During this intestine movement in the water, that which ascends to the surface to be cooled, by losing one degree of its heat, warms more than 500 times its bulk of air one degree, and thus tempers remarkably the air passing over it. Hence places in the vicinity of the sea and of lakes are warmer in winter than places further inland, although nearer to the equator. England is much warmer in winter than central Germany, which lies south of England; and the coast of Scotland and of the north of Ireland are warmer than London:—snow never lies long upon these coasts. As continental or inland countries have thus in winter an extreme of cold, so they have in summer an extreme of heat. Water admits the rays of the sun, and absorbs the heat into the whole thickness of its mass, and therefore is warmed very slowly; but the dry earth retains all the heat near its surface, and is therefore soon heated to excess.

The ventilation of our dwellings and halls of assembly (as explained previously) is owing to the motion produced by the changed specific gravity of air when heated. The air which is within the house becomes warmer than the external air, and the latter then presses in at every opening or crevice to displace the other. The ventilation of the person by the slow passage of air through the texture of our clothing is a phenomenon of the same kind; and thicker clothing acts chiefly by diminishing the rapidity of this passage. Hence an oiled-silk or other air-tight covering laid on a bed has greater influence in preserving warmth than one or two additional blankets, and is not generally used, only because it prevents ventilation, and by shutting in the insensible perspiration, soon produces dampness. From the part of bedclothes immediately over the person there is a constant outward oozing of warm air, and there is an oozing inward of cold air in lower situations around. In many persons the circulation of the blood is so feeble that in winter, they have great difficulty in keeping their feet warm, even in bed, unless with the assistance of a bottle of hot water or some such means, and in consequence they often pass sleepless nights, and suffer in their general health. In such cases, at the suggestion of the author, a long flexible tube has been used,—as of spiral wire, covered with leather or varnished cloth, by which a person can send down to his feet his hot breath, and thus apply them effectually a natural animal warmth, as in a cold day he does to his hands by blowing upon them through his gloves.

The power of fluids to diffuse heat being due to their power of *carrying*, and not of *conducting* it, the consequence should follow, that any circumstance which impedes the internal motion of the fluid particles, should diminish the suffusing power. Accordingly we find, that fluids in general transfer heat less readily in proportion as they are more viscid. Water, for instance, transfers less quickly than spirits; oil than water; molasses or syrup than oil; and water thickened by starch dissolved in it, or which has its internal motion impeded by feathers or thread immersed in it, less quickly than where it is pure and at liberty. Cooling being merely a motion the reverse of heating, it is influenced by the same law. Hence the reason why thick soups, pies, puddings, and all semifluid masses, retain their heat so long—so much longer than equal bulks of mere fluid. The same law affords explanation of the facts, that very porus masses and powders, as charcoal, metal filings, saw dust, sand, &c., conduct heat more slowly than denser masses,—their interstices being filled with air, which scarcely *conducts* heat, and which, by the structure of the substance, has no freedom of motion or circulation by which it might *carry* the heat.

"*Heat spreads, also, partly by being radiated or shot light from one body to another, through transparent media or space with readiness affected by the material and the state of the giving and receiving surfaces.*" (Read the Analysis, page 256.)

If a heated ball of metal be suspended in the air, a hand brought in any direction near to it, will experience the sensation of heat; and beneath it the sensation will be as strong as on the sides, although the heat has to shoot down through an opposing current of air approaching the heated ball, to rise from it, as explained in a previous section. A delicate thermometer substituted for the hand, will equally detect the spreading heat, and if held at different distances, will prove it to diminish in the same ratio as light diminishes in spreading from any luminious centre, *viz.*, to be only a fourth part as intense at a double distance, in a corresponding proportion for other distances. If the heated body be enclosed in a vacuum, a thermometer placed near it will still be affected in the same manner. If a screen be interposed between the body and the thermometer, the latter will not be affected at all, proving the heat to spread in straight lines. Heat, when diffusing itself in this way, to distinguish it from heat passing by contact and communication, as described in the last section, is called *radiant* heat; that is to say, spreading in radii or rays all around its source as light spreads.

Radiant heat resembles light yet in other respects. It as rapidly permeates certain transparent substances, and its course suffers in them a degree of the bending, termed refraction by opticians. It is reflected from many kinds of polished surfaces, just as light is reflected from a common mirror; and many such surfaces directed to one point or centre (as when Archimedes made the sun his assistant to burn the Roman ships) or a single concave surface, having its own centre or focus, will concentrate heat just as light. Its motion in the sun-beam is so rapid, as for any distance at which men can try the experiment, to appear instantaneous; and the rays of heat from hot iron or burning charcoal concentrated at great distances by suitable mirrors, affect a thermometer as quickly as the heat of the sun similarly reflected. Although light and heat are united in the sun's ray, they are still separable by our glass prisims or lenses; and the focus of heat behind a burning glass is not precisely the focus of light. Heat, in radiating through air, does not warm the air, and its passage is not sensibly affected by winds or any other motion of the air.—These resemblances in the phenomena of light and heat have by some inquirers been held to prove that the two classes of appearances are only different modifications of action in the same subtile substance or either,

The diffusion of heat by radiation, as it takes place in an instant to any distance, and begins whenever there is any inequality of temperature between bodies exposed to each other, would produce instant balance of temperature throughout nature, but that heat leaves and enters bodies with readiness, depending on the condition of their surfaces, and on their internal conducting powers. A black stone-ware ter-pot, for instance, will radiate away 100 degrees of its heat in the same time that a pot of polished metal will radiate only 12 degrees.

Professor Leslie was the first to see the importance of investigating this subject, and he had the merit of contriving well-adapted means, and of detecting many of the important facts. As common thermometers are not sufficiently delicate to determine very sudden changes of temperature, where the influence is so light as in many cases of radiant heat, he used the beautiful *differential thermometer* contrived by himself, in conjunction with concave mirrors, (as represented on next page,) to concentrate the heat, and accumulate its energy. Then taking as his heated body a cubical tin vessel filled with boiling water, and covering it successively with plates or layers of different substances and with different colors, and exposing the thermometer to it for a given time under all the changes, he noted the number of degrees which the thermometer rose, (as seen in the table which here follows) and thus ascertained the radiating power of each sort of covering.

Lamp black	100°
Writing paper	98
Crown glass	90
Ice	87
Isinglass	75
Tarnished lead	45
Clean lead	19
Iron polished	15
Tin plate	12
Gold, silver and copper	12

He next reversed the experiments by using his hot-water vessel always in the same state, and covering the thermometer bulb with the different substances and colours, and thus he ascertained that the comparative *absorbing* powers of the substances and colours were very nearly proportioned to their *radiating* powers: lamp-black, for instance, absorbed or was heated 100°, while the polished metals absorbed or were heated only 12°, and so for the others. And, lastly, the absorbing powers being an indication of the opposite or *reflecting* powers (for a body absorbing only a given proportion of the heat which falls on it, must reflect the remainder,) he, by the same experiments, ascertained the radiating, absorbing, and reflective or mirror powers of the bodies, and therefore all the important points respecting radiant heat in its relation to them.

It seems paradoxical that the putting a clothing of a thin cotton or woolen fabric upon the polished tin vessel, should cause the heat to be received by it or dissipated from it much sooner than if the vessel were naked, but such is the fact. And metal with a scratched or roughened surface radiates or receives much more rapidly than polished metal.

The property of absorbing heat depends much upon the colour of the substance, and, as a general rule, the dark colours, *viz.*, those which absorb most light, absorb also most heat. Dr. Franklin proved this by laying pieces of cloth of different colours on snow, and exposing them during a given period to the sun's rays: while he noted the different depths to which by the melting of the snow under them, the pieces sank. Hence comes the importance of having a white dress in summer, that by it, with the sun's light, the heat also may be repelled. And a white dress in winter is good because it radiates little. Polar animals have generally white furs. White horses ars both less heated in the sun, and less chilled in winter, than those of darker hues.

The rate of cooling in heated bodies must be influenced by all the particulars noted above *viz.*, substance, surface, colour, and by the excess of heat in the cooling body as compared with those around it.

The concentrating apparatus used for experiments on the radiation of heat consists of two concave tin mirrors, here represented at *a* and *b*, so formed and placed in relation to each other that all the rays of light or heat issuing from the focus of one, as at *c*, shall, after a double reflection, be collected in the focus of the other, *d*. A stand under one focus *c* is intended

Fig. 124.

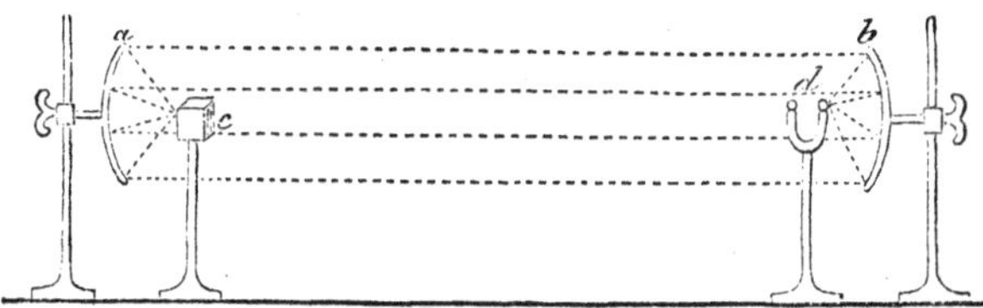

to support the body giving out or receiving heat, and a stand under the other *d* is meant to support the thermometer. For farther explanation of the action of such mirrors, we may refer to what was said of the concentration of sound in the section on *Acoustics*, or to what follows in the section on *Optics*, on the concentration of light. The general rationale of such facts is, that heat, light, sound, and elastic ball, &c., when reflected from any point of a surface, returns, if it fall perpendicularly to that point, in the same line by which it approached; but if it fall obliquely, or from one side of the perpenpicular, it returns in a line deviating as much on the other side. Now the surfaces of concave mirrors are so formed, that every ray issuing from the focus shall, when reflected, become parallel to every other ray—as represented by the dotted lines in the figure; and it is the property of a similar mirror receiving parrallel rays to make them all meet in its focus:—thus, any influence radiating from *c* towards the mirror *d*, will again, after two reflections, be collected at *d*. The purpose and effect of such mirrors in experiments on heat, are merely to concentrate feeble influences, so that they may be more accurately estimated. To show the effect and mode of action of such mirrors, they may be placed exactly facing each other at any convenient distance, and then a hot body of any kind, as a metallic ball or a canister of boiling water, being placed in one focus while a thermometer stands in the other, the thermometer will instantly rise; although if left in any intermediate situation nearer to the hot body, and therefore not in the focus, it will not be affected. If burning charcoal be placed in one focus, and a readily combustible substances in the other, the latter may be set fire to, at the distance of thirty feet or more.

If, in one focus of the mirror apparatus described above, there be placed, instead of the canister of hot water, a piece of ice, the thermometer in the other focus immediately falls. This has been called the radiation of cold, and persons were at one time disposed to think that it proved cold to have a positive existence distinct from heat. The case, however, is merely that the thermometer happens then to be the hotter body in one focus of the mirrors, placed in close relation with a colder body, the ice, in the other, and consequently by the law of equal diffusion, it must share its heat with the ice, and will fall. The mirrors in any case have merely the effect, by preventing the spreading and dissipation of the radiant heat from either focus except towards the other of making two distant bodies act upon each other as if they were very near. All the heat that seeks to radiate from the thermometer *d* in the direction of the surface of the mirror *b*, if not met by an equal tension or force of temperature in the other mirror or focus, to which they are directed at *a* and *c*, will radiate away to *c*, and become deficient at *d*. Some inquirers have believed that heat was constantly radiating in exchange from substance to substance (as light radiates between opposed bodies) only more copiously from the side where the temperature was highest: others have held that motion took place only where there was excess of heat; that is, when the balances of temperature was destroyed; and this is the simplest view.

There is a remarkable difference in one respect between the heat of the sun and that radiated from any other source, namely that the first passes through air, glass, water, and transparent bodies generally, very readily, while the latter, although not obstructed by air, is almost totally intercepted or absorbed, in passing through any of the other substances named. In our drawing-rooms it is common to have plate.glass fire-screens, which, while they allow the light to pass, defend the face from the heat; but all persons know that the heat of the sunbeams, as well as their light, enters our green-houses through the glass which covers them. A glass screen interposed between the concave mirrors in the apparatus above described, destroys almost entirely the effect of the heated body placed in one focus, on the thermometer in the other, and the trifling effect really produced has appeared to some to be owing to the heat that is absorved by the screen on one of its sides, and then after passng through it by conduction, is radiated from the other. This conclusion seemed to be supported by the fact that screens of metal or of glass, covered with lamp black, paper, &c., allow transmission nearly in proportion to their several absorbent and radiant powers. More careful experiments, however, have seemed to prove that, even at a low temperature, a certain portion of the heat is suddenly radiated through the glass, and at a high temperature, a much larger portion. A glass mirror reflects the light of a fire, but at first retains nearly all the heat, and only radiates it afterwards as a hot body.

The doctrines of radiant heat make us aware of the importance of having vessels of polished metal for containing liquids or other things which we desire to keep warm; hence, tea and coffee-pots, dishes for soup, &c., should be polished. As a black earthen tea-pot loses heat by radiation nearly in proportion to the number 100, while one of silver or other polished metal loses only as 12, there will be a corresponding difference in their aptitude for extracting the virtues of any substance infused in them. Pipes for the conveyance of steam or hot air, if left naked, should be of polished metal; but after arriving at a place where they have to give out their heat, their surface should be blackened and rough. . A coat of polished mail is not a cold covering. A mirror intended to reflect heat should be of highly polished metal, and such, for an obvious reason, the interior of a screen behind roasting meat is attempted to be made. A fireman's mark is usually covered externally with smooth tin foil. It is of advantage that the bottom of a tea-kettle or other cooking vessel be externally black, because the bottom has to absorb heat, but the top should be polished because it has to confine.

The interesting phenomenon of dew was not at all understood until lately, since the laws of radiant heat have been investigated. At sun-rise, in particular states of the sky, every blade of grass and leaflet is found not wetted, as if by a shower, but studded with a row of distinct globules most transparent and beautiful, bending it down by their weight, and falling like pearls when the blade is shaken. These are formed in the course of the night by a gradual deposition on bodies rendered by radiation colder than the air around them, of part of the moisture which rises invisibly from water surfaces into the air during the heat of the day. In a clear night the objects on the surface of the earth radiate heat to the sky through the air which impedes not, while there is nothing nearer than the stars to return the radiation; they consequently soon become colder, and if the air around has its usual moisture, part of this will be deposited on them, in the form of dew, exactly as the invisible moisture in the air of a room is deposited on a cold bottle of wine when first brought from the cellar. Clouds, by obstructing the radiation spoken of, obstruct the formation of dew. Air itself seems not to lose heat by radiation. A thermometer placed upon the earth any time between sunset and sunrise, generally stands considerably lower than another suspended in the air a few feet above it; owing to the radiation of heat upwards from it and from the earth, while the surrounding air remains nearly in the same state. During the day, while the sun shines, the earth is warmer than the air. The reason why the dew falls, or is formed so much more copiously upon the soft spongy surface of leaves and flowers, where it is wanted, than on the hard surface of stones and sand where it would be of no use, is the difference of their radiating powers, There is no state of the atmosphere in which artificial dew may not be made to form on a body, by sufficiently cooling it and the degree of heat at which the dew begins to appear is called the *dew-point* being an important particular in the meteorological report of the day. In cloudy nights heat is radiated back from the clouds, and the earth below not being so much cooled, the dew is scanty or deficient. And it is, when uninformed persons would least expect the dew, *viz.:* in very warm clear nights, and perhaps when the beautiful moon invites to walking, as in some of the evenings of autumn with the harvest moon and harvest occupations—that the dew is more abundant, and the danger greater to delicate persons of taking harm by walking among the grass.

"*Heat by entering bodies expands them, and through a range which includes as three successive stages the forms of solid, liquid, and air or gas; becoming thus, in nature, the grand antagonist and modifier of the effects of that attraction which holds corporeal atoms together, and which, if acting alone, would reduce the whole material universe to one solid, lifeless mass.* (Read Analysis, page 256.)

If an experimenter take a body which is as free from heat as human art can obtain it—a bar of solid mercury, for instance, as it exists in a polar winter—and if he then gradually heat such body, it will acquire an increase of bulk with every increase of temperature; first, for a time, there will be simple enlargement or expansion in every direction; then the mass will in addition be softened; then it will be melted or fused; that is to say, in the case supposed, the solid bar will be reduced to the state of liquid mercury, with the cohesive attraction of the atoms nearly overcome: if the mass be still farther heated, it will gain bulk until at a certain point, the atoms will be repelled from one another to much greater distances, constituting then a very elastic fluid called an air or gas, many hundred times more bulky than the same matter in the solid or liquid state, and forcibly distending an appropriate vessel as common air distends a bladder; susceptible, moreover, of dilating indefinitely farther, by farther additions of heat, or by diminution of the atmospheric, or other pressure, against which it had to rise during its formation. A subsequent removal of the heat from the gaseous mercury, will cause a progress of contraction corresponding to the previous progress of expansion, and the various conditions or forms of the substance above enumerated, will be reproduced in a reversed order, until the solid mass re-appear, as at first.

What is thus true of mercury is proved, by modern chemical art, to be true also of all the ponderable elements of our globe, and of many of the combinations of these elements,—as water, for instance, familiarly known in its three forms of *ice, water* and *steam;* although compound substances generally, by great changes of temperature, are decomposed into their elements.

A student might at first have difficulty in believing that the beautiful variety of soil, liquid, and air, found among natural bodies could depend upon the quantities of heat in them, because these forms are all seen existing at the same common temperature; but he afterwards learns that each substance has its peculiar relation or affinity to heat, and that hence, while at the medium temperature of the earth, some bodies contain so little as to be solids—like the metals, stones, earths, &c.; others have enough to be liquids—as mercury, water, oils, &c., and others have enough to be airs—as oxygen, nitrogen, hydrogen, &c. Men, until better informed, are prone to deem the tastes in which bodies are most frequently observed by them, the natural or essential states of such bodies; and the Indian king reasoned but in the usual way when he held the Dutch navigators, newly arrived on his shores, to be gross imposters, because they said that in their country, at one time of the year, water became so hard that they could walk upon it, and drive their carriages upon it, and shape it into solid blocks. All persons err like this king, who in thinking of the different substances familiarly known to them, regard their accidental state of solid, liquid or gas, which state is really dependent on the temperature of the bodies, and therefore on the particular climate or situation on earth where they are found, to be in them an essential natural character. As well might a person who had never seen silk, but as a delicate gauze or satin enveloping some lovely human form, refuse to recognize it in the unsightly coil of the worm which produces it.

The degree in a general state of temperature at which the substances most important to man change their states from solid to liquid, or from liquid to air, will be noticed in a future page. Here we have only to remark, that the differences are very great. Mercury melts at about 80 degrees below the melting point of ice, and porcelain at about 30,000 degrees above. There are some substances which require so high a temperature for their fusion or for their conversion into gas, that human art has difficulty, or even finds it impossible, to produce the changes by simple concentration of heat; but all such substances are quickly reducible to the liquid form when placed in contact with others for which they have a chemical affinity, and which possess already the form of liquid or air; as when gold or platinum are dissloved in nitro-muriatic acid—flint in the fluoric acid—carbon in hydrogen gas. Now many persons may not have reflected that the dissolving a solid in any fluid menstruum is merely another mode of melting it by heat; yet this is the truth, for the menstruum is itself fluid, only because of the much heat which it contains, and in dissolving the more obdurate substances, it does so merely because its atraction for the substance brings the particles into union with the heat which already exists in itself. Heat, then, is the one universal solvent, or cause of fluidity. Its influence in this view is interestingly seen in the fact, that a fluid when heated can dissolve much more of solid than when cold. Water, while hot, keeps dissolved twice as much of many salts as it can when its temperature has fallen, as is proved by the crystals of salt formed in any saturated solution as it cools.—There are, again, in nature many substances having such an affinity for heat, that, until lately, they have been known only as airs; and even in the present advanced state of art, they cannot, by any degree of mere cooling, be reduced to the liquid or solid form; yet all such, when pressure is added to the cooling, or when the chemical attraction for them of some other substances which already exist in the liquid or solid state is made to co-operate, may be reduced. An instance is afforded when oxygen is made part of a liquid acid, or of a solid ore.

Of solids, some on receiving heat become very soft before they are liquefied, as pitch, glue, iron, &c.; others change completely at once, as ice in becoming water: and some pass at once to the state of air, without, therefore, having assumed at all the intermediate state of liquid—they are *sublimed*, as it is called, and on cooling again may be caught in a powdery state, as seen in that form of sulphur or of benzoin, termed the *flower* of the substance. Of the latter class, also, are camphor, arsenic, corrosive sublimate, and the substance called iodine, which last, from the state of rich ruby crystals, becomes at once, on being heated, a dense transparent gas of the same hue, and in cooling resumes its crystalline form.

The reader having arrived at this place, may peruse again with advantage the five pages near the beginning of this work, which treat of the influence of heat on the *constitution of masses*.

"*Each particular substance, according to the nature, proximity, &c., of its ultimate particles, takes a certain quantity of heat, (said to mark its* CAPACITY) *to produce in it a given change of temperature or caloric tension.*" (Read the Analysis, page 256.)

A pound of water, for instance, that its temperature may be raised one degree, takes thirty times as much heat as a pound of mercury. This may be proved in various ways. First, if the heat be derived from any uniform source, the water must remain exposed to it thirty times as long as the mercury. Second, if both substances, after being equally heated, be placed in ice until cooled to the freezing point, the heat which escapes from the water will melt thirty times as much ice as that which escapes from the mercury, instead of the two becoming of a middle temperature, as in the case when equal quantities of hot and cold water are mixed, and every degree of heat lost by the one quantity becomes just a degree gained by the other—the pound of hot water, by giving up one degree to the pound of cold mercury, raises the temperature of the latter thirty degrees; and in the same proportion for other differences:—or on reversing the experiment, a pound of hot mercury will be cooled thirty degress by warming a pound of water one degree.

Now each particular substance in nature, just as water or mercury, has its peculiar capacity for heat; and experiments made by the modes of mixture and of melting ice above described have led to the construction of tables which exhibit the relations. The following short table is an abstract, showing the comparative capacities of equal weights of some common substances. Water, for reasons of convenience, has been chosen as the standard of comparison. It appears that a pound of hydrogen gas takes about twenty times more heat to produce it in a given change of temperature, than a pound of water, while a pound of gold takes about twenty times less, and, therefore, four hundred times less than the hydrogen. The figures in the table, by marking the comparative capacities for heat of various substances necessarily indicate, also, the comparative quantities of ice melted by equal weights of the substances in cooling through an equal number of degrees.—A pound of water, the standard, must cool 140 degrees, that is, must give up 140 degrees of its heat, to melt one pound of ice.

Gases.	
Hydrogen	$21\frac{1}{2}$
Atmospheric air	$1\frac{3}{4}$
Carbonic acid gas	$1\frac{3}{5}$
Common steam	$1\frac{1}{2}$
Liquids.	
Solution of carbonate of ammonia	2
Alcohol	$1\frac{1}{10}$
Water	1
Milk	1
Olive oil	$\frac{3}{4}$
Linseed oil	$\frac{1}{2}$
Sulphuric acid	$\frac{1}{3}$
Quicksilver	$\frac{1}{30}$
Solids.	
Ice	$\frac{9}{10}$
Wheat	$\frac{1}{2}$
Charcoal	$\frac{1}{3}$
Chalk	$\frac{1}{4}$
Glass	$\frac{1}{5}$
Iron	$\frac{1}{8}$
Zinc	$\frac{1}{10}$
Gold	$\frac{1}{20}$

We may remark here that some late researches, by another mode of trial, make the capacity of air to be only a quarter that of water, although in the preceding table it appears to be one and three-quarters. Now as the other aeriform fluids have been compared with water through the medium of atmospheric air, if there be an error with respect to this, it must run through all the figures noting the capacity of other aeriform substances.

If we seek a reason or reasons why there should be among bodies the differences of capacity here stated, the circumstances chiefly attracting attention are the following. 1st. Equal weights of the various substances have very different bulks or volumes, and therefore have different room in which the heat may lodge;—as a pound of mercury, for instance, is only one-fourteenth part as bulky as a pound of water. That the bulk, however, is not the only influencing circumstance appears in the fact, that mercury only has one thirtieth of the capacity of water. 2d. In equal bulks of different substances, the space may be more completely occupied by the particles of one than of another—as is probably true of the particles of mercury compared with those of water. 3d. But as the facts are not fully accounted for by these two circumstances, we must infer that there is some difference in the ultimate particles of bodies affecting their relations to heat. We shall now review more particularly the various circumstances mentioned.

First. The influence of bulk or volume, in determining the capacity for heat, is proved by the facts stated in the preceding table, and by many others. In the table, for instance, it is seen that hydrogen and the gases generally, with their great comparative bulk, have also great capacity; that liquids have less capacity than gases; that solids have less than liquids—but the capacity, as already stated, is not in strict proportion to bulk; for hydrogen which is many thousand times more bulky than an equal weight of water, has only twenty-one times the capacity. Again, if any body whatever be suddenly compressed into less bulk, heat will issue from it as if squeezed out. Thus iron or other metal, suddenly condensed by the heavy blow of a hammer, is thereby rendered hotter, that is expelled heat gradually spreads from it. Because water and spirit, on being mixed, occupy less space than when separate, there is from the mixture a corresponding discharge of heat. But the truth is most remarkably exemplified in airs or gases, owing to their great range of elasticity. They may be condensed or dilated a hundred-fold or more, and there will be a simultaneous concentration or diffusion of their heat, that is to say, the production, in the space occupied by them of intense heat or cold. The heat of air just condensed, or the cold of that which has just expanded, is much greater than even the most delicate thermometer can indicate, for there is so little heat altogether even in a considerable volume of air, that the mass of a mercurial thermometer, although absorbing a great part of it, would be little affected. The extent, however, of the change of temperature is seen in the facts, that, by the sudden condensation of air we may inflame tinder immersed in it, and by allowing air suddenly to expand, we may convert any watery vapour diffused through it into ice or snow. Nay, air, containing carbon in perfect solution, as is true of the common coal gas, if first condensed to expel heat, and then allowed suddenly to expand, will be so cooled that the carbon will be separated like a black cloud, as snow is separated in the case before described. The cold which separates or freezes carbon from a gas holding it in solution, must be very intense. It might be expected that air suddenly compressed into half its previous volume, should become just twice as hot as before, or if suddenly dilated to double volume, should be only half as hot, thus enabling us to ascertain the whole quantity of heat contained in it; but the facts are not so; the temperature changes, near the middle degrees of the scale at least, much less than the density. Air in doubling its volume from a common density, becomes colder only by about 50° of Farenheit's thermometer.

The different capacity for heat of air in different states of dilation, produces effects of great importance in nature as well as in the arts. Thus,

On the surface of the earth near the sea-shore, the air of the atmosphere has a certain density (a cubic foot weighs about one ounce and a quarter) dependent on the weight and pressure of the superincumbent mass; but on a mountain top 15,000 feet high, as half the mass of the atmosphere is below that level (see "Pneumatics,") the air is bearing but half the pressure, and consequently any quantity of it has twice the volume of an equal quantity at the sea-side, with a temperature consequently many degrees inferior. The air which is at any time on a mountain-top, may, a little while before, have been on an adjoining plane or shore, and in gradually climbing the mountain side as a wind, it must have been gradually expanding and becoming cooler in proportion to the diminishing pressure. It is found that air, on rising from the sea-shore, becomes one degree colder nearly, for the first 200 feet of perpendicular ascent, and that air becomes altogether about 50° colder in rising 15,000 feet; so that at this latter elevation, water exposed to the air is frozen even near the equator, where the temperature of low plains is at least 80°. It thus appears that if a man could travel with the wind so as to remain always surrounded by the same air, he might begin his journey with it from the summer vineyard of the Rhine, might soon after find it the piercing blast of the Alpine summits; and again, a little after, without any change having occurred in the absolute quantity of its heat, might feel it as the warm breath of the flowers on the plains of Italy.

The explanation is thus given of why very elevated mountains in all parts of the earth are hooded in perpetual snows. We have just said that even at the equator, where the average temperature near the sea is 84°, water will be frozen when carried to an elevation of 15,000 feet. A line, therefore, traced round a mountain at this level would divide the portion of it destined to sleep under lasting ice and snow from the portion below covered with green herbage. This line, wherever found, is called the *snow line, or line of perpetual congelation.* At the equator it is high in the atmosphere, because there is a difference of about 40° between the average temperature of the country and the freezing point of water, *viz.*, the difference between 84° and 32°, and an elevation of 15,000 feet corresponds to this difference; but in a progress towards the poles, the line is met with gradually nearer to the earth, as the difference between the average temperature and the freezing point is less. In Switzerland, the snow line is at 6,500 feet above the sea; in Norway, it is below 5,000. With respect to the line of congelation, it is farther to be remarked, that in tropical countries, because the temperature of the air is nearly uniform during the whole year, the line or limit of frost and snow is distinct and unvarying, that is to say, is narrow, particularly where the acclivity is considerable; but in countries to the north and south, which have strong contrast of summer and winter, the line rises in summer and falls in winter, and thus becomes broad and less evident; in the hot season much snow is melted or half melted above the middle of the line or belt, while in winter much snow and ice is accumulated below that, to be melted again when summer returns.

In the breadth of the line of congelation for changeable climates, we have the reason of the formation of what are called *glaciers* around snow-capped mountains situated in such climates, and around such only. The snow near the upper part of ths broad line having been only softened or half thawed in the preceding summer, becomes in winter almost as solid as ice, and in the succeeding summer vast masses of it, detached by the action of the sun and of the internal heat of the earth, and loaded with more recently deposited snow, are constantly falling down into the neighbouring valleys within the broad line of congelation; where, being accumulated, and the crevices filled up with snow or with water which hardens to ice, they form at last the huge glaciers or seas of ice, *mers de glace,* which renders certain regions so remarkable. The falling of the masses above described (called in Switzerland *avalanches,*) is what renders the ascent to snow-clad mountains so terrific and dangerous. Around Mount Blanc, in the awful solitudes of the elevated valleys, the avalanches are thundering down almost without interruption during the whole summer,—in which season only the attempt to ascend the mountain can be made: and a pistol shot, or any considerable agitation of the air, may suffice to set loose masses that would sweep away a whole convoy. Beneath glaciers there is always going on a melting of that part of the ice which is in contact with the earth, and hence a stream of water constantly issues from the bed of every glacier. These streams in Switzerland are the beginnings of the magnificent rivers the Rhine and Rhone —Like the avalanches breaking loose in summer among the mountains, there are in polar seas vast masses of ice detached from the shores, and which afterwards drift into warmer seas to be melted. These often become as rafts to the arctic bear, and to his surprise, carry him to new latitudes, and leave him at last to perish in the midst of the wide ocean, when his support has vanished from beneath him.

Although the proofs are not so immediately apparent, the line of congelation exsits as truly every where in the open sky, over sea and plane, as where there are mountain heights to wear its livery; and considerably below the line, the cold, aided by electrical agency, is sufficient to produce in the form of mist or clouds, a deposition from the air of the watery vapour contained in it. There is thus in nature an admirable provision to shade the earth at proper times from the too powerful rays of the sun, or to supply rain as wanted, without the transparency of the inferior regions of the atmosphere being much affected. As the watery vapour rising from sea or lake, and invisibly diffused in the atmosphere, can only reach to the height where the cold is great enough to condense it, the clouds may in general be regarded as the top of that atmosphere of watery vapour or aeriform water, which is always mixed more or less with the atmosphere of mere air; and as the quantity of watery vapour which can exist invisibly in a given space, depends altogether on the intensity of heat present, the clouds in a cold or humid atmosphere will be low, and in a warm or a dry atmosphere will be high, or there may be none. An aëronaut mounting in his balloon through a clear sky, often enters a dense cloudy stratum, and for a time is surrounded by the gloom almost of night, the face of earth being hidden from him below, while the heavenly bodies are equally veiled from him above; but rising still higher, he again emerges to brightness, and looks down upon the fleecy ocean rolling beneath him, as the climber to a lofty peak looks down from the ever pure atmosphere around on the inferior region of clouds and storms.

The diminished temperature of air in the higher regions of the atmosphere, often enables the natives of temperate climates, when forced to reside in hot tropical countries, inimical to their health, to find near at hand, on some mountain height, the congenial temperature of their early homes. The author of this work, during a visit to the then not long inhabited Island of Penang in the strait of Malacca, examined this fact with pleasure not readily forgotten. The centre of the island is occupied by a lofty mountain ridge thickly wooded, on the northern summit of which a few residences visible from the sea-shore like eagles' nests on a cliff, had just been constructed. Towards these one morning at sunrise, on an active little horse of the country, and along a tolerable road, he began to climb from the hot plain below. At first there were around him purely tropical objects, inspiring tropical feelings,—the latter modified indeed by the reflection that his track lay through a forest, in which until lately the foot of man never penetrated, and where the trees nursed through ages to their greatest growth, and the stupendous precipices and the sublime waterfall had so recently been exposed to human observation; but as he gradually ascended, he perceived the character of the vegetation to be changing and the air to be becoming so light and cool as strongly to awaken in him thoughts of distant England—nay, almost the illusion that he was there. When he had reached the summit, however, and a clear space opened to view the whole country around, his attention was soon recalled to the fervid land of the sun. At first, from the elevation being so great, the eye took account only of the grander features of the scene, and which were such nearly as might be met with on a Grecian or Italian shore: the expanse of sunny water in that beautiful strait, the opposite continent with its river winding seaward across the plain, the town and the roadstead near it crowded with ships, which appeared only as specks on a wide-spread map; but, on closer inspection, and particularly with the aid of the telescope, were described the rich groves or cocoa-nut and banana, the plantations of spice, and cotton, and sugar-cane, the tawny labourers, the bamboo dwellings, the fanciful canoes, or prows, and other objects of the like character. And such was the scene which even under the equator, a person could place under his eye, while the thermometer near him stood as in an English month of May.

The interiors of the islands of Jamacia and Hayti have many situations of great extent, which combine, as above described, the advantages of tropical situation and temperate climate, and which might well be inhabited by English labouring colonists. The vast plain of Mexico, and much of the central land of South America, are similarly circumstanced: and it is not uncommon, where the ascent to the gigantic Andes is gradual, to find at the bottom of the ridge a town, whose markets are stored only with the productions of the equator, while in a town higher up will be seen only what belongs to the temperate skies of Europe:—climates of the earth naturally distant, thus meeting, as it were, in amicable vicinity, on the same rising plain.

The facts detailed in the preceding paragraphs are intended to illustrate the subject, of the relation of *volume* in a body to the capacity for heat. We now proceed to speak of *density* in the same respect.

Second. It might be anticipated that a dense body, or one in which the constituent particles may be supposed to fill more completely the space occupied by it than the particles do in a rarer body, would have smaller capacity for heat, in proportion to the smaller space left vacant in its mass: and in a general comparison of the capacities of *equal bulks* of different substances, such anticipation is partly verified,—as when a pint of dense mercury is found to have only about half the capacity which a pint of lighter water has. The relation, however, is by no means universal, nor at all in proportion to the differences of density. Water, which is denser than oil, and according to the hypothesis should have less capacity, yet has nearly double the capacity; and mercury, which being nearly fourteen times denser than water, might be expected to have only a fourteenth of the capacity, has really for equal volumes a half, or, as formerly stated, for equal weights, a thirtieth.

Third. We are at last, therefore compelled to admit that the relation between various substances and heat, which we call capacity for heat, depends much more on the nature of the ultimate particles of the substances than either on the absolute bulk or comparative density of the masses. Throwing much light on this subject, it has been ascertained in late times, that all material substances are composed of extremely minute unchangeable atoms, of which, in different substances, the comparative weights have been determined, although not the absolute weights; that is to say, for example, the atom of gold is known to weigh four times as much as the atom of iron, although we do not know how many thousands or millions of atoms are required to form a grain of either. Now very recent researches seem to prove that for each ultimate atom, no matter of what substance, nearly the same quantity of heat is required to produce in a mass of the atoms a given change of temperature. Thus an ounce of iron, which has four times as many atoms as an ounce of gold, has four times the capacity for heat. This law seems to hold for all simple substances; but for compounds there seems to be another law not yet ascertained.

Instead of the term *capacity for heat* used in the preceding pages, with respect to particular substances, that of *specific* heat has by some authors been preferred; but as the latter gives to a commencing student the idea rather of *kinds* of heat than of *quantities*, the term capacity has been retained.

"*Each substance in nature, for a given change of temperature, undergoes expansion in a degree proper to itself, the expansion generally increasing more rapidly than the temperature, as the cohesion of the particles becomes weaker from increased distance; being remarkably greater, therefore, in liquids than in solids, and in airs than in liquids; the rate being quickened, moreover, near the points of change.*" (See the Analysis, page 256.)

The following table, containing the names of some common substances, solid, liquid, and aeriform, shows, by the figures following each name, how much the substance increases in bulk, by having its temperature raised from that of freezing to that of boiling water. A lump of glass, for instance, would gain one cubic inch for every 416 cubic inches contained in it; while a mass of water would gain one inch for twenty-three, dilating thus for the same range of temperature eighteen times more than glass.

Solids.	
Glass gains one part in	416
Deal	416
Steel	283
Iron	271
Brass	177
Silver	175
Lead	117
Liquids.	
Mercury gains one part in	55
Water	23
Fixed oils	12
Alcohol	9

Airs.

Common air, all gases and vapours	gain one part in - -	3

We have to warn readers here not to confound the increase by heat of the general *bulk* of a solid body with the increase of its *length*. The latter is only one-third as great as the former. This will be understood by considering that the increase of bulk is divided between the *length*, *breadth*, and *depth* (or *thickness*. If the substance of a metalic square rod or wire, be dilated by heat a one-hundreth part of its bulk, it does not gain all that hundreth at its end, becoming 101 inches (or other measure) long, instead of 100; but every part becomes deeper and broader in the same proportion as it becomes longer (we may suppose it divided into a row of equal little cubes,) and the rod gains in length only the third part of an inch. A fluid enclosed in a tube unchangeable by heat (if such tube there were) would show its whole dilatation in an increase of length, because there could be no swelling laterally, and its extremity, therefore, from any variation of temperature, would have a triple extent of motion. A degree of this consequence is obtained in our common thermometers, because the containing glass, although dilatable by heat, is so much less dilatable than the fluid within. As regards solids, we have to inspire so much more frequently respecting the dilatation in length, breadth, &c., that is, to say, the *linear dilatation*, in one direction, than the increase of general bulk, that tables are frequently made stating only the linear dilatation. It may be found at once from the above table, by recollecting that it is one-third of the increase of bulk:—thus, as glass, in passing from the freezing to boiling heat of water, dilates one part in 416 of its bulk, it will dilate only one-third of a part in length, or a whole part in an extent of three times 416 or 1,248.

The expansion of solids by heat has been ascertained by bringing microscope instruments to bear on rods of the different substances heated to various degrees, in troughs of oil or water. The expansion of fluids, again, is found by filling a glass vessel with a known weight of any fluid, and then ascertaining how much is made to run over or escape by a given increase of heat; or how much the fluid rises into a long tubular neck like the stock of a thermometer. This quantity, added to what is required to fill the increased dimensions of the heated glass vessel, (which from the ascertained expansion of glass is known,) forms the whole of the increase

The general and comparative expansion of solids by heat is exemplified in the following cases:

An iron bullet, when heated, cannot be made to enter an opening, through which when cold it passes readily.

A glass stopper sticking in the neck of the bottle, often may be released by surrounding the neck with a cloth taken out of warm water; or by immersing the bottle in the water up to the neck; or by making strong friction on the neck by a tape or any soft rope put around it, and then pulled backwards and forwards. By any one of these means the binding ring of the neck is heated and expanded sooner than the stopper, and so becomes for a short time slack or loose upon it.

Pipes of cast-iron for conveying hot water, steam, &c., if of considerable length, must have joinings which allow a degree of shortening and lengthening, otherwise a change of temperature may destroy them. An incompetent person undertook to warm a large manufactory by steam from one boiler. He laid a rigid main pipe along a passage, with lateral branches passing through holes into the several apartments; but on his first admitting steam, the expansion of the main pipe tore it away from all its branches.

In an iron railing, a gate which during a cold day may be loose and easily shut or opened, in a warm day may stick, owing to their being greater expansion of it and of the neighbouring railing than of the earth on which they are placed. Thus also the centre of the arch of an iron bridge is higher in warm than in cold weather; while on the contrary, in a suspension or chain bridge, the centre is lowered.

The iron pillars now so commonly used to support the front walls of those houses of which the ground stories meant to serve as shops have spacious windows, in warm weather really lift up the wall which rests upon them, and in cold weather allow it again to sink or subside considerably more than if the wall were brick from top to bottom.

In some situations (as lately was seen in the beautiful steeple of Bow-church, in London) where the stones of a building are held together by clamps or bars of iron driven into the stones, the expansion in summer of these clamps will force the stones apart sufficiently for dust or sandy particles to lodge between them: and then, on the return of winter, the stones not being at liberty to close as before, will cause the ends of the shortened clamps to be drawn out, and the effect increasing with each revolving year, the structure will at last be loosened and may fall.

The pitch of a piano-forte or harp is lowered in a warm day or in a warm room, owing to the expansion of the strings being greater than of the wooden frame-work; and in cold the reverse will happen. A harp or piano which is well tuned in a morning drawing-room cannot be perfectly in tune when the crowded evening party has heated the room.

Bell-wires too slack in summer, may be of the proper length in winter.

One admirable contrivance for keeping the pendulum of a clock always of the same length, by making the greater expansion by heat of a middle bar of brass counteract the smaller expansion of two side-rods of steel, was explained under the head of "*Pendulum*," as was also the construction of a balance-wheel having a corresponding property. A difference of a 100th of an inch in the length of a common pendulum, causes a clock to err ten seconds in twenty-four hours, and a rise or fall of 25° of Farenheit's thermometer produces this difference. Another kind of compensation pendulum, not less admirable, distinguished by the name of its inventor *Graham*, is obtained by substituting for the solid bob or ball at the bottom, a glass vessel containing mercury. The mercury on expanding by heat, has its centre of gravity raised just enough to compensate for the lengthening of the rod of the pendulum.

Crystals when heated, do not expand quite equally in breadth and in length. The same is true of fibrous substances, as wood which expands and contracts more in breadth than in length. This is instanced in the leaking, during cold weather, of a ship's deck which, in warm weather, is tight;—an occurrence which the author, in rounding the Cape of Good Hope, had to regret as the cause of destruction to some valuable specimens of natural history which he had collected among the Eastern Islands.

Bodies expanded by heat, unless when their intimate composition is changed by it, regain exactly their former dimensions on being cooled.

As is seen in the preceding table, the expansion of liquids by heat is much greater than of solids.

A cask quite filled with liquid in the winter, must, in summer, force its plug or burst; and a vessel which has been filled to the lip with warm liquid, will not be full when the liquid has cooled. Hence a cunning dealer in liquids has tried to make his chief purchases in very cold weather, and his chief sales in warm weather.

There exists in the case of water, an extraordinary exception, already mentioned to the law of expansion by heat and contraction by cold, producing unspeakable benefits in nature. Water contracts only down to the temperature of 40°, while, from that to 32°, which is its freezing point, it again dilates. One curious consequence of this peculiarity is exhibited when a pool or well happens to be formed on the upper surface of a mass of ice, as on one of the glaciers of Switzerland and elsewhere, namely, that the well goes on quickly deepening itself, until it penetrates to the earth beneath. Supposing the surface of the water originally to have nearly the temperature of the melting ice, or 32°, but to be afterwards heated by the air and sun, instead of the water being thereby dilated or rendered specifically lighter, and detained at the surface, it becomes heavier the more nearly it is heated to 40°, and therefore sinks down to the bottom of the pit or well; but there, by dissolving some of the ice, and being consequently cooled, it is again rendered lighter, and rises to be heated as before, again to descend; and this circulation and digging cease only when the water has bored its way quite through.

Airs are expanded by heat still more than liquids.

The expansion of aeriform bodies by heat produces many important effects in nature. Some of these have already been considered in the preceding parts of this work, as the rising of heated air in the atmosphere, causing the winds all over the earth; the same in our fires and chimneys supporting combustion and ventilating and purifying our houses; the same again from around animal bodies, removing the poisonous or contaminated air which issues from the lungs, and insuring a constant supply of fresh air for the support of life, &c.

It is remarkable with respect to aeriform bodies, that, unlike solids and liquids, *they* are all *equally* dilated by the same change of temperature, receiving an increase of about a third part of their bulk (37½ parts in 100) on being heated from the freezing to the boiling point of water, *viz.*, 180°,—their bulk being therefore doubled from the same standard point by about 500°. This general truth holds, not only with respect to the more permanent airs or gases, but also with respect to all steams or vapours in the dry state, that is, when not in contact with the liquid producing them. The probable reason of this uniformity is, that cohesive attraction which varies so much in different solids and liquids, modifying the effects of heat upon them in aeriform fluids does not exist at all.

The extent of this dilatation for airs is so much greater than for liquids or solids, that it forces itself much more strikingly upon the common attention. Thus, a bladder containing considerably less than its fill of air, becomes tense immediately on being held to the fire. The air in a balloon just escaping from a cloud, has been so suddenly expanded by the direct rays of the sun, as to injure the texture of the balloon; and probably some of the fatal accidents among aeronauts have been owing to this occurrence. Burning fuel conveyed into a vessel or case which can be suddenly and strongly closed, will produce an expansion of the air confined with it, capable of bursting any vessel of ordinary strength—in short, will produce an explosion.

Now, if not before, at any rate soon after steam-engines began to be used, and had so strikingly shown to what important purposes the force of an expanding aeriform fluid might be applied the thought would naturally occurr that the force of common air dilating by heat might also be rendered useful. Accordingly a variety of air-expansion engines have been proposed, but as yet no one has been reduced to profitable practice. Had the truth been generally known, which very recent investigations have proved, that a given quantity of heat, when used to dilate air, produces several times as much expansive power as when used to form steam, the attempts to bring such an application of heat under control would probably have been more numerous, and possibly, by this time, more successful. The subject is so interesting that we shall subjoin a few remarks upon it.

To produce a cubic foot of common steam from water originally cold, about 1,160 degrees of heat are required, as will be explained a few pages hence. The same quantity of heat would double the volume of about five cubic feet of atmospheric air,—as is known from the comparative capacities for heat of the two substances, and the rate of dilatation of air when heated. Now the value for work of the foot of steam passing from the boiler into a working cylinder would be, to press up the piston of the steam-engine through a foot, as from *c d* to *a b*, with a force all the way of 15 lbs. per inch of the piston surface; while the working valve of the five feet of air in dilating to double bulk, would be to lift up the piston five times as far as the steam, *viz.*, from *g h* to *e f*, but with a force gradually diminishing (represented here by the shaded part of the figure) as the expansion went on, from 15 lbs per inch at the begining until the air had dilated to its destined volume, when the force would altogether cease; its whole effect, therefore, would be five feet impulsion of the piston with a pressure averaging between 15 lbs. and nothing, *viz.*, 7½ lbs. per inch; and the friction in the two cases and the varying intensity of the latter pressure being neglected, the force of the air would be 2½ times as great as that of the steam. But it is farther to be considered, that only about half the heat of a fire is applied to use in a steam-engine, *viz.*, that part which enters the boiler, while the remainder passed up the chimney; and in an air-engine probably the whole might be applied. In an air-engine, moreover, there might be a great increase of power from the combustion, or semi-explosion of the inflammable gas evolved from the fuel. We see from this of what importance the discovery would be of a means enabling us effectually to apply the force of expanding air.

Fig. 125.

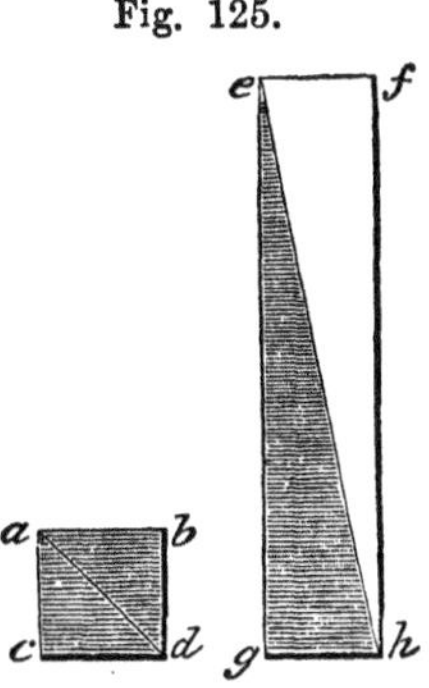

If we suppose a fire *a* to be placed on a grate near the bottom of a close cylinder, *d a*, and the cylinder to be full of fresh air recently admitted, and if we farther suppose the loose piston *g d* to be pulled upwards, it is evident that all the air in the cylinder above *d* will be made to pass by the tube *e* through the fire, and will receive an increased elasticity tending to the expansion or increase of volume, which the fire is capable of giving it. If there were only the single close vessel *d a*, the expansion might be so strong as to burst it; but if another vessel *b c* of equal size were provided, com-

Fig. 126.

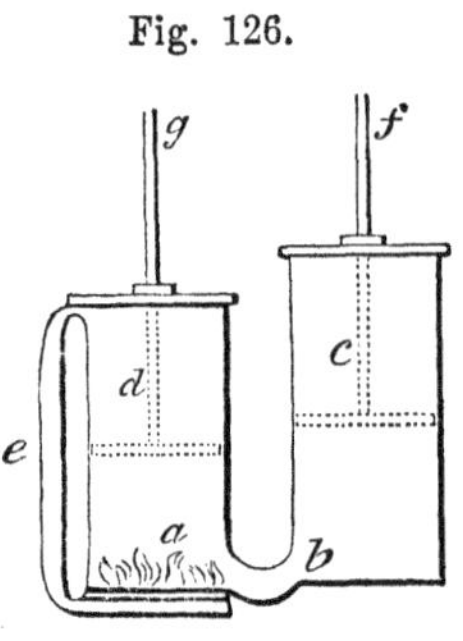

municating with the first through the passage *b*, and containing a *close*-fitting piston *e f*, like that of a steam-engine, the expansion of the air in the first cylinder would act to lift the said piston, and so might work water-pumps, or do any other service which a steam-engine can perform, At the end of the lifting stroke of the piston *f c*, it might be made to open an escape-valve for the hot air, placed in any convenient part of the apparatus, and to cause the descent of the blowing piston *d* to expel that air, while a new supply of fresh air would enter by another valve into the cylinder above *d*. The engine would then be ready to repeat its stroke as before, and the working would be continued as in a steam-engine.

The preceding simple conception of an air-engine occurred to the author's thoughts while considering the application of a condensed air-furnace to some chemical purposes. It appeared to him that, in applying any such engine to use, the chief difficulties to be surmounted would be, to prevent the very heated air and dust from injuring the valves and other working parts of the engine, and to obviate the inconvenience of the inequality of power at different parts of the stroke. Various expedients occurred to him. The overheating might be prevented by surrounding the cylinder, &c., with water; and both cylinder and piston would suffer less from dust, if, instead of the common piston *c*. represented above, a great hollow plunger *a* were used, (such as is here represented, and is now common in water-pumps for mines) embraced by an air-tight neck or collar at *b c*, which neck would be the only part of the cylinder requiring to be made with nicety. But a more complete security would be obtained by interposing water between the hot air and the piston, as represented in this other sketch, where the working cylinder *d* has a water-vessel *b* connected with it, and the heated air is admitted to *b* to press upon a float on the water-surface, to lift the working piston *d e*. This construction, too, if desired, would allow the fire chamber *a* to be made larger than the cylinder, and to be kept constantly filled with highly expansive air, each discharge of which into the space *b* would be replaced by cold air either from the space above the piston *d*, driven in through a tube as the piston ascended, or from a distinct blowing cylinder worked by the beam. And if it were wished to apply the same principle to an engine working with double strokes, that is, forcing the piston alternately up and down, as in the double stroke steam-engine, the object might be attained, by having a second water-vessel *f* communicating with the

Fig. 127

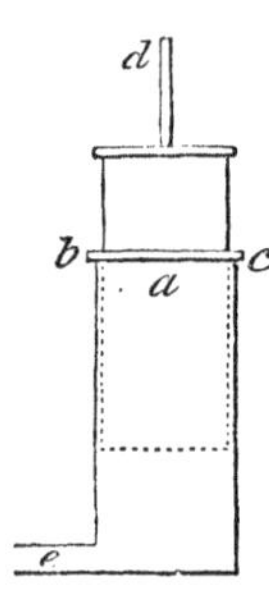

Fig. 128.

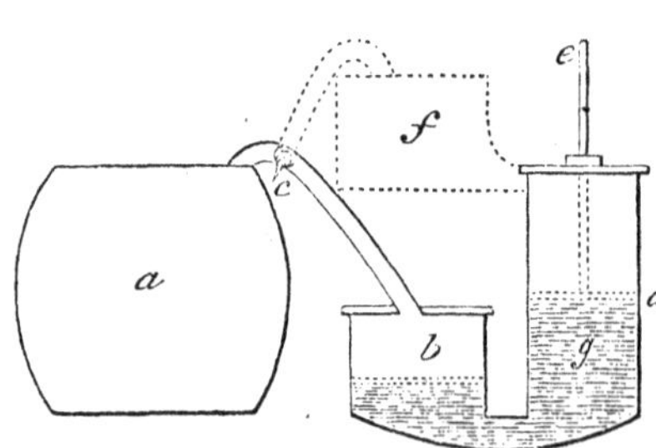

part of the working cylinder above the piston *d;* and the air would pass alternately to the one or the other vessel *b* or *f*, by the operation of the cock *c*, as steam passes in a steam-engine; the supply of fresh air to the chamber *a* would be given by a blowing cylinder worked through a connection with the engine, as the air-pump of a steam-engine is worked.

The sketch of an air-engine, as here given, was included in the specification of a patent for another object engaged some years ago by a friend of the author's; but that friend being almost immediately called to other business, and the author's professional engagements forbidding his attention to the subject, it was not prosecuted. In the specification, drawn up by an engineer in the town, some minor adaptations were described. One experiment has lately been made by a Swedish engineer with the simple form of dry apparatus described at page 285, for the purpose of ascertaining its power, and the effect was found to be several fold greater than of steam from the same quantity of fuel; but the apparatus was rude, and only calculated to prove in a short trial, the existence of the power, but not the fitness of the machine to endure uninjured, or to be rendered easily obedient to control; a complete experiment, therefore, remains still to be made. Could an obedient and durable engine be contrived, at all approaching in simplicity to the plan given above, its advantages over the steam-engine would be very considerable. First, its original cost would be much less, by reason of its small comparative size, its simplicity, and the little nicety of workmanship required. Secondly, it would require much less room, and would be very light; hence its peculiar fitness for purposes of propelling ships and wheel-carriages. Thirdly, the quantity of fuel required being so much less, would not load the ship or carriage leaving little room for anything else. Fourthly, the expense of fuel and repairing would be little. Fifthly, the engine could be set to work in a few minutes, where a steam-engine might require hours. Sixthly, little or no water would be required for it.

Fig. 129.

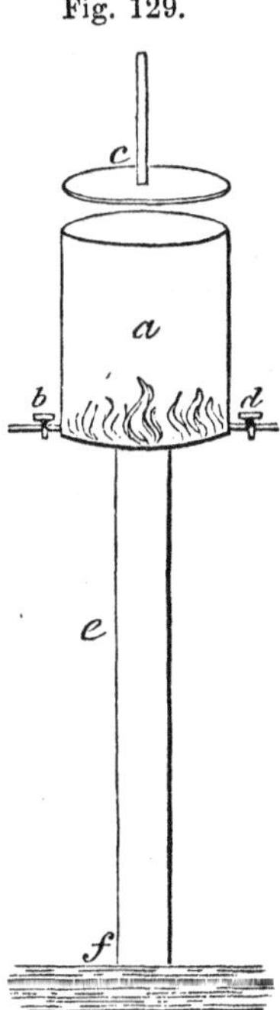

Another modification of air-engine, called a *gas vacuum engine*, has lately been proposed, and many expensive trials have been made of it; but it is in its nature a most wasteful machine, evidently throwing away at least nine-tenths of the power which the combustion generates, It was of this nature in an experiment which the author witnessed. A little of the common coal-gas was admitted by the cock *b* at the bottom of the cylinder *a*, and was there inflamed, the lid *e* being at the time raised. The combustion rarified the lower stratum of air, so that the air above was expelled, and about one-fifth of the original contents of the cylinder was caused to occupy the whole. The lid was shut down, as nearly as could be judged, at the moment of greatest expansion, so that when the small portion of air and vapour remaining within was again cooled, the interior of the cylinder approached nearly to the state of vacuum. It, in fact, retained only a fifth of the air. A communication being then opened from the vacuous cylinder by the tube *e* to a water reservoir ten feet below, the water was driven up by the atmospheric pressure, until it filled more than half of the cylinder. The water so raised was then made to turn a common water-wheel, and to do work. A much larger quantity of water, however, could be raised to the same height at less expense by a steam-engine. The proposer also hoped that he would be able to make the atmosphere pressing into its imperfect vacuum, act directly upon a piston as steam does, and with power cheaper than that of steam; but in this anticipation too he was completely in error. To produce his imperfect vacuum cost him very nearly at the same rate as it costs to produce the perfect vacuum in a steam-engine, and his vacuum for equal bulks was worth as a working power, only about one-fourth as much as the steam vacuum. This may be understood by considering, that in a perfect vacuum a piston rises all the way with the same force, which if common steam be used, is 15 lbs. per inch, the piston may be supposed to rise from *c d* to *a b*, but if the vacuum were only three-fourths towards being complete, the pressure on the piston would be only three-fourths of 15 lbs. at the commencement of the stroke, and then rapidly diminishing, would have ceased altogether when the piston had made three-quarters of its journey, or to *f*. The force in the first case would be represented by the whole line *c d* and the whole space *c d b a*, and in the second by the shortening lines and the triangular space *c e f*.

Fig. 130.

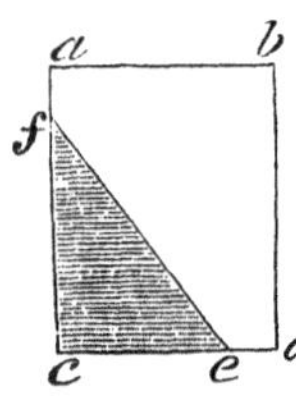

On considering the foregoing diagrams, we may perceive that in the vacuum engine, by far the greater part of the force produced by the combustion of the gas is absolutely wasted, or put to no use, namely, the whole expansive force during the sudden combustion or explosion. It is evident that if a tenth part of the aeriform contents of a cylinder acquire elasticity enough, (a fourteenth part of a nice experiment does so) to be able afterwards to occupy the whole cylinder, that tenth must begin its expansion with the force of a tenfold atmospheric condensation, or pressure of 150 lbs. on the square inch of a piston withstanding it, which pressure will then gradually diminish as the piston rises, but will amount to an average of five times the atmospheric pressure, or 75 lbs. per inch all the way; being therefore quadruple, or more, that of steam against a perfect vacuum, and, therefore, again, by our former calculation, more than twelve times greater than the force obtained from the imperfect vacuum of the engine under consideration.

It is a question which the author thinks will one day be answered in the affirmative, whether nearly the whole force of exploding gas may not be converted into a calmly working power, producing from a given expenditure, ten times or more the effect obtained in the vacuum-engine described above, and, therefore, an effect more than equal to that of a steam-engine incurring the same expense. There are probably various ways in which the object may be obtained. The following sketch is offered merely to give the reader an idea of a machine for such a purpose.

Fig. 131.

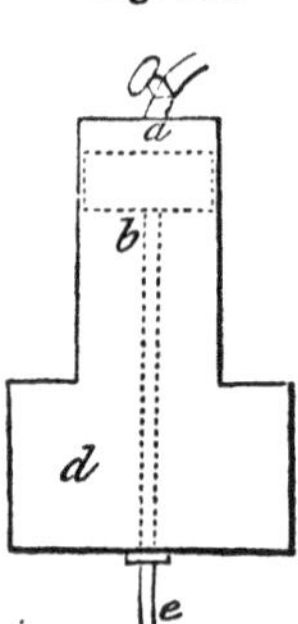

Suppose *b* to be a very heavy close-fitting piston sliding in the cylinder containing it; and suppose the space *d* open to the cylinder, to be filled with atmospheric air of double or greater density; then if a mixture of explosive gases admitted by a cock to the chamber *a* (formed between the piston and end of the cylinder) be inflamed, the heavy piston will be shot forward like a cannon-ball, against the condensed air in *d*; and owing to the momentum acquired in the first instance, it will advance much beyond the point where the exploded gas and air in *d* would balance each other at rest. The quantity of gases admitted would be just such as to carry it to the end of the cylinder. The piston rod *e* would then by a catch or racket, be connected with the work to be done, and after the condensation of the exploded gases in a cylinder, would be pressed back again, with the greater than atmospheric force in *d*, as if urged by high pressure steam. Figure 127 at page 286 represents a form of cylinder which might also answer for this purpose, the heavy plunger being thrown up, to work by its weight in descending.

It is to be remarked that the first modification of air-engine described at page 285, is partly an explosive engine such as contemplated above, for the gas separated from the coal during the moment of slackened combustion while the lately used air is passing out, becomes an explosive accumulation for the fresh air about to enter, The trial alluded to above proved this to be the fact.

" *The expansion of bodies by heat increases more rapidly than the temperature, and particularly near the melting and boiling points, that is, their points of changing into liquid or air being, however, exactly proportioned to the temperature after the change into air.* (See Analysis, page 256.)

If a given quantity of heat, that, for instance, contained in some measure of boiling water or of common steam, be added to a mass of cool water, it will produce in this a certain increment of bulk; and if other equal quantities of heat be afterwards successively added under the nice management which such an experiment requires, each new addition will produce a greater increment of bulk than the preceding, particularly when the water approaches to boiling; but after the water is converted into steam, any farther increase of bulk will be exactly proportioned to the increase of temperature. The same truths may be proved by the converse experiment of abstracting successively equal quantities of heat from steam or water (as by making it melt equal quantities of ice,) and noting the rate of contraction. What is thus true of water in relation to heat, is true also of bodies generally, each, however, having a rate of expansion and temperature for melting and boiling proper to itself. The quickened rate of expansion in solids and liquids might have been anticipated from reflecting, that each successive quantity of heat added to a mass, meets with less resistance to its expanding power than the preceding quantity, owing to the diminishing force of cohesion of the particles as the mass enlarges; while in an air or gas, again, as cohesion has altogether ceased, each addition of heat is at liberty to produce its full and equal effect.—If the capacity of substances for heat did not increase with their bulk, the terms "increase of heat" and "increase of temperature" would have the same meaning, and the subject would be more simple.

The reflection will naturally occur here, that as in the common thermometer, the mercury must rise or expand more for a given quantity of heat added at a high than at a low temperature, the scale should be divided to correspond with the inequality. Now this reasoning is good, but the difficulty of complying with it in practice is such, that the inconvenience of the slight error arising from an equal division is commonly submitted to. An air thermometer with equal divisions is very correct, but from wanting many of the advantages of the mercurial thermometer, is little employed; and fortunately in the mecurial thermometer there is such a counterbalancing relation between the expansion of the mercury and of the containing glass, as to render the error alluded to, at least for any middle range of temperature, very trifling. The subject of unequal thermometric dilatation in the same liquid, and of the differences in that respect in different liquids, depending on the proximity to their boiling points, &c., is well illustrated by Du Luc's experiment of filling various thermometer-glasses with different liquids, and while they are being heated through the same range of temperature, noting their comparative indications. He marked on each tube the points at which the liquid in it stood when the bulb was placed, first in freezing and afterwards in boiling water, and he then divided the intervening space into eighty parts or degrees. The discordance of the dilatations in the different tubes when the instruments were afterwards placed together, and heated from the freezing to the boiling degrees of water, was as here detailed.

Mercury.	Olive Oil.	Alcohol.	Water.
0	0	0	0
10	9.5	7.9	0.2
20	19.3	16.5	4.1
30	29.3	25.6	11.2
40	39.2	35.1	20.5
50	49.2	45.3	32
60	59.3	56.2	45.8
70	69.4	67 8	62
80	80	80	80

The singular discrepancy in the case of water is owing to the peculiarity described in former pages, of its contracting by cold only down to 40° of Fahrenheit, and then dilating again until it freezes.

Laborious investigations have been made by the French chemists to discover a comprehensive law determining the rate of expansion in all bodies, but the object is not yet satisfactorily accomplished.

" *To melt a solid body, or to vaporize a liquid, a large quantity of heat enters it, but in the new arrangement of the particles and generally increased volume of the mass, the heat becomes hidden from the thermometer and is called* LATENT HEAT. *It reappears during the contrary changes, after whatever interval.*" (See the Analysis, page 256.)

The expansion of bodies by heat, instead of proceeding throughout in some nearly uniform or gradual manner, exhibits in its course two singular transformations of the body: the first, when the solid breaks down into a liquid; the second, when the liquid swells out into an air or gas; so that there are, in all, three very distinct modifications or states of existence for the body dependent on the agency of heat. The substance of water, for instance when at a low temperature, exists in the solid form called *ice;* but at 32° of Fahrenheit, on receiving more heat it gradually becomes liquid or *water*, and on receiving more at 215°, even under the resisting pressure of the atmosphere, it acquires a bulk nearly 2,000 times greater than it had as a liquid, (gradually as regards the whole, but suddenly as regards each separate portion,) being then called *steam* or aeriform water. And other bodies under analogous circumstances, undergo similar changes. It is farther remarkable, that although during the changes a large quantity of heat enters the mass, producing in one case liquidity, in the other the form of air, the temperature is the very same, immediately after, as immediately before the change, the last received heat becoming hidden or latent in the mass :—thus water running from melting ice affects the thermometer but as the ice does, and steam over boiling water appears no hotter than the water. The glory of originally discovering the facts, to recall which the terms *latent heat or caloric of fluidity*, have since been used, belongs to the illustrious Dr. Black. The construction of the modern steam-engine was an early result of kindred investigations made by his friend, James Watt.

We select the following instances as serving to display the subject of *latent heat* in its various bearings.

A mass of ice brought into a warm room, and there receiving heat from every object around it, will soon reach the temperature of melting or 32° but afterwards both the ice and the water formed from it will continue at that temperature until all be melted;—the heat which continues to enter, effecting a change only in the form of the mass. And in the case supposed, whatever time was required for heating the mass of ice *one degree*, just one hundred and forty times as much will be required for melting it; proving that 140° is the latent heat of water.

If two similar flasks, one filled with ice at 32°, and the other with water at 32° be placed in the same oven or over the same flame, the water will gain 140 degrees of heat while the ice is nearly being melted into water at 32 : and in the course of the experiment, a correspondence will always exist between the phenomena; for instance, when the water has gained 14° of heat, it will be found that just a tenth part of the ice is melted.

If equal quantities of hot and cold water be mixed together, the whole acquires a middle temperature, each degree lost by the hot water becoming a degree gained by the cold; but if a pound of ice at 32°, and a pound of water 140° hotter be mixed together, the 140° of heat will go merely to melt the ice, for there will result two pounds of water at 32°.

If a flask of water at 32°, or its freezing point, and a similar flask of strong brine (which does not freeze until cooled to near *zero*) also at 32°, be exposed together in the same cold place, it will be found that when the brine has lost 10° of its heat the water flask will still exhibit an undiminished temperature, but a fourteenth part of its contents will be converted into ice. Now as in such a case the water flask must continue to radiate away heat just as much as the other, it can maintain its temperature by absorbing into its general mass the heat which was latent in the portion of water frozen.

It is possible, by slowly cooling water which is kept in perfect repose, to lower its temperature, while yet liquid, ten degrees below its ordinary freezing point; but then on the slightest agitation, ice will be formed. It might be expected, in such a case, that the whole water will instantly freeze, because all is colder than common ice; but no, only a fourteenth part freezes; and singularly, both that fourteenth and the remaining liquid are rendered in the moment ten degrees warmer—rising to 32°. Here the 140° of latent heat escaping from the fourteenth part of the water which freezes, become 10° of sensible heat for the whole mass, so that the remaining water has the temperature at which it only begins to freeze.

Strong solutions in hot water of various neutral salts, if allowed to cool while exposed to atmospheric pressure, soon deposit crystals of the salts; but in a close vessel, which protects them from such pressure, they will remain liquid even when cold. Now at the moment of opening such a vessel to admit the pressure, the salt immediately crystallizes, and the latent heat given out by the solidifying particles warms very sensibly the remaining liquid and the vessel.

From the preceding facts it may be perceived, that the quantity of ice formed or melted in any case, becomes a correct measure of the quantity of heat transferred. From this consideration, the illustrious Lavoisier constructed his calorimeter, or heat measure. It is a case or vessel lined with ice, and the quantity of heat given out by any body placed in it is indicated by the quantity of water collected from the melted ice.

Had the latent heat of water been only 1° or 2° instead of 140°, the earth, except in its tropical regions, would have been scarcely habitable. The cold of a single night might have frozen an ocean, and the heat of a single day might have converted the accumulated snows of a winter into one sudden and frightful inundation. As the fact is, however, both changes are beautifully gradual, and easily controlled or prepared for.

The fact of latent heat in other liquids than water is familiarly exhibited in the slow melting of various substances as—of the metals; lead or pig-iron for instance—of butter or oils—of glass, &c.; and on the other hand, in the slow solidification of any melted masses when heat is again abstracted.

The substances below enumerated, while passing from the solid to the liquid state, absorb and render latent the quantities of heat here noted; which quantities are therefore called the latent heats of the liquids.

Ice - - - -	140°
Mercury - - -	142
Bees' wax - - -	170
Tin - - - -	442
Zinc - - - -	492

If a piece of frozen mercury (the temperature of which is at least 40° below zero) be thrown into a little water at 32°, the latent heat of the water immediately passes into the mercury and melts it; but, singularly, the water in the act of melting the mercury, is itself frozen.

"Latent Heat of *aeriform fluids*."

Water in a vessel placed over a fire gradually attains the boiling temperature or 212°, but afterwards its temperature rises no more, for the farther addition of heat becomes *latent* in the steam escaping during the ebulition. One way of determining the quantity of heat which becomes latent in steam is to note how much more time is required for boiling a quantity of water to dryness, than for merely heating it to the boiling point, or through any certain number of degrees. The experiment indicates about 1,000°; that is to say, 1,000 times as much heat is latent in any quantity of water formed into steam, as would raise the temperatnre of the liquid water one degree. Watt had found that water in a vessel placed over a lamp was about six times as long in being completely evaporated, as in being originally heated from an ordinary temperature to that of boiling.

If we place in the same oven, or over similar flames, two like vessels containing water, one of which is open at the end and the other is strongly closed, the two will gain heat equally up to the boiling point, but afterwards the open vessel from giving out steam will remain at the same temperature, while the other, by confining the heat which enters, will show the temperature continuing to rise as before, until the increasing tendency of the water to dilate forces the vessel open. Supposing the water in the latter vessel, before vent is given, to have become 100° hotter than common boiling water, instead of the whole, when at liberty, being immediately converted into steam, as might be expected, only a tenth part will be so changed (the same quantity as will be found to have already escaped from the other vessel,) for the tenth part requiring in the form of steam 1000° of latent heat, will take the excess of 100° from the other nine parts, and will leave them as common boiling water. If, however, water heated considerably beyond the boiling point be allowed to expand *very suddenly*, the whole is blown out of the vessel as a mist, by the steam formed at the same instant through every part of the mass; but the whole mass in such a case is no more converted into true steam than the whole of very brisk *soda-water* is converted into gas when similarly thrown out by the sudden extrication of the carbonic acid gas, on uncorking the bottle. Misconception of this matter has led to most wasteful experiments on steam-engines of very high pressure. Mr. Perkins, for instance, thought he truly described what was accomplished by saying of the water that it had "flashed into steam."

The same indication of the latent heat of steam is obtained by the converse experiment of first converting a quantity of water into steam, and then admitting it to cold water or to ice. A pound of steam will raise the temperature of ten pounds of cold water 100 degrees, or will melt about 8⅓ pounds of ice.

In the great quantity of heat which becomes latent in steam, we perceive the reason why water projected upon a raging fire so powerfully represses it, and hence, again, why *fire* and *water* are so often adduced proverbially as exemplifying a fierce antagonism.

It was when Watt had discovered how much heat was lost when steam was lost, that he contrived the separate condenser for his steam-engine, by which he at once saved three-fourths of the fuel formerly used

Substances differ among themselves in regard to the latent heat of their vapours as much as in their other relations to heat. Thus the latent heat of the vapor or steam of—

Water	is 1,000°
Vinegar	900
Alcohol	442
Ether	300
Oil of turpentine	177

From the less latent heat in these last-mentioned vapours than in that of water, we might at first suppose that there would be great advantage from using them in steam-engines. Accordingly numerous experiments have been made, and patents secured under this idea; but the fact is, that in the same proportion as the heat is less, the volume of the vapour is less, and therefore no mechanical advantage is obtainable.

The influence of external pressure in keeping the particles of liquids together, in opposition to the repulsion of heat seeking to render their mass aeriform, was considered in the chapter on "*Pneumatics*;" but to make the present section complete, the subject must be shortly resumed.

Because water or any liquid, under the pressure of the atmosphere, while receiving heat, remains tranquil, and apparently unchanged, until it reaches what is called its boiling point, at which a bubbling or conversion into vapour takes place, we might suppose its ordinary boiling temperature necessary to enable it under any circumstances, to assume or to maintain the form of air. But this is no more true than that a common spring compressed against any obstacle or force, has no tendency to expand or recover itself till the moment when at last it overcomes the obstacle. Liquid water with its heat is really a spring compressed by the powerful weight of the atmosphere, and seeking to expand itself into steam with force proportionate to its temperature. Even at 32°, or its freezing point, as is found by placing it in a vacuum, it seeks to assume the form of air, with a force of pressure 1½ ounce on each square inch of its surface, and can be restrained only by a counter-pressure of that amount; and at any higher temperature, to correspond with the greater dilating tendency, the restraining force must also be greater: at 100°, for instance, it must be 13 ounces; at 150°, 14 lbs.; at 212°, 15 lbs.; at 250°, 30 lbs., and so on as stated in the preceding part of this work:—and whenever the restraining force is much weaker than the expansive tendency the formation of steam takes place rapidly and far below the surface of the liquid, so as to produce the bubbling and agitation called boiling. Now it is because the atmosphere or ocean of air which surrounds the earth happens to have in it 15 lbs. weight of air over every square inch of the earth's surface, and presses on all things there accordingly, that 212° happens to be called the boiling point of water. An atmosphere less heavy would have allowed liquid to burst into vapour at lower temperatures, and one more heavy would have had a contrary effect.—The exact degree of expansive force for every degree of temperature in water and other liquids, has been ascertained by heating them in vessels furnished either with properly loaded valves, as at *f* in this figure, or with a tall upright tube, as *d b*, into which the liquid *c* may force a column of mercury to an elevation marking the expansive tendency; the valve and mercury being, of course, protected from the external atmospheric pressure, or the necessary allowance being made for that pressure. Boiling at the bottom of a deep vessel is resisted by the weight of the liquid in addition to that of the atmosphere, as already explained, and consequently the temperature at which it occurs there, is higher than near the surface of the vessel. Boiling heat is greater, also,—in a deep mine, where of course there is additional depth and weight of atmosphere over any exposed liquid,—at times when the barometer is unusually high, that is to say, when the atmosphere is unusually heavy—in cases where air or steam is confined over the boiling surface so as to press more upon it, as when brewers for a time shut the lid or valve of their great boilers, &c. Water placed on the fire in a strong vessel, from which steam cannot at all escape, may be rendered even red-hot, without a bubble forming or one particle being dissipated; but the tendency to expand into steam is then great enough to burst any known material of moderate thickness. The Marquis of Worcester exploded a cannon by shutting up water in it and then surrounding it with fire.—Boiling temperature is lower again when the experiment is made on mountains or in other situations above the level of the sea, where there is less height of air resting over the boiler. In the city of Mexico, which is 7,000 feet above the sea, water boils before it reaches the heat of 200°, instead of, as in places

Fig. 132.

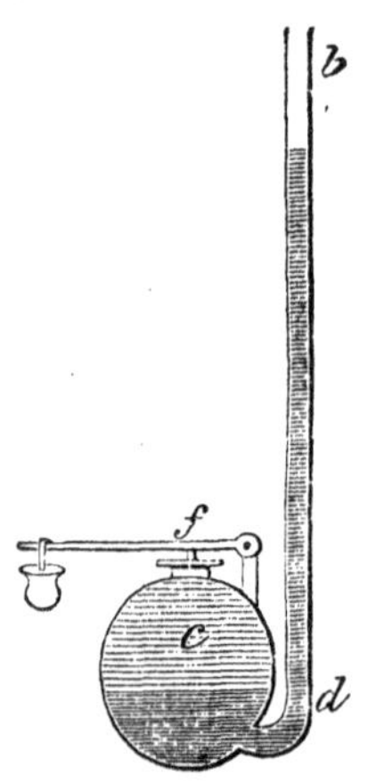

near the sea-level, at 212°. Wollaston's thermometer, beautifully adapted for determining the height of mountains, balloon ascents, &c., by merely indicating the heat of boiling water in any situation, is a fine illustration of this truth. If in any place we take off the atmospheric pressure from a liquid, as by placing it in the receiver of an air-pump, it will boil at very low temperatures indeed. Water thus treated boils at a temperature many degrees below the heat of English summer days; and ether boils when colder than common ice.—Generally, in a vacuum, substances boil at a temperature 124° lower than while restrained by the atmospheric pressure.

Consequences of these truths respecting the boiling temperature are the following.

As water at any temperature is tending to dilate itself into steam, with force proportioned to the temperature, the steam rising from any mass of water presses on the surface of the vessel containing it with that force; and in a steam-engine, therefore, the temperature of the water in the boiler tells the degree of force with which the steam is acting on the piston.

Because in the case of steam the same law holds as for aeriform fluids generally, *viz.*, that the outward elasticity or spring increases in proportion as the fluid is more condensed—high pressure steam is merely condensed steam, just as high-pressure air is condensed air; and to obtain a double or triple pressure, we must have twice or thrice the quantity of steam under the same volume.

The reason that high pressure steam issuing from a boiler heated to 300° or more is not hotter than low-pressure steam from a boiler at 212°, is, that in the instant when the high-pressure or condensed steam escapes into the air, it expands until balanced by the pressure of the atmosphere, that is, until it becomes low-pressure steam, and it is cooled by the expansion, as air is cooled on escaping from any condensation.

The vessel called *Papin's Digester*, is merely a metallic pot or boiler, which can be kept closed in spite of the force of the steam formed within it; and in such a vessel, water can be heated far beyond the ordinary boiling point,—sufficiently for instance, to dissolve and extract all the gluten or jelly of bones, and to form from them a rich soup where common boiling would procure nothing;—or even to melt lead lying in water.

The person who increases the fire under a boiling pot with the hope of making the water hotter, is foolishly wasting the fuel, for the water can only boil, and it does boil at 212° of the thermometer.

As different substances under any given pressure, become aeriform at different temperatures, mixtures of such may be decomposed by heat. If a mixture of spirit and water, for instance, be placed over a fire, the spirit will boil off long before the water. If the spirituous vapour be caught apart and condensed, the operation is called *distillation*. All distillations are of the same nature.

The instrument here represented consists of a glass tube blown into bulbs at the two ends *a* and *b*, and hermetically sealed after receiving into it some water, but no air. There will always be in the apparently empty part a stream or aeriform water of density proportioned to the temperature. If one of the bulbs be heated more than the other, the steam or vapour in that one will, for the reasons stated above be denser and stronger than in the other, and will

Fig. 133.

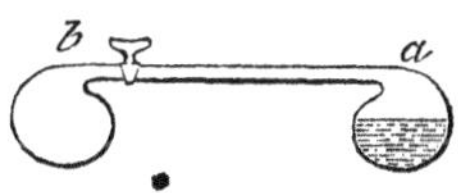

therefore be forcing its way into the other; where, owing to the lower temperature, a part of it will be relapsing into the state of water, making room for more. Hence, if the difference of temperature between the bulbs be long maintained, the whole water will, by a sort of distillation, gradually pass into the colder bulb. If the difference of temperature become at any time considerable, the liquid will boil in the warmer bulb, even although the source of heat be only the living hand grasping it.

To the author of this work it appears that by a larger apparatus made on this principle, fresh water might be conveniently obtained from salt water on board ship, or on islands deficient in fresh springs. Suppose any two air-tight vessels like *a* and *b*, of large size, communicating by a tube furnished with a stop-cock near *b*: then if the vessel *a* were filled with salt water, and were heated by being exposed to the sun, (its surface being blackened to absorb heat, and protected by glass from the cooling effect of the air,) and if the other vessel *b* were made a vacuum by pumping out from its bottom the water with which it had been previously filled, and were then kept as cold as possible by wetted coverings and a current of air,—on opening the cock at *b*, vapour would pass over from the warmer vessel to be condensed in the colder, and there would be a distillation from sea-water by the natural action of the sun alone, of a water naturally fresh and pure. Cases have occurred where a knowledge of this fact would have saved shipwrecked crews from perishing by thirst; and there are rocky islands in the ocean where there is no supply of fresh water but from precarious rains or importation from abroad, but which might be rendered pleasantly habitable by the adoption of such a means.

When a substance has reached the temperature at which it boils, that is to say, at which its power of emitting vapour becomes rather more than a balance to the atmospheric pressure, its dilating force is strong indeed. Persons may not reflect that 15 lbs. on a square inch is about a ton on a square foot.—and such is the power with which the vapour of all boiling substances rises from them—sufficient in a single Cornish steam-engine to urge the piston with the force of 600 horses! But even at temperature much below boiling, the tendency to expand, as already stated, is still very great, and although not attracting common attention, is silently working many beautiful and important ends in the economy of nature.—As in a perfect vacuum, freezing water gives out a steam or vâpour that would lift an opposing weight with force of 1½ ounce per inch, or 16 lbs. on a square foot: and even solid ice gives out its vapour of nearly equal strength,—so also do many other liquids and solids give out their vapours. Thus in the apparently empty space called the Terricellian vacuum, over the mercury in a barometer tube, there is always an aeriform mercury, dense in proportion to the temperature; and around camphor, and the essential or volatile oils, &c., there is similarly an atmosphere of the substance in the form of air.

It had for a considerable time been known that into a perfect vacuum many bodies emitted almost instantly in the form of air, a quantity of their substances proportioned to their temperature; but it was reserved for Mr. Dalton to make the admirable discovery, that even into any space filled with air, these vapours arise in quantity and density the same as if air were not present—the two fluids seeming to be independent of each other, with the exception that in a vacuum the equal diffusion of vapour takes place at once, while in a situation already occupied by air, it proceeds more slowly as the vapour can force its way through the particles of the air, and in general takes place by a tranquil evaporation from the surface instead of the agitation of ebullition. In an apartment with an open vessel of water in it, there is soon, although invisible, a steam of watery vapour mingled with the air, as dense as if the room were a vacuum at the same temperature.

Consequences of this important truth are the following:

That it is only an atmosphere of the substance of each body, which, by pressing on the body, can prevent its farther dissipation by heat. Thus we can save camphor, musk, smelling oils, spirits, water, &c., only by placing them in closed bottles or vessels, in which, additionally to the air present, an atmosphere of their own substance is formed, involving the remaining masses with pressure proportioned to their temperature and its density.

The important process of drying things is merely the placing them under an elevated temperature if attainable, and in an atmosphere not containing so much of the liquid as to be saturated at the temperature. The effect of wind or motion of the air in quickening evaporation, is owing to its removing air saturated with the moisture, and substituting air which is not—thus producing nearly the case of the substance placed in a vacuum.

If air at a certain temperature, contain mixed with it as much water as can be sustained in the form of invisible vapour at that temperature, and if then, by any cause, as by rising in the atmosphere, the air be cooled, it will abstract heat from the vapour, and cause a portion to be precipitated or visibly condensed into a fog or rain. Water rising as invisible vapour from the surface of a lake or river, often, when it has reached a certain height, is condensed into the stratum of clouds which there appears, and which for a time may remain usefully protecting the fields from the intense meridian sun, or may fall again as refreshing showers over the country.

It is the tranquil and invisible evaporation of which we are now speaking, which lifts from the surface of the wide ocean all the water which, after condensation, returns to the ocean in the form of the myriads of river streams which give life and beauty to the face of nature.

In warm climates there are inlets of the sea, occasionally shut off from the parent ocean, and where, after the sun's rays have drunk up all the water, the deposited salt remains to be carried away in loads for the uses of man, as sand is carried away from any ordinary shore. There are in the bowels of the earth prodigious accumulations of salt, some of which may have been formed in this way, during the revolutions of the world in remote past time, and which are now turned to man's account as salt-mines. When the Nile overflows its banks with water holding in solution, although in almost imperceptible proportion, mineral substances brought from the interior of Africa, some of that water admitted into reservoirs, and afterwards dried up by the sun's heat, leaves a rich store chiefly of chrystallized natron or soda.

The following are other instances of vapour which is invisible while at a higher temperature, but is thickly precipitated when air, with which it is mixed, is cooled, or when it touches a colder solid body :—the steam observed at night and morning hovering over brooks and marshes heated by the sun during the day :—the frost-smoke, as it is called, which lies on the whole face of the greenland seas in the beginning of winter, where the water, warmed by the long day of the polar summer, continues to emit its vapour for a considerable time after summer is past, into an atmosphere, become too cold to preserve it invisible :—the breath or perspiration of animals, of horses in particular, after strong exertion, becoming so strikingly visible in cold and damp weather, or even in warm weather, when the air is already charged with moisture:—in cities where there are deep drains communicating with kitchens, manufactories, &c., and constantly filled with moist warm air, the vapour-loaded air, although clear or transparent in the drain, immediately on escaping into a frosty atmosphere lets go its moisture, with the appearance of steam issuing from a great subterranean caldron. Steam over water in any boiler, is transparent or perfectly aeriform—as may be seen when water is made to boil in a vessel of glass, but as soon as it is cooled by contact or admixture of colder air, it ceases to be true steam, and is condensed into small particles of water suspended in the air. Many persons while thinking of steam, figure it only in this latter state, as particles of water mixed with air nearly as a subtile powder might be mixed, and its substance occupying really no more space than the original water did. Now until steam is cool and condensed, it is of a nature to fill alone any appropriate vessel and powerfully distend it, just as air fills and distends a bladder. Steam issuing from the spout of a kettle is hardly seen near the mouth, but as its distance from the spout increases, it is cooled into a thick cloud or vapour.

In a vessel from which air and atmospheric pressure are excluded, even the temperature of freezing air being sufficient to maintain permanently in the state of gas or air, many substances which exist as liquid under the atmospheric pressure—and the whole mass of such a substance, when placed in a vacuum, not being instantly converted into gas, because the portion which first rises becomes an atmosphere weighing upon the remaining mass, and because, moreover, that portion, by absorbing from the mass much heat into the latent state, cools the mass much below the freezing point; we see why the liquids now spoken of are so rapidly cooled to at least the freezing point if placed where a vacuum can be maintained, that is to say, where, after common air has been removed, the aeriform matter arising from them, and absorbing their heat, is also promptly, and in a continued manner, abstracted. It is thus that water placed in an exhausted receiver of an air-pump is so rapidly cooled, and that when there is beside it a vessel of concentrated sulphuric acid, or other substance capable of absorbing the watery vapour as formed, it is soon reduced to the state of ice; or again, that water, or even mercury, surrounded by ether evaporating in a vacuum, is so quickly frozen. It is thus, also, that if one bulb of the instrument described at page 296, be immersed in a freezing mixture, the water in the other and distant bulb will soon become ice; for the vapor arising from that water in the vacuum maintained throughout the apparatus by the freezing mixture, is immediately condensed again in the immersed bulb, and leaves the vacuum still free for the ascent of more vapour, to carry away more heat from the water as latent heat, and to make it freeze.

As we have explained, also, that in a liquid there is the same tendency to evaporation, whether it be or be not exposed to the air, we see the reason why all evaporation is a very cooling process. The effect, however, in air is neither so rapid nor so great as in a vacuum: first, because the presence of the air impedes the spreading of the newly-formed vapour from the liquid surface, and keeps it where its pressure resists the formation of more vapour; and, secondly, because the air in contact with the liquid, shares its higher temperature with the liquid. Still, in India, flat dishes of water, placed during the night on beds of twigs and straw kept wet and in a current of air, soon exhibit thin cakes of ice—and thus, ice is procured in India for purposes of luxury.

The absorption of latent heat in the evaporation which goes on from the sea and earth in all warm climates, greatly tempers the heat of these climates, and the vapour afterwards spreading to the poles, as explained in "*Pneumatics*," under the head of *Winds*, carries warmth thither to be given out when it is recondensed into the form of rain, or is solidified as snow. The formation any where of mist or rain warms the air most sensibly, by the liberation of the latent heat from the precipitated vapour. Again, the liquid water, which, during winter, is converted into snow or ice, had been a reservoir of latent heat stored to temper the frosty air of the commencing cold season; and in the following spring, such ice and snow serve as empty receptacles, in which the first violence of the returning sun hides or expends itself; allowing the temperature to change more gradually, and for many living beings, therefore, more safely. The vast stores of ice and snow among high mountains, as among the Alps and Pyrenees, are often, during the summer, stores of mild temperature to regions around: for besides cooling the air near them, they are the never-failing sources of the rivers which run from them during the whole of summer, carrying freshness through distant lands:—from the Alps, for instance, proceed the Rhine and the Rhone —most romantic and beautiful of European streams; and from the Pyrenees, the rapid Gave, &c., which, while channels around from lower regions are almost dried up by the summer heat, flows only the more freshly as the heat is greater, and the feeding snows are more abundantly dissolved.

Men in artificially raising temperature, are generally causing the liberation of heat which had been previously latent; and in lowering temperature or producing cold, they affect their purpose almost solely by rendering a quantity of heat latent.

Lavoisier thought that the heat of all combustion was merely the latent heat of the oxygen gas concerned in the combustion, given out during its combination with the burning body. It is so in part, but we now know that it depends more on the intensity of the chemical action between the combining substances. The water thrown upon quick-lime to slake it, becomes solid in combination with the lime, and gives out its latent heat so remarkably as often to set fire to a wooden vessel or ship containing it.

When dwelling-houses, green-houses, manufactories, &c., are warmed, as is now common, by the admission of steam into systems of pipes which branch over them, the heat is chiefly that previously latent in the steam, and which spreads around as soon as the steam, by touching pipes of a lower temperature, is condensed to a state of water. The modes of most profitably effecting these purposes have to be considered in a future chapter.

For producing artificial cold, our processes generally involve the circumstance either of a solid changing into a liquid, during which it absorbs, and hides in its new constitution much of the heat previously sensible in it and in the liquid dissolving it; or of a liquid changing into vapour, during which heat equally becomes latent. Thus by dissolving a salt, nitre, for instance, in water, we obtain a solution which is very cold.

In India, the common mode of cooling wine for table is to surround the bottles with nitre thus melting; and the water of the solution being evaporated again before next day, the salt is left ready for use as before. Such is the mutual attraction of water and many salts, that they readily combine, assuming the liquid form, even when water is used in the solid state of ice; and as, in that case, both the water and the salt render heat latent, the fall of temperature is very great. Thus common salt and snow (or ice) when mixed, dissolve into liquid brine 37° colder than freezing water, or 5° below the zero of Fahrenheit.—The last mentioned fact explains the common practice of sprinkling salt on an ice-covered pavement before a street door to clear away the ice. The salt and ice quickly combine and form liquid brine, which either of itself runs off into the gutter and disappears, or is easily swept of, or its water evaporates, leaving only the salt behind. It is true that the brine is at first a refrigerating mixture, which cools still more the pavement and the neighboring ice, but all which is touched by the salt is melted. Servants usually err in using a pickaxe or spade immediately after the sprinkling, instead of waiting, and with a broom spreading the melting salt compleatly over the place.

Frigorific Mixtures.

Substances mixed.		Thermometer sinks.
Common salt	1 part	From any temperature to 5° below zero.
Snow or pounded ice	2 —	
Common salt	5 —	From any temperature to 25° below zero.
Snow or ice	12 —	
Nitratė of ammonia	5 —	
Snow	3 —	From 32° above to 23° below zero.
Diluted sulphuric acid	2 —	
Fused Potass	4 —	From 32° above to 51° below zero.
Snow	3 —	
Nitrate of ammonia	1 —	From 50° to 4° above zero.
Water	1 —	
Sulphate of soda	8 —	From 50° to 0° or zero.
Muriatic acid	5 —	

We have already described under other heads the frigorific effect of evaporating in a vacuum or in the air, and of the operation of condensing a gas to squeeze the heat out of it before letting it expand again to a great volume.

For any given substance, the changes of state from solid to liquid, and from liquid to air, happen under similar circumstances, so precisely at the same temperature, that they mark fixed points in a general scale of temperature, and enable us to regulate and compare our various thermometers. (See Analysis, page 256.)

As we can neither weigh heat, nor measure its bulk, nor see it, and as, even if our sense of touch were a correct judge in the matter, which it is not, we dare not touch things which are very hot or cold, some other means were wanted for estimating the presence in bodies of this very subtile principle;—and a mean has been found in the measurement of its most obvious and constant effect, namely that dilatation or expansion of bodies, which again ceases when the heat is withdrawn. Any substance so circumstanced as to allow this expansion to be accurately measured, becomes to us a *thermometer* or *measure of heat.*

In *solid* substances the direct expansion by heat is so small as to be seen or measured with diffculty. In *airs*, again, the expansion is very extensive; but there is the objection that in any apparatus yet contrived which will allow their expansion completely to appear, they cannot be protected from the varying pressure of the atmosphere—an influence which affects their volume even more than common changes of temperature. But *liquids* are free from both disadvantages, and when placed in a glass bulb, as *a*, having a long neck or stalk $a\ b$, into which the liquid may rise when expanded by heat, to be measured, they form the most generally convenient of thermometers. Then, among liquids, mercury is, on several accounts, singularly pre-eminent. In mercury, the range of temperature between freezing and boiling reaches a higher point than in any other liquid, and a lower point than in all others except alcohol; its little capacity for heat and ready conducting power, causes it to be very quickly affected by change of temperature; its expansion is singularly equable for equal increase of heat through the important middle part of the scale, which includes the common temperature on earth, namely, from freezing to the boiling heat of water; and it is easy to proportion the bulb and the stalk to each other, so that a small difference of temperature shall cause the mercurial column in the stalk to rise or fall very conspicuously.

Fig. 134.

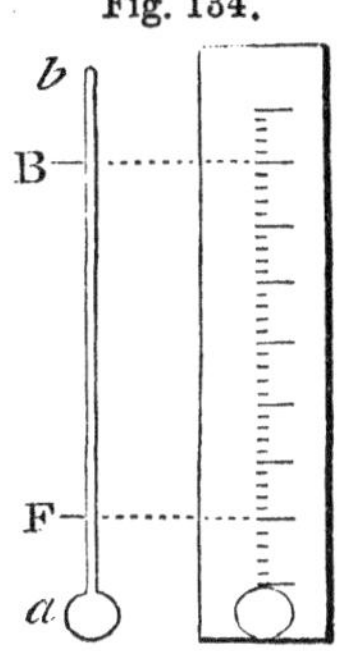

Now, when the important fact was ascertained that solid water or ice melts in every case at precisely the same temperature, and that pure liquid water in a metallic vessel, and under a given atmospheric pressure, boils always at the same temperature, it followed that by placing such a thermometer as above described, first in melting ice and then in boiling water, and marking upon the stalk the two points at which the mercury stood, as represented here by F and B, two fixed or variable points would be obtained, and the interval between them might be divided on the glass, or on a suitable scale to be attached to the glass, into any convenient number of parts to be called degrees; it followed farther, that by continuing the division to any extent both above and below the fixed points, a general scale of temperature would be obtained, with respect to which all thermometers made on the same principle would perfectly agree, although the size of the divisions on the stalks would vary according to the comparative capacities of the bulk and stalk in the different instruments. Our Newton had the honor first to propose the regulating points of freezing and boiling, and they are now universally adopted, but the interval between them has been variously subdivided;—that is to say, there has not been agreement among philosophers as to what they would call a degree of heat. In the Centigrade thermometer, which is the most simple, the division is into 100 equal parts; in Reaumur's, which is commonly used in France, it is into 80 parts; and in Fahrenheit's, which is used in England, it is into 180°. In Fahrenheit's, moreover, the freezing point, instead of being called *zero*, as in the others, is called 32°, because the maker choose to begin counting from the lowest heat which he met in Iceland, and which was 32° below freezing of his scale.—To turn the degrees of any one of these thermometers into degrees of any other, we have only to recollect that 9° of Fahrenheit are equal to 5° of the Centigrade, and to 4° of Reaumer. Therefore, multiplying by 9 and dividing by 5 or 4, or the reverse, and adding or subtracting the 32° of Fahrenheit, gives, as the result, the degree desired.

The bulb of a mercurial thermometer is formed by heating to fusion in the flame of a lamp, the end of a glass tube, which has a very small and equable bore, and then blowing into the tube until the softened end swells like a soap-bubble, to the size desired. The mercury is forced into such a bulb through its long stalk by the pressure of the atmosphere,—thus. First, a portion of the air originally in the bulb being expelled by warming the bulb, the open end of the stalk is immersed in mercury, and when the air remaining in the bulb cools and contracts, a little mercury enters. Secondly this admitted mercury having been made to boil, so as to fill with its vapour the whole capacity of the bulb and tube, and to expel the air, on the open end being again immersed in mercury, and the mercurial vapour within being condensed, the atmosphere presses in fresh mercury to fill the whole vacuum. To complete the making of the thermometer, the bulb is again heated to expel so much of the mercury as that when cold, the tube shall be about one-third full of it, and then, before the heated mercury begins to recede, the end or opening is hermetically closed by directing upon it the point of a blow-pipe flame which fuses the glass.

Although the *direct* expansion of any solid body by a moderate change of temperature is so inconsiderable as to be with difficulty measured, M. Breguet, of Paris, lately with much ingenuity contrived a thermometer which makes it very evident. Having soldered side by side two very small flattened wires of silver and platinum, or of any other metals having different expansibility by heat, he found that all changes of temperature made such compound wires bend to a great extent, as a sheet of damp paper curls on being held before the fire. The metal most shortened or least lengthened acting like a bow-string to pull the other into the arched form; and he then found, on giving to the compound wire a spiral or cork-screw form, and securing the upper end of it to a fixed stand, while the lower was left free to move, that an index like the hand of a watch attached to the lower end was turned completely round by a certain change of temperature, and afforded on a circle of degrees placed like a watch face below it, indications which perfectly agreed with those of good mercurial thermometers. Other modifications of the same principle have since been successfully tried, so simplified and reduced in bulk as to be introduced into the structure of a pocket watch.

Air is a substance on several accounts admirably adapted to the formation of a thermometer; for it has great extent of dilatation from small increase of heat; it quickly receives impressions, and its dilatation is equal for equal increments of heat at all temperatures;—but, as already stated, there is the strong objection that the pressure of the atmosphere cannot be excluded, without at the same time confining the air, and effecting its expansion. Mr. Leslie, however, has used for particular purposes an air thermometer, which he calls the differential thermometer. It consists of two bulbs *a* and *b*, filled with air and connected by a bent tube *d c*, containing liquid, the instrument being hermetically sealed, so that the atmosphere cannot affect the air within. The greater heat in the bulb *b* than in the other, as when that bulb is touched by the warm hand or is exposed to the sun's ray, is marked by the descending of the liquid in the tube *d*, which has a scale attached to it.—We may observe that equal divisions or degrees marked on the scale of this thermometer, do not mark equal changes of temperature, for the increasing condensation and resistance of the air in the cold bulb require the force overcoming it progressively to increase. If the resistance, on the contrary, were unvarying, as in an air thermometer open to a steady atmosphere, equal extent of motion in the fluids would mark equal increments of heat. An air-thermometer made of a simple bulb and long stalk of semi-transparent porcelain, with the mouth downwards, and containing in its neck melted lead or other fusible metal instead of mercury, is well adapted for measuring very high temperatures.

Fig. 135.

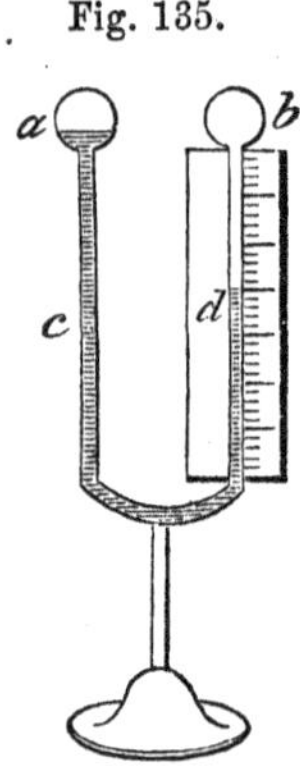

Temperatures, below that of freezing mercury, are usually measured by alcohol which substance has not yet been frozen; and temperatures higher than of boiling mercury, are measured by the expansion of air or of metals, as above described, or by the contraction of pieces of baked clay, which, when highly heated, lose water and become semi-vitrified, The use of baked clay was proposed by Wedgewood, and the apparatus has been called Wedgewood's *Pyrometer*, or fire measure. All contrivances for measuring heat may be graduated so as to correspond with the scale adopted for the mercurial thermometer.

It is most interesting, while considering the vast number and importance of the phenomena produced by heat, to observe the degrees in the general scale of temperature at which these severally take place. In the following table, a selection of the facts is classified, the temperatures being all referred to the scale of Fahrenheit's thermometer.

Table of facts connected with the influence of heat corresponding to certain temperatures.

	Wedgewood.	Fahrenheit.
Highest temperature measured	240°	32,277°
Chinese porcelain softened	156	21,357
Cast-iron thoroughly melted	150	20,577
Greatest heat of a commen smith's forge	125	17,327
Flint glass surface	114	15,897
Stone ware baked in	102	14,337
Welding heat of iron	92 to 95	13,427
Delft ware baked in	41	6,407
Fine gold melts	32	5,237
Settling heat of flint glass	29	4,847
Fine silver melts	28	4,717
Brass melts	21	3,807
Full red heat (*the beginning of Wedgewood's Pyrometer*)	0	1,077
Heat of a common fire		790
Iron red in the dark		750
Quicksilver boils		660
Linseed oil boils		600
Lead melts		594
Sulphur melts		226
Water boils		212
A compound of three parts tin, five of lead, and eight of bismuth melts		210
Alcohol boils		174
Bees'-wax melts		142
Ether boils		98
The present medium temperature of the globe		50
Water freezes		32
Milk freezes		30
Vinegar freezes		28
Strong wine freezes		20
Weak brine freezes	*zero*	0
Quicksilver freezes	*below zero*	40
The air sometimes at Hudson's Bay		50
Greatest artificial cold yet measured		91

There is reason for thinking that the higher temperature noted in this table are considerably too high, owing to the insufficiency of the thermometer or Pyrometer (Wedgewood's) by which they were estimated.

It is a curious inquiry, suggested by contemplating the preceeding table, how much heat may yet remain in bodies at the lowest temperature which we know? No conjecture was hazarded on the subject until Dr. Irving thought it might be elucidated by comparing the quantity of heat which becomes latent in a body on changing form, with the capacity of the body before and after the change. For instance, with respect to water, he said: as it requries one-tenth more heat to make a certain change in the temperature of water than of an equal quantity of ice, probably ice-cold water contains altogether just one-tenth more heat than of an equal quantity of ice at the melting point: then as we know the water to contain exactly 140° more heat than the ice, *viz.*, its latent heat, the whole or absolute quantity of heat in it will be ten times 140°, or 1,400°. By applying this reasoning, however, to other substances than water, it evidently is fallacious; and the conclusion follows that we have as yet no means of solving the question;—the thermometer no more telling us the absolute quantity of heat in any body than the rising and falling of the tide between any two rocks tells us the total depth of the rocky chasm.

From what is said in the last and in preceding paragraphs, it is evident that the thermometer gives very limited information with respect to heat: it merely indicates, in fact, what may be called the tension of heat in bodies, or the tendency of the heat to spread from them. Thus it does not discover that a pound of water takes thirty times as much heat to raise its temperature one degree as a pound of mercury; nor does it discover the caloric of fluidity absorbed when bodies change their form, and which is called "latent heat" only because hidden from the thermometer; nor does it tell that there is more heat in a gallon of water than in a pint; and if an observer did not make allowance for the increasing rate of expansion with increasing temperature, in the substance used as a thermometer, he would believe the increase of heat to be greater than it is; and lastly, when a fluid is used as a thermometer, the expansion observed is only the excess of the expansion in a fluid over that in the containing solid, and subject to the irregularities of expansion in both substances;—all proving that the indications of the thermometer, unless interpreted by other circumstances and our knowledge of the general laws of heat, no more disclose the true relations of heat to bodies, than the money accidentally in a man's pocket tells his rank and riches.

"*Heat, by its different relations to different substances, has a powerful influence on their chemical combinations.*" (See Anaylsis, page 266.)

By observations made and recorded through past ages, man has now come to know that the substances constituting the world around him although appearing to differ in their nature, almost to infinity, are yet all made up of a few simple elements variously combined; and he has discovered that the peculiar relations of these elements to heat, particularly their being unequally expanded by it, and their undergoing fusion and vaporization at different temperatures, furnish him with ready means of separating, combining, and new-modifying them to serve to him most useful purposes. Where the primitive savage, looking around on rocks and soils, saw in their diversified aspect almost as little meaning as did the inferior animals which participated with him the shelter of the wood or cave, his descendant, with penetration sharpened by science, descries at once the treasures of the mine, and aided by heat, whose wonderful energies he has learned to control, pursues through all the Protean disguises of ores, and salts, and solutions, each of the wished-for substance, until he secures it apart. For instance, in what to his forefathers for thousands of years appeared but a red dross, he knows that there lies concealed the precious iron—king of metals! and soon forcing this in his ardent furnace to assume the metallic form, he afterwards, with implements made of it, moulds all other bodies to his will: the trees from the forest, and the rocks from the quarry, in obedience to these, are fashioned by him as if they were of soft clay, and at his command rise into magnificent structures of palaces and ships, with which the earth and the ocean are now so thickly covered.—The minute detail of the relations to heat of particular substances forms a great portion of the department of science called *chemistry* (a name taken from an Arabic word signifying *fire;*) but a general review of the subject belongs to this work.

The most common ores of metals are combinations of the metals with oxygen, carbonic acid, or sulphur, substances all of which are volatilized at much lower temperatures than the metals. Now simply roasting, as it is called, or strongly heating the ores, suffices often to drive away great part of these adjuncts; and where additional assistance is required, it is obtained by mixing with the ore something which when heated attracts the substance to be expelled more strongly than the metal does. Charcoal, for instance, heated with oxide-ore, takes the oxygen, and flying off with it as carbonic acid, leaves at the bottom of the furnace or crucible the vivified or pure metal.

Mercury mixed with the dross of a mine, dissolves any particles of gold or of silver existing in the dross, and the ingredients of the solution may afterwards be obtained separate by mere heating—the mercury passing away as vapour to where it is cooled and again condensed for subsequent use, and the more fixed gold or silver remaining in the vessel—and just as in all other distillations, like that of spirit from wine, or of essential oils from water, &c., there is the separation by heat of a more volatile from a less volatile substance. The only difference between what is called drying by heat and distilling, is that in the one case the substance vaporized, being of no use, is allowed to escape or become dissipated in the atmosphere; while in the other, being the precious part, it is caught and condensed into the liquid form. The vapour from drying linen, if caught would be distilled water. The abundant vapour from wheaten bread while being baken, is chiefly a spirit like what is obtained from malt, and by an ingenious apparatus lately contrived, may be caught and preserved.

A piece of cold charcoal lies in the air for any length of time without change: but if heated to a certain degree, the mutual cohesion of its particles is so weakened, in other words, the particles are so repelled and separated from each other, that their attraction for the oxygen in the air around is allowed to operate, and they combine with that oxygen, so as to produce the phenomenon of combustion. The same is true, under similar circumstances, of almost any dry vegetable or animal substances, and of several of the metals.

Nitre, sulphur and charcoal, while cold, may be mixed together most intimately without any change taking place; but if the mixture or any part of it, be heated to a certain degree, the whole explodes with extreme violence, for it is gunpowder. By the change of temperature, and the consequently altered relative attractions of the different substances, a new chemical arrangement of them then takes place with the intense combustion and expansion, which constitute the explosion.

Sea sand and soda very intimately mixed, and even ground together, if remaining cold, remain also merely an opaque and useless powder; but if the mixture be heated, to diminish the cohesion of the particles of each substance to those of its own kind, so that the mutual attraction of the two substances may come into play, the two substances melt together, and unite chemically into the beautiful compound called glass; a product than which art has formed none more admirable—which, in domestic use alone, is fashioned into the brilliant chandelier and lustre, into the sparkling furniture of the side-board, into the magnificent mirror plate, and when extended across the window opening, admits the light while it repels the storm.

Perhaps the influence of temperature on chemical union is nowhere more remarkably exhibited, than in retarding or hastening the decomposition of dead vegetable and animal substances. The functions of life bring into combination to form the various textures of these organic or living bodies, chiefly four substances, *viz.*, *carbon*, or coal; the ingredients of water, or *oxygen* and *hydrogen*; and lastly, *nitrogen*—which substances, when in the proportions found in such bodies, have but slight attractions for each other, and all of which, except the *carbon*, usually exist as airs. Their connection, therefore, is easily subverted; and particularly by a slight change of temperature, which either so weakens their mutual hold as to allow new arrangements to be formed, or altogether disengages the more volatile of them.—At a certain temperature, a solution of sugar (which consists of the three substances first mentioned, carbon, oxygen, and hydrogen) undergoes a change into a spirituous wash, from which spirit or alcohol may then be obtained by distillation: but if the heat be continued under certain circumstances the liquid undergoes a second change, or new arrangement of constituent particles, and becomes vinegar: under still other circumstances, it undergoes a third change, which is a destructive decomposition, or rotting, as we call it, and the oxygen and hydrogen ascends away as airs. But sugar and many similar vegetable compounds, preserved at a low temperature, remain unchanged for ages.

Again, as regards dead animal substances, we find that although at a certain, not very elevated temperature, they undergo that change in the relations of their elements which we call putrefaction, during which nearly their whole substances rise again to form part of the atmosphere, still at or below the temperature of freezing water, they remain unaltered for any length of time.

In the middle of summer, recently caught salmon, or other fish, packed in boxes with ice, may be conveyed fresh from the most remote parts of Britain to the capital. In our warmest weather, any meat or game may be long preserved in an ice-house. In Russia, Canada, and other northern countries, on the setting in of the hard frosts, when food for the cattle and poultry is with difficulty procured, the inhabitants kill their winter supply, and store up their provender of frozen flesh or fowl, as in other countries men store that which is salted or pickled.—The most striking illustration which we can adduce of this kind is the fact that on the shore of Siberia, in 1801, in a vast block or island of ice, then accidentally broken and partially melted, the carcass of what has been called the antediluvian elephant was found, perfectly preserved—an elephant differing materially from those now existing on earth, and having a skelton exactly similar to the fossil specimens found deep buried in various countries. The carcass was soon discovered by the hungry bears of the district, which were seen eagerly feeding on its flesh, as if it had died but yesterday, although it must have been of an era long anterior to that of any existing monument on earth, of human art, or even of human being. After it had fallen from the ice to the sandy beach, and its tusks had been carried away for sale by a Tunguisian Fisherman, and much of its flesh had been devoured, a naturalist from St. Petersburg who visited it foumd an ear still perfect, and its long mane, and a part of its upper lip, and an eye with the pupil, which had opened on the glories of a former or younger world! About 30 lbs. weight of its hair, which had been trodden into the sand by the bears while eating the carcass, and part of the skin, were preserved, and are now distributed in different museums of natural curiosities. A piece of the skin with the hair upon it is to be seen in the museum of the London College of Surgeons.

"*Heat has powerful influence also on animated nature, both vegetable and animal.*" (Read the Analysis, page 256.)

As the detail of the relations of heat to particular inanimate substances belongs to the province of chemistry, so does the detail of its relations to particular living vegetables and animals belong to the department of Physiology; but a general review of the subject is required in a treatise on Natural Philosophy.

The influence which heat exerts on inanimate nature, is, by the common mind, more immediately and completely perceived than its influence on beings which have life. Thus, to all it is obvious, that the contrast between a winter and summer landscape, is owing chiefly to the effect of heat on the water of the landscape;—that during the absence of heat, there is the dry barren deformity of accumulated ice and snow, covering every thing, the roads impassable, the rivers bound up, perhaps hidden, the air deprived of moisture, and loaded often with powdery drift:—but that on heat returning, the gliding streams again appear, the cascades pour, the rills murmur, the canals once more offer their bosom to the boats of commerce, the lake and pool again show their level face, reflecting the glories of the heavens, and the genial shower falls upon the bosom of the softened earth, become ready to receive the spade or the ploughshare. But this change is not at all greater than what happens to a winter tree acted upon by the warmth of spring.—Again, it may be said with truth, that heat applied to the cold boiler of a steam-engine, is the cause of all its succeeding motions; of the heaving of its beam and pumps, the opening and shutting of its valves, the turning of its wheels, and its ultimate performance of any work, as of spinning, or weaving, or grinding, or propeling vehicles by land and water; but as truly may it be said, that heat coming to a seed which has lain cold for ages, is the cause of its immediate germination and growth; or coming to a lately frozen tree is the cause of the rising of its sap, the new budding and unfolding of its leaves and blossoms, the ripening of its fruits. And what is true of one seed or tree, is true of the whole of the vegetable creation. When the warm gales of spring have once breathed on the earth, it soon becomes covered, in field and in forest, with its thick garb of green, and soon opening flowers or blossoms are every where breathing back again a fragrance to heaven,—among these the heliotrope is seen always turning its beautiful disc to the sun, and many delicate flowers which open their leaves only to catch the direct solar ray, closing them often even when a cloud intervenes, and certainly when the chills of night approach. On the sunny side of a hill, or in the sheltered crevice or a rock, or on a garden wall, with warm exposure, there may be produced grapes, peaches, and other delicious fruits, which will not grow in situations of an opposite character—all acknowledging heat as the immediate cause, or indispensable condition, of vegetable life.

And among animals, too, the effects of heat are equally remarkable. The dread silence of winter, for instance, is succeeded in spring by one general cry of joy. Aloft in the air the lark is every where caroling; and in the shrubberies and woods, a thousand little throats are similarly pouring forth their songs of gladness—during the day, the thrush and blackbird are heard above the rest, and in the evening the sweet nightingale: for all birds it is the season of love and of exquisite enjoyment. And it is equally so for animals of other kinds: in favoured England, for instance, in April and May, the whole face of the country resounds with lowings, and bleatings, and barkings of joy. And even man, the master of the whole, whose mind embraces all times and places, is far from being insensible to the change of season. His far-seeing reason of course draws delight from the anticipation of autumn, with its fruits; and his benevolence rejoices in the happiness observed among all inferior creatures; but independently of these considerations, on his own frame the returning warmth exerts a direct influence. In his early life, when the natural sensibilities are yet fresh and unaltered by the habits of artificial society, spring to man is always a season of delight. The eyes brighten, the whole countenance is animated, and the heart feels as if new life were come, and has longings for fresh objects of endearment. Of those who have passed their early years in the country, there are few who, in their morning walks in spring, have not experienced, without very definite cause, a kind of tumultuous joy of which the natural expression would have been, how good the God of nature is to us. Spring, thus, is a time when sleeping sensibility is roused to feel that there lies in nature more than the grosser sense perceives. The heart is then thrilled with sudden ecstacy, and wakes to aspirations of sweet acknowledgment.

Besides the effects of heat now mentioned, and which are comparatively transient as being connected with the seasons, there are other effects on animated nature of a more prominent character. Certain species of vegetables and animals, by their relations to heat, are confined to certain latitudes or climates; as the orange tree and bird of paradise, to warm regions; the fir tree, and arctic bear, to those that are colder:—and when individuals of either class can support diversity of climate, they acquire a certain character according to the climate—as seen in the sheep and dogs of the various regions of the earth. In this latter respect there is no instance more interesting than that furnished by the varieties of the human race. If we assume that the whole sprung from one stock, what a contrast is there between the native of equatorial Africa, of temperate Europe, and of the Polar zone: between the Negro, the Greek, and the Esquimaux: or, again, between the dark slender children of Hindostan, the strongly-knit fairer Roman or Spaniard, and the taller, ruddy, powerful Briton. And in the female sex of the last-named countries, we may remark the gentleness and singular devotedness of the Indian woman, the more commanding dark eye and gesture of the graceful nymph of Italy or Spain, and the happily attempered mixture of qualities in the fair and much-favoured daughters of Britain.

The very important influence of heat upon the temporary bodily state of animals, becomes an object of much study to the physician. It explains among many other facts, the connection of temperature with the rise of fevers and other pestilences, the powerful remedial efficacy of hot and cold bathing, of changes of climates, of regulating the temperature of air breathed by invalids, the protection from clothes, houses, &c.

"*The great natural source of heat is the sun.*" (See Analysis, page 256.)

To admit this, it is only necessary to think of the comparative temperatures of night and day, of seasons and of climates, and to reflect that the sun is the sole cause of the differences. We need not wonder, then, that to many savage nations, seeking the course of their life and happiness, the sun has been the object, not only of admiration, but of worship.

The heat comes from the sun with all his light. If a sunbeam enter by a small opening an apartment otherwise closed and dark, it illuminates intensely the spot or object on which it first falls, and its light being then scattered around, all the objects in the room become feebly visible. Again, a cold thermometer, held to receive the direct ray, rises much, while, in any other situation, it is less affected; proving the heat to be, like the light, at first concentrated, and then widely diffused, losing proportionally of its intensity. Light passes from the sun to the earth in about eight minutes of time, as will be fully explained in a future chapter; and there is every reason to conclude that heat travels at the same rate.

Human art can gather the sunbeams together, and the intense heat existing in the focus of their meeting, is another proof that the sun is the great source of heat. A pane of glass in a window, or a small mirror will reflect the sun's rays so as strongly to affect the eye at a distance of miles—and the heat accompanies the ray, for by many such mirrors directed towards one point, a combustible object placed there may be inflamed Archimedes set fire to the Roman ships by sunbeams, returning from many points to one, his God-like genius thus rivaling by natural means, the supposed feats of fabled Jupiter with his thunderbolts. Again, when the light of a broad sunbeam is made by a convex glass or lens to converge to one point or focus, the concentrated heat is also there—for a piece of metal held in the focus will drop like melting wax: and if the glass be purposely advanced, its focus will similarly advance, and will pierce through the most obdurate substances, as red-hot wire pierces through paper or wood. A hunter on the hill, and travelling hordes on the plains, often conveniently light their fires at the sun himself, by directing his energies through a burning glass.

The direct ray of the sun, simply received into a box which is covered with glass to exclude the cold air, and is lined with charcoal or burned cork to absorb heat, and to prevent the escape of heat once received, will raise a thermometer in the box to the temperature of 230° of Fahrenheit, a temperature considerably above that of boiling water. And the experiment succeeds in any part of the earth where there is a clear atmosphere, and where the sun attains considerable apparent altitude. We see, therefore, that a solar oven might in some cases be used. In operating with the apparatus suggested by the author, and described at page 295, for distilling water by the heat of the sun, the vessel intended to absorb the heat, and to act as the still, should be enclosed in a case covered and lined as above described.

Reflecting on such facts as now recorded, and on the globular form and the motions of our earth, we have a reason and the measure of the differences of climate and of season found upon the earth. It is evident that the part of the globe turned directly to the sun, receives his rays as abundantly as if it were a perfect plane, directly facing him, while on parts, which, as viewed from the sun, would be called the sides of the globe, with the increasing obliquity of aspect, an equal breadth or quantity of rays is spread over a larger and a larger surface; and at the very edge the light passes level with the surface, and altogether without touching. The sunny side of many a steep hill in

England receives the sun's rays in summer as perpendicularly as the plains about the equator; and such hill-sides are not heated like these plains, only because the air over them is colder—just as mountain tops, even at the equator, owing to the rarified, and, therefore, cold air around them, remain for ever hooded in snow. In England, at the time of equinoxes, a level plain receives only about half as much of the sun's light and heat as an equal extent of level surface near the equator: and in the short days of winter it receives considerably less than a third of its summer allowance.

There are few contrasts in nature more striking than between the consequences of different intensity of the sun's influence: for instance, the inhabitants of India, at midday, with the thermometer at 120°, are running to the shade of their bangalows, darkening their windows, hanging wetted mats upon the walls and roofs, sprinkling water on the floors, fanning themselves with ever moving punkas, and feeling the slightest covering or exertion too much—while, on the other hand, the dwellers in Greenland, with the thermometer below zero, are loaded with furs, and are seeking the direct sunshine or heat from a fire, as their life and comfort. Again, there is the contrast observed on passing, as the author once did, very rapidly, from such a paradise as Rio de Janeiro, with all its vegetable riches, to Tristan de Cunha, and the Isle of Desolation in the Southern Ocean, which exhibit only cold and naked rocks; but yet, where the scene swarms with its appropriate inhabitants—the sea with seals, and the air with clouds of sea-fowl, playing over the never-resting waves like flakes of eddying snow. Were a person for a moment to doubt whether the sun be the real cause of such differences, and of the fact that certain creatures are found only in certain zones of the earth, let him reflect on the extraordinary migration of animals, which have their home, not in any fixed region, but wherever the sun has for a time particular degree of influence, and which acccordingly follow the sun in the changes of season. We have the swallow in vast numbers coming to visit the British isles in the spring, to play over our woods and waters, in pursuit of the insects which the heat breeds to fill the air,—welcome harbingers of the coming summer and its riches; and in autumn, the same creatures are seen congregating on our shores, to wing their flight back in united multitudes to more southern countries, where, in turn, there is a temperate influence of the sun. The same reason brings to England the nightingale, and makes our woodlands resound with the notes of the cuckoo. In the waters of our bays and coasts, again, there appear with the seasons the vast shoals of fish, as of the herring and mackeral, which prove such abundant food for millions of human beings; and we have salmon, at stated times, penetrating from the ocean far up the mountain streams, to deposit its spawn for further supply; all by their movements contributing to the harmonious and beneficent system of the universe.

With respect to the sun as a source of heat, there have been two opinions among philosophers, one class believing that the sun is an intensely heated mass, which radiates its heat and light around, like a mass of intensely heated iron; and another class holding that heat is merely an affection or state of an ethereal fluid, which occupies all space, as sound is an affection or motion of air, and that the sun may produce the phenomena of light and heat without waste of its temperature or substance, as a bell may produce the phenomenon of sound; holding, farther, that the sun, below its luminous atmosphere may be habitable, even by such animals as live on this earth. Those who take the first view, are awakened to the dread contemplation of a universe carrying in itself the seeds of its own decay, or at least of great periodical revolutions: the others may view the universe as destined to last nearly unchanged, until a new act or will of its Creator shall again alter or destroy it.

Of one fact there can be no doubt, namely, that the present temperature of the surface of the earth is much lower than the temperature in remote past time. The rocks called primitive, as granite and gneiss, constituting the interiors of our great mountain masses and the substrata of our plains, bear evident marks of there having been at one period a molten state, from which they have been solidified by a very gradual cooling: and even the whole mass of the earth at some time must have been so fluid or soft, as, in obedience to gravity, to have assumed its rounded form, and in obedience to the centrifugal force of its whirling, to have bulged out, at its great circumference or equator, the thirty-four miles which its equatorial diameter exceeds the polar; the same, by the by, in degrees corresponding to the various speed of rotation, being true of all the other planets belonging to the solar system. Again, while in excavating below the surface of the globe, or in examining its structure as exposed to view by volcanic or other convulsions, men encounter, in very many situations, a thickness of more than a mile, of the wreck and remains of former states of the world—as, on digging eighty feet under vineyards near Mount Vesuvius, they encounter the more recently buried cities of Herculaneum and Pompeii—they farther discover that the animal and vegetable remains buried, without number, in the present cold climates of the earth, and evidently near where the creatures lived, are all of kinds now inhabiting only the warmer or tropical regions. Lastly, in the operations of mining, the deeper men go, the higher they find the temperature to be, at the rate of a degree for about 200 feet of descent; which fact as heat tends to equable diffusion, proves both that the central heat of our earth must have had another source than a radiation from the sun of the present intensity; and that the surface of the earth is now radiating away more heat than it receives from the sun. The conclusion then follows, that the temperature of the world is still falling although perhaps so slowly that a change may not be detected even within centuries. Possibly, in very remote antiquity, that may have been true which the early Greeks erroneously thought true in their day, *viz.*, that the equator of the earth, by reason of its great heat, the sun's influence there being joined to the heat from within, was a barrier, impassable by man, between the northern and southern hemispheres.

"*Electricity a source of heat.*" (See the Analysis.)

This subject can only be satisfactorily entered upon in the chapter devoted exclusively to electricity, and is, therefore deferred. Suffice it here to say, that while an electrical discharge of current passes from one situation to another, the substance serving as a conductor is often heated, melted, or dissipated, in such a manner as to make it doubtful whether human art has any more powerful means of producing these effects. We may remark, too, that in certain cases of the electrical current, the heat is accompanied by as intense a light as art can exhibit.

"*Combustion and other chemical actions as sources of heat.*" (See Analysis, page 256.)

Of the phenomena of nature there is perhaps none which, to the uninstructed, appears so inexplicable and so wonderful as that of *fire* or *combus-*

tion—whether contemplated in its beauty or in its terrors. Fire is seen in its beauty when used by man for his domestic purposes, as when it blazes cheerfully over his parlour hearth, or beams a steady light around from his lamps and chandeliers. It is seen again in its terrors, when spreading by accident from some focus, it envelopes in sudden flame and quickly consumes the surrounding objects, perhaps the draperies and other furniture of a single apartment; or wider spread, the valuable contents of a spacious mansion; or still wider spread, and deafening uproar, a whole town or a forest:—and it is fire which, labouring within the bowels of the earth, first prepares and then urges up to heaven the volcanic eruptions of flames and red-hot rocks, during which the region around often quakes and is uptorn, so that the cities are demolished into the sudden tombs of the inhabitants, the course of the rivers is changed, the plains are converted into lakes, or the lakes'-beds into dry land. Fire is awfully seen also in some meteors, and when, intentionally lighted by human hands, it bursts from the cannon to produce the carnage of battle. Fire among many nations of antiquity was regarded with awe and holy reverence, the sun himself being honoured chiefly as its concentration or supposed abode. There were sacred fires in many of the temples, and fire was used to complete the splendour of the most august ceremonies. Nay, even Moses, a worshipper of the one true God, has given records of the *Burning Bush* and of burnt-offerings made to that God: and at the present day, in many Christian churches there are ever-burning lamps and frequent magnificent illuminations. Now this principle of *fire*, which, when the savage man first saw it spreading after the thunder-clap or the rubbing of forest branches in a storm, so as to threaten universal destruction, he so natuarally accounted the demon, if not the God of nature; this principle man's art has now tamed to be a most obedient and by far the most useful of his servants. *Fire* being in truth but a concentration of the element of *heat*, which in its tranquil and invisible diffusion we have already contemplated as the beneficent life or soul of the universe—the cause of seasons and climates and of all the changes or activity which distinguish a living world from a dead and frozen mass; man, by acquiring command over it, commands heat when and where he wills, and thus truly becomes in a second degree the ruler of nature. Fire in man's service may be figured as a legion of serving spirits to whom no labour is difficult, who, in any particular case have power or magnitude exactly proportioned to the quantity of food or fuel afforded; of whom, moreover, man can, at any moment, conjure up one or many by the magic stroke of his flint and steel. In every private dwelling he has of these fiery spirits as domestic slaves—in the kitchen and in the parlour. In his manufactories they are melting glass for him, and reducing ores, and boiling and evaporating for a hundred purposes. But it is chiefly when chained to the steam-engine, that they show their prodigious powers:—as when, putting forth a giant's strength, they heave a river from the bottom of a mine, or urge up a vast ship through the winter storm; or when equaling, if not surpassing, in nice dexterity, the work of human hands, they twist the silken or cotton threads, and weave them into most delicate fabrics. Men, grown familiar with such prodigies, have almost ceased to be moved by them; but even now few persons can resist a feeling of wonder and admiration when chemistry is calling forth the hidden spirit of combustion in some new and less familiar guise:—as, for instance, when a piece of iron wire, lighted as a taper in oxygen gas, burns with such resplendent brilliancy; or when phosphorus similarly placed, throws around its overpowering flood of flame; or when small portions of the metal called potassium, cast upon the surface of water, become as beads of most intense light running about there, and crossing as in a merry dance;—or, lastly, when flames produced from particular substances are seen rising deep-tinged with most vived and beautiful colours.

Singularly interesting, then, to philosophers, as in such particulars the phenomena of combustion must always have appeared, one may wonder that its true nature could remain to them so long a mystery; but until the admirable researches of Davy, made only a few years ago, their conjectures had scarcely approached the truth. An opinion long prevailed, that in every combustible substance there was present a certain quantity of a something denominated *phlogiston*, which, on being disengaged or separated, became obvious to human sense as light and heat. The white oxide of zinc, for instance, named the flowers of zinc, and into which the metal is changed by burning, was supposed to be the metal deprived of its phlogiston; and when on this oxide being heated with charcoal, the metal again appeared, it was supposed simply to have recovered phlogiston from the charcoal. The illustrious Lavoisier had the merit of most clearly disproving this hypothesis, by showing that the flowers or powders obtained from a metal by burning it, were heavier than the piece of metal from which they were produced, by the exact weight of the oxygen gas, which disappeared in the combustion, &c.; and he showed farther, that in this and many other cases, combustion was merely the act of two substances combining chemically; but he fell into an error almost as great as that which he overthrew, by supposing that for combustion, oxygen had always to be one of the combining substances, and that the heat and light given out in every case had been previously latent in the oxygen.

When Sir Humphrey Davy began his labours on the subject, than which labours there is not, perhaps on record a more perfect specimen of scientific research, it was already known, first,—that bodies when compressed or by any means reduced in bulk, generally give out a part of their heat, as—when air condensed in the match-syringe lights tinder,—or when water and sulphuric acid, uniting into a compound of smaller volume than the separate ingredients, becoming very hot,—or when water, poured upon quick-lime to slake it, and becoming solid with it, produces heat sufficient to inflame wood, as has been fatally proved by the burning of many lime-loaded ships:—and that in such cases, the heat produced during the chemical union depended more upon the energy of the action which united the substances than upon the change of volume produced.

Farther, it was known that any substance having its temperature raised, by whatever means, to 800° or more of Fahrenheit's thermometer, became incandescent or luminous—as when iron, or stone, or any substance not dissipated by heat, is placed in a common fire; in the first degree the substance being said to be red-hot, and at higher temperatures to be white-hot.

Out of these two truths Davy constructed his explanation. He asserted—that in any case, combustion is merely the appearance produced when substances, having a still stronger attraction for each other than quick-lime and water, are, with intense energy, combining chemically, so as to become heated at least to the degree of incandescence, and that during the phenomenon there is not, as was formerly supposed, something altogether consumed or destroyed, or something called *phlogiston* escaping, but that the substances concerned are only assuming a new form or arrangement. Thus, if a piece of charcoal be enclosed in a glass vessel filled with air, of which vessel the mouth dips into a liquid to confine the air, and if the charcoal be then heated to a certain degree, by means of a burning-glass or otherwise, the cohesion of its particles yields to their attraction for the oxygen of the air around them, and they immediately begin to combine with the oxygen so energetically as to produce a heat still much greater, accompanied by the light or incandescence of combustion. The charcoal, under these circumstances, soon entirely disappears, or is dissolved in the air, as sugar may be dissolved in water; but if the air be afterwards weighed, it is found to have gained in weight the exact weight of the charcoal which has disappeared; and a chemist can again separate the charcoal from the air, and use either for any purpose as before. In like manner, if a piece of iron wire be heated at one end, which is then plunged into a jar of oxygen gas, it will burn as a most brilliant taper, and will gradually fall in the form of oxidized drops or scales of iron, to the bottom of the vessel. Now during this process, the quantity of oxygen will be diminished, but if the scales mentioned be collected, they will be found to weigh just as much more than the original wire expended, as there is of oxygen lost or combined with them. A chemist can separate this iron and oxygen, and exhibit them apart as before, without loss. Again, if iron and sulphur, in certain proportions, be heated together, they unite with vivid combustion, but the product weighs exactly as much as the original ingredients.

While every instance of combustion is thus only a case of chemical union going on with such intensity of action as to produce incandescence, still, according to the nature of the substances combining, the appearance varies much. It may be, for instance, with flame or without flame. The great combining substance in nature, that is to say, the most universally distributed, is oxygen, of which the name is now become familiar even to the ears of the unlearned. It forms four-fifths of the substance of water and one-fifth of our atmosphere, being on the latter account present wherever man can be, and ready to unite itself with any matter exposed to it at the necessary temperature. Now, of substances burning in air, those which are originally aeriform, as coal-gas, or which on being heated, are rendered aeriform, or vapourized before the union takes place as oil or wax, assume the appearance of flame—which means that the aeriform particles usually invisible are raised to the incandescent temperature; but when the substance combining with the oxygen remains solid, while its particles are gradually lifted away by the oxygen acting only at the surface of the mass, it appears during the whole time, only as a red-hot stone. The latter is the case of charcoal, coke, Welch stone-coal, &c., while in the case of wood, common coal, &c., a greater or less portion of the inflammable matter is by the heat of the combustion converted into vapour, and produces the beautiful appearance of flame.

Of the substances called combustible, and thus called because they combine with oxygen so energetically as to become incandescent, there are only a few, as the metals called potassium, sodium, &c., which will begin to unite with oxygen, or to burn at the common temperature of our globe, the others requiring to be at some higher temperature. Thus phosphorous begins to burn at 150°, sulphur at 550°, charcoal at 750°, hydrogen at 800°, &c., it appearing that up to these temperatures the attraction of the atoms of the substances among themselves is sufficient to resist the other attraction or that of oxygen. But when the combustion once begins, the temperature, from the effect of the combustion itself, rises instantly much beyond the degree necessary for the commencement of the process. Oxygen and hydrogen, which begin to burn or combine at 800°, produce a flame of as intense heat as human art can excite.

On the circumstance that bodies require to have certain preparatory temperature before beginning thus to combine with oxygen, depend many important facts of nature and art. Hence the safety with which most combustibles may be exposed at ordinary temperatures to the contact of atmospheric air: otherwise, coal, wood, &c., in the moment of being exposed to the air, would catch fire, as really happens to phosphorated hydrogen gas; or to the metal called potassium, even when thrown into cold water, the metal attracting the oxygen from the water instantly, and with intense combustion. If a fire or flame be so small that, from the rapid absorption or heat by bodies around, it does not produce heat enough to maintain the inflaming temperature of the substance, the combustion will soon be extinguished. Thus a common coal fire, if not watched, and the remaining fuel occasionally gathered together to reduce the surface of wasteful radiation, will be extinguished long before the whole fuel is expended:—but not so with a fire of wood or of paper, which substances burn more readily than coal. The Welch stone-coal can only be made to burn in very large masses, or when mixed with a more inflammable coal or other fuel, or when fed by air already heated. Some of our manufactures have lately been improved by causing the air which feeds the working fire to pass through metal tubes heated by lying in another fire. In common fires much coal is burned at temperatures so low as to be nearly useless. A substance placed in pure oxygen gas burns with much greater intensity, and will begin burning at a lower temperature than if placed in atmospheric air, which contains only one-fifth of oxygen and four-fifths of another substance, nitrogen, which does not aid the combustion—because, in the latter case, the nitrogen by absorbing much of the heat of the combustion, lowers the temperature. Iron wire will burn as a taper in oxygen, but not in common air; and a common taper or flaming piece of wood just extinguished by blowing on it, will immediately be rekindled if placed in oxygen. Again, a lamp with a very small wick, as of one thread, and producing, therefore, very little heat, will not burn in cold weather, or in an ice-house, and at any time will be extinguished by a foreign body brought near it so as to cool it—a piece of ice, for instance, or a small metallic nob, presented to it on the end of a wire, or a metallic ring let down over it; but if the ball or ring be hot, the effect will not follow. By more powerful refrigerating processes, even a considerable lamp or candle may be put out. These discoveries led Davy to the construction of his miner's safety-lamp, which is merely a lamp surrounded by a wire gauze, of which the meshes are of such size that a flame of hydrogen gas attempting to pass through is so cooled by the heated absorbing and heat-conducting power of the metal as to be extinguished. A wire gauze gradually let down upon any common flame, annihilates the part of it which should appear above the gauze; but the combustible vapour passing invisibly through the gauze, may be lighted afresh on its upper side. Oxygen and hydrogen, which are the constituents of water, when uniting, produce such intense heat that the momentary expansion of the newly formed water, then in the state of steam, is such as to constitute a violent explosion; and when a jet of the two gases mixed gives a continued flame, the most refractory substances melt in it like wax in a common taper—yet these gases may be kept mixed together in the cold reservoir of a blow pipe without combining, and when they are set on fire, while issuing as a jet, the flame does not travel inwards through the opening, as might be feared, because it is cooled by the metal of the orifice. A mixture of oxygen and hydrogen passing through the small channels left in a tube filled with fire, may be lighted at the end of the tube without danger of explosion.

While solid bodies become very visible or incandescent at about 100° of Fahrenheit, airs, owing to their tenuity of condition, require to be heated much farther before they take on the vivid appearance of flame; and airs of light atoms, like hydrogen, require to be heated still more than heavier airs. Thus the flame of pure hydrogen is pale and blue, but a wire held in it becomes much more luminous than the flame itself; and the flame of mixed oxygen and hydrogen escaping from a very minute orifice in a glass tube, may itself be scarcely visible, while the extremity of the tube heated by it becomes like a brilliant star. Hence the light of many flames may be increased by placing a wire gauze or other solid body in the flame; as is seen when a piece of lime is placed in a flame of mixed oxygen and hydrogen. Consideration of this subject enables us to explain why common coal gas, which consists of hydrogen holding a quantity of carbon in solution, gives a stronger light than pure hydrogen; and why oil gas, which contains about twice as much carbon as the coal gas, gives also about twice as much light; —or it appears that the atmospheric air, which first mixes with these gases as they issue to burn, is sufficient to combine with all their hydrogen (which it most strongly attracts) but not at the same time with all their carbon: the particles of the carbon, therefore, are first separated or precipitated in the flame, and become so many solid particles most intensely heated and luminous; and afterwards, when they have ascended a little higher, they meet with new oxygen and burn in their turn, giving a second quantity of light. That this decomposition of the gas really occurs is proved by placing a wire gauze in the flame, for then we find that if it be held near the middle of the flame, it becomes immediately loaded with particles of charcoal separated there, and cooled by it so as to cohere; while if held at the bottom of the flame where the carbon is not yet separated, it retains none, and if held at the top of the flame, where they are already burned, it retains none. A candle or lamp is said to smoke when the heat produced by it, or the quantity of oxygen allowed to approach the flame, is not sufficient to effect the total combustion of the carbon which rises in the flame. The hollow or circular wick of the common Argand lamp, and the similar form given to gas flames is useful, by admitting air to the inside as well as to the outside of the flame, and the lofty glass chimney is to quicken the currents of air.

When oxygen mixed with certain of the inflammable gases or vapours is heated, although only to a temperature considerably below that of common burning or explosion, a union still takes place, but very slowly, so that the temperature never rises to what is necessary to exhibit flame. This phenomenon has been called invisible combustion. It is remarkably exemplified on plunging plantinum or gold wire moderately heated into such a mixture: the combination then goes on in the immediate vicinity of the hot wire; and although without flame, still with sufficient disengagement of heat to maintain the wire in an incandescent or luminous state, so long as there are gases left to combine. Thus the vapour always arising at a common temperature from the mouth of a phial of ether, (ether consists chiefly of hydrogen and carbon), if made to pass through a coil of heated plantinum wire, while combining, by this slow combustion, with the oxygen of the air around it, gives out heat enough to keep the wire so luminous as to serve as a little lamp by which to read from the dial-plate of a watch through the night. A beautiful modification of this principle has been adopted in the miner's safety-lamp; and when the air of the mine is too impure to maintain the flame, it still suffices thus to produce a continued light from the incandescent metal.

"*Fuel.*"

Heat being, in the sense already explained, the life of the universe, and man having command over nature chiefly by his power of controlling heat, which power again comes to him with the ability to produce combustion, it is of great interest to inquire what substances he can most easily procure as food for combustion, or *fuel*, and how these may be most advantageously employed. To speak on this subject at all fully in reference to the various arts of life would be to compose an extensive work, but an interesting sketch may be comprised within narrow limits.

Although there are a great number of substances, which, in the act of their chemical union, occasion the heat and light which constitute the combustion, still by far the greater part of these, in an uncombined state, are so sparingly distributed in nature, and are therefore procurable with such difficulty, that heat obtained by sacrificing them would be much too expensive to be within common means. Providence, however, has willed that the elementary substance in nature which has the most energetic attraction for almost all other substances, and which therefore produces in uniting with them the most intense heat, is also of all the most universally distributed. This substance is oxygen. It forms part of our atmosphere, and therefore penetrates, and is present wherever man can breathe, offering itself at once at his service. Then for the purpose of combining with the oxygen, there are chiefly two other substances also very widely scattered, and therefore easily procurable and cheap. These are carbon and hydrogen, the great materials of all vegetable bodies, and therefore of our forest trees, and of coal beds, many of which are evidently the remains of antediluvian forests. Carbon is found nearly alone in hard coal, but it is united with a large proportion of hydrogen in caking coal, and in wood, wax, resins, tallow and oils. The gases obtained from these last-mentioned substances and now used for illumination are merely hydrogen, holding in solution certain proportions of carbon; and all bodies which burn with flame give out such gases in the act of combustion. In the mass of the earth, as far as known to man, the stones, earths and water, forming its surface, are already combinations of oxygen with other substances, and are therefore not in a state to produce fresh combustion; but carbon and hydrogen, by various processes of vegetable and animal life, are in numberless situations becoming accumulated, so as to be fit for fuel;—as by other processes the atmosphere is always preserved with its due proportion of oxygen.

The name fuel is given only to the substances which combine with oxygen, and not to the oxygen itself, probably because the former being solid or liquid, and therefore more obvious to sense, had attracted human notice as producers of combustion long before the existence of the aeriform agent, oxygen, was even suspected.

Oils, fat, wax, &c., from becoming aeriform in their combustion, exhibit the appearance of flame, as already explained, and hence these substances are chiefly used for the purpose of giving light; while wood and coal are more frequently used for mere heating. But the chemist's lamp, by which he distills and evaporates, and his common blow-pipe for directing the point of any flame upon a substance to melt it, and his blow-pipe fed with mixed oxygen and hydrogen, whose flame is capable of melting the most refractory substances—prove that it is chiefly the expense of the former kinds of fuel which has limited them so much to the office of light-giving. Lately an important application of coal-gas and of oil or fat as heat-giving fuel has been made in a general cooking apparatus, which promises to effect a considerable diminution of house-keeping expense.

Wood was the common fuel of the early world when coal-mines were not yet known, and still in many countries it is so abundant as tó be the cheapest fuel. Charcoal is the name given to what remains of wood after it has been heated in a close place; during which operation the hydrogen and other minor ingredients are driven away in the form of vapour. Charcoal is nearly pure carbon. Coke, again, is the charcoal obtained by a similar process from coal. The wood and coal, if much heated in the air, would burn or combine with the oxygen of the air; but heated in a vessel or place which excludes air, they merely give out their more volatile parts.

Good coal, where it abounds, is now for ordinary purposes by much the cheapest kind of fuel; and since within a few years men have learned to separate from it, and to use, instead of oil and wax, its illuminating gas, namely, its hydrogen holding in solution a little carbon, it has become doubly precious to them. A person reflecting that heat is the magic power which vivifies nature, and that coal is what best gives heat for the endless purposes of human society, cannot without admiration think of the rich stores of coal which exists treasured up in the bowels of the earth for man's use. And Britain, in this respect, is singularly favored. Her extensive coal mines are in effect mines of latent labour or power vastly more precious than the mines of gold and silver in Peru,—for a hundred men in England, of whom a part dig coal, and the remainder apply it in manufactures, can get in exchange for the merchandize they produce more of either gold or silver than a hundred men working in the American mines are able to extract. These coal-mines, therefore, may be said to produce abundantly every thing which labour and ingenuity can produce, or which our money can buy, and they have essentially contributed to render Britain the mistress of the industry and commerce of the earth. Britain has become to the civilized world around, nearly what an ordinary town is to the rural district in which it stands, and of this vast and glorious city the mines in question are the coal-cellars, stored at the present rate of consumption for above 1,000 years; a supply which, as coming improvements in the arts of life will naturally bring economy of fuel, or substitution of other means to effect similar purposes or changes in the channels of industry,—may be regarded as exhaustless.

The coal in many mines is evidently the remains of forests, which have existed in very remote time, swept together either during convulsions of nature or by the more gradual operation of rivers, (as may be seen at present where the great Mississippi is carrying an uninterrupted stream of floating timber to the sea, and depositing it at the bottom of the Mexican gulf) the accumulated wood being afterwards compressed and solidified by superincumbent deposit sof earthy matters, aided probably by the action of heat. In many coal-beds the trees yet retain their form, so that their species can be easily distinguished, and there are buried among them other vegetable and animal remains of cotemporaneous inhabitants of the earth. Coal is found of different qualities. In some places it is almost unmixed carbon, and exceeding solid, as if it had been coked by subterranean heat. Such is the stone-coal of Wales, which in 100 parts contains 97 of pure carbon, with only three of hydrogen and earthy matter. In other places the coal contains hydrogen in nearly as large proportion as wood does, so combined with part of the carbon as to form the oily or pitchy substances existing in the coal which, when burning, produce flame, and when rising unburned, constitute smoke.

The comparative values, as fuel, of different kinds of carbonaceous matter, have been found to be as in the following tables.

One pound of	Melts of ice
Good coal	90 lbs.
Coke	84
Charcoal of Wood	95
Wood	32
Peat	19

Lavoisier, in making experiments on combustibles generally, to ascertain the quantities of oxygen expended, and of heat given out during the combustion of a given quantity of each, obtained the following results:

One pound of	Melts of ice	Takes of oxygen
Hydogen gas	370 lbs.	7½ lbs.
Carburetted hydrogen	85	4
Olive oil	120	3
Wax	110	3
Tallow	105	3
Charcoal	95	2⅔
Phosphorus	100	1⅓
Sulphur	25	1

The following remarks with respect to fuel, and the modes of using it, seem to demand a place here.

A pound of coke produces nearly as much heat as a pound of coal; but a pound of coke is the produce of a pound and a third of coal, although the coke is more bulky than the coal.

It is wasteful to wet fuel, because the moisture in being evaporated carries off with it as latent, and therefore useless heat, a considerable proportion of what the combustion produces. It is a very common prejudice, that the wetting of coal, by making it last longer, is effecting a great saving; but it does so merely by restraining the combustion or producing a smaller fire, and with the bad fire, there is also much waste of heat. To illustrate the influence of watery vapour upon combustion, we may mention the fact, that a manufacturer who tried to blow his fire by forcing steam into the furnace with the air, extinguished his fire; and the analogous fact, that ordinary fires burn better in winter than in summer, although the temperature be lower, because cold air is generally much drier than that which is warm.

Coal which contains much hydrogen, as all flaming coal does, is used wastefully when any of the hydrogen is allowed to escape unburned; for, first, the great heat which the combustion of such hydrogen would produce is not obtained; and, secondly, the hydrogen while becoming gas, absorbs into the latent state still more heat than an equal weight of water would. Now the smoke of a fire is the hydrogen of the coal rising in combination with a portion of carbon. We see, therefore, that by causing a fire to destroy or burn its smoke, we not only prevent a nuisance, but effect a great saving. The reason that common fires give out so much smoke is, either that the smoke or vaporized pitch is not sufficiently heated to burn, or that the air mixed with it as it ascends in the chimney, has already, by passing through the fire, been deprived of its free oxygen. If the pitch be very much heated, its ingredients assume a new arrangement, becoming transparent, and constituting the common coal-gas of our lamps; but at lower temperatures, it is seen jetting, from cracks or openings in the coal, as a dense smoke—a smoke, however, which immediately becomes a brilliant flame if lighted by a piece of burning paper or the approximation of the combustion. The alternate bursting out and extinction of these burning jets of pitchy vapour, contribute to render a common fire the lively and agreeable object which it is in the winter evenings. When coal is first thrown upon a fire, a great quantity of vapourized pitch escapes as a dense smoke, too cold to burn, and for a time the flame is smothered, or there is none; but as the fresh coal becomes heated, its vapour reproduces the flame as before. In close fire-places, as those of steam-engines, brewing and dyeing apparatus, &c., all the air which enters after the furnace-door is shut, must pass through the grate and the burning fuel, so that its oxygen is consumed by the red-hot coal before ascending to where the smoke is. The smoke, therefore, however hot, passes away unburnt; unless, sometimes, as over foundry furnaces, where the heat is very great indeed; and it burns as a flame or great gas lamp at the chimney-top on reaching the oxygen of the open atmosphere.

There have been many modes proposed of destroying smoke: one has been to admit to the space above the fire, by a suitable opening, a certain quantity of fresh air, the oxygen of which may inflame the smoke. This plan is efficient at a certain point of time after the addition of fresh fuel, and then for a time effects the saving of fuel; but the difficulty of admitting just the quantity of air required to suit the varying demand for it, has not been overcome, and hence, from there being no saving on the whole, the plan has been abandoned. When just enough air entered, the flame produced gave so intense a heat as in several cases to have burned or destroyed the parts of valuable boilers exposed to it; and when, on the contrary, too much air entered, it injuriously cooled the boiler. A contrivance at present commonly adopted for burning smoke, is that of Mr. Brunton, namely, a circular fire-grate, kept turning like a horizontal wheel, on which, by machinery, coal is made to fall in a gradual manner, so as to be uniformly spread over it. The coal falls so gradually, that although there is generally a little smoke from it, there is never much,—the oxygen which finds entrance through and around the grate, being always in quantity the same, and nearly sufficient. A smoke-consuming fire would be perfect, in which the fuel were made to burn only at the upper surface of its mass, so that the pitch and gas disengaged from it, as the heat spread downwards, might have to pass through the burning coals where fresh air were mixing with them; and thus the gas and smoke, being the most inflammable parts, would burn first and be all consumed. This was the principle proposed in a fire-place suggested by the author for the great brewery of Mr. Meux in his neighbourhood, and tried at the time when attempts were extensively made to abate the nuisance of smoke in towns. The experiment proved the theoretical perfection of the method, and that it would produce a saving of 15 or 20 per cent. on the expenditure of coal; but before a durable grate of the kind was completed, the Welch stone-coal was introduced, which has 97 per cent. of pure carbon, and therefore has no pitch to evaporate, and no smoke,—and it was at once adopted there and in many other places. A little common coal is added to it to make it more easily combustible.

Coal in a deep narrow trough, as *a b c d*, if lighted at its surface *a b*, burns with a lofty flame as if it were the wick of a large oil-lamp; for all the gas given out from the coal below, as that is gradually heated, passes through the burning fuel and becomes a flame. Now, if we suppose several such troughs placed together with intervals between them, like the fire bars of a common grate or furnace, there would be a perfect non-smoking fire-place. Such was that made on the occasion mentioned; and although, as a mere experimental apparatus, it was flimsy and imperfect, it put beyond a doubt the possibility of accomplishing its object. The reason of the vast saving of fuel by such a grate is, that the smoke instead of stealing away latent heat—being itself also the most combustible and precious part of the fuel, here gives all its powers and worth to the purpose of the combustion. The coal rested on moveable bottoms in the troughs (here shown by dotted lines) and was moved up like the wick of a lamp by its screw. The bottoms may be lifted in many ways. Or, without bottoms the coal may be raised in each trough successively by an iron lever or bar introduced by the fire-man under the coal, fresh coal being at the same time put into the trough to fill the space under the lever. The author believes that this principle of construction will still be extensively adopted for the Newcastle or flaming coal,—the consequences would be so important. The principle has been already introduced for common parlour fires by Mr. Cuttler in his stove, which is merely a common grate, having, instead of bottom-bars, a deep box to hold the coal for a whole day, with a moveable bottom, which lifts the coal up as wanted. From such a fire there is always ascending a long beautiful flame; and much more heat is given out than from the same quantity of coal burned in the common way; the chimney never requires sweeping, for there is absolutely no smoke and therefore no soot.

Fig. 136.

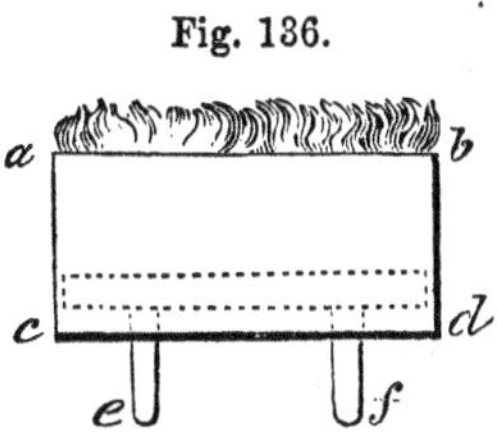

Smoke is effectually consumed also in a fire-place in which the air feeding the combustion is caused to pass downwards through the burning fuel, so as to carry the smoke with it, instead of upwards as usual. This result is attained by having the chimney to communicate with the ash-pit. The chief objection is, that the fire-bars are quickly destroyed by the intense heat to which they are exposed; and to obviate this instead of solid bars, tubes have been used with water passing through them, and admitting the feeding air only above.

It is evident that if a house or apartment with the air in it, were once warmed to a certain degree, it would for ever retain its temperature, but for the escape of heat through the walls and windows, or with the air from within, whether passing away as necessary ventilation or as waste. A perfect system of heating, therefore, would consist in diminishing as much as possible these causes of loss, with reference, however, both to the expense of the means and the salubrity of the dwelling, and in producing and distributing the heat judiciously. It may be asserted that a fourth part of the fuel generally expanded in English houses, if more skilfully used, would better secure comfort and health than all that is now expanded. But it does not accord with the character of this general work to enter into minute detail on the subject. Remarks were made upon it in the chapter on "*Pneumatics*," under the head of "warming and ventilating," and more minute information may be obtained from Mr. Tredgold's work, expressly devoted to it.

The detailed consideration of furnaces, blow-pipes, &c., may appear to some so closely connected with our present subject as to demand a place here, but by so treating of them we should be encroaching on the province of the chemist, &c. We may state generally, that furnaces are merely structures by which coal or other fuel heated to the degree at which it combines rapidly with the oxygen of the atmospheric air, is placed in circumstances favourable to the rapid renewal of the air,—and a common blow-pipe is merely a jet of air thrown from a minute opening into any flame, so as with great decision to direct the point of the flame upon the body to be heated. The sand-bath and water-bath of the chemist are merely means of insuring a more uniform or steady temperature:—a vessel imbedded in sand, so that heat can reach it only through the sand, cannot be very suddenly heated or cooled, because sand is a slow conductor; and a vessel immersed in boiling water, can never have greater heat than 212°, or the boiling heat of water. For certain purposes, hotter baths, as of high-pressure steam, or of solutions of certain salts, or of vapour of oil of turpentine, or of boiling whale oil, have been used. On such subjects, readers may consult works on "chemistry applied to the arts."

"*Condensation and friction as causes of heat.*" (Read Analysis, page 256.)

A soft iron-nail laid upon an anvil, and receiving in rapid succession three or four powerful blows of a hammer becomes hot enough to light a match, and if longer hammered, will become incandescent or red hot,—partly from the diminished volume or condensation of the iron, on the principle already explained, and partly from the percussion or friction, in a way not yet well understood, but probably electrical

In the familiar case of the mutual percussion of flint and steel, small portions of one or both are struck off by the violence of the collision, in a state of white heat, and the particles of the iron burn in passing through the air; in a vacuum, the heated particles are equally produced, but are scarcely visible from this combustion not occurring. In both cases, they suffice to inflame gunpowder, or to light tinder. When the materials are good the shower or burst of sparks from the sudden blow is most brilliant.

The heat produced by friction alone without perceivable condensation of the bodies concerned, is exemplified in many facts. Two dry branches of a tree kept strongly rubbing against each other by the wind, have sometimes set a wood on fire. Savages light their fires by analogous means. Men warm their cold hands in winter, by rubbing them against each other, or against their coat sleeves. Again, the axle-tree of a heavily laden wagon or other carriage, if left without oil, is often heated so as to inflame wood near it. The line attached to a whale-harpoon as it runs over the side of the boat when the whale dives after the harpoon has entered his flesh, requires that water be constantly thrown upon it to prevent its setting fire to the boat. The cable of a ship drawn very rapidly through the hawse-hole by the falling anchor produces there intense heat and smoke. When a great ship is launched from the builder's yard into the deep, and glides along the sloping beams, a dense smoke usually rises from the points of rubbing contact.

"*The functions of animal life a source of heat.*" (Read the Analysis, p. 256.)

It is one of the remarkable facts in nature, that living animal bodies, and to a certain degree living vegetables also, have the property of maintaining in themselves a peculiar temperature, whether surrounded by bodies that are hotter or colder than they. Captain Parry's sailors, during the polar winter, where they were breathing air that froze mercury, still had in them their natural warmth of 98° of Fahrenheit: and the inhabitants of India, where the same thermometer stands sometimes at 115° in the shade have their blood at no higher temperature.

It was at one time the favourite explanation of this, that animal heat was produced in the lungs, during respiration, from the oxygen then admitted. This oxygen combines with carbon from the blood, and becomes carbonic acid as in combustion, and it was supposed to give out a portion of its latent heat, as in actual combustion; which heat, being then spread over the body by the circulating blood, maintained the temperature. We now know that if such a process assist, which it probably does,—for the animal heat has generally a relation to the quantity of oxygen required in any particular case, and when an animal being already much heated needs no increase, very little oxygen disappears,—still much of the effect is dependent on the influence of the nerves either directly or indirectly, through the vital functions governed by them. Mr. Brodie, in his important experiments upon the subject found that, although in animals apparently dead from injury done to the nervous system, he could artificially continue the action of respiration, with the usual formation of carbonic acid, still the temperature fell very quickly. Again, the maintenance of low temperature in an animal immersed in air hotter than itself, is partly attributable to the copious perspiration and evaporation which then take place, and which absorb into the latent form the excess of heat then existing. Perspiration, both from the skin and internal surface of the lungs, occurs generaly in proportion to the excess of heat. Dogs and other animals when much heated, as they cannot throw off or diminish their natural covering, increase the evaporating surface by protruding a long, humid tongue.

The power in animals of preserving their peculiar temperature has its limits. Intense cold coming suddenly upon a man who has not sufficient protection, first causes a sensation of pain, and then brings on an almost irresistible sleepiness, which if indulged in proves fatal. Sir Joseph Banks having gone on shore one day near the cold Cape Horn, and being fatigued, was so overcome by the feeling mentioned, that he entreated his companions to let him sleep but for a little while. His prayer, if granted, might have allowed to come upon him that sleep which ends not—as, under similar circumstances, it came upon so many thousands of the army which Buonaparte led into Russia, and lost there during the disastrous retreat through the snows. The celebrated bulletin which allowed that, in one night, when the thermometer of Reaumur stood at 19° below zero, 30,000 horses died, declared not the number of human victims—tenderly loved husbands and brothers, and children of thousands anxiously waiting their return, but doomed never to see them more. Cold in inferior degrees, and longer continued. acting on persons imperfectly protected by clothing, &c., induces a variety of diseases which destroy more slowly; as many of the winter diseases of England. A great excess of heat, again, may at once excite a fatal appoplexy, and heat in inferior degrees, but long continued, may cause those fevers, &c., which prevail in warm climates, and which are so destructive to strangers.

Each species of animal has a temperature natural to it, and in the diversity are found creatures fitted to live in all parts of the earth, what is wanting in internal bodily constitution being found in the admirable covering which has been provided to protect them—covering which grows from their bodies, with form of fur or feather, in the exact degree required, and even so as in the same animal to vary with climate and season. Such covering, however, has been denied to man; but the denial is not one of unkindness,—it is an indication of his superior nature and destinies. God-like reason was bestowed on him, by which he subjects all nature to his use, and he was left to clothe himself.

The human race is naturally inhabitant of a warm climate, and the paradise described as Adam's first abode, may be said still to exist over vast regions about the equator. There the sun's influence is strong and uniform, producing a rich and warm garden, in which human beings, however ignorant of the world which they had come to inhabit, would have their necessities at once supplied. The ripe fruit is there always hanging from the branches; of clothing there is required only what moral feelings may dictate, or what may be supposed to add grace to the form; and as shelter from the weather, a few broad leaves spread on connected reeds, complete the Indian hut. The human family, in multiplying and spreading in all directions from such a centre, would find to the east and west, only the lengthened paradise, with slightly varying features of beauty; but to the north and south, the changes of season, which make the bee of high lattitudes lay up its winter store of honey, and send migrating birds from country to country in search of warmth and food, would also rouse man's energies to protect himself. His faculties of foresight and contrivance would come into play, awakening industry: and as their fruits, he would soon possess the knowledge and the arts which secure to him a happy existence in all climates, from the equator almost to the pole. And it is chiefly because man has learned to produce at will, and to control, the wonder-working principle of heat, that in the rude winter, which seems the death of nature, he, and other tropical animals and plants which he protects, do not in reality perish—as exemplified when a Canary bird escapes from its cage, or an infant is exposed among the snow-hills. By producing heat from his fire, he then obtains a novel but most pleasurable sort of existence: and in the night, while the dark and freezing winds are howling over his roof, he basks in the presence of his mimic sun, surrounded by his friends and all the delights of society; while in his store-rooms, or in those of merchants at his command, he has the treasured delicacies of every season and clime. He soon becomes aware, too, that the dreary winter, instead of being a curse, is really in many respects a bleseing, by arousing from the apathy to which the eternal serenity of a tropical sky so much disposes. He sees that in climates where labour and ingenuity must precede enjoyment, every faculty of mind and body is invigorated; and that hence the sterner climates produce the perfect man;—that in them the arts and sciences have reached their present advancement, and the brightest examples have arisen of intellectual and moral excellence: while from them, as centres, knowledge and example are spreading over all the earth and promising soon to render the whole of human kind but one large and happy family.

PART IV.

(CONTINUED.)

SECTION II.—ON LIGHT, OR OPTICS.

ANALYSIS OF THE SECTION.

Light is an emanation from the sun and other self-luminous bodies, becoming less intense as it spreads, and which, by falling on other bodies, and being reflected from them to the eye, renders them visible. Its absence is called darkness. It moves with great velocity, and in straight lines where there is no obstacle—leaving shadows where it cannot fall. It passes readily through some bodies—which are therefore called transparent—but when it enters or leaves their surfaces obliquely, it suffers at them a degree of bending or refraction proportioned to the obliquity. And a beam of white light thus refracted or bent, does nót all bend equally, but is divided or resolved into beams of what are called the elementary colours, which colours, on being again blended, become the white light as before.

Transparent bodies, as glass, may be made of such form as, by the powers of refraction thence received, to cause all the rays which pass through them from any given point to bend and meet again in another point beyond them;—the body, then, because usually in form somewhat resembling a flat bean or lentil, being called a LENS. *And when the light thus proceeding from every point of an object placed before a lens is collected at corresponding points behind it, a perfect image of the object is there produced, which may be seen from any situation on a white screen placed to receive it, or in the air, if viewed from behind. Now the most important optical instruments, and even the living eye, are merely various arrangements af parts for producing and examining such images as now described. When this image is received upon a suitable white surface or screen in a dark room, the arrangement is called, according to minor circumstances, a* CAMERA OBSCURA, *a* MAGIC LANTERN, *or a* SOLAR MICROSCOPE. *And the* EYE *itself is, in fact, but a small camera obscura, enabling the mind to judge of external objects, by the size, brightness, colour, &c., of the very minute but most perfect images or pictures formed at its back part, on the smooth screen of nerve called the retina. The art of painting aims at producing on a larger scale such a picture as is formed on the retina, and which, when afterwards held before the eye, and reproducing itself in miniature upon the retina, may excite the same impression as the original objects.—When the image beyond a lens, formed as above described, is viewed in the air, by looking at it from behind, that is, from a situation where the light continued from it passes, then there is exhibited the arrangement of parts constituting the* TELESCOPE *or* COMMON MICROSCOPE.

Ligh falling on very smooth or polished surfaces, is reflected so nearly in the order in which it falls, as to appear to the eye receiving it as if coming directly from the objects originally emitting it—and such surfaces are called mirrors. Mirrors may be plane, convex, or concave; and certain forms will concentrate light, to produce images by reflection, just as lenses produce them by refraction; so that there are reflecting telescopes, &c., as there are refracting instruments of the same name. Light, again, falling on bodies of rougher or irregular surface, or which have other peculiarities, is so modified as to produce all those phenomena of colour and varied brightness seen among natural bodies, and giving them their distinctive characters and beauty.

"*Light.*" (See the Analysis.)

The phenomena of *light* and *vision* have always been held to constitute a most interesting branch of natural science; whether in regard to the beauty of light, or its utility. The beauty is seen spread over a varied landscape—among the beds of the flower-gardens, on the spangled meads, in the plumage of birds, in the clouds around the rising and setting sun, in the circles of the rainbow. And the utility may be judged of by the reflection, that if man had been compelled to supply his wants by groping in utter and unchangeable darkness, even if originally created with all the knowledge now existing in the world, he could scarcely have secured his existence for one day. Indeed, without light, the earth would have been an unfit abode, even for grubs, generated and living always amidst their food. Eternal night would have been universal death. Light, then, while the beauteous garb of nature, clothing the garden and the meadow—glowing in the ruby—sparkling in the diamond, is also the absolutely necessary medium of communication between living creatures and the universe around them. The rising sun is what converts the wilderness of darkness which night covered, and which, to the young mind not yet aware of the regularity of nature's changes, is so full of horror, into a visible and lovely paradise;—no wonder, then, if in early ages of the world, man has often been seen bending the knee before the glorious luminary, and worshipping it as the God of Nature. When a mariner, who has been toiling in midnight gloom and tempest, at last perceives the dawn of day, or even the rising of the moon, the waves seem to him less lofty, the wind is only half as fierce, and hope and gladness beam on him with the light of heaven. A man, wherever placed in light, receives by the eye from every object around, nay, from every point in every object and at every moment of time, a messenger of light to tell him what is there, and in what condition. Were he omnipresent, or had he the power of flitting from place to place with the speed of the wind, he could scarcely be more promptly informed. Then, in many cases where distance intervenes not, light can impart at once knowledge, which, by any other conceivable means, could come only tediously, or not at all. For example, when the illuminated countenance is revealing the secret workings of the heart, the tongue would in vain try to speak, even in long phrases, what one smile of friendship or affection can in an instant convey;—and had there been no light, man never could have suspected the existence of the miniature worlds of life and activity which, even in a drop of water, the microscope discovers to him; nor could he have formed any idea of the admirable structure of many minute objects.

It is light, again, which gives the telegraph, by which men readily converse from hill to hill, or across an extent of raging sea—and it is light which, pouring upon the eye through the optic tube, brings intelligence of events passing in the remotest regions of space.

"*Emanation from the sun,*" *&c.* (See the Analysis, page 325.)

The relation of the sun to light is most strikingly marked in the contrast between night and day. In tropical countries where the sun rises and sets almost perpendicularly, and allows not the long dawn and twilight of temperate latitudes, the change from perfect darkness to the overpowering effulgence of day, and the contrary change, are so sudden as to be most impressive. An eye turned in the morning to the east has scarcely noted a commencing brightness there, when that brightness has already become a glow; and the clouds floating near so as to meet the upward rays, appear like masses of gold fleece suspended in the sky: a little after the whole atmosphere is bright, and as the stream of light reaches the lofty mountain-tops, it makes them shine like burnished pinnacles; then as that stream descends to lower and lower levels, the inhabitants in succession see the radiant orb first rising above the horizon like a tip of flame, and soon displaying all its breadth and glory, too bright for the eye to dwell upon. With evening the same appearances recur in a reversed order, ending, as in the morning they began by complete darkness.

Light emanates also from the stars, but they are so distant as in that respect to be of little importance to this earth. And all bodies in combustion are self-luminious, as exemplified in our common fires and lamps. And there are still other transient sources in animal and vegetable nature, and among solar phosphori.

There have been two opinions respecting the nature of light; one, that it consists of extremely minute particles darting all around from the luminious body; the other, that the phenomenon is altogether dependent on an undulation among the particles of a very subtile elastic fluid diffused through space—as sound is dependent on an undulation among air-particles. To admit the first opinion, the particles of light must be held to be most wonderfully minute, for a common taper can fill with them during hours a space of four miles in diameter; and with the extreme velocity of light, if its particle possessed at all the property of matter called inertia, their momentum should be very remarkable;—yet, even a large sunbeam collected by a burning-glass, and with the precautions necessary in the case, thrown upon the scale of a most delicate balance, has not the slightest effect upon the equilibrium. Such, and many other facts to be treated of in subsequent parts of this work, lead to the opinion that there is an undulation of an elastic fluid concerned in producing the phenomenon of light,—although the fact of light spreading so nearly in straight lines, as if only the crown of the wave had existence, instead of being diffused like sound, is an important difference.

"*Becoming less intense as it spreads.*" (See the Analysis, page 325.)

Any emanation from a central point in spreading through water space, becomes proportionally thinner or less intense. Thus, if a taper be placed in the centre of a box, each side of which is a foot square, the light falling on the sides of the box will have a certain intensity there:—if the taper be then placed in a box with sides of two feet square, there will be only the same quantity of light, but it will be spread over four times the surface, (a square of two feet is made up of four squares of one foot,) and will, therefore, on any part of that surface, be only one-fourth part as strong or intense as in the first box:—and so for any other size of box or space, the intensity will diminish as the square of the distance increases.

Hence four times as much light and heat fall upon a foot of this earth's surface as upon a foot of the surface of the planet Mars, which is twice as distant from the sun:—as four times as much light and heat fall on a man who is at one yard from the fire, as on another who is distant two yards.

"*Falling on other bodies makes them visible.*" (Read the Analysis, page 325.)

If the window-shutter of an apartment be perfectly closed, an eye there turns upon an absolute blank: it perceives nothing.

If a ray of the sun be then admitted, and made to fall upon any object, that object becomes bright, and affects the eye as if it were itself luminious. It returns a part of the light which falls upon it, and it is visible in all directions, proving that it scatters the received light all around. This scattered light, again falling on other objects, and reflected from and among them until absorbed, like echo repeated many times and lost between perpendicular rocks, makes all of them visible, although in a less degree, and the whole apartment is said to be lighted. If the sun's ray be made to fall upon a thing which, from its nature, reflects much of the light, as a sheet of white paper, the apartment will be well lighted:—if, on the contrary, it be received on black velvet, which returns hardly any light, the apartment will remain dark;—and, again, if received on a polished surface or mirror, which returns nearly the whole light, but in one direction only, and therefore throws it upon some other single object, the effect will be according to the nature of that object, and nearly as if the ray had fallen directly upon it.

Now all bodies on earth, and among these the constituent particles of the mass of atmosphere surrounding the earth, retain and diffuse among themselves, for a time, the light received directly from the sun, and by so doing, maintain every where that milder radiance so agreeable to the sight, which renders objects visible when the sun's direct ray does not fall upon them. But for this fact indeed, all bodies shadowed from the sun, whether by intervening clouds or by any other more opaque masses on earth, would be perfectly black or dark; that is, totally invisible. And without an atmosphere, the sun would appear a round luminious mass in a perfectly black sky. On lofty mountain summits, where half the atmosphere is below the level, the direct rays of the sun are painfully intense, and the sky is of darkest blue.

A shadow is the name given to the comparative darkness of places or objects which are prevented by intervening things from receiving the direct rays of some luminious body shining on the things around. The apparent darkness of a shadow, however, is not proportioned to its real darkness, but the intensity of the surrounding lights. A landscape may be very bright, even when the sun is veiled by a cloud, and then little or no shadow is perceived; but as soon as the cloud passes away, deep shadows are cast behind or beyond every projecting object. Yet the objects and places then appearing so dark, are in reality more illuminated than before the shadow existed, for they are receiving, and again scattering new light from all the more intensely illuminated objects around them. A finger held between a candle and the wall casts a shadow of a certain intensity; if another candle be then placed in the same line from the shadow, the shadow will appear doubly dark, although in fact more light will be reaching it and reaching the eye from it than before; it will be more dark only by comparison. If the candles be separated laterally, so as to produce two shadows of the finger, but which coincide or overlap in one part, that part will be of double darkness, as compared with the remainder, The most accurate mode of comparing lights is to place them at such distances from a screen or wall, as to make them at the same time throw equally dark shadows of the same object; and then according to the law of decreasing intensity explained above, to calculate the intensities of the sources of light by the difference of their distances from the wall. The eye judges very easily of the equal intensity of compared shadows of the same object.

The real darkness of a shadow, then, depends on the number and nature of the light reflecting objects around it. Thus shadows are less remarkable opposite to any white surface, as that of a recently painted wall, than in other situations. The reason why the moon, when eclipsed, that is, as will be afterwards explained, when passing behind the earth, or through the shadow cast by the earth in a direction away from the sun, becomes almost, if not quite invisible, is that there are no other moons or bodies bearing laterally on the moon to share their light with it. And the reason why our nights on earth are darker than the shadows behind a house or rock in the sunshine of day, is merely that there are not other earths near us to reflect light into the great night-shadow of the earth, as there are other houses and rocks to illuminate the day-shadow of these. The moon is the only light-reflecting body which the earth has near it; and we perceive how much less dark the earth's night-shadow is when the moon is so placed as to bear upon it. The eclipsed moon, again, is invisible to men on earth, because it receives neither sunshine nor reflected light from this earth, for the side turned towards us faces the shadowed part of the earth; but when the moon is near the situation in which it is called new moon, or between us and the sun, the shaded side of the moon is then, in a degree, visible to us because facing the enlightened side of the earth; the bright crescent, or part of the moon illuminated by the sun, appearing to embrace the non-illuminated part, and giving occasion to the popular saying, that the new moon holds the old moon in its arms.

Many persons have doubted whether the light of the moon could be altogether reflected light of the sun; the moon appearing to them more luminous than any body on earth merely exposed to the sunshine. Their error has arisen from contrasting the moon while returning direct sunshine with the shadows of night on the earth around them. But could they at night see on a hill near them, a white tower or other object scattering light as when it receives the rays of the sun, that object being nearer than the moon, would appear to them almost to be on fire, and much brighter than the moon. The moon when above the horizon in the day time, is perfectly visible on earth, and is then throwing towards the earth just as much light as during the night; but the day moon does not appear more luminous than any small white cloud: and although visible every day, except near the change, many persons have passed their lives without ever observing it. The full moon gives to the earth only about a one-hundred-thousandth part as much light as the sun.

"Light moves with great velocity" (See Analysis, page 325.)

The extraordinary precision with which the astronomical skill of modern days enables men to foretell the times of remarkable appearances or changes among the heavenly bodies has served for the detection of the fact, that light is not an instantaneous communication between distant objects and the eye, as was formerly believed, but is a messenger which requires time to travel: and the rate of travelling has been ascertained.

The eclipses of the satellites or moons of the planet Jupiter had been carefully observed for some time, and a rule was obtained which foretold the instants in all future time when the satellites were to glide into the shadow of the planet and disappear, or when again to immerge into view. Now it was found, that these appearances took place 16½ minutes sooner when Jupiter was near the earth, or on the same side of the sun with the earth than when it was on the other side, that is to say more distant from the earth by one diameter on the earth's orbit; and at all intermediate stations, the difference diminished from the 16½ minutes, in exact proportion to the less distance from the earth. This proves, then, that light takes 16½ minutes to travel across the earth's orbit, and 8¼ minutes for half that distance, or to come to us from the sun.

The velocity of light, ascertained in this way, is such, that in one second of time, *viz:* during a single vibration of a common clock pendulum, it would go and come from London to Edinburgh 200 times, the distance between these being 400 miles. This velocity is so surprising that the philosophic Dr. Hooke, when it was first asserted that light was thus progressive, said he could more easily believe the passage to be absolutely instantaneous, even for any distance, than that there should be a progressive moment so prodigiously rapid. The truth, however, is now put quite beyond a doubt by any collateral facts bearing upon it.

As regards all phenomena upon earth, they may be considered as happening at the very instant when the eye perceives them; the difference of time being too small to be appreciated; for, as shown in the preceding paragraph, if our sight could reach from London to Edinburgh, we should perceive a phenomenon there in the four-hundredth part of a second after its occurrence.

It is hence usual and not sensibly incorrect, when we are measuring the velocity of sound, as when a cannon is fired, by observing the time between the flash and the report, to suppose that the event takes places at the very moment when it is perceived by the eye.

In using a telegraph, no sensible time is lost on account of light requiring time to travel. A message can be sent from London to Portsmouth in a minute and a half; and at the same rate a communication might pass to Rome in about half an hour, to Constantinople in forty minutes, to Calcutta in a few hours, and so on. A telegraph is any object that can be made to assume different forms or appearances at the will of an attendant, and so that the changes may be distinguished at a distance. A pole with movable arms is the common construction, each position of the arms standing for a letter, or cypher, or word, or sentence, as may be agreed upon. Telegraphic signals between ships at sea are generally made by a few flags, the meanings of each being varied by the mast on which it is hoisted, and by its combination with others.

"*Light proceeds in straight lines,*" &c. (Read the Analysis, page 325.)

We have scarcely a clearer notion of a straight line than that received from the direction in which light moves:—but we can verify a line so obtained by other means, as by stretching a cord between the two extremes, or by suspending a weight by a cord, and making a moveable solid measure to correspond with the cord, which standard may be used in any other case.

We can see through a straight tube, but not through a crooked one. The vista through a long straight tunnel is striking as an illustration of this fact and of the diminution of the apparent size of objects as they are more distant. If a person enter one end of the canal tunnel two miles long, cut through the chalk-hills near Rochester as part of the canal which joins the Thames and Medway rivers, the opening at the distant end is seen as a minute luminous speck, having the form of a general arch; and a person who has advanced half way through the tunnel may see the luminous speck at each end, then appearing a little larger than in the former case.

In taking aim with gun or arrow, we are merely trying to make the projectile go to the desired object nearly by the path along which the light comes from the object to the eye.

A carpenter looks along the edge of a plank, &c., to see whether it be straight.

Because light moves in straight lines, if a number of similar objects be placed in a row from the eye, the nearest one hides the others. In a wood or city, a person sees only the trees or houses that are next him.

He who believes that a squinting person can see round a corner, may also believe that a crooked gun can shoot round a corner.

All astronomical and trigonometrical observations are made on the faith of this property of light, the observer holding that any object is situated from him in the direction in which the light comes to him from it. When the mariner, after watching for hours in cloudy weather has caught a glimpse of the sun or star through his sextant-glass, he has ascertained his place among the trackless waves, and boldly advances through the mist of hidden dangers. And the beam darting from the light-house across the stormy sea, would be useless if the light moved not in a straight line.

"Leaving shadows where it cannot fall." (See the Analysis, page 325.)

The form of shadows proves that light moves in straight lines, for the outline of the shadow is always correctly that of the object as seen from the luminous body. If the light bent round the body, this could not be.

The shadow of a face on the wall is a correct profile.

As a wheel presented edgeways to the eye appears only like a broad line, becomes oval or round as it is more turned, so a wheel presented edgeways to the sun or other light casts a linear shadow on the wall behind it, the shadow becoming oval or round as the position is changed.

A globe, a cylinder, a cone, and a flat circle, will all throw the same round shadow if held with their axes pointing to the luminous body, and, therefore, by the shadow only, these objects could not be distinguished.

The figure of a rabbit cut in paste-board, will throw the same shadow on the wall as the animal itself; and, again, that shadow may be well imitated by a certain position of the two hands joined, as is known to those who find pleasure in witnessing the surprise and delight of a child who beholds such a shadow made to mimic the actions of life.

A man under the vertical sun stands upon his little round shadow; but as the sun declines in the afternoon, the shadow juts out on the opposite side, and at last may extend across a whole field.

A distant cloud which appears to the eye of an observer only as a streak along the sky, may yet be broad enough to shadow a whole region; for clouds generally form in level strata, and when viewed by a spectator on earth at a distance are seen nearly edgeways.

The velocity of the wind may be ascertained by marking the time which the shadow of a cloud takes to pass over a plain or other space of known dimension.

A body held between a candle and the wall darkens a portion of the wall, or casts its shadow there; and the whole space between it and the wall is a shadowed space, for anything introduced there is as much shadowed as the portion of the wall. Thus, also, all the heavenly bodies which revolve about the sun cast a shadow beyond them or away from the sun, as is seen when one of them, before brightly visible, passes where the shadow of another is. The satellites or moons of Jupiter, when they suddenly disappear to our glasses, or are eclipsed as we term it, have generally only plunged into the shadow of the planet, and are not hidden by being then on the other side of the planet, as many suppose. When our own moon is eclipsed, that phenomenon so awful in the early ages of the world, she is only passing through the long shadow which the earth casts beyond it.

When in the case of a luminous centre and a body casting a shadow, the centre is larger than the body, then the cross section of the shadowed space, or the shadow as thrown on a plane surface, will be less than the body, and less, moreover, the farther the surface is from the body, for the shadowed space terminates in a point. This is true of the shadows of all the planets and of the earth, because they are less than the sun. On the contrary, if the light-giving surface is smaller than the opaque body, the shadow will be larger than the body. The shadow of a hand held between a candle and the wall is gigantic; and a small pasteboard figure of a man placed near a narrow centre of light, throws a shadow as big as a real man. The latter fact has been amusingly illustrated by the art of making phantasmagoric shadows.

When the surface which receives a shadow is not directly exposed to the light, the shadow may be much larger than the object, even although the sun himself be throwing the light;—as is seen when a slightly projecting roof, or a veranda, shadows from the high sun of summer noon the whole front of a house; or, as is proved by the long evening shadows of all countries, a low wall will shadow from the setting sun a whole field.

"*Light passes readily through some bodies—which are, therefore, called transparent; but when it enters, or leaves their surfaces obliquely, its course is bent.*" (Read the Analysis, page 325.)

It may well excite the surprise of inquirers that light, of which the constitution is so fine or flimsy, should still be able to dart readily and in every direction through great masses of solid matter, but such is the truth. Thick plates of solid glass, blocks of rock crystal, mountains of ice, &c., are instantly pervaded by the beam of the sun.

What it is in the constitution of one mass as compared with another, which fits the one to transmit light, and the other to obstruct it, we cannot clearly explain, but we perceive that the arrangement of the particles has more influence than their peculiar nature. Nothing is more opaque than thck masses of the metals, but nothing is more transparent than equally thick masses of the same metals in solution, nor than the glasses of which a metal forms a large proportion. The thousand salts formed by the union of the metals or earths with the diluted acids, are all transparent, when in cooling from the fluid to the solid state, their particles have been allowed to arrange themselves according to the laws of their mutual attraction, that is to say, to form crystals; but the same substances in other states, as when reduced to powder, are opaque. Even the pure metals themselves, when reduced to leaves of great thinness, are transparent, as may be perceived by looking at a lamp through a fine gold leaf. It is to be remarked, however, that even the most transparent bodies intercept a considerable part of the light which enters them: a depth of seven feet of pure water intercepts about one-half, so that the bottom of the sea is very dark. And of the sun's light, when passing obliquely through the atmosphere towards the earth, as when the sun has lately risen or is about to set, only a small part arrives.

Light having once entered a transparent mass of uniform nature passes forward in it as straightly as in a vacuum; but at the surface, whether on entering or leaving it, if the passage be oblique, and if the mass be of a different density from the transparent medium around it, a very curious and most important phenomenon occurs, namely, the light suffers a degree of bending from its antecedent direction, or a *refraction*, proportioned to the obliquity.

But for this fact, which to many persons might at first appear a subject of regret, as preventing the distinct vision of objects through all transparent media; light could have been of little utility to man. There could have been neither converging lenses as now, nor any optical instruments, of which lenses form a part, as telescopes and microscopes; nor even the eye itself, which has its crystalline lens.

Fig. 137.

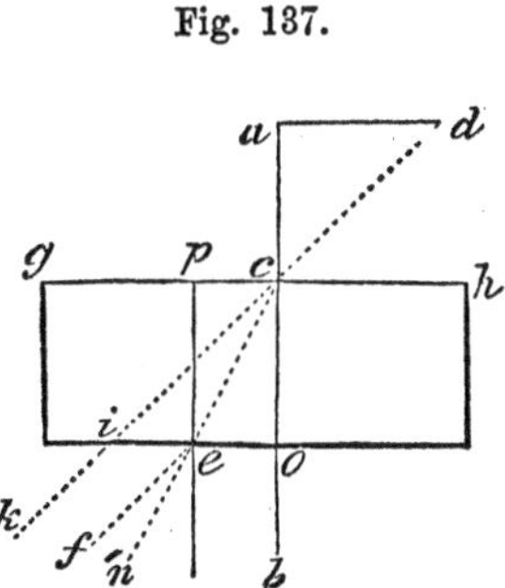

Light falling from the air directly or perpendicularly upon a surface of water, glass, or any such transparent body, passes through without suffering the least bending;—a ray for instance, shot from *a* to the point *c*, in the surface of a piece of glass *g h*, would reach directly across to *o* and *b*; but if the ray fell obliquely, as from *d* to *c*, then, instead of continuing in its first direction to *i* and, *k*, it would at the moment of its entrance be bent downwards in the path *c e*, nearer to a line *c o*, called the perpendicular to the surface at the point of entrance,—and the moving straightly while in the substance of the glass it would, when it passed out again at *e*, in the opposite surface, be bent just as much as at first, but in the contrary direction, or away from a similar perpendicular at that surface, *viz.*, into the line *e f*, instead of *e n*. A ray, therefore, passing obliquely through a transparent body with parallel surfaces, has its course shifted a little to one side of the original course, but still proceeds in the same direction, or in a line parallel to the first—as here shown in the line *e f*, parallel and near to the line *i k*; if the surfaces of the body are not parallel, the ray is ultimately bent as will be explained some pages hence.

The degree of bending or refraction of light in traversing a single transparent surface is measured by comparing the obliquity of its approach to the surface with the obliquity of its departure after passing; and for this purpose a line is supposed to be drawn perpendicularly through the surface of the point where the ray passes (as *a b* in the above figure drawn through *c*, where the ray *d c* passes) and the relative positions of the ray to this line on both sides of the surface, are easily ascertained. Thus the line *a d*, drawn from any point of the ray before passing to such perpendicular, is a measure of the original obliquity or angular distance of the ray, and is called the *sine of the angle of incidence*, and the other line *o c* drawn from a corresponding point of the ray after passing to the perpendicular, is a measure of the obliquity after refraction, and is called the *sine of the angle of the refraction:*—by comparing these two lines in any case, the problem is solved.

When light passes obliquely from air into water, the refraction or bending produced is such, that the line *a d* measuring the obliquity before refraction, is always longer than the line *o e* measuring it after refraction, by nearly one-third of the latter, and the refractive power of water is, therefore, signified by the index 1⅓ or 1,33; in like manner the greater refractive power of common glass has the index ½, of diamond the index 1½, and so on. And it is important to remark, that for the same substance the same relation holds, whatever the obliquity of the incidence ray may be. If, for instance, where the obliquity, as measured by its *sine*, is 40, and the refraction, is half or 20, then in the same substance an obliquity of 10 will occasion a refraction of 5, and obliquity of 4 will occasion a refraction of 2, and so on.

As a general rule, the refractive power of transparent substances of media is proportioned to their densities. It increases, for instance, through the list of air, water, salt, glass, &c. But Newton, while engaged in his experiments upon this subject, observed that inflammable bodies had greater refractive powers than others, and he then hazarded the conjecture, almost of inspired sagacity, which chemistry has since so remarkably verified, that diamond and water contained inflammable ingredients. We now know that diamond is merely crystallized carbon, and that water consists of hydrogen or inflammable air and oxygen. Diamond has nearly the greatest light-bending power of any known substances, and hence comes in part its brilliancy as a jewel.

No good explanation has been given of the singular fact of refraction; but to facilitate the conception and remembrance of it, we say that it happens as if it were owing to an attraction between the light and the refracting body or medium. The light approaching from *d* to *c*, for instance (in the last figure,) may be supposed to be attracted by the solid body below it, so as at the surface to be bent into the direction *c e*; and, again, on leaving the body to be still equally attracted and bent back, so as to take the direction *e f*, instead of *e n*; and we see why the attraction and bending should be greater, the greater the obliquity.

The following are familiar examples of this bending of light in passing from one medium to another.

If an empty basin or other vessel *b c f e*, be in the sun's light, so that the rays falling within it may reach low on the side as to *d*, but not to the bottom, then, on filling the vessel with water, the sun will be found to be shining on the bottom or down to *e*, as well as on the side. The reason of this phenomenon is, that water being a denser medium than air, the light, on entering it at *c*, is bent towards the perpendicular (*c f*,) at the point of incidence, and so reaches the bottom. Again, if a coin or metal, were laid on the bottom of such a vessel at *e*, it would not, while the vessel were empty, be seen by an eye at *a*, but would be visible there immediately on the vessel being filled with water;—because then, the light leaving the coin in the direction *e c*,

Fig. 138.

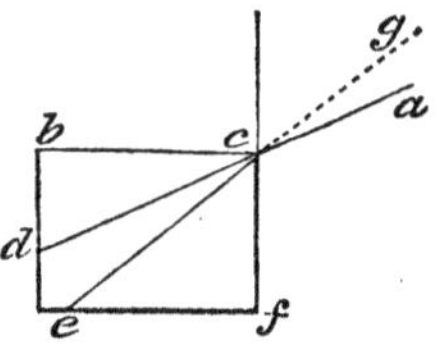

towards the edge of the vessel, would at *c*, on passing from the water into air, be bent away from the perpendicular, and instead of going to *g*, would reach the eye at *a*. The coin, moreover, would appear to the eye to be in the direction of *c d*, instead of the true direction *c e*: for the eye not being able to discover that the light had been bent in its course, would judge the object to be in the line by which the light came from it.

It is thus because objects at the bottom of water, when viewed obliquely, do not appear so low as they really are, that a person examining a river or pond, or any clear water, from its bank, naturally judges its depth to be less than it is. Many a young life has been sacrificed to this error. A person looking from a boat directly down upon the objects at the bottom of water, sees them in their true directions, but even then not in their true distances, as will be afterwards explained; and if he view them more and more obliquely, the appearance becomes more and more deceiving, until at last it represents them as at much less than half of their true depth.

The ship in which the author sailed, once in the middle of the China Sea, where no danger was apprehended, entered by a narrow passage a large hore-shoe enclosure of coral rocks. When the looker-out gave the alarm, the predicament had become truly terrific. On every side, in water most singularly transparent, the rocks appeared to be almost at the surface of the water, and the anchor, which in the first moment had been let go to arrest the ship, appeared to have been dragged to a shallow place. It seemed that if the ship, then drawing 24 feet, or the depth of a two-storied house, moved but a little way in almost any direction, she must inevitably meet her destruction On sending boats around to sound and to search, the place of entrance was again discovered, and was safely traversed a second time as an outlet from that terrible prison.

On account of this bending of light from objects under water, there is more difficulty in hitting them with a bullet or spear. The aim by a person not directly over a fish, must be made at a point apparently below it, otherwise the weapon will miss by flying too high. The spear, sometimes used in this country for killing salmon, is a common weapon among the islanders of the Atlantic and Pacific Oceans for killing the albacore; the use of it, like that of the fly-hook in England, affording to the fishermen, sport as well as profit. The author once witnessed at St. Helena, this employment of the spear. A small fish previously stunned, that it might not try to escape, was every minute or two thrown upon the water as a bait, in the sight of perhaps a hundred great albacores, greedily waiting for it at one side below, and knowing the danger to which they exposed themselves by darting across to seize it. Some albacore bold enough, soon made at the mouthful, apparently with the speed of lightning, but yet with speed which did not save him, for every now and then the thrown spear met him, and held him writhing there in a cloud of his death-blood After a victim so destroyed, the scene of action was changed.

The bending of light when passing obliquely from water, is also the reason of the following facts. A straight rod or stick, of which a portion is immersed in water, appears crooked or broken at the surface of the water, the portion immersed seeming to be bent upwards. That part of a ship or boat visible under water, appears much flatter and shallower than it really is. A deep-bodied fish seen near the surface of water, appears almost a flat fish. A round body there appears oval. A gold fish in a vase may appear as two fishes, being seen as well by light bent through the upper surface of the water, as by straight rays passing through the side of the glass. To see bodies under water, in their true directions and nearly of their true proportions, the eye must view them through a tube, of which the lower end closed with a plate glass, is held in the water.

As light is bent on entering from air into water, glass, or other substance denser than air, so it is also bent on coming from void space into the ocean of our atmosphere. Hence none of the heavenly bodies, except when directly over our heads, are seen by us in their true situations. They all appear a little higher than they really are, and more so the nearer they are to the horizon; as when to a spectator at *d*, suppose on the surface of the earth, a star really at A appears to be at *a*, because its ray, on reaching the atmosphere at *c*, is bent downwards. In astronomical books there is always introduced a table of refraction as it is called, showing what correction must be made on this account for different apparent altitudes. This effect of our atmosphere so bends the rays of the sun, that we see him in the morning before he is really above the horizon, and we see him in the evening after he is really below it—for the ray coming horizontally from *e* to *d*, appears to come from *b*, although in truth it really comes from the lower situation B, and is bent into the level line only at *e*. Our atmosphere thus, by the bending of light as well as by itself becoming luminous, lengthens at dawn and twilight the durations of the precious day. As the atmosphere is denser near the surface of the earth than higher up, the light is more and more bent as it descends, and hence describes a course which is sensibly curved, and therefore unlike the course of light in water.

Fig. 139.

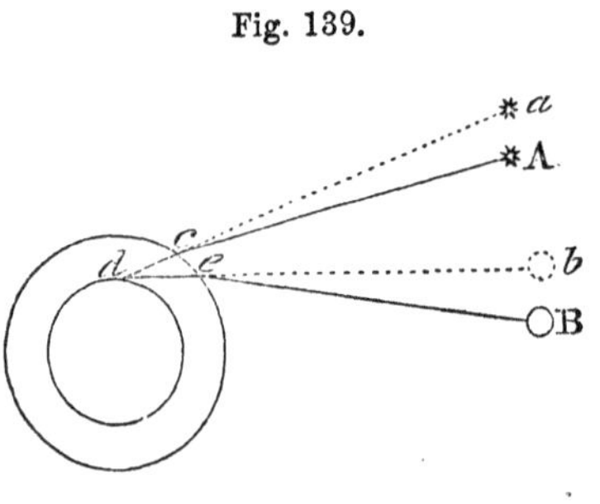

Certain states of the atmosphere, depending chiefly on its humidity and warmth, change very considerably its ordinary refractive power; hence, in one state, a certain hill or island may appear low and scarcely rising above the intervening heights or ocean, while in another state, the same object will be seen towering above; and from a certain station, a city in a neighbouring valley may be either entirely visible, or it may show only the tops of its steeples, as if the bed on which it rested had sunk deeper into the earth. In days of ignorance and superstition, such appearances occasionally excited a strange interest.

Owing to the bending of light in passing through the media of different densities, a beautiful phenomena is often observable in a day of warm sunshine. Black or dark-colored substances, by absorbing much light and heat from the sun's rays, and warming the air in contact with them, until it dilates and rises in the surrounding air, as oil rises in water, cause the light from more distant objects, reaching the eye through the rarefied medium, to be bent a little;—and owing to the heated air rising irregularly under the influence of the wind and other causes, these objects acquire the appearance of having a tremulous or a dancing motion. In a warm, clear day, the whole landscape at last appears to be thus dancing.

The same phenomenon is to be observed at any time, by looking at an object beyond the top of a chimney from which hot air is rising. An illicit distillery has been discovered by the exciseman happening thus to look across a hole used as the chimney, although charcoal was the fuel, and there was no vestige of smoke.

This bending of light by the varying states of the atmosphere renders precaution necessary in making very nice geometrical observations:—as in measuring base lines for the construction of maps or charts.

As it is the obliquity with which a ray traverses the surface, which, in any case of refraction, determines the degree of bending, a body seen through a medium of irregular surface appears distorted according to the nature of that surface. It is because the two surfaces of common window-glass are not as in the case of plate-glass perfect planes, and perfectly parallel to each other, that objects seen through a common window appear generally more or less out of a shape; and hence come the elegance and beauty of plate-glass window: and hence the singular distortion of things viewed through that swelling or lump of glass, which appears at the centre of certain very coarse panes and which remains where the glass-blower's instrument was attached.

The refraction or bending of light is interestingly exemplified in the effect of the glass called a prism, *viz.*, a wedge or three-sided rod of glass such as that of which the end is here represented at *b c*. A ray from *a* falling on the surface at *b* is bent *towards* the internal perpendicular, and therefore reaches *c*, but on escaping again at *c*, it is bent *away* from the external perpendicular and thus with its original deviation doubled, goes on to *d*.

Fig. 140.

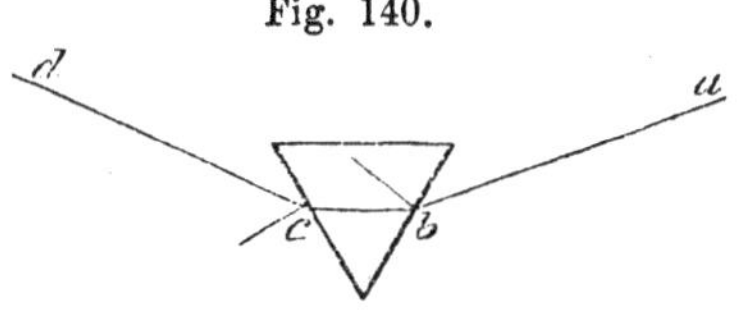

The law of lights bending, according to the obliquity with which it traverses the surfaces of a transparent body, is well elucidated by the effect of what is called a multiplying glass; that is to say, a piece of glass like *a b c e*, having many distinct faces cut upon it at angles with each other. If a small object, a coloured bead for instance, be placed at *d*, an eye at *e* will see as many beads as there are distinct surfaces or faces at the glass; for first, the ray *d a*, passing perpendicularly, and therefore straight through, will form an image as if no glass intervened; then, the rays from *d* to the surface *b*, will be bent by the oblique surface, and will show the object as if it were in the direction *e b;* and the light falling on the still more oblique surface, *c*, will be still more bent, and will reach the eye in the direction *c e*, exhibiting a similar object also in that direction—and so of all the other surfaces. If the eye were at *d*, and the object at *e*, the result would still be the same. A plate of glass roughened, or cut into cross furrows, becomes a very good screen or window-blind, by disturbing the passage of light through it so that objects beyond it are not distinguishable.

Fig. 141.

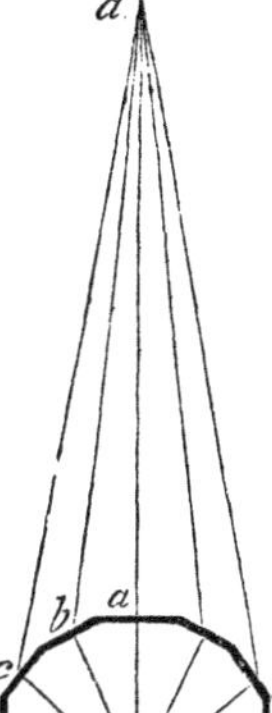

"*And a beam of white light thus made to bend, is resolved into beams of the various primary colours; which beams, however, on being again blended, become white light as before.*" (Read the Analysis, page 325.)

The most extraordinary fact connected with the bending of light is that a pure ray of white light from the sun admitted into a darkened room by a hole

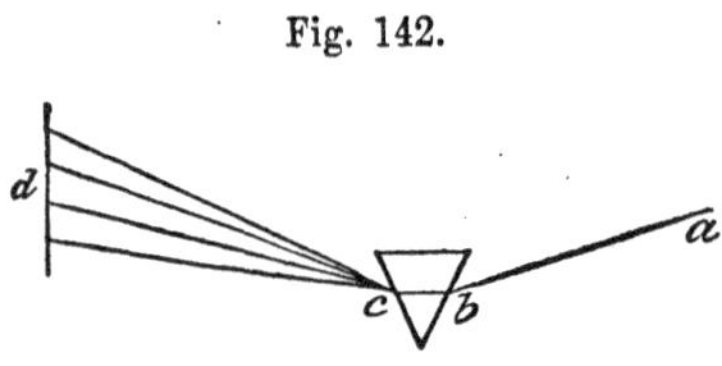

Fig. 142.

in the window-shutter, and made to bend by passing through transparent surfaces which it meets very obliquely (as the ray *a*, admitted and made to bend by passing through the prism of glass *b c*, to fall upon the wall at *d*,) instead of bending altogether and appearing still as the same white ray, is divided into several rays, which falling on the white wall, are seen to be of different most vivid colours. The original white ray is said thus to be analyzed, or divided into its elements.

This solar spectrum, as it is called, formed upon the wall, consists, when the light is admitted by a narrow horizontal slit, of four coloured patches corresponding to the slit, and appearing in the order, from the bottom of red, green, blue, and violet. If the slit be then made a little wider, the patches at their edges overlap each other, and produce by the mixture of their elementary colours, certain new tints. Then the spectrum consists of the seven colours commonly enumerated and seen in the rainbow, *viz.*, red, orange, yellow, green, blue, indigo, and violet.—Had red, yellow, blue, and violet been the four colours obtained in the first experiment, the occurrence of the others, *viz.*, of the orange, from the mixing edges of the red and yellow—of the green, from the mixture of the yellow and blue,—and of the indigo, from the mixture of blue and violet, would have been anticipated. But the facts of the case not being such, we see that they are not yet well understood. When Newton first made known the phenomenon of the many-coloured spectrum, and the extraordinary conclusions to which it led, he excited universal astonishment; for the common idea of purity, the most unmixed was that of white light. In farther corroboration of the notion of the compound nature of light, he mentioned, that if the colours which appear on the spectrum be painted separately around the rim of a wheel, and the wheel be then turned rapidly, the individual colours cease to be distinguished and a white band only appears where they are whirling; also, that if the rays of the spectrum, produced by a prism, be again gathered together by a lens, they reproduce white light. The red is the kind of light which is least bent in refraction, and the violet that which is most bent. It was at one time said, as an explanation, that the differently coloured particles in light had different degrees of gravity or inertia, and were therefore, not all equally bent. It is farther remarkable, with respect to the solar spectrum, that much of the heat in the ray is still less refracted than even the red light, for a thermometer held below the red light rises higher than in any part of the visible spectrum;—and there is an influence or something in the beam more refrangible than even the violet rays, and capable of producing powerful chemical and magnetical effects. The different spots of colour in the spectrum are not all the same size, and there is a difference in this respect according to the refracting substance.

All transparent substances in bending light produce more or less of the separation of colour; but it is an important fact, that the quality of merely bending a beam, or of *refraction*, and that of dividing it into coloured beams, or of *dispersion*, are distinct qualities, and not having the same proportions to each other in different substances. Newton, from not discovering this, concluded that a perfect telescope of refraction could never be made; he supposed that the bent light would always become coloured, and so render the objects indistinct. We now know, however, that by combining two or more media, we may obtain bending of light without dispersion,—thus, by opposing a glass which bends five degrees and disperses one degree, to another glass which bends three degrees and disperses one, the opposing dispersions will just counterbalance or neutralize each other, while the two degrees of excess of bending will remain to be applied to use.

The diversified colours of the substances around us depend merely upon their fitness, from texture or other cause, to reflect or transmit other modifications of common light, and the colour is not a part or property of the body itself. We shall soon find that the vivid colours of the rainbow are merely the white light of the sun, reflected to us after being bent and modified by the colourless drops of falling rain; and that the sparkling with appearance of rubies and emeralds, which we see in cut-glass lustre, is a phenomenon of the same kind:—and that by scratching the surface of a piece of metal so as to have a given number of lines in a given space, we can cause the same substance to appear of any colour we please.

"*Transparent bodies, as glass, may be made of such form as to cause all the rays of light which pass through them from any one point, to bend so as to meet again in another corresponding point beyond them,—the body itself, from the required form generally resembling that of a bean or lentil, being then called a* LENS." (Read the Analysis, page 325.)

The innumerable rays of light (of which five only are here represented,) issuing from any point as *c*, towards any surface in the situation *a b*, are said to form a cone or pencil of diverging light. Now it is evident that to make

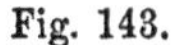

Fig. 143.

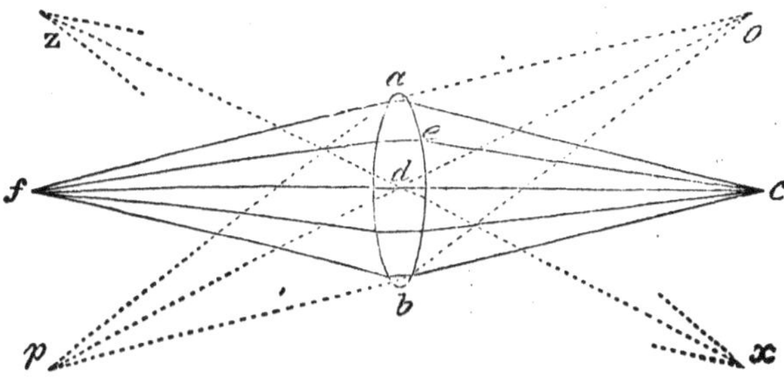

all such rays converge or meet again in one place as *f*, beyond a transparent body place at *a b*, it would be necessary, while the middle ray or axis of the pencil *c d* did not bend at all, for the others to be bent more and more, in proportion as they fell upon the body farther and farther from the centre *d*. Recollecting, then the law of refraction, that light entering from air through the surface of any denser medium, as glass, is bent there towards the perpendicular at the internal surface, in proportion to the obliquity of incidence, and on leaving the opposite surface, is correspondingly bent away from its external perpendicular, (see the case of the prism at p. 337,) we see that if a piece of glass were placed at *a b*, of such form that the rays falling upon it from *c* should meet and leave its surfaces with greater and greater obliquity in some regular proportion, as the points of incidence were more distant from the centre *d*, the purpose would be obtained. And we have the satisfaction of knowing that a glass, of which the surface is ground—which it easily may be—to have a regular convexity or bulging, as if it were a portion cut off from the surface of the globe, can be shown to answer very correctly the required condition. Such a glass similarly ground on both sides, is here represented edgeways between *a* and *b*, where the ray *c d* falling on its middle, or perpendicularly, and similarly leaving it, is seen going straight through to *f*, but the ray *c e* meeting the surface with a certain degree of obliquity is bent down a little, first on entering the surface at *e*, and then as much more on leaving the opposite surface with equal obliquity, and so arrives at *f*; then the ray *c a*, for corresponding reasons is still more bent, and equally arrives at *f*;—and the case would be similar of any other rays that might be examined. The point *f* is usually called a *focus* (meaning a fire-place,) because when the light of the sun is thus gathered, the heat concentrated with it is powerful enough to make combustibles inflame.—We have here to remark farther, that in accordance both with calculation and experiment, the direction in which a pencil of rays falls upon a lens does not affect the result of the covergence to a focus, only the focus is always in the direction of the central ray of the pencil or beam; it will be at *p*, for instance, for light issuing from *o*, and at *z* for light issuing from *x*.

The lens represented at *a b* above or in the annexed diagram, at fig. 1, having both sides convex, is called *a double convex lens.* A glass convex only on one side, and plane or flat on the other, as shown at fig 2, would

Fig. 144.

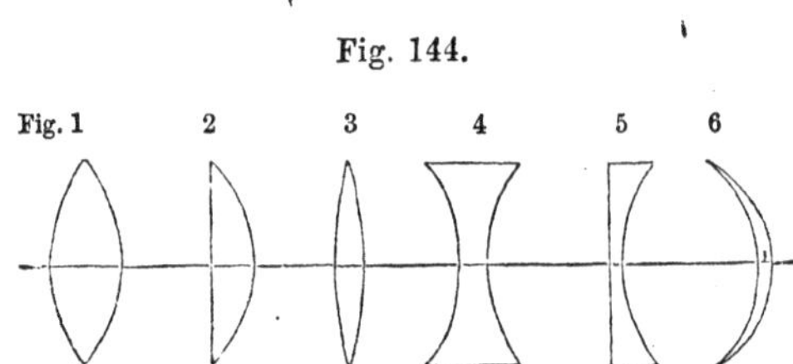

as effectually gather the rays, but with half the power, and the point of meeting or focus would be therefore proportionably more distant. Such a glass is called a *plano-convex lens.* Then the gathering or converging power of any glass, whether doubly or singly convex, is in proportion to the degree of its convexity or bulging of surfaces, for the less it bulges, the more nearly does it approach to a plane glass, and the more it bulges, the more obliquely will the rays, at any distance from the centre, fall upon its surface, and the sooner, therefore, in consequence of their being more bent, will they all meet the axis ray; hence, fig. 1 would converge much more quickly than fig. 3, which represents nearly a common spectacle glass; and a very minute globe is the form most powerfully converging of all. The surfaces of fig. 1 are proportions of a small globe; those of fig. 3, are smaller portions, but of a globe much larger. Concave lens as—fig. 4, a double concave, and at fig. 5, a plano-concave lens, in obedience to the same law of refraction, spread rays, or bend them away from the axis of the pencil, in the same degree that similarly convex lenses gather them. A concave lens, therefore, receiving the converging pencil of rays from a convex lens, might restore them to their former direction. Very useful purposes, as will be afterwards explained, are served in optics, by certain combinations of differently formed lenses. A lens may be convex on one side and concave on the other, as at fig. 6, called a meniscus lens, because it resembles the crescent moon, and its effect will be according to the curve which predominates.

A person collecting the case of the "multiplying glass," described a few pages back, might say,—but is not a convex lens merely a multiplying glass of a much greater number of faces, and if so, why instead of one image, does it not make thousands? The answer is, that the multiplying glass, by every face, bends *a set* of rays, capable of forming a distinct and complete image; but the lens has no surface large enough to bend more than a single ray and it concentrates all the single rays into one place, to form there one image of great vividness and beauty.

"*And when the light proceeding from every point of an object placed before a lens is collected in corresponding points behind it, a perfect image of the object is there produced. When the image is received upon a suitable white surface in a dark place, the arrangement is called, according to minor circumstances, the* CAMERA OBSCURA, SOLAR MICROSCOPE *or* MAGIC LANTERN." (Read the Analysis, page 326.)

Words are wanting to express the admirable consequences to man of the curious property of a lens that it can bring together to focal points behind all the rays of light which traverse it from any points of an object placed before it. The following instance will lead to the understanding of others. If a lens as *a*, be placed so as to fill up an opening made in the window-shutter of a darkened room, then, from any object before that opening—as the cross here presented, all the light which each point emits towards the lens will be concentrated or gathered together in a corresponding focal point behind the lens or within the room and if a sheet of paper be held there at the distance of the focal points, a beautiful image of the object will be seen upon the paper.

Fig. 145.

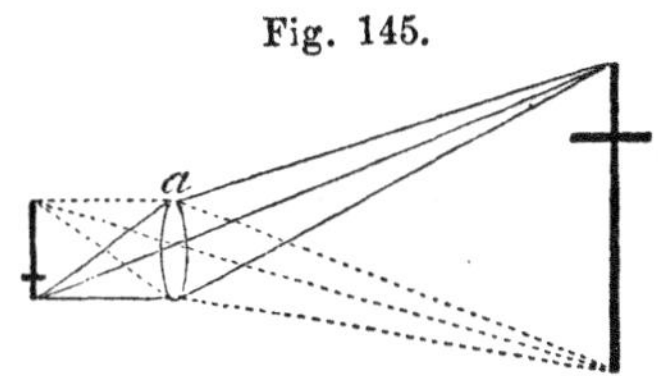

In these few words, we have described the interesting contrivance called the *camera obscura* or *dark chamber;* and when a glass is chosen of proper size and focal distance, and a screen or the wall of the chamber (if at the required distance,) is properly prepared to receive the light, the most enchanting portraiture is instantly produced of the whole scene which the window commands. With what rapture does the school-boy first view this lovely picture drawn by nature's own pencil, and with colors taken directly from the sun's bright ray—with what rapture, as his eyes search over it, does he recognize, perhaps, his playmates there, and the river in which he bathes, and where he sails his boat, and the wood in whose solitudes he loves to wander, and the mountain heights which he climbs to meet the fresh breeze, and at a distance from the world, to allow his young fancy to work, beginning to shoot far into time and space. The great peculiarity of such a picture is, that it does not, like others, portray still-nature, but every thing with appropriate motion or changes: the playmates are all in action: the leafy trees wave in the wind, the clouds sail along, the sun may rise or may set, and even the lightning's gleam may dart across: or, again commenced enterprize may be brought to a close; the traveller may climb the distant hill and disappear; the fisherman may draw his net and secure his prize; the contested race may be won or lost. A Maylan chief in the island of Sumatra, was so surprised and pleased by a small portable camera obscura which the author happened to have among his apparatus, that he seemed disposed to give for it almost any thing he possessed.

It appears in the last diagram that the image formed beyond a lens by the gathered light, is in a contrary position to the object itself,—that is, inverted,—because the light from the top of the object darts through the opening or glass in a descending direction, and that from the bottom rises to the opening, and in the same direction passes beyond it. It is usual, therefore in a camera obscura, to place a small mirror immediately behind the lens, so as to throw all the light which enters downwards, to a whitened table, upon which the picture may be conveniently contemplated.

The camera obscura often gives very useful assistance to young painters, by enabling them to trace correctly the outlines of the objects placed before it, and also to study effects of light, shade and colour, more profitably than they at first can, by looking at the objects themselves. The laws of perspective are most intelligibly illustrated in this most true picture

An effect, approaching in a degree to that of the complete camera obscura now described, is produced by merely making a small hole in the shutter of a dark room, and letting the light which enters by it fall on any white surface beyond. The whole landscape is then dimly portrayed upon the surface. Barry, the painter, while lying on a sick bed, mistook such a scene appearing on the ceiling of his room for a supernatural vision. If a cross be held before the opening as in the last figure, it is evident that from every point of the cross light will enter by the opening, and will fall on corresponding parts of a sheet of paper held behind,—but as the light from each point is not a single ray, but a spreading pencil or cone of light, it will fall on the paper, not on one point, but on a surface at least as large as the opening, and thus the light from adjoining points will mix at the edges, and will render the images misty and indistinct, somewhat like those on the back of tapestry. If the opening be very small, the picture will be well defined, but very feebly illuminated; and if the opening be of considerable size, the mixing of the pencils will be so great as to leave no particular object distinguishable. But, in the latter case, and however large the opening be, if a lens be introduced, it will converge every pencil of light to an exact point, and the picture will instantly be rendered perfectly clear. A lens is never held up in the light without forming beyond it pictures such as now described, of every visible object about it, and the pictures are not seen, only because there are no screens placed to receive them, and because they are so numerous as to confuse one another,—in other words, because they are not admitted singly into a *dark* chamber.

The distance from a lens at which an image is formed or the rays of the light meet, depends first upon the refractive or bending power of the lens, and therefore, on its form and on the nature of its substance; and, secondly, upon the direction of the rays of light when they reach the lens, as to whether they are divergent, parallel or convergent. We have already explained that glass refracts about twice as much as water, and that diamond refracts about twice as much as glass: and we have considered the effect of different degrees of convexity in lenses—arising equally whether the lens be of water enclosed between glasses like watch-glasses, or of solid glass, or of rock-crystal or of diamond itself. We now proceed to consider the joint effect of the refractive power, and of the direction of the incident rays.

Rays falling from *a* on a comparatively flat or weak lens at L, might meet

Fig. 146.

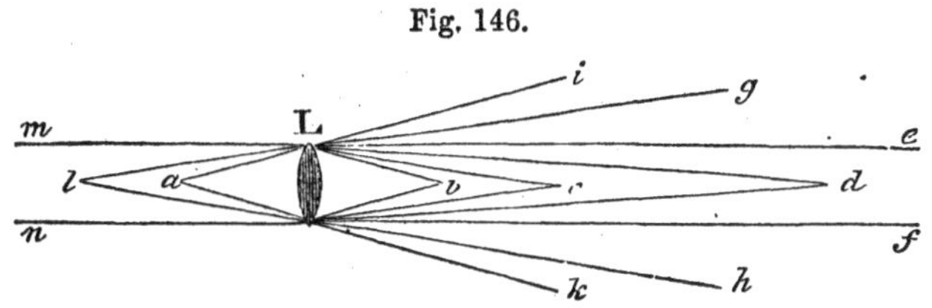

only at *d*, or even farther off: while, with a stronger or more convex lens, they might meet at *c* or at *b*; a lens weaker still might only destroy the divergence of the rays, without being able to give them any convergence or to bend them enough to bring them to a point at all—and then they would proceed all parallel to each other, as seen at *e* and *f*;—and if the lens were yet weaker, it might only destroy a part of the divergence, causing the rays from *a* to go to *g* and *h*, after passing through, instead of to *i* and *k*, in their original direction.

In an analogous manner, light coming to the lens in contrary directions from *b c d*, &c., might, according to the strength of the lens, be all made to come to a focus at *a* or at *l*, or in some more distant point; or the rays might become parallel, as *m* and *n*, and, therefore, never come to a focus, or they might remain divergent.

It may be observed in the figure above, that the farther an object is from the lens, the less divergent are the rays which fall from it upon the lens; or the more nearly do they approach to being parallel. From *b* there is much divergence in the exterior rays, from *c* less, from *d* less still, and rays from a great distance, as those represented by *e* and *f*, appear quite parallel. If the distance of the radiant point be very great, they really are so nearly parallel, that a very nice test is required to detect the non-accordance. Rays for instance, coming from the earth to the sun, do not diverge the millionth of an inch in a thousand miles. Hence, where we wish to make experiments with parallel rays, we take those of the sun.

Any two points so situated on the opposite sides of a lens, as that when either becomes the radiant point of light, the other is the focus of such light, are called *conjugate foci*. An object and the image of it formed by a lens, must always be in *conjugate foci*, and as the one is nearer the lens, the other will be in a certain proportion more distant.

What is called the principal focus of a lens, and by the distance of which from the glass we compare or classify lenses among themselves, is the point at which the sun's rays, or any parallel rays, are made by it to meet; and thus, by holding the glass in the sun, and noting at what distance behind it the little luminous spot or image of the sun is formed, we can at once ascertain the focus of a glass—as at *a* for the rays *e* and *f*.

It is a remarkable coincidence that the bending power of the common glass used for lenses should be such, that the focus of a double lens is just where the centre of the sphere would be, of which the surface of the lens is a portion. This gives us another fact with which to associate the recollection that the focus is nearer as the convexity of the lens is greater, that is to say, as the surface is a portion of a smaller sphere, And such being the law, it may be proved by calculation as well as by the fact, that if a candle be held in relation to a lens at twice the principal focal distance, suppose at *c* for a lens with the focus at *a*, the image of the candle will be formed at *l* just as far on the other side. Thus, then, by trying with the lens until the image of a candle is formed at the same distance from it as the object is, we have a second mode of ascertaining the focal distance of a lens. Other kinds of glass and other substances refract with different power; but the facts now stated should be retained in the memory as standards of comparison.

Because the focal point of light passing through a lens is at the same distance from the centre of the lens, in whatever direction the light passes through, a surface placed to receive the image of any broad object should really be concave, that is to say, all parts of it should be at the same distance from the centre of the lens, otherwise the image will be more perfect either at its middle than towards its edges, or *vice versa*—but it is not found necessary to attend to this in common practice, where the object and its image are not of great extent.

The size of an image formed behind a lens is always proportioned to the distance of the image from the lens, and the image is much larger or smaller than the object as it is farther from or nearer to the lens than the object. This will be evident from considering the annexed figure *c* represents the place of a lens, and the lens, according to its power, will form an image of the cross *a b*, in some situation, as at *d*, *e*, *g*, &c. Now whenever the image is formed and by whatever lens, one end of it must be in contact with the line *a g*, and the other end with the line *b h*; and as these lines cross each other at *c*, and widen regularly afterwards, a line adjoining them (and the image is such a line,) must always be shorter the nearer it is to *c*, that is to say, shorter in proportion to the converging power of the lens.

Fig. 147.

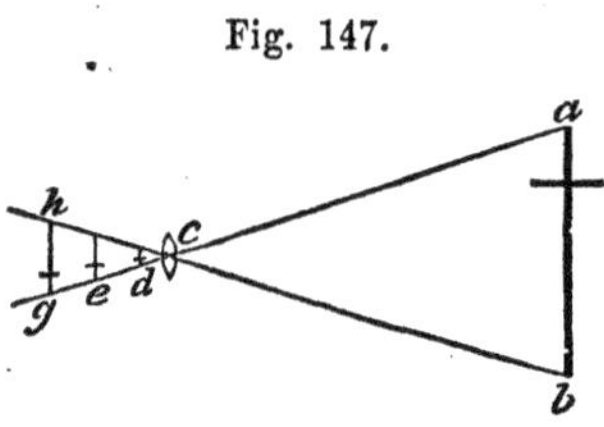

Many persons may not have reflected, that the luminous circle called the focus of a burning glass, is really but the image or picture of the sun formed by the glass or lens. The intensity of the heat and of the light is of course in proportion as the image is smaller than the glass which forms it, and the nearer that the image is formed to the lens, or the more powerfully convergent that the lens is, the smaller will the image be. Mr. Parker's famous burning lens, which cost £700, and is now the property of the Emperor of China, was three feet in diameter, and the diameter of the sun's image formed by it was one inch: it concentrated the light and heat, therefore, about 1,300 times. To render the effect still more powerful, a smaller lens was placed behind the larger, farther reducing the size of the image to one-sixth. Very surprising effects were produced by this lens, in the melting of metals, inflaming of combustibles, &c. The size of burning lenses, until lately, was limited by the difficulty of obtaining the great pieces of glass required to form them: but they are now built up of many pieces suitably united together. Some large lenses have been made of water, that is, of water enclosed between menicus glasses, like watch-glasses. A common goblet of water, or a vase holding gold-fishes, has in some cases acted as a burning glass, setting fire to the curtains, near which it had been left in the sunshine.

And the nearer that an object is brought to a lens, the more distant, and therefore the larger will its image be; for, as the rays falling upon a lens are divergent in proportion to the nearness of the object, and therefore with the same power of lens, must meet farther behind (as seen in the figure at page

343,) then the axis of the rays, as the lines *c a* and *c b* in the last figure, will have separated far before the rays meet, and will have made the image proportionally larger. If we suppose little *d* in the same diagram to be the object, its image would be *a b*. The sun is exactly as much larger than his image formed by a burning-glass, as he is more distant from it than the image; and if we had a screen of sufficient size hung up in a distant space, a very bright object of a quarter of an inch in diameter might be made by a lens to form an image as broad as the sun.

From all these considerations, we see that, in a camera obscura, the screen should be from the lens, at the distance of its principal focus for distant objects, and a little farther than this for near objects Accordingly the lens is generally fixed in a sliding piece, which allows the distance from the screen to be adjusted to circumstances. If the representation be desired large, the lens must be of a long focus; if small the lens must be of a short focus. Again, when the reversed use of the lens, a small object as *d* is to be magnified to such a size as *a b*, then the object must be placed a little beyond the focus of the glass; for if placed nearer, the pencils of rays from it would never be gathered to focal points at all, and no image would be formed at any distance.

When, as alluded to in the last sentence, a small object is placed very near a lens, and the image of it is thrown upon the wall of a dark room, perhaps a hundred times farther from the lens than the object is, the image is a greatly magnified representation of the object, *viz.*, it is a hundred times longer and a hundred times broader, and therefore has ten thousand times as much surface as the object; but if in this experiment the object be illuminated only in an ordinary degree, the light from it is so scattered as not to suffice for distinct division. Hence, to attain fully in this manner the purpose of a microscope, a very strong light, concentrated by a suitable mirror or glass, must be directed upon the object. When the light of the sun is used in such a case, the complete apparatus is called a *solar microscope*, and serves beautifully to display the structure of any minute objects. When artificial light is used, as of a lamp, the apparatus is called the *lucernal microscope* or *magic lantern*.

A good solar microscope becomes one of the most interesting presents which science has made to man, for aiding him in his researches into the secrets of nature. With the late improvements in the construction of lenses by which the dispersion of light or the rainbow-fringe, is prevented, (as will be explained under the head of Telescopes,) objects may be magnified two or three hundred thousand times, and still be so luminous as to be beautifully distinct; thus a cheese-mite will appear of the dimensions of a hog, and creatures altogether invisible to the naked eye, or perceived by it only as minute white points are discovered to be animated beings, having the perfect proportions, and often the beauty of larger animals, and endowed with similar appetites, passions, and apparent ingenuity, but with an activity far surpassing that met with in the more bulky creation. A judicious selection of objects for the solar microscope is calculated exceedingly to surprise the mind on its first attending to them, and to fill it with high conceptions of the infinity of creation. With the common microscope only one person at a time can feast his wonder; but with a solar, a whole roomfull of company may at once contemplate the same objects and witness the same actions, and thus have their admiration increased by the consciousness of sympathy.

The magic lantern, we have said, consists of a powerful lens, with objects, highly illuminated by lamp-light, placed so near it that their images are formed far off, and are therefore proportionally larger. For the magic lantern the objects are generally paintings made on thin plates of glass with transparent colours; and each plate is formed to slide through a slit or passage behind the lens. The lens itself, or what may be called half of it, (for there are often two lenses joined to give greater power,) is moveable with the tube which is seen projecting from the lantern, so that its distance from the object may be varied, and thus a corresponding approach to or receding from the screen may be allowed, which will produce an increase or lessening of the magnitude of the visible picture on the wall.

Some public lectures on astronomy and other branches of natural history, have had the drawings and paintings required for the elucidation of their subjects made in miniature upon glass, to be magnified afterwards to the degree desired, and shown upon any part of the lecture-room by the magic lantern.

A thick fog or smoke at night will sometimes reflect the images of a magic lantern so as to make them distinctly visible; and there are several cases on record, where persons, wickedly ingenious in this way, have terrified ignorant individuals almost to death, by throwing spectres from a concealed lantern. Some years ago a sentinel in St. James' Park was thus persuaded that he had seen supernatural beings near him among the trees.

A very charming illusion is produced by a magic lantern manœuvred on one side of a thin screen, while the spectators not aware of the existence of a screen, are sitting on the other side. The image—let us suppose it that of a genius flying in the air—may be first thrown upon the screen from the lantern while very near, and then it will be small, and, if desired, exceedingly bright because the light is much concentrated. If the exhibitor then gradually recede from the screen, adjusting at the same time the distance of the lens from the picture, the image will become progressively larger, and to the spectators will appear to be soaring and approaching, until at last the expanded wings and limbs seem hovering almost over their heads. An endless variety of most ingenious and beautiful exhibitions of this kind have been made under the name of *phantasmagoria* or *raising of spectres*.

"*The* EYE *itself is, in fact, but a small camera obscura.*" (Read the Analysis, page 325.)

Who could at first believe that in describing the camera obscura, as we have now done, we had in reality been describing only a large model of that most interesting of the objects of creation, the living eye itself, the inlet of man's knowledge,—or what may be called the visible dwelling of the soul—that from which the fire of passion darts, through which the languor of exhaustion is perceived, and in which life and thought seem concentrated! Yet the eye is nothing but a simple camera obscura: formed of the parts described above as essential to the camera obscura: but in its simplicity so perfect that they who delight to find around them tangible evidences of the existence of an all-wise and good Creator, point to this in the midst of thousands, as one of the most undeniable and triumphant proofs. We shall now describe the eye and its actions: and keeping present to us the idea of the camera obscura, as already treated of, we shall find that the use of the various parts will be declared by merely enumerating them. This paragraph should be perused while the reader has the opportunity of observing either his own eye reflected in a mirror, or the eye of some companion near him.

The *human eye*, then is a globular chamber of the size of a large walnut, having for its outer wall a very tough membrane called, from its hardness, the *sclerotic coat*, in the front of which there is one round opening or window, named, because of its horny texture; the *cornea*. The chamber is lined with a finer membrane or web, the *choroid* (having relation to colour,) which, to insure the internal darkness of the place, is covered with a black paint, the *pigmentum nigrum*. This lining is bordered at the edge of the round window by a folded drapery, the *ciliary processes*, hidden from without by being behind the curious contractile window-curtain the *iris*, (so named for its rainbow variety of colour in different persons,) through the central opening of which, called the *pupil*, the light enters. Immediately behind the pupil is suspended, by attachments among the ciliary processes, the *crystalline lens*, a double convex, most transparent body of considerable hardness, which so influences the light passing through it from external objects, as to form perfect images of these objects, in the way already described, on the back wall of the eye, over which the optic nerve, there called the *retina*, is spread as a second lining. The eye is maintained in its globular condition by a watery liquid which distends its external coverings, and which, in the space before the lens, or *the anterior chamber of the eye*, being perfectly limpid, is called the *aquous humour*, and in the remainder or larger *posterior chamber*, being enclosed in a pellucid spongy structure, so as to acquire somewhat of the appearance of melted glass, is called the *vitreous humour*.

The annexed figure represents an eye of the common dimensions, supposed to be cut through its middle, from above downwards, so as to show the edges of the coats, &c. C is the outer or *sclerotic* coat, known popularly, where most exposed in front, as the *white of the eye*. A is the transparent cornea joined to the edge of the round opening of the sclerotic; it is more bulging than the sclerotic, or forms a portion of a smaller sphere than the general eyeball, so that, while it may be truly called a *bow-window*, it, or rather the convex surface of its contained water, is also a powerful lens for acting on the pencils of entering light. At B, the similarity all round the edge of the cornea, is attached to the window-curtain or *iris*, shown here edgeways immersed in the aqueous humour, and hanging inwards from above and below towards its central opening or *pupil*, through which the rays of light are passing to the lens. The iris has in its structure two sets of fibres, the circular and the radiating, which cross and act in opposition to each other;—when the circular fibres contract, the pupil is lessened, when the radiating contract, it is enlarged; and the changes happen according to the intensity of light and the state of sensibility of the retina, as may at any time be proved by closing the eyelids for a moment to make the pupil dilate, and then opening them towards a strong light, to make it contract. Behind the pupil is seen the *lens* D

Fig. 148.

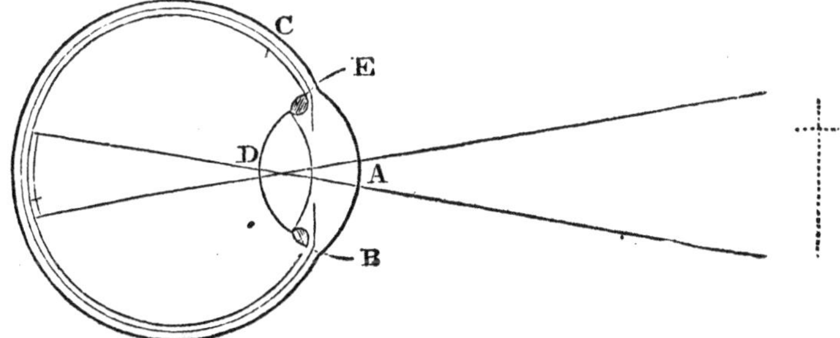

with its circumference attached to the *ciliary processes* E: it is more convex behind than before. The disease of the eye called *cataract* (from the Greek word implying *obstruction*,) is the circumstance of the lens become opaque, and the cure is to extract the lens entirely, or to depress it to the bottom of the eye, and then to substitute for it externally a powerful artifical lens or spectacle-glass. The three lines marking here the boundary of the eye stand for its three coats as they have been called, the strong *sclerotic*, and the double lining of the *choroid* and *retina*. The figure of a cross is represented upon the retina as formed by the light entering from the cross without (which cross has to appear here small and near, although supposed to be large and distant.) The image of the cross is inverted, as explained for the camera obscura: but we shall learn below that the perception of an object may be equally distinct in whatever position the image fall on the retina. It has been explained above, that a lens can form a perfect image of considerable extent only on a concave surface,—and the retina is such a surface. The present diagram further explains what is meant by the *anterior* and *posterior chambers* of the eye, namely, the compartments which are before and behind the crystalline lens D.

The nature of the eye as a camera obscura is beautifully exhibited by taking the eye of a recently killed bullock, and after carefully cutting away the back part of the two outer coats, by going with it to a dark place and directing the pupil towards any brightly illuminated objects; there may then be seen though the semi-transparent retina, left as a screen at the back of the eye, a minute but perfect picture of all such objects—a picture, therefore, formed on the back of the little apartment or camera obscura by the agency of the convex cornea and lens in front. The picture is inverted, for reasons explained above.

Understanding from all this, that when a man is said to be looking at an object, his mind is in truth only taking cognizance of the picture or impression made on his retina, it excites admiration in us to think of the exquisite delicacy of texture and of sensibility which the retina must possess, that there may be the perfect perception which really occurs of even the separate parts of the minute images there formed. A whole printed sheet of newspaper, for instance, may be portrayed on the retina on less space than the surface of a finger-nail, and yet not only shall every word and letter be separately perceivable, but in the centre of the picture at least, even an imperfection of a single letter. Or, more wonderful still, when at night an eye is turned up to the blue vault of heaven, there is portrayed on the little concave of the retina the boundless concave of the sky, with every object in its just proportions. There a moon in beautiful miniature may be sailing among white-edged clouds, and surrounded by a thousand twinkling stars, all in just proportion, so that to an animalcule within and near the pupil, the retina might appear another starry firmanent decked in its glory. If the image in the human eye be thus minute, what must they be in the little eye of a wren, or of other animals smaller still! How wonderful are the works of nature!

Because the images formed on the retina are always inverted as respects the true position of the objects producing them—just as happens in a simple camera obscura—persons have wondered that things should appear upright, or in their true situations. The explanation is not difficult. It is known that a man with wry neck judges as correctly of the position of the objects around him as any other person—never deeming them to be inclined or crooked, because their images are inclined in relation to the natural perpendicularity of his retina; and that a bed-ridden person, obliged to keep his head upon his pillow, soon acquires the faculty of the person with wry neck; and that an affected girl inclining her head while trying her various attitudes, learns from much practice, to judge of the manœuvres of a beau as conveniently in that way as in any other; and that boys who at play bend themselves down to look backwards through their legs, although a little puzzled at first, because the usual position of the images on the retina is reversed, soon see as well in that way as in any other. It appears, therefore, that while the mind studies the form, colour, &c., of external objects in their images projected on the retina, it judges of their position, not by the accidental position of the images on the retina, but by the direction in which the light comes from the object towards the eye—no more deeming an object to be placed low because its image is low in the eye, than a man in a room into which a sunbeam enters by a hole in the window-shutter, deems the sun low because its image is on the floor. A candle carried past a key-hole, throws its light on the opposite wall, so as to cause the luminous spot there to move in a direction the opposite of that in which the candle is carried; but a child is very young, indeed who has not learned to judge at once of the true motion of the candle by the contrary apparent motion of the image. A boatman, who, being accustomed to his oar, can direct its point against any object with great certainty, has long ceased to reflect, that to move the point of the oar in some one direction, his hand must move in the contrary direction. Now the seeing things upright by images which are inverted, is a phenomena akin to those which we have here reviewed.

Another question somewhat allied to the last is, why, as we have two eyes, and an image of any object placed before them is formed in each—why the object does not appear to us to be double. In answer to this, again, we shall only state the simple facts of the case. As in two chess-boards there are corresponding squares, so in the two eyes there must be corresponding points, and when on those points a similar impression is made at the same time, the sensation or vision is single; but if the impression be made on points which do not correspond, owing to some disturbance of the natural position of the eyes, the vision becomes double. Healthy eyes are so wonderfully associated, that from earliest infancy they constantly move in perfect unison. By slightly pressing a finger on the ball of either eye, so as to prevents its following the motion of the other, there is immediately produced the double vision; and tumours about the eye often have the same effect. Persons who squint have always double vision: but they acquire the power of attending to the sensation in one eye at a time. Animals which have the eyes placed on opposite sides of the head, so that the two can never be directed to the same point, must have in a more remarkable degree the faculty of thus attending to one eye at a time.

The corresponding points in the two eyes are equidistant and in similar directions from the centres of the retinæ, which centres are called the points of distinct vision, and at them the imaginary lines named the axes of the eye terminate—but it is worthy of remark that these points, in being both to the right or both to the left of the centres, must be one of them on the inside of the centre as regards the nose, and the other on the outside—that is to say, a point of the left eye between the centre and nose, has its corresponding point in the right eye between the centre and the cheek—and from this fact arises consequences meriting attention. When the two eyes were directed to any object, their axes meet at it, and the centres of the two retinæ are opposite to it, and all the other points of the eyes have perfect mutual correspondence as regards that object, giving the sensation of single vision; but the images formed at the same time, of an object nearer to or farther from the eye than the first supposed, cannot fall on *corresponding* points for an object nearer than where the axes meet would have both its images on the outsides of the centres, and an object more distant would have both its images on the insides of the centres, and in either case the vision would be double. Thus if a person hold up one thumb before his nose, and the other in the same direction, but farther off, by then looking at the nearest, the more distant will appear double, and by looking at the more distant, the nearest will appear double.

The reason for applying the term "point of distinct vision" to the centre of the retina, is felt at once by looking at a printed page, and observing that only the one letter to which the axes of the eye is directed, is distinctly seen; and, consequently, that although the whole page be depicted on the retina at once, the eye, in reading has to direct its centre successively to every part.

On examining a dead eye, the point of distinct vision is distinguishable from the retina around by being more transparent. It might have been expected that this point would have been where the optic nerve enters the eye: but, in fact, the optic nerve enters considerably nearer to the nose than the point of distinct vision is; and singularly, where it enters, the part is altogether blind or insensible. Had the two optic nerves, therefore, entered, at *corresponding* points of the retina, (in the sense explained above,) there would have appeared a black spot on every object opposite to the insensible points; but as the case really stands, the part of any object from which the light passes to the insensible or blind part of one eye must be opposite to a sensible part of the other. The existence of the blind spot, where the nerve of the eye enters, is discoverable by placing in a row three objects—wafers, for instance—across a table, with intervals of about two inches between them. and then looking with one eye, (the other being shut) from a distance of about eight inches, at the wafer which is on the side of the nose;—the middle wafer will be invisible, although the eye will see that on each side of it; and if the eye be then directed to the middle wafer, the external one will disappear. Another proof is obtained by shutting one eye and looking with the other at the points of two fingers held together before it;—if one of the fingers be then gradually moved away laterally, its point when at a certain distance from the other will disappear, but will be seen again when its distance is still increased.

It appearing, from the explanations now given, that there canot be perfect sight unless where a perfect image is formed on the retina, and the truth having been formerly explained, that images behind any lens will be at different distances from it, according to the various distances of the objects in front, that is to say, according as the pencils of light which fall upon it have more or less of divergence in them, it follows, that the eye in being able, as it is, to see distinctly objects at different distances, (the nearest is about five inches,) possesses a power of altering the relation of its parts to accommodate itself to the circumstances. We do not yet perfectly know whether it does this by lengthening or changing the form of the ball through the action of the surrounding muscles, or by changing the place or the form of the lens, but that one or more of these events occurs there can be no doubt.

Among the eyes of the myriads of mankind, however, it happens that all do not originally possess these powers exactly in the requisite degree, and that many lose them, as life advances, from a natural and usual decay.

Persons are called *short-sighted* whose eyes from too great convexity of the cornea or lens, have so strong a bending or converging power, that the rays of light entering them are brought to a focus before reaching the retina—at *a*, for instance, instead of at *b*: so that the rays, by spreading again beyond the focus, produce on the retina that sort of indistinct image which is seen in the camera obscura, of which the screen is too distant from the lens. This defect of sight obliges the individual when using the naked eye to hold objects very near it, that the consequent greater divergence of the rays may be proportioned to the unusual refracting power of the eye—or the person may find a remedy in placing concave lenses between the object and the eyes, which lenses, by rendering light from objects at a usual distance more divergent, (as explained page 340,) cause the perfect images in the eye to be formed farther from the lens, and thereby on the retina itself. Without concave *spectacles*—as the lenses are called when fixed together in a frame—persons with the defect now under consideration cannot see distinctly any object that is distant, for the rays, coming nearly parallel, are quickly gathered to a focus. This defect often diminishes with years, and the person who in youth needed spectacles, in old age sees well without them.

Fig. 149.

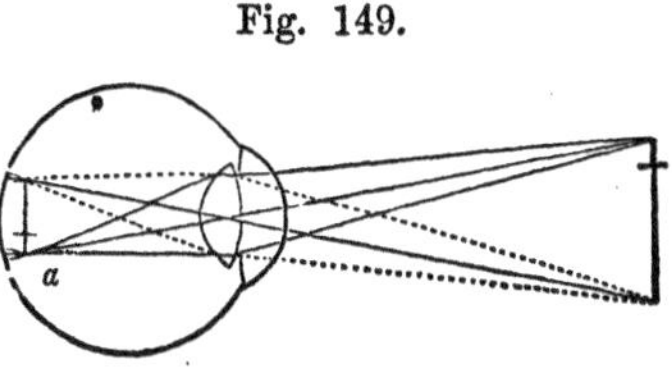

There is an opposite defect of deficient converging power in the eye, dependent on a too great flatness of the cornea or lens, and which is much more common than the last-mentioned defect; indeed, the great majority of persons after middle age sooner or later begin to experience it. In this case, the rays of light are not yet collected into a focus when they reach the retina: they would only meet at *b*, for instance, instead of as they should do at *c*, and hence the image is indistinct, in the same manner as in a camera obscura, when the screen is held too near the lens. Persons suffering this defect cannot, when using the naked eye, see distinctly to it, because the gathering or converging power of the eye cannot conquer the great divergence of rays coming from a near point; and hence such persons always remove objects under consideration to a considerable distance, often to that of arm's length, so as to receive from them only the rays nearly parallel. These persons in contradistinction to the last described, are called *long-sighted* persons; and after middle age, most persons become more or less long-sighted. Their defect is remedied by the common convex spectacles, which do part of the converging work, so to express ourselves, before the light enters the eye, leaving undone only that which the eye can easily accomplish. As this defect, like the last, is met with in all degrees, spectacles must be chosen accordingly. Certain curvatures or strengths of these have been particularized and numbered as naturally belong to different ages or periods of life, but each person should choose under the direction of an experienced judge, until that strength be found which enables him to read, without any straining of the eyes, at the common distance of from twelve to eighteen inches. We cannot apply the mind to this part of our subject without feeling admiration at what science has accomplished for man in assisting and restoring his sight. Now that in civilized

Fig. 150.

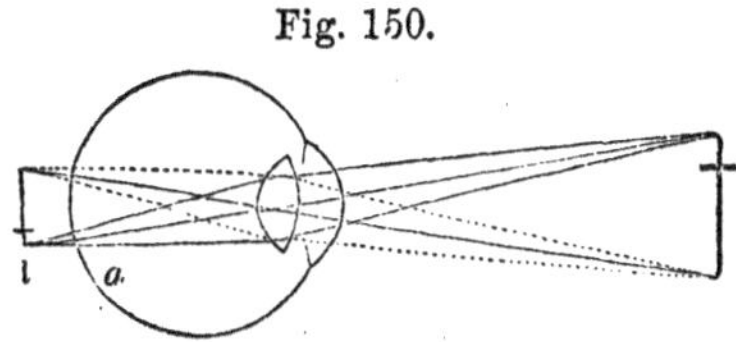

society, the common employments and enjoyments of life require a visual power capable of distinguishing such minute objects as written or printed characters, to deprive old men of their spectacles, would be to condemn many of them to useless inactivity and a listless blank of mind for the remainder of their lives.

An eye much accustomed to examine near and minute objects, often loses something of its pliancy, and becomes defective when tried at a distant thing, as the watchmaker's eye, the engraver's, &c. On the other hand, the old seamen's, which has so often and uninterruptedly been bent on the distant horizon, straining to catch the view of an expected sail, or of land, has a power of discovering distant things which is wonderful; but it often experiences deficiency in regard to near things.

A man who uses his eyes under water sees very indistinctly, because the difference of density between water and the eye not being so great as between air and the eye, the bending or refraction of light entering from the water is not so great as of light entering from air, and the internal structure of the human eye being adapted to the greater refraction, perfect images are not formed on the retina. A man to see well under water, therefore, requires to aid the usual power of his eyes by strong convex spectacles. It is to meet the necessity now explained, that the lens of a fish's eye is extremely convex, or almost round, as is every day seen in the white round bead which issues from the eye of a boiled fish—that little globe being the crystalline lens of the fish coagulated or hardened like the white of an egg during cooking.

There are many important considerations connected with the sensibility of the retina, which regard rather the laws of life than of light, but we must here glance at a few of them.

Any impression of light made upon the retina lasts for about the sixth of a second. Hence when the burning end of a stick is made to sweep rapidly across the view, its path appears to the eye a line of light: and if it be made to revolve in a circle six times in a second, as when moved by the hand or fixed to a turning wheel, that circle will appear to the eye a complete ring of fire. The polished end of an elastic wire, of which the other end is fixed in a block of wood, when caused to vibrate, similarly forms a line or a curve of light. A harp-string, while vibrating as it sounds, appears like a flat transparent riband. Lightning or other meteor darting across the sky, although in fact but a moving luminous point, is generally thought of as a long line of light: the term forked-lightning has reference to this prejudice. The same remark applies in a degree to a sky-rocket in its rapid ascent. Two or more colours painted separately on the rim of a wheel which is made to turn rapidly, appear to a spectator to be as completely united as if they were really mixed;—it has been already explained how patches of all the colours of the rainbow, when mixed in this way, form white light. If on one side of a card a little bird be painted, and on a corresponding part of the other side a cage, then, on making the card turn rapidly by twisting between the fingers and thumbs two threads fixed to its opposite edges, the bird and cage will be seen at once, and the bird will appear to be imprisoned in the cage;—or, if a pensive Juliet, sitting in her bower, occupy one side of the card, and a longing Romeo the other, by the magic turn of the threads the lovers may instantly be brought together. Dr. Paris displayed taste and an amiable ingenuity in designing this toy with great variety of subjects.

A certain intensity of light is necessary to distinct vision, but the degree varies with the previous state of the organ. A person passing from the bright day into a shaded room, might for a time fancy himself in total darkness, and to persons sitting in the room, and become so accustomed to the less light as to see well with it, he might appear to be almost blind. The dawn of morning after the darkness of night appears much brighter than an equal degree of light in the evening. When, as the night falls, our lamps or candles are first introduced, the glare is often for a time offensive to the eye; and a similar feeling but still stronger is experienced, when in the morning, bed-room window shutters or close drawn curtains are suddenly opened After the repose of night, the sensibility of the eye, when first opened, is often such that the globules of blood moving in the capillary vessels of the retina produce the impression there of little balls of light pursuing one another along the tortuous vessels. To a prisoner after long confinement in a dark dungeon, the light of the sun is almost insupportable. And a dungeon, which to common eyes is utterly dark, still to its long-held inmate has ceased to be so;—there are various instances in the records of the barbarous ages of prisoners confined for years in darkness, deemed absolute, but who after a time could see in it, and make entertaining companions of the mice and spiders which frequented their cells. The darkness of a total eclipse after bright sunshine, appears deeper than that of midnight, because of the sudden contrast. The long polar night of months ceases to appear very dark to the polar inhabitants If an eye be directed for a time to a black wafer laid on a sheet of white paper, and be then turned to another part of the sheet, a portion of the paper at that other part, of the size of the wafer, will appear brilliantly illuminated; for the ordinary degree of light from it appears intense to the part of the retina lately receiving almost none. An eye directed long and intensely upon any minute object—as when a sailor watches a speck in the distant horizon, supposed to be a ship, or when a hunter on the brown heath keeps his eye fixed on a bird nearly of the colour of the heath or when an astronomer gazes long at a little star—has the sensibility of its centre at last exhausted, and ceases to perceive the object; but on directing the axis of the eye a little to one side of the object, so that an image may be formed only *near* the centre, the object may be again perceived, and the centre in the mean time enjoying repose, will recover its power.

But the most extraordinary fact connected with the sensibility of the retina is, that if part of it be strongly exercised for a time by looking at some bright-coloured object, on the eye being then turned away or altogether shut, an impression of spectrum will remain of the same form as the object lately contemplated, but of a perfectly different colour. Thus if an eye be directed for a time to a red wafer laid on white paper, and be then shut or turned to another part of the paper, a beautifully bright green wafer will be seen; and *vice versa*, a green wafer will produce a red spectrum: an orange wafer will similarly produce a blue spectrum; a yellow one a violet spectrum, &c., and a cluster of wafers will produce a similar cluster of opposite colours. Then if the hand be held over the closed eye lids to prevent entirely the approach of light to them, the spectrum of bright objects will appear luminous surrounded by a dark ground, and when the hand is again removed, the contrary will be true. Again, if the eye be in a degree fatigued by looking at the setting sun, or even at a window with a bright sky beyond it, or at any very bright object, on then shutting it, the lately contemplated forms will be perceived, first of one vivid colour, and then of another, until perhaps all the primary colours have passed in review. These extraordinary facts prove that the sensations of light and colour, although excitable by light are also producible without it. This truth gave occasion to Darwin's ingenious theory, that the sensation of any particular colour, of red, for instance, is dependent upon a certain state of contraction of the minute fibres of the retina,—and that the fibres, when fatigued in that condition, seek relief when at liberty, by throwing themselves into an opposite state,—as a man whose back is fatigued by bending forward, relieves himself not by merely standing erect, but by bending the spine backwards—which new condition in the eye, whether produced by light or by any other cause, gives the sensation of green. He applied his explanation similarly to all other cases of colour. It is remarkable that the colours which thus appear opposite to each other in kind are those which, when the solar spectrum produced by a prism, as described a few pages back, is painted round a wheel or circle, are opposite to each other in place.

There are persons who, although having distinct perceptions of form, and of light and shade, have not the power of distinguishing colours. It is common for such persons to deem pink and pea-green (naturally opposites) the same colour, and therefore, not to distinguish difference of colour in a red berry and the leaves around it. A man with this defect trusting to his own judgment, has, without knowing it, dressed himself like a parrot.

' *The mind judges of external objects by the relative size, brightness, colour, &c., of the minute but perfect images or pictures of them formed at the back of the eye on the expansion of nerve called the retina; and the art of painting is successful in proportion as it produces on a larger scale such a picture, which, when afterwards held before the eye to reproduce itself in miniature upon the retina, may excite the same impression as on the original object.*" (Read the Analysis, page 325.)

We now understand how an admirable miniature resemblance of the objects before us is produced upon the retina of the eye, by the light from them refracted in passing through the different parts of the eye; but after all, this is only a picture, and the inquiry remains—which many persons would suppose so simple as to be trifling, but which is in reality most curious and important —how we are thereby enabled to judge of the magnitudes, distances, and other particulars respecting the things examined? Here it will be found, to the surprise of persons first entering upon the subject, that we learn the meaning of a scene or pictorial signs only gradually, as we do of any other system of signs, and that a person whose eyes, although perfect, had been kept covered from infancy up to maturity, would no more "see," in the complete sense of the word, that is, understood, any sense or prospect on which he first opened his eyes, so as to have a perfect picture of it on his retina, than a child understands or can read a printed page, when he first looks into a book. Most interesting information has been obtained on this subject, by observing the facts where blindness from birth has, by surgical operation, been suddenly cured in persons arrived at maturity.

If a man were placed from infancy in an apartment fitted up as a camera obscura, and had no means of becoming acquainted with external nature, but by watching the images appearing on the screen, he could learn scarcely any thing of the universe around him; but if after a time he were allowed to walk out, and to examine by the touch and by the measurement the different objects whose images he was in the habit of viewing, and to ascertain what size, shape and distance of an object corresponded with a certain magnitude form, position, and brightness of image, the transient imagery might at last be to him a very clear indication of the real particulars: making him in imagination present to the objects, almost as if he went and examined them with his hands. In the same manner, in a degree, the mind may be considered as stationed in or about the little camera obscura of the eye whence it cannot itself escape to examine external nature; but must learn the meaning of the images formed on the retina, by commanding the services of the bodily limbs or members, and the other organs of sense.—The judging of things by sight, then, is merely the interpreting one set of signs, as judging by sounds or language is interpreting another, and judging by hieroglyphics or any written character is interpreting a third. The common visual signs on the retina, however, are of all signs the most readily learned or understood, from having certain fixed relations in form, magnitude and position to the things signified; while words, hieroglyphics, and written characters, are quite arbitrary, and have no such relations.

Bodies differ and are distinguished among themselves chiefly by their comparative dimensions, that is, their form and magnitude, or shape and size: and to ascertain these and the relative distances and positions, are the great objects which by the eye the mind seeks to accomplish. Now it effects its ends by considering collectively.

1st. *The space and place* occupied by objects in the field of view, measured by what is called the *visual angle.*

2*d*. *The intensity of light, shade, and colour.*

3*d*. *The divergence of the rays of light.*

4*th*. *The convergence of the axes of the eyes.*

We shall treat of these particulars separately in the order now stated.

1*st*. *The space and place occupied in the field of view, measured by the visual angle.*

The term field of view is used to designate that open or visible space before the eyes, in which objects are seen: and it may mean either the small field visible in one position of the eyes, or that which is perceived on direct-

Fig. 150.

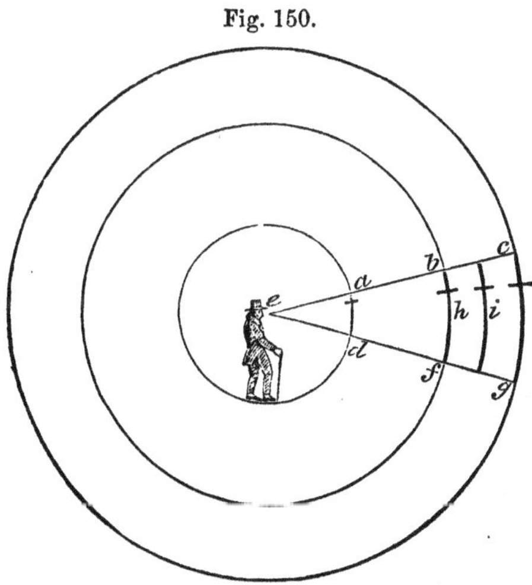

ing them all round. If a man as at *e*, were surrounded by a large globe or sphere of glass as *a*, through which his eye at the centre might view the several objects around occupying certain situations and certain proportions of the circumference, and if the sphere had any equal divisions or degrees marked upon it all round, he would be able at once to say exactly what portion of his sphere or field of view was shadowed or occupied by any single object, as the cross here shown at *i*, and thus to describe very intelligibly either for his own recollection, or to inform others, its relative magnitude and situation as then appearing to him,—just as he might say, on looking at a tree in the garden through a common window which is a portion of the field of view really divided by the cross bars, whether he saw the whole tree through one pane or through several, and through which pane or panes he saw it. It may be remarked farther, that whether the supposed sphere of glass were large or small, viz., were as *b* or *c*, the part of its surface apparently occupied by any object either beyond or within it, would bear the same proportion to the whole surface;—if *a d* were a tenth of the small circle or globe, *c g* would be a tenth of a larger. Now as men have found it convenient to consider a circle (and every circle) as divisible into 360 degrees, (which are smaller, therefore, in a small than in a larger circle, although in each having the same relation to the whole,) the ready mode of comparing the apparent magnitude of objects is to say how many of these degrees of the field of view each object occupies: and this is really what is meant by the apparent size of an object. And because the most convenient way of measuring a portion of a circle, of which the whole is not seen, is to measure by a fit instrument the angle or corner formed at its centre by lines drawn from the extremities of the portion,—as here the angle at *e* formed by the lines *c e* and *g e*, the object is said either to occupy a certain number of degrees of the circumference of the circle, or to subtend an angle of the same number of degrees at its centre, and this angle is called the *visual angle*, the subject of our present disquisition.

The visual angle, then, in regard to any object, is that included between the lines or rays, as *a u* and *d i*, which, from the extreme points of the object,

Fig. 151.

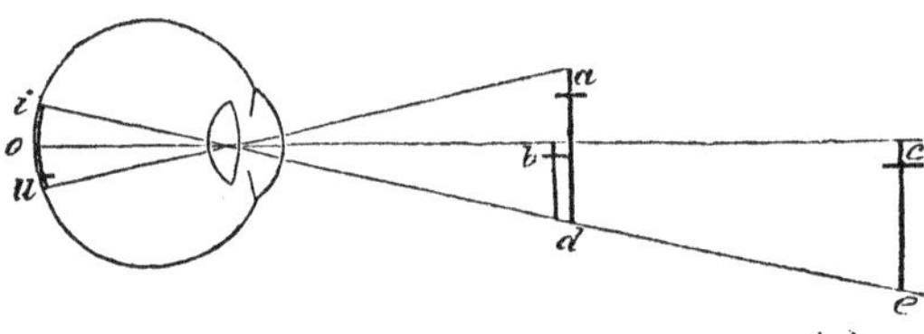

as *a d*, meet and cross in the lens of the eye, and go afterwards to form the extremes of the image on the retina, and, as formerly explained, the angle is the same on each side of the lens, *viz*, towards the object or towards the image.

Now if all bodies were at the same distance from the eye, the magnitude of their images formed on the retina, or in other words, of the visual angles subtended by them, would be an exact measure of their comparative real magnitudes, as is seen in *i u*, the image of the great cross *a d*, and in *i o* the image of the small cross *b d*: but it is evident here, that the cross *c e*, which is twice as large as *b d*, makes, because twice as far off, an image of only the same size as *b d*, and an image therefore only half as large as that of a cross *a d* equal in size with itself: and the same rule of proportion holds for all other comparative distances—at a hundred times the distance, an object appearing only the hundreth part as tall, and so forth. To judge, therefore by the eye, of the true size of an object, we must know its distance as well as its apparent size or visual angle.

Many familiar facts receive their explanation from the law of the visual angle or apparent size being less always in proportion as the distance of an object is greater.

A man (instead of the cross here shown) at *d*, standing near the outside of a window, as *b c* (here represented edgeways) may to the eye of a spectator within the window at *h*, subtend the same visual angle, or appear as tall as the window, the light from the man's head passing through the top of the window, and that from his feet passing through the bottom; but if the man then moves away from the window, the eye of the spectator will be able to see his whole body through a smaller and a smaller extent of the window,—as through half its height or *a c*, when he is twice as distant, or at *f*, and through the third or *o c*, when he shall be three times as distant, or at *g*, and so forth, for any other distance; so that soon a small figure of a man cut in paper, if applied upon the glass, would exactly cover the part of it through which the light from him entered the spectator's eye, and would then, by completely hiding him from view, be an exact measure of his apparent size: and at last a fly passing over the pane might equally hide him, and the fly then would subtend a larger visual angle than he, that is to say, would be forming on the retina a larger image than the man. Thus it often happens in reality, that a person sitting near a window, and intent upon some object of study or of conversation, mistakes a fly on the glass for a man at a distance; or, on the contrary, a man for a fly. It is ascertained that the eye, with an ordinary degree of light, can see an object which in the field of view occupies only the sixtieth of a degree (or one minute.) This space is about the 100th of any inch in a circle of twelve inches in diameter, the eye being supposed in the centre of the circle. Now a body smaller than this at six inches from the eye, or any thing, however large, placed so far from the eye as to occupy in the field of view less space than this, is invisible to ordinary sight. At four miles off, a man becomes thus invisible. A pin-head near will hide a house on a distant hill—nay, will hide even the planet Jupiter, although 1,000 times bigger than this earth.

Fig. 152.

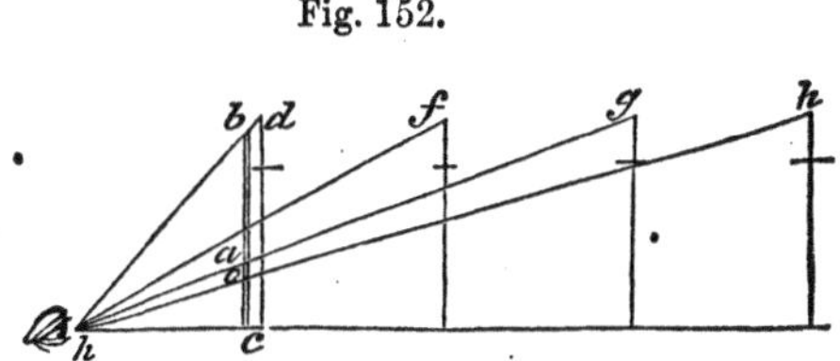

In accordance with the principle now explained, a marine telescope has been constructed, in which the field of view is divided by fine cross wires, or otherwise, so that the person using it can say at once how much of its field any object occupies. When ships are in chase, it is common, by this instrument, or some other which will detect a change of visual angle, or apparent size, to view the fleeing or pursuing ship; and if the apparent size be observed to increase, the conclusion follows that the ships are nearing each other; if, on the contrary, the size diminishes, the chased ship is escaping.

By applying this rule, whenever the real size of a distant object is known, the distance is ascertainable, and, *vice versa*, where the distance is exactly known, the size is determinable:—for it is evident that if a body, as a ship, known to be 100 feet tall, occupy or subtend in the field of vision the 360th part of a whole circle, or one degree, the whole circle must be in circumference 360 times 100 (hundred) feet, or 36,000; and the diameter of any circle being nearly one-third of its circumference, while, in the case supposed, the distance of the ship is the half-diameter, we learn that distance. Again, if we know the distance of a ship or other object to be a mile, and if we then find the visual angle subtended by the object to be the 1,000th part of a circle, we know its true size to be the 1,000th part of a circle, of which the half-diameter or radius is one mile. It is by applying this rule in a manner to be afterwards explained, that we determine the size of the heavenly bodies.

We now perceive that if the rays of light coming to the eye through a plate of glass, from objects seen beyond it, could leave marks in the glass, at the points where they passed, and marks capable of giving out the same kind of light as the objects, there would be formed upon the glass such a representation or picture of the objects formed or viewed *through* it, that when held before the eye, it would form on the retina, the image or images the same in almost all respects as the objects themselves; for from the different points of the glass, light could dart to the eye of the same kinds and in the very same directions as that originally coming from the objects. Now the art of painting seeks so to dispose lights, shades and colours on any plane surface, as to produce the sort of representation of objects here contemplated, while the picture-frame has to recall the window-frame, or edge of the plate of glass through which the true scene is supposed to be viewed. It is admirable how perfectly this art now accomplishes its ends; and although there are still trifling differences between the effect upon the eye, of the picture and of the realities—which peculiarities we shall consider presently, and how they may be combated so as to render the illusion almost perfect,—it is not one of them, as might be supposed from the small extent of the canvass, that the picture appears to the retina smaller than the objects themselves. Few people, before studying this subject, are aware that in a good picture the size of the figures is always made exactly such, that at the distance from the eye at which the picture is meant to be viewed, they produce on the retina the very same size of image as would be produced by the realities seen under the aspect represented in the picture. To become sensible of this, let a person look through a window-pane, with the eye at the distance of eight inches from it, and let him trace with a sharp point upon the glass, previously coated with gum, the outline of the scene beyond—perhaps a street or square,—he will find that the outline of a man seen there at the distance of twenty paces, and appearing perfectly to coincide with the boundaries of the person, so that, if opaque, it would just hide the person, will be scarcely half an inch tall, while the figure of the man a few hundred paces off, will appear so small, that the minuter fractures could not be distinguished, even if they could be drawn.

Now as a person who reads the description of an elephant, does not deem the animal larger or smaller because of the size of letter used in the printing, or in the size of the accompanying engraved representation; and as a man in a picture gallery viewing minatures and larger portraits, does not conceive of the originals according to the size of the representations: and as a man viewing a well-executed picture of a Grecian temple, never dreams, unless his attention be particularly directed to the fact, that upon the canvass the distant pillars of the rows are much smaller than the near ones; but in all such cases the mind merely uses the *signs* to help it to conceive of the *things* according to previous knowledge, or to other principles of judging:—so in any common case of seeing, the mind takes little account of the *apparent* size of objects, but passes instantly from the types to the realities, which are, generally, more or less known, and it soon ceases to be aware that the apparent size of the same object ever changes. Few persons, for instance, are aware that when two friends shake hands, each appears to the mere eye of the other ten times taller than when he has walked ten paces away; or that a chair at one end of a room appears to a person sitting at the other, only half as large as a chair in the middle of the room, but such are the facts; and they may be immediately proved by holding a common eye-glass or ring at a certain distance from the eye, and then looking through it at any similar objects placed at different distances; then, while of a chair standing near, only a small part will be visible through the ring—of a distant chair the whole may be seen; and so of any other case. At five miles distance, the fleets which met on the great day of Trafalgar might have been seen through a marriage-ring as the picture-frame. There are occasions, however, where the usual collateral helps to the immediate recognition of objects being wanting, the observer's attention is strongly aroused to the facts of their diminutive appearance produced by distance; for instance, when a man on a long sea-voyage first approaches a land of which the features are in a degree new to him; as when an Englishman arriving in India, scarcely believes that the little specks which he sees scattered along the shore are commodious dwellings, or that what seem to him only luxuriant herbs or bushes, are magnificent palm-trees.

For the same reason that a distant body to the mere eye appears diminutive, namely, the smallness of the visual angle subtended by it, so does a distant motion to the mere eye appear slow. A carriage dashing past a pedestrian in the street, may surprise him by its speed; but if viewed at the same time by a spectator at the top of a lofty tower near, it seems to be but crawling along the pavement. A ship driven before a tempest, seems to a sailor on board almost to fly through the white foam which surrounds her; but if then observed by a spectator on shore, as an object on the distant horizon, she is scarcely perceived to change her place. A balloon high in the air, and borne along on the wings of the wind at the rate of seventy or eighty miles an hour, may still for a time leave a spectator on earth doubtful as to whether it be in motion, or in what direction it moves. The moon in her orbit wheels round the earth at the astonishing rate of about 2,000 miles an hour, yet, owing to her distance from it, her motion is not visible to the naked eye of the inhabitants of the earth, except by comparing her place at considerable intervals. In respect to bodies still more distant than the moon, the truth at present under consideration is still more striking.

Having now explained how the apparent transverse measures or breadth of bodies and of space, in other words, the visual angle subtended by them, is affected by their distance from the eye, we proceed to show how it is affected also by their position.

A globe at a certain distance from the eye, however turned, preserves the same appearance in the field of view, and its outline traced upon a plate of glass held across between it and the eye, is, like its direct shadow upon a wall, always a circle; but an egg which, held in one position, produces a circular outline or image, when held in another, produces an image nearly oval. A wheel when viewed sideways appears a perfect circle, when viewed edgeways it appears a broad straight band or line, and in any intermediate position it appears oval. The *apparent* form, then, of a body, is only a hint to the mind from which, by former experience or instruction, it guesses at the true form. If a man had never seen an egg but endways, he never could have known that it was not a sphere.

If any long, straight object, as a beam, be placed with one of its ends directly to the eye, that end only can be seen, and according to the case, must appear a square or circle of the diameter of the beam; if it then be placed with its side directly to the eye, its whole length will be seen; and if placed in any intermediate position, it will appear more or less shortened; —in all cases, its outline on the retina being similar to that of its shadow on a wall directly behind the person. A man has advanced on a spear pointed directly to his eye without seeing, or on the end of a bar of iron carried on the shoulder of a porter in the street. A common telescope held with its end to the eye, appears a perfect circle, if then inclined a little, it seems to jut out on one side, and as the inclination is increased, it juts out more and more, until it displays its whole length. A great ship of war whose stern is towards a spectator, appears a rounded building with its rows of windows like those of a peaceful habitation; but as it turns, it gradually reveals the long batteries of bristling cannon. A straight row of a thousand similar objects, as of soldiers in rank, pillars, trees, &c., may appear to a person at the extremity as only one object of the kind, the nearest individual completely hiding all the others; but if viewed from the side and at a certain distance, the individuals may be counted.

The appearance now treated of is called *foreshortening*, and is to be noted wherever surfaces or lines are not placed so as directly to face the spectator.

Perhaps the most important case of foreshortening is when the eye looks more or less obliquely along an extensive plane surface, the general surface of the earth, for instance, or of the sea, by estimating aright the foreshortening of which, we judge of the distance or situation of the objects placed upon it. It will be readily perceived, that in all such cases the more distant portions of the surface are progressively more foreshortened than the nearer;—for a man standing at *a* on the plane as *a b*, and with his eye at *c*, looking down before him, sees a portion of the surface *a d* almost directly, or with a little foreshortening, and an extent, as *a d*, equal to the height of the eye, will subtend in his eye, an angle of 45°, or half a right angle, *viz.*, the angle *a c d*, and therefore rather more than half of all that can be subtended by a straight line or space from his feet to the horizon, however distant; the next equal spaces, *viz.*, *d f*, will subtend an angle of only 18°, *viz.*, *d c f*, the next of 8°, *viz.*, *f c g*, and so on; and as he carries his view more and more forward, the surface becomes to it more and more oblique, until at last the light coming seems more to skim along the level than to rise. This explains why a person having a side view of a row of separate objects,

Fig. 153.

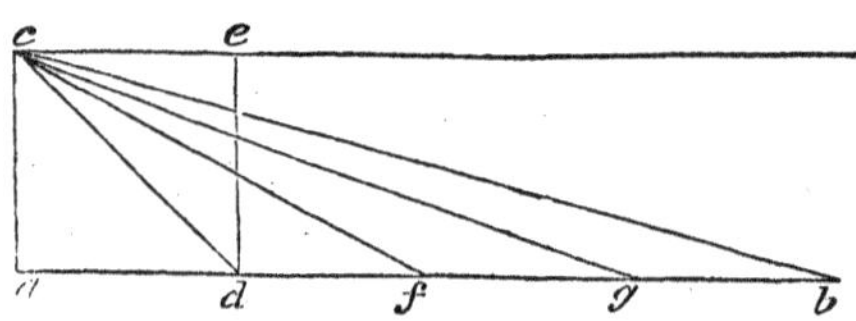

as of men in line, trees, pillars, &c., may see through or between the nearest of them, but towards the extremes sees them as if standing in closest possible array, or as if forming a continued surface. The same remark explains why masses of cloud scattered uniformly over the sky, may allow a spectator to see wide intervals of the blue heaven over head, while all around there is a dense cloudy mass appearing to rest on the horizon.

If a man standing on a hill look down upon a field or plain which is well known to him, and if he see some objects near its side, and some near its middle, and some near its distant border, he knows at once how far they are from him and from one another. Similarly, when viewing the ocean from a lofty cliff, and seeing ships scattered over its face, he may judge correctly of their distance, for he can see only a certain extent of ocean which becomes to him as a known field. The man stationed at the flag-staff on the High Knowl of St. Helena, looks down upon the circular field of the Atlantic a hundred miles broad, and can tell the distance of a sail in sight to within a mile or two. Now, although the ground-plan of an extensive landscape may not be so level as the face of the ocean, there is still an approximation, which very considerably assists a spectator's judgment of distances.

Painters are not only careful to foreshorten, according to the proportion explained above all, the objects which they portray, but they often avail themselves of the principles to produce most striking effects. For instance, Martin, in many of his beautiful designs, by judicious foreshortening, has exhibited miles in extent of gorgeous architecture and of armed men, on a space of canvass that would seem scarcely more than sufficient to receive a very few figures; he has made a single magnificent pillar or accoutred warrior in the foreground, become the type which first fills the mind with admiration, and then sends it along the retiring lines of beautiful perspective, where every tip or edge renews the first impression. A man lying on a high table or bed, with his feet towards the spectator, is foreshortened into a roundish heap, of which the soles of the feet hide the greater part. This is the description of the painting which has been called the "Miraculous Entombment," and it is because an unreflecting spectator moving sideways with the expectation of seeing more of the body, still sees only the soles of the feet, and may suppose the body turned round so as to keep the feet towards him, that the painting has received its appellation. For nearly the same reason, the eye of a common full-faced portrait may seem to follow a spectator to whatever part of the room he goes,—for by moving to one side he cannot see the side of the eye-balls. It is related of a murderer, that he was impelled to commit suicide by observing that the eyes of the portrait of his victim were always fixed upon him. A rifleman portrayed as if taking aim directly in front of the picture, will appear to every spectator in the room to be pointing at him especially. To terrify young ladies, a little arch Cupid has been similarly represented with his arrow pointed directly at them, and just ready to let it slip from his bended bow.

As the painter, availing himself of a knowledge of the principles now explained, by which the eye usually judges of size and distance, may produce on his canvass the most charming illusions, so may the tasteful landlord, in his ornamental gardens and pleasure-grounds, by working his levels into artificial undulation of hill and dale, and clothing these with tree and edifice of magnitudes to correspond—make the eye of a spectator luxuriate in the contemplation of the supposed extensive plains, lofty mountains, widespread lakes, and distant padogas—all within the narrow space of an acre or two; thus, by other means, producing on the retina the same impressions as Claude Poussin, or Wilson has done by the finest pictures.

When any objects or mass of objects is foreshortened, by one part being farther from the eye than another, that part appears also in a proportion smaller than the other. For example, in a straight row of similar houses, pillars, trees, &c., (see the next cut,) those nearest to the eye will, on a glass held before the eye to receive their images, from the largest images, and there will be a gradual diminution from the largest to the least, so that lines drawn upon the glass along the tops and bottoms of the images would tend to a point, called for a reason to be explained below, the *vanishing point.* Thus a person looking from a window along a straight street, must, to see the chimney of the nearest house, look through the top of the window, and to see the street door must look through the bottom; but the most distant house, both top and bottom, may be concealed from view by a little spot upon the glass at the height of the eye. This remarkable tapering of foreshortened objects may of course be strikingly observed on looking at any correctly made drawing or engraving meant to represent a retiring row of similar objects;—such drawing being in truth an attempt to realize by art, on the surface of a sheet of paper, the appearance of the objects as seen through a window or aperture the size of the paper.

The art which gives rules for tracing objects on a plane surface, as they would appear to an eye looking at them through that surface, if transparent, with their various degrees; first, of apparent diminution on account of distance; and, secondly, of foreshortening on account of the obliquity of view, is called, from the Latin word signifying *to look through the art of perspective.* It regards entirely the two particulars now mentioned; and notwithstanding the terror with which, in the imaginations of many young painters, the study of it is clothed, by reason of the mathematical difficulties with which it has usually been mixed up, it is in itself exceedingly simple. We hope that a person capable of ordinary attention, will, after what we have already said, and after the few additional remarks which we have still to make on the appearances of nature, be able readily to understand the great rules of perspective. Although, without a knowledge of these rules, a quick eye soon enables its possessor to sketch from nature with much truth; and although the two instruments, the *camera obscura,* already described, and *camera lucida,* to be described in a future page, give almost mathematical accuracy to drawings from nature, without requiring other skill in the draughtsman than to trace and make permanent, with ink or pencil, the lines of light which he sees on the paper; still the subject is so interesting to all who attempt to sketch, and indeed to all who wish to look intelligently either at nature or at the works of art, that none who have the opportunity of studying it should neglect it.

Fig. 154.

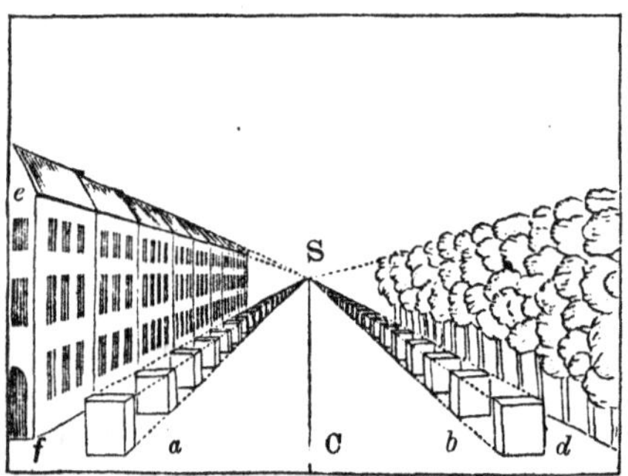

Supposing a straight row of similar objects, as of the stone blocks or pillars represented here from a or b to S, to be viewed by a person stationed near the side and end of the row, as over the point C, then, because, as already explained, objects to the eye appear smaller in exact proportion to their increased distance from it, the second block, if twice as far off as the first, would appear only half as large; the third if three times as far, would be only one-third as large, and so on to any extent and for any other proportions; and if the 1,000th of any other block, owing to its distance, subtended to the eye an angle less than the sixtieth of a degree of the field of view, it would be altogether invisible (as explained at page 333,) even if nothing intervened between it and the eye. Then, where the row ceased to be visible from the minuteness of the parts, or from the fact of the nearer objects concealing the more remote, it might be said to have reached its *vanishing point.* When a student of perspective has learned all that regards the vanishing *point* in relation to a line, and the corresponding vanishing *line*, in relation to a surface, he has learned half of his art. The above cut considered as the representation of a street running directly south to *S*, sketched from a window looking along its centre, will serve as a useful illustration.

It is important, first, to remark, that in any case of a straight line, or a row of objects thus vanishing from sight, as here the line or row *a S*, in whatever direction it points, whether east, west, north, or south, &c., in that direction, too, will its remote or vanishing extremity appear to be from the eye. In this sketch the row *a S* is supposed to run directly south; and although the eye to see the near end of it, would have to look towards the left hand, or in a degree east, still every successive pillar would be more and more nearly south, and the point in the heavens, or in the picture, or in a transparent plane before the eye, where the line would vanish, would be so nearly south from the eye, and not to the east, because the pillars happened to the east of the individual, that no ordinary measure would detect the little want of correspondence; then similarly, if there were more rows of objects, as of pillars, houses, trees, &c., parallel to the first, but considerably apart from each, as the lines here *a S*, *b S*, *d S*, &c., still all would vanish or seem to terminate in the very same point of the field of view. The reason of this is easily understood. Let us suppose a line drawn directly south from the eye to the point S, between the parallel lines of pillars, houses, and trees, *a* S, *b S*, *d S*, &c., also pointing directly south; and let us suppose the two rows of pillars to be 100 feet apart, then evidently for the same reason as the space between the top and bottom of the pillars, that is to say, their height becomes apparently less and less as their distance from the eye increases, so will the space between each pillar and its opposite, or between it and the point corresponding to it in the visual ray along which the eye looks, become apparently less, and therefore the lines of pillars really 100 feet apart from each other, and 50 feet from the visual ray, will, at a certain distance from the eye, (*viz.*, where a space of 50 or 100 feet is apparently reduced to a point,) appear to join, and the three lines will appear to meet in that point, beyond which none of them can be visible, and which is therefore the vanishing point of all. The conception of this truth may be facilitated by supposing a planet to be visible to the exact point of the heavens at the moment of observation; then, if the three parallel lines were continued on to the planet, and were visible all the way, they would arrive there with the interval between them just as when they left the earth; but as a planet, although thousands of miles in diameter, owing to its distance from the earth, appears on earth only as a point, much more would two lines only 100 feet apart be there undistinguishable in place by human sight. And what is true of space of 100 feet between parallel lines, is equally true of a mile or of thousands of miles. As a general rule, therefore, it holds, that all lines in nature parallel to each other, when represented in perspective, tend towards an end in the same vanishing point; and that point is the situation where the line terminates, along which the eye looks when directed parallel to any one of the real lines. And this is true only of lines all in the same level or horizontal plane, *viz.*, such as might lie along the surface of the sea, but also of lines placed one above another, as those running along the tops and bottoms of the pillars here, or along the walls, roofs, and windows of the houses, or along the roots and summits of the trees, and indeed of all lines in whatever situation, provided they are parallel to the visual ray, and therefore to one another. And the truth holds equally with respect to lines which do not vanish at the "point of sight," or centre of the picture, as with respect to those which do. When it is ascertained, therefore, that a line or boundary of any natural or artificial object has a certain inclination to the axis of the picture, or to what we have described as the principal visual ray, then also is it known that all parallels to that line have their vanishing point in the same spot of the field of view, and a line supposed to be drawn from the eye to the heavens, or really drawn from the eye to the picture in that direction, marks upon the picture the true vanishing point.

It will be understood why, in a long arched tunnel or a cathedral, with many longitudinal lines on its floor, walls, roof, &c., all such lines seen by an eye looking along from one end, appear to converge to a point at the other, like the radii of a spider's web; and why similarly in the representation of a common room, viewed from one end, all the lines of the corners, tops and bottoms of windows, floor, stripes, on a carpet, corners of tables, &c., being parallel to each other, tend to the same vanishing point as V. The appearance of the lines in the floor of this room may recall that of the furrows in a ploughed field as seen from one end, when they appear like the ribs of a fan spread out towards the spectator. The same considerations will explain the phenomenon often to be observed, of two little clouds seen near each other, and almost motionless for a time in the distant sky, but which on approaching the spectator with the wind, appear gradually to separate, and in a corresponding degree to enlarge, until one of them sweeps past considerably to the right hand, and the other considerably to the left; after this, they lessen and approximate as they before enlarged and separated, and at last beyond the spectator, appear as small and as near as when first observed. Clouds being so mutable and uncertain in their forms, persons have been led to deem all apparent changes in them, of form, size and place, to be real changes, and not, as they generally are, mere optical or perspective illusion.

Fig. 155.

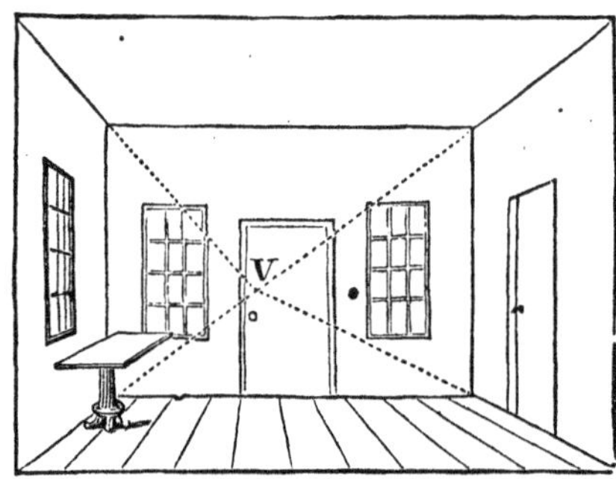

By far the most important vanishing point in common scenes is the middle of the horizon or level line, and in a picture properly placed, it is at the exact height of the eye. It is marked S in the figure before the last, and V in the last figure. Because in houses, the roofs, foundations, floors, windows, &c., are all horizontal, the vanishing points of their lines must be somewhere in the horizon, and if the spectator be in the middle of a street or of a building, and be looking in the direction of its walls, their vanishing point will be in the centre of the scene or picture; if he be elsewhere, it will be to one side. In holding up a picture-frame, through which to view a scene suitable for a picture, it would be found most generally befitting to raise it until the line of the horizon appeared to cross it at about one-third from the bottom :—this fact becomes the reason of the rule in painting, so to place the horizontal line. In beginning a picture, this line is usually the first line drawn on the canvas, marking the place of the vanishing points of all level line and surfaces. And the eye of the spectator is supposed to be placed before the middle of it and generally about as far from the picture as the picture is itself long, such being the extent of view which the eye at one time most conviently commands.

Understanding now that the apparent or perspective direction of all lines in a scene is towards their vanishing points as above discovered, and the rule having been given for determining these points in a drawing, we proceed to show how much of a line drawn to any vanishing point belongs to the known magnitude of any object through which it passes; in other words, how much an object is in perspective foreshortened in consequence of its distance and obliquity of position.

If we suppose A S P to represent a plate of glass seen edgeways, and that towards the point S in it, an eye is looking from the point D, evidently then, a line from P continued in the direction of R and beyond until vanished from sight, would have as its perspective image or representation on the glass a line reaching from P to S,—S being, moreover, the *point of sight* here, and the pictorial vanishing point of the line. Now to divide the *representative* line P S so as to correspond with any given portions of the *original* line P R, &c., it would only be necessary to draw other lines from the place of the eye D to cut or touch the original line in the situations desired, and these lines would cut the perspective line S P as required : for instance, the portion of true line *a b* would be represented by that portion of the image-line S P included between the two lines *a* D, and *b* D, and so of any other portions. There are figures drawn on many mathematical scales by which such problems as this can be at once approximatively solved; and it would be possible, by trigonometrical calculation, to solve them exactly in all cases : but the most generally convenient mode in practice is to sketch on the intended drawing (as that of which the boundaries are given in the next cut)the kind of measure shown above, by setting off from the point of sight S, a distance on the horizontal line, as at D, equal to the distance of the eye from the picture, and then by oblique lines from D drawn upon the base line P R, to cut the perpendicular line P S in the situations desired—as is seen in the last figure, which differs from fig. 161 chiefly in having the *point* of distance marked *before* its point of sight, instead of, as here, *laterally*. And the line P S being always cut by the oblique line from D in proportion to the length of base-line between P and the extremity of the oblique line, a horizontal line drawn through any point in it,

Fig. 156.

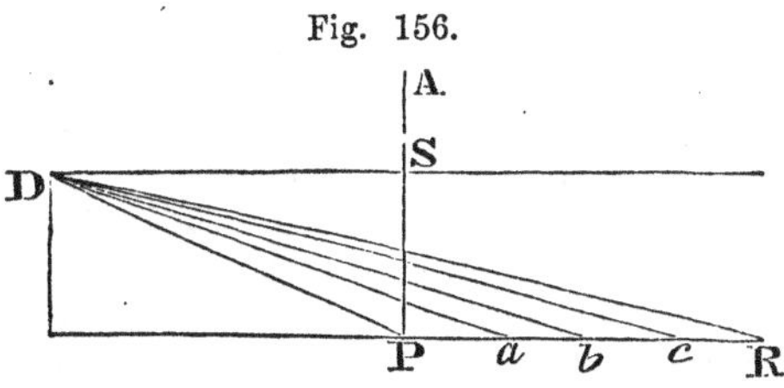

cuts in corresponding proportions all the other lines which have their vanishing points in the horizontal line at S, for instance *a* S, *b* S, &c. Thus, to draw in perspective, on the surface above represented and prepared, a chess-board or board of squares, it is necessary to set off the breadth of the board on the

Fig. 157.

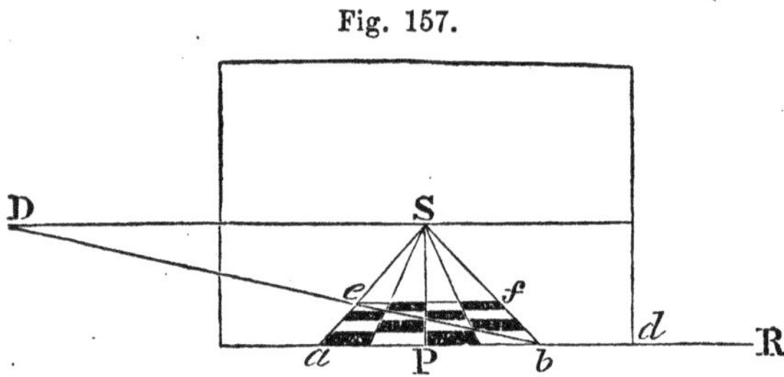

base-line to the right and left of P, *viz.*, at *b* and *a*, and then to draw to the point of sight as a vanishing point, the lines *a* S and *b* S, part of which lines will, therefore, represent the sides of the board, and then to draw the diagonal *b* D, which, for the reason above stated, will cut the lines P S and *a* S in proportion to the length of base-line to the right of their extremities; *a e f b*, therefore, is a square seen in perspective, and any number of smaller included squares are made by drawing lines from the vanishing point to equal divisions on the base, and making cross lines where the diagonal cuts these.

Much of the delight which the art of painting is calculated to afford is lost to the world, because persons in general know not how to look at a picture. Unless a spectator places himself where he can see the objects in true perspective, so that he may fancy himself looking at them through a window or opening, every thing must appear to him false and distorted. The eye should be opposite the *point of sight* of the picture, and, therefore, on a level with the line of the *horizon*, and it should be at the required distance, which is generally at least as great as the length of the picture. But blame not unfrequently rests also with the artist, from his having neglected the study of perspective. It is very common, for instance, to see miniature resemblances of architectural structures so foreshortened and tapered, that the eye, to see them in true perspective, would require to be within an inch of the paper; whence at the usual distance of ten or twelve inches they are seen as hideous distortions. The specimens, in the few preceding pages, necessarily exemplify in a degree this error, because the *point of distance* had to be marked where there was but a small page. These figures, therefore, by any person studying the subject particularly, should be drawn on a scale so much larger as to allow the eye really to view them at the distance supposed.

A means of judging of the dimensions of the bodies by the visual angle, but which depends neither on the absolute size of the image, nor on the foreshortening of the ground plane on which the body stands, is to use known objects in view as measures for others near them which are unknown.

If any person of our acquaintance be standing at some distance from us near another person who is a stranger, we know how tall the stranger is by taking the acquaintance as a measure.

In pictorial representations of objects little familiar as to many people are the Egyptian pyramids, the bodies of the whale, the elephant, the camel, &c., human beings may be represented around them to serve as measures for the less known object. The Colossus of Rhodes seen from afar, might to a stranger have appeared but an ordinary statue of a man, but the exact magnitude would have been known as soon as a ship of known dimensions were seen sailing into port between his gigantic limbs.

When an unpracticed eye is first directed from a distance to a great ship of war, it will on many accounts dwell upon the object with wonder and admiration; but it may not judge truly of the enormous magnitude until it be near enough to perceive the sailors climbing on the rigging, and appearing there, by comparison, as flies or little birds appear among the branches of a majestic tree.

By having a measure of this kind presented to us, the magnitude and elevation of great edifices are rendered more obvious. The magnificient pile of St. Paul's in London becomes still more striking, when we discover visitors looking from the balconies near the summit-cross. They appear so minute among the surrounding huge masses that a person is for a while disposed to doubt whether they be men: but the fact once ascertained, the grandeur of the temple is most impressive.

Manny persons cannot distinguish between the little pilot balloon (sometimes dispatched before the great one to show the direction of the wind) and the great balloon itself, until with the last they perceive the aeronauts as little black points suspended under the globular clouds.

Strangers visiting Switzerland, on first entering the valleys there, are often much deceived as to their extent. Being familiar generally with more lowly hills and shorter valleys at home, which, however; from being near to the eyes, form bulky images, and having at first no other measure, they almost universally underrate the Alpine dimensions:—they will wonder, for instance, in the valley of Chamouny, that they should be travelling swiftly for hours without reaching the end, where on entering they did not anticipate a drive of more than half an honr.

The author once sailed through the Canary Islands, and passed in view of the far-famed Peak of Teneriffe. It had been in sight during the afternoon of the preceding day, at a distance of more than 100 miles, disappointing general expectation by appearing then only as an ordinary distant hill rising out of the ocean, but next morning, when the ship had arrived within about twenty miles of it, and while another ship of the fleet, holding her course six miles nearer to the land, served as a measure, it stood displayed as one of the most stupendous single objects which on earth, and at one view, human vision can command. The ship in question, whose side, showing its tiers of cannon, equalled in extent the front of ten large houses in a street, and whose masts shot up like lofty steeples, still appeared but as a speck rising from the sea, when compared with the huge prominence beyond it, towering sublimely to heaven, and around which the masses of cloud, although as lofty as those which sail over the fields of Britain, seemed still to be hanging low on its sides. Teneriffe alone of very high mountains, rises directly and steeply out of the bosom of the ocean, to an elevation of 13,000 feet, and as an object of contemplation, therefore is more impressive than even the still loftier summits of Chimborazo or the Himalayas, which rise from elevated plains, and in the midst of surrounding hills.

It is because objects which are nearly on a level with us, as contrasted with such as are either much above or much below, are in general more numerously surrounded by other known objects which serve as measures of comparison, that we judge so much more correctly of the size and distance of those near our level than of others.

A man walking like ourselves on the sea-shore or other level, is at once fully recognized; and probably it may not occur to us, that he appears smaller on account of the distance; but if the same man be seen afterwards at an equal distance above us, collecting the sea-fowl's eggs on the face of a cliff, or below us, gathering shells on the beach when we ourselves have reached the height, he appears no bigger than a crow; yet in all the cases he is where the same bulk forms the same magnitude of image on the retina.

Even on a horizontal plane, if the general surface be bare and uniform, single distant objects appear very diminutive. This is true, for instance, of a man seen apart from his caravan, while journeying across a sandy desert; but a man viewed at an equal distance, in the midst of a cultivated landscape, appears of his natural size. The same is true of a boat or ship seen out on the high sea, as constrasted with the like viewed in a harbor where other known objects are near them.

We may now understand why the sun and moon, at rising or setting, appear to us much larger than when they have attained meridian height—although, if we examine them by any measure of the visual angle, as simply by looking at them through the same ring or tube, we find that there is scarcely a difference; and singularly we find the difference to be, that the orbs appear horizontally even less than when seen on the meridian, owing to our being then about 4,000 miles more distant from them. The sun and moon as they appear from this earth are nearly of the same size, each occupying in the field of view about the half of a degree, or as much as is occupied by a circle of a foot in diameter when held 125 feet from the eye—which circle, therefore, at that distance, and at any time, would just hide either of them. Now when a man sees the rising moon apparently filling up the end of a street, which he knows to be 100 feet wide, he very naturally believes that the moon then subtends a greater angle than usual, until the reflection occurs to him, that he is using as a measure, a street known indeed to be 100 feet wide, but of which the part concerned, owing to its distance, occupies in his eye a very small space. The width of the street near him may occupy 60° of his field of view, and he might see from between the houses many broad constellation instead of the moon only, but the width of the street far off may not occupy, in the same field of view, the twentieth part of a degree, and the moon, which always occupies half a degree will there appear comparatively large. The kind of illusion now spoken of is yet more remarkable when the moon is seen rising near still larger known objects, for instance, beyond a town, or a hill which then appears within the luminous circle. Any person who from the river-side terraces of Greenwich has observed the sun setting beyond London, with St. Paul's Cathedral included in the glorious picture, will recollect a most interesting example of our present subject.—That our ocular judgment of the size of the sun or moon is thus influenced by the presence or absence of objects of comparison, and not by the place of the bodies in the sky, is proved by the fact, that a person viewing these bodies at any elevation from the bottom of some of the Swiss valleys, where he might almost suppose himself placed at the centre of the earth, and looking abroad along an endless extent of precipices—if he can closely compare them with certain known magnitudes of ridge or forest bounding his view, sees them as large as they appear from other situations when rising beyond a low horizon. Another proof is afforded by the case of a balloon at a great elevation seen crossing the disc of the sun or moon, and then appearing, however large in reality, as an absolute speck within the vast luminous area. In a future paragraph it will be explained, that another circumstance contributes to cause the sun and moon, when low, to appear larger than when high, namely, their apparent dimness, owing to the obstruction of their light in traversing the low dense atmosphere.

It may be remarked here, that the visual estimate formed of the great size of the sun and moon when they are seen on the horizon, is not an illusion, as is popularly supposed, but an approximation to truth, still prodigiously short of the reality. When we see a distant tree, or a house, or a hill, apparently within the circumference of one of these orbs, it is really true that the orb is larger than the tree, or house, or hill, just as another more distant hill would be larger than nearer objects similarly surrounded by its outline; but the celestial body is so much larger, that even if the whole British Isles could be lifted away from the earth, and suspended near the moon, as a map in the sky, they would hide from a spectator on earth but a small part of the disc of the moon.

Having now shown that the visual angle or apparent size can be a measure of the distance of any object only when the true size also is known, or of the true size when the distance is known, we proceed to examine other means which the eye commands for guessing at distances.

2*d*. *Intensity of light, shade and colour.* (See the Analysis, pages 325 and 355.)

It has already been explained that light, like every other influence, radiating from a centre, becomes rapidly weaker as the distance from the centre increases, being, for instance, only one-fourth part as intense, at double distance, and in a corresponding proportion for other distances; while it is still farther weakened by the obstacle of any transparent medium through which it passes. Now the eye soon becomes sufficiently familiar with these truths, to judge from them, with considerable accuracy, of the comparative distances of objects.

The fine Gothic pile of Westminster Abbey may break upon the view in some situations where nearer edifices, and perhaps some minor imitations of its beauties, already fill the eye with their strong lights, but the misty or less distinct outline of the venerable pile may warn the approaching stranger of its true magnitude, and prepare him for the enjoyment which a nearer inspection of its grandeur and perfection is to afford.

A small yacht or pleasure-boat may be built according to the same model or with the same comparative dimensions as a first-rate vessel of war, and may be in view from the shore at the same time, only so much nearer than the ship, that both shall form images of the same magnitude on the retina of a spectator. In such a case, an unpractised eye might have difficulty to discriminate, but to an old seamen, the bright lights of the little vessel, contrasted with the softer or more misty appearance of the larger, would declare the truth at once. A haziness occurring in the atmosphere between the little vessel and the eye might considerably favour the illusion.

In a fleet of ships, if the sun's direct rays fall upon one here and there through openings among the clouds, while the others remain in shade, the former, in appearance, starts towards the spectator. In like manner, the mountains of an unknown coast, if the sunshine fall upon them, appear comparatively near, but if the clouds again intervene, they recede and mock the awakened hope of the approaching mariner.

A conflagration at night, however distant, appears to spectators, generally, as if very near, and inexperienced persons often run towards it with the hope of soon arriving, but find, after miles travelled, that they have made but a little part of the way.

A person ignorant of astronomy deems the heavenly bodies so much nearer to the earth than they are, merely because of their being so bright or luminous. The evening star, for instance, seen in a clear sky, over some distant hill-top, appears as if a dweller on the hill might almost reach it—for the most intense artificial light which could be placed on the height would be dim in comparison with the beauteous star; yet to a dweller on the hill it appears just as distant as to one on a plain; nay, at a thousand miles farther west, and, therefore, nearer, the appearance would still be nearly the same.

The concave of the starry heavens appears flattened above, or as if its zenith were nearer to the earth than its sides or horizon, because the light from above having to pass through only the depth or thickness of the atmosphere, is little obstructed, while of that which darts towards any place horizontally through hundreds of miles of dense vapour-loaded air, only a small part arrives.

The sun and moon appear larger at rising and setting than when midway in heaven, partly, as already explained, because they can then be easily compared with other objects, of which the size is known, but partly, also, because of the much less light arriving from them in the former situation, while their diameters are nearly the same.

A fog or mist is said to magnify objects seen through it. The truth is, that because it diminishes the intensity of the light from them, it makes them appear farther distant without lessening the visual angles subtended by them; and because an object at two miles, subtending the same angle as an object at one mile is twice as large, the conclusion is drawn that the dim object is large. Thus, a person in a fog may believe that he is approaching a great tree, fifty yards distant, when the next step throws him into the bush which had deceived him.—Two friends meeting in a fog have often mutually mistaken each other for persons of much greater stature.—A row of fox-glove flowers on a neighbouring bank has been mistaken for a company of scarlet-clad soldiers on the more distant face of the hill.—There are, for similar reasons, frequent misjudgings in late twilight and early dawn.—The purpose of a thin gauze screen interposed between the spectators in a theatre and some person or object meant to appear distant, is intelligible on the same principle: a boy near, so screened, appears to be a man at a distance.—The art of the painter uses sombre colours when his object is to produce in his picture the effect of distance. On the alarming occasion of a very dense fog coming on at sea, where the ships of a fleet are near to each other, without wind, and where there is considerable swell or rolling of the sea, much damage is often done: but it is to be remarked in such a case, that the size of each ship approaching to the shock, is always, in the apprehension of the crew of the other ship, exaggerated.

The celebrated *Spectre of the Brocken*, among the Hartz Mountains, is a good illustration of our present subject. On a certain ridge, just at sunrise a gigantic figure of a man had often been observed walking, and extraordinary stories were related of him. About the year 1800 a French philosopher and a friend went to watch the apparition; but for many mornings they paraded on the opposite ridge in vain. At last, however, the monster was seen, but he was not alone; he had a companion, and singularly he and his companion aped all the motions and atitudes of the observer and his companion; in fact the spectres were merely shadows of the observers, formed by the horizontal rays of the rising sun falling on the morning fog which hovered over the valley between the ridges; and because the shadows were very faint, the figures were deemed distant, seeming men walking on the opposite ridge, and because a comparatively small figure seen near, but supposed distant, appears of gigantic dimensions, these shadows were accounted giants.

While the comparative intensities of light coming from bodies considered as wholes, or from their sides similarly exposed to the source of light—furnish an indication of their different distances from the observer, the comparative intensities from their sides dissimilarly or unequally exposed to the source of light, and which, therefore, reflect light to the eye, or are illuminated in different degrees, furnish an indication of the forms and attitudes of the bodies. In observing, for instance, a white house exposed to the sun, it is seen that the side directly receiving the rays is highly illuminated or bright, while the other sides are less so, and are said to be in the shade—a shade which is more or less deep in proportion as there are few or many sources of reflected light near it. The different faces or walls of such a house are to the sense of the observer, as strongly distinguished from each other, by the mere difference of shade, as if they were of different colours, or as if they were examined by the touch, or by walking round them. If the object were a ball instead of a square house, there would still be as great differences of shade in the half not receiving direct rays, but the parts, instead of forming abrupt contrasts like the walls of a house, would appear to melt into each other, marking the beautiful round contour of the object. The consideration of all such cases forms the subject of chiaro-oscuro, so interesting to the painter.

Had there not been in nature the provision of light and shade now described, the sense of sight would have been of comparatively little use, and a mass of things in the light, if of the same colour, would have been as little distinguishable from one another by a person looking directly at them, as things forming a mass, or shadow are distinguishable by a person looking at the shadow. It is this provision, therefore, which enables us, independently of colour, to distinguish the profile or outlines of different bodies placed near to each other, and to distinguish in the same body the protuberent or other form of the surfaces which is towards the observer. But for this, it would have been impossible to distinguish, for instance, between a white wall when naked and when having various white objects placed before it; and it would have been impossible to distinguish between the rounded figures, if similarly coloured, of a flat circle, a sphere and a cone, all directly opposed to the eye; but in reality, by some difference of shade, the white objects are distinguished from the wall, and in the three geometrical figures, the uniformly bright surface of the circle, the soft rounded shadowing of the sphere and the shade coming to a point on the cone, at once declare the true forms. But for the shadowed parts, the facade of a white palace of varied architecture would have been an unmeaning sheet of lights: the lights, however, and shadows produced by the juttings and recesses, mark the variety of surface most completely; and the round pillar is distinguished from the square, and every pediment, and capital, and architectural ornament, stands out pleasingly conspicuous. But for light and shade again, the "human face divine." would have been an unmeaning patch of flesh, for there are few lines in it but those made by different exposures to the lights, and yet its every prominence and depression, and every momentary change, are so truly indicated to the eye that it becomes full of meaning or expression. How well mere light and shade serve to convey what the eye has to learn of a scene or object, may be perceived by examining any of the admirable engravings which now abound, and which, although made up entirely of degrees of shade, or of black and white, are scarcely inferior in expression to finished paintings.

The student of painting soon learns that the lines called outlines, by which he first sketches subjects, do not exist at all in nature, and have to be again effaced in his finished work: for they only mark the place where lights and shades happen to meet. Much may be conveyed to the mind, however, by a mere outline, and particularly of lines if different breadth and thickness are used to mark the situation of the fainter and deeper shadows.

The subject of chiaro-oscuro is not so simple as, from the fact of the sun being the great source of light, might at first be supposed; for although this be true, still every body which reflects the sun's light becomes a new source to those about it, and the shading of a picture must have reference to all such sources, and to the colours of the body itself, and of the neighbouring bodies.

In looking at an extended landscape, it is seen that the near objects considered as wholes, are comparatively bright, that their shadows are strongly marked, and that their peculiar colours are everywhere easily distinguishable —as of flowers, fruit, foliage, &c., but of objects farther off, the colours, with increasing distance become dim, the lights and shadows melt into each other or are confused, and the illumination altogether becomes so faint that the eye at last sees only an extent of distant blue mountain or plain—appearing bluish, partly because the transparent air through which the light must pass has a blue tinge, and partly because the quantity of light which can arrive through the great extent of air is insufficient to exhibit the detail. The ridge called Blue Mountains in Australia, another of the same name in America, and many others elsewhere, are not really blue, for they possess all the diversity of scenery which their climates can give, but to the eye which first discovered them, bent on them generally from a distance: they all at first appeared blue, and they have retained the name.

In a good picture, where, upon canvass stretched on a frame, the artist has disposed the lights, shades and colours in the very situations and with the intensities which they would have had on coming from the real scene to the eyes, through a plate of glass filling up the frame, all that we have now been saying is strictly exemplified. In the foreground, the objects are large and bright, but as the scene is supposed to be gradually more remote, the size and brightness correspondingly diminish, until, at last, there is only a dim mixture of bluish or grayish masses forming the horizon and sky.

A child during what may be called the education of the sense of sight, has a strong perception of the vast differences of appearance which things assume according to their accidental distance from the eye, their position, their exposure to light, &c.; for many of these differences, being at first calculated to deceive the young judgment, have, from time to time, been noted by him with surprise. Thus, a boy when he first discovers that a ship which at the quay, with her white sails spread out, concealed from him half the heavens, is in an hour or two afterwards, seen by him on the distant horizon as a dark speck hardly big enough to hide one star, has his attention strongly awakened, and he feels surprised; or, again, when he discovers that the faint blue unchanging mass which he had always observed bounding in one direction, the view from the house of his infancy, is a distant mountain side, thickly inhabited and corvered with fields and gardens, where in succession, all the bright colours of the different seasons predominate—he is equally struck. But as soon as experience has enabled him to interpret readily and correctly, the visual signs under every variety of circumstance, his attention passes so instantly from them to the realities—which alone are interesting to him—just as it might pass from the paper and printing of a newspaper to the important intelligence communicated by them—that he very soon ceases to be aware that the sign, which, in every case, similarly suggests the object, is not, also, in every case similar to itself, and the very same true and complete representation of the reality. The prejudice that the sign is of this nature becomes quickly so strong, that even a difficult effort has been made by a grown person again to attend to. the mere *appearances*, in any scene of which the *realities* are known.

This attempt to analyze mere appearances, and which, in one sense, is an attempt to unlearn something, or to retrograde, is called, as already stated, the study of *perspective.* When it regards the apparent reduction of size and, the foreshortening of bodies under various circumstances, it is called *linear perspective*—when it regards the fading of light and the modifying of colour it is called *aerial perspective.* As the art of painting depends entirely upon the understanding of these two departments, the gradual progress which it has made in different countries is a measure of the degree in which the common prejudice that things *appear* as they *are* has, in them, been overcome. Where this prejudice exists, any untaught person conceives a good painting to be merely a miniature representation drawn according to a certain reduced scale,—as of an inch to a yard,—and in which all the demensions of things are to be measured as simply as in the reality—while the colour, as to vividness, &c., should perfectly agree with the originals. This statement is remarkably illustrated by the facts, that children in their rude atempts to paint, always aim at realizing the notion of the art above detailed, and that such has been the first stage of painting in every country. In Europe now, owing to the labours of men of genius, art in painting may be said almost to rival nature, producing scenes as lovely as the finest of nature's scenes, and scarcely distinguishable from them : but in other countries, as in China and India, among the native artists, the first stage of the art is still in existénce. In a Chinese picture, owing to the absence of perspective proportions, an extensive subject is only a collection of portaits of men and things drawn all on the same scale, and placed near one another, and where all the colours are as vividly shown as if the objects were only a few feet from the eye; there, the figures at the bottom or fore-ground are supposed to represent the objects nearest to the spectator, while the figures higher up are supposed to be of more remote objects, all appearing as they might be seen in succession by a person who had the power of flying over the country. This kind of picture or representation, although not natural, if all viewed at once, may communicate more information than a single common painting, for it is equivalent to many such. In Europe, lately, the principle has been again usefully acted upon for certain purposes, as for representing on one long sheet or on a succession of sheets, connected in a suitable manner, the banks of a river or a line of a road. The banks of the Rhine particularly have thus been admirably portrayed, so that, the spectator directing his eye along the paper feels almost as if carried in a balloon to view in detail the whole of the real and enchanting scenery. The principle might, perhaps, with advantage, be acted upon still more extensively—for instance, to produce, instead of common maps or charts of countries, true bird's-eye views, over which the eye, moving from place to place, and at every new point of sight, would see a certain portion of the country, as a bird or aeronaut would, the sketch being supposed to be taken from that certain elevation deemed most suitable for the ends in view.

3d *Divergence of the rays of light.* (See Analysis, page 355.)

This is the next circumstance to be mentioned by which the eye judges of distance. Supposing the line E F to mark the place and breadth of the pupil of the eye, the light entering from an object at *a* which is near (it is here placed nearer than an object could be seen in reality,) is very divergent, or is spreading with a large angle; from *b* the pencil of rays is less divergent, or opens with a smaller angle; from *c* it is less divergent still, and so on. Now the eye, to form an image on its retina requires to exert a bending power exactly proportioned to the divergence of the received rays: and it appears to have a sense of the effort made, which becomes to the person a kind of measure of the distance of the object. This divergence of the rays entering the eye, is the chief circumstance in which the most perfect painting must still differ in its effect upon the eye from a natural scene—for, first, in the natural scene, the objects are generally more distant than their representation can be; and secondly, while, in nature, every object, according to its distance, is sending rays which reach the eye with different divergence, and which rays, therefore, can produce distinct images on the retina at any one time, only of the objects which are at the same distance from the eye, the rays from a picture, which is a single plane surface, come from every part with the same divergence, and the eye must feel a disappointment in not having to accommodate its power of bending, to the different distances attempted to be portrayed on the canvass. It might be expected that this kind of disappointment would be more felt on looking at a common picture placed a few feet from the eye, than at the sort of picture called panorama, which is on a larger scale and proportionately more distant, but such is not the case: and the reason seems to be that in the former the illusion is not intended to be complete, the fact of its being but a picture not being at all concealed, and the eye is therefore at once told to expect a difference of feeling;—but in the panorama, the whole circumstances are arranged to deceive the eye entirely, if possible, and to make it believe that the images on the retina are formed by light from the objects themselves,—then to the eye, really deceived in all other particulars, the non-accordance with nature in this one is strongly, and, by some persons, even painfully felt, so as on their first entering the place to cause headache or giddiness.—The illusion and consequently the pleasure from viewing any picture may be made more complete by the spectator using lenses or spectacles, such that the focal distance shall be equal to the distance of the painting from the eye; because such lenses, as was formerly explained, would render all the rays entering the eye nearly parallel, and therefore very nearly such as would arrive from objects at a considerable distance.

Fig. 158.

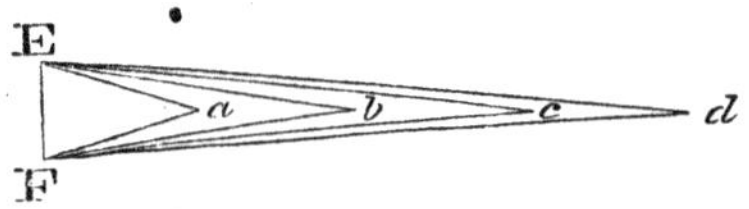

E F a b c d

4th. Convergence of the axes of the eyes. (See the Analysis, page 355.)

This is the last circumstance to be mentioned, by which a person through the eye, judges of the distance of objects. In consequence of there being in the two eyes corresponding parts which must be similarly affected by any object, that the person may have a single vision of it, as was explained in a former page, the axes of both eyes must point to the object, and if it happen to be very near, they will meet and cross each other so near the face as to produce the appearance of squinting,—seen when a person trying to look at the point of his nose,—but if the object be more distant, the obliquity will be less, until at last the eyes directed to a thing at a very great distance, will have their axes almost parallel. The last figure may serve also to explain this subject. Supposing E and F to mark the place of the two eyes if the object looked at be near them, as at *a*, they must be very much turned inwards, that their axes may meet at *a*; if it be at *b*, they will be less turned, if at *c* less still, and so forth.

When the eyes are not directed to any thing in particular, the axes generally become parallel, or as if they were pointed to a very distant object; and because this happens generally when persons are reflecting on things which are absent and seen only by the mind's eye, it is an expression of countenance held to mark contemplation or thoughtfulness.

The direction of the visual axis is another particular, like the divergence of light as to which a mere picture, can never produce upon the eye precisely the effect of the objects themselves. To see a picture, the axes must meet at it, and generally, therefore, at a few feet from the eye; while to see the objects of nature, they often do not meet nearer than at miles. By a glass, however, as will be explained a little farther on, it is possible to correct also this defect, and to render the optical illusion, as regards still objects, absolutely complete.

When a picture has to represent objects supposed far from the eye, the farther the picture itself is placed from the eye supposing the figures to be made proportionately large, the more nearly perfect will the illusion become, because the divergence of rays and convergence of the axes (the two circumstances in which the effect of a mere picture on the eye must always differ from the effect of a real scene) will be in proportion more nearly natural. This explains in part why the picture called panorama (from Greek words signifying a *view* of *the whole*) is an exhibition so charming; for usually the painting is far removed from the eye, and is drawn on a proportionately large scale, and the eyes feel that the light comes from a considerable distance, and that their axes do not need to converge very much; and when in such a case, the first impression of the want of absolute conformity to nature has passed away, the illusion becomes nearly complete. But a not less important peculiarity in the panorama is, that instead of being a painting on a plane surface like common pictures, and embracing only a small part of the field of view, it is on a curved surface which entirely surrounds the spectator, and on which all the objects visible in the various directions from the supposed place are seen in the very situations which in nature they hold; and the spectator is enabled to conceive much more distinctly of each particular by seeing it in relation to others around. Few persons can forget the impression made on them by the first panorama which they may have seen; and after increased maturity of judgment, they will discover still more and stronger reasons for admiring this almost miraculous mode of instantly transporting them to any distance, beyond seas and other dangers, to contemplate at their ease the most interesting scenes on earth, represented under the most favourable circumstances of position, light and weather. Hence few persons of good taste neglect the opportunity now, in most great towns so frequently offered, of obtaining at little cost so high a gratification.

To correct the slight remaining optical defects of a common panorama, a large lens may be used, of which the focal distance is equal to the distance of the picture from the eye. This has the effect of diminishing the divergence of the rays until it becomes exactly that which belongs to the supposed remoteness of the objects, and it also bends the light so that the axes of the eyes may be nearly parallel. The author has found a convenient mode of using the lens for such a purpose to be to cut out two round pieces from opposite sides of it, and to form them into a pair of spectacles:—from one lens three pairs may be formed. Panorama exhibitors should keep such lenses or spectacles for the use of visitors.

The effect of the magnitude and distance of the ordinary large panoramic views might, with the assistance of proper glasses, be had from even the smallest picture on engraved representation embracing the same field; and it is remarkable that some enterprising person has not undertaken to publish sets of interesting views fitted to be used in that way. A common panorama, occupying a circular wall of 150 feet in circumference and twenty feet high, may be reduced—and still retaining the same truth of proportions, to appear on a piece of paper five feet long and eight inches high or broad; and if this were set up in a suitable frame, like a wall, round the head of a spectator, while its edges were concealed by drapery or otherwise, and the eye could only view it through fit glasses placed in its centre and made to turn round so as to command the whole, it could not by any ordinary spectator be distinguished from the large panorama. With the art of lithography, now so well adapted for producing soft representations of scenery, the expense of such views might be very moderate, allowing them to form a common part of library furniture. When we reflect upon the expansion of mind obtained by travelling, and that not a few of the advantages would follow a familiarity with a good selection of panoramic views, it is not perhaps too much to suppose that courses of instruction in geography, history, &c., may before long be illustrated by this most interesting mode of aiding the conception and memory.

Common paintings and prints may be considered as detached parts of a panoramic representation, showing as much of that general field of view which always surrounds a spectator, as can be seen by the eye kept in one place, and looking through a window or other opening of moderate size. The pleasure from contemplating these is much increased by using a lens or such spectacles as above described. There is in the shops such lens, with the title of *optical pillar machine*, or *diagonal mirror*, fitted up so that the print to be viewed is laid upon a table beyond the stand of the lens, and its reflection in a mirror supported diagonally over it, is viewed through the lens. The illusion is rendered more complete in such a case by having a box, as *a b*, on the bottom of which the painting is laid, and at the top of which the lens and mirror, fixed in a smaller box at *a*, are made to slide up and down to allow of a ready adjustment of the focal distance. This box used in a reverse way becomes a perfect camera obscura. The common show-stalls seen in the streets are

Fig. 159.

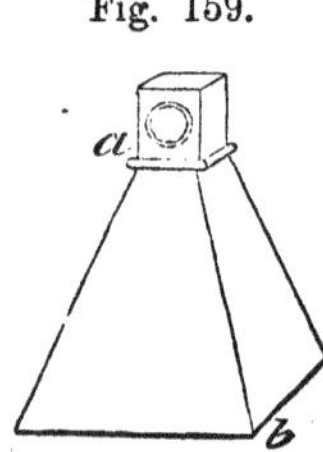

boxes made somewhat on this principle, but without the mirror; and although the drawings or prints in them are generally very course, they are not uninteresting. To children whose eyes are not yet very critical, some of the show-boxes afford an exceeding great treat.

A still more perfect contrivance of the same kind has been exhibited for some time in London and Paris under the title of *Casmoramo,* (from Greek words signifiying *views* of the *world,* because of the great variety of views.) Pictures of moderate size are placed beyond what have the appearance of common windows, but of which the panes are really large convex lenses fitted to correct the errors of appearance which the nearness of the pictures would else produce. Then, by adding various subordinate contrivances, calculated to aid and heighten the effects, even shrewd judges have been led to suppose the small pictures behind the glasses to be very large pictures, while all others have let their eyes dwell upon them with admiration, as magical realizations of the natural scenes and objects. Because this contrivance is cheap and simple, many persons affect to despise it; but they do not thereby show their wisdom: for to have made so perfect a representation of objects, is one of the noblest triumphs of art, whether we regard the pictures as drawn in true perspective and colouring, or the lenses which assist the eye in examining them.

It has already been stated that the effect of looking through such glasses at near pictures, is obtainable, in a considerable degree, without a glass, by having the pictures very large, and placing them at a corresponding distance. The rule of proportion in such a case is, that a picture of one foot square at one foot distance from the eye, appears as large as a picture of 60 feet square at 60 feet distance. The exhibition called the *Diorama* is merely a large painting prepared in accordance with the principle now explained. In principle it has no advantage over the cosmoramo or the show-box, to compensate for the greater expense incurred, but that many persons may stand before it at the same time, all very near the true point of sight, and deriving the pleasure of sympathy in their admiration of it, while a slight motion of the spectator does not make his eye lose the right point of the view.

A round building of prodigious magnitude has lately been erected in the Regent's Park in London, on the walls of which is painted a representation of London and the country around, as seen from the cross on the top of St. Paul's Cathedral. The real scene is unquestionably one of the most extraordinary which the world affords, and this representation of it combines the several advantages of—the circular view of the panorama—the size and distance of the great diorama—and that from the details being so minutely painted, distant objects may be examined by a telescope or opera-glass.

From what has now been said, it may be understood, that for the purpose of representing still-nature, or mere momentary states of moving objects, a picture truly drawn, truly coloured, and which is either very large to correct the divergence of light and convergence of visual axes, or if small, is viewed through a glass, would affect the retina exactly as the realities. But the desideratum still remained of being able to paint motion. Now this, too, has been recently attempted, and in many cases with singular success, chiefly by making the picture transparent, and throwing lights and shadows upon it from behind. In the exhibition of the diorama and cosmorama there have been thus represented with admirable truth and beauty such phenomena as—the sunbeams occasionally interrupted by passing clouds, and occasionally gilding the varied scene: perhaps darting through the windows of a venerable cathedral and illuminating the interesting objects in its interior—the rising and disappearing of mist over a landscape—running water, as, for instance, the cascades among the sublime precipices of Mount St. Gothod, in Switzerland;—and one of the most striking scenes of all, a great fire or conflagration. In the cosmoramo of Regent Street, the great fire of Edinburgh was admirably represented; first, that noble city was seen sleeping in darkness as the fire began, then the conflagration grew and lighted up the sky, and at short intervals, as the wind increased, or as roofs fell in, there were bursts of flame towering to heaven, and vividly illuminating every wall or spire which caught the direct light—then the clouds of smoke were seen rising in rapid succession and sailing northward upon the wind, until they disappeared in the womb of distant darkness. So naturally was all this represented, that no stranger can have viewed the appalling scene with indifference, while on those who knew the city, the effect can scarcely have been weaker, than if they had witnessed the reality. The mechanism for producing such effects is very simple; but spectators, that they may fully enjoy them, need not particularly inquire about it.

It is remarkable, when the imagination is once excited by some beautiful or striking view, how readily any visual hint produces clear and strong impressions. One day in the cosmorama, a school-boy visitor exclaimed with fearful delight that he saw a monstrous tiger coming from its den among the rocks;—it was a kitten belonging to the attendant, which by accident had strayed among the paintings. And another young spectator was heard calling that he saw a horse galloping up the mountain side;—it was a minute fly crawling slowly along the canvas. There is, in this department, a very fine field yet open to the exercise of ingenuity, for the contemplation of pictures representing motion or progressive events, may be made the occasion of mental excitement the most varied and intense. For instance, there are few scenes on earth calculated to awaken more interesting reflections on the condition of human nature than that beheld by a person who sails along the river Thames from London to the sea, a distance of about sixty miles, through the wonders which on every side there crowd on the sight—the forests of masts from all parts of the world—the glorious monuments of industry, of philanthropy, of science—the endless indications of the riches, the high civilization, and progressive happiness of the people. Now this scene was lately in one of our theatres, strikingly portrayed by what was called a *moving panorama* of the southern bank of the Thames. It was a very long painting, of which a part only was seen at a time gliding slowly across the stage, and the impression made on the spectators was that they themselves were sailing down the river in a steamboat, and viewing the fixed realities. In the same manner might be most interestingly represented the whole coast of Britain, or any other coast, or any line or road, or even a line of balloon flight. There was another *moving panorama* exhibited about the same time at Spring Garden, aiming at an effect of still greater difficulty, *viz.*, to depict a course of human life; and the history chosen was that of the latter part of Bonaparte's career. Scenes representing the principal events, were, in succession, made to glide across the field of view, and were so designed that the real motion of the picture gave to the spectator the feeling that the events were then in progress; and with the accompaniments of clear narration and suitable music, they produced on those who viewed them the most complete illusion. The story began by recalling the blow struck at Bonaparte's ambition in the battle of Trafalgar; and to mark how completely, by representations of various movements and situations of the battle, the spectators were in imagination made present to it, the author may mention that on the occasion of his visiting the exhibition, a young man seeing a party of British represented as preparing to board an enemy's ship, started from his seat with a *hurrah*, and seemed quite confounded when he discovered that he was not really in the battle. To the views of Trafalgar succeeded many others, similarly introduced and explained, in each of which the hero himself appeared: there were his defeat, at Waterloo—his subsequent flight—his delivery of himself to the British admiral—his appearing at the gangway of the Bellerophon to thousands of spectators in boats around, while in Plymouh harbour, previous to his departure for ever from the shores of Europe—his house and habits during his exile, with various picturesque views of St. Helena;—and last of all, that solemn procession, in which the bier with his lifeless corpse was moving slowly on its way to the grave under the willow-tree. The exhibition now spoken of might have been made better in all respects, yet in its mediocrity it served to prove how admirably adapted such unions of painting, music, and narration are to affect the mind, and therefore to become the means of conveying most impressive lessons of historical fact and moral principle.

Painting, whether employed to portray scenes of entirely still-nature, or scenes involving some kind of motion as above described, has still, as its great aim or end, merely to represent interesting subjects, and to give to the spectator as much as possible that clear conception of them which is obtained by ocular examination of realities; and thus, as a system of visual signs of thought, it becomes, like language, which is a system of audible signs, a means of expanding the boundaries of individual human existence into wider space and time, and thus of elevating human nature. While it portrays only strict matters of fact, whether of past or present time, as particular human individuals, objects of natural history, the beautiful and magnificent scenes of nature, interesting events which the artist has the means of faithfully representing, &c., it may be called truly historical painting, embodying the materials of true history, both natural and civil, and then it is of singular value. But even when applied to other purposes, it may be fraught with delight; and just as language, of which the grand object or use is to express strict truths, has still been admirably employed in giving a permanent existence to a variety of fictions, from the wildest fables and rhapsodies to the historical plays and novels of modern times, as those of Shakspeare and of Scott—which plays and novels, although not furnishing true portraits of individual human nature, are yet most correct portraits of general human nature—so may painting be employed in embodying fictions adapted to its peculiar powers, and it may do so in a manner to prove the artist endowed with the highest degree of human genius. It should always be recollected, however, that what is usually dignified with the name of historical painting, really bears to historical truths only the kind of relation which novels and plays bear to it, and often approaches even less nearly to the truth; for it pretends to relate a thousand minute circumstances which no history has preserved, and which, therefore, only the imagination of the artist can supply. Thus when a painter, knowing that Lucretia stabbed herself in the presence of her father and others, after the crime of Tarquin, exhibits a woman dying, and a certain number of persons around her in horror and astonishment, he no more represents the real Lucretia and her friends than he represents any other particular young woman and her friends; for he is quite assured that not one of the figures in such a picture is a portrait of the individual whose name it bears; his picture, therefore, in so far, is as untruth or fiction, while it very probably has some of the additional errors and even absurdities so common among historical painters, in respect of national usage in costume, religion, manners, &c, and in respect to the general personal appearance,— as when a Rubens, wishing to represent Sabine or other ladies, gave them the Dutch corpulency deemed comely in his own country, although it strikingly contrasted with the true forms of Italian or Grecian nymphs. From all this it appears that historical pictures may often be regarded as portraitures, not of the realities, but of comedians acting scenes in historical plays intended to represent the realities.

In dealing with the events of ordinary history, there is no strong reason why artists may not please themselves and their spectators as we have now been describing; but it may admit of doubt whether similar liberties should be allowed with respect to religion Yet any painting of *the last supper*, for instance, or of the *ascension*, is not more true than a theatrical representation. To judge of the nature of such a picture we have only to suppose any of the events recorded in the New Testament to be represented by a painter in China with the countenances seen on Chinese tea-boxes; such a representation would appear in Europe revoltingly absurd; but the common practice here is only a degree better, Italian countenances being usually substituted for the Jewish; and twenty painters, undertaking the same subject generally put different persons into all the situations. Then it can produce no pleasing impression on a Christian's mind to be told, that an admired painting of the crucifixion was made chiefly from the body of an executed mur derer; or that for a praised representation of the triumphal entry into Jerusalem, the painter had deemed his own physiognomy the most befitting for the principal figure, while he copied the portrait of Voltaire as a specimen o' the bad Jews, of Newton as a specimen of the good, and of wife, cousins, acquaintances, and old clothes-men, to make up the remaining groups. With the knowledge that such things have often been, it need not surprise that many persons of correct feeling turn with horror from all these mimicries and falsehoods, to seek their idea of God and his providence in the sublime descriptions of his attributes, which written language conveys and which all creation, in a mute language not less impressive, so strongly confirms. When men generally could not read, and as a mass were extremely ignorant, various means of fixing their attention upon religious subjects might be useful, and therefore proper, as sacred plays, certain processions, pictures, &c., which have now in many countries ceased to be either; but a person of good sense will continue to regard with a certain respect whatever at any time may have contributed to reclaim portions of mankind from barbarism and wickedness to the just appreciation as the divine charities of a pure religion.

There are in painting other classes of fictions, which pretend to nothing beyond fiction, and which yet are truly admirable; such are personifications of the virtues and vices, serving to recommend the practice of the former, and to deter from that of the latter—almost all Hogarth's works are of this character, and evince the highest mental acumen and genius:—then may be mentioned the personifications of what have been called the elements and powers of nature, including many of the personages of the heathen mythology—then other generalizations of the characteristics of human or other nature, as scenes of domestic affection, of the play of the passions, &c., &c.; and because many subjects when so sketched, are intelligible to the eye with the suddenness of lightning, where longest verbal description would convey the idea but imperfectly, the art of painting, in regard to them, possesses a truly magical and inestimable power.

As painting, whether employed to represent matters of fact or of fiction, can accomplish its ends only through the means of *drawing* or linear perspective, and of shading and colouring, or aerial perspective, these subjects require to be studied by every artist with great attention; but it is important for all to be aware that the greatest mastery over these, which are merely the mechanical parts of the art, will go a very short way towards producing good performances, unless there be present also the genius to select or to compose subjects worthy of being represented,—indeed, will go little farther to make a painter than the learning of mere penmanship goes to make a historian or a poet. This remark seems the more necessary, because there is in human nature a disposition to value so much the means by which important ends are attained, that often the end itself is forgotten in the contemplation of the means—as when a person, perceiving that money will procure all desirable things, at last becomes the insane miser, and dies from want of the common necessaries of life rather than touch his hoarded treasures:—while among painters, as among persons of other occupations, the talent for the inferior or more mechanical departments of the art, is more common than for the higher. Do we not see the subordinate accomplishments of the painter, by not a few, both artists and pretended connoisseurs, supposed to be the prinpal? But this is evidently to value the dress or clothing instead of the person; or like the bibliomaniac, to regard the type and binding of books more than the subject. To prove how unessential what is called high-finishing in painting, is to the complete attainment of the purposes of the art, we may instance the cartoons of the immortal Raphael, which to the mere mechanic in art appear almost daubs, although exciting such enthusiasm in the superior mind: and many of the mere sketches of genius are to a true taste more precious than some of the most labored pieces in our galleries. As it is of no importance to a man who sees approaching the friend of his heart, whether it be by day-light or candle-light, or with the source of light above or below, &c., provided there is light enough for him to distinguish clearly the friend of his heart, so is it of no importance how any interesting subject is represented, provided the picture vividly excite a true conception of the subject. A painter will discover the difficulties which a brother artist had to surmount in representing an object in some particular predicament, as regards the light, &c., and may estimate the talent accordingly: but the great proportion, even of the most accomplished ordinary spectators, will generally be looking beyond the sign to the thing signified, heedless of the artist's difficulties. In consequence, however, of the prejudice in favor of "a sweet or adorable bit of colouring," as it will sometimes be called—and which in truth may have the merit of most natural colouring, there are preserved in many galleries pictures disgusting in almost all other respects, as of drunken Dutch boors, with fiery noses and physiognomies degrading to human nature, &c., &c., on seeing which the man of taste deplores that the art of painting should so often have been prostituted by clever men to the vile purpose of representing things of worse than no interest.

"*When the image formed, as above described, beyond a lens, is viewed in the air by an eye placed still farther beyond in the same direction, the arrangement, according to minor circumstances, constitutes either the common* TELESCOPE *or* MICROSCOPE." Read the whole second paragraph of the Analysis, page 325.

The name of *telescope* (a compound Greek term, signifying to *see far*, as *microscope* signifies to *see* what is *small*,) applies to that wondrous instrument of modern invention by the use of which the intelligent soul may be said, on the beams of light as its path, to dart widely into space for the purpose of contemplating the distant glories of creation; or again, by which it can command distant objects instantly to approach, for the purpose of convenient inspection. In ancient times, a man, while looking with admiration on the bright face of the moon, might have exclaimed "How pleased would I be, had I the power to fly upwards to that celestial orb, the better to understand its nature and beauties;" but he could little have anticipated that the day was coming when human ingenuity would find means in a great measure to satisfy the wish :—now the telescope is this means, for one which merely doubles apparent magnitudes, shows the moon exactly as she would appear to a person who had ascended towards her from the earth a distance of 120,000 miles, while one of greater power produces effects correspondingly great. —But to examine the heavenly bodies is by no means the only use of the telescope, man being often extremely interested to discover what is passing at a distance on the surface of the earth around him. Thus, by a telescope, the military chief obtains a close view of approaching friends or foes, while they are yet concealed from the naked eye, in the blue mist of distant mountain or plain—and similarly the sea-captain, while persons around him perceive only a little speck on the far horizon, discovers there a ship of class and nation at once evident to him, and with the crew of which, by the additional use of signal flags, he is enabled readily to converse At midnight, a telescope directed to a distant cathedral, may so effectually call it into the presence of the observer, that on the clock-turret may be watched the progress of the slow hands which tell of the unceasing lapse of time. A man, in the midst of a wide plain, or on a lofty hill-top, or far on the face of a lake, who might suppose himself quite alone and unseen, would yet, by a telescope, be instantly placed under the observation of whoever chose to watch him. And the same might happen to a man within the high walls of his own garden, or even within his house near an open window, if a straight line could reach him from some station where an observer was. Some remarkable cases of actions, imagined by the parties to have been done in perfect secrecy, have thus been brought to light.

Now the telescope, with its extraordinary powers, exhibits but another modification of the simple case (described at page 341, and exemplified in the camera obscura, &c.,) of an image formed for visual inspection beyond a lens And we shall here explain that its powers depend altogether on the two circumstances, first of its large lens collecting, for the formation of the image (subsequently transferred to the observer's retina) a thousand times or more the quantity of light which the naked pupil could receive; and second, of its forming by this light an image, to which the eye may be brought very near, so as to examine it with magnifying glasses of any power.

To understand this well, we must recall, that the nature of the bending of light in passing through a lens is such, that all the rays reaching the lens

Fig. 160.

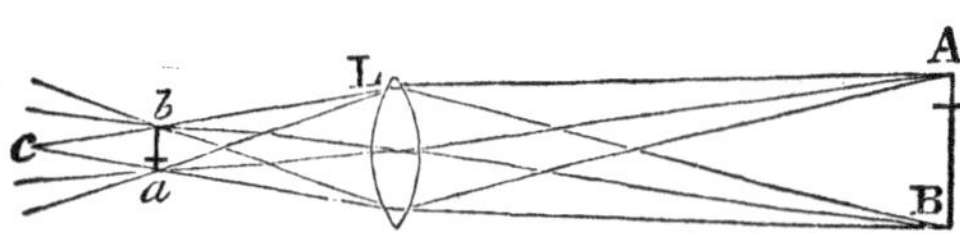

from any point of a visible object in front, and forming what is called a *pencil of light*—as that spreading from the point A of the cross here represented, to the lens L—are collected in a corresponding point, as *a*, at the focal distance beyond the lens, so as always to meet the central ray of the pencil here, (the direct line A *a*,)and, therefore, when the light comes from above the centre of the lens, the focal meeting is below, as shown here; and when it comes from below, the meeting is above; then the same happening as regards every visible point of the object (the rays from only the two extreme points A and B are here represented) at corresponding points beyond the lens in the space between *a* and *b*, the collected light, if received on a white screen placed there, as in the camera obscura, will make apparent to an eye in any direction a beautiful inverted image of the object. Now in the place where the rays meet to form this image, if no screen be interposed, the rays, although not lost or destroyed, but merely cross each other in the air, without interference, nearly as they previously crossed in the lens, and spread again beyond the focal points, or towards *c* as here shown, as they originally spread from the several points of the object itself; an eye, therefore, placed anywhere beyond *c*, must receive portions of the pencil from every point of the image, and may see the image in the air as it would see an object situated where the image is, in the focus of the lens. This may be observed at once by holding a spectacle-glass or any lens at a proper distance between an object and the eye.

Now a telescope is merely a long tube, blackened within to exclude and destroy useless light, and having a large lens, called the *object-glass*, filling its distant end, to gather the light from the objects in front, and with that light to form images towards the other or near end of the tube, where the eye may conveniently inspect them. These images, for a purpose to be immediately explained, are examined through another lens called the *eye-glass*, which is fixed in a small tube made to slide backwards and forwards in the large, so as to admit of the focal distances being adjusted to the power of different eyes, &c. The accompanying sketch shows the progress of the

Fig. 161.

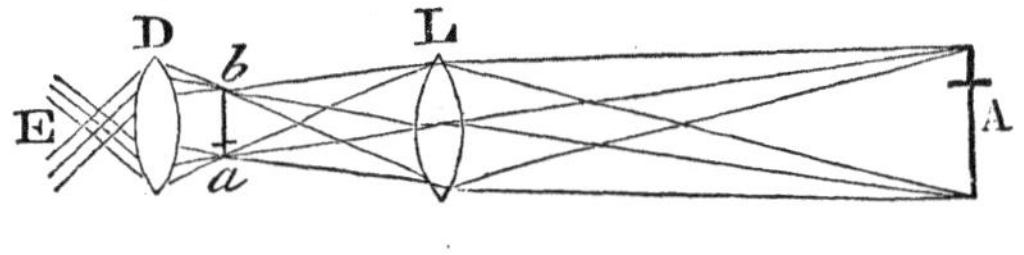

light from the object A, through the object-glass L, to form an image *b a*, and afterwards to be bent by the eye-glass, D, so as to enter the pupil of the eye at E, where the rays cross, to form the last image on the retina.

In the simple telescope with only two lenses, as above represented, called the *astronomical telescope*, or the *night-telescope*, because chiefly used at night, the image is inverted; but this is a circumstance of no importance in viewing the heavenly bodies; to fit the instrument, however, for viewing terrestrial objects, it is necessary to place in the tube another simple or compound lens, which shall form a second image from the first, and by inverting a second time, shall produce an image equally upright.

To determine how much larger an object will appear when viewed through a certain telescope,—for instance, through one with an object-glass of three feet focus,—than when viewed by the naked eye, we must recollect that the image is formed in the focus of the object-glass, or at *b a*, in the last figure, and subtends from the centre of that glass or lens, the same visual angle as the object itself (a fact explained at page 357,) and to an eye placed there would appear of the same size as the object, but if the eye be brought nearer to the image than the centre of the object-glass, the image will appear by so much taller and broader, and thus, as compared with the object, may be called so much magnified. Now as the naked eye cannot see distinctly an object nearer to it than at about six inches, because of the great divergence of light from a nearer radiant point, the telescope in question, without an eye-glass, would allow the eye to come only six times nearer to the image than when at the centre of the object-glass and would only magnify the diameter six times; but if then an eye-glass, as D, of half an inch focus were placed half an inch from the image, so as to render the rays of every pencil parallel, and therefore fitted to the powers of the eye, while the different parcels would cross each other a little way beyond the glass, as shown above, an eye placed to receive in its pupil the crossing parcels, would see the image as large as if at half an inch from it, and therefore 72 times nearer than if viewed from the object-glass, and therefore again as of 72 times greater diameter. Now, as in all cases, the image in a telescope is in the focus both of the object-glass and eye-glass, and is therefore nearer to the latter than to the former in proportion as their focal distances differ, the magnifying power is measured by that difference—in the case at present supposed, the difference is as 72 to 1, and 72 is the magnifying power of the telescope. The rule is generally thus expressed, "divide the focal distance of the object-glass by that of the eye-glass, and the quotient is the magnifying power." It is always to be remembered, that if the diameter of an object be magnified 10 times, the surface is magnified 100 times, and so in proportion for other numbers.

With such means of aiding the sight, then, it is that we discover the mountains of our moon, and can even measure their altitudes; that we can see the four beautiful moons of the planet Jupiter; that we can perceive marks and irregularities on the surfaces of the other planets, enabling us to say at what rate they severally whirl round their axes, experiencing the phenomena of day and night;—and that we can determine many other interesting particulars.

The discovery of the telescope is said to have been first made accidentally by the children of a Dutch spectacle-maker, while playing with their father's work; but it was turned to no use until Galileo, led by science, fell upon it again, and with the knowledge of its worth, obtained from it the most sublime results. The human heart can rarely have throbbed with such delight as Galileo's when he first directed his optic tube to the heavens, and through it contemplated so many glorious objects before unseen by human eye;—as the planet Venus, our beautiful morning and evening star, appearing not a circle, but a crescent like our moon in her quarters—as the satellites of Jupiter—the rings of Saturn—myriads of stars until then invisible to man; and in a word, when he beheld the undoubted proofs of the true system of the universe, as his genius had before conceived it, uniting the greatest simplicity with unspeakable grandeur.

The Galilean telescope was simply a large object-glass to collect much light, with a small concave eye-glass placed so as to intercept the converging rays before they reach their focus, and to change their convergency into the parallelism which the eye could command. This telescope, although magnifying less than that made of two convex glasses, as above described, still, from occasioning no loss of light by the crossing of rays in forming an image, was of considerable power. The common opera-glass is a telescope made on this principle.

It was explained at page 338, that a ray of light, in being bent or refracted by transparent media, as by a lens, is also divided into rays of the different colours seen in the rainbow. Hence an image formed behind a simple lens has coloured edges or fringes. This fact rendered the images of small objects much magnified, in the first made telescopes, very indistinct: and, but for the important discovery made by Dollond, the optician, that different kinds of glass have *dispersive* and *refractive* powers of different relative force, so that a concave lens of a certain curve applied to a convex lens might completely counteract the dispersion of colour by the latter, while it left enough of the convergence of the rays for the formation of an image—refracting telescopes would have always been very imperfect. Dollond called his telescopes *achromatic*, or *not-colouring*. It is very remarkable, that he had the fortune to obtain some glass for his purposes, more suitable than any which has been procured since, or which could be made by known rules, until the late improvements in the manufacture suggested by the ingenuity of Mr. Farraday. The author of this work carried abroad with him a small telescope of old Dollond's, which often gave more correct information respecting minute colored objects at a distance, as signal flags at sea, than much larger glasses of modern make.

The MICROSCOPE of greatest power and with the form called compound, in its structure approaches very closely to the telescope, the chief difference being, that while in the telescope a large distant object forms in the focus of the object-glass an image exactly as much smaller than itself as the distance of the image from the glass is less—in the microscope conversely, a small object placed near the focus of the object-glass produces a more distant image, as much larger than itself as the image is more distant than it—and in the one case as in the other, the image is viewed through an appropriate eye-glass. The object-glass in the telescope is large, in the microscope it is generally very small. If, in the latter, an object-glass be used of one-eighth of an inch focal distance, and the object be so placed that its image is formed at six inches, the image will be of diameter 48 times as great as the object, or will have nearly 2,500 times as much surface: and if that image be viewed through an eye-glass of half an inch focus, the image will appear still twelve times larger, or 30,000 times larger than the object.

A simple convex lens is called a single microscope, and it magnifies, as already explained, chiefly by allowing the eye to be brought much nearer to the object than the distance at which the object could be seen without the glass; but even where the distance of the eye and object is not changed, a lens interposed will still magnify by bending the light, as at *d* and *f*, making that which comes to the eye at *e* from the top of such an object as the little cross *a*, to appear to come from *b*, and that from the bottom to come from *c*, thus magnifying the cross here represented by the black lines to appear of the size represented by the dotted lines. A concave lens minifies for the contrary reason.

Fig. 162.

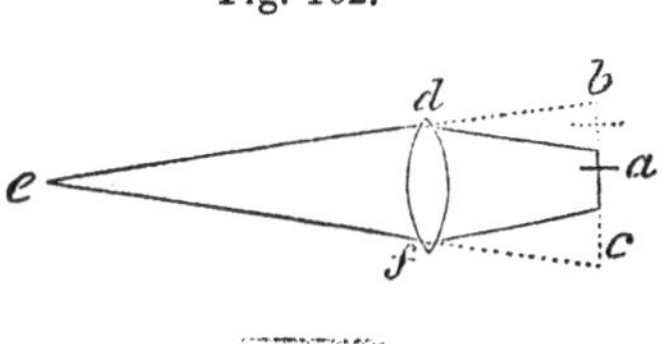

Perhaps there is not a greater treat for a person who has feeling for the beauties of nature, than to explore with the microscope. While the telescope lifts the mind to the contemplation of boundless space occupied by myriads of suns, and exhibits this globe of ours as less, compared with the universe around it, than a leaf is compared with a forest, or one grain of sand compared with all which lies on the ocean's shore, the microscope again excites new astonishment by showing on a leaf, or in a single drop of some water in which the leaf has been infused, thousands of living creatures, and of creatures not imperfect because thus small, but endowed with organs and parts as complex and curious as those of an elephant. And he who admires the curious structure of a honey-comb, may bend his eye through the microscope upon the cut surface of a willow-branch, or of other wood, there to see a similar structure more wonderful still; or he may compare the lace of a fly's wing with the most perfect which human art can weave; or the beautiful proportions and perfection of the limbs and weapons of an insect, invisible, perhaps, to the naked eye, with any larger objects of the kind already known to him.

Telescopes and microscopes might with propriety be both called microscopes, for often the telescopic object subtends to the naked eye even a smaller angle than the objects which the microscope examines. The minutest visible insect at hand may hide from the eye a planet at a distance. The image in the telescope, however, is always much smaller than in the microscope because the rays from a distance being nearly parallel, must form the image nearly in the principal focus of the object glass; while for the microscope, the rays from the near object being very divergent may be made to form the image far beyond that focus, and therefore proportionally larger.

" *Light falling on very smooth or polished surfaces, is reflected so nearly in the order in which it falls, as in many cases to appear to the eye as if coming directly from the objects originally emitting it,—and such surfaces are called* MIRRORS ; *the surface of which is flat as well as polished, is called a plane mirror.*" (Read the Analysis, page 325.)

If, on a marble slab, or other flat surface, (represented here at M R, with the edge supposed towards the spectator,) a ball were projected from A perpendicularly towards D, the ball would rebound directly back to A, but if projected obliquely, as from B to D, it would not return to the first situation B, but to *b*, a corresponding situation on the opposite side of the perpendicular, thus making the *angle* of the return or *reflection* equal to the *angle* of approach or *incidence ;* the same would be true of a ball approaching obliquely from any other point, as C, and rebounding to *c*. Now light is reflected from polished surfaces according to the same law, so that an eye at A would see itself as if placed at *d*, an eye at *b* would see an object really at B, as if it were at *e*, and so forth.

Fig. 163.

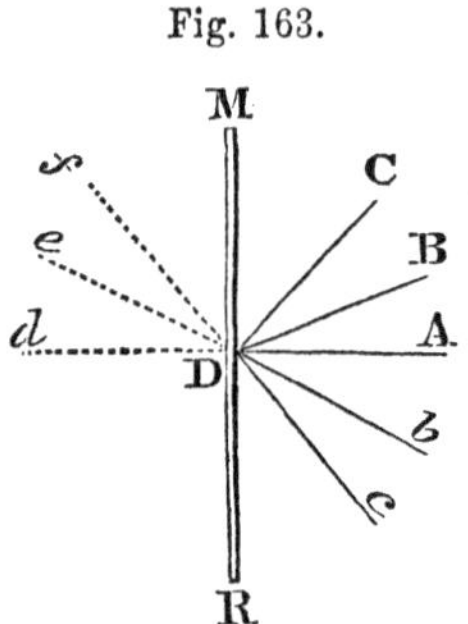

Where the existence of a mirror is not suspected, the objects reflected from it are held to be realities placed beyond where it is. A wild animal will attack its image in a glass: and the fable says that a dog crossing a brook, quitted the piece of meat in its mouth to catch the tempting image which he saw in the water below. The reason that an object seen in a plane mirror appears to be just as far beyond the mirror as its true distance on the side of the spectator, is, that the diverging rays of a pencil of light have the same divergence after as before reflection.

Any plane very smooth surface reflects light as now described, and is a mirror; but different substances send back very different proportions of the light which falls on them. A highly polished metallic surface is the best mirror, often returning three-fourths of the whole light. Hence, in reflecting telescopes, the mirrors are made of polished metals.

Our common looking-glasses are really metallic mirrors, for it is the smooth, clear surface of the quicksilvered tin foil behind the glass which reflects the light, the glass itself merely serving the purpose of preserving the metallic surface perfectly clean and flat. There is always an imperfection in such glass mirrors, when used for viewing objects obliquely, because the external surface of the glass acts also as a mirror, although so much more feebly than the metal behind, and forms a separate image not quite coinciding with the other, and therefore mixing with and confusing it.

The mirror-power of glass unaided is seen from the panes of a plate-glass window, which make objects in front very visible, although by no means with clearness comparable to that from a metallic surface. All common panes of glass in windows, or in print frames, &c., reflect as much light as plate-glass, but the reflection being irregular because the surface is irregular, scarcely attracts notice.

The smooth surface of a fluid is a mirror, which is, moreover, horizontal; and when that surface is metallic, as of mercury, the mirror is most perfect. In water, spirits, oil, or any other liquid, it is also perfect, but feebler.

The mirror of liquid quicksilver is sometimes used by astronomers in observing the apparent altitudes of the heavenly bodies, for the image in the mirror appearing exactly as much below the horizon as the object is really above it, half the distance between them is the true height.

A varnished picture or any japanned surface, is a mirror: nay, even a polished table of mahogany or other wood—as it is well known among playful children. The author, while writing this, has before him a table covered with black leather, and in that covering, as a mirror, he clearly sees the bright objects beyond the table. Polished stones, as marble slabs, &c., reflect as much as glass. Even a surface of air may be a mirror, as where a cold and dense stratum happens to lie in contact with a warmer and rarer stratum. In such cases, where particular causes have unequally heated different levels of the atmosphere, the trees, islands, &c., happening to be below, are reflected from above, and appear as if in the sky. This phenomena is called *mirage*. It is often to be observed over the burning sands of Africa, where the air is much heated; and elsewhere certain kinds of mists and thin clouds produce a similar effect, causing, for instance, a ship to appear as if suspended aloft, with keel uppermost.

In certain cases, an object seen by the light reflected from a mirror appears reversed, as when the right hand of a person standing before a glass becomes the type for the left hand of the image; or when a tree, or rock, or mountain, seen in the mirror of a lake, has its top downwards.

It is on this account, that a man painting his own portrait from a mirror, is apt to reverse all the accidental characteristics of the countenance or person, not the same on both sides; and if, as is generally true, one eye be higher than the other, or the nose be a little to one side, a very incorrect resemblance will be produced. Hence also a person whose countenance is at all thus peculiar never sees himself in a mirror as he appears to others; and a belle or beau, who has decided that a curl is more graceful on the left temple, may unconsciously leave it on the right.

By an image, however reflected from a first mirror to a second, and from that to the eye, persons may see the object, or themselves, if they choose, as others see them. What a pity that there are not some moral mirrors to answer an analogous purpose!

A candle placed between two parallel mirrors fixed on opposite sides of a room, makes visible in either glass to a spectator near the middle of the room an endless straight line of lights. If the glasses be inclined to each other, the lights will appear as if placed in the circumference of a circle, of which the centre is where the prolonged mirrors would meet: this fact is well illustrated in the beautiful toy called the *kaleidoscope*. It is possible to place a few mirrors in such situations around an apartment, that a man entering it, may see himself multiplied into a crowd, and a few ornamental pillars may produce the effect of thousands formed into long colonades of retiring lines.

The sun or moon reflecting in a still lake, appear as they do in the sky; but if the surface of the water become at all ruffled by the breeze, instead of one distinct image, there will be a long line of bright tremulous reflection. The reason of this appearance is, that every little wave, in an extent perhaps of miles, has some part of its rounded surface with the direction or obliquity which according to the required relation of the angles of incidence and reflection, fits it to reflect the light to the eye, and hence every wave in that extent sends its momentary gleam, which is succeeded by others.

Although the external surface of glass reflects but a small part of the light which falls upon it, being, therefore, a feeble mirror, still curiously, if light which has entered a piece of glass, fall very obliquely upon the back or internal surface, instead of passing out there, it is more perfectly reflected than it would be by the best metallic mirror. This light from A entering a piece of glass at B, is entirely reflected at C, the back of the piece, and escapes at D towards E. The back of a wedge of glass, or common prism, thus becomes a perfect mirror.

Fig. 164.

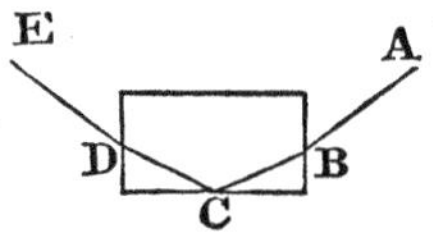

E A D B C

It is this fact which enabled Dr. Wallaston to construct that beautiful little instrument called by him the *Camera Lucida*. The two surfaces at the back of the small prism of glass A become mirrors, the first reflecting to the second, and the second to the eye at E, the objects in the landscape before it, while the eye also sees through the glass to the paper below at B, and may suppose the imagery to be feebly portrayed on the paper: with a pencil that appearance is made permanent, and a correctly-drawn outline of the scene is at once obtained The instrument for assisting draftsmen is still simpler than the camera obscura. Other modifications of the instrument have since been contrived.

Fig. 165.

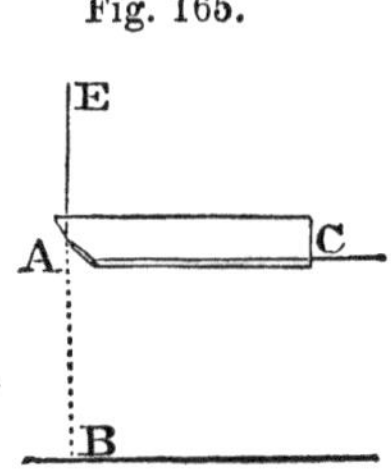

The same fact of the internal surface of a transparent mass becoming a mirror, gives us the explanation of that apparition or phenomenon so admired before it was understood, and not less admired since—the *rainbow*, or *arc in the sky*, as in France and elsewhere it is named—an object which the poets of nature have almost worshipped for its beauty, and which few of us can cease to remember as one of the delights of our boyish days, when we saw it stretching over the haunts of our young pleasures, and may have pursued it in the hope of catching some of the falling rubies and emeralds, or bright coloured dew of which it might be composed.

When a partial shower of rain falls on the side of the landscape opposite to where the sun is shining, there immediately appears in the shower a variegated arch, red at its external border or confine, and then successively orange, yellow, green, &c. (in the order of the colours of the prismatic spectrum described at page 337,) towards its inner border. Its centre is directly opposite to the sun, or at the end of the straight line supposed to be drawn from the sun through the eye of the spectator towards the opposite horizon; and being, therefore, always under the horizon, the bow is less than a semicircle. The diameter of the circle of which the bow is a part, occupies nearly 82° of the field of view, that is to say, the bow always coincides with a hoop of one foot diameter held eight inches from the eye. There is a second bow of much fainter light than the first, and with the colours in reverse order: it is of 108° diameter, and therefore external to the other.

Now the explanation of this miracle of beauty is simply as follows. While the sun shines upon the spherical drops of falling rain, its light falls upon the whole central part of any drop, passes completely through, but that portion which enters near the edge of the drop, as at *a*, is refracted or bent, and reaches the back surface of the drop at *y* so slantingly, or an angle so great, that it suffers there an entire reflection instead of being transmitted; the ray, therefore is turned to *b*, where it escapes from the drop, and as here shown, descends to the earth or eye in the direction *b e*. Thus every drop of rain on which the sun shines is a little mirror suspended in the sky, and is returning at a certain angle all round it, *viz.*, at an angle of 41°, a portion of the light which falls on it; and the eye placed in the required direction receives that reflected light. If in this case, however, there were *reflection* only, and not also *refraction* with *separation of colours*, the rainbow would be only a very narrow resplendent arc of white light formed of millions of little images of the sun; but in truth, because the light, which enters near the edge of the drop, traverses the surface very obliquely, it is much bent or refracted before its reflection, as seen at *a*, and is divided into rays of its seven colours, as it would be on passing through a prism (as explained at page 337;) and this division or separation continuing after the light again escapes from the drop at *b*, instead of one white ray descending from each drop to a certain point of the earth, seven rays descend (here marked by dotted lines from the figure 1 on the left hand to 7, 6, 5, &c., on the right, and with separation greater than occurs in reality to make it very evident,) and of these rays, an eye can only receive one at a time from the same drop, which drop will then appear of the colour of the ray: but for the same reason that seven eyes placed in a line from above downwards, as at 7, 6, 5, &c., on the right would be required to see the seven colours from one drop in the centre of the bow, so one eye looking in the direction of seven drops situated in a corresponding row, as from 1 to 7 on the left, will catch the lower or red ray of the upper, the orange or second ray of the next, the yellow or third ray of that which follows, and so on, while it will lose all the others, and thus will see the several drops as if they were each of one colour only. Of such elements, then found in the same relative directions all around the eye, the glorious arch is formed.

Fig. 166.

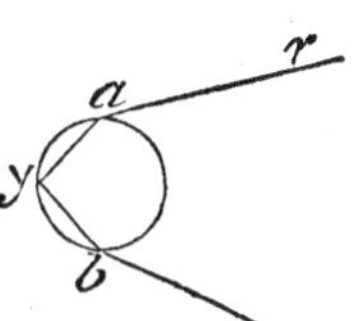

Fig. 167.

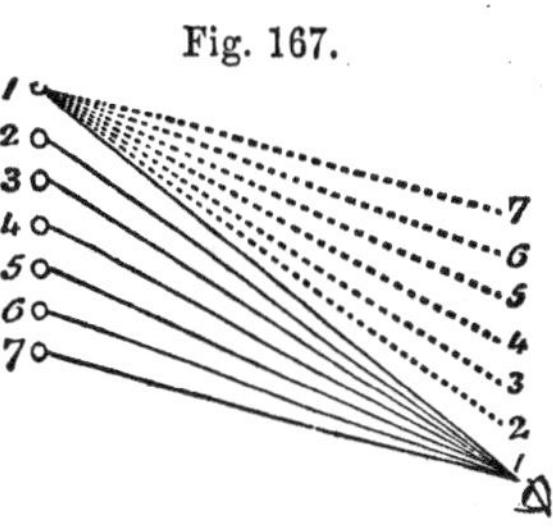

No two eyes can see the same rainbow, that is, can receive light from the same drops at the same time; and the same eye does not for two instants receive light from the same drops. This rainbow can never appear to a person on a plain, unless when the sun is within 41° of the horizon, for otherwise the centre of the rainbow would be more than 41° under the horizon, and therefore the whole circumference would be below it too.

Fig. 168.

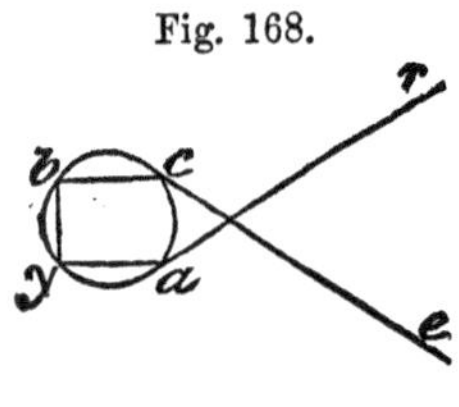

We have described above what is called the principal bow, formed in the drops by two refractions, and one reflection of light. To produce the fainter second or external bow, mentioned above, and of which the colours are in reverse order, the light which enters on the under side of the drop as at *a*, is reflected first at *y*, then again at *b*, and escapes at *c* towards the eye: after two reflections as well as two refractions. As the semi-diameter of the bow is 54°, it may be visible when the internal bow is not.

An artificial rainbow may be produced in sunshine at any time by scattering water-drops from a bush or otherwise; and a rainbow is often seen among the spray of a lofty waterfall, or of a stormy sea. The cut-glass ornaments of chandeliers, &c., produce colours on the same principle as rain-drops; as do also mist and particles of frozen water between a luminous body, and the eye exhibiting the circular coloured *halos* often observed around the sun and moon. A white *halo* is light reflected from the external surfaces of drops or particles.

"*Mirrors may be plane, convex, or concave; and certain curvatures will produce images by reflection, just as lenses produce images by refraction; in consequence, there are reflecting telescopes, microscopes, &c., as there are refracting instruments of the same names.*" (See the Analysis, page 325.)

While a plane surface reflects light, so that what is called the image in it of a known object may readily be mistaken for the reality, convex or concave mirrors reflect as if every distinct point of them were a separate small plane mirror, and their effects on light correspond with a relative inclination of the different parts. The only forms of much importance are the regularly spherical or parobolic concave and convex mirrors. We shall now find that these produce on light similar effects with lenses, only the concave mirror answers to the convex lens, and the convex mirror to the concave lens. It is the concave mirror which gathers the light to form images in the most perfect telescopes that exist, as those of Herschel and others. Admirable as is the refracting telescope, it still falls short in certain respects of the telescope acting by reflection.

Fig. 169.

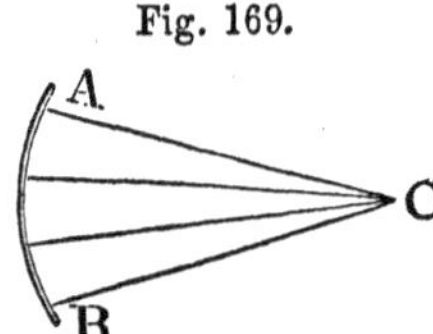

In a hollow sphere, or part of a sphere with polished internal surface, if rays radiate from the centre in all directions, they reach every part perpendicularly, and therefore are thrown back to the centre. Thus if A B were a concave spherical mirror, of which C. were the centre, rays issuing from C. would, in obedience to the law that the angles of incidence and reflection are equal, again meet at C.

It can be proved also, that any ray parallel to

the axis, falling upon such a mirror, will be reflected inwards so as to cut the axis half-way between the mirror and its centre, *viz.*, at D, the centre being C. Then as all parallel rays must meet in the same point, that point becomes a focus, as already explained for lenses, and there an image of the sun will be formed when the mirror is held directly towards the sun. This point is called the principal focus of the mirror.

Fig. 170.

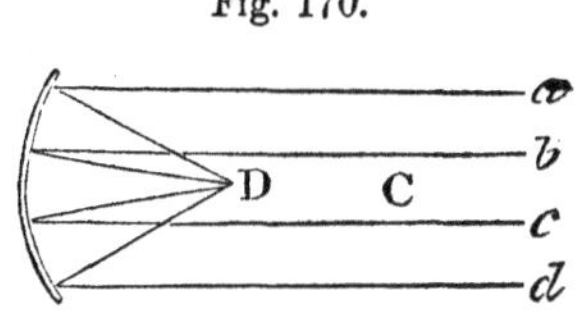

For the same reason that parallel rays meet in the focus, so will rays, issuing from the focus towards the mirror, become parallel, after reflection, as seen above or in the figure at page 271; and if they be then caught in a second and opposite mirror, as also represented at page 271, corresponding effects will follow.

Now, for a concave mirror, as already explained for a lens, when rays fall on it obliquely from one side of the axis, their focus will be on the opposite side, and therefore the mirror will form an inverted image of any body placed before it, just as the lens does; and the image will be near or distant and large or small, according to the divergence of the approaching rays, exactly as happens with lenses; and thus the camera obscura, magic lantern, telescopes and microscopes, may all be formed by mirrors, as they may be by lenses. Moreover, concave mirrors magnify, as concave lenses of the opposite names do. The two subjects of images by refraction and by reflection run so nearly parallel, that it would be useless repetition here to enter upon the detailed consideration of the latter subject, and we shall therefore content ourselves with showing why a concave mirror magnifies and why a convex mirror minifies.

A concave mirror magnifies because the light from the top of the cross at A, reaching the mirror where it can be reflected to an eye placed at F, *viz.*, at E, seems to the eye to come from C, and the light of B similarly appears to come from D, so that the cross A B, by the reflection, seems to the eye to be of the greater dimensions C D.

Fig. 171.

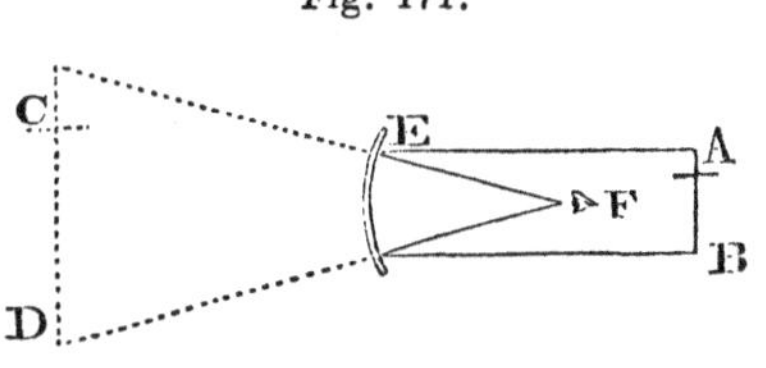

In the convex mirror, again, for corresponding reasons, the cross A B appears only as C D, and therefore much smaller than the reality.

Concave or magnifying mirrors are often used by persons in shaving.

A convex mirror is a common ornament of our apartments, exhibiting a pleasing miniature of the room and its contents.

Fig. 172.

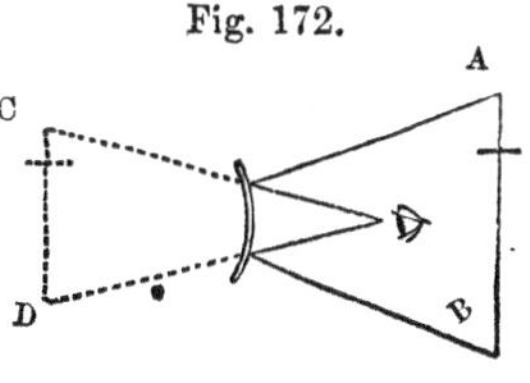

Any polished convex body is a mirror, and therefore the ball of the human eye is one, in which we may contemplate most perfect miniatures of surrounding things. It is the image of the window or of the sun in the convex mirror of the eye, which painters usually represent by a spot of white paint there; and a similar luminous spot or line must be made when they have to represent almost any of the pieces of furniture which have rounded polished surfaces as bottles, glasses, smooth pillars, &c.,

Convex lenses thus are also mirrors to all the objects around them, and very strikingly so, owing to the perfection of the form of a lens. The polished back of a watch, often, in the same way, attracts the attention of a child, who wonders to see there so clearly "the little baby."

It has been mathematical amusement to calculate what kind of distortion mirrors of unusual forms will produce, and then to make distorted drawings, which, when reflected from such mirrors, might produce in the eye the natural image of the objects.

When a concave mirror is used for a telescope, the image formed in front of it, and to be examined through the magnifying eye-glass, may be viewed, —first, as in Herschel's telescope, by the spectator turning his back to the real object, and looking in at the mouth of the telescopic tube, near to the edge of which the image is thrown by a slight inclination of the mirror at its bottom :—or, secondly, as in the *Newtonian* telescope, through an opening in the side of a tube, after being reflected by a small plane mirror, placed diagonally in the centre of the tube :—or, thirdly, as in the *Gregorian* telescope, through an opening cut in the principal mirror or speculum, after being reflected towards that opening by a small mirror placed in the centre of the tube ; this last arrangement is that preferred for smaller telescopes, because the spectator, while seeing the image, is also looking in the direction of the object.

Reflecting telescopes have the advantage of being perfectly *achromatic*, that is, of producing no coloured or rainbow edges to the images ; for compound light is reflected, although not refracted entire, all the colours following the same law of equal angles of incidence and reflection.

Herschel's largest telescope had a mirrow of 48 inches in diameter, and therefore received about 150,000 times more light than an unassisted eye, forming with that light, at a focal distance of 40 feet, a large image admirably distinct. It was with such a telescope that, in the obscurity of remote space, Herschel discovered the immense planet rolling along, which in honour of his royal patron, he called the *Georgium Sidus*, but which now, by the decision of the scientific world, bears his own name ;—and with such he discovered moons before unseen of other planets, and he unravelled the celestial nebulæ and clustered stars of the milky way and, in a word, unveiled vastly more than had before been done, the system of the boundless universe. If this world were to last for millions of years, the discoveries made by Herschel's telescope would mark a memorable epoch of its early history.

"*Light returned from, or passing through bodies of rougher or irregular surface, or which have other peculiarities, is so modified as to produce all those phenomena of colour and varied brightness seen among natural bodies, and giving them their distinctive characters and beauty.*" (See the Analysis, page 325.)

General remarks on this part of our subject were made in the beginning of the section, in the explanations of how objects not self-luminious become visible by reflecting the light of other bodies, and of how the prism separates a ray of white light into rays of the several colours which are seen also in the rainbow—which rays, on being again mixed, become white light as before:—and much beyond these remarks we have not the intention of now proceeding To give a full account of the matters that might come within the scope of this department, would occupy the pages of a large volume, for there would be to pass in review—the various opinions which have existed on the *intimate nature* of light,—the facts connected with what has been called the *polarization* of light,—the relation of light in its *double refraction*, to the ultimate structure of material masses, &c., all which subjects are in certain respects highly interesting, but—as some of them are not yet completely investigated—as respecting others various opinions prevail,—as they involve few matters yet applied to common use,—as the reasonings about them are far removed from ordinary trains of thinking, and refer to facts altogether unknown to common observation,—we hold them not to be fit parts of a popular treatise on light. We may state, however, that persons who have the leisure and mathematical preparation necessary for pursuing the study, will find their labour in it richly rewarded.

What we deem necessary here to add, is, that white light, in falling upon any transparent substance, as air, water, glass, &c., reduced to thin plates of films, is so affected, that for certain degrees of thinness, different for each substance, it is decomposed, and is reflected or is transmitted, not as white light, but as some of the colours of the rainbow, and the colour reflected in any case, is always the opposite or complement of that which is transmitted, that is to say, is such that the two brought together make white light as before. The facts may be studied as Newton originally studied them, in the thin plate of air which occupies the space between a convex lens and a plane surface of glass upon which the lens is laid,—in which plate, as the distance from the point of the apparent contact of the glasses increases, there are all degrees of thinness, and with these appear successive rings of vivid colours. The same truth is exemplified in the colours of a soap-bubble, which brighten as the bubble swells and becomes of thinner substance, and are different as the thickness is different and greater from above downwards;—and it is exemplified also in the colours seen in the fissures of cracked ice or crystalline spars, and in numerous other common facts. Now, whatever be the reasons of such decomposition of light—and the explanation is not yet complete—we cannot doubt that in natural bodies generally, the colours, opacity, transparency, &c., depend entirely upon the volume and arrangement of the minute fibres or plates, with included interstices, which constitute the volume or structure of each mass Accordingly, whatever changes that arrangement may change also the colour of the mass Thus, by drawing a certain number of minute lines on a certain extent of any metallic surface, we may make it of what colour we please; and mother-of-pearl owes its vivid colours and beauty entirely to its furrowed or striated surface, as is proved by our making an impression of that surface on sealing wax and perceiving that the wax exhibits similar colours.

The investigations in progress respecting the phenomena of light, are furnishing new proofs of the extreme simplicity of nature, amidst the boundless extent and most curious variety. When men thought of the sense of touch only as it exists at the tips of the fingers, or on the general surface of the body they were far from suspecting that the sense of hearing had the near relation to it which subsequent discoveries have proved, and still less, that the sense of sight was only yet a finer touch than hearing. But step by step they have ascertained, 1st, in relation to sound, that the air through which it usually reaches the organ of hearing, is a material fluid as much as water, consisting of the same or smaller particles, only more distant among themselves,—and that a motion or trembling in the air, by affecting nerves exposed in the ear, produces the sensation of sound by slight repeated pressures on these nerves, as the trembling in a log of wood caused by the action of a saw produces a peculiar sensation of touch in the nerves of a hand laid on the log;—and, moreover, that sound in all its varieties, is merely such trembling affecting a structure of nerve in the ear, which nerve is made as much more readily excitable than the nerves in the fingers or general cutaneous surface, as the action or impulse of trembling air is more delicate than the stronger pressures of common occurrence.—And, 2dly, in the investigation respecting light, this kind of comparison is carried a step farther, for it is become matter almost of certainty that the sensation of light is produced in the suitable nervous tissue of the eye, called the retina by a trembling motion in another fluid than air, which fluid pervades all space, and in rarity or subtilety of nature surpasses air yet more than air does water or solids;—and that, while in sound, different tones or notes depend on the *number* of vibrations in a given time, so in light do different colours depend on the *number* and *extent* of the vibrations. Can human imagination picture to itself a simplicity more magnificent and fruitful of marvellous beauty and utility than all this?—But yet farther, as air answers in the universe so many important purposes besides that of conveying sounds—although this, alone, comprehends language, which almost means reason and civilization—so also does the material of light minister in numerous ways, in the phenomena of heat, electricity and magnetism.

The truths now positively ascertained with respect to the nature of light and vision, are among those in the wide field of human inquiry, which, acting on ordinary apprehension, most forciby place the student, as it were, in the very presence of Creative Intelligence, awakening in him the most elevated thoughts of which the human mind is capable. Had there been no light in the universe, all its other perfections in regard to man had existed in vain. This earth would have been to its human inhabitants what any unknown shore would be to exiles abandoned upon it after their eyes were put out: every movement might be to their destruction, for their perceptions, being limited by the length of their arms, and of their fearful groping steps, the wretched beings separating when impelled by hunger to search for food, would probably scatter to meet no more. But the material of light exists, pervading all space, and certain impressions made upon it in one place, extend rapidly over the universe, the progressive impression being called a ray, or *beam of light*. The beams of light, then from all parts coming to every individual, may be regarded as millions of supplementary arms or feelers belonging to the individual, and which reach to the end of the universe, so that each person, instead of being as a blind point in space, becomes nearly omnipresent; then these limbs or feelers have no weight, they are never in the way, they impede nothing, and they are only known to exist when their use is required!

But this miracle of LIGHT would have been totally useless, and the paradise of earth would have been to man still a dark and dreary desert, had there not been farther the twin miracle of the EYE, an organ of commensurate delicacy to perceive the light. In the *Eye* we have to admire the round cornea, of such perfect transparency, placed exactly in the anterior of the ball, (and elsewhere it had been useless,) then exactly behind this, the beautiful curtain, the iris, with its opening, called the pupil, dilating and contracting to suit the intensity of light—and exactly behind the iris, again, the crystalline lens possessing important and remarkable properties, and which, by acting on the entering light, forms of it on the retina beautiful pictures or images of the objects in front;—the most sensible part of the retina being where the images fall. Of these parts and conditions, had any one been otherwise than as it is, the whole eye had been useless, and light useless, and the great universe useless to man, for he could not have existed in it.—Then, farther, we find that this precious organ, the eye, is placed in the person, not as if by accident, any where, but aloft on a befitting eminence, where it becomes the glorious watch-tower of the soul; and, again, not so that to alter its direction the whole person must turn, but in the head, which, on a pivot of admirable structure, moves while the body is at rest; besides that, the ball of the eye itself can roll in its place, and is furnished with muscles which, as the will directs, turn it with the rapidity of lightning to sweep along the horizon, or take in the whole heavenly concave;—then is the delicate orb secured in a strong socket of bone, and there is over this the arched and padded eyebrow as a cushion, to mitigate the shock of blows, and with its inclined hairs to turn aside any descending perspiration or other moisture which might incommode;—then is there the soft and pliant eyelid with its beauteous fringes, incessantly wiping the polished surface, and spreading over it the pure moisture poured out from the lachrymal glands above, of which moisture the superfluity, by a fine mechanism, is sent into the nose, there to be evaporated by the current of the breath;—still farther, it is to be noted, that instead of there being only one such precious organ, there are two, lest one, by accident, should be destroyed, but which two have so entire sympathy, that they act together as only one more perfect; then the sense of sight continues perfect during the period of growth from birth to maturity, although because the eye then increases in size, the distance between the lens and the retina is constantly increasing;—and the pure liquid which fills the eye, if rendered turbid by accident or disease, is by the actions of life, although its source be the thick red blood, gradually restored to transparency.—The mind which can suppose or admit that, within any limits of time, one single such apparatus of vision could have been produced by accident, or without design, must surely be of extraordinary character, or must have received unhappy bias in its education; but the mind which can still farther admit that the millions of human eyes which now exist on earth, all equally perfect, can have sprung from accident—and that the millions of millions of other eyes throughout the almost innumerable species of the living creation, where each is adapted to the peculiar nature and circumstances of the animal which bears it, can be accident; and, lastly, that the countless millions of all these well adapted kinds, which have existed in past ages, were all but accidents—the mind which can admit this, must have some of its highest faculties either benumbed or destroyed.

As a concluding reflection with respect to vision, we may remark, that all the provisions above considered have mere utility in view, for any one of them wanting would leave a necessary link in the chain of creation wanting: but we have shown, in a preceding part of the work, that if there had been white light only, susceptible as now of different degrees of intensity and shade, the merely useful purposes of vision should have been answered about as perfectly as with all the colours of the rainbow—a truth instanced in the facts, that many persons do not distinguish colours, and that it imports not whether a person view objects in the morning, or at mid-day, or at even-tide, or through plain glass or coloured glass, provided there be light and shade enough to show them clearly. While, therefore, the existence of light generally, and of the eye, speaks of Creative Power and Intelligence, the existence of colours, or of that lovely variety of hues exhibited in flowers, in the plumage of birds, in the endless aspects of the earth and heavens—because appearing expressly planned to give delight to animated beings, speaks of Creative Benevolence, and may well excite in us towards the Being in whom these attributes reside, the feelings associated in our minds during this earthly scene, with the endearing appellation of "Father."

Fig. 173.

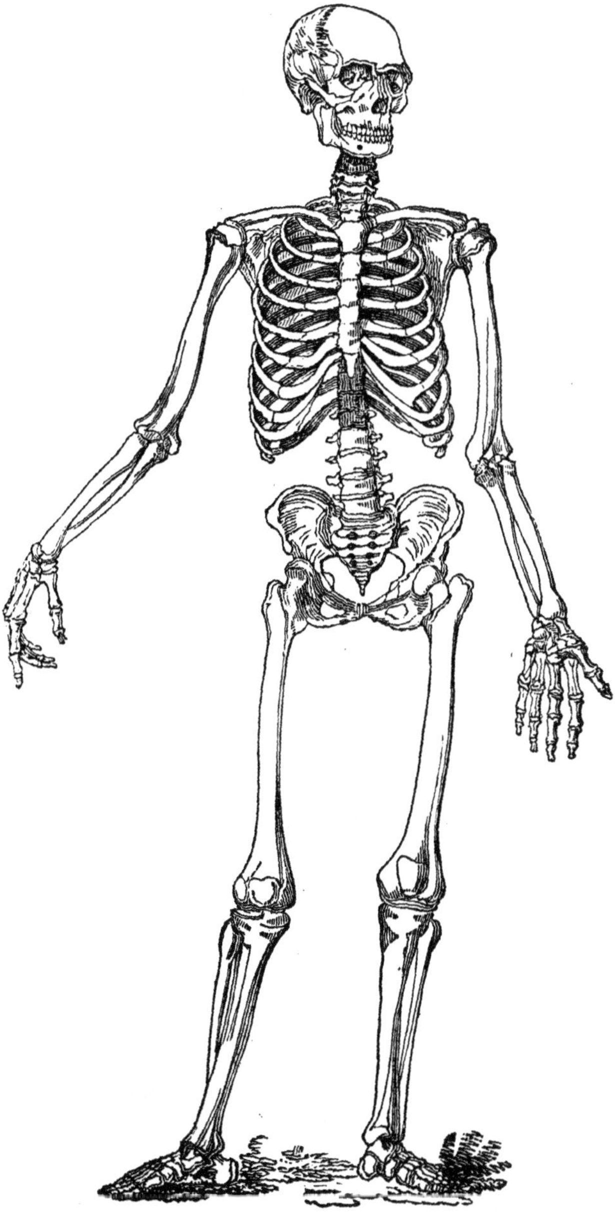

PART V.

ANIMAL AND MEDICAL PHYSICS.

Section I.

Mechanism of the Human Skeleton.

Having now completed our study of general mechanics, we shall proceed, with the light thence derived, to examine that most interesting illustration of many of the truths—the solid frame-work of the human body—a perfect work of an unerring Engineer!

There is scarcely a part of the animal body, or an action which it performs, or an accident that can befall it, or a piece of professional assistance which can be given to it, that does not furnish illustration of some truth of natural philosophy; but were we here to enter into much detail, we should be giving minute lessons in medical science, instead of explaining general laws. We shall therefore only touch upon as many particulars as will make the understanding of all the others easy; trying to conclude, among our illustrations, such matters of importance as would be likely to escape the notice of a hasty student.

The *cranium* or *skull* has been already mentioned as an instance of the arched form answering the purpose of giving strength. The brain, in its nature, is so tender or susceptible of injury, that slight local pressure disturbs its action. Hence a solid covering like the skull was required with those parts made stronger and thicker which are most exposed to injury. An architectural dome is constructed to resist one kind of force only, always acting in one direction, *viz.*, gravity; and therefore its strength increases regularly towards the bottom, where the weight and horizontal thrust of the whole are to be resisted; but in a skull, as in a barrel or egg-shell, the mere tenacity of the substance is many times greater than sufficient to resist gravity, and therefore the form and securities are calculated to resist forces of other kinds operating in all directions. When we reflect on the strength displayed by the arched film of an egg-shell, we need not wonder at the severity of blows which the cranium can withstand.

In the early fœtal state, that which afterwards becomes the strong bony case of the brain exists only as a tough flexible membrane. Ossification commences in this membrane long before birth, at a certain number of points from which it spreads, and the portions of the skull formed around these points soon acquire the appearance of so many scales or shells applied on the surface of the brain, and held together by the remaining membrane not yet ossified. They afterwards become firmly fixed together, by projections of bone from each, shutting in among similar projections of the adjoining ones, until all mutually cohere by perfect dove-tailed joints, like the work of a carpenter. These joints are called sutures of the cranium, and are visible to extreme old age. Through early childhood, the cranium remains to a certain degree yielding and elastic, causing the falls and blows, so frequent during the lessons of walking, &c., to be borne with comparative impunity. The mature skull consists of two *layers* or *tables*, with a soft *diploe* between them; the outer table being very tough, with its parts dove-tailed into each other as tough wood is joined by human artificers; while the inner table is harder and more brittle, (hence called *vitreous*) with its edges merely lying in contact.

A very severe partial blow on the skull generally fractures and depresses the part, as a pistol-bullet would: while one less severe, but with more extended contact, being slowly resisted by the arched form, often injures the skull by what is correspondent to the *horizontal thrust* in a bridge, and causes a crack at a distance from the place struck—generally half way round to the opposite side. The French in speaking of this effect, usė the term *contre-coup*. Sometimes in a fall with the head foremost, the skull would escape injury, but for the trunk which falls upon it, and drives the end of the spine against or even through its base.

In the lower jaw we have to remark the greater mechanical advantage, or lever-power, with which the muscles act, than in other parts of animals. The temporal and masseter muscles pull almost *directly*, or at right angles to the line of the jaw, while in most other cases, as in that of the deltoid muscle lifting the arm, the muscles act very *obliquely*, and with power diminished in proportion to the obliquity. An object placed between the back teeth is compressed with the whole direct power of the strong muscles of the jaw. Hence the human jaw can crush a body which offers great resistance, and the jaws of the lion, tiger, shark, and crocodile, &c., are stronger still.

The *teeth* rank high among those parts of the animal body which appear almost as if they were severally the results of distinct miraculous agencies —so difficult is it to suppose a few simple laws of life capable of producing the variety of form and fitness which they exhibit. They constitute a beautiful set of chisels and wedges so arranged as to be most efficient for cutting and tearing, and grinding the food, with their exterior enamel so hard, that few substances in nature can make an impression upon it. In early states, of society, teeth were used for many purposes for which steel is used now. It seems, however, as if the laws of life, astonishing to human intellect as they are, had still been inadequate to cause teeth cased in their hard and polished enamel, to grow as the softer bones grow; and hence has arisen a provision more extraordinary still. A set of small teeth appear soon after birth, and serve the child until six or seven years of age: these then fall out, and are replaced by larger ones, which endure for life; the number of the latter, however, being completed only when the man or woman is full grown, by the four teeth, called wisdom teeth, from coming with the person's maturity, to fill up the then spacious jaw.

The spine or back bone, in its structure, has as much of beautiful and varied mechanism as any part of our wonderful frame. It is the central pillar of support and great connecting chain of all the other parts; and has at the same time, the office of containing within itself, and of protecting from external injury, a prolongation of the brain, called the spiral marrow, more important to animal life than the greater part of the brain itself. It has united in it the apparent incompatibilities of great elasticity, great flexibility in all directions, and great strength, both to support a load and to defend its important contents,—as we shall now perceive.

Elasticity.—The head rests on the elastic column of the spine, as softly as the body of a carriage rests upon its springs. Between each two of the twenty-four vertebræ or distinct bones of which the spine consists, there is a soft elastic *intervertebral substance*, about half as bulky as a vertebræ, and which yields readily to any sudden jar: then the spine is waved or bent like an italic *f*, as if perceived on viewing it sideways, or in profile, and by this reason, also, it yields to any sudden pressure operating against either end. The bending might seem a defect in a column intended to support weight, but the disposition of the muscles around is such as to leave all the elasticity of that form, and a roomy thorax, without any diminution of strength.

Flexibility.—The spine has been compared to a chain, because it consists of many distinct pieces (twenty-four.) They are in contact by smooth rubbing surfaces, which allow of a degree of motion in all directions; and a little motion comparatively between each two adjoining pieces, becomes a great extent of motion in the whole line.

The *strength*—of the spine as a whole, is shown in the fact of a man's easily carrying upon his head or back a weight heavier than himself; and the strength of each separate vertebræ surrounding the spinal marrow, is evident in its being a double arch, or strong irregular ring. The spine increases in size towards the bottom, in the justest proportion, as it has more weight to bear. The articulating surfaces of the spine are so many, and so exactly fitted to each other, and are connected by such number and strength of ligaments, that the combination of pieces, becomes, in reference to motion, a much stronger column than a single bone of the same size would be.

Considering the great number of parts forming the spine, and their nice mutual adaption, it might be expected that injuries and diseases of the structure would be very frequent. The reverse, however, under natural circumstances, is true; so that while hundreds and thousands of works have been published on the diseases of almost every other part of the body, hardly any have been written on spine-affections, and what have appeared are of very recent date. One reason of this is, that whatever regards health and disease is now much more completely analyzed than formerly; but another and the chief reason is, that from a change in modern times introduced into the system of education for young ladies, a considerable proportion of them having grown to womanhood with weakened and crooked spines.—The subject merits further consideration here.

To the well-being of the higher classes of animals, a certain degree of exercise of their various parts is not less necessary than their nourishment, and if, during the period of growth, such exercise be withheld by any cause, the body never acquires its due proportions and strength. To prompt young creatures to the required exertion, nature has given them an overflow of life and energy, as evinced in the ever-changing occupation of a child in the quick succession of its ideas, in its jumping and skipping, and using all the modes of roundabout action to expand muscular energy, instead of seeking, as in after life, to accomplish its ends in the shortest ways:—and as seen among the inferior animals, in the play of kittens, puppies, lambs, &c. But, strongly as nature has thus expressed herself, tyrant fashion, with a usual perversion of common sense, had of late times, in England, for young women of the higher classes formed a school discipline, directly at war with nature's dictate; so that a stranger arriving from China, might almost suppose it our design to make crooked and weak spines by that discipline, as it is the design in China to make little feet by the iron shoe. The result is the more striking, when the brothers of the female victims, and who of course have similar constitutions, are seen to be robust, healthy, and well-formed. A *peasant-girl,* when her spirits are buoyant, is allowed to obey her natural feeling, and at proper times to dance, and skip, and run, until healthy exhaustion asks that repose which is equally allowed; and thus she grows up strong and straight, but the *young lady* is receiving constant admonition to curb all propensity to such vulgar activity, and often, just in proportion as she subdues nature, she receives the praise of being *well-bred.* The multifarious studies, also, of the latter come powerfully in aid to the admonition, by fixing her for many hours every day to sedentary employment; and the consequences soon follow, of weakness in the body generally from the want of the natural quantity and variety of muscular exertion, but weakness of the back particularly, from the manner in which the sitting is usually performed. It would be accounted great cruelty to make a delicate girl stand all day, because her legs would tire, but this very cruelty is in almost constant operation against her back, as if backs could not tire as well as legs. When she is allowed to sit down because she has been long standing, great care is taken that the muscles of the back, which still remain in action as she sits, shall not be at all relieved; for, from the idea that it is ungraceful to loll, she is either upon a stool which has no back at all, or upon a very narrow chair with a perpendicular back. Now neither of these seats relieve her spine, the stool, however, being less hurtful than the chair, because it allows the spine to bend in different ways so as to rest the different sets of muscles alternately, while the chair keeps the spine constantly upright and nearly unmoved. The excessive fatigue soon causes the spine, somewhere, to give way and to bend, and the curvature often becomes permanent. And, as when a bend takes places in one situation, there immediately follows an opposite bend above or below, to keep the centre of gravity of the body always directly over the base, the curve thus becomes double, like an italic *f*, and the distortion is rendered complete.—In bending the spine is sometimes also partially rotated or twisted, so as to show from behind that waving profile which should be seen only from the side.

When owing to such discipline the inclination of the back has once been begun, it is often rapidly increased by the means used to correct it. Strong stiff stays are put on to support the back, as is said, but which in reality, by superseding the action of the muscles placed there by nature as the supports, cause these to lose their strength, and to be unable, when the stays are withdrawn, to support the body. Longer sittings in the narrow upright chair are then recommended, and sometimes the back is forcibly stretched by pullies, so the patient is kept all day and night lying on an inclined board, losing her health, &c.;—the only things guarded against being, the patient should take due exercise and air, and should rest properly when she is not taking exercise. With many persons the prejudice had at last grown up, that strong stays should be put on at a very early age, to prevent the first approach of the mischief, and that children should always be made to sit on straight-backed chairs, or to lie on hard planes; and it is probable, that if these cures and preventives had been adopted as universally and strictly as many deemed them necessary, we should now scarcely have in England a young lady of healthful form. What would be said of the person who should try to improve the strength and shape of a young race-horse or greyhound, by binding tight splints or stays round its beautiful young body, and then tying it up in a stall! But this is the kind of absurdity and cruelty which has been so commonly practised in this country towards beings than whom, as nature offers them, the universe surely contains none more faultless.

A pernicious prejudice, with respect to such curvature or distortion of the spine, long existed, namely that it was a scrofulous affection; and many mothers concealed it as much as possible, and sought remedy from quacks far from home. In consequence, until within a few years, the management, of spine diseases was chiefly in the hands of some irregular members of the profession,—and a rich source of wealth it became to them, from many of their remedies being calculated rather to prolong than to cure the evil. The practice in such cases, however, has now fallen into the hands of the profession generally; the science having detected the true cause of the evil, its frequency is already diminished. It has been shown that to prevent the disease is easy, and that the best cures are those conducted on the general principles of improving the health of the patient by fit regimen, of prescribing such exercises as may directly strengthen the affected part, and of causing the patient, when reposing to assume positions which directly counteract the morbid tendency.

Some might expect here a long description of machines employed in the treatment of spine affection: but the list of those which are useful or safe is very short:—a sofa to rest upon during the day and a fit bed for the night; (the "hydrostatic bed,") proposed by the author of this work, and described in the next chapter, has certain advantages;) choice of pleasant means of taking exercise, such as the skipping-rope, shuttle-cock, dum-bells, a rope-ladder to climb, a winch to turn, &c. :—and where it is much desired that the young lady should employ herself in the sitting attitude, as in practising music, a chair may be used with crutches rising from its side, or with straps descending from pullies in an overhanging canopy or crane, and kept tight by proper weights at their distant ends, to support the head and shoulders. The author has had a small crane of wood made, which well answers the last mentioned purpose, and may be attached to a common chair. It would be out of place here to detail those particulars of constitutional treatment which, in peculiar habits, may be required to aid the effects of the means above described

The ribs —Attached to twelve vertebræ in the middle of the back, there are the ribs or bony stretchers of the cavity of the chest, constituting a structure which solves, in the most perfect manner, the difficult mechanical problem of making a cavity with solid exterior, which shall yet be capable of dilating and contracting itself. Each pair of corresponding ribs may be considered as constituting a hoop: which hangs obliquely down from the place of attachment behind, so that when the forepart of all the hoops is lifted by the muscles the cavity of the chest is enlarged.

We have to remark the double connection of the rib behind, first to the bodies of two adjoining vertebræ, and then to process or projection from the lower, thus affecting a very steady joint, and yet leaving the necessary freedom of motion: and we observe the forepart of the rib to be joined in the breast-bone by flexible cartilage, which allows the degree of motion required there without the complexity of a joint, and admirably guards, by its elasticity against the effects of sudden blows or shocks.

The muscles, which have their origin on the ribs and their insertion into the bones of the arm, afford us an example worth remembering of action and reaction being equal and contrary. When the ribs are fixed, these muscles move the arm; and when the arm is fixed, as by resting on a chair or other object, they with equal force move the ribs The latter occurrence is seen in fits of asthma and dyspnœa.

The human skeleton, with its naked ribs, is so associated in the common mind, with ideas of death and loss of friends, and all the terrors of doubtful futurity, that to most persons it is an object of abhorrence; but to the philosophic mind, which rises superior to place and time, the so admirable adaptation of all the parts to their purposes, and of parts which, being purely mechanical, are perfectly understood, makes it independently of all professional considerations, an object of the most intense interest. Such mechanism reveals, by intelligible signs, the hand of the Creator; and a man may be said sublimely to commune with his Maker, who contemplates and understands the structure aright.

The *shoulder-joint* is remarkable for combining great extent of motion with great strength. The round head of the shoulder-bone, that it may turn freely in all ways, rests upon a shallow cavity or socket in the shoulder-blade; and the danger of dislocation from this shallowness is guarded against by two strong bony projections above and behind. To increase the range of motion to the greatest possible degree, the bone called the shoulder-blade, which contains the socket of the arm, slides above itself upon the convex exterior of the chest, having its motions limited in certain directions by its connection, through the collar-bone or clavicle, with the sternum.

The *scapula* or *blade-bone* is extraordinary as an illustration of the mechanical rules for combining lightness with strength. It has the strength of the arch from being a little concave, like the dished wheel already described, and its substance is chiefly collected in its borders and spines, with thin plates between, as the strength of a wheel is collected in its rim, and spokes, and nave.

The bones of the arms, considered as levers, have the muscles which move them attached very near to the fulcra, and very obliquely, so that the muscles, from working through a short distance, comparatively with the displacement of the resistances at the extremities require to be of great strength. It has been calculated that the muscles of the shoulder-joint, in the exertion of lifting a man upon the hand, pull with the force of two thousand pounds.

Notwithstanding all the securities of the shoulder-joint now described in the infinite variety of twists, and falls and accidents to which men, in the busy scene of society, are liable, the joint is frequently dislocated, that is, the rounded head of the humerus or arm-bone slips from his socket, with instant lameness as a consequence.

In the treatment of dislocations and fractures of the frame-work of the human body, the surgeon cannot avoid displaying strikingly either his professional skill or ignorance. With what ease does the displaced arm or thigh-bone return to its socket, under the guidance of the skilful hand; and to what horrible, and often unavailing torture, is the patient subjected, when, in such a case, ignorance dares to act! It is very painful to allow the imagination to dwell upen the records of ancient surgery, and to be made present, as it were, to the stretching of patients on the rack with pullies and powerful engines, to do, what better information could have accomplished with such gentleness. And would that the records of modern times contained no instances of individuals crippled for life by bad practice. To a practioner in this branch, impunity and a quiet conscience can now be secured only by his having a perfect knowledge of anatomy, and familiarity with the laws of mechanical philosophy.

With our present information on these subjects, we are suprised at the detail of the practices and errors promulgated in former times, owing to imperfect knowledge of mechanics, even by authors of the highest credit. It would hardly be believed that so distinguished an ornament of English surgery as Mr. Pott, should assign as one reason for not pulling by the hand or foot, in reducing a dislocation of the shoulder or hip, that the intervening joints prevented the strain from reaching the part desired.

Some surgeons, possessing a certain degree of knowledge in mechanics, but only that degree which is dangerous, having heard that the lever was a powerful engine, have tried to replace bones solely by leverage, as it was called. Thus, a man's dislocated arm has been placed over the back of a chair as a fulcrum, or over the top of a door, and while the weight of the suffering body was hanging to it on one side as the resistance, force has been applied to the other side, enough sometimes to break the bone, or to tear away the ligaments and soft parts about the joint.

Other surgeons, after learning in the same way the effects of the pulley, have wished to do all by irresistible extension, and instead of borrowing the moderate assistance which might be useful, have torn muscles and ligaments from their attachments.

It is not the object of this work to enter into an extended examination of the accidents which befall the body requiring mechanical skill for their proper management, for this would be to deliver a course of instruction on practical surgery; but it is wished to awaken the attention of the medical student to those valuable general principles which may furnish direction in most difficulties. Knowing these principles, and possessing good sense, he will often be a more effective minister of his art than a man full of learned precedents, who knows them not. To make this lesson more impressive to his young readers, the author may take the liberty of adducing his own experience. When he was himself very young, and had not yet had extensive practical experience, he was thrown into a situation where a heavy medical charge developed upon him, and where, through accidents among a numerous crew, during a very eventful voyage, which led to intercourse with the savage inhabitants of unfrequented coasts, he had, within twenty-six months, more practice in singular wounds, dislocations, and fractures, than falls to the lot of many practitioners during a life:—in that time he became strongly impressed with the importance to the medical man of such knowledge as he now recommends: and he had reason to rejoice that although Natural Philosphy was not then much insisted upon in the course of professional education, circumstances had led him to look carefully at the body through that medium.

The os humeri, or bone of the upper arm, is not perfectly cylindrical, but like most of the other bones called cylindrical, it has ridges to give strength, on the principle explained in the chapter "on strength of materials."

The elbow joint is a correct hinge, and so strongly secured that it is rarely dislocated without fracture.

The fore-arm consists of two bones with a strong membrane between them. Its great breadth, from this structure, affords abundant space for the origin of the many muscles which go to move the hand and fingers: and the very peculiar mode of connection of the two bones, gives man that most useful faculty of turning the hand round, into what are called the positions of pronation and supination,—exemplified in the action of twisting or of turning a gimblet.

The old surgeons, who acted frequently by rules of routine rather than by reasons, in the accident of fracture to one or both bones of the fore-arm, often applied a tight bandage, which pulled the bones at the fractured part close to each other, and thus injured the future shape and strength of the arm.

The wrist. The many small bones forming the wrist have a signal effect of deadening, in regard to the parts above, the shocks or blows which the hand receives.

The annular ligament is a strong band passing round the joint, and keeping all the tendons which pass from the muscles above to the fingers, close to the joint. It answers the purpose of so many fixed pullies for directing the tendons: without it, they would all, on action, start out like bow-strings, producing deformity and weakness.

The human hand is so admirable, from its numerous mechanical and sensitive capabilities, that one opinion at one time prevailed, that man's superior reason depended on his possessing such an instructor and such a servant. Now, although reason, with hoofs instead of fingers, could never have raised man much above the brutes, and probably could not have secured the continued existence of the species,—still, the hand is no more than a fit instrument of the Godlike mind which directs it.

The pelvis, or strong, irregular ring of bone on the upper edge of which the spine rests, and from the sides of which the legs spring, forms the centre of the skeleton. A broad bone was wanted here to connect the central column of the spine with the lateral columns of the legs, and a circle was the lightest and strongest. If we attempt still farther to conceive how the circle could be modified so as to fit it—for the spine to rest on, for the thighs to roll in, for muscles to spring from, both above and below, for the person to be able to sit, &c., we shall find, on inspection, that all our anticipations are realized in the most perfect manner. In the pelvis, too, there are the thyroid hole and ischiatic notches, furnishing subordinate instances of contrivance to save material and weight:—they are merely deficiencies of bone where solidity could have given no additional strength. The broad ring of the pelvis protects most securely the important organs placed within it.

The hip joint exhibits the perfection of the ball and socket articulation. It allows the feet to move round in a circle, as well as to have the great range of backward and forward motion, exhibited in the action of walking. When we see the elastic, tough, smooth cartilage which lines the deep socket of this joint, and the similar glistening covering of the ball or head of the thigh-bone, and the lubricating synovia poured into the cavity by appropriate secretaries, and the strong ligaments giving strength all around, we feel how far the most perfect of man's work falls short of the mechanism exhibited in nature.

The thigh-bone is remarkable for its two projections near the top, called trochanters, to which the moving muscles are fixed; and which lengthen considerably the lever by which the muscles work. The shaft of the bone is not straight, but has a considerable forward curvature. Short-sightedness might suppose this a weakness, the bone being a pillar to support a weight; but the bend gives it in reality the strength of the arch, to bear the action of the mass of muscles called *vastus,* which lies and swells upon its fore part.

The knee is a hinge joint of complicated structure, claiming the most attentive study of the surgeon. The rubbing parts are flat and shallow, and, therefore, the joint has little strength from form; but it derives security from the numerous and singularly strong ligaments which surround it. The ligaments on the inside of the knees resemble, in two circumstances, the annular ligaments of joints, *viz.,* in having a constant and great strain to bear, and yet in becoming stronger always as the strain increases. The line of the leg, even in the most perfect shapes, bends inward a little at the knee, requiring the support of the ligaments; and in many persons it bends very much; but the inclination does not increase with age. The legs of many weakly in-kneed children become straight by exercise alone. This inclination at the middle joint of the leg, by throwing a certain strain on the ligaments, gives, in such actions as jumping, running, &c., an increase of elasticity to the limb.

In the knee there is a singular provision of loose cartilages between the ends of the bones. They have been called friction-cartilages, from a supposed relation in use to friction-wheels, but their real effect seems to be, to accommodate, in the different positions of the joint, the surfaces of the rubbing bones to each other.

Under the head *Pneumatics*, we shall find that the bones forming the joints, are held together, independently of their ligaments, by a constant pressure of the atmosphere, amounting in the knee, for instance, to upwards of sixty pounds.

The great muscles on the fore-part of the thigh are contracted into a tendon a little above the knee, over and in front of which the tendon has to pass to reach the top of the leg, where its attachment is. The part of the tendon over the joint becomes bony, and forms the patella or knee-pan, often called the pulley of the knee. This peculiarity enables the muscles to act more advantageously, by increasing the distance of the rope from the centre of the motion. The petella is, moreover, a sort of shield or protection to the fore-part of this important joint.

The leg below the knee, like the fore-arm already described, has two bones. They offer spacious surface of origin for the numerous muscles required for the feet, and they form a compound pillar of greater strength than the same quantity of bone as one shaft would have had. The individual bones also are angular instead of round, hence deriving greater power to resist blows, &c.

The ankle-joint is a perfect hinge of great strength. There is in front of it an angular ligament, by which the greater part of the tendons passing downwards to the foot and toes are kept in their places. One of these tendons passes behind and under the bony projection of the inner ankle, in a smooth, appropriate groove, exactly as if a little fixed pulley were there.

The heel, by projecting so far backwards, is a lever for those strong muscles to act by, which form the calf of the leg and terminate in the tendo-achillis. The muscles by drawing at it, lift the body, in the actions of standing on the toes, walking, dancing, &c. In the foot of the negro, the heel is so long as, in European estimation, to appear ugly; and its great length rendering the effort of smaller muscles sufficient for the various purposes, the calf of the negro's leg is smaller than of other races of men.

In a graceful human step the heel is always raised before the foot is lifted from the ground, as if the foot were part of a wheel rolling forward; and the weight of the body, supported by the muscles of the calf of the leg, as just described, rests for the time on the fore-part of the foot and toes. There is at that time a bending of the foot in a certain degree. But where strong wooden shoes are used, or any shoe so stiff that it will not yield and allow this bending of the foot, the heel is not raised at all until the whole foot rises with it, so that the muscles of the calf are scarcely used, and in consequence soon dwindle in size, and almost disappear. Many of the English farm-servants wear heavy stiff shoes, and in London may constantly be seen as the drivers of country wagons, with fine robust body and arms, but with legs which are fleshless spindles, producing a gait most awkward and unmanly. The brothers of these men, otherwise employed, are not so mis-shapen; and even they themselves, when they choose to become soldier, and are trained in military exercises, lose their peculiarity. What a pity that, for the sake of a trifling saving, graceful nature should be thus deformed. An example of an opposite kind is seen in Paris, where, as the streets have no side pavements, and the ladies are obliged consequently to walk almost constantly on tiptoe, the great action of the muscles of the calf has given a conformation of the leg and foot, to match which the Parisian belles proudly challenge all the world,—not aware, probably, that it is a defect of their city to which the boasted peculiarity is mainly due.

A person confined to his bed for a week or two by sickness, has generally to remark a much greater wasting of the legs than of the arms: the reason of which is, that the muscles of the leg, in ordinary cases, being more in use than those of the arms, their ordinary bulk is more dependent on use, and they suffer a corresponding change from inaction.

Such facts as now mentioned, bear directly on the subject so near the hearts of many English mothers, *viz.*, the weak and crooked backs of their daughters. From such they may understand that strong stays, which in part supercede the action of the muscles placed by nature around the spine to support it, cause these muscles to dwindle, and afterwards, when the support of the stays fails or becomes unequal, leave the back to bend or twist. Stays, therefore, can neither help to make strong and well-formed backs. originally, nor can they be a remedy after the weakness has commenced. A healthy young woman from the country, with spine lying deep between the firm cushions of muscles which support it, if, according to town fashion, braced up in tight stays, will frequently, at the end of a short time, exhibit such a wasting of the flesh, that the points of bone in the spine may be counted by the eye, all the way down.

The arch of the foot is to be noticed as another of the many provisions for saving the body from shocks by the elasticity of the supports. The heel and the ball of the toes are the two extremes of the elastic arch, and the leg rests between them.

Connected with elasticity, it is interesting to remark how imperfectly a wooden leg answers the purpose of a natural one. The centre of the body, when supported by the wooden leg, which always remains of the same length, must describe, at each step a portion of a circle of which the bottom nob of the leg is the centre; and the body is, therefore, constantly rising and falling;—but with the natural legs, which, by gentle flexture at the knee, are made shorter or longer in different parts of the step as required, the body is carried along in a manner perfectly or nearly level. In like manner, a man riding on horseback, if he kept his back upright and stiff, has his head jolted by every step of the trotting animal; but the experienced horseman, even without rising in the stirrups, by letting the back yield a little at each movement, as a bent spring yields during the motion of a carriage, can carry his head quite smoothly along.

In a general review of the skeleton, we have to remark, 1st, the nice adaptation of all the parts to one another, and to the strains which they have respectively to bear; as—in the size of the spinal vertebræ increasing from above downwards—the bones of the leg being larger than those of the arm, and so on. 2dly, the objects of strength and lightness combined; as by the hollowness of the long bones—their angular form—their thickening and flextures in particular places where great strain has to be borne—the enlargement of the extremities to which the muscles are attached, lengthening the lever by which these acts, &c. 3dly, we have to remark the nature and strength of material in different parts, so admirably adapted to the purposes which the parts serve: there is bone for instance, in one place nearly as hard as iron, where, covered with enamel, it has the form of teeth, with the office of chewing and tearing all kinds of matter used as food; in the cranium again, bone is softer but tough and resisting; in the middle of the long bones it is compact and little bulky, to leave room for the swelling of the muscles lying there; while at either end, with the same quantity of matter, it is large and spongy, to give a broad surface for articulation; and in the spine, the bodies of the vertebræ, which rest on an elastic bed of intervertebral substance are light and spongy, while their articulating surfaces and processes are very hard. In the joints we see the tough elastic smooth substance called cartilage, covering the ends of the bones, defending and padding them, and destroying friction. In infants we find all the bones soft or gristly, and therefore calculated to bear with impunity the falls and blows incidental to their age; and we see certain parts, where elasticity is necessary or useful, remaining cartilage or gristle for life, as at the anterior extremities of the ribs. About the joints we have to remark the ligaments which bind the bones together, possessing a tenacity scarcely equalled in any other known substance; and we see that the muscular fibres whose contractions move the bones and thereby the body,—because they would have rendered the limbs clumsy even to deformity had they all passed over the joints to the parts which they have to pull,—attach themselves, at convenient distances, to a strong cord called a tendon, by means of which, like a hundred sailors at a rope they make their effort effective at any distance. The tendons are remarkable for the great strength which resides in their slender forms, and for the lubricated smoothness of their surfaces. Many other striking particulars might be enumerated, but these may suffice. Such, then, is the skeleton or general frame-work of the human body; less curious and complicated, perhaps, than some other parts of the system which we have yet to examine, but so perfect and so wonderful, that the mind which can attentively consider it without emotion is in a state not to be envied.*

This living force of man has been used as a working power in various ways, as in turning a winch—pulling a rope—walking in the inside of a large wheel to move it, as a squirrel or turnspit dog moves his little wheel, &c. Each of these has some particular advantage: but the mode in which for many purposes, the greatest effect may be produced, is for a man to carry up to a height his body only, and then to let it work by its weight in descending. A bricklayer's labourer would be less fatigued to lift twice as many bricks to the top of a house in the course of a day, by ascending the ladder without a load, and raising bricks of nearly his own weight over a pully each time in descending, than by carrying fewer bricks and himself up together, and descending again without a load, as is still usually done.

Reflection, independently of experiment, would naturally anticipate the above stated result, for the load which a man should be best able to carry, is surely that from which he can never free himself,—the load of his own body. Accordingly the strength of muscles and disposition of parts are all such as to make his body appear very light to him.

The question which was agitated with such warmth some time ago as to

* In the second and third editions of this work a criticism was introduced at this place of a treatise on "Animal Mechanics," published as a part of the *Library of Useful Knowledge*; in which treatise the author from imperfect acquaintance with Natural Philosophy had fallen into many grave errors. That note, having answered its purpose is not repeated here.

the propriety of condemning men and women to work on the tread-mill receives an easy decision here. They work by climbing on the outside of a large wheel or cylinder, which is turned by their weight, and on which, to avoid falling from their proper situation, they must advance just as fast as it turns. There are, on the outside of the cylinder, projections or steps for the feet, and the action to the workers is exactly that of ascending an acclivity. Now, as nature has fitted the human body for climbing hills as well as for walking on plains, the work of the tread-mill, under proper restrictions as to duration, must be nearly as natural and healthful as any other. Its effects have ultimately proved it to be so.

Animal power being exhausted in proportion, as well to the time during which it is acting, as to the intensity of force exerted, there may often be a great saving of power by doing work quickly, although with a little more exertion during the time. Suppose two men of equal weight to ascend the same stair, one of whom takes only a minute to reach the top, and the other takes four minutes, it will cost the first but a little more than a fourth part of the fatigue which it costs the second, because the exhaustion has relation to the time during which the muscles are acting. The quick mover may have exerted, perhaps, one-twentieth more force in the first instant, to give his body the greater velocity which was afterwards continued, but the sloth supported his load four times as long.

A healthy man will run rapidly up a long stair, and his breathing will scarcely be quickened at the top; but if he walk up slowly, his legs will feel fatigued, and he will have to wait some time before he can speak calmly.

For the same reason coach-horses are much spared by being made to gallop up a short hill, and then being allowed to go more slowly for a little time, so as to rest at the top.

The rapid waste of muscular strength which arises from continued action, is proved by keeping the arm extended horizontally for some time. Few persons can continue the exertion beyond a minute or two. In animals which have long horizontal necks, there is a wonderful provision of nature in a strong elastic substance on the back or upper part of the neck which nearly supports the head independently of muscular exertion.

In farther illustration of the truth that strength is saved in many cases by doing work quickly, we may recall the fact explained at page 60, that a body thrown or shot upwards with double velocity, rises four times as far when that with a single velocity, or half the other.

"*Instruments.*"

The following remarks regard some instruments used by medical men, and which range under the present division of "Mechanics."

The obstetric forceps—As the blades beyond the joint or fulcrum are longer than the handles, the pressure on the head included in them is less than that exerted by the hand that uses them, but its degree should always be kept present to the mind of the operator.

The vectis, or lever used instead of the forceps just mentioned, is a dangerous instrument in unskilful hands. In fact whenever it is used as a lever, in the common acceptation of the term, with some part of the pelvis as the fulcrum, the use of it is a piece of unskilful cruelty; for the soft parts between the bone and the instrument are bruised not only with the whole force of the hand, but with twice or thrice as much, according as the resistance is nearer to the fulcrum than to the hand. The instrument is safely used, only when the operator makes one of his hands the fulcrum, and uses the other as the power, or makes different parts of the same hand answer both purposes; and then there is a resemblance between the action of the vectis and of a hook.

The *levator*, or lever for rising the broken and depressed portion of the skull in trepanning, has a fulcrum attached to it by a joint. Care should be taken to place this fulcrum where its pressure cannot be injurious.

The *circular-saw* or *crown-saw* of the trephine, should be worked with a quick motion and gentle pressure, for the reason given at page 107, when treating of cutting instruments. The purpose is thereby attained, both sooner and better, and the head of the patient is less shaken.

For the same reason, in amputation, a light and quick motion of the straight saw causes less jarring, and effects an easier division of the bone.

In using the amputation knife, the speed, neatness and success of the operation are all favoured by blending the drawing or saw motion of the knife, with the pressure towards the bone

These last observations are of a hundred similar, which might be made to prove the extreme importance to a surgeon of having familiarity with the use of tools and instruments. Perhaps a person cannot better acquire this than by practising, while young, some amusing work of carpentry. Manual dexterity, and a little readiness at mechanical contrivance, so frequently prove of importance to persons in all stations, that it is a great defect in the prevailing system of general education, not to cultivate them with greater attention.

The tooth-key is an instrument found in many hands, and persons who do not pretend to more than the lowest degree of skill in the healing art, are not afraid to use it. The consequence is, that perhaps scarcely a day passes in which teeth are not broken and jaws splintered, and gums bruised even to sloughing, by the unskilful or awkward use of it. The common tooth-key may be compared to wheel and axle, the hand of the operator acting on two spokes of the wheel to work it, while the tooth is fixed to the axle by the claw, and is drawn out as the axle turns. The gum and alveolar process of the jaw form the support on which the axle rolls. The common errors in tooth-drawing by the key, are these:

1st. Turning the key towards that side where the adjoining teeth are so close that the tooth to be drawn cannot pass, without either breaking one of them, or being itself broken. Sometimes two or even three teeth are thus unfixed instead of one.

2d. Neglecting the natural inclination of the tooth. By winding it round in the direction in which it already inclines, and in accordance with a bend which is generally found in it, the operation is easy and safe; but by drawing it in the opposite way, it not unfrequently is broken, or it splinters the part of the jaw-bone in which it was set.

3d. If the tooth-claw be blunt, its point may slip upon the tooth, so as to produce a jar which is very apt to break the tooth

4th. Unless the axle or fulcrum of the key be made to rest as evenly as possible on the gum, it will tear or otherwise injure the gum. It should rest, if possible, over the part of the bone in which the tooth is set, for if not—as when a back tooth is drawn by an instrument resting on a part considerably anterior to it—the twist produced is painful, and there is danger of splintering.

A man who, after making these reflections, operates leisurely a few times on the dead subject, will be able to give instant and safe relief to very common and most intense suffering. And it is hardly excusable in any medical man who may be placed where a professed dentist cannot be procured, to neglect acquiring a talent so easy.

Some dentists, by a strong forceps made for the purpose, pull teeth *directly* out; others by a simple sharp-pointed lever push them out; and others use a forceps in the manner of the tooth-key, by resting one side of it on the gum as a fulcrum, and then giving it a twisting motion: in the latter case, the resting side of the forceps is formed like the bolster of a tooth-key. But much more in all cases depends on the dexterity of the operator than on the form of the instrument.

Steel trusses for ruptures are among the blessings to suffering humanity which modern ingenuity has supplied. From the unhealthy employments of some men, and the early dissipations or unnatural modes of life of others, debilitated constitutions are frequent, and are often transmitted to offspring; and one of the lamentable effects is that weakness of the flesh forming the sides of cavities which, and on occasions of strong effort, allows, at particular points the living parts to protrude from within, so as to form tumours under the skin. A person perfectly healthy may suffer the same injury from a very violent strain or pressure on the abdoman. The occurrence is called *hernia* or *rupture;* the most common hernia is that of the small intestine through the groins.

Formerly this occurrence disabled for life. A man who had hernia was discharged from the army or navy; he could not ride on horseback, or take usual exercise; he could not lift a weight; and in a word he often became a miserable burden to himself and others. Now, by fitting the pad of a good steel truss to the part, the rupture is as perfectly restrained, as if the hand of a skilful surgeon were constantly there. The truss may be put on and off, with as little reflection or trouble, as a part of the ordinary dress, and the man becomes again almost as fit for all the duties of life as if he were without his ailment.

The old and still common form of the steel truss is that of a half or three-quarter hoop, so bent and tempered, that when put upon the patient, one end which has a pad upon it, presses with a given force, over the opening by which the rupture protrudes, and the remainder tightly embraces the body. The objections to this kind of truss are, the difficulty of making it to fit exactly; its being rather troublesome to put on and off; and its pressing disagreeably round the body.

Another kind of truss, free from these objections, consists of a little more than half a hoop, with a pad at each end: one of the pads supports the weakness, and the other rests upon the centre of the back, to bear all the strain there, while the hoop itself passing round the hip farthest from the hernia, reposes loosely on the hip. This truss may be called self-adjusting, for it almost of itself falls into its place, and needs no fastenings; the same truss fits all persons of one size, whatever their shape; and the strength may be adjusted by changing the number of plates in the spring-hoop.

Tourniquets, crutches, splints, &c., &c., are so simple in all respects as not to merit special notice here.

This section contains some of the reflections which, in contemplating the human skeleton, occur to a person familiar with mechanical philosophy: and the more complete such a person's knowledge is of anatomy, physiology, surgery and medicine, the more numerous will be the professional objects on which this philosophy will shed a light, dissipating doubt and error. The author has not entered into more minute detail, because it would have been encroaching upon the office of the teachers of particular departments, and because he thinks that any one who is not enabled, by the examples here given, to make the applications of the general laws to all possible cases, may account the study of the healing art unsuited to the faculties with which he is endowed.

PART V.

(CONTINUED.)

SECTION II.—DOCTRINES OF FLUIDITY IN RELATION TO ANIMALS.

In the preceding sections on the laws of fluidity, occasional illustrations have been taken from the animal economy; but there are many other particulars of the same class and of great interest, which it is convenient to consider apart under the four heads of, 1st. *The circulation of the blood;* 2d. *The respiration and voice;* 3d. *The digestion*, and 4th. *The pelvic phenomena.* It is important to remark here, that this section cannot be understood by a person ignorant of those which precede.

THE CIRCULATION OF THE BLOOD.

PERHAPS there are few points more remarkable in the history of the progress by which man has arrived at his present knowledge of the universe, than that he was so long ignorant of the fact that the blood in his own and in other animal bodies is constantly circulating. England claims as one of her sons, the man whose powerful intellect at last established this truth, in opposition to strong appearances, and to the most fixed prejudices. Dr. Harvey published his proofs in the year 1619. A person who tries to imagine what the science of medicine could have been while it took no account of this fact, on which, as a basis, much of the certain reasoning about the phenomenon of life must rest, is prepared for what old medical books exhibit of the writhings of human reason, in attempts to explain and to form theories, while a fatal error was mixed with every supposition.—The chief circumstances which prevented the earlier discovery of the circulation was, that on examining dead bodies, the arteries were always found empty of blood;—which was the reason, also, of those vessels being called *arteries* or *air-tubes.*

We now know, that as the Thames water spreads over London in pipes to supply the inhabitants generally, and to answer the particular purposes of brewers, bakers, tanners, and others, and is then in great part returned to where the current sweeps away the impurities, so, nearly in the human body, does the blood spread from the centre, through the arteries, to nourish all the parts, and to supply material of secretion to the liver, the kidneys, the stomach, and other viscera; and returns from these by the veins towards the heart and lungs, to be purified and to have its waste replenished, that it may again renew its course.

The circulation may be more particularly described thus: From the left chamber or *ventricle* of the strong muscular mass, the *heart*, a large tube arises, called the *aorta ;* and by a continued division or ramification, opens a way for the bright scarlet blood to the very minutest part of the living frame,—the extreme divisions of twigs being so small that they are called *capillary* or hair-like tubes. At the termination of these vessels, the blood, after answering the purposes of general nutrition, &c., by which it loses its bright colour, enters the commencement of the *venous tree*, or returning channel; and gliding successively from smaller to larger branches, returns towards the right chamber or ventricle of the heart requiring purification and partial renewal. Considering the great arterial and venous systems of the body as *twin trees*—the *scarlet* and the *purple*, with corresponding and meeting branches. and with trunks which touch each other at the heart—it will appear that they again similarly meet or inosculate by their extreme roots, and thus form a continued or circular channel. The root of the venous tree, by which the blood spreads from the right chamber of the heart through the lungs is called the *pulmonary artery*, and that of the arterial tree, by which the blood returns to the left chamber, is called the *pulmonary vein*. Both of these ramify in the spongy masses of the lungs forming a great part of the pulmonary substance. Fresh material for the blood is brought from the *digestive organs* by the *lacteal absorbents* and *thoracic duct*, and is constantly pouring into a large vein near the heart, to be completely mixed with the dark or returning blood by a violent agitation or *churning* during its passage through the heart. The mixture, on leaving the right ventricle, is strained through the minute ramifications, the vessels in the lungs, and at the same time is exposed to the action of the air entering the cells of the lungs in respiration, by which exposure the dark purple blood becomes again pure scarlet, and when it reaches the left chamber or ventricle, is ready to set out on its journey as before, charged with new life and nutriment. The two chambers or ventricles of the heart have each an anti-chamber or *auricle*, (so called from an external resemblance to a dog's ear,) into which the blood is first received from the veins; and there are valvular doors between the auricle and ventricle, which allow the blood to pass readily into the ventricle, but oppose its recoil during the ventricular contraction. Similarly acting valves are placed between the ventricles and great arteries. There are valves, also, in many of the veins, over the body, to secure the natural course of the circulation. Besides the important change of purification which the blood undergoes in passing through the lungs, its composition is much influenced by the action of the kidneys, of the exhalent of the skin, and of the liver,—the two former relieving it of superfluous moisture and salts, the last from a large quantity of matter in the form of bile.

The description given above of the circulation of the blood is only an outline; and yet by showing the manner in which fresh material enters, it contains more than Harvey knew of the subject. In this department of knowledge, as in most others, we have advanced from the very general and vague, to the more particular and precise,—and just as the general nature of steam was known long before it served in steam-engines, and as the period of the moon's revolutions had been accurately observed for thousands of years before the fluctuations in her velocity could be calculated so as to make her the mariner's best guide in his courses across the ocean, so, when Harvey had proved the general fact of the circulation of the blood, he had left much yet to be done, by observing and raising from subordinate facts, to render the knowledge available for the many useful purposes which it is calculated to serve.

Within a few years only, has the importance of the subordinate circumstances been fully appreciated,—as is evinced by the numerous works composed to elucidate them; but many of which works have served only to prove, that if the difficulties were to be solved by natural philosophy, medical men in general had not yet studied it sufficiently to be able to use it successfully. In this section it will be attempted to place certain important points of the subject in a clear light: and by referring directly to the general laws of nature, as explained in the body of the work, to settle existing disputes on some of these points, to remove remaining doubts on others, and to suggest some important new applications.

The fact of the circulation of the blood being once admitted, an inquirer who contemplates the apparatus by which it is effected, is led by the general analogies of nature to conclude—1st. That the ventricle of the heart, at each contraction, empties itself into the great artery, as a forcing-pump at each stroke empties itself into its pipe;—2d. That the consequent jet causes a wave to pass along the extremities of the arterial tree, (accounted simply elastic,) so as to produce every where, what is called the pulse; 3d. That the force of the heart acting along the arteries, forces the blood through their open capillary extremities into the commencing veins, and along the veins back to the heart again. Now these assumptions, which Harvey believed completely to describe the circulation, are all nearly true: but are still so far from being either the exact or the whole truth, that they leave important facts unexplained. Thus:—1st. The pulse, instead of being the wave, as slowly progressive as the view above given anticipates, is almost as instantaneous over the whole body as a shock of electricity; 2d. The arteries are all found empty after death, although they have no power of emptying themselves; and if an artery be tied in the living body, the part beyond the ligature, and cut off, therefore, from the influence of the heart, is equally emptied;—3d. The rapidity of the blood's passage through the capillaries varies very much, but it does not vary in exact accordance with the changes in the rapidity or force of the heart's action.—These and other facts ascertained since Harvey's day, not exactly squaring with his views, have rendered further investigation necessary; and it is now additionally known—1st. That the coats of the arteries are not simply elastic but actively contractile; and 2dly. That the capillary vessels can move the blood independent of the heart. In analyzing this subject, it is convenient to follow the blood round from the heart to the heart again, through the three stages of, 1st. *The arteries*; 2d. *The capillaries*; 3d. *The veins.*

Motion of the blood in the arteries.

The contractions of the heart inject the blood into the arteries with a force maintaining such a tension in them, that according to the interesting experiments of Dr. Hales, recorded in his *Statistical Essays*, if any artery of a large animal like a horse be made to communicate with an upright tube, the blood will ascend in the tube to a height of about ten feet above the level of the heart and will there continue, rising and falling a few inches with each pulsation of the heart. Now a column of ten feet, as explained at page 120, indicates a pressure of about *four and a half* pounds on the square inch of surface: this, therefore, is the force of the heart urging the blood along the arteries into the veins.—The opposing tension of the veins is much less, because, as will be explained under the proper head, the blood readily escapes from them into the heart; Hales found that in a tube communicating with a vein, the blood stood only a few inches higher than the level of the heart.

In small animals he ascertained the tension of artery and vein to be less than in large ones; and the ratios deduced for the human body, under ordinary circumstances, were an eight feet column, or nearly four pounds per inch for the arteries; and half a foot column, or a quarter of a pound per inch for the veins.

Arteries examined after death are found to consist of—1st, an outer coat of strong *elastic substance;* 2d, a middle coat of *circular fibres;* and 3d, an invex inner coat of *smooth lining membrane.* Their elasticity or power of resisting change of dimension, and of returning to a middle state from either dilatation or compression, because remaining in the dead artery, was the most obvious property, and was that first attended to. Minute observation of the phenomena of life has since determined the following facts, proving and illustrating a contractility resident in the fibrous coat.

1. A small living artery, cut across, soon contracts so as to close its canal and arrest hæmorrhage.

2. While an animal is bleeding to death, the arteries, in accommodating themselves to the decreasing quantity of blood, contract far beyond the degree to which their simple elasticity would carry them; and they relax again after death. Dr. Hales took seventeen quarts of blood from a horse before it died, in whose body only three quarts more were found altogether, and yet the moment before death the tension of the arteries sustained a column of two feet of blood in his experimental tube.

3. The artery of a living animal, if exposed by dissection to the air, sometimes will contract in a few minutes to a great degree: and in such a case, only a single fibre of the artery may be affected, narrowing the channel like a thread tied round it. (*See Parry on the Pulse.*)

4. When a living artery is tied, the part between the ligature and the nearest branch on the side of the heart gradually contracts, and becomes at last a solid or impervious cord.

5. Fluctuation in the vital action of parts, is often attended with sudden increase or diminution of caliber in the arteries concerned.

Although these facts prove indubitably a contractility in the coats of arteries distinct from their elasticity, still, because the circular fibres do not resemble common muscles in colour or in chemical composition, or in being immediately obedient to the stimuli of electricity, pricking, great heat, &c., their contractility was by many persons for a long time denied. The dispute, however, was often more about the words *contractility* and *muscularity*, than about facts.

The pulse in the arteries, chiefly as regards its almost instantaneous occurrence over the whole system, in all states of arterial dilatation, and its great strength and sharpness in very small and remote branches, points also to the active contractility of the arterial coats: for,

1. Were the arterial tree in the living body a system of simply elastic tubes as readily admitting of further dilatation as in the dead body, the first part or trunk would affect the motion of the blood beyond it, nearly as the *air-vessel* (see page 126) placed at the commencement of artificial arrangements of water-pipes affects the motion of the water in them;—that is to say, as the air-vessel converts the sudden and interrupted jets of water from pumps of *fire-engines, town supplying pipes*, &c., into a uniform stream with scarcely a remnant of shock, so, in the arterial branches, simple elasticity would cause a more tranquil flow than indicated by the remarkable gushes from a wounded artery, and a quieter beat than that bounding pulse of life felt in the remote artery of the wrist, as sensibly, in proportion, as near the heart itself.

2. Were the pulse a wave advancing in tubes that yielded as readily as the dead arteries in their middle states of dilatation, it would be more slowly progressive from the heart to the extremities; but it is felt so instantly over the whole body, as to be commonly compared to a shock of electricity.

3. A pulse may be produced artificially in the arteries of a body recently dead by filling them with water to the tension of life, and then injecting at intervals, by a syringe, as much water as the heart throws off blood at a beat; but although the artery is then distended nearly to the limit of its dilatability, and is, therefore, rendered rigid, the beats are weaker than those of the living pulse. A similar experiment, tried by connecting the artery of a dead animal with the corresponding artery of a living one, has a similar result.

4. A tube extensively elastic, that it might convey a wave of liquid with a velocity approaching to that of the pulse, would require to be so tense, from fulness, as to be discernible always by the touch, through any imbedding medium, such as flesh, like a hard cylinder or cord; and it would be acting constantly as a spring tending to straighten itself, and, therefore, would be stiffening the parts through which it passed. Now the living arteries, between their pulsations, are almost as soft and compressible as the surrounding flesh, and they offer no perceiveable opposition to bending, in any movement of the parts. This may be verified by examination of the lips, for instance, or of the fingers; but when a person sits cross-legged, the well-known shaking of the suspended foot, in unison with the pulse, shows the recurring efforts of the artery to straighten itself, during the moments of greater tension.

5. A bulky wave in elastic vessels would have to recoil from the extremities, or to pass through them as a gush; and the recoil would be particularly observeable near the ligature of the tied artery; but examination has not detected such effects in the living body. The operation for aneurism,—in which the artery is tied *beyond* the tumour, instead of, as usual on the side next the heart,—if it checked a strong wave, would almost certainly produce bursting; yet Mr. Wardrope and others have lately performed this operation with successful issue.

6. The wave would be more interrupted by the bandage in the operation of bleeding, than the living pulse is.

7. The pulse of a paralytic limb often seems more affected, than mere change of size in the artery will account for. The same is true, in an opposite way, of the pulse in an artery leading to an inflamed part.

8. If the abdomen of a living animal be opened, the mesenteric artery, in all its ramifications, is seen stiffened and raised up suddenly with every pulsation, in a manner which the spreading of newly received blood in a very yielding vessel does not account for.

9. In the interesting experiments of *Bichat*, *Parry* and others, to ascertain the exact extent of the supposed dilatation and contraction of arteries, during a pulse, not the slightest degree of either was discernible, even when sought for with microscopes.

To explain these and other phenomena, then, it seems necessary to admit, as occurring throughout the whole body, and almost simultaneously with the contraction of the heart itself, such an action of the contractile fibres of the arteries, so as to modify the elasticity of the arteries, and to render them rigid enough, in all degrees of dilation, for the heart to produce its effects through them almost as it would through tubes of metal.—Dr. Young, in a paper published in the Philosophical Transactions for 1809, and characterized by the usual elegance and precision of his writings, has adduced experiments and calculations, to show that waves in elastic vessels advance more quickly than was before imagined: but the spreading of the pulse seems to be yet more rapid than his calculation anticipates.—It is evident, that when arteries, in consequence of depletion, are contracted beyond the middle station of their elasticity, their tension and power of quickly conveying the pulse must be dependent altogether on the condition of their contractile fibres.

The careful experiments which could detect no change of size in the arteries during the pulse, while they disprove the ancient belief of a considerable tumefaction or wave passing along, or of a considerable filling and emptying of arteries, like what occurs in the heart, might also be supposed positively to disprove the occurrence of any general constriction of the vessels on their contents—but erroneously:—for if a man's arterial system, considered as one cavity, be supposed to contain five pounds of blood (which is near the truth,) and if the vessels be thought to embrace their contents, even between the pulses, with force enough to have all a rounded or cylindrical form, although remaining soft and yielding to the pressure of the finger; and if we suppose their coats, during the pulse to be thrown into a sudden contraction, as if in obedience to an electrical shock, still, because blood is incompressible, and because just as much enters the arteries with every pulse as escapes from them before the next, their bulk would not sensibly diminish by the strongest conceivable action of their coats; of which action the only sensible effects would be, that the soft, yielding, and, in some places, compressed tubes would be suddenly converted into hard or resisting cylinders; and that wherever, by any accidental pressure, an artery had been flattened, it would, in regaining its cylindrical form, strike or pulsate against the compressing body.—Whether such an action as this contributes to produce the arterial pulse will be considered under the head of "*the pulse,*" after we have seen how the blood moves in the capillaries and veins

In any admissible view, however, of arterial agency, we find that the arteries contribute to the circulation of the blood, but as tubes which convey it, their own permanent tension, and, therefore, the force with which the blood is pressed into the capillaries, being derived from the heart alone. Even if there be a momentary arterial contraction, such as alluded to above, at the instant of the pulse, it is of too short duration to have an appreciable effect, and probably any effect would be counterbalanced by the same action pervading the capillaries. Many physiologists have had a confused belief that the arteries aided very actively in propelling the blood; but a little reflection would have shown, that as they have no vermicular or progressive contraction, like the intestines, they no more *propel* the fluid within them, than any other tubular conduits do.—Although they be thus in no degree instrumental in the propulsion of the blood, still, by more permanently enlarging or diminishing their caliber, that is, by merely becoming larger or smaller conduits, they may much influence its local distribution, and the speed of its transmission.*

* It has long been a subject of dispute whether the arteries exercise any active power in the circulation of the blood, and many ingenious experiments have been instituted to determine the question. Our author, in denying that the arteries are in any degree instrumental in the propulsion of the blood, is not borne out by recent investigations. M. Poiseuille, indeed, seems satisfactorily to have proved that the contractility of the arteries does assist in the propulsion of the blood. By a series of well devised, and apparently accurate experiments, M. Poiseuille arrived at the unexpected result, that the force of the blood in the arteries will support a column of mercury of the same height with whatever part of the course of the arterial circulation the column is placed in connection—whether for example it is connected with the origin of the carotid, or with a branch derived by repeated subdivision from the crural artery: and he, therefore, concludes that the force with which a mole-

The nature of this work does not allow us to record historically the various errors into which even able men have fallen, in attempting to explain the office of the arteries, but we shall glance at the following —*Dr. Monro* and *John Hunter*, two of the most able physiologists that the world has seen, believed that the arteries did almost as much in propelling the blood as the heart itself. We need not repeat the refutation of this opinion. The ingenious *Bichat*, again, unable to detect either momentary contraction or dilatation in the arteries, thought that the blood was pushed along them by

cule of blood moves, is the same throughout the whole arterial circulation. Following out these researches, he has investigated the manner in which the original impulse communicated by the heart, is transmitted unimpared to distant parts of the circulation, notwithstanding the retarding tendency of friction, and the yielding of the parieties of the vessels.

His first object here was to ascertain whether the arteries are dilated by the stroke of the heart, and impulse communicated to the blood, and what the amount of the dilatation may be. By many physiologists a very extravagant idea used to be entertained of the amount of their dilatation; on the other hand, the latter researches of Parry, and other experimentalists have assigned exceedingly narrow limits to it; nay, by one eminent physiologist, Bichat, it has been denied altogether. M. Poiseuille has determined the point by means of a very satisfactory set of experiments with an apparatus of his own invention, and has ascertained that *dilatation is affected;* but that it is so small as certainly to be indistinguishable in an artery by the unaided senses. This apparatus cannot be thoroughly described without a diagram; it will be sufficient, therefore, for us to mention, that it is so contrived as to contain about eight inches of the carotid artery of the horse in a vessel filled with water and made water tight, except at one point, from which a small horizontal glass-tube issues, about an eighth of an inch in diameter. At each contraction of the animal's heart it was found that the water in the small tube advanced two inches and eight-tenths, and that it retired to its former place during the diastole of the heart. The diameter of the artery was seven-twentieths of an inch. Hence it may be calculated that at each pulsation its capacity was increased about a thirtieth part.

Having ascertained this fact, M. Poiseuille goes on to inquire, whether the power which is expended by the blood in causing this dilatation is restored by the subsequent contraction of the artery. For this purpose a portion of the common carotid artery of the horse, ten inches long, and seven-twentieths of an inch in diameter, taken immediately after death, was connected with the end *a*, of the tube, (see Figure;) a stop-cock, however, being previously fitted between *a* and *b*. The other end of the artery was fixed on a tube of analogous construction, different in fact only in so far as the limb *c d* was inclined at about half a right angle instead of being vertical, and the stop-cock was placed near the end *d*. The whole of the first tube, the artery and part of the descending limb *b c* of the second tube was filled with water, a little mercury then filled the curvature of the second tube, and the ascending inclined limb of that tube above the mercury was filled with water. The stop-cock of the last tube being closed, and that on the first tube being opened, mercury was poured into the former by its end *d*, till the pressure on the artery amounted to ninety five millimetres or about 3.8 inches. The stop-cock of the first tube was then closed and that on the second tube was opened; upon which the water rose instaneously in the latter, a portion flowed out at the top, and the remainder then sank a little, and assumed a fixed level. On making the necessary computations, M. Poiseuille found that the point to which the mercury was raised in the second tube at the moment of the contraction of the artery indicated an elevation of one hundred and ten millimetres or 4.4 inches. Hence *the power with which the arterial coats contract upon themselves after being dilated, exceeds that which is expended in dilating them.* In the present experiment the excess was equivalent to six-tenths of an inch of mercury, or three nineteenths more than the dilating force. In three other experiments, the excess of the column of mercury was 9.20, 14.20, 19.20 of an inch. When repeated with the artery of an animal which had been killed four days before, the excess was less than 4.20 It is evident, therefore, that whatever force the blood after issuing from the heart loses in consequence of its acting on yielding vessels, is completly restored by the elastic contraction of their parietes. The memoirs of M. Poiseuille will be found in the *Repertoir General d'Anatomie* et de Physiologie, Tom, VII.

Fig. 174.

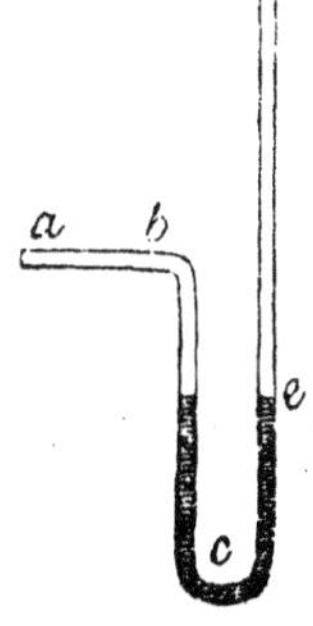

Dr. Badham in an interesting paper in the London Medical Gazette (vol. viii. p. 549) has adduced some strong evidences of the existence of independent arterial action, and gives a sketch of various pathological phenomena, which appears explicable on such an admission, and inexplicable without it.—*Am. Ed.*

the heart instantly through their whole extent, as a solid rod of metal or wood is advanced by an impulse at one end. Dr. Parry took nearly the same view of the subject, and illustrated his idea by referring to the experiment of moving a whole line of billiard-balls by striking the extreme one. Both these authors erred by neglecting the hydrostatical truth, that pressure in a fluid operates equally in all directions, and, therefore, that fluid pressed into a tube tends to dilate the tube, just as powerfully as to drive the fluid forward; and they did not advert to the fact that the progress of the blood in the small arteries is not by waves or successive jets, but is nearly a uniform stream. The blood could only advance, as they supposed, by the arteries becoming, for an instant, absolutely rigid, in consequence of a strong action of their contractile fibres.

It merits notice here, although not strictly a mechanical fact, that arteries permanently increase or diminish in size when a permanent change takes place in the demand for their service. The arteries of the *gravid uterus*, or of an increasing tumour, grow with the part supplied, while on the contrary, those of the stump left after amputation soon remarkably diminish. If the chief artery of a limb be obliterated by any cause, as after the operation for aneurism, the small collateral anastomozing branches increase in size to do its duty.

It is farther remarkable, that when arteries are called upon to carry an increased quantity of blood, they often become tortuous or serpentine, as well as larger; and that arteries leading to parts whose actions are naturally intermitting or fluctuating, have generally the tortuous form. Of these truths, the arteries leading to rapidly growing tumours, or to varicose aneurisms, and the arteries of the uterus and testes, may serve as instances. This bending of arteries, and the very curious divisions into many branches which again re-unite, found in those leading to the brains of some animals, do not seem intended to slacken the rapidity of the sanguineous current, but to give the artery a greater control over the supply.

Passage of the blood through the capillaries.

We have seen that the heart keeps up a tension or pressure in the arteries of about four pounds on the square inch of their surface; and with this force, therefore, is propelling the blood into the capillaries. If these last were passive tubes, constantly open, such force would be sufficient to press the blood through them with a certain uniform velocity; but they are vessels of great and varying activity; it is among them that the nutrition of the different textures of the body takes place, as of *muscle*, *bone*, *membrane*, *&c.*; and that all the secretions from the blood are performed, as of *bile*, *gastric juice*, or *saliva*, *&c.*; and to perform such varied and often fluctuating offices, they require to be able to control, in all ways, the motion of the blood passing through them. The capillaries of the cheek, under the influence of shame, dilate instantly, and admit more blood producing what is called a *blush*;—under the influence of anger or fear, they suddenly empty themselves, and the countenance becomes pallid—tears or saliva, under certain circumstances, gush in a moment, and in a moment again are arrested—if a person having inflammation in one hand be blooded from corresponding veins in both arms at the same time, twice or thrice as much blood will flow from the diseased side as from the other. Similar changes occur in many other instances. Now the only action of cylindrical vessels, capable of causing these phenomena, is contraction and dilatation of their coats; and with reference to such action it merits notice, that arterial branches have more of the fibrous or contractile coat in proportion as they are smaller.

A muscular capillary tube strong enough to shut itself against the arterial current from the heart, is also strong enough to propel the blood to the heart again through the veins, even if the resistance on the side of the veins were as great as the force on the side of the arteries. For if we suppose the first circular fibre of the tube to close itself completely, it would, of course, be exerting the same repellent force on both sides, or as regarded both the artery and vein. If, then, the series of such fibres forming the tube were to contract in succession towards the vein, as the fibres of the intestinal canal contract in propelling the contents of that canal, it is evident that all the blood in the capillary would thereby be pressed into the vein towards the heart. If after this the capillary again relaxed on the side of the artery, so as to admit more blood, and again contracted towards the vein as before, it would produce a forward motion of the blood, first towards the vein, and then in it independently of the heart. As capillary action, however, is not visible, its nature has not yet been positively ascertained :—some persons have deemed it electrical.

It is capillary action which absorbs and moves the fluids of the classes of animals which have no heart. It must also be the power which moves the blood in warm-blooded monsters formed without hearts. There are cases of apparent death among human beings, where the heart remains inactive for days and yet a degree of circulation sufficient to preserve life is carried on by the capillaries. In illustration of capillary action, we have also the absorption, by the lacteals, of nutriment from the alimentary canal; and, perhaps, to a certain extent, the circulation of blood in the livers of animals. In this last case, the blood collected by veins from the abdominal viscera, instead of going directly to the heart, is again distributed through the liver by the branches of the *vena portæ*, and is then again collected by the ordinary veins of the liver, and carried to the heart: it thus moves through two sets of capillaries in passing from the arteries to the heart again.

The action of the capillaries is the cause of that singular fact which prevented the ancients from discovering the circulation of the blood, *viz.*, the empty state of the arteries after death. All the muscular parts of an animal, including, therefore, the contractile coats of vessels, retain their life, or power of contracting, for a considerable time after respiration has ceased—as is seen in the recovery of persons apparently drowned or suffocated : in the leaping of a heart taken from an animal recently killed; in the actions resembling life which can be produced, by the agency of galvanism, in a body recently dead : but the fact is seen still more amply for our purpose, in the total disappearance of a local inflammation after the death of the patient,—for inflammation involves a gorging or over-tension of the capillaries, into which, when the heart has ceased to press blood, the contractile force remaining in them, even under disease and in a dead animal, is sufficient to squeeze the blood out of them, and often to remove all trace of the malady which has been fatal. In ordinary cases, then, the capillaries throughout the body remain alive and active for a considerable time after breathing has ceased, working like innumerable little pumps, and empyting the arteries into the veins. As the red blood is their proper sustenance as well as stimulus, they work as long as there is any of it coming into them from the arteries behind; except, however, the capillaries of the lungs, which soon cease to act, because, after breathing has ceased, they receive only black blood, and are moreover compressed by the collapse of the chest; and all the blood accumulates behind them. The capillaries may continue to be filled from the arteries, either in consequence of their elasticity opening them with what is called a suction power, or of an absorbant power depending on life, like that of the lacteals and of the absorbents all over the body, and, perhaps, of the vessels in the roots of vegetables. When death is produced by lightning, or by the poisons which destroy all muscular irritability, and therefore that of capillaries, the arteries after death are found to contain blood like the veins. In a living body, if an artery be tied, the part beyond the ligature is soon emptied into the veins, and becomes flat.—The experiment has be made even upon the aorta itself.

The empty state of the arteries after death has been ascribed, by some teachers, to the momentum with which they supposed the blood to be thrown out from the heart in its last contraction—sufficient, said they, to squirt it fairly through the most distant capillaries; a doctrine exemplifying the carelessness with which men sometimes receive and repeat opinions, to which their attention has never been fully awakened. Such an effect would not follow, even if the action of a dying heart were the strongest possible; while, in reality, it is in most cases so feeble, that the pulse for sometime ceases to be perceptible at the extremities, and the diminished circulation lets them become cold.—Other physiologists have taught that an artery is capable of contracting directly upon its contents, so as to expel even the last drop;—but large arteries, when emptying, do not contract *roundly* like an intestine; they become *flat* like elastic tubes of leather *sucked* empty, and no contractile action of the vessel itself could bring its sides together in such a manner. If arteries emptied themselves by their own action, the pulmonary artery should be more certainly empty than the aorta, because it is shorter; yet it is always full; for the reason already stated, that the pulmonary capillaries cease to act after respiration has ceased, on account of the blood in them being venous or dark blood, and therefore not life-supporting or stimulant to them.

Passage of blood through the veins.

The veins have much thinner coats than the arteries, and if taken altogether have much greater capacity: for besides being larger than the corresponding arteries, they, exist in many situations, as double sets, an exterior and an interior: they have also very frequent inosculations or communications with each other throughout their whole course.

The simple weight of the column of blood in any descending artery is just sufficient to raise the blood through open capillaries to an equal height in the corresponding vein, according to the hydrostatical law, that fluids attain the same level in all communicating vessels; and therefore, as the arch of the aorta rises considerably above the heart, the gravitating pressure of the descending arterial column of blood would be sufficient to lift that in the veins not only up to the heart, but considerably beyond it. In addition to this influence of gravity on the venous current, the blood is pressed into the arteries, and from them, therefore, towards the veins, with a force from the heart itself, as stated above, of about four pounds to the square inch, or, in other words, as if there were a column of blood eight feet higher than the heart urging the current. It might be expected from the law of equal diffusion of pressure in fluids, that these causes would soon produce a tension in the veins as great as in the arteries: and this does not happen, only because the blood has a ready escape from the veins through the right ventricle of the heart. Under ordinary circumstances, there can be no greater tension in the veins than what is sufficient to lift the blood to the heart and to overcome the friction;—as in an upright leathern tube, open at top, and receiving water at its bottom from a powerful forcing pump, there never can be a greater tension or pressure than what corresponds to the height of the fluid column in the tube, and to the friction between the fluid and tube. In Dr. Hale's experiments, already alluded to, a tube connected with a vein so as to receive its blood, became filled with blood to a height only of about six inches above the level of the heart. As Dr. H. generally cut the vein completely across, and inserted the tube into the portion leading from the capillaries, he would have discovered the whole power with which the blood is pushed along the veins from the capillaries, but for the free lateral communication of veins with each other, which reduces the tension even in an obstruct branch, to the degree existing in the system generally. When, from agitation of the animal, or any straining exertion, the passage of the blood into the heart was impeded, all the veins became tense, and a tube inserted into the returning jugular had blood running over, at a height of three feet above the heart.

If the blood did not escape from the veins, as above described, the only cause which could prevent the venous tension from becoming as great as the arterial, would be obstruction in the connecting capillaries : but the following facts and considerations prove that these vessels, which, in the dead body, allow the passage of injections, in the living body freely allow the passage of blood. 1st. Magendie laid bare the chief artery and vein of a living limb, and at the part, detached them from the flesh underneath, so that he could apply a tight bandage round the limb without including them, and could thus render them the only channels of circulation for the limb beyond the bandage. He then found, that when a separate ligature was put upon the vein, to prevent the return of its blood to the heart, and a puncture was made beyond the ligature, the flux of blood from the puncture was rapid or slow according as the heart was allowed to produce a greater or less degree of tension in the artery :—this tension was regulated by his compressing the artery between the fingers. 2d. After a similar preparation of the parts, the blood will ascend in a tube from the obstructed vein very nearly as high as from the artery 3d. In the common operation of bleeding at the moment of puncturing the vein, the blood often jets from it as from an artery, staining even the top of the bedstead. 4th. The microscope discovers in the capillaries, a uniform forward motion of the blood, as if it were obeying the steady pressure of the arterial tension, and not any intermitting action. 5th. Disturbed action of the heart, by obstructing the passage of the blood through it, is very soon attended with a tumefaction of all the veins leading to the heart : the tumefaction becomes very visible about the neck and head, and in the liver produces swelling and acute pain. 6th. Dr. Young, from experiments made by him, and reported in the philosophical transactions for 1809, concluded that perfectly open capillaries, of the size existing in the living body, should just retard a flow of blood urged by the usual arterial tension, in the degree which really occurs :—a correspondence proving that they must be open; and open vessels, however small, and how slowly soever they transmit the blood, still, if the escape of blood from the veins were arrested, would transmit the arterial tension without diminution. 7th. The fact that after death the capillaries empty the arteries into the veins, proves that, under certain circumstances, the venous tension may become even greater than the arterial.—These facts then and others that might be mentioned, prove incontestably, that the blood is pressed into the veins from the arteries and capillaries, with force sufficient to lift it, not only to the heart again, but many feet farther, *viz.*, about as far as it would ascend in a tube rising from the tense arteries themselves. So little, however, has this important truth been understood, that in elementary works of authority lately published, the venous current is treated of as a very obscure subject; and some authors, in their anxiety to explain it, have assigned causes for it, which, as will appear hereafter, are positive absurdities in physics. The difficulty in the question seems to have arisen from the great disparity observed between the tension in the arteries and in the veins, while the reflection did not occurr, that the disparity was owing to there being a free passage or outlet from the veins through the heart.

The illustrious Bichat, with an inattention to facts, extraordinary in him, persuaded himself that the influence of the heart ceased entirely at the capillaries and that the blood was returned through the veins by the action of the capillaries, alone. How could he avoid the single reflection, that, if the purpose of the arteries had been merely to convey the blood to the capillaries, and not also to bear the force which pressed it into and through them, the extraordinary strength of the arterial coats, and the great power of the heart to fill them and keep up the tension described, would have been quite superfluous?—and he knew that nature does nothing in vain.* The reflection applies strikingly to the pulmonary artery, of which no branch exceeds a few inches in length.

The uniform current of blood along the veins, so apparent in the operation of bleeding, and produced, as now explained, by the combined influence of the heart and capillaries, suffers a considerable disturbance in the neighbourhood of the heart from three causes. 1st. As there is no valve between the veins and the auricles of the heart, each contraction of the right auricle tends to throw the blood back into the veins, as well as forward into the ventricle, and thus produces the venous pulse often felt in the neighbourhood of the chest. 2d. When the chest is expanded by inspiration, it is more roomy than during the collapse of expiration, and the blood then enters it more readily. 3d. While the chest is *inhaling* or *drawing in* air, that is to say, expanding so as to diminish the tension or pressure of the air within it, (see *Pneumatics*,) it is by the same action favouring the entrance of blood through the veins towards the heart placed in it;—on the contrary, while it is *exhaling* or *throwing out* air, it is, with equal force, resisting the entrance of blood, and slackening, or even causing recoil of the inward current. This favouring or resisting force, however, as will be hereafter shown, is only such as to lift or support a column of blood of about half an inch in height. It appears, then, that the entrance of blood into the chest fluctuates by reason of the respiration, &c., as the entrance of a river stream into the sea fluctuates by reason of the ebbing and flowing of the tide. An eye watching the jugular vein, under favourable circumstances, may see it tense or slack in accordance with the opening and shutting of the chest.

It still remains to be ascertained whether or not veins have in themselves any contractile power, such as can partially empty a lower portion into a higher portion beyond an adjoining valve. If so, the valve by then bearing the pressure would let more blood be easily raised from below into the portion so relieved: and the action, without being equal to the office of completely emptying any portion of a vein, would still have the effect of dividing

* This incorrect and inconclusive mode of reasoning is so common that we may be permitted to protest against it. The influence of the heart may cease with the capillaries, and yet nature has done nothing in vain. Before we would be justified in making such a charge against nature we must possess an infinitely more precise knowledge of the circulatory forces and of the functions of the arterial system than we do at present. *Am. Ed.*

a long heavy column into a number of short columns of comparatively little resistance. It is certain, at least, that the valves in the veins, by preventing the falling back of blood which has once passed towards the heart, must affect its flow during bodily exercise; for every time that pressure is made on a vein by a swelling muscle or otherwise, the blood in the part must be forced forward, and cannot return.

The veins which are surrounded by muscles are thinner and weaker than those supported only by the skin. The external veins of the legs are almost as strong as arteries. Proving, however, that the fabric of veins is much weaker than that of arteries, any vein in the living body, made to communicate directly with an artery, soon exhibits what is called a *varicose aneurism*, and swells to bursting. Veins possess power, to a great extent, of adapting themselves to the varying quantity of blood.

Some recent authors, as stated above, either not aware of the facts which prove that the blood is everywhere pressed into the veins with force much more than sufficient to raise it to the heart again; or, being unable, from their little familiarity with mechanical science, to draw exact conclusions from the facts, or to avoid errors in their own hypotheses, have promulgated the opinion that the progression of the blood in the veins is greatly owing to a partial vacuum or a suction power in the heart or chest; that is to say, to the atmospheric pressure remaining constant on the body generally, while it is, at intervals, lessened about the heart. Now the whole influence of this effect or circumstance, as stated above, is merely a slight disturbance of the uniformity of the venous current near the chest. Such a doctrine could not be proposed or entertained for a moment by a person understanding the principle of a common household pump; and its having been published, and tolerated by certain professional men in the present time, will remain a proof to posterity of the deficiency, as regards fundamental science or natural philosophy, now existing in the ordinary medical education. Much ingenuity has been wasted upon it, particularly by Drs. Carson and Barry, the latter of whom, after making laborious experimental investigations on living animals, has even attempted to build upon it a superstructure of medical theory and practice! To say that the influence of the heart or chest is the power which draws the blood to the heart from the general system, is as if one asserted that the rising and falling of the tide at the mouth of a river is the power which collects the tributary streams in the interior country.

We shall enter into a little detail on this subject, because the discussion will elucidate some minor points connected with the circulation.

Presuming, then, that the reader perfectly understands the theory of pumps, and therefore of atmospheric pressure, as explained under *Pneumatics*, he will readily understand the two following propositions, either of which proves it to be a physical impossibility, that a sucking action of the heart or chest can be a cause of the blood's motion along the veins. 1st. The veins are pliant tubes free to collapse, and no pump can lift liquid through such. 2d. The *suction-power* of the chest in healthy respiration is too weak to lift liquid even one inch through tubes of any kind.

A practical illustration of the first proposition is afforded by putting the point of a syringe into a piece of gut, or eel-skin, or vein filled with water, and then trying to pump up the water. The result will be, that the fluid close to the mouth of the syringe will enter it, and then the sides of the pliant tube will collapse against the syringe, making an end of the experiment. In exact proportion to the rigidity of the tube will be the distance to which the influence of the syringe, will extend in it; if, for instance, half an ounce of pressure on the square inch of its surface be required to make it collapse, then the pump will draw up one inch of water, and so for other proportions. If, during the action of the syringe, the tube were allowed to open freely at the bottom into a vessel of water, instead of the syringe then drawing any more water from the vessel into the tube, the original contents of the tube would straightway be discharged downwards into the vessel.

The explanation of all these facts is found in the pressure of the atmosphere, (see from page 153 to page 158) seeking entrance everywhere at the surface of the earth, with a force of fifteen pounds per square inch, and overcoming any opposing force less than this;—a pressure which is sufficient, therefore, to push a column of water thirty-four feet in height, up through a rigid tube into the vacuum of a pump, but will cause the sides of the tubes to collapse, unless able to sustain its force. When nature intends a tube to resist any degree of suction, the tube is made rigid in proportion;—witness the wind-pipe and its branches, which are the only instances in the human body. And if tubes prepared for sucking light air only have received such rigidity, how much stronger would tubes have been made for sucking blood.

Some bad reasoners on this subject have believed, that if a suction power exist, capable of lifting one inch of a column of liquid, any column, however long, must follow the first inch when acted upon by the power; for, say they, the atmospheric pressure, by preventing a vacuum, will prevent separation of the liquid. Now, in the first place, this reasoning is quite inapplicable to pliant tubes, because the ready collapse of their sides will both allow the separation and prevent the vacuum; and, in the second place, with respect to rigid tubes, it is equivalent to asserting that a force just capable of lifting one link of a chain, must, therefore, be able to lift any number of connected links. Water, in a rigid tube, to which the air has no admittance, may truly be considered as a chain, for it is held together by a force of fifteen pounds per inch, pressing inwards at the two ends; and by force inferior to this, cannot lift one portion of it away from another, and, therefore, cannot draw out a drop but by lifting the whole. A man cannot suck any water from a rigid tube which is closed at the bottom; and if the bottom be open, and he has not power to support the whole contained fluid, it will sink from his tantalized lips to stand at an elevation corresponding to his suction power.

To illustrate the second proposition respecting the trifling suction power really residing in the chest, we may state that a person of ordinary strength using the whole power of the chest, (but not of the mouth separated, which is a smaller and much more powerful pump than the chest,) cannot, through a rigid tube, suck water from more than about two feet below his lips, and therefore not half way so far as from the extremities of his body; while, in the opposite action of blowing outwards, as in the attempt to blow through a tube which is dipping into water, he finds nearly the same limit. But in ordinary breathing, instead of force corresponding to a liquid column of two feet, or a *fifteenth* of the atmospheric pressure, the increase and diminution of air-density in the chest are measured by a column of less than one inch or about a *five-hundredth* of the atmospheric pressure. This fact is easily shown by breathing through the nose, while holding in the mouth one end of a glass tube, the other end of which is immersed in water, and then noting how much the water in the tube rises above the surrounding level during *inspiration*, and sinks below it during *expiration*. The mouth, during this experiment, may be considered as a part of the general cavity of the chest, to and from which air is passing by the narrow openings of the nostrils. In tranquil breathing, with both nostrils open, the fluctuation in the tube is less than half an inch each way; with one nostril closed and the other a little compressed, it may amount to a whole inch; and with hurried or convulsive breathing, like that of an animal in terror and in pain, it may exceed twelve inches. Although the measures thus obtained from the mouth are somewhat too small for the changes in the chest itself, because the chest is more remote from the opening by which the external air enters, the difference is very trifling, as may be proved during such experiments, by stopping the nostrils altogether, while the same respiratory efforts are continued; and as is also proved by the agreement of the results with strict calculation founded on the inertia and velocity of the air respired—a calculation similar to that required in adjusting the index to the machine mentioned at page 215, for measuring water-currents. In common healthy breathing, then, while the mouth is open, the fluctuation of pressure in the chest would be measured by less than half an inch motion each way of the liquid column. Dr. Barry, not aware that this point could be so easily determined by the bloodless experiment described above, or even by a simple calculation, sought the solution by numerous trials on living animals, into some part of whose chest he forced a tube. But even if farther experiments had been at all necessary, those of Dr. B. could not have decided the question, first, because the pain and agitation of the dying animals rendered the breathing violent or unnatural; and, secondly, because his experimental tube often or always became a syphon, and he, not adverting to this fact, has not recorded the difference of level in the liquids at the two ends. That the external level was, for the most part, higher than the internal, is proved by his having noticed, almost solely, the *inhaling* action of the chest, although the *exhaling* is often a more powerful effort.

Calling an inch column of blood, then, the measure of the greatest sugescent and repellent powers of the chest during ordinary respiration, we see that the force which really sends the blood from below to the heart, may have to lift a column one inch shorter during *inspiration*, and one inch longer during *expiration:* but this is the full and true measure and nature of the influence of the inspiration on the blood's return to the heart. To say, then, that the atmospheric pressure, modified by respiration, is the great power which moves the venous blood, is just as if we said, that a boy, standing near the fly-wheel of a steam engine of a hundred horses' power, and giving it his Lilliputian thrust, alternately backward and forward, were the prime mover of the machinery.

The truth explained above, that no kind of pump can lift fluid through pliant tubes, free to collapse, like the veins, renders it unnecessary farther to speak here of the pumping action of the heart itself, insisted on by Dr. Carson, or of that other action, mentioned in a subsequent part of this work, to which, also, he attributes great influence, *viz.*, the tendency towards a vacuum external to the lungs and around the heart, produced by the disposition of the lungs to collapse. It may be remarked, however, that this last influence is more considerable than the simple inspiratory action dwelt on by Dr. Barry, and that it operates during expiration nearly as much as during inspiration, varying in force with the degrees of expansion of the chest. It is weaker in the living than in the dead body, because the rigidity of the distended pulmonary arteries helps to support the weight of the living lungs.

Were it necessary to give proofs to persons unable to follow the above argument, that a suction-power in the heart or chest is not the force which draws the blood from the extreme veins, reference might be made to many notorious facts quite incompatible with that supposition; such for instance, as those recorded at page 424, and others. A vein tied, fills tensely below the ligature—a vein cut across bleeds from the orifice which is distant from the heart, and will fill a lofty tube connected with it—the circulation goes on in persons holding their breath—the veins of fishes, which do not breathe, return the blood as well as those of men, &c., &c.*

After the explanations now given, it is almost superfluous to remark that *absorption* in animals cannot depend on atmospheric pressure, and that the effect of cupping-glasses applied to extract blood, or to prevent the absorption of poison in wounds, in no way depends upon the fluctuating density of the air in the chest.† Dr. Barry's reasonings upon these subjects involve the same fallacies as his reasonings on the venous current. With respect to absorption, they neglect the facts of fluids having weight; and with respect to cupping-glasses, of which the true action is explained at page 175, they are equivalent to asserting that the action of pumps drawing water from a river among the hills is influenced by tides, or pumps operating at its mouth in the sea.

If the fluids in animal vessels had no weight, it is true, that in absorption at external atmospheric pressures of fifteen pounds per inch might force new matter into a receiving orifice, at the instant during inspiration, when the opposing pressure in the chest, at the other ends of the vessels, were half an ounce per inch less,—there would be no physical absurdity in opposing this, although there are physiological facts that disprove it—but when we reflect, that in all vessels under the level of the heart, the weight of the contained fluids causes an additional outward pressure of about half an ounce troy for every perpendicular inch of fluid column, making an excess of outward pressure at the toes, for instance, even at the more favourable time of absorption,

* The influence of inspiration of the cavity in the chest exterior of the lungs, and the expansive power of the heart, on the circulation of the blood in the veins, have no doubt been greatly overestimated by Drs. Barry, Carson and others, but our author appears to us to have undervalued their effect. Their joint power is more considerable than the reader might be led to suppose from the perusal of the preceding pages.

The influence of inspiration has been estimated by our author, perhaps justly, as only sufficient to raise a column one inch; if this force acted through *rigid* tubes of the length of the veins, it would produce no movement of the contained fluid; but acting through pliant tubes, it would rise one inch of the blood out of the vein nearest the heart, and if this power acted alone, its effect would here cease. But the vis a tergo, produced by the propulsive power of the capillaries, and, perhaps, also of the heart, prevents the collapse; the vein is kept full, and at every inspiration this power is renewed.

The influence of the tendency towards a vacuum external to the lungs, and around the heart, from the contractile disposition of the resilience of the lungs, is admitted by our author to be more considerable than the inspiratory effort, and it in fact is, we think, greater than is suspected. There are reasons for believing that the lungs do not entirely fill the cavity in which they are contained; the influence of this space is therefore constant, though greater during inspiration, and of course diminished during expiration.

The capillaries, our author has most satisfactorily shown, have a vital expansive power; and though he does not assert that the heart has no such power, he denies that it can have any influence on the movement of the venous blood, since it must act through pliant tubes. This would be the fact if the expansion of the heart were the only moving power, but the vis a tergo prevents their collapse, and the effect of the expansive power of the heart, whatever that may be, is allowed to act.

While, therefore, the action of the capillaries and perhaps of the left ventricle of the heart must be considered as the main forces by which the blood is propelled through the veins, the expansive power of the heart—respiration and the resilience of the lungs, or atmospheric pressure—ought to be viewed as necessary forces, though their precise power cannot readily be estimated.—*Am. Ed.*

† The effect of cupping-glasses in preventing the absorption of poisons has been shown by Dr. Pennock to be owing to mechanical pressure. See the interesting experiments in the *American Journal of the Medical Sciences*, Vol. II.—*Am. Ed.*

of about two pounds per inch, we see that absorption must be a strong *action of life*, able to overcome a great excess of mechanical resistance, instead of a passive phenomenon obeying an excess of mechanical force. If a mere balance of pressures acted at the orifices, as Dr. B. supposes, the blood and other fluids would be constantly oozing out from all orifices below the heart, as blood really does from an artificial opening, with force that would fill a tube reaching as high as the heart. It would be good news for proprietors of mines, and other persons having to raise water, if by taking off an ounce or two per inch of the atmospheric pressure at the top of a full pipe, the atmospheric pressure continuing elsewhere would then force water in at openings below and cause the upward current:—but in truth to make the atmosphere efficient below, powerful steam-engines or other means must be used to take off a pressure above, of at least half an ounce per square inch, for every inch in height the water has to rise.

Another erroneous conception of atmospheric pressure, akin to that which we have been considering, is expressed in the following reasoning on the progress of blood in the veins. "The atmosphere presses 15 lbs. per square inch on all things; the blood therefore, in a vein which has 20 inches of surface is pressed upon through the flesh, with a force of 20 times 15, or 300 lbs., while a cross section of the vein near the heart would measure less than one inch. The blood, therefore, is always running towards the heart, to escape from a powerful excess of atmospheric pressure."—This paradox is solved by the law of fluid pressure, explained at page 131. The same reasoning would prove that an eel-skin suspended by its lip, and filled with water, when exposed to the pressure of the atmosphere, should quickly be emptied; and nearly the same would prove that a long sharp wedge thrown into water, should be always moving in a direction away from its point; and that a ship formed like the wedge, should make quick speed across the sea without either oar or sail.

A knowledge of the facts detailed under the three heads of *arteries*, *capillaries* and *veins*, prepares us for the discussion of the following subjects.

The force of the heart.

The arterial tension of four pounds to the square inch, marked by its supporting in a tube connected with the arteries, a column of blood eight feet high, (see page 417,) is produced by the action of the heart; but as the heart, while injecting the blood against this resistance, has moreover to overcome the *inertia* both of the quantity injected and of the mass in the great artery, first moved by the injection, as also the resisting *elasticity* of the vessel which yields to momentary increase of pressure, the heart must act with a force exceeding four pounds on the inch. And as the left ventricle of the human heart, when distended, has about ten square inches of internal surface, the whole force exerted by it is a matter of simple calculation. It is remarkable, as there is this easy means of solving the question, that the accurate Magendie, in his recent elements of physiology, should speak of it as undetermined; and should cite, as the best approximation, an estimate made from the obscure circumstance of a loaded foot shaking in unison with the pulse, when suspended in the cross-legged sitting attitude.

Some physiologists have expressed surprise that the force of the heart should be so great as it is, remarking that much less would have sufficed to propel the blood to the most distant capillaries; but they did not reflect that the heart, besides carrying on the general circulation, has to force blood into those parts of the flesh which, in the various positions of sitting, lying, standing, &c., are for the time compressed by the whole weight of the body; for that, if it were not strong enough for this purpose, either the compressed parts, deprived of their nourishment, would quickly die, or the person obliged to be every moment changing position, could obtain no lengthened repose. In illustration of this point, we may advert to the frequent occurrence, in diseases where the power of the heart is for the time weakened, for sloughings, or bed-sores in the bearing parts, causing many cases of illness to terminate fatally which would otherwise soon have terminated in health. The author of this work has had great satisfaction in suggesting a means of entirely preventing deplorable termination, namely, that which he is now about to describe under the title of

The HYDROSTATIC BED *for Invalids.*

In many of the diseases which afflict humanity, more than half of the suffering and danger is not really a part of the disease, but the effect or consequence of the confinement to which the patient is subjected. Thus a fracture of a bone of the arm is as serious a local injury as a fracture of one of the bones of the leg; but the former leaves the patient free to go about and amuse himself, or attend to business as he wills, and to eat and drink as usual—in fact, hardly renders him an invalid; while the latter imprisons the patient closely upon his bed, and brings upon him, first, irksomeness of the continued position, and then the pains of the unequal pressures borne by the parts on which the body rests. These, in many cases of confinement, disturb the sleep and the appetite, and excite fever, or such constitutional irritation as much to retard the cure of the original disease, and not unfrequently to produce new and more serious disease. That complete inaction should prove hurtful to the animal system, may by all be at once conceived; the operation of the continued local pressures will be understood from the following statements. The health, and even life, of every part of the animal body, depend on the sufficient circulation through it of fresh blood, driven in by the force of the heart. Now when a man is sitting or lying, the parts of his flesh compressed by the weight of the body, do not receive the blood so readily as at other times; and if from any cause the action of his heart has become weak, the interruption will both follow more quickly and be more complete. A peculiar uneasiness soon arises where the circulation is thus obstructed, impelling the person to change of position; and a healthy person changes as regularly, and with as little reflection, as he winks to wipe and moisten his eyeballs. A person weakened by disease, however, while he generally feels the uneasiness sooner, as explained above, and therefore becomes what is called restless, makes the changes with much fatigue; and should the sensations after a time become indistinct, as in the delirium of fever, in palsy, &c,, or should the patient have become too weak to obey the sensation, the compressed parts are kept so long without their natural supply of blood that they lose their vitality, and become what are called sloughs or mortified parts. These have afterwards to be thrown off, if the patient survive, by the process of ulceration, and they leave deep holes, requiring to be filled up by new flesh during a tedious convalescence. Many a fever, after a favourable crisis, has terminated fatally from this occurrence of sloughing on the back or sacrum; and the same termination is common in lingering consumptions, palsies, spine diseases, &c., and generally in diseases which confine the patients long to bed.

It is to mitigate all, and entirely to prevent some of the evils attendant on the necessity of remaining in a reclining posture, that the hydrostatic bed is intended. It was first used under the following circumstances.

A lady after her confinement, which occurred prematurely, and when her child had been for some time dead, passed through a combination and succession of low fever, jaundice, and slight phlegmasia dolens of one leg. In her state of extreme depression of strength and sensibility, she rested too long in one posture, and the parts of the body on which she had rested all suffered : a slough formed on the sacrum, another on the heel; and in the left hip, on which she had lain much, inflammation began, which terminated in abscess. These evils occurred while she was using preparations of bark, and other means, to invigorate the circulation, and while her ease and comfort were watched over by the affectionate assiduity of her mother with numerous attendants. After the occurrence, she was placed upon the bed contrived for invalids by Mr. Earle, furnished for this case with pillows of down and of air of various sizes, and out of its mattrass portions were cut opposite to the sloughing parts; and Mr. Earle himself soon afforded his valuable aid. Such, however, was the reduction of the power of life, that in spite of all endeavors, the mischief advanced, and about a week later, during one night, the chief slough on the back was much enlarged, another had formed near it, and a new abscess was proceeding in the right hip An air-pillow had pressed where these sloughs appeared. The patient was at that lime so weak that she generally fainted when her wounds were dressed; she was passing days and nights of uninterrupted suffering, and as all known means seemed insufficient to relieve her, her life was in imminent danger.

Under these circumstances, the idea of the hydrostatic bed occurred to me. Even the pressure of an air-pillow had killed her flesh; and it was evident that persons in such condition could not be saved unless they could be supported without sensible inequality of pressure. I then reflected, that the support of water to a floating body is so uniformly diffused, that every thousandth of an inch of the inferior surface has, as it were, its own separate liquid pillar, and no part bears the load of its neighbour—that a person resting in a bath is nearly thus supported—that this patient might be laid upon the surface of a bath over which a large sheet of the water-proof India-rubber cloth were previously thrown, she being rendered sufficiently buoyant by a soft mattress placed beneath her—thus would she repose on the face of the water, like a swan on its plumage, without sensible pressure anywhere, and almost as if the weight of her body were annihilated. The pressure of the atmosphere on our bodies is of fifteen pounds per square inch of its surface, but because uniformly diffused, is not felt. The pressure of a water-bath or depth to cover the body, is less than half a pound per inch, even on the under side where it is greatest, and is similarly unperceived. A bed such as then planned, was immediately made. A trough of convenient dimensions (6 feet long, 2 feet 8 inches wide, and 11 inches deep, are good common dimensions) was lined with metal to make it water-tight; it was about half filled with water, and over it was thrown a sheet of the India-rubber cloth as large as would be a complete lining to it if empty. Of this sheet the edges, touched with spirit varnish to prevent the water creeping round by capillary attraction, were afterwards secured in a water-tight manner all round to the upper border or top of the trough, shutting in the water as closely as if it had been in bottles, the only entrance left being through an opening at one corner, which could be perfectly closed. Upon this beautiful dry sheet a suitable mattress was laid, and constituted a bed ready to receive its pilow and bed-clothes, and not distinguishable from a common bed but by its most surpassing softness or yielding. The bed was carried to the patient's house, and she was laid upon it; she was instantly relieved in a remarkable degree: sweet sleep came to her; she awoke refreshed; she passed the next night much better than usual; and on the following day, Mr. Earle found that all the sores had assumed a healthy appearance; the healing from that time went on rapidly, and no new sloughs were formed. When the patient was first laid upon the bed, her mother asked her where the down pillows, which she before had used, were to be placed; to which she answered, that she knew not, for that she felt no pain to direct; in fact, she needed them no more.

It may be here recalled to mind, that the human body is nearly of the specific gravity of water, or of the weight of its bulk of water, and therefore, as is known to swimmers, is just suspended or upheld in water without exertion when the swimmer rests tranquilly on his back with his face upwards. He then displaces water equal to his own body in weight as well as in bulk, and is supported as the displaced water would have been. If his body be two and a half cubical feet in bulk, (a common size,) he will just displace two and a half cubic feet of water, equal in weight to his body. If, however, instead of displacing the water with his mere body, he choose to have something around or under him which bulky with little weight, as the mattrass of the bed above described, then, after his weight is forced two cubical feet of that under the level of the water around, he will float with four-fifths of his body above the level, and will sink much less into his floating mattrass than a person sinks in an ordinary feather-bed. It thus appears that by choosing a certain thickness of mattrass, and if unusual positions are required by having different thicknesses in different parts, or by placing a bulk of folded blanket or of pillow over or under the mattrass in certain situations, any desired position of the body may be easily obtained. If the water be about six inches deep, which in general will suffice, the person standing upon any part of the bed, or sitting with the knees raised, will cause the part of the mattrass on which he rests gently to touch the bottom, because a narrow end of the body cannot displace water equal to the bulk of the whole, but even then the person is as if standing or sitting on a soft sofa. If it be desired to prevent the mattrass, when used as a seat, from touching the bottom, the object may be attained by having under its middle a broad band or strap fixed to one edge of the trough, and connected with the other by buttons or otherwise, so as to be tightened to allow the mattrass to descend just so far, and no farther.

This bed is a warm bed, owing to water being nearly an absolute non-conductor of heat from above downwards, and owing to its allowing no passage of cold air from below. From this last fact, however, less of the perspiration, sensible and insensible, is carried off by the air than in a common bed, and unless the patient can leave the bed daily to let it be aired like a common bed, there will be a necessity for ventilation to prevent the perspiration from being condensed on the water-sheet below. This ventilation is perfectly obtained by placing under the mattrass, arranged like the bars of a gridiron, small flexible tubes of tinned wire, wound spirally, with their ends open to the atmosphere, either directly or through two larger tubes crossing and connecting their extremities near the ends of the mattrass, and then issuing at the corners of the bed from under the clothes. This bed is in itself as dry as a bed can be, for the India-rubber cloth (of which bottles can be made) is quite impermeable to water, and the maker is now preparing cloth expressly for this purpose. Then, as Sir Humphrey Davy recommended that his safety lamp should be double, some persons may prefer a double sheet, to obviate the possibility of accident. Unlike any other bed that ever was contrived, it allows the patient, when capable of only feeble efforts, to change his position, almost like a person swimming, and so to take a degree of exercise, affording the kind of relief which, in constrained positions, is obtained by occasional stretching, or which an invalid seeks by driving out in a soft-springed carriage. It exceedingly facilitates turning for the purpose of dressing wounds, for by raising one side of the mattrass or depressing the other, or merely by the patient's extending a limb to one side, he is gently rolled over, nearly as if he were simply suspended in water; and it is possible even to dress wounds, apply poultcies, or place vessels under any part of the body, without moving the body at all: for there are some inches of yielding water under the body, and the elastic mattrass may at any part be pushed down, leaving vacant space there, without the support being lessened for the other parts. Then, with all the advantages which other invalid beds posses, and with those which are entirely its own, it may yet be made so cheaply, that even in hospitals, where economy must prevail, it may at once be adopted for many of the bed-ridden. Mr. Earle within a few days of seeing the first one, had others made for patients in St. Bartholomew's Hospital, and has been as much pleased with the results of them as of the first. The bed has since been introduced into St. George's Hospital by Mr. Keate, and elsewhere.—The author has now seen enough of the effects of this bed to make him feel it a duty at once to publish a notice of it. With it, evidently, the fatal termination called sloughing, now so common, of fevers, and other diseases need never occur again. And not only will it prevent that termination, but by alleviating the distress through the earlier stages, it may prevent many cases from even reaching the degree of danger. Then it is peculiarly applicable to cases of fractured bones, and other surgical injuries; to palsies, diseases of the hip joint, and spine; and universally, where persons are obliged to pass much time in bed. And in all cases of curvature of the spine, either actually existing or threatened, it affords a means of laying a patient in any desired position, and with any degree of pressure incessantly urging any part of the spine back to its place. If used without the mattrass, it becomes a warm or a cold bath, not allowing the body, however, to be touched by the water, and in India, it might be made a cool bed for persons sick or sound, during the heats which there prevent sleep and endanger health. There are numerous other professional adaptions and modifications of it, which will readily occur to practitioners sufficiently versed in the department of natural philosophy (hydrostatics) to which it belongs. Before reflection, a person might suppose a resemblance between it and an air-bed or pillow, calling this a water-bed or pillow; but the principles of the two are perfectly distinct or opposite. An air-pillow supports by the *tension of the surface* which encloses the air, and is therefore like a hammock or the tight sacking under the straw mattrass of a common bed, and really is a hard pillow; but in the hydrostatic bed, there is no tense surface or web at all; the patient is floating upon the water, on which a loose sheet is lying, merely to keep the mattrass dry, and every point of his body is supported by the water immediately beneath it. To recall the difference here described, and which is of great importance, the bed is better described by the appellations of *hydrostatic-bed*, or *floating-bed*, than of *water-bed*.

The author has given no exclusive right or privilege to any person to make this bed. He has hitherto employed the carpenter nearest to him, Mr. Smith, 253 Tottenham-court Road, at the back of Bedford Square; Mackintosh & Co., 58 Charing Cross, the manufacturers of the cloth; and Mr. Williams, 25 Cleveland Street, Fitzroy Square; but any carpenter or upholsterer may learn to supply them, and he gives free permission to all. He hopes, for the sake of the poor, that a trough, without metallic lining, and with a cheaper water-proof cloth, may be found to answer satisfactorily.

The principle of the hydrostatic bed, is applicable, also, to couches for invalids, and with certain considerable modifications, to the construction, also, of chairs; and there are other means than the water-proof sheet of adapting the hydrostatic principle for all—but the subject has already occupied its full share of this volume.

The preceding paragraphs are intended as much to direct in the choice and use of common beds for the sick, as to announce and describe the hydrostatic bed for the cases in which it may be required. At present, the medical attendant generally leaves whatever regards the bed to the judgment of friends or nurses; but evidently, he who has been led to reflect how much the course and event of a maladay may depend on the patient's being supported, so that no pain shall arise from local pressure, and as little muscular weariness as possible from constrained position, will deem the bed-management worthy of his own attention, and will be able more judiciously both to choose and to use beds. There is a bed constructed of spiral springs, which may be made so as to diffuse the support more equally than any except the hydrostatic bed; and had professional men generally been acquainted with it, it would have been more used than it is, and would have received various modifications, of which it is susceptible, for medical purposes. It has long been known, chiefly, however, as a mechanical curiosity, or an object of luxury, and was introduced into this country about seventy years ago by Mr. Merlin; but it has been so little known, that a few years ago an English tradesman thought he might appropriate the manufacture by taking a patent for it. It is now made by upholsterers generally, and the same principle is applied in the construction of sofas, chairs, and carriage cushions.

The velocity of the circulating blood.

This has been much overrated. 1st. By assuming that the ventricles of the heart are both completely filled from the auricles and emptied towards the arteries at each pulsation:—an assumption disproved by inspections of the exposed heart of a living body, and by the fact of the valves between the auricles and ventricles not closing so perfectly as quite to prevent regurgitation. 2d. By supposing the issue of blood from a wounded artery or vein to be the measure of the usual velocity. Now it would be as reasonable to suppose the issue of water from a wounded pipe connected with any reservoir to be the measure of a continued current in that pipe, although, in truth, the issue would be the same even if the water in the pipe were usually at rest. 3d. By supposing the *frequency* of the pulse to be a measure. Now we know, that in diseases of debility, and in animals bleeding to death, the pulse usually becomes more frequent as it becomes more feeble, and as there is less blood moving: *viz.*, the heart very partially discharging its contents at each contraction. 4th, and lastly. By supposing the *strength* of the pulse to be the measure. Now we find that the pulse in an artery just tied, and in which, consequently, there is no current at all, is scarcely weaker than in an open artery. The common fact of a person's feet remaining stone-cold for hours, although the arteries leading to them pulsate nearly as usual, is a proof that exceedingly little blood is passing through the capillaries at the time, and that the strength of the pulse, therefore, is no measure of the speed of the blood.

The ventricles of the heart appear, under common circumstances, to throw out about an ounce and a half of blood at every contraction—or about seven pounds per minute. Now if the body contain about twenty pounds altogether, as seems to be the case, the whole would circulate twenty times in an hour. This would give an average velocity of about eight inches per second in the aorta, but gradually less in the smaller arteries, because whenever a vascular channel subdivides, the branches taken collectively have considerably greater area than the trunk from which they arise, and the current diminishes in a corresponding proportion, just as the speed of a river is always less in the parts of the channel which are deeper and broader. The velocity in the extreme capillaries is found to be often less than one inch per minute. In the veins, the blood must move more slowly than in corresponding arteries, in proportion as the veins are more capacious than the arteries.

The pulse.

The opinion which the ancients held, that the arteries contained *vital spirits* or *air* and not *blood*, rendered the pulse, to them, a very mysterious phenomenon; and many curious hypotheses were framed to explain it. These it would now be unprofitable to detail. Even Harvey's grand discovery of the circulation, however, has not rendered the subject so simple as might have been anticipated. The following opinions now exist, or have lately existed with respect to the pulse.

1st. The great majority of physiologists have believed that a tumefaction is produced in the aorta by each jet of blood from the heart, and spreads afterwards as a wave into all the arterial branches. 2d. Many have supposed an extensive contractile action of the arteries themselves, corresponding to that of the heart. 3d. Bichat, unable by any means to detect the slightest change of diameter in the arteries during pulsation, but perceiving that in many situations they were at the time somewhat lengthened, so that straight portions became bent, and portions originally bent, were bent still more, held that this locomotion or changing of place in the arteries was the cause. 4th. Others have supposed the impulse of the heart's contraction to be transmitted through the fluid blood, somewhat as sound is transmitted through bodies generally, or as a blow struck on one end of a log of wood, is felt distinctly by a hand applied to the other, although there be no visible locomotion. 5th. Dr. Young, in the paper in the Philosophical Transactions already alluded to, has shown that a sudden rush forward of the blood in the artery, such as might be produced by injection at one end of a rigid tube, would be felt by a finger applied to the artery, quite as distinctly as a tumefaction; and he deems this occurrence to be a chief cause of the pulse. Dr. Parry, in his work on the pulse, points to this almost as exclusively the cause.

Now the truth is, that the pulse in the living body does not depend exclusively upon any one of the particulars just noticed, but has all of them as elements; and its fluctuations and varieties depend upon proportions in which these elements are combined. We shall review them again to prove this.

1st. At each jet of blood thrown into the aorta, a tumefaction or wave *must* spread from the heart to the extremities; for it is evident, that if blood be at all pushed into the arterial system, it either must dilate it, or cause an equal quantity to be expelled at the same instant from the distant extremities: now as the passage of blood through the capillaries appears perfectly uniform, there must be an intermediate dilatation. Dr. Parry and others should not have denied this dilatation because they could not see it: for even if its advancing front were more considerable than it is, it passes with such velocity that like a cannon-ball crossing before the face, it would not be perceived.

2d. Contraction of the arterial coats certainly does not take place in the manner and to the extent supposed by some, who have spoken of it as resembling the contraction of the heart itself, and as what might be a substitute for the action of the heart in propelling the blood; but as shown at page 414, the rigidity of tube which in all degrees of arterial dilatation causes the pulse to be transmitted so quickly, can depend on nothing but a contractile action of the fibres. There are some reasons for doubting whether this rigidity may not increase at the moment of the pulse.

3d. Unless the arterial tubes were absolutely inelastic, which they are far from being, they *must* be lengthened a little by a sudden injection of blood, and, therefore, at all the curvatures particularly, there *must* be a degree of the locomotion described by Bichat, often sensible to a finger applied.

4th. That a tangible shock is conveyed through a fluid without any apparent accumulation of the fluid or change of velocity, and much in the manner of sound, is proved by the facts, that we may discover the working of a water-pump at very great distances, through iron pipes connected with it, and even through elastic pipes of leather, as those of a common fire-engine, from which the water is spouting, nevertheless, in a uniform stream. The pulse in a tied artery, in which there is no current or rushing wave, must be chiefly from this cause and from the locomotion of the artery.

5th. That any additional quantity of fluid injected into elastic vessels already full, *must* spread all over with a *forward rush*, affecting the finger of an examiner, as described above, is also most certain. As the heart, however, often beats without discharging much of its blood, and as in many arteries from inaction of the capillaries, or pressure of the blood for a time makes little or no progress, while the pulse, however, remains very distinct, the pulse in such cases must be produced independently of the forward rush. An animal intestine prepared, and filled with water or air, and laid upon a table—or a full vein in the living body, carries a rapid and distinct pulse to a great distance when gently tapped by the finger. The cause of the sensation, then, cannot be the simple forward *rush* without tumefaction, described by Dr. Young and Dr. Parry.

In whatever proportions these particulars combine to form the pulse, its force will be proportioned to the size of the artery. Hence as an artery leading to an inflamed part becomes of greater calibre, its pulse also becomes stronger.

It is a remark respecting the pulse, appearing to the author worthy of deep consideration, that if the purpose of the heart and arteries were merely the propulsion and conveyance of the blood, their structure and action would form most signal deviations from the ascertained rules of fitness in mechanics. In machines of human contrivance, it is one of the most important maxims "to avoid shocks, or jerking motions;" and in former parts of this work, we have described fly-wheels, air-vessels, springs, &c., as means of accomplishing this object, and thereby of preventing the tearing and straining of parts which would else happen. In the human body, also we have to describe the admirable elasticity of the spine, of the arch of the foot, of the cartilages of joints,

&c., as contrivances answering the same end; and to remark that, in other cavities than the heart, which are alternately filled and emptied like it, as the stomach, bladder, uterus, &c., there is a smooth and gradual action. The heart alone is the rugged anomaly, which, from before birth unto the dying moment, throbs unceasingly, and sends the bounding pulse of life to every part; and which, moreover, instead of being secured and tied down to its place, is attached at the extremity of the aorta, like a weight at the end of an elastic branch of a tree, and every time that it fills the aorta, is thrown with violence, by the consequent sudden tendency of that vessel to become straighter, against the ribs, in the place where the hand applied, feels it so distinctly beating.

Now one use of the pulsation of the heart probably is, by the *agitation* and *churning* which the blood suffers in passing through it, to keep in complete mixture all the heterogeneous parts of the blood, which so readily separate when left to repose:—but this cannot be the only use, for the object might have been more simply detained; and we may conclude that the phenomenon has relation to some important laws of life still hidden from us. The cause commonly assigned for the heart's contraction is the peculiar stimulus of the blood; yet if we reflect that the heart will beat after removal from the body and when it contains only air, and that during life it beats with extraordinary regularity, whether the state of the circulation allow it to empty itself at each beat or not, we perceive that the case is more obscure. We cannot contemplate this subject attentively without perceiving a strong analogy between the action of the heart and some electrical phenomena in which there are successive accumulations and exhaustion of power; and, recollecting the important relations which late researches have shown to exist between electricity and certain other actions of life, the inquiry becomes very interesting. Galvanism can excite the muscles to their usual actions; it powerfully affects the secretions and the digestive function; and the breathing in asthma; strong animal passion seems to produce electrical excitement: and certain animals have the faculty of stunning their enemies by an electrical discharge. The pulse, then, in its sudden, strong, and regular recurrence, may be a kindred phenomenon. In this view, there would be less difficulty in supposing a momentary stiffening or slight contraction of the whole arterial system, such as the sudden rising of the mesenteric arterial tree so readily suggests: if there be such, however, it is still closely connected with, and proportioned to, the action of the heart; for it occurs only with that action, it indicates any disturbance in the action, and as death approaches, it ceases in the remote extremities first.

The preceding considerations exhibit the pulse as a complex subject, and one on which professional opinions are not yet settled. By showing its close relation to the powers of life, they also prove it to be an object of high importance to the medical practitioner. This last truth has scarcely been questioned but by persons either utterly uninformed or singularly deficient in the power of tactile discernment; yet, because no simple and good analysis of the pulse, and detail of its relation to morbid states, has been made and published, the degrees of skill acquired by individual practitioners with respect to it are very various, and in a great measure accidental. Some practitioners try the pulse merely for form's sake, because patients expect it; many examine it only to count its frequency; but others read in it, with confidence, much of the history and probabilities of the disorder, and decide on the treatment accordingly. Few who have attended to the subject at all, can confound the pulses of such diseases, as acute rheumatism, gastric inflammation, the fits of ague, &c. The author remembers to have conversed with a Chinese practitioner who had only the scanty medical information of his countrymen, but who judged by the pulse with a singular penetration.

The changing circumstances in the state of the circulatory system, connected with health and disease, and discoverable by a finger watching the pulse, seem to be chiefly the following; and the epithets added in italics, are those which seem best to indicate the sensations perceived. The artery at the wrist is that generally chosen for examination, because it is not like others imbedded in soft parts, having only the skin over it, and nothing between it and the bone below.

1st. The number of the contractions of the heart in a given time, and the regularity of their recurrence.—Pulse, *frequent, slow, intermittent, equal, regular, of varying force.*

2d. The degree of the heart's contraction, or the quantity of blood ejected at each time; and the corresponding state of the capillaries as to the quantity of blood passing through them.—Pulse, *full, long, labouring, bounding, feeble..*

3d. The force of the heart's action, with the correspondent arterial tension or rigidity.—Pulse, *hard, sharp, strong, wiry, weak, soft, yielding.*

4th. The suddenness of the individual contractions of the heart, and the rigidity of the vessels in conveying the shock.—Pulse, *quick, tardy.*

5th. The size of the artery for the time, whether larger or smaller than usual.—Pulse, *large, small.*

Superficial as is this sketch, it may show that a good treatise on the subject of the pulse, as connected with disease, is yet a desideratum in medicine. The sort of empirical but useful tact which many persons acquire, is not fitted to satisfy the physician who reasons deeply, and whose mind should have always present to it the various constituents of the pulse, and all the important circumstances of health or disease related to its indications. The laboured treatise of *Solano, Bordue, Boerhaave,* &c., may treat of what were clear ideas to their authors, but by not referring the physical causes of many varieties, they become so obscure to others, that many of the divisions and denominations appear altogether fanciful. Dr. Young's excellent paper in the Phiosophical Transactions, details important facts, as far as it goes, but it was not intended to point out all the pathological relations. Dr. Y., guided by general principle, asserted a progressive motion of the pulse, while other authorities were holding it to be quite simultaneous over the whole system. He might have mentioned in proof, that careful examination can practically detect a succession of beats at different distances, particularly at the four stations; 1st, of the heart; 2d, in the lip; 3d, at the wrist; 4th, at the ankle:—but the interval of time, even between the extremes, being only a small part of a second, persons will often fail to make their first experiment satisfactorily. Dr. Parry's treatise on the pulse, which is the last one of note, although having excellencies, errs—in attributing the phenomenon to one cause so exclusively—in denying arterial dilatation, because it was not discovered by his mode of searching for it, in supposing that a liquid column in an elastic tube, can be made to advance like a solid rod, or line of billiard-balls. The too common neglect of mechanical philosophy by medical, men is signally proved, by our finding in works of authority, published at the present day, such statements as that the arterial pulse may be more frequent or less frequent than the beatings of the heart. *Dr. Good* (*Study of Medicine*) says, that there may be various frequency of pulse in various parts of the body at the same time: *Richerand*

(*Physiologie*) says, the pulse is more frequent in the artery leading to a whitlow than at the same time elsewhere; and many practitioners share these notions. What a satire on the medical profession is this disagreement on a point which, to common observers, seems, obove all others, to occupy the attention of the attendant on the sick!

Having now explained the circulation of the blood in general, we proceed to consider some cases where mechanical circumstances modify it.

Circulation in the head.

The head may be considered as an air-tight vessel or cavity of bone, containing chiefly brain and blood, and having openings occupied by blood-vessels, leading to and from the heart. The atmospheric pressure, therefore, always keeps the head full, as it keeps the top of a syphon full; and because the substance of the brain itself does not more than water, sensibly change in bulk by any ordinary degrees of pressure, there must always be the same quantity of blood in the head, how much soever the quantity may vary in the body generally. Regard to this important truth, a knowledge of which has followed the discovery of the true nature of atmospheric pressure, enables us to explain many hitherto obscure facts, both in health and disease; —as the following instances will show.

If, from any cause, the arteries in the head become too full of blood, in the same proportion the veins must become too empty; or, if the veins become too full, the arteries must be too empty; and in either case, the circulation in the head will be in a corresponding degree impeded, because, when one part of a channel is narrowed or diminished, the current throughout the whole is slackened. Now, as insensibility supervenes when the supply of fresh blood to the brain is interrupted, and death follows if the interruption continue long, it seems evident that in many of the cases of apoplexy, where, on inspection, there is found nothing but a fulness of the arterial or of the venous system of the head, death has happened merely because the circulation was arrested in this way. In other parts of the body, not circumstanced like the brain, an excess of blood in one set of vessels may happen without inducing deficiency in another, and therefore with perfect impunity to the individual.

Simple increase of pressure produced by the blood on the brain, provided the proper balance exist between the quantity in veins and arteries, has no injurious effect. This is proved by the safe descent of a person in a diving-bell, where, at thirty-four feet under the surface of the water, the body is bearing an additional pressure of fifteen pounds on a square inch (see page 163,) which pressure through the blood-vessels affects the brain as much as any other part. On the other hand, when a man climbs a mountain, or is lifted up in a balloon, the brain is less pressed than usual; but the proper balance in artery and vein being maintained, no inconvenience is felt. The inhabitants of some of the valleys among the Andes are as far above the sea as they would be at the top of Mont Blanc, where the atmosphere presses only half as much as on the sea-shore; but they enjoy good health.

As the box of the cranium encloses the brain so as to leave no vacant space, it is evident, that when the heart injects blood with unusual violence, the strain at first is borne chiefly by the cranium, and not by the coats of the blood-vessels. Hence, the arteries of the brain need not to be, and are not, nearly so strong as those of other parts of the body.

The veins of the brain are also peculiar. Common veins in the head would, for the reasons above given, collapse by any sudden tension of the arteries there, and if they did, insensibility or death would ensue, on account of the consequent stoppage of the circulation. The chief channels, therefore, for the refluent blood, instead of being common compressible veins, are what have been called *sinuses*, or grooves in the bone itself, with exceedingly strong membranous coverings, supported so that the channels become in strength, and as to maintenance of their capacity, a little inferior to complete channels of bone. This singular deviation in the structure of the cerebral veins from what is found elsewhere, and without which deviation, animal existence could not be continued, is one of those particulars which powerfully affect the contemplative mind, as proofs of the designing intelligence which has planned this glorious universe.

From not adverting sufficiently to the fact now explained, of the cranium being a vessel always full, and which will hold only a certain quantity, misconception has prevailed among medical men with respect to many of the affections of the brain.

It has been said, for instance, that the substance of the brain cannot bear pressure with impunity, for that stupor immediately follows pressure, however produced. Now the truth is, that pressure produces stupor only when it interferes with the circulation. In wounds with loss of a large piece of the cranium, the brain will bear very rough handling, because if compressed at one part, it may bulge at another, and leave the circulation free; but if the wound be small, pressure made through it instantly affects the whole brain, and the blood is prevented from entering from the heart. Let one reflect, for an instant, on what happens to the fœtal head during parturition—how often it escapes elongated and bent, almost as if it were of soft clay—yet the child lives and thrives, and the natural form is soon recovered. The reason is, that the fœtal skull is soft, and pressure in one part is compensated for by a bulging or extension in another, and the blood is not expelled.

Water in the head, again, is said to kill by this fatal pressure on the tender brain; but, in reality, it kills by keeping out of the blood, and so mechanically arresting the circulation. Accordingly we see, that where the *fontanelle* still remains open, or where the *sutures* or joinings of the skull will yield, water may accumulate to a great degree without causing much disturbance.

A tumour in the brain, which would be of no consequence if the brain were unconfined, soon becomes fatal by occupying room in the skull, and to the extent of its size excluding or checking the supply of blood.

If the substance of the brain at all increase and diminish in bulk, as muscles, &c., under certain circumstances, do, in the body below, all such changes must produce a considerable effect on the cerebral circulation and functions.

Effects of position on the circulation.

While a man is in a standing attitude, the heart and arteries have to send the blood up the head against gravity; but in the horizontal position, the blood, if equally propelled, must arrive with greater force, because gravity then does not resist. Hence headache, no other symptom arising from fulness of blood in the arteries of the head, is often relieved by the upright position, and is increased by lying down.

Many people who have had a slight degree of toothache during the day, find it intolerable when they lie down at night, and are relieved again by rising and walking about. Commonly they suppose that it is the cold of the night which then lulls the pain; but it is in fact the change of position. The author knew a lady who was obliged to sleep for months in the sitting posture, because she had a *tic doloureux* in the face whenever she lay down; and another who was under the same necessity for a considerable period after an inflammatory affection of the brain, because if her head fell low during sleep she was immediately assailed by a terrific dream of swords driven into the brain.

Delirium in fever is sometimes checked at once by elevating the head. On account of the great relief thus obtained, some continental practitioners had proposed to support the patients occasionally in an upright posture.

Apoplexy has often been brought on by a man bending his head down in the act of tying his shoe, or of pulling on his boot.

Children and professed tumblers being much in the habit of placing their bodies in all positions, feel no inconvenience from having the head downwards; apparently, because arteries and veins usually become strong enough to bear the pressure to which they are habitually exposed; but to many old people, accustomed to keep the head always up, the attempt would be fatal.

Ulcers on the legs are often obstinate and will bleed, because the veins about them are too weak to support the lofty columns of blood above. Hence the frequent counsel given in such cases to keep the feet raised upon a chair, and the utility of certain modes of bandaging.

Many inflammations of the legs and feet become exceedingly painful when the limbs are in a hanging position, and the pain is relieved by laying them horizontally.

Many anasarcous or dropsical affections of the legs increase towards night, because, during the dependent position of the legs through the day, the absorbents have not power to lift the fluid. The swelling disappears again before morning.

When the heart has to send blood upwards, it requires to act more strongly than when the body is horizontal, and the pulse increases five or six beats in the minute; hence the common rule to make a patient with hæmorrhage lie in the horizontal position, that the heart may become tranquil and allow the bleeding to cease.

Fainting from diminished arterial tension.

Fainting, which is a temporary cessation of the action of the heart, and hence, as explained above, of the action of the brain for want of blood, is produced by several causes, and among others, by any occurrence which renders the blood-vessels about the heart suddenly less full of tense than usual. It would appear that the heart being accustomed, when it contracts, to a certain degree of resistance, has its action disturbed when the resistance is much diminished.

Thus hæmorrhage, from any cause, by lessening the general tension of the sanguiferous system, often causes fainting. The state is relieved by lying down; probably because the still remaining weaker action of the heart is sufficient to send blood to the head along a horizontal course, until the gradual contraction of the whole vascular system reproduces the tension necessary to perfect action. A small quantity of blood taken away *suddenly*, affects the circulation as much as a larger quantity taken *gradually*, apparently because a certain space of time is required for the gradual lessening of the vessels.

The operation of *tapping* for dropsy in the abdomen would often bring on fainting, but for the precaution of tightening a broad bandage upon the body as the water flows. The reason is, that the sudden removal of a large quantity of fluid which had been compressing all the abdominal vessels, and keeping them perhaps only half full of blood, allows them again suddenly to receive their natural quantity, and thus produces a relaxation of the other parts of the vascular system.

Sudden parturition often causes faintness for the same reasons.

Even rising up suddenly from a horizontal position will cause an approach to fainting in weak people, or in those who have been long bed-ridden: probably because the heart having for a time been accustomed to send blood only in a horizontal direction to the head, does not at an instant exert the additional power required to lift an upright column with equal force;—besides, that the blood does not then return to the heart by the veins, from the inferior parts of the body so readily as before.

These various facts, now easily understood, form the reason of a rule which is a great modern improvement in the practice of the healing art, *viz.*, in bleeding for the cure of inflammation, to take the blood away as *quickly* as possible. This subject deserves a little farther consideration.

A great proportion of dangerous diseases involve inflammation of some vital organ; an inflammation consists chiefly, as already stated at page 422, of a gorging or over-distension of the capillary vessels in the part. The nature of the capillaries, again, is such (page 422) that when not maintained constantly full by the pressure of the heart behind them, they gradually by their own action, empty themselves towards the veins—as is seen in the disappearance of a local inflammation soon after the death of the person, or in the fact of the arteries being emptied of blood after breathing ceases, &c. Now ever since medicine deserved the name of an art, practitioners have accounted the lancet their sheet-anchor in inflammatory disease; but it is only in late times, since the circulation of the blood was understood, that they have known the rationale of the remedy, *viz.*, that it acts by diminishing vascular tension, and hence the action of the heart, and so allowing the small vessels to empty themselves by their own force, and to recover sufficiently to resist the return of an excessive load. It is still more lately that they have understood how much more suddenly and completely the disease is cured by abstraction of a small quantity of blood *so rapidly* as to produce fainting, than of a much larger quantity *so slowly* that only weakness follows. Judicious treatment now cures inflammation much more certainly and completely than was done formerly, yet with much smaller loss of the precious blood, and with less danger of those diseases of weakness, or of that complete breaking up of the constitution, which often follow great depletion. To induce faintness, *large* openings are to be made into the veins—sometimes into two veins at once, and the patient is kept in the upright attitude, Often thus an inflamed eye, which was as red as scarlet before bleeding, in a few minutes is rendered nearly of the natural appearance; and intense internal inflammations, as of the brain, lungs, bowels, &c., which, if neglected, would be shortly fatal, are removed in the same manner. In all these cases, the faintness seems to be almost equally efficacious, whether it happens after the loss of ten ounces of blood, or of fifty; or even, as sometimes occurs, when it happens without bleeding at all, after merely tying the arm in preparation.

Reflection upon these circumstances led the author to think that, in certain cases, the beneficial effects of blood-letting might be attainable by the simple means of *extensive dry cupping*, alluded to at page 176; that is to say, by diminishing the atmospherical pressure on a considerable part of the body, on the principle of the cupping-glass used very gently, and thus suddenly removing for a time from about the heart, a quantity of blood, sufficient by its absence to produce faintness. The results of trial have been such as to give great interest to the inquiry, and the author's leisure will be devoted to the prosecution of it.—An air-tight case of copper or tin-plate, or of air-tight cloth kept extended by hoops, being put upon a limb, and made close by a suitable collar tied at the same time round its mouth and the limb,—on part of the air being then extracted by a suitable syringe, in an instant the vessels all over the limb become gently distended with blood; and as the blood is suddenly taken from the centre of the body, faintness is produced, just as by bleeding from a vein. The excess of blood may be retained in the limb as long as desired; for the circulation is not impeded. To produce a powerful effect with a slight diminution of pressure, more than one limb must be operated upon at the same time.

An instrument resembling the contrivance now described, was proposed about twenty years ago by a non-professional person, as a means of drawing all sorts of diseases out of the body through the pores of the skin. He enclosed a leg in an air-tight case; he then admitted steam to heat the limb, and relax the pores of the skin, as he said, and then he worked an air-pump to draw out the disease. He called the engine the *air-pump vapour bath.* In various cases where its true action was desirable, although not understood by the proposer, nor judiciously managed, it proved beneficial.

The operation of applying tourniquets or bandages round the limbs, so as to influence the transmission of the blood, affects the action of the heart. It is said sometimes to have prevented the accession of the ague. It is a means akin to those above described.

Because arteries are stronger and more tense than veins, a bandage may be put round a limb tight enough to close the veins but not the arteries, and the limb will then swell beyond the ligature. By thus putting tight elastic bandages round all the limbs at once, and immersing them in warm water to favour the dilatation of their vessels, so much blood may be suddenly detained in them as to cause the person to faint. Such a means, therefore, might also be used remedially.

In the same way, a tight handkerchief, or stock round the neck, will often retain the venous blood in the head, and cause apoplexy.—Strong pressure made on a jugular vein kills as certainly as if made on the windpipe.

When a *hernia*, or other tumor, is strangulated, it swells in the manner above described, and if not relieved, soon mortifies.

Diffused pressure, like that made by rolling a bandage round a whole limb, or by immersing the limb in fluid, must affect the circulation. The veins will be more compressed than the arteries, by reason of the distending force in them being less. Varicose veins, therefore, are usefully supported by a bandage or laced stocking. The reason why this manner of supporting assists so powerfully in the healing of ulcers on the legs, may be, that the support affects the capillaries and absorbents, as well as the larger vessels.

Poultices, by their weight, produce a soft compression of the parts on which they are applied; and in certain cases, may benefit by mechanically squeezing the excess of blood out of the weakened vessels.

The author has relieved the chronic inflammation of a sprained ankle, by ordering the foot and leg, covered with an oiled-silk stocking, to be enclosed in a boot strong enough to support the pressure of quicksilver, which was then poured into the boot. The effect is a pressure by the fluid metal on the weak vessels, of one pound to the square inch, for every two inches of the depth of metal above the part.—A height of four or five inches gives the relief expected. A much greater elevation would stop the circulation altogether. No bandage can press with uniformity approaching to this action of a fluid.

The effect of continued pressure, in removing tumours of various kinds, is explicable on the same principle. The author doubts not that in such cases, pressure properly managed, would prove a more valuable remedy than is at present generally supposed. The elastic steel half-hoop with one cushion before and another behind, lately introduced for the relief of hernia, affords an admirable mode, in certain cases, of producing a uniform pressure of the nature spoken of, and of any desired force.

When a man stands in a bath, with the water up to his chin, there is a pressure of the water upon his body, proportioned everywhere to the depth, (see page 130.) This pressure must produce a considerable effect on the blood-vessels of the lower parts of the body. We see in this that a bath must propel the blood from all other parts of the body, towards the cavity of the chest, which the pressure cannot reach. It is this effect which in part causes the feeling of thoracic oppression experienced by persons on first plunging into water, which feeling is usually attributed altogether to the cold.

The old practice of placing a patient in a pit, and surrounding his body with earth and sand, must have had a mechanical action of the kind now contemplated, in addition to any other influence.

Transfusion of blood from a vein of a healthy person into that of one fainting or dying from hæmorrhage, is an operation the converse of some of those mentioned above. It has been frequently performed with success. The cases to which it seems best fitted, are those of flooding after parturition, and of wound; and there can be no doubt that many of the lives lost from these so frequently recurring causes, might be saved by its adoption. The blood to be injected is received into a vessel, as in common bleeding, from which vessel, by a fit syringe, (to be described in a future page,) it is transferred, as it flows, into an opened vein of the patient. The admission of air with the blood would be fatal, and has therefore to be most carefully guarded against. The last interesting report upon this subject is that of Dr. Blundell, in his Physiological Essays.

RESPIRATION AND VOICE,

The doctrines of fluidity, illustrating and illustrated by the animal respiration and voice.

As the motion of a windmill depends altogether on the breeze to which its vanes are exposed, so does the motion and the life of that most wonderful of structures, the animal body, depend on the supply of air for its breathing. If this supply be withheld but for a few moments, painful convulsions ensue; and if for a still longer period, the body, however perfect and beautiful, is made a lifeless corpse, soon to putrify and be decomposed.

The mechanical nature of air, as to its lightness, elasticity, &c., and the fact of its forming an ocean around the earth of about fifty miles high, are now well understood, and have been fully explained under *pneumatics;* but the precise nature of its life-sustaining action has yet to be elucidated by farther research of chemists and physiologists. Thus far, however, we know—that the ingredient called *oxygen*, constituting a fifth of the atmosphere, is the most essential part—that air, by being breathed once is rendered unfit for farther respiration at the time—and that a man requires about a gallon per minute. The enterprising Mr. Spalding, who introduced the use of the diving-bell, descended for the last time with a companion on the coast of Ireland, when owing to the signal cord becoming entangled round the great rope of the bell, which had turned in descending, he could not make their want of air known above, and both were found dead when the bell was drawn up soon after, although the water had not touched them. Of a hundred and forty-six Englishmen, who, in the year, 1750, were made prisoners at Calcutta, and where thrown into the close dungeon, since called the *black-hole*, only twenty-three survived the few hours of their confinement, and one of the most appalling recitals of human suffering existing on record, is what these persons had afterwards to make.

We know generally of the life-supporting action of air, that it consists in some change operated by the air on the blood; and we know that the function of respiration has merely to bring air and blood together in the cavity of the chest, that this change may take place. The blood while in the chest is moving along a part of its circle, in vessels of extreme minuteness and thinness, and the air at each inspiration rushes in among these, so that every globule of blood passes within its influence. And the blood, which, after having served the purposes of the body, arrives at this part of its course, black and impure, immediately after its exposure to the air enters the left chamber of the heart, of a beautiful scarlet colour, and thence departs to carry new life to the general system.

The minute vessels through which the circulating blood is strained in the chest, do not hang loose in the cavity, but are supported by running through spungy masses, called the lungs, which consist chiefly of these vessels and of thin membrane formed into cells. The cells at every inspiration, receive fresh air through the cartilaginous windpipe which branches into them, and at every inspiration they return the changed air by the same channels to the atmosphere. The lungs of a child, before birth, are perfectly collapsed, or without the least air in their structure, and hence are dense enough to sink in water; but after breathing, they retain a portion of air, and will float. This fact has been accounted a test of whether a child had been born dead or alive; but because putrefaction, &c., will cause air to be in lungs which have never breathed, the criterion may be fallacious.

The chest is a large cavity, of form approaching to that of a common bee-hive, bounded laterally by the encircling ribs, behind the spine, and before by the sternum, and divided below, from the abdomen or belly, by a strong membranous and muscular expansion, called the diaphragm. The ribs, in the natural state, hang obliquely downwards from their posterior attachments, and on being raised in front, they widen or increase the size of the cavity, as already explained at p. 403. The cavity is farther enlarged by the descent of the diaphragm, which may be regarded as both the floor of the chest and the roof of the abdomen, and which being convex upwards like a dome, by contracting itself to a more flat condition, sinks out of, and enlarges the chest, while it descends into, and diminishes the abdomen, or at least causes protrusion of its sides.

Now on the chest being enlarged by the rising of the ribs and descent of the diaphragm, or by either singly, the air rushes into it through the mouth and windpipe, exactly as air rushes into a common bellows through its pipe, when the valve is shut and the two boards are drawn apart; and air is again expelled from the lungs by the contraction of the chest, as from the bellows by the approximation of the boards. Into both cavities air enter, because with the enlarging dimensions, the air which was within dilates, and becomes less powerfully tense or resisting against the external pressure of the atmosphere, and so allows more air to rush in to restore the equilibrium. The air is expelled again by the contraction of the cavities, because by being compressed, its elastic force or tension becomes greater than that of the external air, which it therefore easily repels, and so in part escapes. By immersing in water an india rubber bottle, and then opening and shutting it, the entrance and exit of fluid in this manner may be rendered very apparent.

That the air admitted into the chest should have the fullest action on the blood passing there, it was necessary that the spongy mass of lungs in which the blood-vessels ramify, should occupy the whole of the cavity, and be equally distributed. Now while the equable distribution is effected by the uniform elasticity or resilience which belongs to the structure of the lung the complete filling of the cavity is obtained, not by general attachment between the lungs and the ribs or sides of the chest, as might be expected, but by the following means, equally simple, and still more perfect. The spongy mass of the lungs, is completely covered by a strong adherent membrane, called the pleura, through which air cannot pass; between this membrane and a similar lining of the chest there is no air or empty space, and therefore in the raising and falling of the ribs during respiration, this membrane remains always in contact with the lining of the ribs, just as a bladder put into a bellows as a lining, with its mouth secured around the nozzel, is filled and emptied, and remains in contact with the interior of the bellows, in all the states of dilatation, as if there were attachments in a thousand places. This construction allows the lungs to have a singular freedom of play during all the motions of the body; a freedom further provided for by their being divided into five portions or lobes, which slide upon one another: of these, three occupy the right side of the chest, and two with the heart occupy the left. The right and left sides of the chest are rendered cavities quite distinct from each other by the *mediastinum*, a strong membranous partition. The mechanical disposition of the contents of the chest, as now described, is productive of certain consequences which it is important to understand;—for instance,

If a wound be made in one side of the chest so as to admit air, the lungs of that side collapse in obedience to their weight and elasticity; and as the chest afterwards enlarges and diminishes in respiration, air more easily enters and leaves the space around the collapsed lung, through the wound, than it can enter or leave the lung itself through the windpipe; because, in the first case, it has no force to overcome, and in the second, the elasticity, weight and inertia of the lung oppose Thus the lungs of the wounded side become collapsed and useless. If such a wound, therefore, were made in both sides of the chest at once, even without hurting any part within, the person, unless assisted, would die of suffocation. The kind of resistance required in such a case, is first to press the ribs down so as to empty the chest of air as much as possible, and then to keep the wounds close or covered while the ribs rise again; the air, of course, will then enter by the natural road, the only one left, to fill the chest, and will distend the lung, and reach the blood in the pulmonary vessels as usual. Then by straining with the muscles or the chest, as in the action of blowing, and at the same time preventing the breath from escaping by the mouth or nose, all the air which had entered by any wound in the chest may be expelled. In Benjamin Bell's system of surgery, which was long the manual practitioners, counsel on this head was given the very contrary of that required, and, of course, any patient treated according to it must have been lost.

In cases of dangerous hæmorrhage from a lung, caused by a wound in the side, the proper practice is to allow the lung to collapse, as now explained, that the hæmorrhage may be checked: and when the danger is past, the treatment above described is to be adopted to restore the natural play of the lung. Life may be supported for a long time by the lung in one side of the chest.

In cases of hæmoptysis, or spontaneous bleeding from the lungs, a disease so often fatal, life might sometime be saved or prolonged by making an opening between two of the ribs, and allowing the lung to collapse. The affected lung is often pointed out by the circumstances; and the opening, when properly made, would be no more dangerous than in the case where, by a similar opening, water or pus is discharged from the chest.

The same operation has been tried as a forlorn hope in pulmonary consumption. This disease is often limited to the lung of one side, and as the alternate stretching and collapse of the deceased lung during respiration, together with the contact of the air, powerfully prevent an ulcer there from healing, or inflammation from subsiding, a new chance of recovery is given by allowing the deceased lung to collapse and remain at rest. Some cases are recorded where cure is said to have followed this operation, and certainly where the circumstances are favourable for it, and where death must ensue unless it can save, it is worth trying.

When ribs are fractured, it is the practice to put a bandage round the chest, so as for the time to prevent almost entirely the respiratory motion of the ribs, and the breathing is then performed chiefly by the rising and falling of the diaphragm or floor of the chest as above described. Although a person with broken ribs is wisely for a time subjected to the unnatural restraint, it is surely the height of folly to inflict the same on healthy beings, as is yet, however, so commonly done among young women, and often to the destruction of their health, by the fashion of bracing the body in tight stays.

The force of a healthy chest's action in blowing is equal, as stated in last section, to be about *one pound* on the inch of its surface; that is to say, the chest can condense its contained air with that force, and can, therefore, blow, through a tube, the mouth of which is two feet under the surface of the water. In the opposite action of sucking or drawing in air, the power, is nearly the same. In both actions it is possible to use the cavity of the mouth separately from that of the chest; and the mouth, being smaller, with stronger muscles about it in proportion to its size, it can act more strongly. Some men can suck with the mouth so as to make nearly a perfect vacuum, or to lift water nearly thirty feet. An expert operator with the blow-pipe can keep up an uninterrupted blast by shutting the mouth behind, while he inhales, and replenishing it as is required in the intervals.

In *coughing*, the *glottis* or top of the windpipe, by a curious sympathy of parts, is first closed for an instant, during which the chest is compressing and condensing its contained air, and on the glottis being then opened, a slight explosion as it were, of the compressed air takes place, and blows out any irritating matter that may be in the air-passages; just, only with inferior force, as the burst from the chamber of an air-gun discharges its bullet. This shutting of the glottis to allow the compression of the air, and the subsequent opening to allow the discharge, may recur at every minute intervals, and many times for one fill of the chest, as is instanced in hooping-cough. The action of coughing is often produced by irritation from a cause which cannot be removed by coughing, as inflammation of the chest, or tubercles; or even by irritation in a distant part, as when children are teething, or when the stomach is overloaded.

Sneezing is a phenomenon resembling cough, only the chest empties itself at one throe, and chiefly through the nose, instead of through the mouth, as in coughing. The irritation that produces sneezing is generally in the nose; but, as in the case of cough, sneezing may occur from distant sympathies: witness that from worms in the bowels.

Laughing consists of quickly repeated expulsions of air from the chest, the glottis being at the time in a condition to produce voice; but there is not between the gusts, as in coughing, complete closure of the glottis.

Crying differs from laughing almost solely in the circumstances of the intervals between the gusts of air being longer. Children laugh and cry in the same breath, and it is often difficult to mark the moment of change.

Hiccup is the sudden stopping, by a closure of the glottis, of a strong inspiration at its commencement. If the inspiratory effort be afterwards continued, it may cause air from the atmosphere, or half-digested food from the stomach, to enter the œsophagus.

In *straining* to lift weights, or to make any powerful effort, the air is shut up in the lungs, that there may be steadiness and firmness of the person. At such a time, by the compression and condensation of air around the heart and large blood-vessels, the blood is determined violently outwards from the chest, and often rises to the head, with force that produces giddiness, or even apoplexy,—and the eye will sometimes become suddenly bloodshot from a small vessel giving way; and leech-bites will break out afresh.—The force of this pressure outwards is measured, as already stated, by a column of about two feet of blood; and this is, therefore, the measure of the additional arterial and venous tension in the body generally.

Suffocation is the name given to what happens when the supply of air to the lungs is, any way, prevented. The blood, not then refreshed by the approach of the air, rises in the brain unfit for its purpose, and confusion of thought is immediately produced, soon followed by convulsion and death.

When this happens from mechanical obstruction at the narrow entrance of the windpipe, as in croup, by the tenacious films thrown off from the inflamed lining of the air-passages, life may be saved by making a new entrance for air through the windpipe, lower down in the neck, and keeping it free by a little tube inserted, until the obstruction above be removed. Where children die with croup, it is frequently not from the violence of the constitutional disease, but from detached films thus accidentally sticking in the narrow entrance of the air-passage.

In the cases of strangling and hanging, the tight binding of the rope or ligature crushes inwards the cartilaginous rings of the windpipe, and shuts the air-passage. It may also cause apoplexy, by arresting the passage of blood to and from the head: and there may be dislocation of the cervical vertebræ of the spine.

In *drowning*, communication with the atmosphere is cut off altogether by the supernatent water. If, during submersion, the chest expands, it can receive water only, instead of air. The nerves and muscles, however, at the entrance of the windpipe, are so irritable, as to be immediately excited by the contact of any unusual matter, and, for a considerable time, they keep the passage shut against the liquid seeking entrance. It is partly on this account that the body of a person, after submersion in water and apparent death, may, often, if recovered within a moderate time, be restored to life.

The apparatus of the Humane Society for the recovery of persons apparently drowned, includes a bellows for producing artificial respiration. This bellows resembles a common bellows, except that its flap or valve, instead of being internal, is external like a large flute-key, and has a spring to close it, obedient to the finger of the operator. The bellows receives its charge of fresh air on being expanded, while the valve is open; it sends the charge into the lungs on being compressed while the valve is shut; it withdraws the charge again on being expanded with the valve shut; and the impure air is thrown out to the atmosphere on its being compressed with the valve open. These changes, repeated and continued, produce the artificial respiration required. It is most important to remark here, that if air be injected into the lungs, either in too large quantity or very suddenly, instead of recalling or sustaining life, it is as certain a means of killing as a dagger driven through the heart. This truth has been but lately known, and ignorance of it has probably decided the fate of many persons, treated with a view to recovery after submersion. The operator should reflect that he is dilating the delicate air-cells of the lungs with the force of an hydraulic press; and that if he does so very suddenly, although to a small extent, he still may rupture many small blood-vessels, before they can empty themselves so as to yield. In a bellows for the purpose of artificial respiration, there should be the means of checking its opening to suit the capacity of the patient's chest, and there should be a cock in the pipe or nozzle to regulate the speed of the passing air.

In addition to the artificial breathing for the recovery of suspended animation, it is often necessary to restore natural warmth of the body, to rub the limbs in aid of the circulation, to administer stimulants by the mouth, to excite by galvanism, &c.

It seems to be an error, and probably often a fatal error, in the present mode of treating persons apparently drowned, to use cold instead of warm air for the artificial respiration. Thus while the important object of restoring the temperature of life is sought by all external means, the great inconsistency is committed of blowing cold air upon an internal surface of the body more extensive than the external; and until that reciprocal action of the air and blood begins, which constitutes the slow combustion of natural respiration, every bellows-full of cold air admitted, brings back with it a portion of the remaining central warmth, and may thus chill so as to make the recovery impossible:—as a fire which has fallen very low may be immediately extinguished by the same action of a bellows, which a little before would have made it blaze. Air might easily be heated for the purpose of respiration by pouring boiling water into a vessel containing it, and then connecting the bellows with that vessel by a fit pipe, or by making the bellows draw through a pipe partially immersed in hot water:—a quart of boiling water has heat enough in it to warm many gallons of air to blood-heat. This plan would not only avoid the mischiefs arising from the cold air, but by affording the means of applying warmth even higher than that of life, it might probably furnish the most useful of all stimulants to the parts about the heart. A healthy man can breathe with impunity air that is much hotter than boiling water.

Late physiological investigations have shown that the breathing, or mechanical action of the chest in respiration, is so dependent upon the influence of the brain, as to be disturbed and even stopped when the brain is embarrassed; they have shown, farther, that the action of the heart is dependent on the breathing, but not on the brain, except as the cause of the breathing—

for that respiration kept up artificially, will preserve the circulation and the life for a considerable time after the brain has altogether ceased to act, or even has been removed from the body. Now, some interesting experiments of Mr. Brodie have proved, that certain poisons are fatal, merely because they suspend for a time the action of the brain—through which suspension, the actions, first of the chest, and then of the heart, cease, and death ensues: but that in such cases, if the action of the chest be maintained artificially, the circulation and life of the body will be for a time continued, and the brain may gradually recover from the effect of the poison, so as to resume its office. Thus certain cases of poisoning, which formerly would have been fatal may now end in recovery.

An important application of this discovery is to the treatment of cases of convulsion, particularly those occurring from teething or other irritations in infancy. The respiration ceases in these cases often only because the action of the brain is suspended; and if the respiration be continued artificially, the circulation and life will also continue for a time, during which the brain may recover itself, either spontaneously, or in consequence of remedies employed, and life may be saved.—The chest of an infant is comparatively so small, that it may be filled from the mouth and windpipe of a grown person, with air which has not descended to that person's lungs, and therefore has not been rendered unfit by respiration; and on the little chest being afterwards compressed by the hand, the air will return. The air may be blown directly into the child's mouth through a thin handkerchief laid over the mouth, or it may pass through a tube inserted into the nostril or treachea:—to prevent it from passing into the stomach, the larynx must be pressed against the œsophagus during its entrance. Let all who try this remedy, keep present to their minds the danger of inflating too much.

Any medicated air is generally inhaled by a patient from an oiled-silk, or other air-tight bag, or from a light gasometer (see page 211.) When the compound nature of our atmosphere was first discovered, great advantages were anticipated to medicine from the use of pneumatic or aerial mixtures. These expectations have not been realized, but the subject still remains highly deserving of research.

THE VOICE AND SPEECH.

The chest and air-passages, with certain additional parts, constitute the organs of voice and speech.

An inquirer into the constitution of the universe around him, meets with few things calculated more to surprise him than that faculty in the human mind by which it can associate the ideas of objects with any arbitrary signs, so closely that the ideas are afterwards excited by the signs almost as vividly as by the objects themselves. The inhabitants of China, for instance, have contrived many thousand grotesque characters, and determined what object each should recall; when a Chinese by study becomes familiar with these, he may have his bodily eye poring over pages of crooked and unseemly scratches, while his mental eye through them sees only a pleasing succession of the most beautiful imagery of nature; and the characters may be rendered intelligible to the deaf and dumb man, as well as to him who speaks; and they serve as a media of thought and communication through many provinces and countries of which the spoken languages have no common resemblance.

But if the ready resemblance of *visible* signs be wonderful, which have a permanent existence, and which often may have some resemblance in form to the things signified, how much more wonderful is it that an *audible* sign, that is, a passing sound or fugitive breath, called by man a word, should serve as well; and that by a succession of mere sounds, having so little natural connection with the things signified, that they are totally different in different countries, and are changing with fashion from age to age, any train of thought may be made to pass through the minds of an audience, so as to excite and to leave impressions almost as vivid as from realities! Such, however, is the fact, and it is greatly owing to this and to a correspondent faculty of producing at will a sufficient number of distinguishable sounds, that man owes his elevation above the brutes of the field. His godlike powers of intellect would have remained dormant and unknown, had he wanted the faculty of comparing invisible thoughts with those of his fellow men, and of arranging and recording them by means of signs.—Written language is a double remove from the objects themselves, being *visible signs* not of things, but of the *audible signs.*

The admirable apparatus by which man is enabled to produce a sufficient variety of sounds to answer his purposes, passes generally under the denomination of *the organs of speech;* because the act of using sounds which have meanings assigned to them is called speech. It consists of the chest for containing air; of the larynx or cartilaginous box, with its narrow aperture called the glottis, at the top of the windpipe, for producing the voice, and varying its pitch; and of the short tube of the mouth, with the tongue and lips, for farther modifying the voice.

In the chapter on acoustics, we explained that sound is the name given to the effect produced upon the ear by certain tremblings conveyed to it generally through the medium of the air; and we explained how air, forced from the human lungs through the opening at the top of the windpipe, causes the elastic lips of that opening to vibrate, and to excite the tremblings. We have now to show that this sound, in passing forward from the top of the windpipe, may be modified at the will of the individual, in a great variety of ways —a variety, however, which is still very simple.

The modifications of voice easily made, and easily distinguishable by the ear, and therefore fit elements of language, are about fifty in number; but no single language contains more than about half of them. They are divisible into two very distinct and nearly equal classes, called, for reasons now to be explained, *vowels* and *consonants.*

Those of the first class are the simple voices issuing through the open mouth, and influenced only by the degrees in which the mouth is opened and elongated. They may be continued as long as there is breath to issue from the chest, and it is for this reason that they are named *vowels* or *calling sounds.* The Roman letters, A, E, I, O, U, as generally pronounced on the continent of Europe, and of which the sounds correspond nearly to those *aw*, *a*, *e*, *o*, and *oo*, of English writing, indicate the most easily distinguishable vowels. Sounds passing through the mouth in its most natural state of relaxation, is heard as the modifications expressed by the Italian E, (or the *a* of the English word *care ;*) if the mouth be then widened, the sound becomes A (as in the English word *bar ;*) if the mouth be narrowed, we hear the I (or the *e* of the English *tedious ;*) if the mouth be elongated and at the same time widened, we hear the O (as in the English word *bore ;*) and if more elongated but narrowed, we hear the U (as in the English *rude.*) The possible number of vowels, however, is as great as the possible degrees in which the dimensions of the mouth may be altered. About twenty of them are sufficiently distinguishable, but few languages comprehend more than twelve. Modern art can produce the vowel sounds mechanically by means of tubes of certain dimensions.

The alphabets of Europe are very faulty in not all using the same characters for the same sounds, and in not having, according to the true intent of an alphabet, a character for each distinct sound. In English, one letter is used for several sounds, as A, in *water*, *far*, *fat*, *fate*, where it indicates four sounds perfectly distinct. In repeating the English alphabet, the A is pronounced as the broad E of the Italians and of continental Europe, and the E as the I; and the I (in *tide*, for instance,) as the dipthong AI of more correct alphabets; and the U (in *muse*,) as the dipthong IU. In consequence of the changes which the English have made in the meaning of the Roman letters, they now experience increased difficulty in learning modern continental languages; and their own pronunciation of the ancient languages, to all but themselves, is ridiculous, and almost unintelligible. The same cause renders the pronunciation of English difficult to foreigners, and thus restricts much in other countries, the cultivation of English literature.

To explain the second class of the modifications of sound, called *consonants*, we may remark, that while any continued or vowel sound is passing through the mouth, if it be interrupted, whether by a complete closure of the mouth, or only by an approximation of parts, the effect on the ear of a listener is very remarkable, and is so exceedingly different, according to the *situation* in the mouth where the interruption occurs, and to the *manner* in which it occurs, that many most distinct modifications thence arise. Thus any continued sound, as A, if arrested by a closure of the mouth at the external confine or lips, is heard to terminate with the modification which we choose to express by the letter P, that is, the syllable AP has been pronounced; but if, under similar circumstances, the closure be made towards the back of the mouth by the tongue rising against the palate, we hear the modification expressed by the letter K, and the syllable AK has been pronounced; and if the closure be made in the middle of the mouth by the tip of the tongue rising against the roof, the sound expressed by T, is produced and the syllable AT is heard—and so of others. It is to be remarked, also, that the ear is equally sensible of the peculiarities whether the closure precedes the continued sound or follows it: that is to say, whether the syllables pronounced are as above, AP, AT, AK, or on the contrary, PA, TA, KA.—The modifications of which we are now speaking, appear, then, not so much to be sounds, as distinguishable manners of beginning and ending sounds; and it is because they are thus only perceivable in connection with vocal sounds that they are called *consonants*.

Now in the mouth, considered as a vocal tube, there are three situations, in which interruption of voice or breath may most conveniently be made, and there are six modes of making it at each; so that eighteen distinct interruptive modifications or consonants hence arise. These we shall now describe.

The three great *oral positions*, as they may be called, are,

1st. At the external confine of the mouth, or lips, giving the *labial* articulations, of which is an example.

2d. In the middle of the mouth, where the tip of the tongue approaches the palate behind the teeth, producing the *palatal* articulations, of which T is an example.

3d. Near the back of the mouth, where the body of the mouth approaches the palate, giving th *guttural* articulations, of which K is an example.

And the *six modes* in which the voice or breath may be affected in passing through each of the three positions of the mouth, are,

1st. A sudden and complete stoppage, producing what may be called a *mute* articulation : *viz.*, P, in the labial position; T, in the palatal; and K, in the guttural. (See here the general table of articulations next page which table may be considered as representing the tube of the mouth, with the letters so placed as to show in what situations in the mouth the sounds represented by them are severally produced.) A mute may also be made by stopping the breath exactly at the teeth, *viz.*, a *dental-mute;* but it is hardly distinguishable from the *palatal mute*, produced just behind it, and being less perfect is not used.—Some awkward speakers, substituting it for the proper mute, are said to *speak thick.*

2d. A sudden shutting, as in the last case, but the voice being allowed to continue until the part of the mouth behind the closure be distended with air.—This produces the *semi-mutes*, B, D, and G, (as heard in the syllables AB, AD, AG,) for the three positions. There might be a dental *half-mute*, but it is not more used than the *dental-mute*, and for the same reasons. If the sides of the tongue be depressed, after it has taken the position required for T or D, the sound L is produced; and the letter is, in the table, placed below D, although the sound, from being continuable, is not in any sense a *mute.*

3d. The positions closed as for the mutes, while sound is allowed to pass by the nose.—Thus arise the *semi-vowels* or *nasals*, M, N, NG, for the three positions,—NG (as in *king*) is a simple sound, although our imperfect alphabet has no single letter for it. The nasal sound of the French language, which gives it so great a peculiarity, approximates to the English NG, but differs from it in sound being allowed to pass by the mouth, as well as by the nose. It is pointed at by the small *n* in the table, and like the other sounds which do not occur in the English language, is here pointed in the Italic character.

4th. Breath only (or whisper) allowed to pass at the three oral positions, nearly closed.—Hence come the sounds which we call *aspirates*, *viz.*, F, for the labial position, TH and S, for the palatal, and CH (heard in the Scottish word *loch*,) for the guttural; the TH and CH, are simple sounds, although each expressed in Britain by two letters. The TH is heard in the word *bath*, and is the sound expressed by the single letter θ of the Greeks. The CH is heard in the German *ich*, and is the χ of the Greeks. The *soft aspirate* TH is more easily made by pressing the tongue gently against the teeth, and allowing the breath to pass all round, than by the true palatal approximation of parts, and the *soft dental aspirate*, therefore, is used in preference to the *palatal.* The letter S is the *hard palatal aspirate*, and differs from the *soft palatal aspirate* TH, in the breath being made to issue with greater force, and only by a narrow space over the centre of a rigid tongue, instead of on all sides of a soft tongue, as for TH. French people on first attempting to pronounce TH, substitute for it the D, or the S, or the Z (which is nearly related to S, as explained below.) The author has found it easy to enable them to pronounce the TH at once, and perfectly, by explaining its nature as above. If we depress the sides of the tongue while pronouncing S, we produce the simple sound expressed by the English double letter S H, just as by depressing the sides, of the tongue while making D, we produce L.

5th. Using *voice* in the same manner as *breath* or *whisper* is used for the aspirates.—This produces the sound called *vocal aspirates, viz.*, V, TH, Z, J, and GH. TH *vocal aspirate,* is heard in *bathe,* as contrasted with the *simple aspirate* in *bath;* Z comes from the S position, only with *sound* instead of *breath;* SH pronounced with *voice,* becomes the J of the French in the word *je,* or the sound heard in the middle of the English word *vision.* GH is a simple sound in German, but not in English.

Table of Articulations.

Mute.	Semimute.	Semivowel or nasal.	Aspirate.	Vocal aspirate.	Vibratory.	
P	B	M	F	V	*pr*	LABIAL.
.	.	.	th	th	.	Dental.
T	D	N	S	Z	R	PALATAL.
.	L	.	sh	J	.	with the edges of the tongue depressed.
K	G	ng	*ch*	*gh*	*ghr*	GUTTURAL.
		n	H			

6th. Shaking the approaching parts in the three positions.—We thus make *vibratory sounds,* of which the middle position gives the common R, the only one of them used in England. Some bad speakers of English, however, make the *labial vibratory* by shaking the P in such words as *property;* and many use the *guttural,* which is the *burr* of Northumberland, and the common affectation in Parisian speech, termed *parler gras,* or *grasseyer.*

Additional Remarks.

The sound of H is an *aspirate* produced even behind the situation of the *guttural aspirate ch;* it is, indeed, merely a forcible passing of the breath through the very back part of the mouth or throat.

CH, in such words as *chain*, means T before *sh*.

J, as heard in the English name *John*, is a compound sound, *viz.*, D before the simple J of the table, which is the S of *vision*.

LL. The liquid or double LL of the French as heard in the word *paille*, is merely L with the letter Y begun to be pronounced after it. It is heard in the English words *billiard* and *halyard*, and would be their terminating liquid were the syllable *ard* not pronounced. The double LL of the Welch, as in the name of *Lloyd*, has the first L pronounced as an *aspirate*, that is, as a whisper, and the second in the ordinary way.

GN. The soft G N of the Italians and French, is the English N with Y begun to be pronounced after it. It is heard in our word *tanyard;* and in the Italian words *pegnio bagnio;* and in the French word *craignent*.

C, in English, stands always either for S or K, as in the words *certain* and *car*, and has no sound proper to itself.

Q, in English, expresses the sound of the letter K, with U following it: and yet, uselessly, U is always written after Q.

X, in English, means either KS, as in the word *axle*, or GZ as in the word *example*.

The consonants are best heard by sounding them with voice before them: that is to say, by making them rather terminate a syllable than begin it: pronouncing B, D, G, thus *eb, ed, eg*, rather than their common alphabetical names *be, de, ge*.

The labial sounds may be made either by the two lips, or by one lip and the opposite teeth. F may be pronounced, for instance, by the lips only, or by the lips and teeth: and some persons awkwardly make it by the under teeth and upper lip.

The letters Y and I, in most modern languages, stand for nearly the same sound. In English, for instance, *bullion* and *minion* might be written *bullyon* and *minyon*, without suggesting a change of pronunciation. In the words, *yard, you, yes*, &c., the Y is a short I, very closely joined to the following sound.—W is also thus a short U, as perceived in the words *war, we*, &c.

The author believes the analysis of articulation to be the best basis for a system of short-hand written characters. He has tried such a system, and found it exceedingly convenient.

Lisping is chiefly the habitual substitution of the aspirate TH for the S and SH.

Whispering is articulation without voice; that is to say, articulation while breath only is passing.

Stuttering, stammering or *hesitation of speech*, are terms implying an interrupted articulation, accompanied generally with more or less of straining and distortion of feature. It is remarkable with respect to this defect, that when the present work was first published, scientific or regular medicine had taught as yet no certain cure for it, although the frequent success of non-professional, and often ignorant individuals, by a mode of treatment which they solemnly bound their patients not to divulge, proved the cure, in certain cases, to be both possible and not difficult.—The author's attention had been drawn to the subject some years before, by an interesting case submitted to him of stuttering connected with other disease; and it was in analysing the subject with a view to the treatment of that case, that he framed the analysis of articulation contained in the preceding pages, and drew up a part of the additional observations which are now to follow. A cure was obtained; but as the case possessed a favourable peculiarity in the powerful mind of the individual, to which the author attributed great importance, as he had little leisure from his ordinary professional duties, to pursue the subject, or to ascertain in what respects his plan might differ from that employed by the most successful of the practitioners who concealed their proceedings, he gave his remarks in former editions of this work, merely as continued elucidation of the subject of speech. He is now, however, enabled to state, that his analysis has completely detected the nature of the morbid affection, and that it directs simple and effectual means of relief. He declined meddling with many cases offered to him after the original publication of his work, from the impression that the cure in the instance mentioned above, was owing at least as much to the ingenuity and perseverance of the patient, as to his suggestions, and, therefore, that his professional superintendence of the discipline required for ordinary cases would demand care and attention which he could not spare; but subsequent experience has proved to him that the business is altogether very simple and easy, and as regards children, may be managed by any intelligent instructor of youth who chooses to devote attention to it, while grown individuals will often be able to relieve themselves by the study of the present section; and he hopes that in very few cases will the counsel of a person familiar with the anatomy and actions of the organs be found to fail.

Command over the organs of speech is acquired in the same way as over all the other muscular organs of the body; those, for instance, used in walking, skating, fencing, performing on musical instruments, &c.: that is to say, at first, a distinct act of volition is required for every individual movement; but the law of association or habit of rendering the actions easier with each successive repetition, they are at last formed into connected tribes or trains, which appear as obedient to a single wish as the separate elements originally were. A child at first exerts as distinct and powerful a volition to pronounce the syllable *pa*, as after some practice to double the syllable and make it *papa;* or after still more practice, to pronounce the longest and hardest word of the language:—nay, at last, where there is strong and healthy power of association, complete sentences, and even rounded periods of eloquence, are poured out like single words, the mind of the speaker seeming at liberty, after each sentence or period is begun, to meditate and prepare that which is next to follow. As the faculties of locomotion and of speech are acquired in infancy and early childhood, persons no more recollect how they gradually acquired them than how their limbs grew; but the progress, described above, may be watched by any individual of mature years in his own person, while he is learning such an art as that of playing on a musical instrument. He will find, that at first, every finger which is moved to produce a note, obeys a distinct thought and volition; that soon short trains of connected notes become obedient to the will almost like a single note; that then by degrees, longer and longer trains or passages become familiar, until at last the instrument is obedient to the practised player, as voice is to the singer or speech to the orator.

There is a great original diversity among individuals as to their powers of muscular association, and therefore, also, as to their aptitude for acquiriug the various faculties of which we have been speaking. Thus some children walk well before a year, others require a much longer time, and some never succeed perfectly until they have had lessons from the dancing-master or drill serjeant. So, again, many people, by ear and imitation alone, learn easily to play on musical instruments; but others must begin by studying the written notes, and the precise *fingering* by which each note is produced on the instrument; and many, unless the notes be constantly before them cannot play at all. So, again, all persons may be said to learn to speak at first by ear and imitation; but many grow up to a certain age with defects, which judicious lessons from parents or other tutors are required to remove; and there are some as stutterers, who, owing to a naturally weak or irregular association, or to some accident in early life, which has strongly affected their nervous system, retain defects which no ordinary teaching can correct. It appears, then, that an analysis and scale of articulate sounds, with minute description of the organic actions required to produce them, like the scale which we possess for music, in the *gamut* and rules for fingering, should give nearly the same assistance to the speaker which the gamut gives to the player. The table and analysis contained in the preceding pages is intended to supply this information. It is constructed from minute consideration of the organs of speech while in action. It agrees in many respects with the common grammatical divisions of elementary sounds, but in others it pursues the analysis in a different way, and considerably farther. A person who understands it well, will have, while he speaks, an intelligent perception of what he is doing, in addition to the parrot-like faculty of habit, or of repeating by rote, and will thus command any desired sound by two powers instead of one. And, as a musician, when his musical memory fails him, finds help by thinking of his written notes and their relation to his instrument, so may a stutterer, when hesitating at any sound, receive benefit by thinking of the letter which represents it, and of the position of the organs required for that letter. Then, by frequent practice in making the particular combinations of sound which are difficult to him, he may strengthen the useful habit, and ultimately overcome his defect.

The most common case of stuttering, however, is not, as has been almost universally believed, where the individual has a difficulty in respect to some particular letter or articulation, by the disobedience of the parts of the mouth which should form it to the will or power of association, but where the spasmodic interruption occurs altogether behind or beyond the mouth, *viz.*, in the glottis, so as to affect all the articulations equally. To a person ignorant of anatomy, and therefore knowing not what or where the glottis is, it may be sufficient explanation to say, that it is the slit or narrow opening at the top of the windpipe, by which the air passes, to and from the lungs, being situated just behind the root of the tongue. It is that which is felt to close suddenly in hiccup, arresting the ingress of air, and that which closes, to prevent the egress of air from the chest of a person lifting a heavy weight or making any straining exertion; it is that, also, by the repeated shutting of which a person divides the sound in pronouncing several times, in distinct and rapid succession, any vowel, as o, o, o, o. Now the glottis, during common speech, needs never to be closed, and an ordinary stutterer is instantly cured, if, by having his attention properly directed to it, he can keep it open. Had the edges or thin lips of the glottis been visible, like the external lips of the mouth, the nature of stuttering would not so long have remained a mystery, and the effort necessary to the cure would have been suggested to the most careless observer: but because they were hidden, and professional men had not detected in how far they were concerned, and the patient himself had only a vague feeling of some difficulty, which, after straining, grimace, gesticulation, and sometimes almost general convulsions of the body, gave way, the uncertainty with respect to the subject has remained. Even many persons who, by attention and much labor, had overcome the defect in themselves, as Demosthenes did, have not been able to describe to others the nature of their efforts, so as to ensure imitation; and evidently the quacks who have succeeded in relieving many cases, but in many also have failed, or have given only temporary relief, have not really understood what precise end in the action of the organs their imperfect directions were accomplishing.

Now, a stutterer, understanding of anatomy only what is stated above, will comprehend what he is to aim at, by being farther told, that when any continued sound is issuing from his mouth, as when he is humming a single note or tune the glottis is necessarily open, and therefore, that when he chooses to begin pronouncing or droning what we have already described to be the simplest of vocal sounds, namely the vowel *e*, and in its less distinct modification, as heard in the English word *certain* or in the French word *que* (to do what at once no stutterer has difficulty,) he thereby opens the glottis, and renders the pronunciation of any other sound easy:—or if, when speaking or reading, he joins his words together, nearly as a person joins them in singing, (and this may be done without its being at all noted as a peculiarity of speech, for many persons do it in their ordinary conversation,) the voice never stops, the glottis never closes, and there is of course no stutter. The author has given merely this explanation or lesson, with examples, to persons who before would have required half an hour to read a page, but who immediately afterwards read it quite smoothly; and who then, on transferring the lesson to the speech, by continued practice and attention, obtained the same facility with respect to it. There are many persons not accounted peculiar in their speech, who in seeking words to express themselves, or while coming to a decision, often rest between their words on the simple sound of *e* mentioned above, saying, for instance, hesitatingly, "e. . . . . . I e . . . . . . think e . . . . . . I shall," the sound never ceasing until the end of a sentence, however long it may be delayed. Now a stutterer, who, to open his glottis at the begining of a phrase, or in the middle after any interruption, uses such a sound, would not even at first be more remarkable than a drawling speaker, and he would only require to drawl for a little while, until practice facilitated his command of the other sounds. Although producing the simple sound mentioned is a means of opening the glottis, which by stutterers is found very generally to answer, there are cases in which such other means may be more suitable, as the intelligent preceptor will soon discover.—Were it possible to divide the nerves of the muscles which close the glottis, without at the same time destroying the faculty of producing voice, such an operation would be an immediate and certain cure of stuttering.

While the spasmodic closure of the glottis, as above described, is the common cause of stuttering, there are also cases in which the cause is a spasmodic prolongation of some of the aspirates or semivowel sounds, as of *s m l*, &c., Fortunately, however, the substitution of the simple sound is equally the cure for all.

While the cure of many stutterers has been accomplished by their own efforts, after the study of what is written in this section, for others, and particularly for young people, the following have been found to be farther useful rules or forms of direction; and a commentary upon them making them perfectly intelligible, would seem to comprehend all that can be communicated upon the subject.—1. Familiarize yourself with the idea of a *continued sound* as of the roar of the sea or waterfall, or the note of an organ-pipe, and feel that your speech is to be as uninterrupted.—2d. Then never stutter more, but substitute always the simple continued sound for any threatened defect, and rest upon it until power be felt to overcome the difficulty.—3d. Never repeat words or syllables.—4. The simple sound must become the first syllable (*closely* joined) of every difficult word, until the morbid habit be weakened. The object of all these directions is to enable the patient, first to substitute universally the *drawl* for the *stutter*, and then, as soon as possible, to discard the *drawl* too.

The view given above of the nature of stuttering and its cure, explains the following facts, which to many persons have hitherto appeared extraordinary. —Stutterers often can sing well, and without the least interruption; for the tune being continued the glottis does not close.—Many stutterers also can read poetry well, or any declamatory composition, in which the uninterrupted tone is almost as remarkable as in singing.—A person who draws a deep breath before beginning to speak, as he cannot long retain the air, and the glottis must be open to let it escape, is to a degree secured against the occurrence of stuttering. The secret remedy of an American quack, who years ago got much money from Englishmen, was the direction thus to fill the chest before beginning to speak. A Dr. McCormac, also, who published a work on this subject, founded on the erroneous idea, that stuttering was an effort to speak while *inhaling* air, instead of while *exhaling*, gave the same direction.—The cause of stuttering being a weak and easily disturbed association of certain muscular actions, we have the reason why any degree of anxiety or dread as to speaking well, exceedingly increases the defect; and why many stutterers, who cannot make themselves intelligible in society, still, when alone can speak and read as perfectly as any other person. This explains also why many stutterers, who have gone to live for a time at the houses of pretended curers of their defect, have felt themselves singularly relieved from the moment of entering the house; because knowing that they were expected to speak ill, they had no fear of disagreeably attracting attention, and therefore had their powers much more at command. These persons, on returning to the world, have generally stuttered as badly as ever, but many of the asserted cures of stuttering, with certificates obtained from the parties at the time have been of the nature now described.—The cause of stuttering being so simple, as above described, one rule given and explained may, in certain cases, instantly cure the defect, however aggravated, as has been observed in not a few instances; and this explains also why an ignorant pretender may occasionally succeed in curing, by a given rule of which he knows not the reason, and which he cannot modify to the peculiarities of other cases.—The same view of the subject explains why the speech of a stutterer has been correctly compared to the escape of liquid from a bottle with a long, narrow neck, coming—" either by hurried gushes or not at all:" for when the glottis is once opened, and the stutterer feels that he has the power of utterance, he is glad to hurry out as many words as he can, before the interruption recurs.

The study of the table of articulations lead to the immediate correction of many minor defects in utterance, and is calculated to facilitate the acquirement of foreign languages. A lisping person, for instance, is cured at once, by being told that the tongue must not touch the teeth in pronouncing the letter S; and a Frenchman who deems it impossible for him to pronounce the English sound of TH, discovers that he cannot avoid doing so if he rests his tongue softly against his teeth, opened a little, and then forces breath or sound to pass between the tongue and teeth.

Several of the modern languages of Europe consist of nearly the same elementary or radical words, and differ among themselves chiefly by the prevalence in each of certain terminations and of one or other of the related and convertible sounds classified in the analysis given above. A student, therefore, who, by analytical investigation, or considerable practice, has become impressed with the peculiar genius of a language, may invent, or determine by analogy, even before minute study, the majority of those words belonging to each which have sprung from a common origin. This remark is so true with respect to the languages of Italy, Spain, Portugal, and even France, that to persons familiar with them, they are at last listened to rather as the same language spoken by different individuals, than as languages in themselves different.

Ventriloquism is the name commonly given to the art by which an individual can assume characters of voice and speech which are not natural to him, and thus, although alone, can imitate closely a conversation held between two or more persons.

The most remarkable diversity is obtained by speaking during inspiration, instead of, as usual, during expiration. The voice so produced is more feeble than the ordinary voice, and when accompanied by other circumstances favouring the illusion, it may suggest very completely the idea of a boy calling from the bottom of a pit, or from the interior of a chimney, &c. An unsuspecting peasant may be tricked into unloading his hay-wagon by an expert ventriloquist, who makes him believe that there is a poor child packed under the heap and ready to be smothered there.

A person, by a little practice, may acquire the power of producing, without the slightest apparent motion of the lips or countenance, all the articulations except the labial, and of them the F, V and M may be tolerably imitated by parts behind; hence by avoiding words in which P and B occur, such person may speak without visible movements of the organs, and if he assume the attitude of a listener, he may make the deception of ventriloquism complete. The idea which some authors have had (see *Good's Study of Medicine, &c.*) that the articulations of the ventriloquist are not produced by the tongue and mouth, as in common speech, is altogether an error. The art, carried to a certain degree, is not very difficult, as any person may ascertain who tries it after considering minutely the nature of common speech.

There are also striking varieties of voice producable by speaking with a more acute or grave pitch than usual, and with different degrees of contraction of the mouth; but these may be more properly called *imitations* than *ventriloquism*.

The variety of effect in sound which the human organs are capable of producing is truly surprising. There are adepts in the art of imitations, who not only mimic the speech of all ages and condition of the human race, but the songs of birds, the cries of animals, and even not a few of the sounds producing among inanimate things. Many of these performances become in the highest degree ludicrous, and furnish favourite amusements in our theatres. A Mr. Henderson, of London, about the end of the eighteenth century, used to *kill his calf*, as he called it, to crowded houses every night. After dropping a screen between him and the audience, he caused to issue from behind it all the sound, even to the minutest particular, which may be heard while a calf is falling a victim in the slaughter-house;—the conversation of the butchers, the struggling and bellowing and quick breathing of the frightened animal, the whetting of the knife, the plunge, the gush, the agony;—and, revolting as the occasion is in itself, the imitation was so true to nature, that thousands eagerly went to see the art of the mimic.

The following cases of inanimate sound may be closely imitated by the mouth: The working of a grindstone, including the noise of the water into which it dips, the rough attrition of the steel upon it, and the various changes occuring with the change of pressure;—the working of a saw cutting wood;

—the uncorking of a bottle, and the gurgling noise of decanting its contents; —the sound of air rushing into a room in a winter night by a crevice or key-hole—and many others.

It has already been explained, that voice depends on the vibration of the two edges or lips of the slit-like opening of the glottis, by which the air passes to and from the chest. The number of vibrations in a given time, or the pitch of voice depends, of course, on the length and tension of these edges. The length is varied by the position sof the arytænoid cartilages, and the tension by the action of small muscles which act on these; and the cavity of the mouth is enlarged or lessened to accord with the number of vibrations, by the rising or falling of the tongue and larynx which form its bottom. The peculiarities of individual voices must depend chiefly on the size and firmness of the cartilaginious box of the larynx, the strength of the muscles of the chest which force the air through the glottis, and the pliancy of the moving parts.

The glottis is smaller in women than in men, and hence their pitch of voice is higher:—with reference to music, the difference is generally of an octave, or eight notes.

The voice of a boy, in regard to pitch, is generally the same as that of a woman; but at the age of puberty, the sounding organs in the male enlarge suddenly, and render the voice stronger than before, and by nearly an octave graver. The voice of a eunuch is the voice of the boy continued, because the change called puberty does not take place in him.

Complete loss of voice, for longer or shorter periods, is often experienced by persons while in feeble states of health. The vibrating, and, therefore, sounding edges of the glottis, which are usually kept tense by the operation of certain muscles, on these ceasing to act, owing to the state of their nerves, will not vibrate as required, and the voice is lost. Slight colds suffice in many people to produce this effect: in others of morbidly sensitive or delicate nervous temperament, it follows fatigue, or any other cause of debility. Articulation is not destroyed by loss of voice; and whispering answers passably the end of vocal speech.

No intelligent mind can meditate on human speech and its influence in the world, without being roused to vivid admiration. But for speech, the most gifted individuals who have lived, had they existed at all, could have been little superior in their worldly state to the leading oxen of our herds, or to leading monkeys in the woods. As regarded the rest of mankind, Homer and Newton would have lived in vain. At the present day, among the natives of Australasia, where language may be said scarcely yet to exist, human nature is seen thus brutishly debased; while, on the other hand, in the history of the world, we may trace, as a consequence of more perfect speech, all the progress which has been made in arts and civilization. By language, fathers have communicated their gathered experience and reflections to their children, who in their turn become fathers, have transmitted them to succeeding children, with new accumulation; and when, in the course of ages, the precious store had increased, until mere memory could retain no more, the art of writing arose, making language visible and permanent, and enlarging without limit the receptacles of knowledge; and then the art of printing came, which now rolls the still swelling flood into every hamlet and every hut. Language thus, at the present moment of the world's existence, may be said to bind the whole human race of uncounted millions into one gigantic rational being, whose memory reaches to the beginnings of written record, and retains imperishably the important events that have occurred; whose judgment, analysing the treasures of memory, has already discovered many of the sublime and unchanging laws of nature, and has built on them the arts of life, and through them, piercing far into futurity, sees distinctly many events that are to come; and whose eyes, and ears, and observant mind are, at this moment, in every corner of the earth, watching and recording new phenomena, for the purpose of still better comprehending the magnificence and simplicity and beauty of creation.

THE DIGESTION.

The doctrines of fluidity, illustrating and illustrated by certain phenomena of digestion.

The animal body may be seen at first, in the maternal ovary, as a single speck of mucus; but from possessing life—wonderful life—the little nucleus, placed in new circumstances, begins to gather itself substance from around, and it increases in bulk. For a certain time it remains attached to the body of its parent, and draws the material of its increase from its parent's blood; but after that time it is alone and entirely dependent on its own resources. Then we see brought into play that extraordinary apparatus now to be described under the name of the *digestive* or *assimilating organs;* which, under the direction of a nervous energy, can, out of almost any kinds of dead animal or vegetable matter build up the beautiful living body to perfect maturity of size, and form, and faculty. And it is not only while their bodies are growing that animals require to take in and assimilate new matter, but also after maturity, in order to repair the waste of constant action. Supply of fuel and water to the steam-engine is not more necessary than of aliments to the living body.

Some of the less perfect animals take in sustenance almost like vegetables, by absorbent tubes that open on their surface; but by far the greater part receive it first into an interior cavity, where it undergoes certain preparation, and is then offered to internal absorbents, which drink up what is required and carry it into the circulating blood. This internal cavity is called *a stomach.* Its form and appendages differ exceedingly in different animals, according to the nature of the substances which serve for their sustenance, and to various other circumstances.

In man, the process of digestion has the following steps. The food is first received by the *mouth.* It is there broken or torn into small portions by the cutting and grinding wedges, called *teeth,* with which the *jaws* are armed; at the same time a fluid called *saliva* is mixed with it, poured out from glands around, so as to reduce it into a pulpy mass: this mass is then pushed backwards by the *tongue* to enter the long tube called the *gullet* or *œsophagus,* which by successive contraction of circular fibres, propels it down to the pouch of the *stomach,* placed under the edge of the left ribs. From the internal surface of the stomach a liquor oozes, called the *gastric juice,* the most general solvent in nature, and which, attacking the received food, soon reduces it, of whatever kind, to the state of a pultaceous mass, named *chyme;* in this state it enters the narrow *intestinal* canal which is continued from the stomach, where it almost immediately receives a mixture of bile and pancreatic juice poured out from the liver and pancreas. After this mixture, as it gradually passes on, a chemical decomposition and separation of parts takes place, and the pure nutriment of the body assumes the state of a milky fluid floating among refuse. This milky fluid, called *chyle*, is taken up all along the canal by the numberless absorbent mouths of the vessels called *lacteals*, and is then carried to the *thoracic duct*, and by it into the blood to supply the waste. The intestinal canal is about six times as long as the body, affording, therefore, a very extensive surface from which absorbtion may take place. That remnant of the chyme which the absorbents refuse, mixed with various depositions or secretions, continues its journey onwards, and is periodically discharged.

Much of the process which we have now described is *mechanical*, as will appear immediately; other parts of it are *chemical*, such as the solution of the food by the gastric juice, the separation of the milky chyle, &c.; and parts are *vital*, such as the afflux, just when wanted, of saliva, gastric juice, bile, &c., and the muscular and absorbent actions. He who neglects the study of any one of these three classes of particulars, must have a very incomplete acquaintance with the function.—We proceed now to explain the mechanical or physical circumstances connected with digestion.

The abdomen may be considered as a vessel full of liquid, in which, therefore, there is pressure in all directions, increasing with the depth, (see hydrostatics,) and increased also by the action of the surrounding muscles which form the sides of the cavity.

The justness of this view of the abdomen becomes evident, when we consider that only moistened or semifluid food descends into the stomach, that drink follows, and that gastric and other juices are poured out to mix with the food as it passes on to occupy the long intestinal canal; and that then the intestines externally are perfectly smooth, and are moistened by the constant secretion of lubricating serum, so that they slide among each other, without sensible impediment from friction. The abdomen, therefore, is in fact a roundish smooth vessel filled with a thick fluid, which is farther contained in a perfectly pliant and smooth-coated tube.

Thus any part of the contents of the stomach and bowels, in a living man, is supported like water in surrounding water, and therefore, if the whole contents be of equal specific gravity, no part can descend or advance by its weight. Neither can any general pressure, nor contraction of the surrounding parietes, hasten, except at the moment of expulsion, the motion of any contained matter—as has, however, often been supposed; nor can it help to empty one part into another—the stomach, for instance, or the gall-bladder, into the small intestine.

For the same reason, however, the very slightest contractile action of any containing part is sufficient to dislodge its contents—gravity as a resistance being neutralized by the surrounding fluid. And when the gall-bladder, or stomach, or any part of the intestinal tube, becomes so full as to put the elasticity of the coats ever so little upon the stretch, this circumstance alone, unless some muscular action oppose, will cause a discharge of the contents. —The natural action of the intestinal canal is a successive contraction of its circular fibres from above downwards propelling the contents, just as if a small ring or tube were put round the canal and pushed forwards.

These considerations make evident the common error of supposing that vomiting can by the sudden compression of the abdominal viscera *mechanically* emulge or clear the obstructed biliary ducts. If general pressure of the abdomen could produce this and similar effects, a descent in the diving-bell should be a powerful remedy in human maladies; for nearly fifteen pounds on the inch are added to the ordinary abdominal pressure, at a depth of thirty feet in water.

We hence see also the kind of error into which our predecessors fell so generally, when they attributed much of the digestive power of the stomach to its simple pressure upon the food. The idea probably arose from the contemplation of the stomach or gizzard of a fowl, which is a powerful gristly substance, answering the purpose almost of a mouth and teeth, as well as of a stomach.

It is an error also to suppose that quicksilver, which is sometimes swallowed to remove obstruction, runs through the bowels simply by its weight. On first entering the loose small intestine, it must drag the part containing it to the bottom of the abdomen, and in that situation, the whole intestine must pass round it, nearly as a rope passes through a ring fixed to the floor. When the mercury arrives at the part of the intestine called the *cæcum*, where the farther course lies upward along the fixed arch of the colon, it probably can be dislodged only by the patient's lying down. Any useful operation of quicksilver in such cases, may be in its stimulating the bowels, by dragging or displacing them, in the manner above described.

When the abdominal muscles, which are the containing sides of the cavity, become tense, whether from unusual fulness of the cavity, or from their own action in any of the straining exertions, a variety of important mechanical effects ensue. Thus,

A full stomach produces—tension and projection of the belly—projection of the diaphragm into the chest, causing hurried breathing, and impeding speech and singing—expulsion of blood from the abdominal vessels, and, therefore, congestions elsewhere, as in the arteries of the head, sometimes producing apoplexy.

Abdominal fulness, as in *dropsy*, *tympanitis*, *corpulency*, *pregnancy*, &c., produces most of the effects now mentioned in an aggravated degree. If dropsy be allowed to proceed too far without tapping, the patient will die of suffocation from the rise of the diaphragm.—The external veins of the legs and abdomen of a dropsical person are generally turgid, because the blood is pressed into them out of the abdominal cavity, and because the passage of blood through the abdomen is impeded. In tympanitis, or windy dropsy, as it has been called, the viscera hang down in the abdominal cavity, while the air occupies the upper part. In common dropsy the viscera float about and are supported.

Straining or strong action of the abdominal muscle, and therefore, also pressure on the abdominal contents, occur with almost every considerable bodily exertion; for the abdominal muscles are the antagonists of the great muscles on the back and about the spine, and must always come into play with them to give firmness and rigidity to the trunk of the body. This may be seen remarkably in the actions of lifting, running, wrestling, &c. As the abdominal muscles cannot act in a continued way and strongly, unless the ribs, from which they arise, become nearly fixed, the ribs are supported during exertion by the intercostal muscles, and by the air in the chest, then confined by the closure of the air-passages: hence there is generally compression in the chest also when the abdomen is compressed, and the blood is squeezed towards the extremities from both cavities at once. It is important to remark also, that in what are called the strong actions of the chest, as *coughing*, *sneezing*, *blowing*, &c., the abdominal muscles are at least as active as the pectoral: by pulling down the ribs to which they are attached, they narrow the chest, and by compressing the abdominal contents, and thus raising up the diaphragm, they shorten the chest.

The following cases exemplify the effects of straining —The lifting of a great weight, or making any great exertion, drives the blood up to the head; as is marked by the sudden redness of the face.—Coughing or vomiting will cause closed leech-bites to bleed afresh, and sometimes will overcome the action of the sphincter of the bladder or rectum: coughing will also produce vomiting,—Straining to empty the bladder, rectum, or womb, or the effort of vomiting, will cause the rupture of a blood-vessel in the white of the eye, with consequent effusion of blood there.—Apoplexy often happens under the same circumstances, from the breaking of a vessel in the brain.—The rupture of a varicose vein, or of aneurism, generally happens during exertion.—And during exertion, the protrusion is likely to occur at any weak part of the abdominal cavity, of some portion of its contents, producing what is called *hernia* or *rupture.*

Vomiting is produced, not by the forcible contraction of the stomach, as was long supposed, but chiefly by the action of the parietes of the abdomen. This is proved by the fact that the stomach has been removed from a living animal, and a sheep's bladder containing liquid has been substituted for it, in connection with the gullet above and the intestines below; and on then injecting an emetic drug into the veins of the animal vomiting has taken place, as if the stomach had been there and unhurt.* From this we see why, to prevent regurgitation of the food, during exertion, the upper orfice of the stomach requires to be almost as strongly closed as the sphincters below.

A small pump—in this application called the *stomach-pump*—has lately been used in medical practice, for removing poisons from the stomach in cases where the action of vomiting could not be excited. It has already saved many lives. It resembles the common small syringe, except that there are two appertures near the end, instead of one, which, owing to the valves in them, opening different ways, become what are called a *sucking* and a *forcing* passage. When the object is to extract from the stomach, the pump is worked while its sucking orifice is in connection with an elastic tube passed into the stomach, and the discharged matter passes by the *forcing* orifice. When it is desired, on the contrary, to throw the cleansing water or liquid into the stomach, the connection of the tube is reversed.

As a pump may not be always procurable when the occasion for it arises, the profession should be aware that in many cases a simple tube will answer the purpose as well, if not better. Such a tube being introduced, and the body of the patient being so placed that the tube forms a downward channel from the stomach, all fluid matter will escape from the stomach by the tube, as water escapes from a funnel by its pipe; and if the outer end of the tube

* The mechanism of vomiting is still a moot point in physiology. Mr. Haighton, a celebrated English physiologist, opened several animals during the effort of vomiting, and he asserts that he distinctly saw the contractions of the stomach. The more recent experiments of M. Magendie, which were repeated in the presence of a committee of the French Institute, are, however, entirely contradictory of those of Mr. Haighton, and seem to show that the stomach is entirely quiescent in the act of vomiting. M. Maingault, nevertheless, has been led to results opposed to those of M. Magendie, and he is supported by Professor Portal and M. Bourdon, both of whom appeal to experiments, and to some pathological facts, which are very imposing. It appears to us probable that vomiting is usually the *joint effect* of the contraction of the stomach, and of the diaphragm and abdominal muscles, though either is of itself occasionally sufficient for that purpose.—*Am. Ed.*

be kept immersed in liquid, there will be during the discharge a syphon action of considerable force. On then changing the posture of the body, water may be poured in through the tube to wash the stomach, and may, by the same channel, be again discharged. Such a tube, made long enough, might, if desired, be rendered a complete bent syphon, the necessary preliminary suction being produced by a syringe, or by an assistant, who acts through an interposed vessel.

But there is still an easier mode than either of these now described, of dislodging poison from a torpid stomach, *viz.*, merely to place the patient so that the mouth shall be considerably lower than the stomach,—as when the body lies across a chair or a sofa, with the face near the floor,—and then, if necessary, to press on the stomach with the hand. The cardiac orifice opens readily in such a case, and the stomach is emptied like any other inverted vessel.

Useful as the pump may prove, upon occasions, in evacuating the stomach, its more ancient office of injecting the enema is still the more important—and recent experience seems to show that such injection may become a remedy of more extensive utility than had yet been suspected. From an erroneous opinion, that what had been called the *valve of the cæcum* acts as a perfect valve, allowing passage downwards only, few practitioners have ventured to order much liquid to be injected, for fear of overstretching the lower part of the intestine; and the possibility of thus, by injection, relieving disease situated above the supposed valve, has scarcely been contemplated. It is now ascertained, however, that fluid may be safely thrown in even until it reach the stomach.—Perhaps few, if any cases of obstruction of the bowels, could resist the gentle force of penetrating water, so that a mechanical remedy of certain effect may, in many cases, be substituted for the drastic purgatives and pernicious bleedings now used, and often used in vain.—From what has been said above of the abdomen and the intestinal canal, it appears that an injection tends to spread itself with singular uniformity over the whole. This tendency may be rendered obvious to sight by throwing a sheep's intestine, recently extracted, into a bucket of water, and then pumping water in at one end; a stream will issue strongly at the other end, although several feet distant, almost immediately, and without any intermediate part having become very sensibly tense.—Of course, in the living body, in cases of spasms or obstruction, the liquid must be thrown in against resistance very gradually.

That case is called *intro-susception* of the bowel, in which an upper portion falls, or is received into a portion below, (as one part of the finger of a glove may be received into another part,) and the receiving portion of the bowel, mistaking the received for descending food, holds it fast. This occurrence forms a complete obstruction, and generally proves fatal. Many infants with irritable bowels, die of it.—Now a copious enema, such as we have described above, is almost a certain cure. The liquid advances until it reaches the part where the portion of gut has been swallowed by gut below; and as it cannot pass without pushing the intro-suscepted portion back to liberty, it effects the cure.*

The *perpetual syringe* or *little valved pump*, of which we have been speaking as lately used in applications to the animal body, can inject or with-

* It should be remarked, however, that this measure can succeed only whilst the intro-susception is recent; at least before inflammation has occurred and adhesions formed between the intro-suscepted portion and that portion of the bowel in which it is received. Common or atmospheric air, from its great elasticity, lightness, &c., is the best fluid for the injection.—*Am. Ed.*

draw any quantity, and is therefore very superior, for almost every purpose, to the large old syringes which had no valves, and which, without being removed could inject only once their fill. With well-adapted additional apparatus, the same instrument will answer for many purposes, as for throwing up the enema, clearing the stomach, transfusion of blood exhausting cupping-glasses, relieving the over-distended breasts; for the lotio vesicæ, vaginæ vel urethræ, &c. No surgical apparatus is now complete without one. The annexed outline represents such a syringe. The aperature *c* is rendered a *sucking orifice*, by a valve at it, opening inwards; and *a* is the forcing orifice, in consequence of having its valve opening outwards: *b* is the piston, with its handle. The valves may be variously made, or a single *double-way cock* may be used instead of both. Convenient dimensions for the syringe are from four to six inches for the length, and from three-quarters of an inch to an inch and a quarter for the diameter.

Fig. 175.

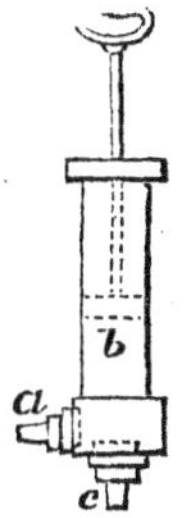

For a case of diseased rectum, where it was necessary to use an enena daily, or oftener, the *enema funnel*, represented here, (from the funnel-shaped mouth *a* downwards, and exclusive of the part above *a*) was contrived, and was found more manageable by the patient than any other instrument. If the tube *a b* be about two feet long, (it may be of metal, oiled silk, caoutchouc cloth, &c.,) the liquid column contained in it suffices to overcome the ordinary abdominal resistance; but if a very short tube be used, the open funnel *a* must be converted into a close vessel, as represented here by the dotted line above the funnel, having a bladder, or other air-tight bag, *d*, connected with it, and a bottle-neck and cork, or a cock, at *c*, for admitting the enema. On pouring in the liquid at *c*, the air in the vessel at *c a* is forced into the bag, and on then closing the opening at *c*, and compressing the bag, it is evident, that any desired degree of injecting pressure may be exerted on the enema. This apparatus is both cheaper and more simple than a syringe, and is equally effectual; and the bag never being wetted, lasts long: *b* is a cock kept shut until the moment of injection.—The principle of substituting, in an injecting apparatus, the pressure of a liquid column for that of a piston, was first suggested in this work; and yet since the publication of the work, more than one patent has been taken for it, for parties seeking to convert it to their profit.

Fig. 176.

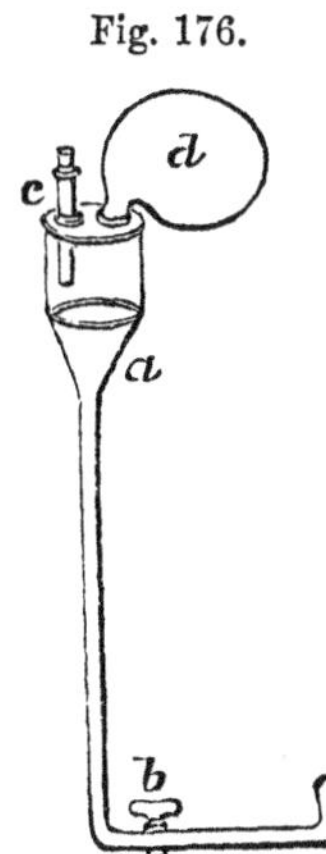

By viewing the abdomen in the true light of a vessel or bag filled with liquid which is seeking to escape in all directions, we have the explanation of several circumstances connected with *hernia* or *rupture*; in which accident, the containing sides of the abdomen in some part have given way, allowing a portion of the viscera to escape, so as to form a tumour under the skin.

Hernia may be produced by all causes which strain or weaken the muscles; as by leaping, lifting, great weights, coughing and sneezing, lying with the belly across a bench or yard, as sailors do on ship-board, over distention of the belly by eating or drinking, corpulency, dropsy, pregnancy; debility of muscle from dissipation, &c.

The reason that a rupture increases so rapidly after it has once begun, is, that the protruding part is truly a fluid wedge, of which, therefore, the opening force increases with the diameter. (See Hydrostatics.) This shows the singular importance of arresting the accident at its very commencement. The truss used to repress rupture were described at page 412.

In attempting to return any part of the abdominal contents which may have escaped as rupture, we should recollect, that a soft uniform compression or squeezing exerted upon the tumour by the hands of the operator, if greater than the internal pressure of the abdomen, is slowly pushing back again any fluid matter that can ooze inward from the tumour; and by thus gradually lessening the size of the tumour, may effect the desired object, without the adoption of the last resource, of cutting parts to widen the inlet. When, in such a case, the operator sees clearly with the mind's eye what is passing under his fingers, his efforts may often be successful, where a less intelligent individual would fail. No man practices medicine long, whatever his nominal department, without having opportunities of saving life, or of preventing a serious operation, by judicious management of recent hernia. The barbarous old fashion of lifting the patient by the heels and shaking him, that the weight of the bowels might drag back again the part which had escaped, was founded on ignorance of the fact, that the weight of the bowels in all positions of the body, is supported almost entirely, not by their attachments, but by the surrounding parts.

The function of digestion or assimilation sketched in the preceding paragraphs, by which the animal body assumes foreign matters from around, and converts them into his own substances, is a subject of study little inviting in some of its details, but taken altogether is one of the most wonderful which can engage the human attention. It points directly to the curious and yet unanswered question—what is LIFE? The student of nature may analyze with all his art those minute portions of matter called *seeds* and *ova*, which he knows to be the rudiments of future creatures, and the links by which endless generations of living creatures hang to existence: but he cannot disentangle and display apart their mysterious LIFE! that something under the influence of which each little germ, when placed in due circumstances, swells out, to fill an invisible mould of maturity which determines its forms and proportions. One such substance thus expands into a beauteous rose-bush; another into an noble oak; a third into an eagle; a fourth into an elephant—yea, in the same way, out of the rude materials of broken seeds and roots, and leaves of plants, and bits of animal flesh, is built up the human frame itself, whether of the active man, combining gracefulness with strength or of the gentler woman, with beauty around her as light. How passing strange, that such should be the origin of the bright eye, whose glance pierces as if the invisible soul were shot with it—of the lips which pour forth sweetest eloquence—of the larynx, whose vibrating fills the surrounding air with music; and, more wonderful than all, of that mass shut up within the body fortress of the skull, whose delicate and curious texture is the abode of the soul, with its reason which contemplates, and its sensibility which delights n these and endless other miracles of creation.

PELVIC APPARATUS.

The Secretion of the Kidneys, &c.

Of the large quantity of fluid daily taken into the human body, much escapes with the breath, as is provided by the visible condensation of it in frosty air, or on any cold polished surface held near the mouth; part escapes by the skin in perspiration; but the greatest part after having answered the purposes of the constitution, is separated from the blood by the two secreting organs, called the kidneys, and from them by fit channels, is carried off, holding in solution various other matters, which the system does not require. The kidneys are situated in the loins, one on each side of the spine; and the constant drain of liquid from them passes down by two membranous canals called *ureters* into the bladder, from which the liquid is again expelled through the urethra, at considerable intervals, according to the rapidity of accumulation.

The bladder is a curious membranous and muscular reservoir, of which the fibres can contract so as to expel the last drop, and yet can yield so as to admit a quart or more.

The passage of fluid downwards through the ureters from the kindeys to the bladder resembles, in some respects, the passage of blood in the veins. Authors have erroneously supposed that the weight of the fluid suffices to cause its descent: but the bladder and ureters are enclosed in a common cavity with the intestinal canal; and while this is full of a semi-fluid mass of greater specific gravity than the urine, the latter is not only supported by the surrounding pressure, as water would be supported by water, but is forced upwards or resisted, as water would be in honey or treacle: in descending, therefore, it obeys some other force than gravity, namely, that of the secreting vessels and heart.

The *ureters, bladder* and *urethra* are the seats of some of the most distressing diseases, to which the human frame is liable. Two classes of these being relievable chiefly by mechanical means, require to be shortly considered here. They are, *obstructions in the urethra;* and *concretions,* or *stones,* as they are called, *in the bladder.*

Obstructions or strictures in the urethra are generally consequences of an inflammation, which has destroyed the dilatability of a part of the canal. They appear as if a thread or a bit of tape were tied round the canal, so as to narrow its caliber. Constant irritation, which destroys the general health, fits of fever, broken rest, and even death from total suppression of urine, have been common consequences of stricture.

Until within a recent period, the treatment of such obstructions was pursued very generally according to a blind routine. The attempt was made either to bore them open by wedges, called bougies, often of doubtful and tedious operation, or to destroy them by caustic passed down to them in the end of a bougie, which caustic often hurt the part of the canal anterior to them, or eat out false passages about the stricture, or open blood-vessels so as to cause dangerous hæmorrhage.

Struck by the defective state of this branch of the healing art, the author of the present work, while abroad, and situated where he had interesting opportunities of observation, had bestowed considerable attention upon it, and he then contrived and tried several new means of relief. These were afterwards brought more extensively into use and improved, and others were added, by his brother Dr. James Arnott, superintendent surgeon in the service of the Hon. East India Company, who gave a minute account of them in a treatise on urethral diseases, and a supplement published in the years 1818 and 1820, They have become, perhaps, still better known in Francee than in England, through the work of Dr. Ducamp, which describes them, and which, having been submitted to the French Institute, and most favourably reported upon by the appointed authorities, soon became a standard treatise in the country;—in France, also, the philosophy of mechanics had been studied by surgeons more generally than in England. It is painful to be obliged to add, that Dr Ducamp as regarded these instruments and the views of disease and treatment which had suggested them, concealed the fact of his being only a translator. The imposition was not discovered at the time of his death, which happened two years afterwards, hastened apparently by the fatigues of the extensive practice which the report of the Acadamy brought upon him. The auther has had so much pleasing intercourse with enlightened and honourable Frenchmen, that it pains him to have this fact to relate.

The objects aimed at by the *new means* were,—to ascertain the exact condition of the diseased canal—to facilitate the passing of instruments in cases of difficulty—and to effect a permanent cure. The following *seven* of these means may here be particularized:

1st. *An examining sound;* being a bougie with the point formed of a softer tenacious material; in which fibres of cotton or silk are mixed to prevent any portion from being broken off or detached during use. This sound, pressed against the obstruction, takes a correct impression of its anterior face, and shows the magnitude and exact position of the still remaining opening.

2nd *An expanding or dilator sound* which is a small tube with a dilatable button at its extremity. The button consists of a little bag, which is passed through the structure empty, and is filled with fluid after it has passed. It readily discovers any other strictures beyond the first, and, to a certain degree, the state of each.

3d. *A conducting canula* or tube, open at both ends. It is passed down to the stricture, for the purpose of supporting and directing small bougies seeking entrances through very narrow strictures, or of guarding the caustic bougie in its approach to the place of its action.

4th. In cases where the attempt to open the passage has failed by all common means, a conducting tube is first introduced, and through it six or more small bougies are passed side by side, so as to probe the whole face of the stricture at the same time. It is thus scarcely possible that the opening should not be found.

5th. Were even this means to fail, the conducting tube may be filled with water, under any degree of pressure, which water will either open the passage for the small bougies, or will itself act as the sharpest and most insinuating of all instruments. The stricture, by which ever means opened will then allow the urine to escape. As patients might fear that water forced towards a bladder already too full would only increase the evil, J. Arnott waited for more numerous proofs of the utility and safety of the practice, before strongly recommending it: Dr. Amussat, of Paris, has since published a statement of numerous cases of retention thus relieved.

6th. A *dilator* for widening the stricture, after a small instrument can be passed through it. It is intended as a substitute for the *bougies* and *sounds* of former times. The chief objections to these last are, the painful friction, the danger of making false passages, the tediousness and imperfection of the cure, and that they cannot dilate any part of the canal beyond the size of its orifice, through which they have to pass, and which during health, is the narrowest part of the canal.

The dilator consists of a tube of thin membrane introduced while empty into the stricture, on a ball-pointed wire and then filled with fluid by a syringe, so to dilate the stricture, with any degree of force, from the mere filling of the part to the strain of the hydrostatic press, sufficient to tear the strongest texture that disease can form. The dilating tube is about two inches long, and its end next to the operator is fixed to the point of a small catheter, through which the distending fluid is injected. The tube is formed of thin silk riband of various sizes, with the edges joined. It is lined with prepared gut of the cat or dog, which is almost as thin as gold-beater's skin, although very strong and water-tight; and it is covered with the same to give the smoothest and softest possible external surface. When complete and enclosing its blunt wire, it is still much less bulky than the bougie which would be required for the same case. Thus, it passes easily; it cannot tear the canal or make false passages; it can enter through a small orifice, and then dilate to any desired extent; and its greatest advantage is, that by swelling so as to follow the yielding of the stricture, it can effect at one application, what only a succession of hard bougies, during long treatment, could accomplish. In one day it has often removed disease which had resisted other means for months or even years.

Some practitioners and critics, not understanding the law of fluid pressure (explained at p. 128,) objected at first to the dilator, that a little water or air pressed into it by a syringe, would be unable to overcome much resistance. Had they seen the instrument lifting so readily as it does, a heavy weight laid upon it, or snapping a strong ligature tied round it they would not have had this prejudice. It was objected, also, that the instrument would do mischief by dilating the urethra before and behind the stricture more than the stricture itself; now its dimensions being determined and fixed by those of its silken tunic, it never can *distend* beyond the diameter chosen, and, therefore, if of the proper size, it can only *press* on the stricture itself. It was also said, that this instrument requires, in the operator, greater manual dexterity and acquaintance with mechanical philosophy than many surgeons possess; but this is merely saying that the arts are progressive, and that the accomplished surgeon of the present day is more dexterous and intelligent than his predecessors of the last centuary. It is not accounted a reason why the delicate apparatus of the oculist should fall into disuse, that all surgeons are not able to apply it.

Some attempts had been made before, to construct a *dilator of fluid pressure*, but they produced nothing of value. For urethral purposes, a simple gut or intestine is worse than useless, for, being yielding in its texture, the surgeon can never know truly the size of his instrument, and therefore may do much mischief by it. Dr. Ducamp, in speaking of the dilator, allows that he did not first invent it, but then, from ignorance of what constitutes its true value, he takes praise to himself for simplifying and improving it, by throwing away the silk, and using the gut only.—A variety of metallic dilators have been contrived and used by English surgeons since the publication of *Arnott's Treatise on Strictures*, but although manageable with less trouble than the fluid dilator, they want its chief merits.

The *dilator* is applicable to many other purposes in surgery besides that now mentioned,—as for removing stricture of the gullet, and of the rectum, for checking hæmorrhage in deep wounds, for dilating wounds as a tent, &c. And the operation of lithotomy was saved to a gentleman, whom Sir Astley Cooper and the auther of this work were attending together, by the dilator opening a *fistula in perineo*, so that a large stone was extracted without cutting. The dilator has also served in removing stones from the female bladder.

7th. Another improved means for the treatment of stricture, described in the *treatise*, is a mode of applying caustic for its entire destruction, but so as not to touch any other part of the canal. Formerly the caustic was applied *to the face* or anterior part of the stricture, and, therefore, had almost always to destroy a portion of the healthy canal before it could reach the narrowest fibres: the extent of such portion depending on the distance from these fibres of the part where the lining of the canal began to be drawn inwards by them. This explains why not unfrequently a hundred applications of caustic were made in a single case, and why, during such treatment, false passages were often bored, and other mischiefs produced. Now by applying the caustic *within* the stricture at once a single application generally suffices. To accomplish this, a ring of caustic is placed (as described in the *Treatise*, and in the *Cases*,) on a bougie of peculiar construction, about an inch from its extremity; and the bougie being then passed down to the stricture through a tube or conductor, and the point being passed beyond the stricture, the caustic is guided to the very spot where it is desired to act.*

* Dr. Ducamp incurred a singular risk in giving himself out as the first proposer of the instruments and practice described above; for he was already known as a translator of English medical books, and the *Treatise on Strictures* of J. Arnott had been held up to public attention two years before by the various medical reviews, in terms such as the following: "We have carefully perused this little volume, and are of opinion that it is by far the best systematic work on the subject in the English language."—It is a judicious compilation, interwoven with much original and acute observation; and it gives publicity to instruments which promise to be of essential benefit to operative surgery."—*Medico-Chirurgical Review, January*, 1819.

Perhaps Dr. Ducamp imagined that the slight alteration proposed by him in the construction of three of the new instruments, might be a shield to him when detected; but as the chief merit was in the analysis of the subject which suggested such instruments, and not in the mere mechanical fulfilment of intentions, even a considerable improvement in the instruments would not have been a sufficient excuse. His changes, however, were either trifling or retrograde. His metallic *dilating sound* is less perfect than metallic sounds contrived by J. A., but not described, because the fluid dilating sound was found to be preferable. His *porte-caustique* is defective in not distending the stricture at the moment of applying the caustic; and his mode of making a *dilator* without the silken tunic, renders it not only a useless, but a dangerous instrument:—indeed, such as obliged him to use the caustic in almost every case. His silence with respect to the *liquid probe* favors the conclusion that he did not understand it, although Dr. Amussat of Paris has since used it with much success:—and the same remark applies to the *double catheter* (see Arnott's cases,) or *sonde a double current*, as it has been called by those who have since used it in Paris.

The following are extracts from the report made by the commissioners of the French Institute, Doctors Deschamps and Percy, in May, 1822, on the subject of Ducamp's work entitled *Traite des retentions d'urine.*

"This treatise concerning a most important malady, because one of the most common and painful which affects humanity, has appeared to us to merit more than ordinary attention.

"When, some years ago, your same commissioners had to express their opinion of another work on this subjct, they commended the zeal and industry of its estimable author (Dr. Petit;) but they could not conceal that there were still imperfections in his modes of treatment; and also that they were almost entirely either borrowed or imitated from the English.

"The work of Dr. Ducamp now leaves us, however, nothing more to desire, and we have no longer reason, as regards this subject, to envy our neighbours. Although a volume of moderate size, it is incomparably more complete and full of matter than the bulky treatises lately published in other countries.

"* * Ducamp leaves all these authors far behind him, whether as to the soundness of his doctrines, the superiority of his trials, or the invention of instruments.

"He takes a print or model of the stricture by an instrument of his invention, called *Sonde Exploratrice.* (Arnott's examining sound, page 471.)

"For introducing bougies in difficult cases, he uses an elastic gum tube, which he calls *conductor.* (*Described above, page* 471.)

"Mr. D. has invented for measuring the length of strictures, &c., an instrument which, when introduced, enlarges beyond the stricture. (*The dilating sound, page* 471.)

"The nitrate of silver, or common caustic, is what he uses for destroying strictures, but he employs it in a new manner, which appears to us to give it new powers, and to deprive it of all its former dangers. * * He carries the caustic *into* the stricture by means of his *porte caustique.* (*See above, page* 473. *No.* 7 *of Mr. Arnott.*)

".........To enlarge the canal at the morbid part of its true caliber, he uses an instrument which he names a *dilatateur.* (*Dilator, page* 471.) He does not conceal that this instru-

Stone in the bladder is another disease relievable chiefly by mechanical means

The urine as secreted in the kidneys, contains dissolved in it, a variety of substances which, under certain circumstances, separate and assume the solid form,—as sugar separates in small crystals from cooling syrup, or salt from cooling brine :—and it is thus that those minute grains are produced which we call *urinary gravel.* A single particle of gravel remaining by an accident in the bladder, soon attracts to itself more matter of the same kind, and becomes the nucleus or centre of an increasing mass, which is what we call the *stone in the bladder.*

In a second Tract by the author's brother, published in 1820,* the follow ing paragraph appears :

" From the severe suffering of the patient labouring under stone in the bladder, and the remedy being an operation so painful and dangerous, that many wear out their lives in certain misery, rather than submit to it, it has arisen that no part of surgery has excited more attention, either in the medical profession or out of it.† No very important change in the treatment of this disease has now been made for upwards of a century; and, indeed, it has appeared to be the opinion of modern surgeons, that the manner of operating practised in Cheselden, about a century ago, and which has been called the 'glory of English surgery' was so nearly perfect as to leave little room for improvement. The hopes which the rapid progress of chemistry, and the grand discoveries relating to stone of Scheele, Wollaston, Fourcroy, and others some time ago gave birth to, that we should be able to dissolve stone by lithontriptics, and thus save the horrors of lithotomy, had again died away, and the researches of many ingenious men who have been, and still are employed about the question, have for their end, more to prevent the formation of stone by remedies and regimen, than to improve the manner of removing it when once formed. I trust, however, notwithstanding the supposed exhausted nature of the subject, that the following essay will prove that much was still possible in the improvement of this department of the healing art."

The publication from which the above paragraph is taken, and the " *Treatise*" which preceded it, in both of which new instruments and new processes were described, and interesting facts were detailed, aroused the public attention in England to the possibility of improving the treatment of stone; and about the same time, a similar spirit awoke with more decided effects in France. The results have now become of great importance to humanity. In the medical publications since that time, cases soon began to be recorded in lithotomy superseded by new means, and lately such cases form the majority. We shall now briefly animadvert to the principal of these

ment had been imagined before him, but he has the merit of perfecting it, and of reducing to practice what before had only existed as a project.

"........In rendering justice to the able men who have preceded Ducamp, we must still say, that no one has displayed so much industry, dexterity, and talent, and we think that he has high claims to the confidence of patients and the gratitude of the profession, and that his work merits the eulogium of the Academy.

(Signed "DESCHAMPS,—PERCY, Reporters.
CUVIER, ... Secretary.

* Cases illustrative of the Treatment of Urethral Obstructions and of Stone. By James Arnott.—Longman and Co., 1820.

† The Catalogue of authors who have written upon stone occupies in Plocquet's *Literatura Medica*, no less than twenty-nine very closely printed quarto pages.

means, intending, however, only to interest the reader in a manner that may lead him to the persual of the original works, where more minute information is to be found. They shall be named in the order in which they have come into use.

The *dilator*, as applied to the treatment of stone, has already been spoken of in the preceding pages.

The *double catheter*. This instrument, with its applications to causes of stone and other affections of the bladder, is described in *Arnott's Cases*. It has two channels, by one of which a fluid may pass into the bladder, while by the other there is a returning current mixed with urine. It is equipped with two pliant tubes, of which one leads, from a *supplying reservoir*, and the other to a *waste vessel*. It will soothe irritation of the bladder, whether arising from stone or not, by keeping the acrid urine in a dilated state, or by applying bland and medicated liquids directly to the internal surface of the bladder. Not being larger than a common catheter, it may be worn for any period as the common catheter now is. It need prevent no sedentary occupation, and may be used during sleep. It will act powerfully to dilate a contracted bladder, if the reservoirs be placed high, and the fluid be caused to distend with the pressure of a lofty column. It also affords by far the best means of admitting to the bladder any solvent of stone. Even pure water is a weak solvent of most animal calculi, as is proved by placing them in a running stream; but the living bladder bears with impunity a diluted acid or alkali.

The *syphon catheter* (also first described in *Arnott's Cases*) is merely a catheter of a length that will allow its external part to descend, so as to constitute the long leg of a syphon. (See *Pneumatics*.) Its outer extremity is turned up a little, or has a portion of soft animal gut tied upon it to act as a valve, for preventing the entrance of air. The most useful application of this instrument is to keep the bladder empty after operations, until the healing process has made a certain advance. The diffusion of urine among the surrounding parts after lithotomy, particularly after the high operation, is often a cause of death; and the syphon catheter, by providing a channel by which the urine must immediately pass away as secreted, obviates the danger. This instrument is sometimes useful in very irritable bladders, by preventing the repeated distensions of the bladder, with the consequent excruciating contraction. It has also relieved in the deplorable case of the bladder torn or opened by sloughing in parturition, as it can keep the unhappy patient quite dry.

A *forceps*, calculated to pass through a tube into the bladder, and to open there, for the purpose of seizing any small stone or other solid object offered to it, was described long ago in the *Armamentum Chirurgicum* of Scultetus, but was again forgotten until John Hunter's investigations led him to a second invention of it. Such an instrument had for a considerable time passed under the appellation of *Hunter's urethra or bladder forceps*, answering for extracting small stones, and therefore, if used in time, preventing occasionally, the necessity of lithotomy. Soon after the publication of Arnott's Essay, it was modified and much more extensively used by Sir Astley Cooper and other surgeons in England.

But a new and intense interest has now been excited with respect to the forceps, as a means for removing stone, by the discovery—also an old discovery revived—that a *straight* tube may be passed to the bladder, as a conductor instead of the *bent* tubes or catheters commonly used. A door is thus, as it were, opened directly into the bladder, through which a stone might even be seen, if light were directed upon it, and through which it easily may be caught and broken to pieces, and brought away without doing injury to the living parts. Dr. Civiale, of Paris, had the merit first of contriving good instruments for this operation, and of himself operating with complete success in many cases.. But the praise of carrying the operation of *Lithotrity* (*stone-wearing-down*,) as it is now named, to its present state of perfection, is shared by various other ingenious surgeons, as Gruithuisen (who first used the straight sound,) Amussat, Leroy, Heurteloup, (who proposed the mode of percussion,) &c. The operator introduces a strong forceps, which seizes and holds fast the stone, he then weakens the stone by boring it in various directions with a simple drill, which passes through the handle of the forceps, and is turned rapidly by a drill-bow acting on its external end, or with a drill of which the point can be bent to one side, so as to excavate to any desired extent; after which weakening, the stone is crushed, either by the forceps which first held it, or by another instrument called *brisecoque*, made on purpose; or without boring at all, Heurteloup and others at once break the stone to pieces by blows of a small hammer acting on a sliding limb of the forceps.

Dr. Darwin, in his *Zoonomia*, published in 1790, proposed to seize stones by forceps passed into the bladder, and then to break them down or destroy them mechanically; but the supposed necessity of working through a long bent tube prevented trials from being made. The author of this work also showed some years ago, before any of the above-described improvements were made (see *Cases*, page 93,) that it was possible to pass a bag into the living bladder, and to enclose a stone there, so that any solvent might be injected into the bag, and again withdrawn without coming into contact with the bladder. This was shown rather to excite attention to the possibility of operating with the living bladder with great precision, than to recommend that precise means of destroying stone.

To all the ingenions instruments above spoken of for breaking down the stone, there is still this objection, that it is broken into such fragments, that many of them require to be afterwards treated as distinct stones, and thus the painful operation has to be repeated again and again, and whole months may pass before the operation be completed.—The author deems it possible to make a forceps of several claws or ribs, which should surround the stone so loosely as to leave it freedom of motion within the claws, like a loose kernal in a shell, and so that on making the forceps itself whirl backwards and forwards, like the drill in Civiale's apparatus, the stone might be quickly rubbed to dust by the friction or file action of the roughened interior of the claws. The bladder would be filled during the operation, with water, or even air, to secure plenty of room for the turning instrument:—or a slender external forceps might be used as a guard, to prevent contact of the bladder with the moving instrument. Out of the body, a stone harder than urinary calculus, placed in such a cage with rough interior, and subjected to the action described, is soon reduced to dust. There are various ways of making a forceps or cage for this operation, which will readily suggest themselves to persons knowing what has already been achieved in this department of practice, and having the ingenuity likely to engage them in such a pursuit.

The *high operation* of lithotomy possesses over the common *lateral operation* such advantages as the following:—thinness of the parts cut through —distance of the knife from important arteries—stones of very large size may be more easily extracted—the prostate gland is not wounded. But the high operation has not become general,—because—there was difficulty in avoiding the peritoneum while making the opening into the bladder—there was danger of effusion of urine among the cut parts, after the operation—and where the bladder was contracted, the incision had to be very deep. Now these objections are obviated by, 1st, the *double catheter*, which will dilate the contracted bladder; 2d, by the *syphon catheter*, which will prevent the effusion of urine; and 3d, by the jointed *sliding sound*, (see *Cases*, page 104,) which will ensure the accurate cutting in the desired place. Had we possessed then, for the removal of stone, no less hazardous means than cutting, the high operations with the new securities might have been best.

When a catheter has to be retained in the bladder after any operation, in cases where, if it slipped out, it might with difficulty be replaced, something should be passed through it like a small spring forceps to expand and become an internal button preventing its escape. (See *Cases*, page 94.)

UTERINE PHENOMENA.

Although so many of the uterine phenomena are mechanical, there are few of them which could be treated of with advantage, except in connection with particulars, of which the consideration does not belong to a work like this. We shall, however, cite the following particulars as examples.

The protection given to the tender fœtus by the *liquor amnii* in which it floats, is such, that a blow from without is expended on the surrounding water, and cannot reach the fœtus.

The head of the fœtus, because ossification begins in it first, becomes of greater specific gravity than the other parts of the body, and therefore generally lies at the bottom of its liquid bed. It is thus ready to appear first in parturition, according to the safest course of delivery.

The membranes distended by the liquor amnii descend before the head, as a soft but powerful wedge preparing the way, according to the principle explained in a previous page.

We have spoken, at page 168, under the name of *pneumatic tractor*, of a circular piece of leather or similar soft substances, kept extended by included solid rings or radii, as being adapted to some purposes of surgery. Now it seems peculiarly adapted to a purpose of obstetric surgery, *viz.*, as a substitute for the steel forceps, in the hands of men who are deficient in manual dexterity, whether from inexperience or natural inaptitude. The forceps, to be well and safely used, requires address, which even the naturally dexterous man cannot possess without a certain degree of continued practical familiarity with it, and except in large towns, a man must be very unfortunate in his practice who often requires it: hence the really small number of persons, who use it well. The consideration of the tractor as a substitute for it belongs properly to the present section: but as the true mode of action of the tractor is not very readily conceived by persons who either have never been instructed in the general laws of physics, or who have ceased to be familiar with them, such persons are advised to read this paragraph in continuation of that at page 168, and to weigh well the following remarks. A tractor of three inches in diameter, would act upon any body, to lift or draw it, with a force of about a hundred pounds—with more therefore, than is ever required or allowable in obstetric practice. In lifting a stone, the tractor does act as if it were glued or nailed to the stone, but merely bears or takes off the atmospheric pressure from one part, and allows the pressure on the opposite side, not then counterbalanced, to push the stone in the direction of the tractor;—so when placed upon the head, it would not pull by the skin, in the manner of a very strong adhesive plaster applied there, as uninformed persons would be apt to suppose, but by taking off a certain atmospheric pressure from the part of the head on which it rested, it would allow the pressure on the other side or behind to urge the head forward on its way. Of course the forwarding pressure in such a case would not operate on the head directly, but through the intervening parietes and contents of the maternal abdomen. It would be much better to have a gentle and diffused action of the tractor over a large surface, than an intense action on a small surface, and therefore a tractor for the purpose now contemplated should not be very small, and should have a little air underneath it in a slight depression or cavity at its centre.—The forceps must be more effective than the tractor for rectifying malposition of the head, and diminishing its transverse diameter; but the tractor will answer both these purposes in a degree greater than many would expect.* The author proposes to publish on this matter, and on some other strictly professional subjects which are lightly touched upon in the present general work, such a practical detail, as for the dilator, syphon catheter, &c., is found in his brother's "*Treatise*" and "*Cases.*"

Conclusion.

It is almost superfluous to remark here, that, for the practice of general and obstetric surgery, learning and judgment are of little avail unless accompanied by manual dexterity: and it is one of the improvements yet to be made in our system of education for various professions, to cultivate more methodically the use of the hands. Children and young people, in obtaining practical familiarity with ingenious toys, tools of carpentry, games of address, musical instruments, &c., are often fitting themselves for the important business of their future life.

* We have been already compelled on one or two occasions to differ from the able author of this work, in relation to the practical application of some of his principles, and we must be again permitted to record our dissent from his opinion that the pneumatic tractor, under certain circumstances is peculiarly adapted as a substitute for the obstetric forceps. Our author cannot be a practical accoucheur or he would at once perceive that the various manœuvres by which labour is assisted with the forceps, cannot be accomplished with the tractor. That address and knowledge are requisite to apply the forceps properly, is no objection to their use; it only shows the necessity of the operators's acquiring this dexterity and knowledge before attempting to apply the instruments; and these acquirments are not so difficult as our author seems to think, nor do we believe that the number who possess them is so very small. It is not contended even by the author that the tractor is superior to the forceps; he only recommends it as being less dangerous in the hands of the unskilful.

Now it might be supposed from this that the tractor is readily applied and cannot effect injury, both of which are erroneous. Every instrument is dangerous in the hands of ignorance. If a person deficient in dexterity could succeed in applying the traction, (of which we have strong doubts, believing it would require, in most instances, even Dr. Arnott's skill and knowledge) it is quite as probable that he would produce injury as benefit. In certain states of labour, the tractor may be applied to the neck of the uterus instead of the head of the child, or to both, drawing out the uterus thus as well as the child; it may be applied before the uterus is sufficiently dilated, or the force may be applied in the wrong direction; indeed, there are but few cases in which force could be applied in the proper direction with the tractor, &c. These accidents cannot happen to the well instructed; but in the hands of such, the forceps are more effectual and equally safe. The tractor, then, requiring skill for its proper application, and being a less efficient instrument than the forceps, ought not, independent of many other reasons, to be recommended. It is not to those who devise imperfect substitutes for valuable instruments, or temporary palliatives for important operations, in order that the awkward and ignorant may imperfectly perform what the skilful or instructed should only attempt, or are capable of accomplishing, that praise is to be awarded. It is the just meed of those who furnish proper instructions for the use of instruments and for performing operations, and present the means of gaining information and skill.—*Am. Ed.*

While the author directs the attention of the profession to the important physical considerations set forth in the preceding pages, he deems it necessary most pointedly to remark, that in the living body *mechanical* principles are generally associated in their operation with the more recondite principles of *chemistry* and of *life*; and that the man who allows his mind to dwell too exclusively on any one of the three classes, must be a very bad reasoner in questions either of health or disease. It is within a very recent period, however, that just views on this subject have begun to prevail, and that the titles of the peculiarly mechanical physician, or chemical physician, or physician attending only to the influence of nerves or life, are likely to be no longer justly applicable. The light of true philosophy is at last breaking in upon the very complex and difficult subjects of medical inquiry; and where formerly keen penetration beheld only confusion, even common minds now begin to see clear divisions and beautiful arrangement.

THE

ANALYTICAL TABLE.

PART II.—DOCTRINES OF SOLIDS or MECHANICS, 84.

PART III.—DOCTRINES OF FLUIDS or HYDRODYNAMICS, 126.

SECT. I.—HYDROSTATICS, or fluids in repose, 126.

THE END.

BLANCHARD & LEA'S MEDICAL AND SURGICAL PUBLICATIONS,

TO THE MEDICAL PROFESSION.

The prices on the present catalogue are those at which our books can generally be furnished by booksellers throughout the United States, who can readily procure any which they may not have on hand. To physicians who have not convenient access to bookstores, we will forward them by mail, at these prices, *free of postage* (as long as the existing facilities are afforded by the post-office), for any distance under 3,000 miles. As we open accounts only with booksellers, the amount must in every case, without exception, accompany the order, and we assume no risks of the mail, either on the money or on the books; and as we deal only in our own publications, we can supply no others. Gentlemen desirous of purchasing will, therefore, find it more advantageous to deal with the nearest booksellers whenever practicable.

BLANCHARD & LEA.

PHILADELPHIA, December, 1860.

*** We have just issued a new edition of our ILLUSTRATED CATALOGUE of Medical and Scientific Publications, forming an octavo pamphlet of 80 large pages, containing specimens of illustrations, notices of the medical press, &c. &c. It has been prepared without regard to expense, and will be found one of the handsomest specimens of typographical execution as yet presented in this country. Copies will be sent to any address, by mail, free of postage, on receipt of nine cents in stamps.

Catalogues of our numerous publications in miscellaneous and educational literature forwarded on application.

☞ The attention of physicians is especially solicited to the following important new works and new editions, just issued or nearly ready:—

TWO MEDICAL PERIODICALS, FREE OF POSTAGE,

Containing over Fifteen Hundred large octavo pages,

FOR FIVE DOLLARS PER ANNUM.

THE AMERICAN JOURNAL OF THE MEDICAL SCIENCES, subject to postage, when not paid for in advance, - - - - - - - - - - $5 00
THE MEDICAL NEWS AND LIBRARY, invariably in advance, - - 1 00
or, BOTH PERIODICALS mailed, FREE OF POSTAGE, for Five Dollars remitted in advance.

THE AMERICAN JOURNAL OF THE MEDICAL SCIENCES,

EDITED BY ISAAC HAYS, M. D.,

is published Quarterly, on the first of January, April, July, and October. Each number contains at least two hundred and eighty large octavo pages, handsomely and appropriately illustrated, wherever necessary. It has now been issued regularly for nearly FORTY years, and it has been under the control of the present editor for more than a quarter of a century. Throughout this long period, it has maintained its position in the highest rank of medical periodicals both at home and abroad, and has received the cordial support of the entire profession in this country. Its list of Collaborators will be found to contain a large number of the most distinguished names of the profession in every section of the United States, rendering the department devoted to

ORIGINAL COMMUNICATIONS

full of varied and important matter, of great interest to all practitioners.

As the aim of the Journal, however, is to combine the advantages presented by all the different varieties of periodicals, in its

REVIEW DEPARTMENT

will be found extended and impartial reviews of all important new works, presenting subjects of novelty and interest, together with very numerous

BIBLIOGRAPHICAL NOTICES,

including nearly all the medical publications of the day, both in this country and Great Britain, with a choice selection of the more important continental works. This is followed by the

QUARTERLY SUMMARY,

being a very full and complete abstract, methodically arranged, of the

IMPROVEMENTS AND DISCOVERIES IN THE MEDICAL SCIENCES.

This department of the Journal, so important to the practising physician, is the object of especial care on the part of the editor. It is classified and arranged under different heads, thus facilitating the researches of the reader in pursuit of particular subjects, and will be found to present a very full and accurate digest of all observations, discoveries, and inventions recorded in every branch of medical science. The very extensive arrangements of the publishers are such as to afford to the editor complete materials for this purpose, as he not only regularly receives

ALL THE AMERICAN MEDICAL AND SCIENTIFIC PERIODICALS,

but also twenty or thirty of the more important Journals issued in Great Britain and on the Continent, thus enabling him to present in a convenient compass a thorough and complete abstract of everything interesting or important to the physician occurring in any part of the civilized world.

To their old subscribers, many of whom have been on their list for twenty or thirty years, the publishers feel that no promises for the future are necessary; but those who may desire for the first time to subscribe, can rest assured that no exertion will be spared to maintain the Journal in the high position which it has occupied for so long a period.

By reference to the terms it will be seen that, in addition to this large amount of valuable and practical information on every branch of medical science, the subscriber, by paying in advance, becomes entitled, without further charge, to

THE MEDICAL NEWS AND LIBRARY,

a monthly periodical of thirty-two large octavo pages. Its "News Department" presents the current information of the day, while the "Library Department" is devoted to presenting standard works on various branches of medicine. Within a few years, subscribers have thus received, without expense, many works of the highest character and practical value, such as "Watson's Practice," "Todd and Bowman's Physiology," "Malgaigne's Surgery," "West on Children," "West on Females, Part I.," "Habershon on the Alimentary Canal," &c.

While in the number for January, 1860, is commenced a new and highly important work,

CLINICAL LECTURES ON THE DISEASES OF WOMEN.

By Professor J. Y. SIMPSON, of Edinburgh.

WITH NUMEROUS HANDSOME ILLUSTRATIONS.

These Lectures, published in England under the supervision of the Author, carry with them all the weight of his wide experience and distinguished reputation. Their eminently practical nature, and the importance of the subject treated, cannot fail to render them in the highest degree satisfactory to subscribers, who can thus secure them without cost. The present is therefore a particularly eligible time for gentlemen to commence their subscriptions.

It will thus be seen that for the small sum of FIVE DOLLARS, paid in advance, the subscriber will obtain a Quarterly and a Monthly periodical,

EMBRACING NEARLY SIXTEEN HUNDRED LARGE OCTAVO PAGES,

mailed to any part of the United States, free of postage.

Those subscribers who do not pay in advance will bear in mind that their subscription of Five Dollars will entitle them to the Journal only, without the News, and that they will be at the expense of their own postage on the receipt of each number. The advantage of a remittance when ordering the Journal will thus be apparent.

As the Medical News and Library is in no case sent without advance payment, its subscribers will always receive it free of postage.

Remittances of subscriptions can be mailed at our risk, when a certificate is taken from the Postmaster that the money is duly inclosed and forwarded.

Address BLANCHARD & LEA, Philadelphia.

ASHTON (T. J.),

Surgeon to the Blenheim Dispensary, &c.

ON THE DISEASES, INJURIES, AND MALFORMATIONS OF THE RECTUM AND ANUS;

with remarks on Habitual Constipation. From the third and enlarged London edition. With handsome illustrations. In one very beautifully printed octavo volume, of about 300 pages. (*Now Ready.*) $2 00.

INTRODUCTION. CHAPTER I. Irritation and Itching of the Anus. II. Inflammation and Excoriation of the Anus. III. Excrescences of the Anal Region. IV. Contraction of the Anus. V. Fissure of the Anus and lower part of the Rectum. VI. Neuralgia of the Anus and extremity of the Rectum. VII. Inflammation of the Rectum. VIII. Ulceration of the Rectum. IX. Hemorrhoidal Affections. X. Enlargement of Hemorrhoidal Veins. XI. Prolapsus of the Rectum. XII. Abscess near the Rectum. XIII. Fistula in Ano. XIV. Polypi of the Rectum. XV. Stricture of the Rectum. XVI. Malignant Diseases of the Rectum. XVII. Injuries of the Rectum. XVIII. Foreign Bodies in the Rectum. XIX. Malformations of the Rectum. XX. Habitual Constipation.

The most complete one we possess on the subject. *Medico-Chirurgical Review.*

Its merits as a practical instructor, well arranged, abundantly furnished with illustrative cases, and clearly and comprehensively, albeit too diffusely, written, are incontestable. They have been sufficiently endorsed by the verdict of his countrymen in the rapid exhaustion of the first edition, and they would certainly meet with a similar reward in the United States were the volume placed within the reach of American practitioners. We are satisfied after a careful examination of the volume, and a comparison of its contents with those of its leading predecessors and contemporaries, that the best way for the reader to avail himself of the excellent advice given in the concluding paragraph above, would be to provide himself with a copy of the book from which it has been taken, and diligently to con its instructive pages. They may secure to him many a triumph and fervent blessing.—*Am. Journal Med. Sciences*, April, 1858.

ALLEN (J. M.), M. D.,

Professor of Anatomy in the Pennsylvania Medical College, &c.

THE PRACTICAL ANATOMIST; or, The Student's Guide in the Dissecting-ROOM.

With 266 illustrations. In one handsome royal 12mo. volume, of over 600 pages, leather. $2 25.

However valuable may be the "Dissector's Guides" which we, of late, have had occasion to notice, we feel confident that the work of Dr. Allen is superior to any of them. We believe with the author, that none is so fully illustrated as this, and the arrangement of the work is such as to facilitate the labors of the student in acquiring a thorough practical knowledge of Anatomy. We most cordially recommend it to their attention.—*Western Lancet.*

We believe it to be one of the most useful works upon the subject ever written. It is handsomely illustrated, well printed, and will be found of convenient size for use in the dissecting-room.—*Med. Examiner.*

ANATOMICAL ATLAS.

By Professors H. H. SMITH and W. E. HORNER, of the University of Pennsylvania.

1 vol. 8vo., extra cloth, with nearly 650 illustrations. ☞ See SMITH, p. 27.

ABEL (F. A.), F. C. S. AND C. L. BLOXAM.

HANDBOOK OF CHEMISTRY, Theoretical, Practical, and Technical;

with a Recommendatory Preface by Dr. HOFMANN. In one large octavo volume, extra cloth, of 662 pages, with illustrations. $3 25.

ASHWELL (SAMUEL), M. D.,

Obstetric Physician and Lecturer to Guy's Hospital, London.

A PRACTICAL TREATISE ON THE DISEASES PECULIAR TO WOMEN.

Illustrated by Cases derived from Hospital and Private Practice. Third American, from the Third and revised London edition. In one octavo volume, extra cloth, of 528 pages. $3 00.

The most useful practical work on the subject in the English language.—*Boston Med. and Surg. Journal.*

The most able, and certainly the most standard and practical, work on female diseases that we have yet seen.—*Medico-Chirurgical Review.*

ARNOTT (NEILL), M. D.

ELEMENTS OF PHYSICS; or Natural Philosophy, General and Medical.

Written for universal use, in plain or non-technical language. A new edition, by ISAAC HAYS, M. D. Complete in one octavo volume, leather, of 484 pages, with about two hundred illustrations. $2 50.

BIRD (GOLDING), A. M., M. D., &c.

URINARY DEPOSITS: THEIR DIAGNOSIS, PATHOLOGY, AND THERAPEUTICAL INDICATIONS.

Edited by EDMUND LLOYD BIRKETT, M. D. A new American, from the fifth and enlarged London edition. With eighty illustrations on wood. In one handsome octavo volume, of about 400 pages, extra cloth. $2 00. (*Just Issued.*)

The death of Dr. Bird has rendered it necessary to entrust the revision of the present edition to other hands, and in his performance of the duty thus devolving on him, Dr. Birkett has sedulously endeavored to carry out the author's plan by introducing such new matter and modifications of the text as the progress of science has called for. Notwithstanding the utmost care to keep the work within a reasonable compass, these additions have resulted in a considerable enlargement. It is, therefore, hoped that it will be found fully up to the present condition of the subject, and that the reputation of the volume as a clear, complete, and compendious manual, will be fully maintained.

BUDD (GEORGE), M. D., F. R. S.,

Professor of Medicine in King's College, London.

ON DISEASES OF THE LIVER. Third American, from the third and enlarged London edition. In one very handsome octavo volume, extra cloth, with four beautifully colored plates, and numerous wood-cuts. pp. 500. $3 00.

Has fairly established for itself a place among the classical medical literature of England.—*British and Foreign Medico-Chir. Review*, July, 1857.

Dr. Budd's Treatise on Diseases of the Liver is now a standard work in Medical literature, and during the intervals which have elapsed between the successive editions, the author has incorporated into the text the most striking novelties which have characterized the recent progress of hepatic physiology and pathology; so that although the size of the book is not perceptibly changed, the history of liver diseases is made more complete, and is kept upon a level with the progress of modern science. It is the best work on Diseases of the Liver in any language.—*London Med. Times and Gazette*, June 27, 1857.

This work, now the standard book of reference on the diseases of which it treats, has been carefully revised, and many new illustrations of the views of the learned author added in the present edition.—*Dublin Quarterly Journal*, Aug. 1857.

BY THE SAME AUTHOR.

ON THE ORGANIC DISEASES AND FUNCTIONAL DISORDERS OF THE STOMACH. In one neat octavo volume, extra cloth. $1 50.

BUCKNILL (J. C.), M. D.,

Medical Superintendent of the Devon County Lunatic Asylum; and

DANIEL H. TUKE, M. D.,

Visiting Medical Officer to the York Retreat.

A MANUAL OF PSYCHOLOGICAL MEDICINE; containing the History, Nosology, Description, Statistics, Diagnosis, Pathology, and Treatment of INSANITY. With a Plate. In one handsome octavo volume, of 536 pages. $3 00.

The increase of mental disease in its various forms, and the difficult questions to which it is constantly giving rise, render the subject one of daily enhanced interest, requiring on the part of the physician a constantly greater familiarity with this, the most perplexing branch of his profession. At the same time there has been for some years no work accessible in this country, presenting the results of recent investigations in the Diagnosis and Prognosis of Insanity, and the greatly improved methods of treatment which have done so much in alleviating the condition or restoring the health of the insane. To fill this vacancy the publishers present this volume, assured that the distinguished reputation and experience of the authors will entitle it at once to the confidence of both student and practitioner. Its scope may be gathered from the declaration of the authors that "their aim has been to supply a text book which may serve as a guide in the acquisition of such knowledge, sufficiently elementary to be adapted to the wants of the student, and sufficiently modern in its views and explicit in its teaching to suffice for the demands of the practitioner."

BENNETT (J. HUGHES), M. D., F. R. S. E.,

Professor of Clinical Medicine in the University of Edinburgh, &c.

THE PATHOLOGY AND TREATMENT OF PULMONARY TUBERCULOSIS, and on the Local Medication of Pharyngeal and Laryngeal Diseases frequently mistaken for or associated with, Phthisis. One vol. 8vo., extra cloth, with wood-cuts. pp. 130. $1 25.

BENNETT (HENRY), M. D.

A PRACTICAL TREATISE ON INFLAMMATION OF THE UTERUS, ITS CERVIX AND APPENDAGES, and on its connection with Uterine Disease. To which is added, a Review of the present state of Uterine Pathology. Fifth American, from the third English edition. In one octavo volume, of about 500 pages, extra cloth. $2 00. (*Now Ready.*)

The ill health of the author having prevented the promised revision of this work, the present edition is a reprint of the last, without alteration. As the volume has been for some time out of print, gentlemen desiring copies can now procure them.

BOWMAN (JOHN E.), M. D.

PRACTICAL HANDBOOK OF MEDICAL CHEMISTRY. Second American, from the third and revised English Edition. In one neat volume, royal 12mo., extra cloth, with numerous illustrations. pp. 288. $1 25.

BY THE SAME AUTHOR.

INTRODUCTION TO PRACTICAL CHEMISTRY, INCLUDING ANALYSIS. Second American, from the second and revised London edition. With numerous illustrations. In one neat vol., royal 12mo., extra cloth. pp. 350. $1 25.

BEALE ON THE LAWS OF HEALTH IN RELATION TO MIND AND BODY. A Series of Letters from an old Practitioner to a Patient. In one volume, royal 12mo., extra cloth. pp. 296. 80 cents.

BUSHNAN'S PHYSIOLOGY OF ANIMAL AND VEGETABLE LIFE; a Popular Treatise on the Functions and Phenomena of Organic Life. In one handsome royal 12mo. volume, extra cloth, with over 100 illustrations. pp. 234. 80 cents.

BUCKLER ON THE ETIOLOGY, PATHOLOGY, AND TREATMENT OF FIBRO-BRONCHITIS AND RHEUMATIC PNEUMONIA. In one 8vo. volume, extra cloth. pp. 150. $1 25.

BLOOD AND URINE (MANUALS ON). BY JOHN WILLIAM GRIFFITH, G. OWEN REESE, AND ALFRED MARKWICK. One thick volume, royal 12mo., extra cloth, with plates. pp. 460. $1 25.

BRODIE'S CLINICAL LECTURES ON SURGERY. 1 vol. 8vo. cloth. 350 pp. $1 25.

BARCLAY (A. W.), M. D.,

Assistant Physician to St. George's Hospital, &c.

A MANUAL OF MEDICAL DIAGNOSIS; being an Analysis of the Signs and Symptoms of Disease. In one neat octavo volume, extra cloth, of 424 pages. $2 00. (*Lately issued.*)

Of works exclusively devoted to this important branch, our profession has at command, comparatively, but few, and, therefore, in the publication of the present work, Messrs. Blanchard & Lea have conferred a great favor upon us. Dr. Barclay, from having occupied, for a long period, the position of Medical Registrar at St. George's Hospital, possessed advantages for correct observation and reliable conclusions, as to the significance of symptoms, which have fallen to the lot of but few, either in his own or any other country. He has carefully systematized the results of his observation of over twelve thousand patients, and by his diligence and judicious classification, the profession has been presented with the most convenient and reliable work on the subject of Diagnosis that it has been our good fortune ever to examine; we can, therefore, say of Dr. Barclay's work, that, from his systematic manner of arrangement, his work is one of the best works "for reference" in the daily emergencies of the practitioner, with which we are acquainted; but, at the same time, we would recommend our readers, especially the younger ones, to read thoroughly and study diligently the *whole* work, and the "emergencies" will not occur so often.—*Southern Med. and Surg. Journ.*, March, 1858.

To give this information, to supply this admitted deficiency, is the object of Dr. Barclay's Manual. The task of composing such a work is neither an easy nor a light one; but Dr. Barclay has performed it in a manner which meets our most unqualified approbation. He is no mere theorist; he knows his work thoroughly, and in attempting to perform it, has not exceeded his powers.—*British Med. Journal*, Dec. 5, 1857.

We venture to predict that the work will be deservedly popular, and soon become, like Watson's Practice, an indispensable necessity to the practitioner.—*N. A. Med. Journal*, April, 1858.

An inestimable work of reference for the young practitioner and student.—*Nashville Med. Journal*, May, 1858.

We hope the volume will have an extensive circulation, not among students of medicine only, but practitioners also. They will never regret a faithful study of its pages.—*Cincinnati Lancet*, Mar. '58.

An important acquisition to medical literature. It is a work of high merit, both from the vast importance of the subject upon which it treats, and also from the real ability displayed in its elaboration. In conclusion, let us bespeak for this volume that attention of every student of our art which it so richly deserves — that place in every medical library which it can so well adorn.—*Peninsular Medical Journal*, Sept. 1858.

BARLOW (GEORGE H.), M. D.

Physician to Guy's Hospital, London, &c.

A MANUAL OF THE PRACTICE OF MEDICINE. With Additions by D. F. Condie, M. D., author of "A Practical Treatise on Diseases of Children," &c. In one handsome octavo volume, leather, of over 600 pages. $2 75.

We recommend Dr. Barlow's Manual in the warmest manner as a most valuable vade-mecum. We have had frequent occasion to consult it, and have found it clear, concise, practical, and sound. It is eminently a practical work, containing all that is essential, and avoiding useless theoretical discussion. The work supplies what has been for some time wanting, a manual of practice based upon modern discoveries in pathology and rational views of treatment of disease. It is especially intended for the use of students and junior practitioners, but it will be found hardly less useful to the experienced physician. The American editor has added to the work three chapters—on Cholera Infantum, Yellow Fever, and Cerebro-spinal Meningitis. These additions, the two first of which are indispensable to a work on practice destined for the profession in this country, are executed with great judgment and fidelity, by Dr. Condie, who has also succeeded happily in imitating the conciseness and clearness of style which are such agreeable characteristics of the original book.—*Boston Med. and Surg. Journal.*

BARTLETT (ELISHA), M. D.

THE HISTORY, DIAGNOSIS, AND TREATMENT OF THE FEVERS OF THE UNITED STATES. A new and revised edition. By Alonzo Clark, M. D., Prof. of Pathology and Practical Medicine in the N. Y. College of Physicians and Surgeons, &c. In one octavo volume, of six hundred pages, extra cloth. Price $3 00.

It is the best work on fevers which has emanated from the American press, and the present editor has carefully availed himself of all information existing upon the subject in the Old and New World, so that the doctrines advanced are brought down to the latest date in the progress of this department of Medical Science.—*London Med. Times and Gazette*, May 2, 1857.

This excellent monograph on febrile disease, has stood deservedly high since its first publication. It will be seen that it has now reached its fourth edition under the supervision of Prof. A. Clark, a gentleman who, from the nature of his studies and pursuits, is well calculated to appreciate and discuss the many intricate and difficult questions in pathology. His annotations add much to the interest of the work, and have brought it well up to the condition of the science as it exists at the present day in regard to this class of diseases.—*Southern Med. and Surg. Journal*, Mar. 1857.

It is a work of great practical value and interest, containing much that is new relative to the several diseases of which it treats, and, with the additions of the editor, is fully up to the times. The distinctive features of the different forms of fever are plainly and forcibly portrayed, and the lines of demarcation carefully and accurately drawn, and to the American practitioner is a more valuable and safe guide than any work on fever extant.—*Ohio Med. and Surg. Journal*, May, 1857.

BROWN (ISAAC BAKER),

Surgeon-Accoucheur to St. Mary's Hospital, &c.

ON SOME DISEASES OF WOMEN ADMITTING OF SURGICAL TREATMENT. With handsome illustrations. One vol. 8vo., extra cloth, pp. 276. $1 60.

Mr. Brown has earned for himself a high reputation in the operative treatment of sundry diseases and injuries to which females are peculiarly subject. We can truly say of his work that it is an important addition to obstetrical literature. The operative suggestions and contrivances which Mr. Brown describes, exhibit much practical sagacity and skill, and merit the careful attention of every surgeon-accoucheur.—*Association Journal.*

We have no hesitation in recommending this book to the careful attention of all surgeons who make female complaints a part of their study and practice. —*Dublin Quarterly Journal.*

CARPENTER (WILLIAM B.), M. D., F. R. S., &c.,

Examiner in Physiology and Comparative Anatomy in the University of London.

PRINCIPLES OF HUMAN PHYSIOLOGY; with their chief applications to Psychology, Pathology, Therapeutics, Hygiene, and Forensic Medicine. A new American, from the last and revised London edition. With nearly three hundred illustrations. Edited, with additions, by FRANCIS GURNEY SMITH, M. D., Professor of the Institutes of Medicine in the Pennsylvania Medical College, &c. In one very large and beautiful octavo volume, of about nine hundred large pages, handsomely printed and strongly bound in leather, with raised bands. $4 25.

In the preparation of this new edition, the author has spared no labor to render it, as heretofore, a complete and lucid exposition of the most advanced condition of its important subject. The amount of the additions required to effect this object thoroughly, joined to the former large size of the volume, presenting objections arising from the unwieldy bulk of the work, he has omitted all those portions not bearing directly upon HUMAN PHYSIOLOGY, designing to incorporate them in his forthcoming Treatise on GENERAL PHYSIOLOGY. As a full and accurate text-book on the Physiology of Man, the work in its present condition therefore presents even greater claims upon the student and physician than those which have heretofore won for it the very wide and distinguished favor which it has so long enjoyed. The additions of Prof. Smith will be found to supply whatever may have been wanting to the American student, while the introduction of many new illustrations, and the most careful mechanical execution, render the volume one of the most attractive as yet issued.

For upwards of thirteen years Dr. Carpenter's work has been considered by the profession generally, both in this country and England, as the most valuable compendium on the subject of physiology in our language. This distinction it owes to the high attainments and unwearied industry of its accomplished author. The present edition (which, like the last American one, was prepared by the author himself), is the result of such extensive revision, that it may almost be considered a new work. We need hardly say, in concluding this brief notice, that while the work is indispensable to every student of medicine in this country, it will amply repay the practitioner for its perusal by the interest and value of its contents.—*Boston Med. and Surg. Journal.*

This is a standard work—the text-book used by all medical students who read the English language. It has passed through several editions in order to keep pace with the rapidly growing science of Physiology. Nothing need be said in its praise, for its merits are universally known; we have nothing to say of its defects, for they only appear where the science of which it treats is incomplete.—*Western Lancet.*

The most complete exposition of physiology which any language can at present give.—*Brit. and For. Med.-Chirurg. Review.*

The greatest, the most reliable, and the best book on the subject which we know of in the English language.—*Stethoscope.*

To eulogize this great work would be superfluous. We should observe, however, that in this edition the author has remodelled a large portion of the former, and the editor has added much matter of interest, especially in the form of illustrations. We may confidently recommend it as the most complete work on Human Physiology in our language.—*Southern Med. and Surg. Journal.*

The most complete work on the science in our language.—*Am. Med. Journal.*

The most complete work now extant in our language.—*N. O. Med. Register.*

The best text-book in the language on this extensive subject.—*London Med. Times.*

A complete cyclopædia of this branch of science. —*N. Y. Med. Times.*

The profession of this country, and perhaps also of Europe, have anxiously and for some time awaited the announcement of this new edition of Carpenter's Human Physiology. His former editions have for many years been almost the only text-book on Physiology in all our medical schools, and its circulation among the profession has been unsurpassed by any work in any department of medical science.

It is quite unnecessary for us to speak of this work as its merits would justify. The mere announcement of its appearance will afford the highest pleasure to every student of Physiology, while its perusal will be of infinite service in advancing physiological science.—*Ohio Med. and Surg. Journ.*

BY THE SAME AUTHOR.

PRINCIPLES OF COMPARATIVE PHYSIOLOGY. New American, from the Fourth and Revised London edition. In one large and handsome octavo volume, with over three hundred beautiful illustrations. pp. 752. Extra cloth, $4 80; leather, raised bands, $5 25.

The delay which has existed in the appearance of this work has been caused by the very thorough revision and remodelling which it has undergone at the hands of the author, and the large number of new illustrations which have been prepared for it. It will, therefore, be found almost a new work, and fully up to the day in every department of the subject, rendering it a reliable text-book for all students engaged in this branch of science. Every effort has been made to render its typographical finish and mechanical execution worthy of its exalted reputation, and creditable to the mechanical arts of this country.

This book should not only be read but thoroughly studied by every member of the profession. None are too wise or old, to be benefited thereby. But especially to the younger class would we cordially commend it as best fitted of any work in the English language to qualify them for the reception and comprehension of those truths which are daily being developed in physiology.—*Medical Counsellor.*

Without pretending to it, it is an encyclopedia of the subject, accurate and complete in all respects—a truthful reflection of the advanced state at which the science has now arrived.—*Dublin Quarterly Journal of Medical Science.*

A truly magnificent work—in itself a perfect physiological study.—*Ranking's Abstract.*

This work stands without its fellow. It is one few men in Europe could have undertaken; it is one no man, we believe, could have brought to so successful an issue as Dr. Carpenter. It required for its production a physiologist at once deeply read in the labors of others, capable of taking a general, critical, and unprejudiced view of those labors, and of combining the varied, heterogeneous materials at his disposal, so as to form an harmonious whole. We feel that this abstract can give the reader a very imperfect idea of the fulness of this work, and no idea of its unity, of the admirable manner in which material has been brought, from the most various sources, to conduce to its completeness, of the lucidity of the reasoning it contains, or of the clearness of language in which the whole is clothed. Not the profession only, but the scientific world at large, must feel deeply indebted to Dr. Carpenter for this great work. It must, indeed, add largely even to his high reputation.—*Medical Times.*

CARPENTER (WILLIAM B.), M. D., F. R. S.,

Examiner in Physiology and Comparative Anatomy in the University of London.

THE MICROSCOPE AND ITS REVELATIONS. With an Appendix containing the Applications of the Microscope to Clinical Medicine, &c. By F. G. SMITH, M. D. Illustrated by four hundred and thirty-four beautiful engravings on wood. In one large and very handsome octavo volume, of 724 pages, extra cloth, $4 00; leather, $4 50.

Dr. Carpenter's position as a microscopist and physiologist, and his great experience as a teacher, eminently qualify him to produce what has long been wanted—a good text-book on the practical use of the microscope. In the present volume his object has been, as stated in his Preface, "to combine, within a moderate compass, that information with regard to the use of his 'tools,' which is most essential to the working microscopist, with such an account of the objects best fitted for his study, as might qualify him to comprehend what he observes, and might thus prepare him to benefit science, whilst expanding and refreshing his own mind" That he has succeeded in accomplishing this, no one acquainted with his previous labors can doubt.

The great importance of the microscope as a means of diagnosis, and the number of microscopists who are also physicians, have induced the American publishers, with the author's approval, to add an Appendix, carefully prepared by Professor Smith, on the applications of the instrument to clinical medicine, together with an account of American Microscopes, their modifications and accessories. This portion of the work is illustrated with nearly one hundred wood-cuts, and, it is hoped, will adapt the volume more particularly to the use of the American student.

Every care has been taken in the mechanical execution of the work, which is confidently presented as in no respect inferior to the choicest productions of the London press.

The mode in which the author has executed his intentions may be gathered from the following condensed synopsis of the

CONTENTS.

INTRODUCTION—History of the Microscope. CHAP. I. Optical Principles of the Microscope. CHAP. II. Construction of the Microscope. CHAP. III. Accessory Apparatus. CHAP. IV. Management of the Microscope CHAP. V. Preparation, Mounting, and Collection of Objects. CHAP. VI. Microscopic Forms of Vegetable Life—Protophytes. CHAP. VII. Higher Cryptogamia. CHAP. VIII. Phanerogamic Plants. CHAP. IX. Microscopic Forms of Animal Life—Protozoa—Animalcules. CHAP. X. Foraminifera, Polycystina, and Sponges. CHAP. XI. Zoophytes. CHAP. XII. Echinodermata. CHAP. XIII. Polyzoa and Compound Tunicata. CHAP. XIV. Molluscous Animals Generally. CHAP. XV. Annulosa. CHAP. XVI. Crustacea. CHAP. XVII. Insects and Arachnida. CHAP. XVIII. Vertebrated Animals. CHAP. XIX. Applications of the Microscope to Geology. CHAP. XX. Inorganic or Mineral Kingdom—Polarization. APPENDIX. Microscope as a means of Diagnosis—Injections—Microscopes of American Manufacture.

Those who are acquainted with Dr. Carpenter's previous writings on Animal and Vegetable Physiology, will fully understand how vast a store of knowledge he is able to bring to bear upon so comprehensive a subject as the revelations of the microscope; and even those who have no previous acquaintance with the construction or uses of this instrument, will find abundance of information conveyed in clear and simple language.—*Med. Times and Gazette.*

Although originally not intended as a strictly medical work, the additions by Prof. Smith give it a positive claim upon the profession, for which we doubt not he will receive their sincere thanks. Indeed, we know not where the student of medicine will find such a complete and satisfactory collection of microscopic facts bearing upon physiology and practical medicine as is contained in Prof. Smith's appendix; and this of itself, it seems to us, is fully worth the cost of the volume.—*Louisville Medical Review*, Nov. 1856.

BY THE SAME AUTHOR.

ELEMENTS (OR MANUAL) OF PHYSIOLOGY, INCLUDING PHYSIOLOGICAL ANATOMY. Second American, from a new and revised London edition. With one hundred and ninety illustrations. In one very handsome octavo volume, leather. pp. 566. $3 00.

In publishing the first edition of this work, its title was altered from that of the London volume, by the substitution of the word "Elements" for that of "Manual," and with the author's sanction the title of "Elements" is still retained as being more expressive of the scope of the treatise.

To say that it is the best manual of Physiology now before the public, would not do sufficient justice to the author.—*Buffalo Medical Journal.*

In his former works it would seem that he had exhausted the subject of Physiology. In the present, he gives the essence, as it were, of the whole.—*N. Y. Journal of Medicine.*

Those who have occasion for an elementary treatise on Physiology, cannot do better than to possess themselves of the manual of Dr. Carpenter.—*Medical Examiner.*

The best and most complete exposé of modern Physiology, in one volume, extant in the English language.—*St. Louis Medical Journal.*

BY THE SAME AUTHOR. (*Preparing.*)

PRINCIPLES OF GENERAL PHYSIOLOGY, INCLUDING ORGANIC CHEMISTRY AND HISTOLOGY. With a General Sketch of the Vegetable and Animal Kingdom. In one large and very handsome octavo volume, with several hundred illustrations.

The subject of general physiology having been omitted in the last editions of the author's "Comparative Physiology" and "Human Physiology," he has undertaken to prepare a volume which shall present it more thoroughly and fully than has yet been attempted, and which may be regarded as an introduction to his other works.

BY THE SAME AUTHOR.

A PRIZE ESSAY ON THE USE OF ALCOHOLIC LIQUORS IN HEALTH AND DISEASE. New edition, with a Preface by D. F. CONDIE, M. D., and explanations of scientific words. In one neat 12mo. volume, extra cloth. pp. 178. 50 cents.

CONDIE (D. F.), M. D., &c.

A PRACTICAL TREATISE ON THE DISEASES OF CHILDREN. Fifth edition, revised and augmented. In one large volume, 8vo., leather, of over 750 pages. $3 25. (*Just Issued*, 1859.)

In presenting a new and revised edition of this favorite work, the publishers have only to state that the author has endeavored to render it in every respect "a complete and faithful exposition of the pathology and therapeutics of the maladies incident to the earlier stages of existence—a full and exact account of the diseases of infancy and childhood." To accomplish this he has subjected the whole work to a careful and thorough revision, rewriting a considerable portion, and adding several new chapters. In this manner it is hoped that any deficiencies which may have previously existed have been supplied, that the recent labors of practitioners and observers have been thoroughly incorporated, and that in every point the work will be found to maintain the high reputation it has enjoyed as a complete and thoroughly practical book of reference in infantile affections.

A few notices of previous editions are subjoined.

Dr. Condie's scholarship, acumen, industry, and practical sense are manifested in this, as in all his numerous contributions to science.—*Dr. Holmes's Report to the American Medical Association.*

Taken as a whole, in our judgment, Dr. Condie's Treatise is the one from the perusal of which the practitioner in this country will rise with the greatest satisfaction.—*Western Journal of Medicine and Surgery.*

One of the best works upon the Diseases of Children in the English language.—*Western Lancet.*

We feel assured from actual experience that no physician's library can be complete without a copy of this work.—*N. Y. Journal of Medicine.*

A veritable pædiatric encyclopædia, and an honor to American medical literature.—*Ohio Medical and Surgical Journal.*

We feel persuaded that the American medical profession will soon regard it not only as a very good, but as the VERY BEST "Practical Treatise on the Diseases of Children."—*American Medical Journal.*

In the department of infantile therapeutics, the work of Dr. Condie is considered one of the best which has been published in the English language. —*The Stethoscope.*

We pronounced the first edition to be the best work on the diseases of children in the English language, and, notwithstanding all that has been published, we still regard it in that light.—*Medical Examiner.*

The value of works by native authors on the diseases which the physician is called upon to combat, will be appreciated by all; and the work of Dr. Condie has gained for itself the character of a safe guide for students, and a useful work for consultation by those engaged in practice.—*N. Y. Med. Times.*

This is the fourth edition of this deservedly popular treatise. During the interval since the last edition, it has been subjected to a thorough revision by the author; and all new observations in the pathology and therapeutics of children have been included in the present volume. As we said before, we do not know of a better book on diseases of children, and to a large part of its recommendations we yield an unhesitating concurrence.—*Buffalo Med. Journal.*

Perhaps the most full and complete work now before the profession of the United States; indeed, we may say in the English language. It is vastly superior to most of its predecessors.—*Transylvania Med. Journal.*

CHRISTISON (ROBERT), M. D., V. P. R. S. E., &c.

A DISPENSATORY; or, Commentary on the Pharmacopœias of Great Britain and the United States; comprising the Natural History, Description, Chemistry, Pharmacy, Actions, Uses, and Doses of the Articles of the Materia Medica. Second edition, revised and improved, with a Supplement containing the most important New Remedies. With copious Additions, and two hundred and thirteen large wood-engravings. By R. EGLESFELD GRIFFITH, M. D. In one very large and handsome octavo volume, leather, raised bands, of over 1000 pages. $3 50.

COOPER (BRANSBY B.), F. R. S.

LECTURES ON THE PRINCIPLES AND PRACTICE OF SURGERY. In one very large octavo volume, extra cloth, of 750 pages. $3 00.

COOPER ON DISLOCATIONS AND FRACTURES OF THE JOINTS.—Edited by BRANSBY B. COOPER, F. R. S., &c. With additional Observations by Prof. J. C. WARREN. A new American edition. In one handsome octavo volume, extra cloth, of about 500 pages, with numerous illustrations on wood. $3 25.

COOPER ON THE ANATOMY AND DISEASES OF THE BREAST, with twenty-five Miscellaneous and Surgical Papers. One large volume, imperial 8vo., extra cloth, with 252 figures, on 36 plates. $2 50.

COOPER ON THE STRUCTURE AND DISEASES OF THE TESTIS, AND ON THE THYMUS GLAND. One vol. imperial 8vo., extra cloth, with 177 figures on 29 plates. $2 00.

COPLAND ON THE CAUSES, NATURE, AND TREATMENT OF PALSY AND APOPLEXY. In one volume, royal 12mo., extra cloth. pp. 326. 80 cents.

CLYMER ON FEVERS; THEIR DIAGNOSIS, PATHOLOGY, AND TREATMENT In one octavo volume, leather, of 600 pages. $1 50.

COLOMBAT DE L'ISERE ON THE DISEASES OF FEMALES, and on the special Hygiene of their Sex. Translated, with many Notes and Additions, by C. D. MEIGS, M. D. Second edition, revised and improved. In one large volume, octavo, leather, with numerous wood-cuts. pp. 720. $3 50.

CARSON (JOSEPH), M. D.,

Professor of Materia Medica and Pharmacy in the University of Pennsylvania.

SYNOPSIS OF THE COURSE OF LECTURES ON MATERIA MEDICA AND PHARMACY, delivered in the University of Pennsylvania. Second and revised edition. In one very neat octavo volume, extra cloth, of 208 pages. $1 50.

CURLING (T. B.), F. R. S.,

Surgeon to the London Hospital, President of the Hunterian Society, &c.

A PRACTICAL TREATISE ON DISEASES OF THE TESTIS, SPERMATIC CORD, AND SCROTUM. Second American, from the second and enlarged English edition. In one handsome octavo volume, extra cloth, with numerous illustrations. pp. 420. $2 00.

CHURCHILL (FLEETWOOD), M. D., M. R. I. A.

ON THE THEORY AND PRACTICE OF MIDWIFERY. A new American from the fourth revised and enlarged London edition. With Notes and Additions, by D. Francis Condie, M. D., author of a "Practical Treatise on the Diseases of Children," &c. With 194 illustrations. In one very handsome octavo volume, leather, of nearly 700 large pages. $3 50. (*Now Ready*, October, 1860.)

This work has been so long an established favorite, both as a text-book for the learner and as a reliable aid in consultation for the practitioner, that in presenting a new edition it is only necessary to call attention to the very extended improvements which it has received. Having had the benefit of two revisions by the author since the last American reprint, it has been materially enlarged, and Dr. Churchill's well-known conscientious industry is a guarantee that every portion has been thoroughly brought up with the latest results of European investigation in all departments of the science and art of obstetrics. The recent date of the last Dublin edition has not left much of novelty for the American editor to introduce, but he has endeavored to insert whatever has since appeared, together with such matters as his experience has shown him would be desirable for the American student, including a large number of illustrations. With the sanction of the author he has added in the form of an appendix, some chapters from a little "Manual for Midwives and Nurses," recently issued by Dr. Churchill, believing that the details there presented can hardly fail to prove of advantage to the junior practitioner. The result of all these additions is that the work now contains fully one-half more matter than the last American edition, with nearly one-half more illustrations, so that notwithstanding the use of a smaller type, the volume contains almost two hundred pages more than before.

No effort has been spared to secure an improvement in the mechanical execution of the work equal to that which the text has received, and the volume is confidently presented as one of the handsomest that has thus far been laid before the American profession; while the very low price at which it is offered should secure for it a place in every lecture-room and on every office table.

A better book in which to learn these important points we have not met than Dr. Churchill's. Every page of it is full of instruction; the opinion of all writers of authority is given on questions of difficulty, as well as the directions and advice of the learned author himself, to which he adds the result of statistical inquiry, putting statistics in their proper place and giving them their due weight, and no more. We have never read a book more free from professional jealousy than Dr. Churchill's. It appears to be written with the true design of a book on medicine, viz: to give all that is known on the subject of which he treats, both theoretically and practically, and to advance such opinions of his own as he believes will benefit medical science, and insure the safety of the patient. We have said enough to convey to the profession that this book of Dr. Churchill's is admirably suited for a book of reference for the practitioner, as well as a text-book for the student, and we hope it may be extensively purchased amongst our readers. To them we most strongly recommend it.—*Dublin Medical Press*, June 20, 1860.

To bestow praise on a book that has received such marked approbation would be superfluous. We need only say, therefore, that if the first edition was thought worthy of a favorable reception by the medical public, we can confidently affirm that this will be found much more so. The lecturer, the practitioner, and the student, may all have recourse to its pages, and derive from their perusal much interest and instruction in everything relating to theoretical and practical midwifery.—*Dublin Quarterly Journal of Medical Science.*

A work of very great merit, and such as we can confidently recommend to the study of every obstetric practitioner.—*London Medical Gazette.*

This is certainly the most perfect system extant. It is the best adapted for the purposes of a text-book, and that which he whose necessities confine him to one book, should select in preference to all others.—*Southern Medical and Surgical Journal.*

The most popular work on midwifery ever issued from the American press.—*Charleston Med. Journal.*

Were we reduced to the necessity of having but one work on midwifery, and *permitted to choose*, we would unhesitatingly take Churchill.—*Western Med. and Surg. Journal.*

It is impossible to conceive a more useful and elegant manual than Dr. Churchill's Practice of Midwifery.—*Provincial Medical Journal.*

Certainly, in our opinion, the very best work on the subject which exists.—*N. Y. Annalist.*

No work holds a higher position, or is more deserving of being placed in the hands of the tyro, the advanced student, or the practitioner.—*Medical Examiner.*

Previous editions, under the editorial supervision of Prof R. M. Huston, have been received with marked favor, and they deserved it; but this, reprinted from a very late Dublin edition, carefully revised and brought up by the author to the present time, does present an unusually accurate and able exposition of every important particular embraced in the department of midwifery. * * The clearness, directness, and precision of its teachings, together with the great amount of statistical research which its text exhibits, have served to place it already in the foremost rank of works in this department of remedial science.—*N. O. Med. and Surg. Journal.*

In our opinion, it forms one of the best if not the very best text-book and epitome of obstetric science which we at present possess in the English language.—*Monthly Journal of Medical Science.*

The clearness and precision of style in which it is written, and the great amount of statistical research which it contains, have served to place it in the first rank of works in this departmentof medical science.—*N. Y. Journal of Medicine.*

Few treatises will be found better adapted as a text-book for the student, or as a manual for the frequent consultation of the young practitioner.—*American Medical Journal.*

BY THE SAME AUTHOR. (*Lately Published.*)

ON THE DISEASES OF INFANTS AND CHILDREN. Second American Edition, revised and enlarged by the author. Edited, with Notes, by W. V. Keating, M. D. In one large and handsome volume, extra cloth, of over 700 pages. $3 00, or in leather, $3 25.

In preparing this work a second time for the American profession, the author has spared no labor in giving it a very thorough revision, introducing several new chapters, and rewriting others, while every portion of the volume has been subjected to a severe scrutiny. The efforts of the American editor have been directed to supplying such information relative to matters peculiar to this country as might have escaped the attention of the author, and the whole may, therefore, be safely pronounced one of the most complete works on the subject accessible to the American Profession. By an alteration in the size of the page, these very extensive additions have been accommodated without unduly increasing the size of the work.

BY THE SAME AUTHOR.

ESSAYS ON THE PUERPERAL FEVER, AND OTHER DISEASES PECULIAR TO WOMEN. Selected from the writings of British Authors previous to the close of the Eighteenth Century. In one neat octavo volume, extra cloth, of about 450 pages. $2 50.

CHURCHILL (FLEETWOOD), M. D., M. R. I. A., &c.

ON THE DISEASES OF WOMEN; including those of Pregnancy and Childbed. A new American edition, revised by the Author. With Notes and Additions, by D. FRANCIS CONDIE, M. D., author of "A Practical Treatise on the Diseases of Children." With numerous illustrations. In one large and handsome octavo volume, leather, of 768 pages. $3 00.

This edition of Dr. Churchill's very popular treatise may almost be termed a new work, so thoroughly has he revised it in every portion. It will be found greatly enlarged, and completely brought up to the most recent condition of the subject, while the very handsome series of illustrations introduced, representing such pathological conditions as can be accurately portrayed, present a novel feature, and afford valuable assistance to the young practitioner. Such additions as appeared desirable for the American student have been made by the editor, Dr. Condie, while a marked improvement in the mechanical execution keeps pace with the advance in all other respects which the volume has undergone, while the price has been kept at the former very moderate rate.

It comprises, unquestionably, one of the most exact and comprehensive expositions of the present state of medical knowledge in respect to the diseases of women that has yet been published.—*Am. Journ. Med. Sciences*, July, 1857.

This work is the most reliable which we possess on this subject; and is deservedly popular with the profession.—*Charleston Med. Journal*, July, 1857.

We know of no author who deserves that approbation, on "the diseases of females," to the same extent that Dr. Churchill does. His, indeed, is the only thorough treatise we know of on the subject; and it may be commended to practitioners and students as a masterpiece in its particular department.—*The Western Journal of Medicine and Surgery*.

As a comprehensive manual for students, or a work of reference for practitioners, it surpasses any other that has ever issued on the same subject from the British press.—*Dublin Quart. Journal*.

DICKSON (S. H.), M. D.,

Professor of Practice of Medicine in the Jefferson Medical College, Philadelphia.

ELEMENTS OF MEDICINE; a Compendious View of Pathology and Therapeutics, or the History and Treatment of Diseases. Second edition, revised. In one large and handsome octavo volume, of 750 pages, leather. $3 75. (*Just Issued.*)

The steady demand which has so soon exhausted the first edition of this work, sufficiently shows that the author was not mistaken in supposing that a volume of this character was needed—an elementary manual of practice, which should present the leading principles of medicine with the practical results, in a condensed and perspicuous manner. Disencumbered of unnecessary detail and fruitless speculations, it embodies what is most requisite for the student to learn, and at the same time what the active practitioner wants when obliged, in the daily calls of his profession, to refresh his memory on special points. The clear and attractive style of the author renders the whole easy of comprehension, while his long experience gives to his teachings an authority everywhere acknowledged. Few physicians, indeed, have had wider opportunities for observation and experience, and few, perhaps, have used them to better purpose. As the result of a long life devoted to study and practice, the present edition, revised and brought up to the date of publication, will doubtless maintain the reputation already acquired as a condensed and convenient American text-book on the Practice of Medicine.

DRUITT (ROBERT), M. R. C. S., &c.

THE PRINCIPLES AND PRACTICE OF MODERN SURGERY. A new and revised American from the eighth enlarged and improved London edition. Illustrated with four hundred and thirty-two wood-engravings. In one very handsomely printed octavo volume, leather, of nearly 700 large pages. $3 50. (*Now Ready*, October, 1860.)

A work which like DRUITT'S SURGERY has for so many years maintained the position of a leading favorite with all classes of the profession, needs no special recommendation to attract attention to a revised edition. It is only necessary to state that the author has spared no pains to keep the work up to its well earned reputation of presenting in a small and convenient compass the latest condition of every department of surgery, considered both as a science and as an art; and that the services of a competent American editor have been employed to introduce whatever novelties may have escaped the author's attention, or may prove of service to the American practitioner. As several editions have appeared in London since the issue of the last American reprint, the volume has had the benefit of repeated revisions by the author, resulting in a very thorough alteration and improvement. The extent of these additions may be estimated from the fact that it now contains about one-third more matter than the previous American edition, and that notwithstanding the adoption of a smaller type, the pages have been increased by about one hundred, while nearly two hundred and fifty wood-cuts have been added to the former list of illustrations.

A marked improvement will also be perceived in the mechanical and artistical execution of the work, which, printed in the best style, on new type, and fine paper, leaves little to be desired as regards external finish; while at the very low price affixed it will be found one of the cheapest volumes accessible to the profession.

This popular volume, now a most comprehensive work on surgery, has undergone many corrections, improvements, and additions, and the principles and the practice of the art have been brought down to the latest record and observation. Of the operations in surgery it is impossible to speak too highly. The descriptions are so clear and concise, and the illustrations so accurate and numerous, that the student can have no difficulty, with instrument in hand, and book by his side, over the dead body, in obtaining a proper knowledge and sufficient tact in this much neglected department of medical education.—*British and Foreign Medico-Chirurg. Review*, Jan. 1860.

In the present edition the author has entirely rewritten many of the chapters, and has incorporated the various improvements and additions in modern surgery. On carefully going over it, we find that nothing of real practical importance has been omitted; it presents a faithful epitome of everything relating to surgery up to the present hour. It is deservedly a popular manual, both with the student and practitioner.—*London Lancet*, Nov. 19, 1859.

In closing this brief notice, we recommend as cordially as ever this most useful and comprehensive hand-book. It must prove a vast assistance, not only to the student of surgery, but also to the busy practitioner who may not have the leisure to devote himself to the study of more lengthy volumes.—*London Med. Times and Gazette*, Oct. 22, 1859.

In a word, this eighth edition of Dr. Druitt's Manual of Surgery is all that the surgical student or practitioner could desire.—*Dublin Quarterly Journal of Med. Sciences*, Nov. 1859.

DALTON, JR. (J. C.), M. D.

Professor of Physiology in the College of Physicians, New York.

A TREATISE ON HUMAN PHYSIOLOGY, designed for the use of Students and Practitioners of Medicine. With two hundred and fifty-four illustrations on wood. In one very beautiful octavo volume, of over 600 pages, extra cloth, $4 00; leather, raised bands, $4 25. (*Just Issued.*)

This system of Physiology, both from the excellence of the arrangement studiously observed throughout every page. and the clear, lucid, and instructive manner in which each subject is treated, promises to form one of the most generally received class-books in the English language. It is, in fact, a most admirable epitome of all the really important discoveries that have always been received as incontestable truths, as well as of those which have been recently added to our stock of knowledge on this subject. We will, however, proceed to give a few extracts from the book itself, as a specimen of its style and composition, and this, we conceive, will be quite sufficient to awaken a general interest in a work which is immeasurably superior in its details to the majority of those of the same class to which it belongs. In its purity of style and elegance of composition it may safely take its place with the very best of our English classics; while in accuracy of description it is impossible that it could be surpassed. In every line is beautifully shadowed forth the emanations of the polished scholar, whose reflections are clothed in a garb as interesting as they are impressive; with the one predominant feeling appearing to pervade the whole—an anxious desire to please and at the same time to instruct.—*Dublin Quarterly Journ. of Med. Sciences*, Nov. 1859.

The work before us, however, in our humble judgment, is precisely what it purports to be, and will answer admirably the purpose for which it is intended. It is *par excellence*, a text-book; and the best text-book in this department that we have ever seen. We have carefully read the book, and speak of its merits from a more than cursory perusal. Looking back upon the work we have just finished, we must say a word concerning the excellence of its illustrations. No department is so dependent upon good illustrations, and those which keep pace with our knowledge of the subject, as that of physiology. The wood-cuts in the work before us are the best we have ever seen, and, being original, serve to illustrate precisely what is desired —*Buffalo Med. Journal*, March, 1859.

A book of genuine merit like this deserves hearty praise before subjecting it to any minute criticism. We are not prepared to find any fault with its design until we have had more time to appreciate its merits as a manual for daily consultation, and to weigh its statements and conclusions more deliberately. Its excellences we are sure of; its defects we have yet to discover. It is a work highly honorable to its author; to his talents, his industry, his training; to the institution with which he is connected, and to American science.—*Boston Med. and Surgical Journal*, Feb. 24, 1859.

A NEW book and a first rate one; an original book, and one which cannot be too highly appreciated, and which we are proud to see emanating from our country's press. It is by an author who, though young, is considerably famous for physiological research, and who in this work has erected for himself an enduring monument, a token at once of his labor and his success.—*Nashville Medical Journal*, March, 1859.

Throughout the entire work, the definitions are clear and precise, the arrangement admirable, the argument briefly and well stated, and the style nervous, simple, and concise. Section third, treating of Reproduction, is a monograph of unapproached excellence upon this subject, in the English tongue. For precision, elegance and force of style, exhaustive method and extent of treatment, fulness of illustration and weight of personal research, we know of no American contribution to medical science which surpasses it, and the day is far distant when its claims to the respectful attention of even the best informed scholars will not be cheerfully conceded by all acquainted with its range and depth.—*Charleston Med. Journal*, May, 1859.

A new elementary work on Human Physiology lifting up its voice in the presence of late and sturdy editions of Kirke's, Carpenter's, Todd and Bowman's, to say nothing of Dunglison's and Draper's, should have something superior in the matter or the manner of its utterance in order to win for itself deserved attention and a name. That matter and that manner, after a candid perusal, we think distinguish this work, and we are proud to welcome it not merely for its nativity's sake, but for its own intrinsic excellence. Its language we find to be plain, direct, unambitious, and falling with a just conciseness on hypothetical or unsettled questions, and yet with sufficient fulness on those living topics already understood, or the path to whose solution is definitely marked out. It does not speak exhaustively upon every subject that it notices, but it does speak suggestively, experimentally, and to their main utilities. Into the subject of Reproduction our author plunges with a kind of loving spirit. Throughout this interesting and obscure department he is a clear and admirable teacher, sometimes a brilliant leader.—*Am. Med. Monthly*, May, 1859.

DUNGLISON, FORBES, TWEEDIE, AND CONOLLY.

THE CYCLOPÆDIA OF PRACTICAL MEDICINE: comprising Treatises on the Nature and Treatment of Diseases, Materia Medica, and Therapeutics, Diseases of Women and Children, Medical Jurisprudence, &c. &c. In four large super-royal octavo volumes, of 3254 double-columned pages, strongly and handsomely bound, with raised bands. $12 00.

*** This work contains no less than four hundred and eighteen distinct treatises, contributed by sixty-eight distinguished physicians, rendering it a complete library of reference for the country practitioner.

The most complete work on Practical Medicine extant; or, at least, in our language.—*Buffalo Medical and Surgical Journal.*

For reference, it is above all price to every practitioner.—*Western Lancet.*

One of the most valuable medical publications of the day—as a work of reference it is invaluable.—*Western Journal of Medicine and Surgery.*

It has been to us, both as learner and teacher, a work for ready and frequent reference, one in which modern English medicine is exhibited in the most advantageous light.—*Medical Examiner.*

We rejoice that this work is to be placed within the reach of the profession in this country, it being unquestionably one of very great value to the practitioner. This estimate of it has not been formed from a hasty examination, but after an intimate acquaintance derived from frequent consultation of it during the past nine or ten years. The editors are practitioners of established reputation, and the list of contributors embraces many of the most eminent professors and teachers of London, Edinburgh, Dublin, and Glasgow. It is, indeed, the great merit of this work that the principal articles have been furnished by practitioners who have not only devoted especial attention to the diseases about which they have written, but have also enjoyed opportunities for an extensive practical acquaintance with them, and whose reputation carries the assurance of their competency justly to appreciate the opinions of others, while it stamps their own doctrines with high and just authority.—*American Medical Journ.*

DEWEES'S COMPREHENSIVE SYSTEM OF MIDWIFERY. Illustrated by occasional cases and many engravings. Twelfth edition, with the author's last improvements and corrections In one octavo volume, extra cloth, of 600 pages. $3 20.

DEWEES'S TREATISE ON THE PHYSICAL AND MEDICAL TREATMENT OF CHILDREN. The last edition. In one volume, octavo, extra cloth, 548 pages. $2 80

DEWEES'S TREATISE ON THE DISEASES OF FEMALES. Tenth edition. In one volume, octavo extra cloth, 532 pages, with plates. $3 00

DUNGLISON (ROBLEY), M. D.,

Professor of Institutes of Medicine in the Jefferson Medical College, Philadelphia.

NEW AND ENLARGED EDITION.

MEDICAL LEXICON; a Dictionary of Medical Science, containing a concise Explanation of the various Subjects and Terms of Anatomy, Physiology, Pathology, Hygiene, Therapeutics, Pharmacology, Pharmacy, Surgery, Obstetrics, Medical Jurisprudence, Dentistry, &c. Notices of Climate and of Mineral Waters; Formulæ for Officinal, Empirical, and Dietetic Preparations, &c. With French and other Synonymes. Revised and very greatly enlarged. In one very large and handsome octavo volume, of 992 double-columned pages, in small type; strongly bound in leather, with raised bands. Price $4 00.

Especial care has been devoted in the preparation of this edition to render it in every respect worthy a continuance of the very remarkable favor which it has hitherto enjoyed. The rapid sale of FIFTEEN large editions, and the constantly increasing demand, show that it is regarded by the profession as the standard authority. Stimulated by this fact, the author has endeavored in the present revision to introduce whatever might be necessary "to make it a satisfactory and desirable—if not indispensable—lexicon, in which the student may search without disappointment for every term that has been legitimated in the nomenclature of the science." To accomplish this, large additions have been found requisite, and the extent of the author's labors may be estimated from the fact that about SIX THOUSAND subjects and terms have been introduced throughout, rendering the whole number of definitions about SIXTY THOUSAND, to accommodate which, the number of pages has been increased by nearly a hundred, notwithstanding an enlargement in the size of the page. The medical press, both in this country and in England, has pronounced the work indispensable to all medical students and practitioners, and the present improved edition will not lose that enviable reputation.

The publishers have endeavored to render the mechanical execution worthy of a volume of such universal use in daily reference. The greatest care has been exercised to obtain the typographical accuracy so necessary in a work of the kind. By the small but exceedingly clear type employed, an immense amount of matter is condensed in its thousand ample pages, while the binding will be found strong and durable. With all these improvements and enlargements, the price has been kept at the former very moderate rate, placing it within the reach of all.

This work, the appearance of the fifteenth edition of which, it has become our duty and pleasure to announce, is perhaps the most stupendous monument of labor and erudition in medical literature. One would hardly suppose after constant use of the preceding editions, where we have never failed to find a sufficiently full explanation of every medical term, that in this edition "*about six thousand subjects and terms have been added*," with a careful revision and correction of the entire work. It is only necessary to announce the advent of this edition to make it occupy the place of the preceding one on the table of every medical man, as it is without doubt the best and most comprehensive work of the kind which has ever appeared.—*Buffalo Med. Journ.*, Jan. 1858.

The work is a monument of patient research, skilful judgment, and vast physical labor, that will perpetuate the name of the author more effectually than any possible device of stone or metal. Dr. Dunglison deserves the thanks not only of the American profession, but of the whole medical world.—*North Am. Medico-Chir. Review*, Jan. 1858.

A Medical Dictionary better adapted for the wants of the profession than any other with which we are acquainted, and of a character which places it far above comparison and competition.—*Am. Journ. Med. Sciences*, Jan. 1858.

We need only say, that the addition of 6,000 new terms, with their accompanying definitions, may be said to constitute a new work, by itself. We have examined the Dictionary attentively, and are most happy to pronounce it unrivalled of its kind. The erudition displayed, and the extraordinary industry which must have been demanded, in its preparation and perfection, redound to the lasting credit of its author, and have furnished us with a volume *indispensable* at the present day, to all who would find themselves *au niveau* with the highest standards of medical information.—*Boston Medical and Surgical Journal*, Dec. 31, 1857.

Good lexicons and encyclopedic works generally, are the most labor-saving contrivances which literary men enjoy; and the labor which is required to produce them in the perfect manner of this example is something appalling to contemplate. The author tells us in his preface that he has added about six thousand terms and subjects to this edition, which, before, was considered universally as the best work of the kind in any language.—*Silliman's Journal*, March, 1858.

He has razed his gigantic structure to the foundations, and remodelled and reconstructed the entire pile. No less than *six thousand* additional subjects and terms are illustrated and analyzed in this new edition, swelling the grand aggregate to beyond sixty thousand! Thus is placed before the profession a complete and thorough exponent of medical terminology, without rival or possibility of rivalry.—*Nashville Journ. of Med. and Surg.*, Jan. 1858.

It is universally acknowledged, we believe, that this work is incomparably the best and most complete Medical Lexicon in the English language. The amount of labor which the distinguished author has bestowed upon it is truly wonderful, and the learning and research displayed in its preparation are equally remarkable. Comment and commendation are unnecessary, as no one at the present day thinks of purchasing any other Medical Dictionary than this.—*St. Louis Med. and Surg. Journ.*, Jan. 1858.

It is the foundation stone of a good medical library, and should always be included in the first list of books purchased by the medical student.—*Am. Med. Monthly*, Jan. 1858.

A very perfect work of the kind, undoubtedly the most perfect in the English language.—*Med. and Surg. Reporter*, Jan. 1858.

It is now emphatically *the* Medical Dictionary of the English language, and for it there is no substitute.—*N. H. Med. Journ.*, Jan. 1858.

It is scarcely necessary to remark that any medical library wanting a copy of Dunglison's Lexicon must be imperfect.—*Cin. Lancet*, Jan. 1858.

We have ever considered it the best authority published, and the present edition we may safely say has no equal in the world.—*Peninsular Med. Journal*, Jan. 1858.

The most complete authority on the subject to be found in any language.—*Va. Med. Journal*, Feb. '58.

BY THE SAME AUTHOR.

THE PRACTICE OF MEDICINE. A Treatise on Special Pathology and Therapeutics. Third Edition. In two large octavo volumes, leather, of 1,500 pages. $6 25.

DUNGLISON (ROBLEY), M. D.,

Professor of Institutes of Medicine in the Jefferson Medical College, Philadelphia.

HUMAN PHYSIOLOGY. Eighth edition. Thoroughly revised and extensively modified and enlarged, with five hundred and thirty-two illustrations. In two large and handsomely printed octavo volumes, leather, of about 1500 pages. $7 00.

In revising this work for its eighth appearance, the author has spared no labor to render it worthy a continuance of the very great favor which has been extended to it by the profession. The whole contents have been rearranged, and to a great extent remodelled; the investigations which of late years have been so numerous and so important, have been carefully examined and incorporated, and the work in every respect has been brought up to a level with the present state of the subject. The object of the author has been to render it a concise but comprehensive treatise, containing the whole body of physiological science, to which the student and man of science can at all times refer with the certainty of finding whatever they are in search of, fully presented in all its aspects; and on no former edition has the author bestowed more labor to secure this result.

We believe that it can truly be said, no more complete repertory of facts upon the subject treated, can anywhere be found. The author has, moreover, that enviable tact at description and that facility and ease of expression which render him peculiarly acceptable to the casual, or the studious reader. This faculty, so requisite in setting forth many graver and less attractive subjects, lends additional charms to one always fascinating.—*Boston Med. and Surg. Journal.*

The most complete and satisfactory system of Physiology in the English language.—*Amer. Med Journal.*

The best work of the kind in the English language.—*Silliman's Journal.*

The present edition the author has made a perfect mirror of the science as it is at the present hour. As a work upon physiology proper, the science of the functions performed by the body, the student will find it all he wishes.—*Nashville Journ. of Med.*

That he has succeeded, most admirably succeeded in his purpose, is apparent from the appearance of an eighth edition. It is now the great encyclopædia on the subject, and worthy of a place in every physician's library.—*Western Lancet.*

BY THE SAME AUTHOR. (*A new edition.*)

GENERAL THERAPEUTICS AND MATERIA MEDICA; adapted for a Medical Text-book. With Indexes of Remedies and of Diseases and their Remedies. SIXTH EDITION, revised and improved. With one hundred and ninety-three illustrations. In two large and handsomely printed octavo vols., leather, of about 1100 pages. $6 00.

In announcing a new edition of Dr. Dunglison's General Therapeutics and Materia Medica, we have no words of commendation to bestow upon a work whose merits have been heretofore so often and so justly extolled. It must not be supposed, however, that the present is a mere reprint of the previous edition; the character of the author for laborious research, judicious analysis, and clearness of expression, is fully sustained by the numerous additions he has made to the work, and the careful revision to which he has subjected the whole.—*N. A. Medico-Chir. Review*, Jan. 1858.

The work will, we have little doubt, be bought and read by the majority of medical students; its size, arrangement, and reliability recommend it to all; no one, we venture to predict, will study it without profit, and there are few to whom it will not be in some measure useful as a work of reference. The young practitioner, more especially, will find the copious indexes appended to this edition of great assistance in the selection and preparation of suitable formulæ.—*Charleston Med. Journ. and Review*, Jan. 1858.

BY THE SAME AUTHOR. (*A new Edition.*)

NEW REMEDIES, WITH FORMULÆ FOR THEIR PREPARATION AND ADMINISTRATION. Seventh edition, with extensive Additions. In one very large octavo volume, leather, of 770 pages. $3 75.

Another edition of the "New Remedies" having been called for, the author has endeavored to add everything of moment that has appeared since the publication of the last edition.

The articles treated of in the former editions will be found to have undergone considerable expansion in this, in order that the author might be enabled to introduce, as far as practicable, the results of the subsequent experience of others, as well as of his own observation and reflection; and to make the work still more deserving of the extended circulation with which the preceding editions have been favored by the profession. By an enlargement of the page, the numerous additions have been incorporated without greatly increasing the bulk of the volume.—*Preface.*

One of the most useful of the author's works.—*Southern Medical and Surgical Journal.*

This elaborate and useful volume should be found in every medical library, for as a book of reference, for physicians, it is unsurpassed by any other work in existence, and the double index for diseases and for remedies, will be found greatly to enhance its value.—*New York Med. Gazette.*

The great learning of the author, and his remarkable industry in pushing his researches into every source whence information is derivable, have enabled him to throw together an extensive mass of facts and statements, accompanied by full reference to authorities; which last feature renders the work practically valuable to investigators who desire to examine the original papers.—*The American Journal of Pharmacy.*

ELLIS (BENJAMIN), M. D.

THE MEDICAL FORMULARY: being a Collection of Prescriptions, derived from the writings and practice of many of the most eminent physicians of America and Europe. Together with the usual Dietetic Preparations and Antidotes for Poisons. To which is added an Appendix, on the Endermic use of Medicines, and on the use of Ether and Chloroform. The whole accompanied with a few brief Pharmaceutic and Medical Observations. Tenth edition, revised and much extended by ROBERT P. THOMAS, M. D., Professor of Materia Medica in the Philadelphia College of Pharmacy. In one neat octavo volume, extra cloth, of 296 pages. $1 75.

ERICHSEN (JOHN),

Professor of Surgery in University College, London, &c.

THE SCIENCE AND ART OF SURGERY; BEING A TREATISE ON SURGICAL INJURIES, DISEASES, AND OPERATIONS.

New and improved American, from the second enlarged and carefully revised London edition. Illustrated with over four hundred engravings on wood. In one large and handsome octavo volume, of one thousand closely printed pages, leather, raised bands. $4 50. (*Just Issued.*)

The very distinguished favor with which this work has been received on both sides of the Atlantic has stimulated the author to render it even more worthy of the position which it has so rapidly attained as a standard authority. Every portion has been carefully revised, numerous additions have been made, and the most watchful care has been exercised to render it a complete exponent of the most advanced condition of surgical science. In this manner the work has been enlarged by about a hundred pages, while the series of engravings has been increased by more than a hundred, rendering it one of the most thoroughly illustrated volumes before the profession. The additions of the author having rendered unnecessary most of the notes of the former American editor, but little has been added in this country; some few notes and occasional illustrations have, however, been introduced to elucidate American modes of practice.

It is, in our humble judgment, decidedly the best book of the kind in the English language. Strange that just such books are not oftener produced by public teachers of surgery in this country and Great Britain. Indeed, it is a matter of great astonishment, but no less true than astonishing, that of the many works on surgery republished in this country within the last fifteen or twenty years as text-books for medical students, this is the only one that even approximates to the fulfilment of the peculiar wants of young men just entering upon the study of this branch of the profession.—*Western Jour. of Med. and Surgery.*

Its value is greatly enhanced by a very copious well-arranged index. We regard this as one of the most valuable contributions to modern surgery. To one entering his novitiate of practice, we regard it the most serviceable guide which he can consult. He will find a fulness of detail leading him through every step of the operation, and not deserting him until the final issue of the case is decided.—*Sethoscope.*

Embracing, as will be perceived, the whole surgical domain, and each division of itself almost complete and perfect, each chapter full and explicit, each subject faithfully exhibited, we can only express our estimate of it in the aggregate. We consider it an excellent contribution to surgery, as probably the best single volume now extant on the subject, and with great pleasure we add it to our text-books.—*Nashville Journal of Medicine and Surgery.*

Prof. Erichsen's work, for its size, has not been surpassed; his nine hundred and eight pages, profusely illustrated, are rich in physiological, pathological, and operative suggestions, doctrines, details, and processes; and will prove a reliable resource for information, both to physician and surgeon, in the hour of peril.—*N. O. Med. and Surg. Journal.*

FLINT (AUSTIN), M. D.,

Professor of the Theory and Practice of Medicine in the University of Louisville, &c.

PHYSICAL EXPLORATION AND DIAGNOSIS OF DISEASES AFFECTING THE RESPIRATORY ORGANS.

In one large and handsome octavo volume, extra cloth, 636 pages. $3 00.

We regard it, in point both of arrangement and of the marked ability of its treatment of the subjects, as destined to take the first rank in works of this class. So far as our information extends, it has at present no equal. To the practitioner, as well as the student, it will be invaluable in clearing up the diagnosis of doubtful cases, and in shedding light upon difficult phenomena.—*Buffalo Med. Journal.*

A work of original observation of the highest merit. We recommend the treatise to every one who wishes to become a correct auscultator. Based to a very large extent upon cases numerically examined, it carries the evidence of careful study and discrimination upon every page. It does credit to the author, and through him, to the profession in this country. It is, what we cannot call every book upon auscultation, a readable book.—*Am. Jour. Med. Sciences.*

BY THE SAME AUTHOR. (*Now Ready.*)

A PRACTICAL TREATISE ON THE DIAGNOSIS, PATHOLOGY, AND TREATMENT OF DISEASES OF THE HEART.

In one neat octavo volume, of about 500 pages, extra cloth. $2 75.

We do not know that Dr. Flint has written anything which is not first rate; but this, his latest contribution to medical literature, in our opinion, surpasses all the others. The work is most comprehensive in its scope, and most sound in the views it enunciates. The descriptions are clear and methodical; the statements are substantiated by facts, and are made with such simplicity and sincerity, that without them they would carry conviction. The style is admirably clear, direct, and free from dryness. With Dr. Walshe's excellent treatise before us, we have no hesitation in saying that Dr. Flint's book is the best work on the heart in the English language. —*Boston Med. and Surg. Journal*, Dec. 15, 1859.

We have thus endeavored to present our readers with a fair analysis of this remarkable work. Preferring to employ the very words of the distinguished author, wherever it was possible, we have essayed to condense into the briefest space a general view of his observations and suggestions, and to direct the attention of our brethren to the abounding stores of valuable matter here collected and arranged for their use and instruction. No medical library will hereafter be considered complete without this volume; and we trust it will promptly find its way into the hands of every American student and physician.—*N. Am. Med. Chir. Review*, Jan. 1860.

This last work of Prof. Flint will add much to his previous well-earned celebrity, as a writer of great force and beauty, and, with his previous work, places him at the head of American writers upon diseases of the chest. We have adopted his work upon the heart as a text-book, believing it to be more valuable for that purpose than any work of the kind that has yet appeared.—*Nashville Med. Journ.*, Dec. 1859.

With more than pleasure do we hail the advent of this work, for it fills a wide gap on the list of text-books for our schools, and is, for the practitioner, the most valuable practical work of its kind.—*N. O. Med. News*, Nov. 1859.

In regard to the merits of the work, we have no hesitation in pronouncing it full, accurate, and judicious. Considering the present state of science, such a work was much needed. It should be in the hands of every practitioner.—*Chicago Med. Journal*, April, 1860.

But these are very trivial spots, and in no wise prevent us from declaring our most hearty approval of the author's ability, industry, and conscientiousness.—*Dublin Quarterly Journal of Med. Sciences*, Feb. 1860.

He has labored on with the same industry and care, and his place among the *first* authors of our country is becoming fully established. To this end, the work whose title is given above, contributes in no small degree. Our space will not admit of an extended analysis, and we will close this brief notice by commending it without reserve to every class of readers in the profession.—*Peninsular Med. Journ.*, Feb. 1860.

FOWNES (GEORGE), PH. D., &c.

A MANUAL OF ELEMENTARY CHEMISTRY; Theoretical and Practical. From the seventh revised and corrected London edition. With one hundred and ninety-seven illustrations. Edited by Robert Bridges, M. D. In one large royal 12mo. volume, of 600 pages. In leather, $1 65; extra cloth, $1 50. (*Just Issued.*)

The death of the author having placed the editorial care of this work in the practised hands of Drs. Bence Jones and A. W. Hoffman, everything has been done in its revision which experience could suggest to keep it on a level with the rapid advance of chemical science. The additions requisite to this purpose have necessitated an enlargement of the page, notwithstanding which the work has been increased by about fifty pages. At the same time every care has been used to maintain its distinctive character as a condensed manual for the student, divested of all unnecessary detail or mere theoretical speculation. The additions have, of course, been mainly in the department of Organic Chemistry, which has made such rapid progress within the last few years, but yet equal attention has been bestowed on the other branches of the subject—Chemical Physics and Inorganic Chemistry—to present all investigations and discoveries of importance, and to keep up the reputation of the volume as a complete manual of the whole science, admirably adapted for the learner. By the use of a small but exceedingly clear type the matter of a large octavo is compressed within the convenient and portable limits of a moderate sized duodecimo, and at the very low price affixed, it is offered as one of the cheapest volumes before the profession.

Dr. Fownes' excellent work has been universally recognized everywhere in his own and this country, as the best elementary treatise on chemistry in the English tongue, and is very generally adopted, we believe, as the standard text-book in all our colleges, both literary and scientific.—*Charleston Med Journ. and Review*, Sept. 1859.

A standard manual, which has long enjoyed the reputation of embodying much knowledge in a small space. The author has achieved the difficult task of condensation with masterly tact. His book is concise without being dry, and brief without being too dogmatical or general.—*Virginia Med. and Surgical Journal.*

The work of Dr. Fownes has long been before the public, and its merits have been fully appreciated as the best text-book on chemistry now in existence. We do not, of course, place it in a rank superior to the works of Brande, Graham, Turner, Gregory, or Gmelin, but we say that, as a work for students, it is preferable to any of them.—*London Journal of Medicine.*

A work well adapted to the wants of the student It is an excellent exposition of the chief doctrines and facts of modern chemistry. The size of the work. and still more the condensed yet perspicuous style in which it is written, absolve it from the charges very properly urged against most manuals termed popular.—*Edinburgh Journal of Medical Science.*

FISKE FUND PRIZE ESSAYS — THE EFFECTS OF CLIMATE ON TUBERCULOUS DISEASE. By Edwin Lee, M. R. C. S., London, and THE INFLUENCE OF PREGNANCY ON THE DEVELOPMENT OF TUBERCLES By Edward Warren, M. D., of Edenton, N. C. Together in one neat 8vo volume, extra cloth. $1 00.

FRICK ON RENAL AFFECTIONS; their Diagnosis and Pathology. With illustrations. One volume, royal 12mo., extra cloth. 75 cents

FERGUSSON (WILLIAM), F. R. S.,

Professor of Surgery in King's College, London, &c.

A SYSTEM OF PRACTICAL SURGERY. Fourth American, from the third and enlarged London edition. In one large and beautifully printed octavo volume, of about 700 pages, with 393 handsome illustrations, leather. $3 00.

GRAHAM (THOMAS), F. R. S.

THE ELEMENTS OF INORGANIC CHEMISTRY, including the Applications of the Science in the Arts. New and much enlarged edition, by Henry Watts and Robert Bridges, M. D. Complete in one large and handsome octavo volume, of over 800 very large pages, with two hundred and thirty-two wood-cuts, extra cloth. $4 00.

*** Part II., completing the work from p. 431 to end, with Index, Title Matter, &c., may be had separate, cloth backs and paper sides. Price $2 50.

From Prof. E. N. Horsford, Harvard College.

It has, in its earlier and less perfect editions, been familiar to me, and the excellence of its plan and the clearness and completeness of its discussions, have long been my admiration.

No reader of English works on this science can afford to be without this edition of Prof. Graham's Elements.—*Silliman's Journal*, March, 1858.

From Prof. Wolcott Gibbs, N. Y. Free Academy.

The work is an admirable one in all respects, and its republication here cannot fail to exert a positive influence upon the progress of science in this country.

GRIFFITH (ROBERT E.), M. D., &c.

A UNIVERSAL FORMULARY, containing the methods of Preparing and Administering Officinal and other Medicines. The whole adapted to Physicians and Pharmaceutists. Second Edition, thoroughly revised, with numerous additions, by Robert P. Thomas. M. D., Professor of Materia Medica in the Philadelphia College of Pharmacy. In one large and handsome octavo volume, extra cloth, of 650 pages, double columns. $3 00; or in sheep, $3 25.

It was a work requiring much perseverance, and when published was looked upon as by far the best work of its kind that had issued from the American press. Prof Thomas has certainly "improved," as well as added to this Formulary, and has rendered it additionally deserving of the confidence of pharmaceutists and physicians.—*Am. Journal of Pharmacy.*

We are happy to announce a new and improved edition of this, one of the most valuable and useful works that have emanated from an American pen. It would do credit to any country, and will be found of daily usefulness to practitioners of medicine; it is better adapted to their purposes than the dispensatories.—*Southern Med. and Surg. Journal.*

It is one of the most useful books a country practitioner can possibly have.—*Medical Chronicle.*

This is a work of six hundred and fifty-one pages, embracing all on the subject of preparing and administering medicines that can be desired by the physician and pharmaceutist.—*Western Lancet.*

The amount of useful, every-day matter, for a practicing physician, is really immense.—*Boston Med. and Surg. Journal.*

This edition has been greatly improved by the revision and ample additions of Dr Thomas, and is now, we believe, one of the most complete works of its kind in any language. The additions amount to about seventy pages, and no effort has been spared to include in them all the recent improvements. A work of this kind appears to us indispensable to the physician, and there is none we can more cordially recommend.—*N. Y. Journal of Medicine.*

GROSS (SAMUEL D.), M. D.,

Professor of Surgery in the Jefferson Medical College of Philadelphia, &c.

Just Issued.

A SYSTEM OF SURGERY: Pathological, Diagnostic, Therapeutic, and Operative. Illustrated by NINE HUNDRED AND THIRTY-SIX ENGRAVINGS. In two large and beautifully printed octavo volumes, of nearly twenty-four hundred pages; strongly bound in leather, with raised bands. Price $12.

FROM THE AUTHOR'S PREFACE.

"The object of this work is to furnish a systematic and comprehensive treatise on the science and practice of surgery, considered in the broadest sense; one that shall serve the practitioner as a faithful and available guide in his daily routine of duty. It has been too much the custom of modern writers on this department of the healing art to omit certain topics altogether, and to speak of others at undue length, evidently assuming that their readers could readily supply the deficiencies from other sources, or that what has been thus slighted is of no particular practical value. My aim has been to embrace the whole domain of surgery, and to allot to every subject its legitimate claim to notice in the great family of external diseases and accidents. How far this object has been accomplished, it is not for me to determine. It may safely be affirmed, however, that there is no topic, properly appertaining to surgery, that will not be found to be discussed, to a greater or less extent, in these volumes. If a larger space than is customary has been devoted to the consideration of inflammation and its results, or the great principles of surgery, it is because of the conviction, grounded upon long and close observation, that there are no subjects so little understood by the general practitioner. Special attention has also been bestowed upon the discrimination of diseases; and an elaborate chapter has been introduced on general diagnosis."

That these intentions have been carried out in the fullest and most elaborate manner is sufficiently shown by the great extent of the work, and the length of time during which the author has been concentrating on the task his studies and his experience, guided by the knowledge which twenty years of lecturing on surgical topics have given him of the wants of the profession.

Of Dr. Gross's treatise on Surgery we can say no more than that it is the most elaborate and complete work on this branch of the healing art which has ever been published in any country. A systematic work, it admits of no analytical review; but, did our space permit, we should gladly give some extracts from it, to enable our readers to judge of the classical style of the author, and the exhausting way in which each subject is treated.—*Dublin Quarterly Journal of Med. Science*, Nov. 1859.

The work is so superior to its predecessors in matter and extent, as well as in illustrations and style of publication, that we can honestly recommend it as the best work of the kind to be taken home by the young practitioner.—*Am. Med. Journ.*, Jan. 1860.

The treatise of Prof. Gross is not, therefore, a mere text-book for undergraduates, but a systematic record of more than thirty years' experience, reading, and reflection by a man of observation, sound judgment, and rare practical tact, and as such deserves to take rank with the renowned productions of a similar character, by Vidal and Boyer, of France, or those of Chelius, Blasius, and Langenbeck, of Germany. Hence, we do not hesitate to express the opinion that it will speedily take the same elevated position in regard to surgery that has been given by common consent to the masterly work of Pereira in Materia Medica, or to Todd and Bowman in Physiology.—*N. O. Med. and Surg. Journal*, Jan. 1860.

At present, however, our object is not to review the work (this we purpose doing hereafter), but simply to announce its appearance, that in the meantime our readers may procure and examine it for themselves. But even this much we cannot do without expressing the opinion that, in putting forth these two volumes, Dr. Gross has reared for himself a lasting monument to his skill as a surgeon, and to his industry and learning as an author.—*St. Louis Med. and Surg. Journal*, Nov. 1859.

With pleasure we record the completion of this long-anticipated work. The reputation which the author has for many years sustained, both as a surgeon and as a writer, had prepared us to expect a treatise of great excellence and originality; but we confess we were by no means prepared for the work which is before us—the most complete treatise upon surgery ever published, either in this or any other country, and we might, perhaps, safely say, the most original. There is no subject belonging properly to surgery which has not received from the author a due share of attention. Dr. Gross has supplied a want in surgical literature which has long been felt by practitioners; he has furnished us with a complete practical treatise upon surgery in all its departments. As Americans, we are proud of the achievement; as surgeons, we are most sincerely thankful to him for his extraordinary labors in our behalf.—*N. Y. Monthly Review and Buffalo Med. Journal*, Oct. 1850.

BY THE SAME AUTHOR.

ELEMENTS OF PATHOLOGICAL ANATOMY. Third edition, thoroughly revised and greatly improved. In one large and very handsome octavo volume, with about three hundred and fifty beautiful illustrations, of which a large number are from original drawings. Price in extra cloth, $4 75; leather, raised bands, $5 25. (*Lately Published.*)

The very rapid advances in the Science of Pathological Anatomy during the last few years have rendered essential a thorough modification of this work, with a view of making it a correct exponent of the present state of the subject. The very careful manner in which this task has been executed, and the amount of alteration which it has undergone, have enabled the author to say that "with the many changes and improvements now introduced, the work may be regarded almost as a new treatise," while the efforts of the author have been seconded as regards the mechanical execution of the volume, rendering it one of the handsomest productions of the American press.

We most sincerely congratulate the author on the successful manner in which he has accomplished his proposed object. His book is most admirably calculated to fill up a blank which has long been felt to exist in this department of medical literature, and as such must become very widely circulated amongst all classes of the profession.—*Dublin Quarterly Journ. of Med. Science*, Nov. 1857.

We have been favorably impressed with the general manner in which Dr. Gross has executed his task of affording a comprehensive digest of the present state of the literature of Pathological Anatomy, and have much pleasure in recommending his work to our readers, as we believe one well deserving of diligent perusal and careful study.—*Montreal Med. Chron.*, Sept. 1857.

BY THE SAME AUTHOR.

A PRACTICAL TREATISE ON FOREIGN BODIES IN THE AIR-PASSAGES. In one handsome octavo volume, extra cloth, with illustrations. pp. 468. $2 75.

GROSS (SAMUEL D.), M. D.,

Professor of Surgery in the Jefferson Medical College of Philadelphia, &c.

A PRACTICAL TREATISE ON THE DISEASES, INJURIES, AND MALFORMATIONS OF THE URINARY BLADDER, THE PROSTATE GLAND, AND THE URETHRA.

Second Edition, revised and much enlarged, with one hundred and eighty-four illustrations. In one large and very handsome octavo volume, of over nine hundred pages. In leather, raised bands, $5 25; extra cloth, $4 75.

Philosophical in its design, methodical in its arrangement, ample and sound in its practical details, it may in truth be said to leave scarcely anything to be desired on so important a subject.—*Boston Med. and Surg Journal.*

Whoever will peruse the vast amount of valuable practical information it contains, will, we think, agree with us, that there is no work in the English language which can make any just pretensions to be its equal.—*N. Y. Journal of Medicine.*

A volume replete with truths and principles of the utmost value in the investigation of these diseases. -*American Medical Journal.*

GRAY (HENRY), F. R. S.,

Lecturer on Anatomy at St. George's Hospital, London, &c.

ANATOMY, DESCRIPTIVE AND SURGICAL.

The Drawings by H. V. Carter, M. D., late Demonstrator on Anatomy at St. George's Hospital; the Dissections jointly by the Author and Dr. Carter. In one magnificent imperial octavo volume, of nearly 800 pages, with 363 large and elaborate engravings on wood. Price in extra cloth, $6 25; leather raised bands, $7 00. (*Just Issued.*)

The author has endeavored in this work to cover a more extended range of subjects than is customary in the ordinary text-books, by giving not only the details necessary for the student, but also the application of those details in the practice of medicine and surgery, thus rendering it both a guide for the learner, and an admirable work of reference for the active practitioner. The engravings form a special feature in the work, many of them being the size of nature, nearly all original, and having the names of the various parts printed on the body of the cut, in place of figures of reference with descriptions at the foot. They thus form a complete and splendid series, which will greatly assist the student in obtaining a clear idea of Anatomy, and will also serve to refresh the memory of those who may find in the exigencies of practice the necessity of recalling the details of the dissecting room; while combining, as it does, a complete Atlas of Anatomy, with a thorough treatise on systematic, descriptive, and applied Anatomy, the work will be found of essential use to all physicians who receive students in their offices, relieving both preceptor and pupil of much labor in laying the groundwork of a thorough medical education.

The work before us is one entitled to the highest praise, and we accordingly welcome it as a valuable addition to medical literature. Intermediate in fulness of detail between the treatises of Sharpey and of Wilson, its characteristic merit lies in the number and excellence of the engravings it contains. Most of these are original, of much larger than ordinary size, and admirably executed. The various parts are also lettered after the plan adopted in Holden's Osteology. It would be difficult to over-estimate the advantages offered by this mode of pictorial illustration. Bones, ligaments, muscles, bloodvessels, and nerves are each in turn figured, and marked with their appropriate names; thus enabling the student to comprehend, at a glance, what would otherwise often be ignored, or at any rate, acquired only by prolonged and irksome application. In conclusion, we heartily commend the work of Mr. Gray to the attention of the medical profession, feeling certain that it should be regarded as one of the most valuable contributions ever made to educational literature.—*N. Y. Monthly Review.* Dec. 1859.

In this view, we regard the work of Mr. Gray as far better adapted to the wants of the profession, and especially of the student, than any treatise on anatomy yet published in this country. It is destined. we believe, to supersede all others, both as a manual of dissections, and a standard of reference to the student of general or relative anatomy.—*N. Y. Journal of Medicine*, Nov. 1859.

This is by all comparison the most excellent work on Anatomy extant. It is just the thing that has been long desired by the profession. With such a guide as this, the student of anatomy, the practitioner of medicine, and the surgical devotee have all a newer, clearer, and more radiant light thrown upon the intricacies and mysteries of this wonderful science, and are thus enabled to accomplish results which hitherto seemed possible only to the specialist. The plates, which are copied from recent dissections, are so well executed, that the most superficial observer cannot fail to perceive the positions, relations, and distinctive features of the various parts, and to take in more of anatomy at a glance, than by many long hours of diligent study over the most erudite treatise, or, perhaps, at the dissecting table itself.—*Med. Journ. of N. Carolina*, Oct. 1859.

For this truly admirable work the profession is indebted to the distinguished author of "Gray on the Spleen." The vacancy it fills has been long felt to exist in this country. Mr. Gray writes throughout with both branches of his subject in view. His description of each particular part is followed by a notice of its relations to the parts with which it is connected, and this, too, sufficiently ample for all the purposes of the operative surgeon. After describing the bones and muscles, he gives a concise statement of the fractures to which the bones of the extremities are most liable, together with the amount and direction of the displacement to which the fragments are subjected by muscular action. The section on arteries is remarkably full and accurate. Not only is the surgical anatomy given to every important vessel, with directions for its ligation, but at the end of the description of each arterial trunk we have a useful summary of the irregularities which may occur in its origin, course, and termination.—*N. A. Med. Chir. Review*, Mar. 1859.

Mr. Gray's book, in excellency of arrangement and completeness of execution, exceeds any work on anatomy hitherto published in the English language, affording a complete view of the structure of the human body, with especial reference to practical surgery. Thus the volume constitutes a perfect book of reference for the practitioner, demanding a place in even the most limited library of the physician or surgeon, and a work of necessity for the student to fix in his mind what he has learned by the dissecting knife from the book of nature.—*The Dublin Quarterly Journal of Med. Sciences*, Nov. 1858.

In our judgment, the mode of illustration adopted in the present volume cannot but present many advantages to the student of anatomy. To the zealous disciple of Vesalius, earnestly desirous of real improvement, the book will certainly be of immense value; but, at the same time, we must also confess that to those simply desirous of "cramming" it will be an undoubted godsend. The peculiar value of Mr. Gray's mode of illustration is nowhere more markedly evident than in the chapter on osteology, and especially in those portions which treat of the bones of the head and of their development. The study of these parts is thus made one of comparative ease, if not of positive pleasure; and those bugbears of the student, the temporal and sphenoid bones, are shorn of half their terrors. It is, in our estimation, an admirable and complete text-book for the student, and a useful work of reference for the practitioner; its pictorial character forming a novel element, to which we have already sufficiently alluded.—*Am. Journ. Med. Sci.*, July, 1859.

GIBSON'S INSTITUTES AND PRACTICE OF SURGERY. Eighth edition, improved and altered. With thirty-four plates. In two handsome octavo volumes, containing about 1,000 pages, leather, raised bands. $6 50.

GARDNER'S MEDICAL CHEMISTRY, for the use of Students and the Profession. In one royal 12mo. vol., cloth, pp. 396, with wood-cuts. $1.

GLUGE'S ATLAS OF PATHOLOGICAL HISTOLOGY. Translated, with Notes and Additions. by JOSEPH LEIDY, M. D. In one volume, very large imperial quarto, extra cloth, with 320 copper-plate figures, plain and colored, $5 00.

HUGHES' INTRODUCTION TO THE PRACTICE OF AUSCULTATION AND OTHER MODES OF PHYSICAL DIAGNOSIS IN DISEASES OF THE LUNGS AND HEART. Second edition 1 vol. royal 12mo., ex. cloth, pp. 304. $1 00.

HAMILTON (FRANK H.), M. D.,

Professor of Surgery in the University of Buffalo, &c.

A PRACTICAL TREATISE ON FRACTURES AND DISLOCATIONS. In

one large and handsome octavo volume, of over 750 pages, with 289 illustrations. $4 25. (*Now Ready*, January, 1860.)

This is a valuable contribution to the surgery of most important affections, and is the more welcome, inasmuch as at the present time we do not possess a single complete treatise on Fractures and Dislocations in the English language. It has remained for our American brother to produce a complete treatise upon the subject, and bring together in a convenient form those alterations and improvements that have been made from time to time in the treatment of these affections. One great and valuable feature in the work before us is the fact that it comprises all the improvements introduced into the practice of both English and American surgery, and though far from omitting mention of our continental neighbors, the author by no means encourages the notion—but too prevalent in some quarters—that nothing is good unless imported from France or Germany. The latter half of the work is devoted to the consideration of the various dislocations and their appropriate treatment. and its merit is fully equal to that of the preceding portion.—*The London Lancet*, May 5, 1860.

It is emphatically *the* book upon the subjects of which it treats, and we cannot doubt that it will continue so to be for an indefinite period of time. When we say, however, that we believe it will at once take its place as the best book for consultation by the practitioner; and that it will form the most complete, available, and reliable guide in emergencies of every nature connected with its subjects; and also that the student of surgery may make it his text-book with entire confidence, and with pleasure also, from its agreeable and easy style—we think our own opinion may be gathered as to its value.—*Boston Medical and Surgical Journal*, March 1, 1860.

The work is concise, judicious, and accurate, and adapted to the wants of the student, practitioner, and investigator, honorable to the author and to the profession.—*Chicago Med. Journal*, March, 1860.

We venture to say that this is not alone the only complete treatise on the subject in the language, but the best and most practical we have ever read. The arrangement is simple and systematic, the diction clear and graphic, and the illustrations numerous and remarkable for accuracy of delineation. The various mechanical appliances are faithfully illustrated, which will be a desideratum for those practitioners who cannot conveniently see the models applied.—*New York Med. Press*, Feb. 4, 1860.

We regard this work as an honor not only to its author, but to the profession of our country. Were we to review it thoroughly, we could not convey to the mind of the reader more forcibly our honest opinion expressed in the few words—we think it the best book of its kind extant. Every man interested in surgery will soon have this work on his desk. He who does not, will be the loser.—*New Orleans Medical News*, March, 1860.

Now that it is before us, we feel bound to say that much as was expected from it, and onerous as was the undertaking, it has surpassed expectation, and achieved more than was pledged in its behalf; for its title does not express in full the richness of its contents. On the whole, we are prouder of this work than of any which has for years emanated from the American medical press; its sale will certainly be very large in this country, and we anticipate its eliciting much attention in Europe.—*Nashville Medical Record*, Mar. 1860.

Every surgeon, young and old, should possess himself of it, and give it a careful perusal, in doing which he will be richly repaid.—*St. Louis Med. and Surg. Journal*, March, 1860.

Dr. Hamilton is fortunate in having succeeded in filling the void, so long felt, with what cannot fail to be at once accepted as a model monograph in some respects, and a work of classical authority. We sincerely congratulate the profession of the United States on the appearance of such a publication from one of their number. We have reason to be proud of it as an original work, both in a literary and scientific point of view. and to esteem it as a valuable guide in a most difficult and important branch of study and practice. On every account, therefore, we hope that it may soon be widely known abroad as an evidence of genuine progress on this side of the Atlantic, and further, that it may be still more widely known at home as an authoritative teacher from which every one may profitably learn, and as affording an example of honest, well-directed, and untiring industry in authorship which every surgeon may emulate.– *Am. Med. Journal*, April, 1860.

HOBLYN (RICHARD D.), M. D.

A DICTIONARY OF THE TERMS USED IN MEDICINE AND THE

COLLATERAL SCIENCES. A new American edition. Revised, with numerous Additions, by ISAAC HAYS, M. D., editor of the "American Journal of the Medical Sciences." In one large royal 12mo. volume, leather, of over 500 double columned pages. $1 50.

To both practitioner and student, we recommend this dictionary as being convenient in size, accurate in definition, and sufficiently full and complete for ordinary consultation.—*Charleston Med. Journ.*

We know of no dictionary better arranged and adapted. It is not encumbered with the obsolete terms of a bygone age, but it contains all that are now in use; embracing every department of medical science down to the very latest date.—*Western Lancet.*

Hoblyn's Dictionary has long been a favorite with us. It is the best book of definitions we have, and ought always to be upon the student's table.—*Southern Med. and Surg. Journal.*

HOLLAND'S MEDICAL NOTES AND REFLECTIONS. From the third London edition. In one handsome octavo volume, extra cloth. $3.

HORNER'S SPECIAL ANATOMY AND HISTOLOGY. Eighth edition. Extensively revised and modified. In two large octavo volumes, extra cloth, of more than 1000 pages, with over 300 illustrations. $6 00.

HABERSHON (S. O.), M. D.,

Assistant Physician to and Lecturer on Materia Medica and Therapeutics at Guy's Hospital, &c.

PATHOLOGICAL AND PRACTICAL OBSERVATIONS ON DISEASES

OF THE ALIMENTARY CANAL, ŒSOPHAGUS, STOMACH, CÆCUM, AND INTESTINES. With illustrations on wood. In one handsome octavo volume of 312 pages, extra cloth $1 75. (*Now Ready.*)

HODGE (HUGH L.), M. D.,

Professor of Midwifery and the Diseases of Women and Children in the University of Pennsylvania, &c.

ON DISEASES PECULIAR TO WOMEN, including Displacements of the Uterus. With original illustrations. In one beautifully printed octavo volume, of nearly 500 pages. (*Now Ready.*)

The profession will look with much interest on a volume embodying the long and extensive experience of Professor Hodge on an important branch of practice in which his opportunities for investigation have been so extensive. A short summary of the contents will show the scope of the work, and the manner in which the subject is presented. It will be seen that, with the exception of Displacements of the Uterus, he divides the Diseases peculiar to Women into two great constitutional classes—those arising from irritation, and those arising from sedation.

CONTENTS.

PART I. DISEASES OF IRRITATION.—CHAPTER I. Nervous Irritation, and its Consequences —II. Irritable Uterus.—III. Local Symptoms of Irritable Uterus: Menorrhagia and Hæmorrhagia; Leucorrhœa; Dysmenorrhœa —IV. Local Symptoms of Irritable Uterus; Complications.—V. General Symptoms of Irritable Uterus: Cerebro-spinal Irritations.—VI. General Symptoms of Irritable Uterus.—VII. Progress and Results of Irritable Uterus.—VIII. Causes and Pathology of Irritable Diseases —IX. Treatment of Irritable Uterus; Removal or Palliation of the Cause. —X. Treatment of Irritable Uterus: To Diminish or Destroy the Morbid Irritability —XI. Treatment of the Complications of Irritable Uterus.—XII. Treatment of the Complications of Irritable Uterus.

PART II. DISPLACEMENTS OF THE UTERUS.—CHAPTER I. Natural Position and Supports of the Uterus.—II. Varieties of Displacements of the Uterus, and their Causes.—III. Symptoms of Displacements of the Uterus.—IV. Treatment of Displacements of the Uterus.—V. Treatment of Displacements; Internal Supports.—VI. Treatment of Displacements; Lever Pessaries.—VII. Treatment of the Varieties of Displacements.—VIII. Treatment of Complications of Displacements of the Uterus.—IX. Treatment of Enlargements and Displacements of the Ovaries, &c.

PART III. DISEASES OF SEDATION.—CHAPTER I. Sedation and its Consequences: Organic and Nervous Sedation; Passive Congestion; Reaction; Treatment —II. Sedation of the Uterus; Amenorrhœa: Sedation of the Uterus from Moral Causes; Sedation of the Uterus from Physical Causes.—III. Diagnosis and Treatment of Sedation of the Uterus.

The illustrations, which are all original, are drawn to a uniform scale of one-half the natural size.

JONES (T. WHARTON), F. R. S.,

Professor of Ophthalmic Medicine and Surgery in University College, London, &c.

THE PRINCIPLES AND PRACTICE OF OPHTHALMIC MEDICINE AND SURGERY. With one hundred and ten illustrations. Second American from the second and revised London edition, with additions by EDWARD HARTSHORNE, M. D., Surgeon to Wills' Hospital, &c. In one large, handsome royal 12mo. volume, extra cloth, of 500 pages. $1 50.

JONES (C. HANDFIELD), F. R. S., & EDWARD H. SIEVEKING, M.D.,

Assistant Physicians and Lecturers in St. Mary's Hospital, London.

A MANUAL OF PATHOLOGICAL ANATOMY. First American Edition, Revised. With three hundred and ninety-seven handsome wood engravings. In one large and beautiful octavo volume of nearly 750 pages, leather. $3 75.

As a concise text-book, containing, in a condensed form, a complete outline of what is known in the domain of Pathological Anatomy, it is perhaps the best work in the English language. Its great merit consists in its completeness and brevity, and in this respect it supplies a great desideratum in our literature. Heretofore the student of pathology was obliged to glean from a great number of monographs, and the field was so extensive that but few cultivated it with any degree of success. As a simple work of reference, therefore, it is of great value to the student of pathological anatomy, and should be in every physician's library.—*Western Lancet.*

KIRKES (WILLIAM SENHOUSE), M. D.,

Demonstrator of Morbid Anatomy at St. Bartholomew's Hospital, &c.

A MANUAL OF PHYSIOLOGY. A new American, from the third and improved London edition. With two hundred illustrations. In one large and handsome royal 12mo. volume, leather. pp. 586. $2 00. (*Lately Published.*)

This is a new and very much improved edition of Dr. Kirkes' well-known Handbook of Physiology. It combines conciseness with completeness, and is, therefore, admirably adapted for consultation by the busy practitioner.—*Dublin Quarterly Journal.*

Its excellence is in its compactness, its clearness, and its carefully cited authorities. It is the most convenient of text-books. These gentlemen, Messrs. Kirkes and Paget, have really an immense talent for silence, which is not so common or so cheap as prating people fancy. They have the gift of telling us what we want to know, without thinking it necessary to tell us all they know.—*Boston Med and Surg. Journal.*

One of the very best handbooks of Physiology we possess—presenting just such an outline of the science as the student requires during his attendance upon a course of lectures, or for reference whilst preparing for examination.—*Am. Medical Journal.*

For the student beginning this study, and the practitioner who has but leisure to refresh his memory, this book is invaluable, as it contains all that it is important to know, without special details, which are read with interest only by those who would make a specialty, or desire to possess a critical knowledge of the subject.—*Charleston Med. Journal.*

KNAPP'S TECHNOLOGY; or, Chemistry applied to the Arts and to Manufactures. Edited by Dr. RONALDS, Dr. RICHARDSON, and Prof. W. R. JOHNSON. In two handsome 8vo. vols., with about 500 wood-engravings. $6 00.

LAYCOCK'S LECTURES ON THE PRINCIPLES AND METHODS OF MEDICAL OBSERVATION AND RESEARCH. For the Use of Advanced Students and Junior Practitioners. In one royal 12mo. volume, extra cloth. Price $1.

LALLEMAND AND WILSON.

A PRACTICAL TREATISE ON THE CAUSES, SYMPTOMS, AND TREATMENT OF SPERMATORRHŒA. By M. LALLEMAND. Translated and edited by HENRY J. MCDOUGALL. Third American edition. To which is added —— ON DISEASES OF THE VESICULÆ SEMINALES; AND THEIR ASSOCIATED ORGANS. With special reference to the Morbid Secretions of the Prostatic and Urethral Mucous Membrane. By MARRIS WILSON, M. D. In one neat octavo volume, of about 400 pp., extra cloth. $2 00. (*Just Issued.*)

LA ROCHE (R.), M. D., &c.

YELLOW FEVER, considered in its Historical, Pathological, Etiological, and Therapeutical Relations. Including a Sketch of the Disease as it has occurred in Philadelphia from 1699 to 1854, with an examination of the connections between it and the fevers known under the same name in other parts of temperate as well as in tropical regions. In two large and handsome octavo volumes of nearly 1500 pages, extra cloth. $7 00.

From Professor S. H. Dickson, Charleston, S. C., September 18, 1855.

A monument of intelligent and well applied research, almost without example. It is, indeed, in itself, a large library, and is destined to constitute the special resort as a book of reference, in the subject of which it treats, to all future time.

We have not time at present, engaged as we are, by day and by night, in the work of combating this very disease, now prevailing in our city, to do more than give this cursory notice of what we consider as undoubtedly the most able and erudite medical publication our country has yet produced. But in view of the startling fact, that this, the most malignant and unmanageable disease of modern times, has for several years been prevailing in our country to a greater extent than ever before; that it is no longer confined to either large or small cities, but penetrates country villages, plantations, and farmhouses; that it is treated with scarcely better success now than thirty or forty years ago; that there is vast mischief done by ignorant pretenders to knowledge in regard to the disease, and in view of the probability that a majority of southern physicians will be called upon to treat the disease, we trust that this able and comprehensive treatise will be very generally read in the south.—*Memphis Med. Recorder.*

BY THE SAME AUTHOR.

PNEUMONIA; its Supposed Connection, Pathological and Etiological, with Au- tumnal Fevers, including an Inquiry into the Existence and Morbid Agency of Malaria. In one handsome octavo volume, extra cloth, of 500 pages. $3 00.

LUDLOW (J. L.), M. D.

A MANUAL OF EXAMINATIONS upon Anatomy, Physiology, Surgery, Practice of Medicine, Obstetrics, Materia Medica, Chemistry, Pharmacy, and Therapeutics. To which is added a Medical Formulary. Third edition, thoroughly revised and greatly extended and enlarged. With 370 illustrations. In one handsome royal 12mo. volume, leather, of 816 large pages. $2 50.

The great popularity of this volume, and the numerous demands for it during the two years in which it has been out of print, have induced the author in its revision to spare no pains to render it a correct and accurate digest of the most recent condition of all the branches of medical science. In many respects it may, therefore, be regarded rather as a new book than a new edition, an entire section on Physiology having been added, as also one on Organic Chemistry, and many portions having been rewritten. A very complete series of illustrations has been introduced, and every care has been taken in the mechanical execution to render it a convenient and satisfactory book for study or reference. The arrangement of the volume in the form of question and answer renders it especially suited for the office examination of students and for those preparing for graduation.

We know of no better companion for the student during the hours spent in the lecture room, or to refresh, at a glance, his memory of the various topics *crammed* into his head by the various professors to whom he is compelled to listen.—*Western Lancet,* May, 1857.

LEHMANN (C. G.)

PHYSIOLOGICAL CHEMISTRY. Translated from the second edition by GEORGE E. DAY, M. D., F. R. S., &c., edited by R. E. ROGERS, M. D., Professor of Chemistry in the Medical Department of the University of Pennsylvania, with illustrations selected from Funke's Atlas of Physiological Chemistry, and an Appendix of plates. Complete in two large and handsome octavo volumes, extra cloth, containing 1200 pages, with nearly two hundred illustrations. $6 00.

The work of Lehmann stands unrivalled as the most comprehensive book of reference and information extant on every branch of the subject on which it treats.—*Edinburgh Journal of Medical Science.*

The most important contribution as yet made to Physiological Chemistry.—*Am. Journal Med. Sciences,* Jan. 1856.

BY THE SAME AUTHOR. (*Lately Published.*)

MANUAL OF CHEMICAL PHYSIOLOGY. Translated from the German, with Notes and Additions, by J. CHESTON MORRIS, M. D., with an Introductory Essay on Vital Force, by Professor SAMUEL JACKSON, M. D., of the University of Pennsylvania. With illustrations on wood. In one very handsome octavo volume, extra cloth, of 336 pages. $2 25.

From Prof. Jackson's Introductory Essay.

In adopting the handbook of Dr. Lehmann as a manual of Organic Chemistry for the use of the students of the University, and in recommending his original work of PHYSIOLOGICAL CHEMISTRY for their more mature studies, the high value of his researches, and the great weight of his authority in that important department of medical science are fully recognized.

LAWRENCE (W.), F. R. S., &c.

A TREATISE ON DISEASES OF THE EYE. A new edition, edited, with numerous additions, and 243 illustrations, by ISAAC HAYS, M. D., Surgeon to Will's Hospital, &c. In one very large and handsome octavo volume, of 950 pages, strongly bound in leather with raised bands. $5 00.

MEIGS (CHARLES D.), M. D.,

Professor of Obstetrics, &c. in the Jefferson Medical College, Philadelphia.

OBSTETRICS: THE SCIENCE AND THE ART. Third edition, revised and improved. With one hundred and twenty-nine illustrations. In one beautifully printed octavo volume, leather, of seven hundred and fifty-two large pages. $3 75.

The rapid demand for another edition of this work is a sufficient expression of the favorable verdict of the profession. In thus preparing it a third time for the press, the author has endeavored to render it in every respect worthy of the favor which it has received. To accomplish this he has thoroughly revised it in every part. Some portions have been rewritten, others added, new illustrations have been in many instances substituted for such as were not deemed satisfactory, while, by an alteration in the typographical arrangement, the size of the work has not been increased, and the price remains unaltered. In its present improved form, it is, therefore, hoped that the work will continue to meet the wants of the American profession as a sound, practical, and extended SYSTEM OF MIDWIFERY.

Though the work has received only five pages of enlargement, its chapters throughout wear the impress of careful revision. Expunging and rewriting, remodelling its sentences, with occasional new material, all evince a lively desire that it shall deserve to be regarded as improved in *manner* as well as *matter*. In the *matter*, every stroke of the pen has increased the value of the book, both in expungings and additions —*Western Lancet*, Jan. 1857.

The best American work on Midwifery that is accessible to the student and practitioner—*N. W. Med. and Surg. Journal*, Jan. 1857.

This is a standard work by a great American Obstetrician. It is the third and last edition, and, in the language of the preface, the author has "brought the subject up to the latest dates of real improvement in our art and Science."—*Nashville Journ. of Med. and Surg.*, May, 1857.

BY THE SAME AUTHOR. (*Just Issued.*)

WOMAN: HER DISEASES AND THEIR REMEDIES. A Series of Lectures to his Class. Fourth and Improved edition. In one large and beautifully printed octavo volume, leather, of over 700 pages. $3 60.

In other respects, in our estimation, too much cannot be said in praise of this work. It abounds with beautiful passages, and for conciseness, for originality, and for all that is commendable in a work on the diseases of females, it is not excelled, and probably not equalled in the English language. On the whole, we know of no work on the diseases of women which we can so cordially commend to the student and practitioner as the one before us.—*Ohio Med. and Surg. Journal.*

The body of the book is worthy of attentive consideration, and is evidently the production of a clever, thoughtful, and sagacious physician. Dr. Meigs's letters on the diseases of the external organs, contain many interesting and rare cases, and many instructive observations. We take our leave of Dr. Meigs, with a high opinion of his talents and originality.—*The British and Foreign Medico-Chirurgical Review.*

Every chapter is replete with practical instruction, and bears the impress of being the composition of an acute and experienced mind. There is a terseness, and at the same time an accuracy in his description of symptoms, and in the rules for diagnosis, which cannot fail to recommend the volume to the attention of the reader.—*Ranking's Abstract.*

It contains a vast amount of practical knowledge, by one who has accurately observed and retained the experience of many years.—*Dublin Quarterly Journal.*

Full of important matter, conveyed in a ready and agreeable manner.—*St. Louis Med. and Surg. Jour.*

There is an off-hand fervor, a glow, and a warm-heartedness infecting the effort of Dr. Meigs, which is entirely captivating, and which absolutely hurries the reader through from beginning to end. Besides, the book teems with solid instruction, and it shows the very highest evidence of ability, viz., the clearness with which the information is presented. We know of no better test of one's understanding a subject than the evidence of the power of lucidly explaining it. The most elementary, as well as the obscurest subjects, under the pencil of Prof. Meigs, are isolated and made to stand out in such bold relief, as to produce distinct impressions upon the mind and memory of the reader.—*The Charleston Med. Journal.*

Professor Meigs has enlarged and amended this great work, for such it unquestionably is, having passed the ordeal of criticism at home and abroad, but been improved thereby; for in this new edition the author has introduced real improvements, and increased the value and utility of the book immeasurably. It presents so many novel, bright, and sparkling thoughts; such an exuberance of new ideas on almost every page, that we confess ourselves to have become enamored with the book and its author; and cannot withhold our congratulations from our Philadelphia confreres, that such a teacher is in their service.—*N. Y. Med. Gazette.*

BY THE SAME AUTHOR.

ON THE NATURE, SIGNS, AND TREATMENT OF CHILDBED FEVER. In a Series of Letters addressed to the Students of his Class. In one handsome octavo volume, extra cloth, of 365 pages. $2 50.

The instructive and interesting author of this work, whose previous labors have placed his countrymen under deep and abiding obligations, again challenges their admiration in the fresh and vigorous, attractive and racy pages before us. It is a delectable book. * * * This treatise upon childbed fevers will have an extensive sale, being destined, as it deserves, to find a place in the library of every practitioner who scorns to lag in the rear.—*Nashville Journal of Medicine and Surgery.*

BY THE SAME AUTHOR; WITH COLORED PLATES.

A TREATISE ON ACUTE AND CHRONIC DISEASES OF THE NECK OF THE UTERUS. With numerous plates, drawn and colored from nature in the highest style of art. In one handsome octavo volume, extra cloth. $4 50.

MAYNE'S DISPENSATORY AND THERAPEUTICAL REMEMBRANCER. With every Practical Formula contained in the three British Pharmacopœias. Edited, with the addition of the Formulæ of the U. S. Pharmacopœia, by R. E. GRIFFITH, M. D. 1 12mo. vol. ex. cl., 300 pp. 75 c.

MALGAIGNE'S OPERATIVE SURGERY, based on Normal and Pathological Anatomy. Translated from the French by FREDERICK BRITTAN, A. B., M. D. With numerous illustrations on wood. In one handsome octavo volume, extra cloth, of nearly six hundred pages. $2 25.

MACLISE (JOSEPH), SURGEON.

SURGICAL ANATOMY. Forming one volume, very large imperial quarto. With sixty-eight large and splendid Plates, drawn in the best style and beautifully colored. Containing one hundred and ninety Figures, many of them the size of life. Together with copious and explanatory letter-press. Strongly and handsomely bound in extra cloth, being one of the cheapest and best executed Surgical works as yet issued in this country. $11 00.

*** The size of this work prevents its transmission through the post-office as a whole, but those who desire to have copies forwarded by mail, can receive them in five parts, done up in stout wrappers. Price $9 00.

One of the greatest artistic triumphs of the age in Surgical Anatomy.—*British American Medical Journal.*

No practitioner whose means will admit should fail to possess it.—*Ranking's Abstract.*

Too much cannot be said in its praise; indeed, we have not language to do it justice.—*Ohio Medical and Surgical Journal.*

The most accurately engraved and beautifully colored plates we have ever seen in an American book—one of the best and cheapest surgical works ever published.—*Buffalo Medical Journal.*

It is very rare that so elegantly printed, so well illustrated, and so useful a work, is offered at so moderate a price.—*Charleston Medical Journal.*

Its plates can boast a superiority which places them almost beyond the reach of competition.—*Medical Examiner.*

Country practitioners will find these plates of immense value.—*N. Y. Medical Gazette.*

A work which has no parallel in point of accuracy and cheapness in the English language.—*N. Y. Journal of Medicine.*

We are extremely gratified to announce to the profession the completion of this truly magnificent work, which, as a whole, certainly stands unrivalled, both for accuracy of drawing, beauty of coloring, and all the requisite explanations of the subject in hand.—*The New Orleans Medical and Surgical Journal.*

This is by far the ablest work on Surgical Anatomy that has come under our observation. We know of no other work that would justify a student, in any degree, for neglect of actual dissection. In those sudden emergencies that so often arise, and which require the instantaneous command of minute anatomical knowledge, a work of this kind keeps the details of the dissecting-room perpetually fresh in the memory.—*The Western Journal of Medicine and Surgery.*

MILLER (HENRY), M. D.,

Professor of Obstetrics and Diseases of Women and Children in the University of Louisville.

PRINCIPLES AND PRACTICE OF OBSTETRICS, &c.; including the Treatment of Chronic Inflammation of the Cervix and Body of the Uterus considered as a frequent cause of Abortion. With about one hundred illustrations on wood. In one very handsome octavo volume, of over 600 pages. (*Lately Published.*) $3 75.

The reputation of Dr. Miller as an obstetrician is too widely spread to require the attention of the profession to be specially called to a volume containing the experience of his long and extensive practice. The very favorable reception accorded to his "Treatise on Human Parturition," issued some years since, is an earnest that the present work will fulfil the author's intention of providing within a moderate compass a complete and trustworthy text-book for the student, and book of reference for the practitioner.

We congratulate the author that the task is done. We congratulate him that he has given to the medical public a work which will secure for him a high and permanent position among the standard authorities on the principles and practice of obstetrics. Congratulations are not less due to the medical profession of this country, on the acquisition of a treatise embodying the results of the studies, reflections, and experience of Prof. Miller. Few men, if any, in this country, are more competent than he to write on this department of medicine. Engaged for thirty-five years in an extended practice of obstetrics, for many years a teacher of this branch of instruction in one of the largest of our institutions, a diligent student as well as a careful observer, an original and independent thinker, wedded to no hobbies, ever ready to consider without prejudice new views, and to adopt innovations if they are really improvements, and withal a clear, agreeable writer, a practical treatise from his pen could not fail to possess great value.—*Buffalo Med Journal*, Mar. 1858.

In fact, this volume must take its place among the standard systematic treatises on obstetrics; a position to which its merits justly entitle it. The style is such that the descriptions are clear, and each subject is discussed and elucidated with due regard to its practical bearings, which cannot fail to make it acceptable and valuable to both students and practitioners. We cannot, however, close this brief notice without congratulating the author and the profession on the production of such an excellent treatise. The author is a western man of whom we feel proud, and we cannot but think that his book will find many readers and warm admirers wherever obstetrics is taught and studied as a science and an art.—*The Cincinnati Lancet and Observer*, Feb. 1858.

A most respectable and valuable addition to our home medical literature, and one reflecting credit alike on the author and the institution to which he is attached. The student will find in this work a most useful guide to his studies; the country practitioner, rusty in his reading, can obtain from its pages a fair résumé of the modern literature of the science; and we hope to see this American production generally consulted by the profession.—*Va. Med. Journal*, Feb. 1858.

MACKENZIE (W.), M. D.,

Surgeon Oculist in Scotland in ordinary to Her Majesty, &c. &c.

A PRACTICAL TREATISE ON DISEASES AND INJURIES OF THE EYE. To which is prefixed an Anatomical Introduction explanatory of a Horizontal Section of the Human Eyeball, by THOMAS WHARTON JONES, F. R. S. From the Fourth Revised and Enlarged London Edition. With Notes and Additions by ADDINELL HEWSON, M. D., Surgeon to Wills Hospital, &c. &c. In one very large and handsome octavo volume, leather, raised bands, with plates and numerous wood-cuts. $5 25.

The treatise of Dr. Mackenzie indisputably holds the first place, and forms, in respect of learning and research, an Encyclopædia unequalled in extent by any other work of the kind, either English or foreign. —*Dixon on Diseases of the Eye.*

Few modern books on any department of medicine or surgery have met with such extended circulation, or have procured for their authors a like amount of European celebrity. The immense research which it displayed, the thorough acquaintance with the subject, practically as well as theoretically, and the able manner in which the author's stores of learning and experience were rendered available for general use, at once procured for the first edition, as well on the continent as in this country, that high position as a standard work which each successive edition has more firmly established. We consider it the duty of every one who has the love of his profession and the welfare of his patient at heart, to make himself familiar with this the most complete work in the English language upon the diseases of the eye. —*Med. Times and Gazette.*

MILLER (JAMES), F. R. S. E.,

Professor of Surgery in the University of Edinburgh, &c.

PRINCIPLES OF SURGERY. Fourth American, from the third and revised Edinburgh edition. In one large and very beautiful volume, leather, of 700 pages, with two hundred and forty illustrations on wood. $3 75.

The work of Mr. Miller is too well and too favorably known among us, as one of our best text-books, to render any further notice of it necessary than the announcement of a new edition, the *fourth* in our country, a proof of its extensive circulation among us. As a concise and reliable exposition of the science of modern surgery, it stands deservedly high—we know not its superior.—*Boston Med. and Surg. Journal.*

The work takes rank with Watson's Practice of Physic; it certainly does not fall behind that great work in soundness of principle or depth of reasoning and research. No physician who values his reputation, or seeks the interests of his clients, can acquit himself before his God and the world without making himself familiar with the sound and philosophical views developed in the foregoing book.—*New Orleans Med. and Surg. Journal.*

BY THE SAME AUTHOR. (*Just Issued.*)

THE PRACTICE OF SURGERY. Fourth American from the last Edinburgh edition. Revised by the American editor. Illustrated by three hundred and sixty-four engravings on wood. In one large octavo volume, leather, of nearly 700 pages. $3 75.

No encomium of ours could add to the popularity of Miller's Surgery. Its reputation in this country is unsurpassed by that of any other work, and, when taken in connection with the author's *Principles of Surgery*, constitutes a whole, without reference to which no conscientious surgeon would be willing practice his art.—*Southern Med. and Surg. Journal.*

It is seldom that two volumes have ever made so profound an impression in so short a time as the "Principles" and the "Practice" of Surgery by Mr. Miller—or so richly merited the reputation they have acquired. The author is an eminently sensible, practical, and well-informed man, who knows exactly what he is talking about and exactly how to talk it.—*Kentucky Medical Recorder.*

By the almost unanimous voice of the profession, his works, both on the principles and practice of surgery have been assigned the highest rank. If we were limited to but one work on surgery, that one should be Miller's, as we regard it as superior to all others.—*St. Louis Med. and Surg. Journal.*

The author has in this and his "Principles," presented to the profession one of the most complete and reliable systems of Surgery extant. His style of writing is original, impressive, and engaging, energetic, concise, and lucid. Few have the faculty of condensing so much in small space, and at the same time so persistently holding the attention. Whether as a text-book for students or a book of reference for practitioners, it cannot be too strongly recommended.—*Southern Journal of Med. and Physical Sciences.*

MORLAND (W. W.), M. D.,

Fellow of the Massachusetts Medical Society, &c.

DISEASES OF THE URINARY ORGANS; a Compendium of their Diagnosis, Pathology, and Treatment. With illustrations. In one large and handsome octavo volume, of about 600 pages, extra cloth. (*Just Issued.*) $3 50.

Taken as a whole, we can recommend Dr. Morland's compendium as a very desirable addition to the library of every medical or surgical practitioner.—*Brit. and For. Med.-Chir. Rev.*, April, 1859.

Every medical practitioner whose attention has been to any extent attracted towards the class of diseases to which this treatise relates, must have often and sorely experienced the want of some full, yet concise recent compendium to which he could refer. This desideratum has been supplied by Dr. Morland, and it has been ably done. He has placed before us a full, judicious, and reliable digest. Each subject is treated with sufficient minuteness, yet in a succinct, narrational style, such as to render the work one of great interest, and one which will prove in the highest degree useful to the general practitioner. To the members of the profession in the country it will be peculiarly valuable, on account of the characteristics which we have mentioned, and the one broad aim of practical utility which is kept in view, and which shines out upon every page, together with the skill which is evinced in the combination of this grand requisite with the utmost brevity which a just treatment of the subjects would admit.—*N. Y. Journ. of Medicine*, Nov. 1858.

MONTGOMERY (W. F.), M. D., M. R. I. A., &c.,

Professor of Midwifery in the King and Queen's College of Physicians in Ireland, &c.

AN EXPOSITION OF THE SIGNS AND SYMPTOMS OF PREGNANCY. With some other Papers on Subjects connected with Midwifery. From the second and enlarged English edition. With two exquisite colored plates, and numerous wood-cuts. In one very handsome octavo volume, extra cloth, of nearly 600 pages. (*Lately Published.*) $3 75.

A book unusually rich in practical suggestions.—*Am. Journal Med. Sciences*, Jan. 1857.

These several subjects so interesting in themselves, and so important, every one of them, to the most delicate and precious of social relations, controlling often the honor and domestic peace of a family, the legitimacy of offspring, or the life of its parent, are all treated with an elegance of diction, fulness of illustrations, acuteness and justice of reasoning, unparalleled in obstetrics, and unsurpassed in medicine. The reader's interest can never flag, so fresh, and vigorous, and classical is our author's style; and one forgets, in the renewed charm of every page, that it, and every line, and every word has been weighed and reweighed through years of preparation; that this is of all others the book of Obstetric Law, on each of its several topics; on all points connected with pregnancy, to be everywhere received as a manual of special jurisprudence, at once announcing fact, affording argument, establishing precedent, and governing alike the juryman, advocate, and judge. It is not merely in its legal relations that we find this work so interesting. Hardly a page but that has its hints or facts important to the general practitioner; and not a chapter without especial matter for the anatomist, physiologist, or pathologist.—*N. A. Med.-Chir. Review*, March, 1857.

MOHR (FRANCIS), PH. D., AND REDWOOD (THEOPHILUS).

PRACTICAL PHARMACY. Comprising the Arrangements, Apparatus, and Manipulations of the Pharmaceutical Shop and Laboratory. Edited, with extensive Additions, by Prof. WILLIAM PROCTER, of the Philadelphia College of Pharmacy. In one handsomely printed octavo volume, extra cloth, of 570 pages, with over 500 engravings on wood. $2 75.

NEILL (JOHN), M. D.,

Surgeon to the Pennsylvania Hospital, &c.; and

FRANCIS GURNEY SMITH, M. D.,

Professor of Institutes of Medicine in the Pennsylvania Medical College.

AN ANALYTICAL COMPENDIUM OF THE VARIOUS BRANCHES

OF MEDICAL SCIENCE; for the Use and Examination of Students. A new edition, revised and improved. In one very large and handsomely printed royal 12mo. volume, of about one thousand pages, with 374 wood-cuts. Strongly bound in leather, with raised bands. $3 00.

The very flattering reception which has been accorded to this work, and the high estimate placed upon it by the profession, as evinced by the constant and increasing demand which has rapidly exhausted two large editions, have stimulated the authors to render the volume in its present revision more worthy of the success which has attended it. It has accordingly been thoroughly examined, and such errors as had on former occasions escaped observation have been corrected, and whatever additions were necessary to maintain it on a level with the advance of science have been introduced. The extended series of illustrations has been still further increased and much improved, while, by a slight enlargement of the page, these various additions have been incorporated without increasing the bulk of the volume.

The work is, therefore, again presented as eminently worthy of the favor with which it has hitherto been received. As a book for daily reference by the student requiring a guide to his more elaborate text-books, as a manual for preceptors desiring to stimulate their students by frequent and accurate examination, or as a source from which the practitioners of older date may easily and cheaply acquire a knowledge of the changes and improvement in professional science, its reputation is permanently established.

The best work of the kind with which we are acquainted.—*Med. Examiner.*

Having made free use of this volume in our examinations of pupils, we can speak from experience in recommending it as an admirable compend for students, and as especially useful to preceptors who examine their pupils. It will save the teacher much labor by enabling him readily to recall all of the points upon which his pupils should be examined. A work of this sort should be in the hands of every one who takes pupils into his office with a view of examining them; and this is unquestionably the best of its class.—*Transylvania Med. Journal.*

In the rapid course of lectures, where work for the students is heavy, and review necessary for an examination, a compend is not only valuable, but it is almost a *sine qua non.* The one before us is, in most of the divisions, the most unexceptionable of all books of the kind that we know of. The newest and soundest doctrines and the latest improvements and discoveries are explicitly, though concisely, laid before the student. There is a class to whom we very sincerely commend this cheap book as worth its weight in silver—that class is the graduates in medicine of more than ten years' standing, who have not studied medicine since. They will perhaps find out from it that the science is not exactly now what it was when they left it off.—*The Stethoscope.*

NELIGAN (J. MOORE), M. D., M. R. I. A., &c.

(A splendid work. Just Issued.)

ATLAS OF CUTANEOUS DISEASES.

In one beautiful quarto volume, extra cloth, with splendid colored plates, presenting nearly one hundred elaborate representations of disease. $4 50.

This beautiful volume is intended as a complete and accurate representation of all the varieties of Diseases of the Skin. While it can be consulted in conjunction with any work on Practice, it has especial reference to the author's "Treatise on Diseases of the Skin," so favorably received by the profession some years since. The publishers feel justified in saying that few more beautifully executed plates have ever been presented to the profession of this country.

Neligan's Atlas of Cutaneous Diseases supplies a long existent desideratum much felt by the largest class of our profession. It presents, in quarto size, 16 plates, each containing from 3 to 6 figures, and forming in all a total of 90 distinct representations of the different species of skin affections, grouped together in genera or families. The illustrations have been taken from nature, and have been copied with such fidelity that they present a striking picture of life; in which the reduced scale aptly serves to give, at a *coup d'œil*, the remarkable peculiarities of each individual variety. And while thus the disease is rendered more definable, there is yet no loss of proportion incurred by the necessary concentration. Each figure is highly colored, and so truthful has the artist been that the most fastidious observer could not justly take exception to the correctness of the execution of the pictures under his scrutiny.—*Montreal Med. Chronicle.*

BY THE SAME AUTHOR.

A PRACTICAL TREATISE ON DISEASES OF THE SKIN.

Third American edition. In one neat royal 12mo. volume, extra cloth, of 334 pages. $1 00.

☞ The two volumes will be sent by mail on receipt of *Five Dollars.*

OWEN ON THE DIFFERENT FORMS OF THE SKELETON, AND OF THE TEETH. One vol. royal 12mo., extra cloth with numerous illustrations. $1 25.

PIRRIE (WILLIAM), F. R. S. E.,

Professor of Surgery in the University of Aberdeen.

THE PRINCIPLES AND PRACTICE OF SURGERY.

Edited by John Neill, M. D., Professor of Surgery in the Penna. Medical College, Surgeon to the Pennsylvania Hospital, &c. In one very handsome octavo volume, leather, of 780 pages, with 316 illustrations. $3 75.

We know of no other surgical work of a reasonable size, wherein there is so much theory and practice, or where subjects are more soundly or clearly taught.—*The Stethoscope.*

Prof. Pirrie, in the work before us, has elaborately discussed the principles of surgery, and a safe and effectual practice predicated upon them. Perhaps no work upon this subject heretofore issued is so full upon the science of the art of surgery.—*Nashville Journal of Medicine and Surgery.*

PARRISH (EDWARD),

Lecturer on Practical Pharmacy and Materia Medica in the Pennsylvania Academy of Medicine, &c.

AN INTRODUCTION TO PRACTICAL PHARMACY. Designed as a Text-Book for the Student, and as a Guide for the Physician and Pharmaceutist. With many Formulæ and Prescriptions. Second edition, greatly enlarged and improved. In one handsome octavo volume of 720 pages, with several hundred Illustrations, extra cloth. $3 50. (*Now Ready.*)

During the short time in which this work has been before the profession, it has been received with very great favor, and in assuming the position of a standard authority, it has filled a vacancy which had been severely felt. Stimulated by this encouragement, the author, in availing himself of the opportunity of revision, has spared no pains to render it more worthy of the confidence bestowed upon it, and his assiduous labors have made it rather a new book than a new edition, many portions having been rewritten, and much new and important matter added. These alterations and improvements have been rendered necessary by the rapid progress made by pharmaceutical science during the last few years, and by the additional experience obtained in the practical use of the volume as a text-book and work of reference. To accommodate these improvements, the size of the page has been materially enlarged, and the number of pages considerably increased, presenting in all nearly *one-half more* matter than the last edition. The work is therefore now presented as a complete exponent of the subject in its most advanced condition. From the most ordinary matters in the dispensing office, to the most complicated details of the vegetable alkaloids, it is hoped that everything requisite to the practising physician, and to the apothecary, will be found fully and clearly set forth, and that the new matter alone will be worth more than the very moderate cost of the work to those who have been consulting the previous edition.

That Edward Parrish, in writing a book upon *practical* Pharmacy some few years ago—one eminently original and unique—did the medical and pharmaceutical professions a great and valuable service, no one, we think, who has had access to its pages will deny; doubly welcome, then, is this new edition, containing the added results of his recent and rich experience as an observer, teacher, and practical operator in the pharmaceutical laboratory. The excellent plan of the first is more thoroughly, and in detail, carried out in this edition.—*Peninsular Med. Journal*, Jan. 1860.

We know of no work on the subject which would be more indispensable to the physician or student desiring information on the subject of which it treats. With Griffith's "Medical Formulary" and this, the practising physician would be supplied with nearly or quite all the most useful information on the subject.—*Charleston Med. Journal and Review*, Jan. 1860.

This edition, now much enlarged, is one of the most useful works of the past year.—*N. O. Med. and Surg. Journal*, Jan. 1860.

The whole treatise is eminently practical; and there is no production of the kind in the English language so well adapted to the wants of the pharmaceutist and druggist. To physicians, also, it cannot fail to be highly valuable, especially to those who are obliged to prepare and compound many of their own medicines.—*N. Am. Med. Chir. Review*, Jan. 1860.

Of course, all apothecaries who have not already a copy of the first edition will procure one of this; it is, therefore, to physicians residing in the country and in small towns, who cannot avail themselves of the skill of an educated pharmaceutist, that we would especially commend this work. In it they will find all that they desire to know, and should know, but very little of which they do really know in reference to this important collateral branch of their profession; for it is a well established fact, that, in the education of physicians, while the science of medicine is generally well taught, very little attention is paid to the art of preparing them for use, and we know not how this defect can be so well remedied as by procuring and consulting Dr. Parrish's excellent work.—*St. Louis Med. Journal*, Jan. 1860.

PEASLEE (E. R.), M. D.,

Professor of Physiology and General Pathology in the New York Medical College.

HUMAN HISTOLOGY, in its relations to Anatomy, Physiology, and Pathology; for the use of Medical Students. With four hundred and thirty-four illustrations. In one handsome octavo volume, of over 600 pages. (*Lately Published.*) $3 75.

It embraces a library upon the topics discussed within itself, and is just what the teacher and learner need. Another advantage, by no means to be overlooked, everything of real value in the wide range which it embraces, is with great skill compressed into an octavo volume of but little more than six hundred pages. We have not only the whole subject of Histology, interesting in itself, ably and fully discussed, but what is of infinitely greater interest to the student, because of greater practical value, are its relations to Anatomy, Physiology, and Pathology, which are here fully and satisfactorily set forth.—*Nashville Journ. of Med. and Surgery*, Dec. 1857.

We would recommend it to the medical student and practitioner, as containing a summary of all that is known of the important subjects which it treats; of all that is contained in the great works of Simon and Lehmann, and the organic chemists in general. Master this one volume, we would say to the medical student and practitioner—master this book and you know all that is known of the great fundamental principles of medicine, and we have no hesitation in saying that it is an honor to the American medical profession that one of its members should have produced it.—*St. Louis Med. and Surg. Journal*, March, 1858.

PEREIRA (JONATHAN), M. D., F. R. S., AND L. S.

THE ELEMENTS OF MATERIA MEDICA AND THERAPEUTICS. Third American edition, enlarged and improved by the author; including Notices of most of the Medicinal Substances in use in the civilized world, and forming an Encyclopædia of Materia Medica. Edited, with Additions, by JOSEPH CARSON, M. D., Professor of Materia Medica and Pharmacy in the University of Pennsylvania. In two very large octavo volumes of 2100 pages, on small type, with about 500 illustrations on stone and wood, strongly bound in leather, with raised bands. $9 00.

*** Vol. II. will no longer be sold separate.

PARKER (LANGSTON),

Surgeon to the Queen's Hospital, Birmingham.

THE MODERN TREATMENT OF SYPHILITIC DISEASES, BOTH PRIMARY AND SECONDARY; comprising the Treatment of Constitutional and Confirmed Syphilis, by a safe and successful method. With numerous Cases, Formulæ, and Clinical Observations. From the Third and entirely rewritten London edition. In one neat octavo volume, extra cloth, of 316 pages. $1 75.

RAMSBOTHAM (FRANCIS H.), M.D.

THE PRINCIPLES AND PRACTICE OF OBSTETRIC MEDICINE AND SURGERY,

in reference to the Process of Parturition. A new and enlarged edition, thoroughly revised by the Author. With Additions by W. V. Keating, M. D. In one large and handsome imperial octavo volume, of 650 pages, strongly bound in leather, with raised bands; with sixty-four beautiful Plates, and numerous Wood-cuts in the text, containing in all nearly two hundred large and beautiful figures. $5 00.

From Prof. Hodge, of the University of Pa.

To the American public, it is most valuable, from its intrinsic undoubted excellence, and as being the best authorized exponent of British Midwifery. Its circulation will, I trust, be extensive throughout our country.

It is unnecessary to say anything in regard to the utility of this work. It is already appreciated in our country for the value of the matter, the clearness of its style, and the fulness of its illustrations. To the physician's library it is indispensable, while to the student as a text-book, from which to extract the material for laying the foundation of an education on obstetrical science, it has no superior.—*Ohio Med. and Surg. Journal.*

The publishers have secured its success by the truly elegant style in which they have brought it out, excelling themselves in its production, especially in its plates. It is dedicated to Prof. Meigs, and has the emphatic endorsement of Prof. Hodge, as the best exponent of British Midwifery. We know of no text-book which deserves in all respects to be more highly recommended to students, and we could wish to see it in the hands of every practitioner, for they will find it invaluable for reference.—*Med. Gazette.*

RICORD (P.), M.D.

A TREATISE ON THE VENEREAL DISEASE. By John Hunter, F.R.S.

With copious Additions, by Ph. Ricord, M.D. Translated and Edited, with Notes, by Freeman J. Bumstead, M.D., Lecturer on Venereal at the College of Physicians and Surgeons, New York. Second edition, revised, containing a *résumé* of Ricord's Recent Lectures on Chancre. In one handsome octavo volume, extra cloth, of 550 pages, with eight plates. $3 25. (*Just Issued.*)

In revising this work, the editor has endeavored to introduce whatever matter of interest the recent investigations of syphilographers have added to our knowledge of the subject. The principal source from which this has been derived is the volume of "Lectures on Chancre," published a few months since by M. Ricord, which affords a large amount of new and instructive material on many controverted points. In the previous edition, M. Ricord's additions amounted to nearly one-third of the whole, and with the matter now introduced, the work may be considered to present his views and experience more thoroughly and completely than any other.

Every one will recognize the attractiveness and value which this work derives from thus presenting the opinions of these two masters side by side. But, it must be admitted, what has made the fortune of the book, is the fact that it contains the "most complete embodiment of the veritable doctrines of the Hôpital du Midi," which has ever been made public. The doctrinal ideas of M. Ricord, ideas which, if not universally adopted, are incontestably dominant, have heretofore only been interpreted by more or less skilful secretaries, sometimes accredited and sometimes not. In the notes to Hunter, the master substitutes himself for his interpreters, and gives his original thoughts to the world in a lucid and perfectly intelligible manner. In conclusion we can say that this is incontestably the best treatise on syphilis with which we are acquainted, and, as we do not often employ the phrase, we may be excused for expressing the hope that it may find a place in the library of every physician.—*Virginia Med. and Surg. Journal.*

BY THE SAME AUTHOR.

RICORD'S LETTERS ON SYPHILIS. Translated by W. P. Lattimore, M. D.

In one neat octavo volume, of 270 pages, extra cloth. $2 00.

ROYLE'S MATERIA MEDICA AND THERAPEUTICS; including the

Preparations of the Pharmacopœias of London, Edinburgh, Dublin, and of the United States. With many new medicines. Edited by Joseph Carson, M. D. With ninety-eight illustrations. In one large octavo volume, extra cloth, of about 700 pages. $3 00.

ROKITANSKY (CARL), M.D.,

Curator of the Imperial Pathological Museum, and Professor at the University of Vienna, &c.

A MANUAL OF PATHOLOGICAL ANATOMY. Four volumes, octavo,

bound in two, extra cloth, of about 1200 pages. Translated by W. E. Swaine, Edward Sieveking, C. H. Moore, and G. E. Day. $5 50

The profession is too well acquainted with the reputation of Rokitansky's work to need our assurance that this is one of the most profound, thorough, and valuable books ever issued from the medical press. It is *sui generis*, and has no standard of comparison. It is only necessary to announce that it is issued in a form as cheap as is compatible with its size and preservation, and its sale follows as a matter of course. No library can be called complete without it.—*Buffalo Med. Journal.*

An attempt to give our readers any adequate idea of the vast amount of instruction accumulated in these volumes, would be feeble and hopeless. The effort of the distinguished author to concentrate in a small space his great fund of knowledge, has so charged his text with valuable truths, that any attempt of a reviewer to epitomize is at once paralyzed, and must end in a failure.—*Western Lancet.*

As this is the highest source of knowledge upon the important subject of which it treats, no real student can afford to be without it. The American publishers have entitled themselves to the thanks of the profession of their country, for this timeous and beautiful edition.—*Nashville Journal of Medicine.*

As a book of reference, therefore, this work must prove of inestimable value, and we cannot too highly recommend it to the profession.—*Charleston Med. Journal and Review.*

This book is a necessity to every practitioner.—*Am. Med. Monthly.*

RIGBY (EDWARD), M.D.,

Senior Physician to the General Lying-in Hospital, &c.

A SYSTEM OF MIDWIFERY. With Notes and Additional Illustrations.

Second American Edition. One volume octavo, extra cloth, 422 pages. $2 50.

BY THE SAME AUTHOR. (*Lately Published.*)

ON THE CONSTITUTIONAL TREATMENT OF FEMALE DISEASES.

In one neat royal 12mo. volume, extra cloth, of about 250 pages. $1 00.

STILLE (ALFRED), M.D.

THERAPEUTICS AND MATERIA MEDICA; a Systematic Treatise on the Action, and Uses of Medicinal Agents, including their Description and History. In two large and handsome octavo volumes, of 1789 pages. (*Now Ready*, 1860.) $8 00.

This work is designed especially for the student and practitioner of medicine, and treats the various articles of the Materia Medica from the point of view of the bedside, and not of the shop or of the lecture-room. While thus endeavoring to give all practical information likely to be useful with respect to the employment of special remedies in special affections, and the results to be anticipated from their administration, a copious Index of Diseases and their Remedies renders the work eminently fitted for reference by showing at a glance the different means which have been employed, and enabling the practitioner to extend his resources in difficult cases with all that the experience of the profession has suggested. At the same time particular care has been given to the subject of General Therapeutics, and at the commencement of each class of medicines there is a chapter devoted to the consideration of their common influence upon morbid conditions. The action of remedial agents upon the healthy economy and on animals has likewise received particular notice, from the conviction that their physiological effects will afford frequent explanations of their pathological influence, and in many cases lead to new and important suggestions as to their practical use in disease. Within the scope thus designed by the author, no labor has been spared to accumulate all the facts which have accrued from the experience of the profession in all ages and all countries; and the vast amount of recent researches recorded in the periodical literature of both hemispheres has been zealously laid under contribution, resulting in a mass of practical information scarcely attempted hitherto in any similar work in the language.

Our expectations of the value of this work were based on the well-known reputation and character of the author as a man of scholarly attainments, an elegant writer, a candid inquirer after truth, and a philosophical thinker; we knew that the task would be conscientiously performed, and that few, if any, among the distinguished medical teachers in this country are better qualified than he to prepare a systematic treatise on therapeutics in accordance with the present requirements of medical science. Our preliminary examination of the work has satisfied us that we were not mistaken in our anticipations. In congratulating the author on the completion of the great labor which such a work involves, we are happy in expressing the conviction that its merits will receive that reward which is above all price—the grateful appreciation of his medical brethren.—*New Orleans Medical News*, March, 1860.

We think this work will do much to obviate the reluctance to a thorough investigation of this branch of scientific study, for in the wide range of medical literature treasured in the English tongue, we shall hardly find a work written in a style more clear and simple, conveying forcibly the facts taught, and yet free from turgidity and redundancy. There is a fascination in its pages that will insure to it a wide popularity and attentive perusal, and a degree of usefulness not often attained through the influence of a single work. The author has much enhanced the practical utility of his book by passing briefly over the physical, botanical, and commercial history of medicines, and directing attention chiefly to their physiological action, and their application for the amelioration or cure of disease. He ignores hypothesis and theory which are so alluring to many medical writers, and so liable to lead them astray, and confines himself to such facts as have been tried in the crucible of experience.—*Chicago Medical Journal*, March, 1860.

The plan pursued by the author in these very elaborate volumes is not strictly one of scientific unity and precision; he has rather subordinated these to practical utility. Dr. Stillé has produced a work which will be valuable equally to the student of medicine and the busy practitioner.—*London Lancet*, March 10, 1860.

With Pereira, Dunglison, Mitchell, and Wood before us, we may well ask if there was a necessity for a new book on the subject. After examining this work with some care, we can answer affirmatively. Dr. Wood's book is well adapted for students, while Dr. Stillé's will be more satisfactory to the practitioner, who desires to study the action of medicines. The author needs no encomiums from us, for he is well known as a ripe scholar and a man of the most extensive reading in his profession. This work bears evidence of this fact on every page.—*Cincinnati Lancet*, April, 1860.

SMITH (HENRY H.), M.D.

MINOR SURGERY; or, Hints on the Every-day Duties of the Surgeon. With 247 illustrations. Third edition. 1 vol. royal 12mo., pp. 456. In leather, $2 25; cloth, $2 00.

BY THE SAME AUTHOR, AND

HORNER (WILLIAM E.), M.D.,

Late Professor of Anatomy in the University of Pennsylvania.

AN ANATOMICAL ATLAS, illustrative of the Structure of the Human Body. In one volume, large imperial octavo, extra cloth, with about six hundred and fifty beautiful figures. $3 00.

These figures are well selected, and present a complete and accurate representation of that wonderful fabric, the human body. The plan of this Atlas, which renders it so peculiarly convenient for the student, and its superb artistical execution, have been already pointed out. We must congratulate the student upon the completion of this Atlas, as it is the most convenient work of the kind that has yet appeared; and we must add, the very beautiful manner in which it is "got up" is so creditable to the country as to be flattering to our national pride.—*American Medical Journal*.

SHARPEY (WILLIAM), M.D., JONES QUAIN, M.D., AND RICHARD QUAIN, F.R.S., &c.

HUMAN ANATOMY. Revised, with Notes and Additions, by JOSEPH LEIDY, M.D., Professor of Anatomy in the University of Pennsylvania. Complete in two large octavo volumes, leather, of about thirteen hundred pages. Beautifully illustrated with over five hundred engravings on wood. $6 00.

SIMPSON (J. Y.), M.D.,

Professor of Midwifery, &c., in the University of Edinburgh, &c.

CLINICAL LECTURES ON THE DISEASES OF FEMALES. With numerous illustrations.

This valuable series of practical Lectures is now appearing in the "MEDICAL NEWS AND LIBRARY" for 1860, and can thus be had without cost by subscribers to the "AMERICAN JOURNAL OF THE MEDICAL SCIENCES." See p. 2.

SARGENT (F. W.), M. D.

ON BANDAGING AND OTHER OPERATIONS OF MINOR SURGERY.

Second edition, enlarged. One handsome royal 12mo. vol., of nearly 400 pages, with 182 wood-cuts. Extra cloth, $1 40; leather, $1 50.

Sargent's Minor Surgery has always been popular, and deservedly so. It furnishes that knowledge of the most frequently requisite performances of surgical art which cannot be entirely understood by attending clinical lectures. The art of bandaging, which is regularly taught in Europe, is very frequently overlooked by teachers in this country; the student and junior practitioner, therefore, may often require that knowledge which this little volume so tersely and happily supplies.—*Charleston Med. Journ. and Review*, March, 1856.

A work that has been so long and favorably known to the profession as Dr. Sargent's Minor Surgery, needs no commendation from us. We would remark, however, in this connection, that minor surgery seldom gets that attention in our schools that its importance deserves. Our larger works are also very defective in their teaching on these small practical points. This little book will supply the void which all must feel who have not studied its pages.—*Western Lancet*, March, 1856.

SMITH (W. TYLER), M. D.,

Physician Accoucheur to St. Mary's Hospital, &c.

ON PARTURITION, AND THE PRINCIPLES AND PRACTICE OF OBSTETRICS.

In one royal 12mo. volume, extra cloth, of 400 pages. $1 25.

BY THE SAME AUTHOR.

A PRACTICAL TREATISE ON THE PATHOLOGY AND TREATMENT OF LEUCORRHŒA.

With numerous illustrations. In one very handsome octavo volume, extra cloth, of about 250 pages. $1 50.

SOLLY ON THE HUMAN BRAIN; its Structure, Physiology, and Diseases. From the Second and much enlarged London edition. In one octavo volume, extra cloth, of 500 pages, with 120 wood-cuts. $2 00.

SKEY'S OPERATIVE SURGERY. In one very handsome octavo volume, extra cloth, of over 650 pages, with about one hundred wood-cuts. $3 25.

SIMON'S GENERAL PATHOLOGY, as conducive to the Establishment of Rational Principles for the prevention and Cure of Disease. In one octavo volume, extra cloth, of 212 pages. $1 25.

TODD (R. B.), M. D., F. R. S., &c.

CLINICAL LECTURES ON CERTAIN DISEASES OF THE URINARY ORGANS AND ON DROPSIES.

In one octavo volume, 284 pages. $1 50.

BY THE SAME AUTHOR. (*Now Ready.*)

CLINICAL LECTURES ON CERTAIN ACUTE DISEASES.

In one neat octavo volume, of 320 pages, extra cloth. $1 75.

The subjects treated in this volume are—RHEUMATIC FEVER, CONTINUED FEVER, ERYSIPELAS, ACUTE INTERNAL INFLAMMATION, PYÆMIA, PNEUMONIA, and the THERAPEUTICAL ACTION OF ALCOHOL. The importance of these matters in the daily practice of every physician, and the sound practical nature of Dr. Todd's writings, can hardly fail to attract to this work the general attention that it merits.

TANNER (T. H.), M. D.,

Physician to the Hospital for Women, &c.

A MANUAL OF CLINICAL MEDICINE AND PHYSICAL DIAGNOSIS.

To which is added The Code of Ethics of the American Medical Association. Second American Edition. In one neat volume, small 12mo., extra cloth, 87½ cents.

TAYLOR (ALFRED S.), M. D., F. R. S.,

Lecturer on Medical Jurisprudence and Chemistry in Guy's Hospital.

MEDICAL JURISPRUDENCE.

Fourth American Edition. With Notes and References to American Decisions, by EDWARD HARTSHORNE, M. D. In one large octavo volume, leather, of over seven hundred pages. $3 00.

No work upon the subject can be put into the hands of students either of law or medicine which will engage them more closely or profitably; and none could be offered to the busy practitioner of either calling, for the purpose of casual or hasty reference, that would be more likely to afford the aid desired. We therefore recommend it as the best and safest manual for daily use.—*American Journal of Medical Sciences.*

It is not excess of praise to say that the volume before us is the very best treatise extant on Medical Jurisprudence. In saying this, we do not wish to be understood as detracting from the merits of the excellent works of Beck, Ryan, Traill, Guy, and others; but in interest and value we think it must be conceded that Taylor is superior to anything that has preceded it.—*N. W. Medical and Surg. Journal.*

BY THE SAME AUTHOR. (*New Edition, just issued.*)

ON POISONS, IN RELATION TO MEDICAL JURISPRUDENCE AND MEDICINE.

Second American, from a second and revised London edition. In one large octavo volume, of 755 pages, leather. $3 50.

Since the first appearance of this work, the rapid advance of Chemistry has introduced into use many new substances which may become fatal through accident or design—while at the same time it has likewise designated new and more exact modes of counteracting or detecting those previously treated of. Mr. Taylor's position as the leading medical jurist of England, has during this period conferred on him extraordinary advantages in acquiring experience on these subjects, nearly all cases of moment being referred to him for examination, as an expert whose testimony is generally accepted as final. The results of his labors, therefore, as gathered together in this volume, carefully weighed and sifted, and presented in the clear and intelligible style for which he is noted, may be received as an acknowledged authority, and as a guide to be followed with implicit confidence.

New and much enlarged edition—(Just Issued.)

WATSON (THOMAS), M. D., &c.,

Late Physician to the Middlesex Hospital, &c.

LECTURES ON THE PRINCIPLES AND PRACTICE OF PHYSIC.

Delivered at King's College, London. A new American, from the last revised and enlarged English edition, with Additions, by D. FRANCIS CONDIE, M. D., author of "A Practical Treatise on the Diseases of Children," &c. With one hundred and eighty-five illustrations on wood. In one very large and handsome volume, imperial octavo, of over 1200 closely printed pages in small type; the whole strongly bound in leather, with raised bands. Price $4 25.

That the high reputation of this work might be fully maintained, the author has subjected it to a thorough revision; every portion has been examined with the aid of the most recent researches in pathology, and the results of modern investigations in both theoretical and practical subjects have been carefully weighed and embodied throughout its pages. The watchful scrutiny of the editor has likewise introduced whatever possesses immediate importance to the American physician in relation to diseases incident to our climate which are little known in England, as well as those points in which experience here has led to different modes of practice; and he has also added largely to the series of illustrations, believing that in this manner valuable assistance may be conveyed to the student in elucidating the text. The work will, therefore, be found thoroughly on a level with the most advanced state of medical science on both sides of the Atlantic.

The additions which the work has received are shown by the fact that notwithstanding an enlargement in the size of the page, more than two hundred additional pages have been necessary to accommodate the two large volumes of the London edition (which sells at ten dollars), within the compass of a single volume, and in its present form it contains the matter of at least three ordinary octavos. Believing it to be a work which should lie on the table of every physician, and be in the hands of every student, the publishers have put it at a price within the reach of all, making it one of the cheapest books as yet presented to the American profession, while at the same time the beauty of its mechanical execution renders it an exceedingly attractive volume.

The fourth edition now appears, so carefully revised, as to add considerably to the value of a book already acknowledged, wherever the English language is read, to be beyond all comparison the best systematic work on the Principles and Practice of Physic in the whole range of medical literature. Every lecture contains proof of the extreme anxiety of the author to keep pace with the advancing knowledge of the day, and to bring the results of the labors, not only of physicians, but of chemists and histologists, before his readers, wherever they can be turned to useful account. And this is done with such a cordial appreciation of the merit due to the industrious observer, such a generous desire to encourage younger and rising men, and such a candid acknowledgment of his own obligations to them, that one scarcely knows whether to admire most the pure, simple, forcible English—the vast amount of useful practical information condensed into the Lectures—or the manly, kind-hearted, unassuming character of the lecturer shining through his work. —*London Med. Times and Gazette*, Oct. 31, 1857.

Thus these admirable volumes come before the profession in their fourth edition, abounding in those distinguished attributes of moderation, judgment, erudite cultivation, clearness, and eloquence, with which they were from the first invested, but yet richer than before in the results of more prolonged observation, and in the able appreciation of the latest advances in pathology and medicine by one of the most profound medical thinkers of the day.—*London Lancet*, Nov. 14, 1857.

The lecturer's skill, his wisdom, his learning, are equalled by the ease of his graceful diction, his eloquence, and the far higher qualities of candor, of courtesy, of modesty, and of generous appreciation of merit in others. May he long remain to instruct us, and to enjoy, in the glorious sunset of his declining years, the honors, the confidence and love gained during his useful life.—*N. A. Med.-Chir. Review*, July, 1858.

Watson's unrivalled, perhaps unapproachable work on Practice—the copious additions made to which (the fourth edition) have given it all the novelty and much of the interest of a new book.—*Charleston Med. Journal*, July, 1858.

Lecturers, practitioners, and students of medicine will equally hail the reappearance of the work of Dr. Watson in the form of a new—a fourth—edition. We merely do justice to our own feelings, and, we are sure, of the whole profession, if we thank him for having, in the trouble and turmoil of a large practice, made leisure to supply the hiatus caused by the exhaustion of the publisher's stock of the third edition, which has been severely felt for the last three years. For Dr. Watson has not merely caused the lectures to be reprinted, but scattered through the whole work we find additions or alterations which prove that the author has in every way sought to bring up his teaching to the level of the most recent acquisitions in science.—*Brit. and For. Medico-Chir. Review*, Jan. 1858.

WALSHE (W. H.), M. D.,

Professor of the Principles and Practice of Medicine in University College, London, &c.

A PRACTICAL TREATISE ON DISEASES OF THE LUNGS; including

the Principles of Physical Diagnosis. A new American, from the third revised and much enlarged London edition. In one vol. octavo, of 468 pages. (*Just Issued*, June, 1860.) $2 25.

The present edition has been carefully revised and much enlarged, and may be said in the main to be rewritten. Descriptions of several diseases, previously omitted, are now introduced; the causes and mode of production of the more important affections, so far as they possess direct practical significance, are succinctly inquired into; an effort has been made to bring the description of anatomical characters to the level of the wants of the practical physician; and the diagnosis and prognosis of each complaint are more completely considered. The sections on TREATMENT and the Appendix (concerning the influence of climate on pulmonary disorders), have, especially, been largely extended.—*Author's Preface.*

*** To be followed by a similar volume on Diseases of the Heart and Aorta.

WILSON (ERASMUS), F. R. S.,

Lecturer on Anatomy, London.

THE DISSECTOR'S MANUAL; or, Practical and Surgical Anatomy. Third

American, from the last revised and enlarged English edition. Modified and rearranged, by WILLIAM HUNT, M. D., Demonstrator of Anatomy in the University of Pennsylvania. In one large and handsome royal 12mo. volume, leather, of 582 pages, with 154 illustrations. $2 00.

New and much enlarged edition—(Just Issued.)

WILSON (ERASMUS), F. R. S.

A SYSTEM OF HUMAN ANATOMY, General and Special. A new and revised American, from the last and enlarged English Edition. Edited by W. H. GOBRECHT, M. D., Professor of Anatomy in the Pennsylvania Medical College, &c. Illustrated with three hundred and ninety-seven engravings on wood. In one large and exquisitely printed octavo volume, of over 600 large pages; leather. $3 25.

The publishers trust that the well earned reputation so long enjoyed by this work will be more than maintained by the present edition. Besides a very thorough revision by the author, it has been most carefully examined by the editor, and the efforts of both have been directed to introducing everything which increased experience in its use has suggested as desirable to render it a complete text-book for those seeking to obtain or to renew an acquaintance with Human Anatomy. The amount of additions which it has thus received may be estimated from the fact that the present edition contains over one-fourth more matter than the last, rendering a smaller type and an enlarged page requisite to keep the volume within a convenient size. The author has not only thus added largely to the work, but he has also made alterations throughout, wherever there appeared the opportunity of improving the arrangement or style, so as to present every fact in its most appropriate manner, and to render the whole as clear and intelligible as possible. The editor has exercised the utmost caution to obtain entire accuracy in the text, and has largely increased the number of illustrations, of which there are about one hundred and fifty more in this edition than in the last, thus bringing distinctly before the eye of the student everything of interest or importance.

It may be recommended to the student as no less distinguished by its accuracy and clearness of description than by its typographical elegance. The wood-cuts are exquisite.—*Brit. and For. Medical Review.*

An elegant edition of one of the most useful and accurate systems of anatomical science which has been issued from the press The illustrations are really beautiful. In its style the work is extremely concise and intelligible. No one can possibly take up this volume without being struck with the great beauty of its mechanical execution, and the clearness of the descriptions which it contains is equally evident. Let students, by all means examine the claims of this work on their notice, before they purchase a text-book of the vitally important science which this volume so fully and easily unfolds.—*Lancet.*

We regard it as the best system now extant for students.—*Western Lancet.*

It therefore receives our highest commendation.—*Southern Med. and Surg. Journal.*

BY THE SAME AUTHOR. (*Just Issued.*)

ON DISEASES OF THE SKIN. Fourth and enlarged American, from the last and improved London edition. In one large octavo volume, of 650 pages, extra cloth, $2 75.

The writings of Wilson, upon diseases of the skin, are by far the most scientific and practical that have ever been presented to the medical world on this subject. The present edition is a great improvement on all its predecessors. To dwell upon all the great merits and high claims of the work before us, *seriatim*, would indeed be an agreeable service; it would be a mental homage which we could freely offer, but we should thus occupy an undue amount of space in this *Journal*. We will, however, look at some of the more salient points with which it abounds, and which make it incomparably superior in excellence to all other treatises on the subject of dermatology. No mere speculative views are allowed a place in this volume, which, without a doubt, will, for a very long period, be acknowledged as the chief standard work on dermatology. The principles of an enlightened and rational therapeia are introduced on every appropriate occasion.—*Am. Jour. Med. Science*, Oct. 1857.

ALSO, NOW READY,

A SERIES OF PLATES ILLUSTRATING WILSON ON DISEASES OF THE SKIN; consisting of nineteen beautifully executed plates, of which twelve are exquisitely colored, presenting the Normal Anatomy and Pathology of the Skin, and containing accurate representations of about one hundred varieties of disease, most of them the size of nature. Price in cloth $4 25.

In beauty of drawing and accuracy and finish of coloring these plates will be found equal to anything of the kind as yet issued in this country.

The plates by which this edition is accompanied leave nothing to be desired, so far as excellence of delineation and perfect accuracy of illustration are concerned.—*Medico-Chirurgical Review.*

Of these plates it is impossible to speak too highly The representations of the various forms of cutaneous disease are singularly accurate, and the coloring exceeds almost anything we have met with in point of delicacy and finish.—*British and Foreign Medical Review.*

We have already expressed our high appreciation of Mr. Wilson's treatise on Diseases of the Skin. The plates are comprised in a separate volume, which we counsel all those who possess the text to purchase. It is a beautiful specimen of color printing, and the representations of the various forms of skin disease are as faithful as is possible in plates of the size.—*Boston Med. and Surg. Journal*, April 8, 1858.

BY THE SAME AUTHOR.

ON CONSTITUTIONAL AND HEREDITARY SYPHILIS, AND ON SYPHILITIC ERUPTIONS. In one small octavo volume, extra cloth, beautifully printed, with four exquisite colored plates, presenting more than thirty varieties of syphilitic eruptions. $2 25.

BY THE SAME AUTHOR.

HEALTHY SKIN; A Popular Treatise on the Skin and Hair, their Preservation and Management. Second American, from the fourth London edition. One neat volume, royal 12mo., extra cloth, of about 300 pages, with numerous illustrations. $1 00; paper cover, 75 cents.

WHITEHEAD ON THE CAUSES AND TREATMENT OF ABORTION AND STERILITY. **Second American Edition. In one volume, octavo extra cloth, pp. 308. $1 75.**

WINSLOW (FORBES), M. D., D. C. L., &c.

ON OBSCURE DISEASES OF THE BRAIN AND DISORDERS OF THE MIND;

their incipient Symptoms, Pathology, Diagnosis, Treatment, and Prophylaxis. In one handsome octavo volume, of nearly 600 pages. (*Just Issued*, June, 1860.) $3 00.

The momentous questions discussed in this volume have perhaps not hitherto been so ably and elaborately treated. Dr. Winslow's distinguished reputation and long experience in everything relating to insanity invest his teachings with the highest authority, and in this carefully considered volume he has drawn upon the accumulated resources of a life of observation. His deductions are founded on a vast number of cases, the peculiarities of which are related in detail, rendering the work not only one of sound instruction, but of lively interest; the author's main object being to point out the connection between organic disease and insanity, tracing the latter through all its stages from mere eccentricity to mania, and urging the necessity of early measures of prophylaxis and appropriate treatment. A subject of greater importance to society at large could scarcely be named; while to the physician who may at any moment be called upon for interference in the most delicate relations of life, or for an opinion in a court of justice, a work like the present may be considered indispensable.

The treatment of the subject may be gathered from the following summary of the contents:—

CHAPTER I. Introduction.—II. Morbid Phenomena of Intelligence. III. Premonitory Symptoms of Insanity.—IV. Confessions of Patients after Recovery.—V. State of the Mind during Recovery.—VI. Anomalous and Masked Affections of the Mind.—VII. The Stage of Consciousness.—VIII. Stage of Exaltation.—IX. Stage of Mental Depression.—X. Stage of Aberration.—XI. Impairment of Mind.—XII. Morbid Phenomena of Attention.—XIII. Morbid Phenomena of Memory.—XIV. Acute Disorders of Memory.—XV. Chronic Affections of Memory.—XVI. Perversion and Exaltation of Memory.—XVII. Psychology and Pathology of Memory.—XVIII. Morbid Phenomena of Motion.—XIX. Morbid Phenomena of Speech.—XX. Morbid Phenomena of Sensation.—XXI. Morbid Phenomena of the Special Senses.—XXII. Morbid Phenomena of Vision, Hearing, Taste, Touch, and Smell.—XXIII. Morbid Phenomena of Sleep and Dreaming.—XXIV. Morbid Phenomena of Organic and Nutritive Life.—XXV. General Principles of Pathology, Diagnosis, Treatment, and Prophylaxis.

WEST (CHARLES), M. D.,

Accoucheur to and Lecturer on Midwifery at St. Bartholomew's Hospital, Physician to the Hospital for Sick Children, &c.

LECTURES ON THE DISEASES OF WOMEN.

Now complete in one handsome octavo volume, extra cloth, of about 500 pages; price $2 50.

Also, for sale separate, PART II, being pp. 309 to end, with Index, Title matter, &c., 8vo., cloth, price $1.

We must now conclude this hastily written sketch with the confident assurance to our readers that the work will well repay perusal. The conscientious, painstaking, practical physician is apparent on every page.—*N. Y. Journal of Medicine*, March, 1858.

We know of no treatise of the kind so complete and yet so compact.—*Chicago Med. Journal*, January, 1858.

A fairer, more honest, more earnest, and more reliable investigator of the many diseases of women and children is not to be found in any country.—*Southern Med. and Surg. Journal*, January 1858.

We gladly recommend his Lectures as in the highest degree instructive to all who are interested in obstetric practice.—*London Lancet*.

We have to say of it, briefly and decidedly, that it is the best work on the subject in any language; and that it stamps Dr. West as the *facile princeps* of British obstetric authors.—*Edinb. Med. Journ.*

BY THE SAME AUTHOR. (*Now Ready.*)

LECTURES ON THE DISEASES OF INFANCY AND CHILDHOOD.

Third American, from the fourth enlarged and improved London edition. In one handsome octavo volume, extra cloth, of about six hundred and fifty pages. $2 75.

The continued favor with which this work has been received has stimulated the author to render it in every respect more complete and more worthy the confidence of the profession. Containing nearly two hundred pages more than the last American edition, with several additional Lectures and a careful revision and enlargement of those formerly comprised in it, it can hardly fail to maintain its reputation as a clear and judicious text-book for the student, and a safe and reliable guide for the practitioner. The fact stated by the author that these Lectures "now embody the results of 900 observations and 288 post-mortem examinations made among nearly 30,000 children, who, during the past twenty years, have come under my care," is sufficient to show their high practical value as the result of an amount of experience which few physicians enjoy.

The three former editions of the work now before us have placed the author in the foremost rank of those physicians who have devoted special attention to the diseases of early life We attempt no analysis of this edition, but may refer the reader to some of the chapters to which the largest additions have been made—those on Diphtheria, Disorders of the Mind, and Idiocy, for instance—as a proof that the work is really a new edition; not a mere reprint. In its present shape it will be found of the greatest possible service in the every-day practice of nine-tenths of the profession.—*Med. Times and Gazette*, London, Dec. 10, 1859.

All things considered, this book of Dr. West is by far the best treatise in our language upon such modifications of morbid action and disease as are witnessed when we have to deal with infancy and childhood. It is true that it confines itself to such disorders as come within the province of the *physician*, and even with respect to these it is unequal as regards minuteness of consideration, and some diseases it omits to notice altogether. But those who know anything of the present condition of pædiatrics will readily admit that it would be next to impossible to effect more, or effect it better, than the accoucheur of St. Bartholomew's has done in a single volume. The lecture (XVI.) upon Disorders of the Mind in children is an admirable specimen of the value of the later information conveyed in the Lectures of Dr. Charles West.—*London Lancet*, Oct. 22, 1859.

Since the appearance of the first edition, about eleven years ago, the experience of the author has doubled; so that, whereas the lectures at first were founded on six hundred observations, and one hundred and eighty dissections made among nearly fourteen thousand children, they now embody the results of nine hundred observations, and two hundred and eighty-eight post-mortem examinations made among nearly thirty thousand children, who, during the past twenty years, have been under his care.—*British Med. Journal*, Oct. 1, 1859.

BY THE SAME AUTHOR.

AN ENQUIRY INTO THE PATHOLOGICAL IMPORTANCE OF ULCERATION OF THE OS UTERI.

In one neat octavo volume, extra cloth. $1 00.

www.ingramcontent.com/pod-product-compliance
Lightning Source LLC
LaVergne TN
LVHW021314110826
845150LV00003B/568

* 9 7 8 1 4 2 5 5 5 8 3 4 5 *